Test Item File

Precalculus
Enhanced with Graphing Utilities
Fourth Edition

Sullivan
Sullivan

Upper Saddle River, NJ 07458

Editor-in-Chief: Sally Yagan
Acquisitions Editor: Adam Jaworski
Editorial Assistant: Christopher Truchan
Executive Managing Editor: Kathleen Schiaparelli
Assistant Managing Editor: Becca Richter
Production Editor: Diane Hernandez
Supplement Cover Manager: Paul Gourhan
Supplement Cover Designer: Joanne Alexandris
Manufacturing Buyer: Ilene Kahn

© 2006 Pearson Education, Inc.
Pearson Prentice Hall
Pearson Education, Inc.
Upper Saddle River, NJ 07458

All rights reserved. No part of this book may be reproduced in any form or by any means, without permission in writing from the publisher.

Pearson Prentice Hall™ is a trademark of Pearson Education, Inc.

The author and publisher of this book have used their best efforts in preparing this book. These efforts include the development, research, and testing of the theories and programs to determine their effectiveness. The author and publisher make no warranty of any kind, expressed or implied, with regard to these programs or the documentation contained in this book. The author and publisher shall not be liable in any event for incidental or consequential damages in connection with, or arising out of, the furnishing, performance, or use of these programs.

> This work is protected by United States copyright laws and is provided solely for the use of instructors in teaching their courses and assessing student learning. Dissemination or sale of any part of this work (including on the World Wide Web) will destroy the integrity of the work and is not permitted. The work and materials from it should never be made available to students except by instructors using the accompanying text in their classes. All recipients of this work are expected to abide by these restrictions and to honor the intended pedagogical purposes and the needs of other instructors who rely on these materials.

Printed in the United States of America

10 9 8 7 6 5 4 3 2 1

ISBN 0-13-149094-X

Pearson Education Ltd., *London*
Pearson Education Australia Pty. Ltd., *Sydney*
Pearson Education Singapore, Pte. Ltd.
Pearson Education North Asia Ltd., *Hong Kong*
Pearson Education Canada, Inc., *Toronto*
Pearson Educación de Mexico, S.A. de C.V.
Pearson Education—Japan, *Tokyo*
Pearson Education Malaysia, Pte. Ltd.

CONTENTS

Chapter	1	Graphs	1
Chapter	2	Functions and Their Graphs	76
Chapter	3	Polynomial and Rational Functions	198
Chapter	4	Exponential and Logarithmic Functions	320
Chapter	5	Trigonometric Functions	415
Chapter	6	Analytic Trigonometry	538
Chapter	7	Applications of Trigonometric Functions	580
Chapter	8	Polar Coordinates; Vectors	601
Chapter	9	Analytic Geometry	678
Chapter	10	Systems of Equations and Inequalities	770
Chapter	11	Sequences; Induction; the Binomial Theorem	826
Chapter	12	Counting and Probability	859
Chapter	13	A Preview of Calculus: The Limit, Derivative, and Integral of a Function	881
Chapter	14	(Appendix) Review	903

Ch. 1 Graphs

1.1 Rectangular Coordinates; Graphing Utilities

1 Use the Distance Formula

Find the distance $d(P_1, P_2)$ between the points P_1 and P_2.

1) $P_1 = (7, 4)$; $P_2 = (-1, -5)$
   A) $\sqrt{145}$ B) 1 C) $\sqrt{17}$ D) 72

2) $P_1 = (7, -1)$; $P_2 = (3, -3)$
   A) 2 B) 12 C) $2\sqrt{5}$ D) $12\sqrt{3}$

3) $P_1 = (-1, -3)$; $P_2 = (7, -7)$
   A) 48 B) 12 C) $4\sqrt{5}$ D) $48\sqrt{3}$

4) $P_1 = (-1, 4)$; $P_2 = (-7, 12)$
   A) 100 B) 10 C) 20 D) 11

5) $P_1 = (0, 0)$; $P_2 = (3, -9)$
   A) $3\sqrt{10}$ B) 90 C) 3 D) 6

6) $P_1 = (-1, -1)$; $P_2 = (-1, 3)$
   A) 5 B) 4 C) 2 D) 3

7) $P_1 = (0.8, -0.1)$; $P_2 = (2.2, 2.3)$ Round to three decimal places, if necessary.
   A) 19 B) 8.786 C) 2.878 D) 2.778

Solve the problem.

8) Find the length of the line segment. Assume that the endpoints of the line segment have integer coordinates. If necessary, round to two decimal places.

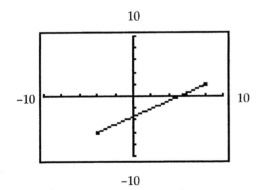

   A) 8.94 B) 14.42 C) 21.63 D) 13.41

9) Find the length of the line segment.

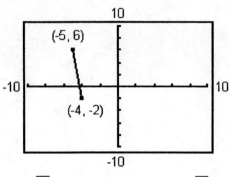

A) $\sqrt{63}$ B) $\sqrt{97}$ C) $\sqrt{65}$ D) 8

Decide whether or not the points are the vertices of a right triangle.

10) (9, 2), (14, 2), (14, 6)
   A) No   B) Yes

11) (−4, 6), (−2, 10), (0, 9)
   A) Yes   B) No

12) (3, 11), (9, 13), (8, 8)
   A) No   B) Yes

13) (−1, −5), (5, −3), (11, −10)
   A) No   B) Yes

Solve the problem.

14) Find all values of k so that the given points are $\sqrt{29}$ units apart.
   (−5, 5), (k, 0)
   A) 7   B) 3, 7   C) −3, −7   D) −7

15) Find all the points having an x-coordinate of 9 whose distance from the point (3, −2) is 10.
   A) (9, 13), (9, −7)   B) (9, −12), (9, 8)   C) (9, 2), (9, −4)   D) (9, 6), (9, −10)

16) Find the area of the right triangle ABC with A = (−2, 7), B = (7, −1), C = (3, 9).
   A) 58 square units   B) 29 square units   C) $\frac{\sqrt{58}}{2}$ square units   D) $\frac{\sqrt{29}}{2}$ square units

17) Find the length of each side of the triangle determined by the three points P_1, P_2, and P_3. State whether the triangle is an isosceles triangle, a right triangle, neither of these, or both.
$P_1 = (-5, -4)$, $P_2 = (-3, 4)$, $P_3 = (0, -1)$

A) $d(P_1, P_2) = 2\sqrt{17}$; $d(P_2, P_3) = \sqrt{34}$; $d(P_1, P_3) = \sqrt{34}$
isosceles triangle

B) $d(P_1, P_2) = 2\sqrt{17}$; $d(P_2, P_3) = \sqrt{34}$; $d(P_1, P_3) = 5\sqrt{2}$
neither

C) $d(P_1, P_2) = 2\sqrt{17}$; $d(P_2, P_3) = \sqrt{34}$; $d(P_1, P_3) = 5\sqrt{2}$
right triangle

D) $d(P_1, P_2) = 2\sqrt{17}$; $d(P_2, P_3) = \sqrt{34}$; $d(P_1, P_3) = \sqrt{34}$
both

18) A middle school's baseball playing field is a square, 80 feet on a side. How far is it directly from home plate to second base (the diagonal of the square)? If necessary, round to the nearest foot.
   A) 114 feet   B) 113 feet   C) 112 feet   D) 120 feet

19) A motorcycle and a car leave an intersection at the same time. The motorcycle heads north at an average speed of 20 miles per hour, while the car heads east at an average speed of 48 miles per hour. Find an expression for their distance apart in miles at the end of t hours.
   A) $52\sqrt{t}$ miles   B) $t\sqrt{68}$ miles   C) $52t$ miles   D) $2t\sqrt{13}$ miles

20) A rectangular city park has a jogging loop that goes along a length, width, and diagonal of the park. To the nearest yard, find the length of the jogging loop, if the length of the park is 125 yards and its width is 75 yards.
   A) 145 yards   B) 345 yards   C) 346 yards   D) 146 yards

2   Use the Midpoint Formula

Find the midpoint of the line segment joining the points P_1 and P_2.

1) $P_1 = (5, 5)$; $P_2 = (3, 1)$
   A) (8, 6)   B) (4, 3)   C) (2, 4)   D) (3, 4)

2) $P_1 = (8, -3)$; $P_2 = (7, 7)$
   A) (1, -10)   B) $(\frac{1}{2}, -5)$ C) $(\frac{15}{2}, 2)$ D) (15, 4)

3) $P_1 = (7, 1)$; $P_2 = (-16, -16)$
   A) (-9, -15)   B) $(-\frac{9}{2}, -\frac{15}{2})$ C) $(\frac{23}{2}, \frac{17}{2})$ D) (9, 15)

4) $P_1 = (0.2, -0.9)$; $P_2 = (-2.6, -2.3)$
   A) (-1.4, -0.7)   B) (-0.7, -1.4)   C) (-1.6, -1.2)   D) (-1.2, -1.6)

5) $P_1 = (y, 8)$; $P_2 = (0, 9)$
   A) (y, 17)   B) $(y, \frac{17}{2})$ C) $(\frac{y}{2}, \frac{17}{2})$ D) $(-\frac{y}{2}, 1)$

6) $P_1 = (9y, 2)$; $P_2 = (10y, 7)$
   A) $(\frac{19y}{2}, \frac{9}{2})$
   B) $(y, 5)$
   C) $(\frac{9y}{2}, \frac{19}{2})$
   D) $(19y, 9)$

Solve the problem.

7) Find the midpoint of the line segment shown on the graphing utility.

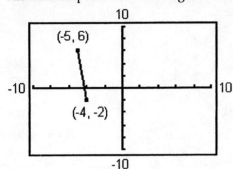

   A) $(-1, 8)$
   B) $(\frac{9}{2}, 2)$
   C) $(-\frac{9}{2}, 2)$
   D) $(-9, 4)$

8) The medians of a triangle intersect at a point. The distance from the vertex to the point is exactly two-thirds of the distance from the vertex to the midpoint of the opposite side. Find the exact distance of that point from the vertex A(3, 4) of a triangle, given that the other two vertices are at (0, 0) and (8, 0).
   A) $\frac{\sqrt{17}}{3}$
   B) 2
   C) $\frac{8}{3}$
   D) $\frac{2\sqrt{17}}{3}$

9) If (−2, −1) is the endpoint of a line segment, and (1, 3) is its midpoint, find the other endpoint.
   A) (6, 5)
   B) (−8, −9)
   C) (4, −5)
   D) (4, 7)

10) If (3, −3) is the endpoint of a line segment, and (2, −8) is its midpoint, find the other endpoint.
    A) (5, 7)
    B) (−7, −5)
    C) (1, −13)
    D) (1, 2)

11) If (−9, 3) is the endpoint of a line segment, and (−8, 1) is its midpoint, find the other endpoint.
    A) (−11, 7)
    B) (−7, 5)
    C) (−13, 5)
    D) (−7, −1)

12) If (−6, −2) is the endpoint of a line segment, and (−8, 3) is its midpoint, find the other endpoint.
    A) (−2, −12)
    B) (−10, −7)
    C) (−10, 8)
    D) (4, −6)

3 Demonstrate Additional Skills and Understanding

Name the quadrant in which the point is located.

1) (17, 10)
   A) II
   B) III
   C) I
   D) IV

2) (−14, 9)
   A) IV
   B) III
   C) II
   D) I

3) (−2, −15)
   A) I
   B) IV
   C) II
   D) III

Precalculus Enhanced with Graphing Utilities 4

4) (10, −17)

   A) III B) I C) II D) IV

Give the coordinates of the points shown on the graph.

5)
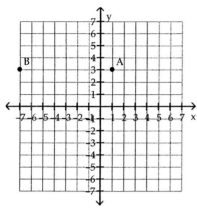

A) A = (1, 3), B = (3, −7)
B) A = (1, 3), B = (3, 3)
C) A = (1, 3), B = (−7, 3)
D) A = (3, 18), B = (3, −7)

6)
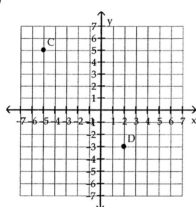

A) C = (−5, 5), D = (2, −3)
B) C = (5, 6), D = (−3, 2)
C) C = (−5, 5), D = (−3, 2)
D) C = (−5, −3), D = (5, −3)

7)
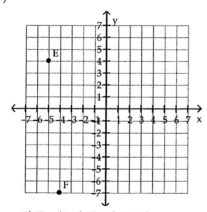

A) E = (4, 6), F = (−7, −4)
B) E = (−5, 4), F = (−4, −7)
C) E = (−5, −7), F = (4, −7)
D) E = (−5, 4), F = (−7, −4)

Determine the coordinates of the points shown. Tell in which quadrant the point lies. Assume the coordinates are integers.

8)

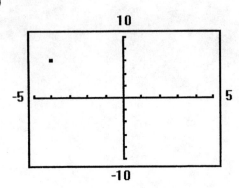

A) (–4, 6); quadrant I   B) (–4, 6); quadrant II   C) (–4, 3); quadrant II   D) (–4, 3); quadrant I

Select a setting so that each of the given points will lie in the viewing window.

9) (8, 3), (5, 1), (9, 19)

A)
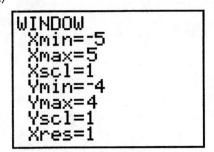

B)
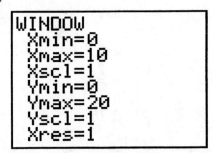

C)

```
WINDOW
 Xmin=-5
 Xmax=5
 Xscl=1
 Ymin=-1
 Ymax=10
 Yscl=1
 Xres=1
```

D)

```
WINDOW
 Xmin=0
 Xmax=20
 Xscl=1
 Ymin=0
 Ymax=10
 Yscl=1
 Xres=1
```

Determine the viewing window used.

10)

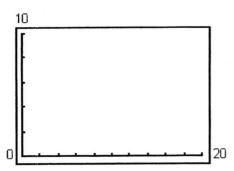

A)
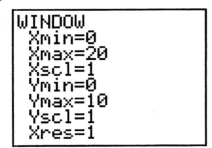
WINDOW
 Xmin=0
 Xmax=20
 Xscl=1
 Ymin=0
 Ymax=10
 Yscl=1
 Xres=1

B)
WINDOW
 Xmin=0
 Xmax=10
 Xscl=1
 Ymin=0
 Ymax=20
 Yscl=1
 Xres=1

C)
WINDOW
 Xmin=0
 Xmax=20
 Xscl=2
 Ymin=0
 Ymax=10
 Yscl=2
 Xres=1

D)
WINDOW
 Xmin=0
 Xmax=10
 Xscl=2
 Ymin=0
 Ymax=20
 Yscl=2
 Xres=1

1.2 Graphs of Equations in Two Variables

1 Graph Equations by Hand by Plotting Points

Graph the equation by plotting points.

1) $y = x + 5$

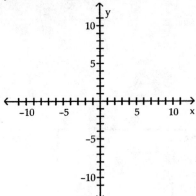

A)
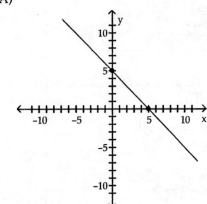

B)
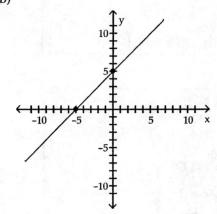

C)
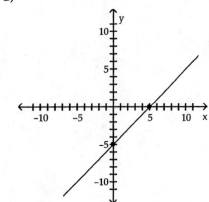

D)
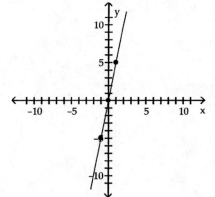

2) $y = 2x + 6$

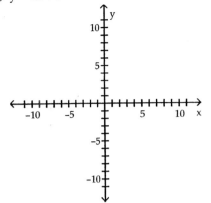

A)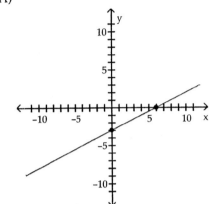

B)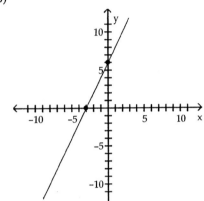

C)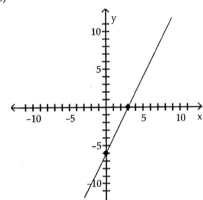

D)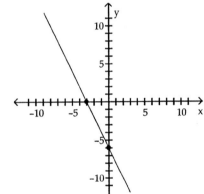

3) $y = x^2 - 4$

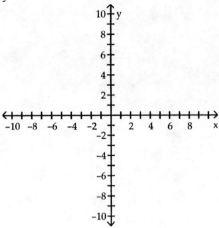

A)
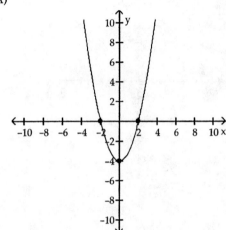

B)
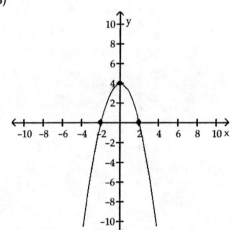

4) $3x + 4y = 12$

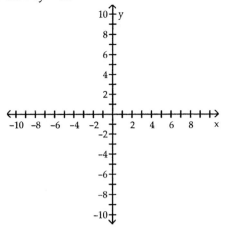

A)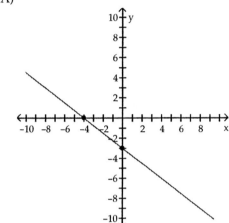

B)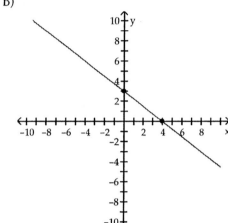

5) $x^2 + 9y = 9$

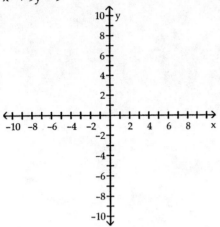

A)

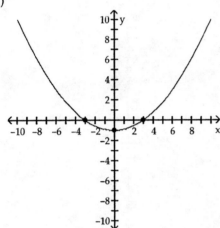

B)

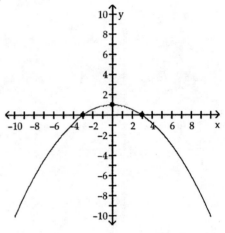

Determine whether the given point is on the graph of the equation.

6) Equation: $y = x^3 - \sqrt{x}$
   Point: $(1, 0)$

   A) No　　　　　　　　　　　　　　　　　B) Yes

7) Equation: $x^2 - y^2 = 64$
   Point: $(8, 8)$

   A) Yes　　　　　　　　　　　　　　　　　B) No

Solve the problem.

8) If $(a, 3)$ is a point on the graph of $y = 2x - 5$, what is a?
   A) 4　　　　　B) -4　　　　　C) -1　　　　　D) 1

9) If $(3, b)$ is a point on the graph of $3x - 2y = 17$, what is b?
   A) $\frac{23}{3}$　　　　　B) -4　　　　　C) 4　　　　　D) $\frac{11}{3}$

2 Graph Equations Using a Graphing Utility

1) There are no exercises for this objective.

3 Use a Graphing Utility to Create Tables

  1) There are no exercises for this objective.

4 Find Intercepts from a Graph

List the intercepts of the graph.

1)
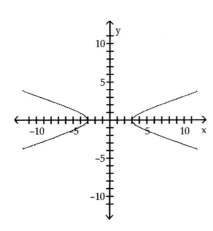

  A) (-3, 0), (3, 0)   B) (-3, 0), (0, 3)   C) (0, -3), (0, 3)   D) (0, -3), (3, 0)

2)
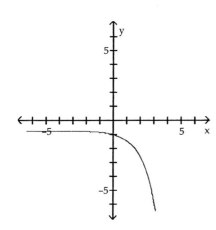

  A) (-1, -1)   B) (0, -1)   C) (0, 0)   D) (-1, 0)

3)

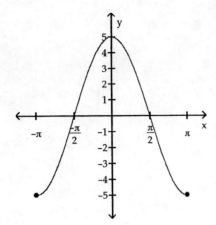

A) $(0, -\frac{\pi}{2})$, $(0, 5)$, $(0, \frac{\pi}{2})$

B) $(-\frac{\pi}{2}, 0)$, $(0, 5)$, $(\frac{\pi}{2}, 0)$

C) $(-\frac{\pi}{2}, 0)$, $(5, 0)$, $(\frac{\pi}{2}, 0)$

D) $(0, -\frac{\pi}{2})$, $(5, 0)$, $(0, \frac{\pi}{2})$

4)

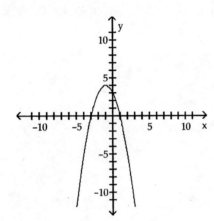

A) (−3, 0), (0, 3), (0, 1)   B) (0, −3), (0, 3), (1, 0)   C) (0, −3), (3, 0), (0, 1)   D) (−3, 0), (0, 3), (1, 0)

5)

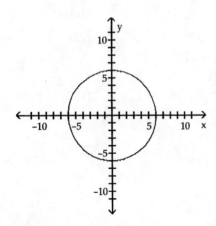

A) (0, 6), (6, 0)

B) (−6, 0), (0, 6)

C) (−6, 0), (0, −6), (0, 0), (0, 6), (6, 0)

D) (−6, 0), (0, −6), (0, 6), (6, 0)

6)

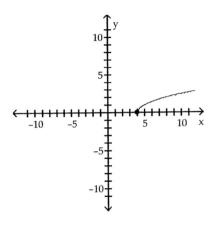

A) (-4, 0)   B) (4, 0)   C) (0, -4)   D) (0, 4)

7)

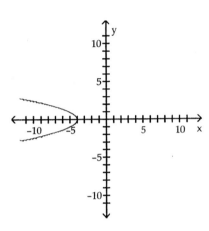

A) (4, 0)   B) (0, -4)   C) (0, 4)   D) (-4, 0)

8)

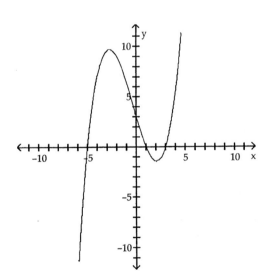

A) (-3, 0), (1, 0), (5, 0), (0, 3)
B) (3, 0), (1, 0) (-5, 0), (0, 3)
C) (3, 0), (0, -3), (0, 1), (0, 5)
D) (3, 0), (0, 3), (0, 1), (0, -5)

9)

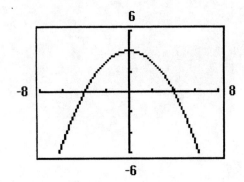

A) (–4, 0), (0, 4), (4, 0) B) (–2, 0), (0, 2), (2, 0) C) (–2, 0), (2, 0) D) (–2, 0), (0, 4), (2, 0)

5 Find Intercepts from an Equation

List the intercepts for the graph of the equation.

1) $y = x + 4$
   A) (4, 0), (0, 4) B) (4, 0), (0, –4) C) (–4, 0), (0, –4) D) (–4, 0), (0, 4)

2) $y = -5x$
   A) (0, 0) B) (–5, 0) C) (0, –5) D) (–5, –5)

3) $x^2 = y$
   A) (0, 0) B) (0, 1) C) (–1, 0), (1, 0) D) (–1, 0), (0, –1), (1, 0)

4) $y^2 = x$
   A) (1, 0) B) (0, 0) C) (–1, 0), (0, 1), (0, –1) D) (0, 1), (0, –1)

5) $x^2 + y - 36 = 0$
   A) (–6, 0), (0, 36), (6, 0)
   C) (0, –6), (36, 0), (0, 6)
   B) (6, 0), (0, 36), (0, –36)
   D) (–6, 0), (0, –36), (6, 0)

6) $y^2 - x - 1 = 0$
   A) (1, 0), (0, 1), (0, –1) B) (–1, 0), (0, –1), (1, 0) C) (0, –1), (1, 0), (0, 1) D) (0, –1), (–1, 0), (0, 1)

7) $4x^2 + 16y^2 = 64$
   A) (–4, 0), (0, –2), (0, 2), (4, 0)
   C) (–2, 0), (–4, 0), (4, 0), (2, 0)
   B) (–16, 0), (0, –4), (0, 4), (16, 0)
   D) (–4, 0), (–16, 0), (16, 0), (4, 0)

8) $9x^2 + y^2 = 9$
   A) (–9, 0), (0, –1), (0, 1), (9, 0)
   C) (–1, 0), (0, –3), (0, 3), (1, 0)
   B) (–3, 0), (0, –1), (0, 1), (3, 0)
   D) (–1, 0), (0, –9), (0, 9), (1, 0)

9) $y = x^2 + 17x + 72$
   A) (–8, 0), (–9, 0), (0, 72)
   C) (0, 8), (0, 9), (72, 0)
   B) (8, 0), (9, 0), (0, 72)
   D) (0, –8), (0, –9), (72, 0)

10) $y = \dfrac{8x}{x^2 + 64}$

    A) (–8, 0), (0, 0), (8, 0)   B) (0, 0)

    C) (0, –8), (0, 0), (0, 8)   D) (–64, 0), (0, 0), (64, 0)

11) $y = \dfrac{x^2 - 64}{8x^4}$

    A) (–8, 0), (8, 0)   B) (0, –8), (0, 8)

    C) (0, 0)   D) (–64, 0), (0, 0), (64, 0)

12) $y = x^2 + 1$

    A) (0, 1)   B) (1, 0)   C) (0, 1), (–1, 0), (1, 0)   D) (1, 0), (0, –1), (0, 1)

13) $y = x^3 - 8$

    A) (0, –2), (–2, 0)   B) (–8, 0), (0, 2)   C) (0, –2), (0, 2)   D) (0, –8), (2, 0)

14) $y = (x - 2)^2 - 1$

    A) (0, 3), (3, 0) (1, 0)   B) (0, 3), (2, 0)   C) (0, 2), (3, 0)   D) (0, 3), (0, 1), (3, 0)

6 Use a Graphing Utility to Approximate Intercepts

Graph the equation using a graphing utility. Use a graphing utility to approximate the intercepts rounded to two decimal places, if necessary. Use the TABLE feature to help establish the viewing window.

1) $y = -2x + 15$

    A) (0, –15), (7.5, 0)   B) (0, 15), (7.5, 0)

    C) (0, 15), (–7.5, 0), (7.5, 0)   D) (0, 15), (–7.5, 0)

2) $y = -4x + 15$

    A) (0, –3.75), (15, 0)   B) (0, 15), (–3.75, 0)   C) (0, 3.75), (15, 0)   D) (0, 15), (3.75, 0)

3) $y = 3x^2 - 19$

    A) (0, –19), (2.52, 0)   B) (0, 19), (–2.52, 0), (2.52, 0)

    C) (0, –19), (–2.52, 0), (2.52, 0)   D) (0, –19), (–2.51, 0), (2.51, 0)

4) $y = 5x^2 - 13$

    A) (0, –13), (1.61, 0), (–1.61, 0)   B) (0, –13), (2.60, 0), (–2.60, 0)

    C) (0, 2.60), (0, –2.60), (–13, 0)   D) (0, 1.61), (0, –1.61), (–13, 0)

5) $3x - 4y = 56$

    A) (0, 14), (18.67, 0)   B) (0, –14), (–18.67, 0), (18.67, 0)

    C) (0, –14), (18.67, 0)   D) (0, –14), (18.66, 0)

6) $6x - 5y = 67$

    A) (0, 11.17), (–13.40, 0)   B) (0, –13.40), (11.17, 0)

    C) (0, 13.40), (–11.17, 0)   D) (0, –13.41), (11.18, 0)

7) $3x^2 - 5y = 34$

    A) (0, 6.8), (−3.37, 0), (3.37, 0)   B) (0, −6.8), (−3.36, 0), (3.36, 0)

    C) (0, −6.8), (−3.37, 0), (3.37, 0)   D) (0, −6.8), (3.37, 0)

8) $4x^2 - 5y = 68$

    A) (0, −13.60), (4.12, 0), (−4.12, 0)   B) (0, −13.59), (4.13, 0), (−4.13, 0)

    C) (0, 13.60), (4.12, 0), (−4.12, 0)   D) (0, 4.12), (0, −4.12), (−13.60, 0)

7 Demonstrate Additional Skills and Understanding

Name the quadrant in which the point is located.

1) (2, 9)

    A) IV   B) II   C) I   D) III

2) (−10, 12)

    A) IV   B) III   C) II   D) I

3) (−2, −4)

    A) II   B) III   C) IV   D) I

4) (13, −13)

    A) I   B) IV   C) II   D) III

Give the coordinates of the points shown on the graph.

5)

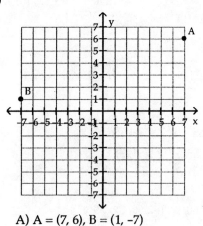

    A) A = (7, 6), B = (1, −7)   B) A = (7, 1), B = (6, 1)

    C) A = (6, 30), B = (1, −7)   D) A = (7, 6), B = (−7, 1)

6)

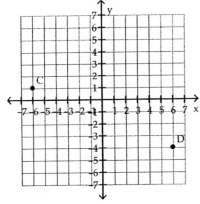

A) C = (1, 4), D = (-4, 6)
B) C = (-6, -4), D = (1, -4)
C) C = (-6, 1), D = (6, -4)
D) C = (-6, 1), D = (-4, 6)

7)

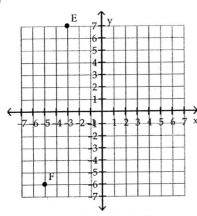

A) E = (7, 10), F = (-6, -5)
B) E = (-3, 7), F = (-5, -6)
C) E = (-3, -6), F = (7, -6)
D) E = (-3, 7), F = (-6, -5)

Determine the coordinates of the points shown. Tell in which quadrant the point lies. Assume the coordinates are integers.

8)

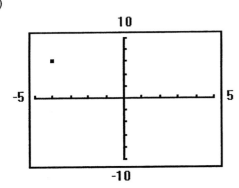

A) (-4, 3); quadrant II   B) (-4, 6); quadrant II   C) (-4, 6); quadrant I   D) (-4, 3); quadrant I

Select a setting so that each of the given points will lie in the viewing window.

9) (8, 3), (5, 1), (9, 19)

A)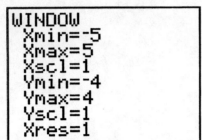

B)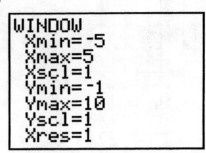

C)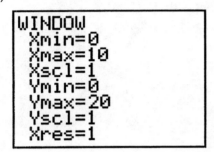

D)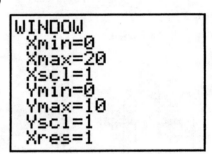

Determine the viewing window used.

10)

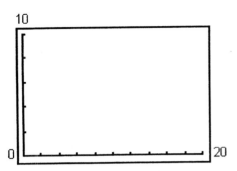

A)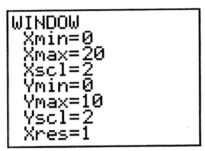

B) WINDOW
Xmin=0
Xmax=10
Xscl=1
Ymin=0
Ymax=20
Yscl=1
Xres=1

C) WINDOW
Xmin=0
Xmax=10
Xscl=2
Ymin=0
Ymax=20
Yscl=2
Xres=1

D) WINDOW
Xmin=0
Xmax=20
Xscl=1
Ymin=0
Ymax=10
Yscl=1
Xres=1

8 Know How to Graph Key Equations

Graph the function.

1) $y = x^2$

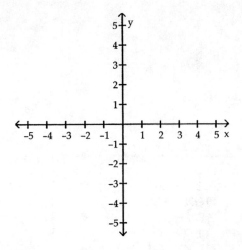

A)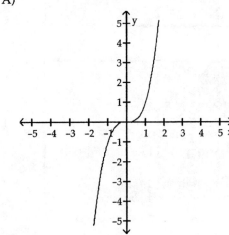

B)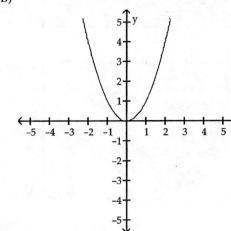

C)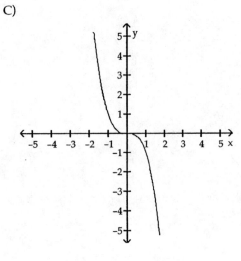

D)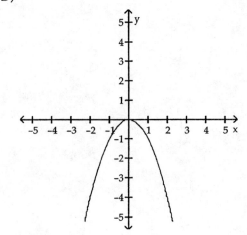

2) $y = x^3$

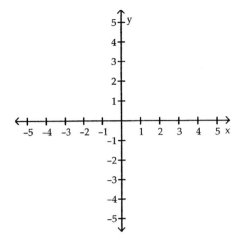

A)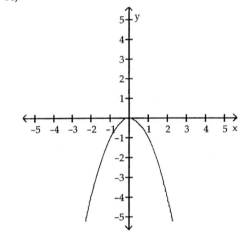

B)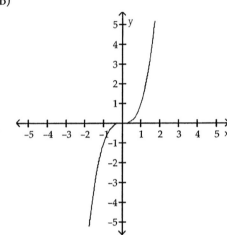

C)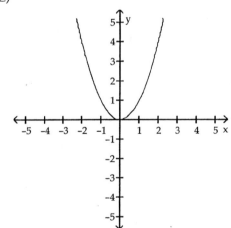

D)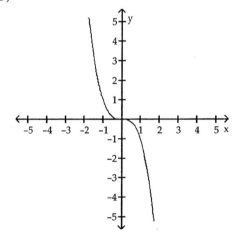

3) $y = \sqrt{x}$

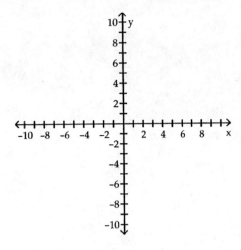

A)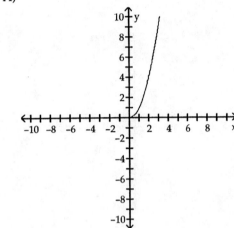

B)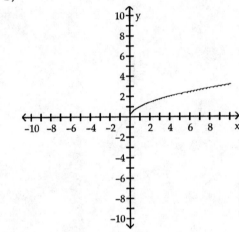

C)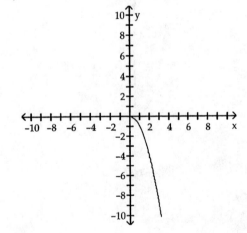

D)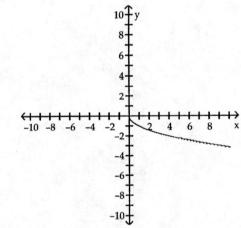

4) $y = \dfrac{1}{x}$

A)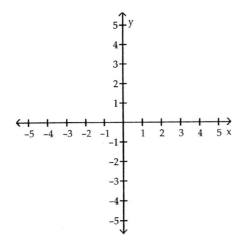

B)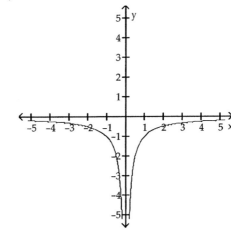

C)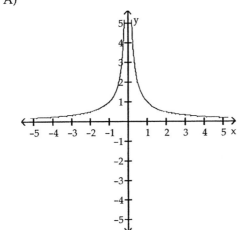

D)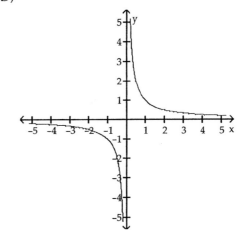

1.3 Solving Equations in One Variable Using a Graphing Utility

1 Solve Equations in One Variable Using a Graphing Utility

Use a graphing utility to approximate the real solutions, if any, of the equation rounded to two decimal places.

1) $x^3 - 6x + 3 = 0$
   A) {-0.48}
   B) {2.15, 0.52, -2.67}
   C) {2.67, -0.52, -2.15}
   D) no solution

2) $x^4 - 3x^2 + 4x + 15 = 0$
   A) {3.94, -1.27}
   B) {2.11, -2.60}
   C) {-0.84, -1.93}
   D) no solution

3) $2x^4 - 5x^2 + 7x = 14$
   A) {-2.31, 1.70}
   B) {-2.32, 1.70}
   C) {-2.31, 1.69}
   D) {-2.30, 1.69}

4) $x^4 - 5x^3 + 6x - 2 = 0$
   A) {4.71, 1.44, -0.38, -0.77}
   B) {4.75, 1, 0.38, -1.13}
   C) {0.31, -5.23}
   D) no solution

5) $-x^4 + 3x^3 + \frac{4}{3}x^2 = \frac{9}{2}x + 2$
   A) {1.09, -0.44}
   B) {-3.34}
   C) {2.82, 1.61, -0.46, -0.97}
   D) no solution

Solve the equation algebraically. Verify the solution using a graphing utility.

6) $14(3x - 9) = 9x - 2$
   A) $\{\frac{124}{51}\}$
   B) $\{-\frac{124}{33}\}$
   C) $\{\frac{124}{33}\}$
   D) $\{\frac{128}{33}\}$

7) $4(x + 5) = 5(x - 8)$
   A) {-60}
   B) {-20}
   C) {5}
   D) {60}

8) $3(2x - 3) = 5(x + 5)$
   A) {19}
   B) {-16}
   C) {-34}
   D) {34}

9) $6x + 3 + 4(x + 1) = 7x - 1$
   A) {7}
   B) $\{\frac{17}{8}\}$
   C) $\{-\frac{1}{5}\}$
   D) $\{-\frac{8}{3}\}$

10) $\frac{-2x + 7}{4} + \frac{5x}{3} = -\frac{3}{2}$
    A) $\{\frac{3}{2}\}$
    B) $\{-\frac{3}{14}\}$
    C) $\{\frac{3}{14}\}$
    D) $\{-\frac{39}{14}\}$

11) $1 - \frac{8}{9x} = \frac{4}{3}$
    A) $\{-\frac{48}{7}\}$
    B) $\{-\frac{7}{2}\}$
    C) $\{-\frac{8}{3}\}$
    D) $\{\frac{8}{3}\}$

Precalculus Enhanced with Graphing Utilities

12) $\dfrac{2}{x} - \dfrac{1}{3} = \dfrac{7}{x}$

   A) $\{\dfrac{7}{2}\}$ B) {-15}   C) {15}   D) {5}

13) $(x+6)(x-1) = (x+1)^2$

   A) {2}   B) {7}   C) $\{\dfrac{7}{3}\}$ D) $\{\dfrac{7}{6}\}$

14) $x^2 - 6x + 5 = 0$

   A) {1, 5}   B) {-1, 5}   C) {-1, -5}   D) {1, -5}

15) $x^2 - 6x - 27 = 0$

   A) {3, -9}   B) {3, 9}   C) {-3, -9}   D) {-3, 9}

16) $4x^2 = -3x + 1$

   A) $\{-\dfrac{1}{4}, 1\}$ B) $\{\dfrac{1}{4}, 1\}$ C) $\{-\dfrac{1}{4}, -1\}$ D) $\{\dfrac{1}{4}, -1\}$

17) $x^3 + 6x^2 - 25x - 150 = 0$

   A) {-5, 5, -6}   B) {5, -6}   C) {25, -6}   D) {-5, 5, 6}

18) $x^3 + 6x^2 + 16x + 96 = 0$

   A) {6}   B) {-4, 4, -6}   C) {-6}   D) no real solution

19) $\sqrt{x+2} = 3$

   A) {11}   B) {25}   C) {9}   D) {7}

20) $\sqrt{x-4} = 4$

   A) {20}   B) {12}   C) {64}   D) {16}

1.4 Lines

1 Calculate and Interpret the Slope of a Line

Find the slope of the line through the points and interpret the slope.

1)
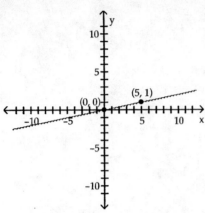

A) 5; for every 1-unit increase in x, y will increase by 5 units

B) -5; for every 1-unit increase in x, y will decrease by 5 units

C) $\frac{1}{5}$; for every 5-unit increase in x, y will increase by 1 unit

D) $-\frac{1}{5}$; for every 5-unit increase in x, y will decrease by 1 unit

Find the slope of the line.

2)
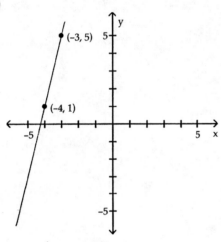

A) $-\frac{1}{4}$ B) 4 C) -4 D) $\frac{1}{4}$

3)

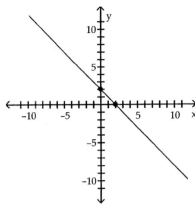

A) 1   B) 2   C) −2   D) −1

4)

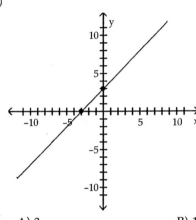

A) 3   B) 1   C) −3   D) −1

5)

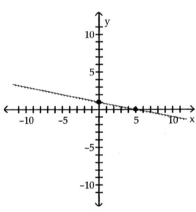

A) −5   B) 5   C) $\frac{1}{5}$ D) $-\frac{1}{5}$

Find the slope of the line containing the two points.

6) (5, −5); (−3, 8)

A) $\frac{13}{8}$ B) $-\frac{13}{8}$ C) $-\frac{8}{13}$ D) $\frac{8}{13}$

7) (2, 0); (0, 5)

   A) $-\dfrac{5}{2}$ B) $\dfrac{2}{5}$ C) $-\dfrac{2}{5}$ D) $\dfrac{5}{2}$

8) (−2, 4); (8, 8)

   A) $\dfrac{2}{5}$ B) $-\dfrac{5}{2}$ C) $-\dfrac{2}{5}$ D) $\dfrac{5}{2}$

9) (−2, −3); (3, 6)

   A) $\dfrac{9}{4}$ B) 3   C) $\dfrac{9}{5}$ D) undefined

10) (2, −6); (5, 2)

   A) 1   B) $-\dfrac{8}{3}$ C) $\dfrac{8}{3}$ D) undefined

11) (−2, 2); (−2, 1)

   A) 1   B) −1   C) 0   D) undefined

12) (4, 5); (−9, 5)

   A) −13   B) 0   C) $\dfrac{1}{13}$ D) undefined

2 Graph Lines Given a Point and the Slope

Graph the line containing the point P and having slope m.

1) P = (-7, 5); m = -3

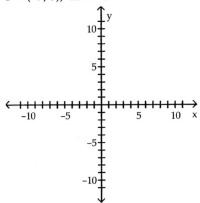

A)
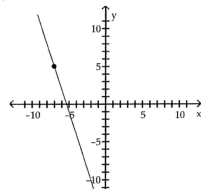

B)
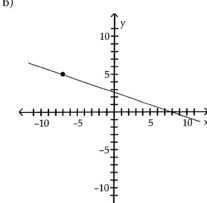

C)
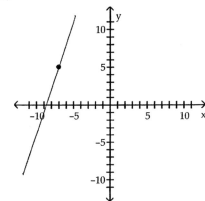

D)
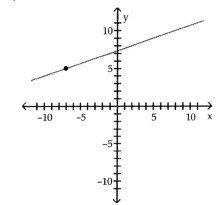

2) $P = (-3, -3)$; $m = \dfrac{2}{3}$

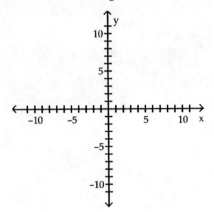

A)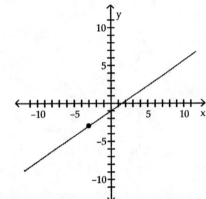

B)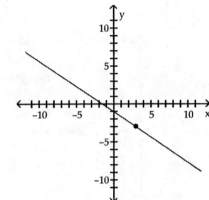

C)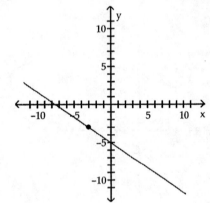

D)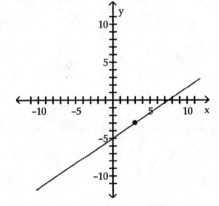

3) $P = (-4, -10)$; $m = -\dfrac{1}{2}$

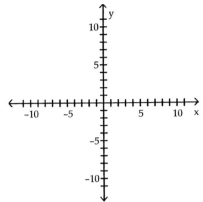

A)
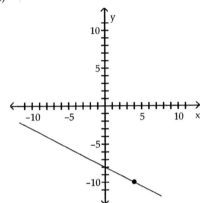

B)
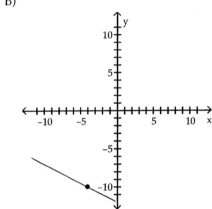

C)
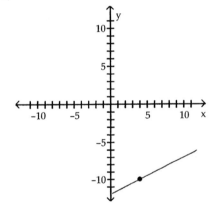

D)
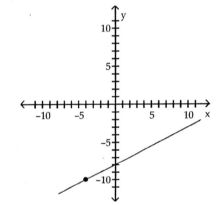

4) $P = (0, 6)$; $m = \dfrac{1}{2}$

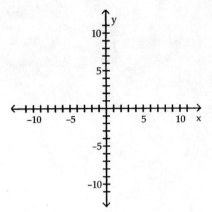

A)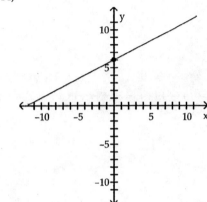

B)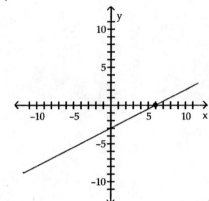

C)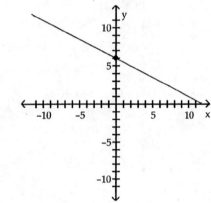

D)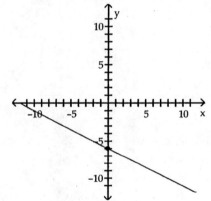

5) $P = (0, 4)$; $m = -\dfrac{3}{5}$

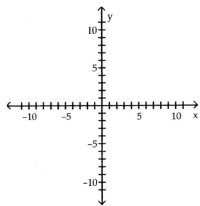

A)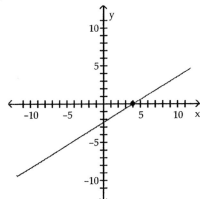

B)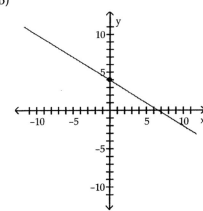

C)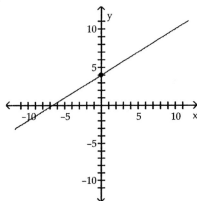

D)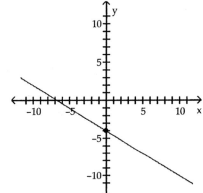

6) $P = (-4, 0);\ m = \dfrac{3}{2}$

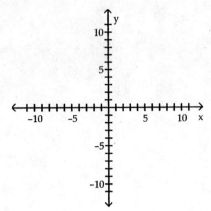

A)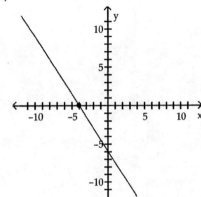

B)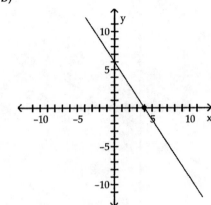

C)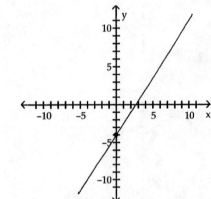

D)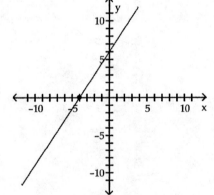

7) $P = (5, 0)$; $m = -\dfrac{2}{3}$

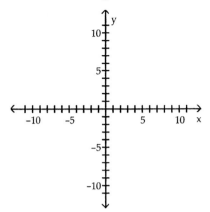

A)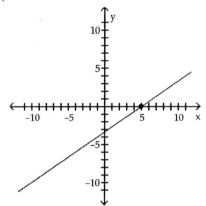

B)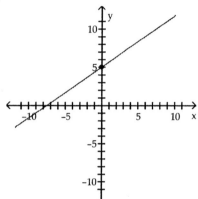

C)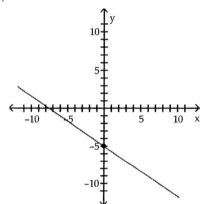

D)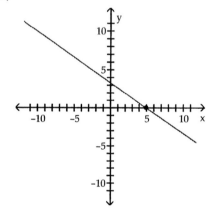

8) P = (7, -10); m = 0

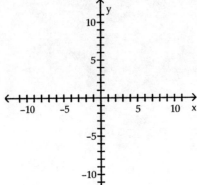

A)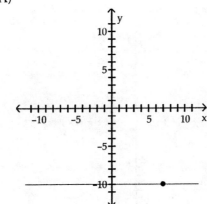

B)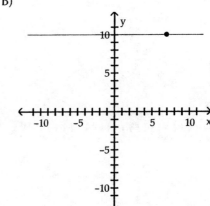

C)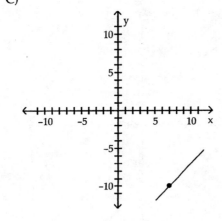

D)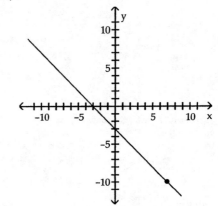

9) P = (1, 2); slope undefined

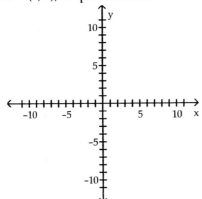

A)

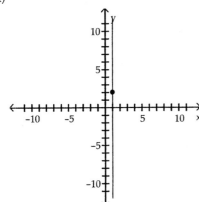

B)

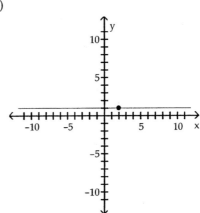

C)

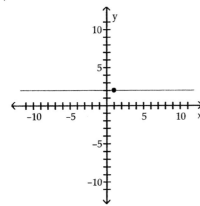

D)

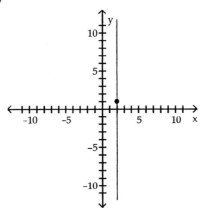

3 Find the Equation of a Vertical Line

Solve the problem.

1) Find an equation of the line with slope undefined and containing the point (9, −3).
   A) y = −3   B) y = 9   C) x = −3   D) x = 9

2) Find an equation of the vertical line containing the point (8, 10).
   A) y = 8   B) x = 8   C) y = 10   D) x = 10

3) Find an equation of the y-axis.
   A) y = 0   B) y = 1   C) x = 0   D) x = 1

4) Find an equation of the line with slope undefined and containing the point $(-\frac{3}{8}, 6)$.

  A) $y = 6$ B) $x = -\frac{3}{8}$ C) $x = 6$ D) $y = -\frac{3}{8}$

5) Find an equation of the vertical line through the point (4.9, 8.1).

  A) $x = 8.1$ B) $x = 13$ C) $x = 4.9$ D) $x = 0$

6) Find an equation of the line through the point $(-\frac{4}{5}, 6)$ with undefined slope.

  A) $x = -\frac{4}{5}$ B) $x = 0$ C) $x = 6$ D) $x = \frac{4}{5}$

4 Use the Point–Slope Form of a Line; Identify Horizontal Lines

Find the point-slope equation for the line with the given properties.

1) slope $= -\frac{2}{5}$; containing the point (-1, -3)

  A) $x - 3 = -\frac{2}{5}(y - 1)$ B) $y - 3 = -\frac{2}{5}(x - 1)$ C) $y + 3 = -\frac{2}{5}(x + 1)$ D) $y + 1 = -\frac{2}{5}(x + 3)$

Find an equation for the line with the given properties. Express the answer using the slope–intercept form of the equation of a line.

2) horizontal; containing the point (-9, -7)

  A) $y = -9$ B) $x = -9$ C) $y = -7$ D) $x = -7$

3) slope $= 0$; containing the point (-10, 7)

  A) $y = 7$ B) $x = 7$ C) $y = -10$ D) $x = -10$

4) horizontal; containing the point (-2.3, -7.6)

  A) $y = -2.3$ B) $y = 9.9$ C) $y = 0$ D) $y = -7.6$

5) horizontal; containing the point $(-\frac{1}{8}, 9)$

  A) $y = 9$ B) $y = -\frac{1}{8}$ C) $y = -9$ D) $y = 0$

5 Find the Equation of a Line Given Two Points

Find the equation of the line in slope-intercept form.

1)
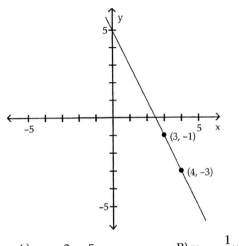

A) $y = -2x - 5$
B) $y = -\frac{1}{2}x - \frac{4}{5}$
C) $y = -2x + 1$
D) $y = -2x + 5$

6 Write the Equation of a Line in Slope-Intercept Form

Find an equation for the line with the given properties. Express the answer using the slope-intercept form of the equation of a line.

1) Slope = 2; containing the point (-3, -4)
   A) $y = -2x - 2$
   B) $y = -2x + 2$
   C) $y = 2x - 2$
   D) $y = 2x + 2$

2) Slope = $-\frac{16}{7}$; containing the point (2, -8)
   A) $y = -\frac{7}{16}x - \frac{24}{7}$
   B) $y = \frac{16}{7}x - \frac{24}{7}$
   C) $y = -\frac{16}{7}x - \frac{24}{7}$
   D) $y + 8 = -\frac{16}{7}(x - 2)$

3) Slope = $-\frac{3}{5}$; containing the point (-8, 4)
   A) $y = \frac{3}{5}x - \frac{4}{5}$
   B) $y = -\frac{5}{3}x - \frac{4}{5}$
   C) $y + 2 = -\frac{3}{5}(x - 2)$
   D) $y = -\frac{3}{5}x - \frac{4}{5}$

4) Containing the points (-4, -7) and (-8, 5)
   A) $y = -3x - 19$
   B) $y = 3x - 19$
   C) $y = -\frac{1}{3}x - 19$
   D) $y + 7 = -3(x + 4)$

5) Slope = -5; y-intercept = 6
   A) $y = -5x + 6$
   B) $y = -5x - 6$
   C) $y = 6x - 5$
   D) $y = 6x + 5$

6) x-intercept = 5; y-intercept = 6
   A) $y = -\frac{6}{5}x + 5$
   B) $y = -\frac{5}{6}x + 5$
   C) $y = \frac{6}{5}x + 6$
   D) $y = -\frac{6}{5}x + 6$

7) x-intercept = 6; y-intercept = -7
   A) $y = \frac{7}{6}x - 7$
   B) $y = \frac{7}{6}x + 7$
   C) $y = -\frac{7}{6}x - 7$
   D) $y = \frac{7}{6}x + 6$

8) Slope undefined; containing the point (5, –2)

   A) y = –2   B) y = 5   C) x = –2   D) x = 5

9) Slope = 0; containing the point (–7, 4)

   A) x = 4   B) y = –7   C) x = –7   D) y = 4

7 Identify the Slope and y-Intercept of a Line from Its Equation

Find the slope and y-intercept of the line.

1) $y = -\dfrac{8}{3}x + 8$

   A) slope = $-\dfrac{8}{3}$; y-intercept = 8
   B) slope = $\dfrac{8}{3}$; y-intercept = –8
   C) slope = 8; y-intercept = $-\dfrac{8}{3}$
   D) slope = $-\dfrac{3}{8}$; y-intercept = –8

2) $\dfrac{1}{11}x + y = -5$

   A) slope = $-\dfrac{1}{11}$; y-intercept = –5
   B) slope = $-\dfrac{1}{11}$; y-intercept = $-\dfrac{1}{5}$
   C) slope = $\dfrac{1}{11}$; y-intercept = –5
   D) slope = –11; y-intercept = –5

3) $-\dfrac{1}{5}y = x - 9$

   A) slope = $-\dfrac{1}{5}$; y-intercept = –9
   B) slope = 5; y-intercept = 45
   C) slope = –5; y-intercept = 45
   D) slope = –5; y-intercept = –9

4) $x + 11y = 1$

   A) slope = $\dfrac{1}{11}$; y-intercept = $\dfrac{1}{11}$
   B) slope = 1; y-intercept = 1
   C) slope = $-\dfrac{1}{11}$; y-intercept = $\dfrac{1}{11}$
   D) slope = –11; y-intercept = 11

5) $-x + 10y = 50$

   A) slope = $-\dfrac{1}{10}$; y-intercept = 5
   B) slope = –1; y-intercept = 50
   C) slope = $\dfrac{1}{10}$; y-intercept = 5
   D) slope = 10; y-intercept = –50

6) $3x - 5y = 15$

   A) slope = $\dfrac{3}{5}$; y-intercept = –3
   B) slope = 3; y-intercept = 15
   C) slope = $\dfrac{5}{3}$; y-intercept = 5
   D) slope = $-\dfrac{3}{5}$; y-intercept = 3

7) $4x + 11y = 44$

   A) slope = $\frac{4}{11}$; y-intercept = -4
   B) slope = $\frac{11}{4}$; y-intercept = 4
   C) slope = $-\frac{4}{11}$; y-intercept = 44
   D) slope = $-\frac{4}{11}$; y-intercept = 4

8) $x + y = -5$

   A) slope = 0; y-intercept = -5
   B) slope = -1; y-intercept = 5
   C) slope = -1; y-intercept = -5
   D) slope = 1; y-intercept = -5

9) $x - y = -2$

   A) slope = 1; y-intercept = 2
   B) slope = 1; y-intercept = -2
   C) slope = -1; y-intercept = -2
   D) slope = -1; y-intercept = 2

10) $y = 11$

    A) slope = 1; y-intercept = 11
    B) slope = 0; no y-intercept
    C) slope = 0; y-intercept = 11
    D) slope = 11; y-intercept = 0

11) $x = 5$

    A) slope = 5; y-intercept = 0
    B) slope undefined; no y-intercept
    C) slope = 0; y-intercept = 5
    D) slope undefined; y-intercept = 5

12) $y = 6x$

    A) slope = 0; y-intercept = 6
    B) slope = 6; y-intercept = 0
    C) slope = -6; y-intercept = 0
    D) slope = $\frac{1}{6}$; y-intercept = 0

13) $x - y = 0$

    A) slope = 1; y-intercept = -1
    B) slope = 1; y-intercept = 0
    C) slope = -1; y-intercept = 0
    D) slope = 1; y-intercept = 1

14) $6x - 5y = 0$

    A) slope = $-\frac{6}{5}$; y-intercept = 0
    B) slope = $\frac{6}{5}$; y-intercept = 0
    C) slope = $\frac{6}{5}$; y-intercept = 6
    D) slope = $\frac{5}{6}$; y-intercept = 0

15) $9x + 3y = 0$

    A) slope = 3; y-intercept = 0
    B) slope = $-\frac{1}{3}$; y-intercept = 0
    C) slope = -3; y-intercept = 0
    D) slope = -3; y-intercept = 9

8 Graph Lines Written in General Form Using Intercepts

Find the intercepts of the graph of the equation, then graph the equation.

1) $6x + 7y = 42$

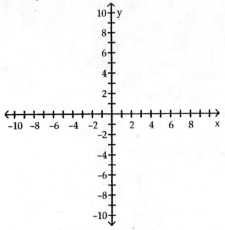

A) x-intercept: 7; y-intercept: 6

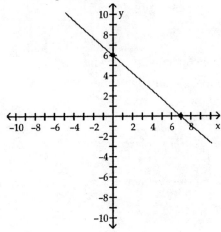

B) x-intercept: -6; y-intercept: 7

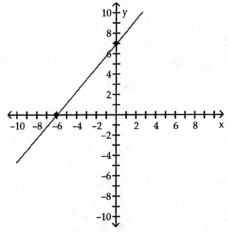

C) x-intercept: -7; y-intercept: 6

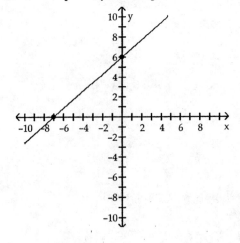

D) x-intercept: -7; y-intercept: -6

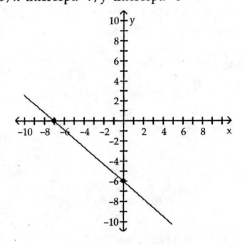

2) −5x + 3y = 15

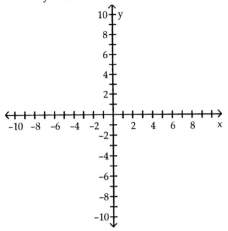

A) x-intercept: 3; y-intercept: 5

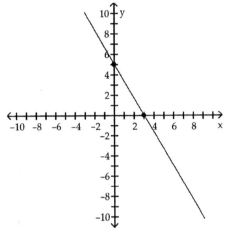

B) x-intercept: 3; y-intercept: −5

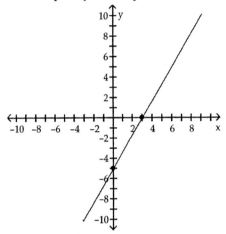

C) x-intercept: 5; y-intercept: −3

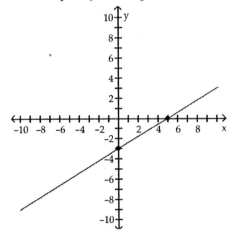

D) x-intercept: −3; y-intercept: 5

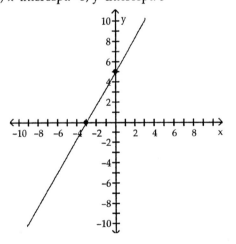

3) $2x + 5y = 8$

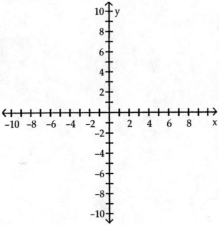

A) x-intercept: $\frac{8}{5}$; y-intercept: 4

B) x-intercept: 4; y-intercept: $-\frac{8}{5}$

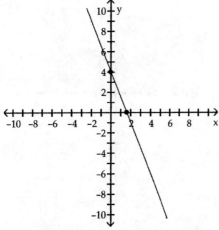

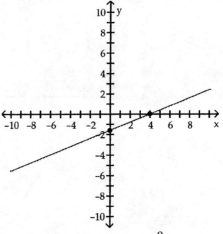

C) x-intercept: 4; y-intercept: $\frac{8}{5}$

D) x-intercept: -4; y-intercept: $\frac{8}{5}$

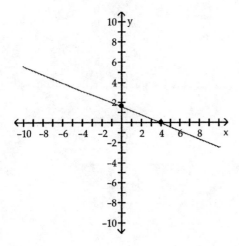

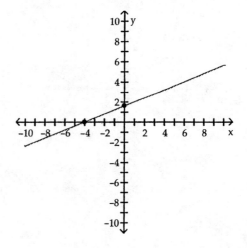

4) $5x + 4y = 8$

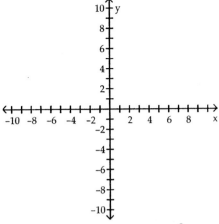

A) x-intercept: 2; y-intercept: $\frac{8}{5}$

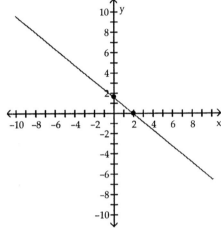

B) x-intercept: $\frac{8}{5}$; y-intercept: 2

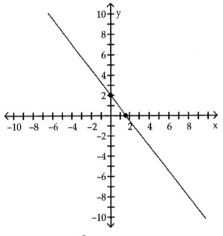

C) x-intercept: $-\frac{8}{5}$; y-intercept: -2

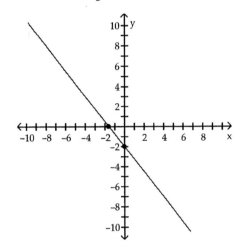

D) x-intercept: $-\frac{8}{5}$; y-intercept: 2

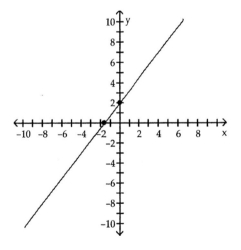

5) $\frac{1}{5}x + \frac{1}{3}y = 1$

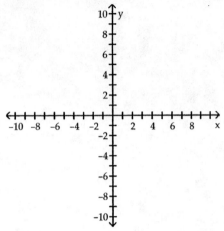

A) x-intercept: 3; y-intercept: 5

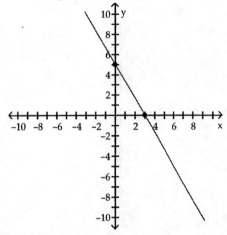

B) x-intercept: -5; y-intercept: 3

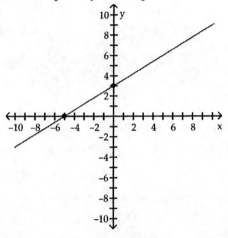

C) x-intercept: -5; y-intercept: -3

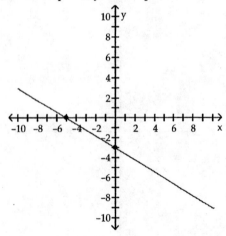

D) x-intercept: 5; y-intercept: 3

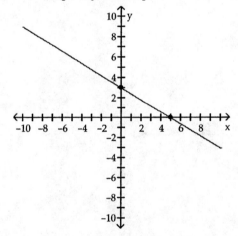

6) $0.4x + 0.8y = 3.2$

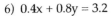

A) x-intercept: −8; y-intercept: −4

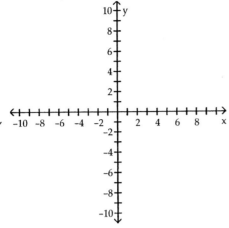

B) x-intercept: 8; y-intercept: 4

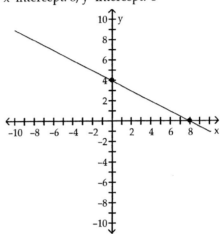

C) x-intercept: −8; y-intercept: 4

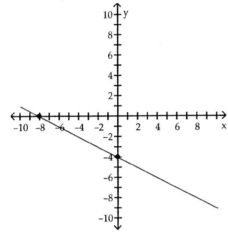

D) x-intercept: −4; y-intercept: 8

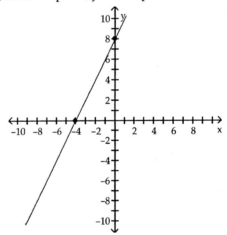

7) $-0.7x + 0.3y = 2.1$

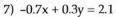

A) x-intercept: 7; y-intercept: –3

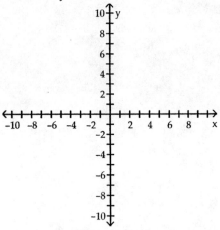

B) x-intercept: –3; y-intercept: 7

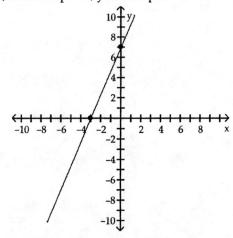

C) x-intercept: 3; y-intercept: 7

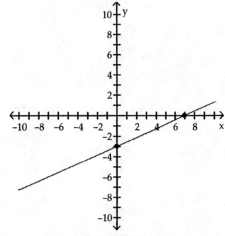

D) x-intercept: 3; y-intercept: –7

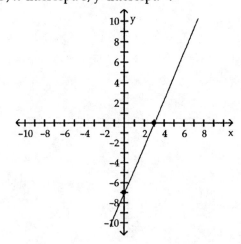

9 Find Equations of Parallel Lines

Find an equation for the line with the given properties. Express the answer using the slope–intercept form of the equation of a line.

1) The solid line L contains the point (3, 1) and is parallel to the dotted line whose equation is y = 2x.

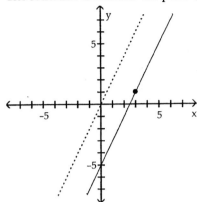

   A) $y = 2x + b$
   B) $y = 2x - 5$
   C) $y = 2x - 2$
   D) $y - 1 = 2(x - 3)$

2) Parallel to the line y = −3x; containing the point (4, 2)
   A) $y = -3x + 14$
   B) $y - 2 = -3x - 4$
   C) $y = -3x - 14$
   D) $y = -3x$

3) Parallel to the line x + 5y = 7; containing the point (0, 0)
   A) $y = -\frac{1}{5}x + 7$
   B) $y = \frac{6}{5}$
   C) $y = -\frac{1}{5}x$
   D) $y = \frac{1}{5}x$

4) Parallel to the line 4x − y = 7; containing the point (0, 0)
   A) $y = -\frac{1}{4}x$
   B) $y = 4x$
   C) $y = \frac{1}{4}x$
   D) $y = -\frac{1}{4}x + 7$

5) Parallel to the line y = 6; containing the point (9, 4)
   A) $y = 6$
   B) $y = -4$
   C) $y = 4$
   D) $y = 9$

6) Parallel to the line x = −6; containing the point (3, 4)
   A) $x = 3$
   B) $y = -6$
   C) $y = 4$
   D) $x = 4$

Find an equation for the line with the given properties. Express the answer using the general form of the equation of a line.

7) Parallel to the line 7x + 6y = 7; containing the point (7, −15)
   A) $7x + 6y = -41$
   B) $6x + 7y = -15$
   C) $7x - 6y = -41$
   D) $7x + 6y = 7$

8) Parallel to the line 4x + 3y = 3; containing the point (6, 0)
   A) $3x - 4y = -24$
   B) $4x + 3y = 24$
   C) $4x + 3y = 18$
   D) $3x - 4y = 18$

10 Find Equations of Perpendicular Lines

Find an equation for the line with the given properties. Express the answer using the slope–intercept form of the equation of a line.

1) The solid line L contains the point (3, 4) and is perpendicular to the dotted line whose equation is $y = 2x$.

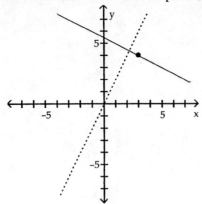

A) $y - 4 = 2(x - 3)$ B) $y = -\frac{1}{2}x + \frac{11}{2}$ C) $y = \frac{1}{2}x + \frac{11}{2}$ D) $y - 4 = -\frac{1}{2}(x - 3)$

2) Perpendicular to the line $y = 2x - 1$; containing the point (-1, 1)

A) $y = \frac{1}{2}x + \frac{1}{2}$ B) $y = -2x + \frac{1}{2}$ C) $y = -\frac{1}{2}x + \frac{1}{2}$ D) $y = 2x + \frac{1}{2}$

3) Perpendicular to the line $y = \frac{1}{8}x + 5$; containing the point (2, -3)

A) $y = 8x - 13$ B) $y = -8x - 13$ C) $y = -\frac{1}{8}x - \frac{13}{8}$ D) $y = -8x + 13$

4) Perpendicular to the line $5x - y = 2$; containing the point $(0, \frac{2}{5})$

A) $y = -\frac{1}{5}x + 2$ B) $y = \frac{1}{5}$ C) $y = -\frac{1}{5}x + \frac{2}{5}$ D) $y = \frac{1}{5}x + \frac{2}{5}$

5) Perpendicular to the line $x - 6y = 6$; containing the point (4, 3)

A) $y = -6x - 27$ B) $y = -\frac{1}{6}x - \frac{9}{2}$ C) $y = 6x - 27$ D) $y = -6x + 27$

6) Perpendicular to the line $y = -6$; containing the point (1, 5)

A) $y = 5$ B) $x = 5$ C) $x = 1$ D) $y = 1$

7) Perpendicular to the line $x = 5$; containing the point (3, 6)

A) $x = 3$ B) $y = 3$ C) $y = 6$ D) $x = 6$

Find an equation for the line with the given properties. Express the answer using the general form of the equation of a line.

8) Perpendicular to the line $8x - 7y = -77$; containing the point (-7, -3)

A) $7x + 8y = -73$ B) $8x + 7 = 8$ C) $-7x + 7y = -77$ D) $7x - 8y = -73$

9) Perpendicular to the line $-2x + 9y = -66$; containing the point $(6, -13)$

   A) $-2x - 9y = 28$ B) $9x + 2y = 28$ C) $9x - 2y = -66$ D) $9x - 2y = 28$

10) Perpendicular to the line $3x - 2y = 6$; containing the point $(0, 6)$

   A) $-2x - 3y = -12$ B) $3x - 2y = 18$ C) $-2x - 3y = -18$ D) $3x - 2y = -12$

11 Demonstrate Additional Skills and Understanding

Find an equation for the line with the given properties. Express the answer using the general form of the equation of a line.

1) Slope = 2; containing the point $(-6, -10)$

   A) $2x + y = -2$ B) $2x - y = 2$ C) $2x - y = -2$ D) $2x + y = 2$

2) Slope = $\frac{3}{4}$; y-intercept = $\frac{1}{2}$

   A) $3x + 4y = -2$ B) $y = \frac{3}{4}x + \frac{1}{2}$ C) $3x - 4y = -2$ D) $y = \frac{3}{4}x - \frac{1}{2}$

3) Slope = $-\frac{6}{7}$; containing the point $(5, 5)$

   A) $6x - 7y = 65$ B) $6x + 7y = 65$ C) $6x + 7y = -65$ D) $7x + 6y = -65$

4) Slope = $-\frac{5}{7}$; containing the point $(0, 4)$

   A) $5x - 7y = 28$ B) $5x + 7y = 28$ C) $5x + 7y = -28$ D) $7x + 5y = -28$

5) Slope = $\frac{6}{7}$; containing the point $(0, 4)$

   A) $-6x - 7y = 28$ B) $-6x + 7y = -28$ C) $-6x + 7y = 28$ D) $7x - 6y = -28$

6) Containing the points $(5, 0)$ and $(0, -8)$

   A) $y = -\frac{8}{5}x + 5$ B) $8x + 5y = 40$ C) $8x - 5y = 40$ D) $y = -\frac{8}{5}x - 8$

7) Containing the points $(-8, 6)$ and $(2, -5)$

   A) $14x - 7y = -7$ B) $-14x + 7y = -7$ C) $-11x - 10y = 28$ D) $11x - 10y = 28$

8) Containing the points $(-3, 2)$ and $(0, -3)$

   A) $5x - 3y = -9$ B) $5x - 3y = 9$ C) $-5x + 3y = -9$ D) $-5x - 3y = 9$

9) Containing the points $(6, 0)$ and $(1, -6)$

   A) $-6x - 7y = -48$ B) $-6x + 5y = -36$ C) $6x + 5y = -36$ D) $6x + 7y = -48$

10) Containing the points $(9, -2)$ and $(-1, 1)$

   A) $3x + 10y = 7$ B) $-3x + 10y = 7$ C) $11x - 2y = 9$ D) $-11x + 2y = 9$

Precalculus Enhanced with Graphing Utilities

Solve the problem.

11) Find an equation for the line graphed on a graphing utility.

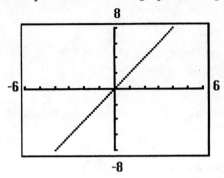

A) $y = \dfrac{3}{4}x$
B) $y = 2x$
C) $y = x$
D) $y = \dfrac{4}{3}x$

12) Find an equation in slope-intercept form for the line graphed on a graphing utility.

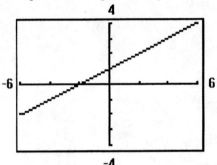

A) $y = x + 1$
B) $y = 2x + 1$
C) $y = \dfrac{1}{2}x + 1$
D) $x - 2y = -2$

13) Find an equation in general form for the line graphed on a graphing utility.

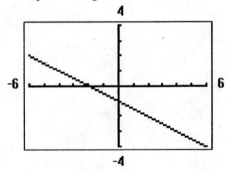

A) $x + 2y = -2$
B) $y = -2x - 1$
C) $2x + y = -1$
D) $y = -\dfrac{1}{2}x - 1$

14) A truck rental company rents a moving truck one day by charging $27 plus $0.09 per mile. Write a linear equation that relates the cost C, in dollars, of renting the truck to the number x of miles driven. What is the cost of renting the truck if the truck is driven 140 miles?

   A) C = 27x + 0.09; $3780.09
   B) C = 0.09x + 27; $28.26
   C) C = 0.09x + 27; $39.60
   D) C = 0.09x − 27; $14.40

15) The relationship between Celsius (°C) and Fahrenheit (°F) degrees of measuring temperature is linear. Find an equation relating °C and °F if 10°C corresponds to 50°F and 30°C corresponds to 86°F. Use the equation to find the Celsius measure of 39° F.

   A) $C = \frac{9}{5}F - 80$; $-\frac{49}{5}$ °C
   B) $C = \frac{5}{9}F - \frac{160}{9}$; $\frac{35}{9}$ °C
   C) $C = \frac{5}{9}F + \frac{160}{9}$; $\frac{355}{9}$ °C
   D) $C = \frac{5}{9}F - 10$; $\frac{35}{3}$ °C

16) Each week a maid service cleans x houses at $60/house. The cost to the maid service for cleaning each house is $40.00. Write an equation that relates the monthly profit P, in dollars, to the number of houses that are cleaned each month. Then use the equation to find the monthly profit when 60 houses are cleaned in a month.

   A) P = 30x; $1800
   B) P = 20x; $1200
   C) P = 60x; $3600
   D) P = 40x; $2400

17) Each week a soft drink machine sells x cans of soda for $0.75/soda. The cost to the owner of the soda machine for each soda is $0.10. The weekly fixed cost for maintaining the soda machine is $25/week. Write an equation that relates the weekly profit P, in dollars, to the number of cans sold each week. Then use the equation to find the weekly profit when 92 cans of soda are sold in a week.

   A) P = 0.65x; $59.80
   B) P = 0.75x − 25; $44.00
   C) P = 0.65x − 25; $34.80
   D) P = 0.65x + 25; $84.80

18) Each day the commuter train transports x passengers to or from the city at $1.75/passenger. The daily fixed cost for running the train is $1200. Write an equation that relates the daily profit P, in dollars, to the number of passengers each day. Then use the equation to find the daily profit when the train has 920 passengers in a day.

   A) P = 1.75x − 1200; $1610
   B) P = 1.75x; $1610
   C) P = 1.75x + 1200; $2810
   D) P = 1.75x − 1200; $410

19) Each month a beauty salon gives x manicures for $12.00/manicure. The cost to the owner of the beauty salon for each manicure is $7.35. The monthly fixed cost to maintain a manicure station is $120.00. Write an equation that relates the monthly profit P, in dollars, to the number of manicures given each month. Then use the equation to find the monthly profit when 200 manicures are given in a month.

   A) P = 4.65x; $930
   B) P = 4.65x − 120; $810
   C) P = 4.65x + 120; $1050
   D) P = 7.35x − 120; $1350

20) Each month a gas station sells x gallons of gas at $1.92/gallon. The cost to the owner of the gas station for each gallon of gas is $1.32. The monthly fixed cost for running the gas station is $37,000. Write an equation that relates the monthly profit P, in dollars, to the number of gallons of gasoline sold. Then use the equation to find the monthly profit when 75,000 gallons of gas are sold in a month.

   A) P = 0.60x − 37,000; $8000
   B) P = 0.60x + 37,000; $82,000
   C) P = 0.60x; $45,000
   D) P = 1.32x − 37,000; $62,000

1.5 Circles

1 Write the Standard Form of the Equation of a Circle

Write the standard form of the equation of the circle.

1)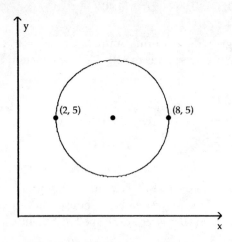

A) $(x + 5)^2 + (y + 5)^2 = 9$
B) $(x - 5)^2 + (y - 5)^2 = 9$
C) $(x + 5)^2 + (y + 5)^2 = 3$
D) $(x - 5)^2 + (y - 5)^2 = 3$

2)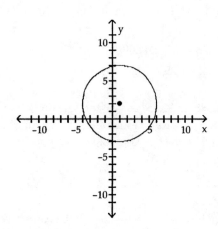

A) $(x + 1)^2 + (y + 2)^2 = 25$
B) $(x + 2)^2 + (y + 1)^2 = 25$
C) $(x - 1)^2 + (y - 2)^2 = 25$
D) $(x - 2)^2 + (y - 1)^2 = 25$

Write the standard form of the equation of the circle with radius r and center (h, k).

3) $r = 4$; $(h, k) = (0, 0)$
   A) $(x - 4)^2 + (y - 4)^2 = 16$
   B) $x^2 + y^2 = 4$
   C) $(x - 4)^2 + (y - 4)^2 = 4$
   D) $x^2 + y^2 = 16$

4) $r = 5$; $(h, k) = (-3, 4)$
   A) $(x - 3)^2 + (y + 4)^2 = 5$
   B) $(x - 3)^2 + (y + 4)^2 = 25$
   C) $(x + 3)^2 + (y - 4)^2 = 25$
   D) $(x + 3)^2 + (y - 4)^2 = 5$

5) $r = 3$; $(h, k) = (-6, 0)$
   A) $(x - 6)^2 + y^2 = 9$
   B) $(x + 6)^2 + y^2 = 9$
   C) $x^2 + (y - 6)^2 = 3$
   D) $x^2 + (y + 6)^2 = 3$

6) r = 11; (h, k) = (0, 3)
   A) $(x - 3)^2 + y^2 = 121$
   B) $x^2 + (y + 3)^2 = 11$
   C) $(x + 3)^2 + y^2 = 121$
   D) $x^2 + (y - 3)^2 = 121$

7) r = $\sqrt{14}$; (h, k) = (7, -9)
   A) $(x - 9)^2 + (y + 7)^2 = 196$
   B) $(x + 7)^2 + (y - 9)^2 = 14$
   C) $(x - 7)^2 + (y + 9)^2 = 14$
   D) $(x + 9)^2 + (y - 7)^2 = 196$

8) r = $\frac{1}{2}$; (h, k) = $(-\frac{1}{2}, 0)$
   A) $x^2 + (y + \frac{1}{2})^2 = \frac{1}{4}$
   B) $(x + \frac{1}{2})^2 + y^2 = \frac{1}{4}$
   C) $(x - \frac{1}{2})^2 + y^2 = \frac{1}{4}$
   D) $(x + \frac{1}{2})^2 + y^2 = \frac{1}{2}$

9) r = $\frac{1}{4}$; (h, k) = $(0, \frac{1}{4})$
   A) $(x - \frac{1}{4})^2 + y^2 = \frac{1}{16}$
   B) $x^2 + (y - \frac{1}{4})^2 = \frac{1}{16}$
   C) $x^2 + (y + \frac{1}{4})^2 = \frac{1}{16}$
   D) $x^2 + (y - \frac{1}{4})^2 = \frac{1}{4}$

Solve the problem.

10) Find the standard form of the equation of the circle. Assume that the center has integer coordinates and the radius is an integer.

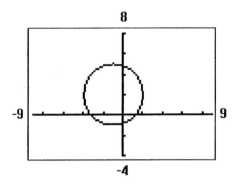

   A) $x^2 + y^2 + 2x - 4y - 4 = 0$
   B) $(x - 1)^2 + (y + 2)^2 = 9$
   C) $x^2 + y^2 - 2x + 4y - 4 = 0$
   D) $(x + 1)^2 + (y - 2)^2 = 9$

2 Graph a Circle by Hand and by Using a Graphing Utility

Graph the circle with radius r and center (h, k).

1) $r = 4$; $(h, k) = (0, 0)$

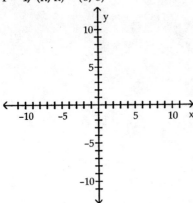

A)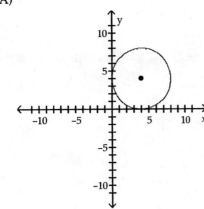

B)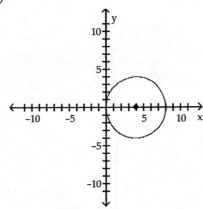

C)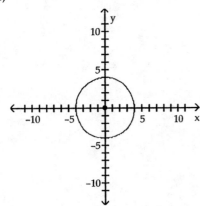

D)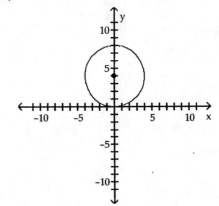

2) r = 4; (h, k) = (0, 4)

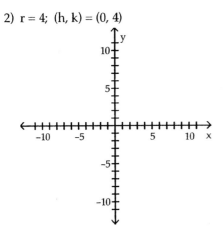

A)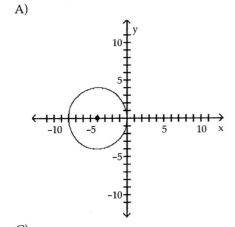

B)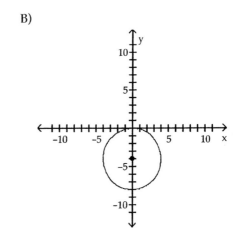

C)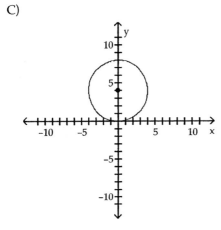

D)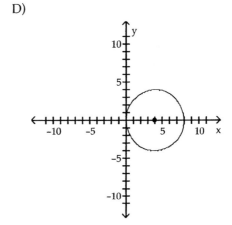

3) r = 2; (h, k) = (4, 0)

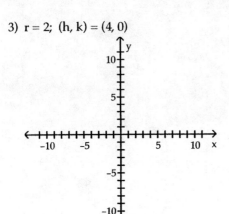

A)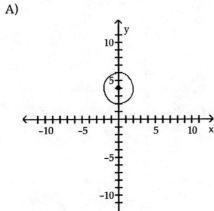

B)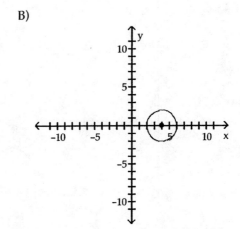

C)

D)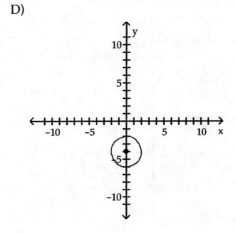

4) r = 4; (h, k) = (−2, −2)

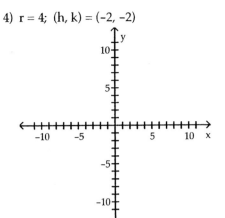

A)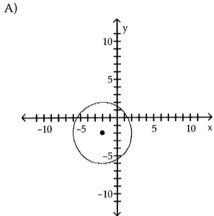

B)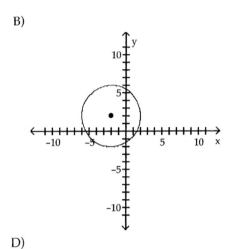

C)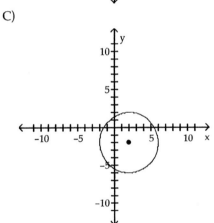

D)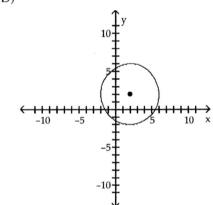

Graph the circle.

5) $x^2 + y^2 = 9$

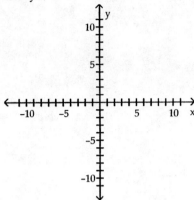

A)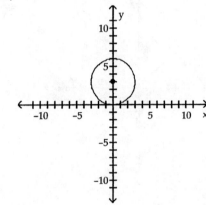

B)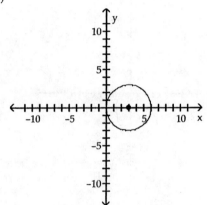

C)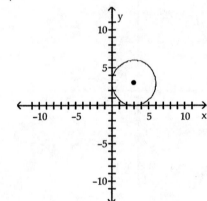

D)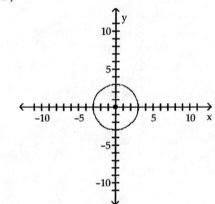

6) $(x-3)^2 + (y-5)^2 = 4$

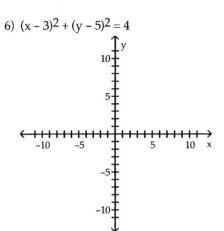

A)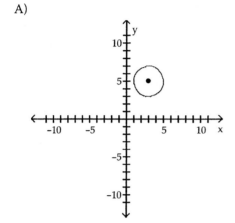

B)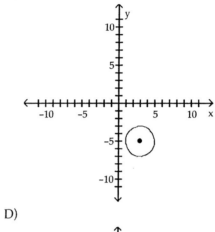

C)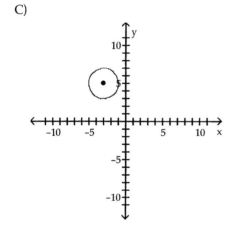

D)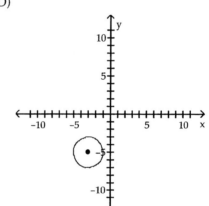

7) $x^2 + (y-1)^2 = 16$

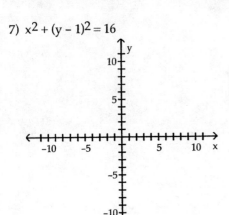

A)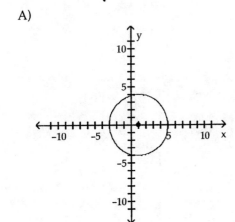

B)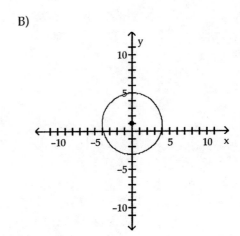

C)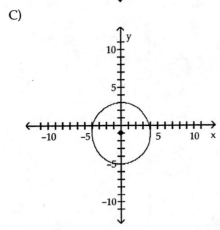

D)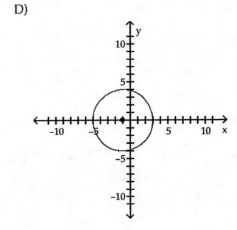

8) $(x-4)^2 + y^2 = 9$

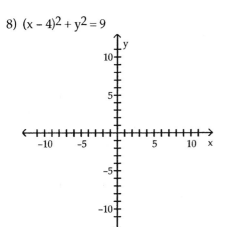

A)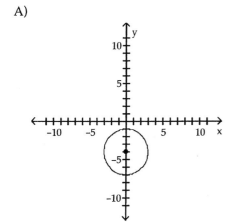

B)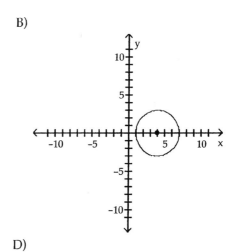

C)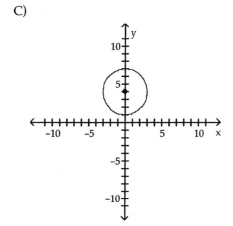

D)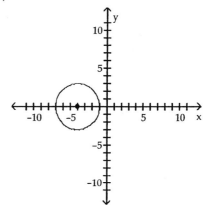

9) $x^2 + y^2 - 2x - 10y + 1 = 0$

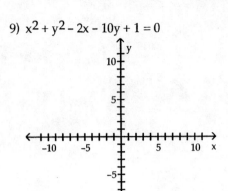

A)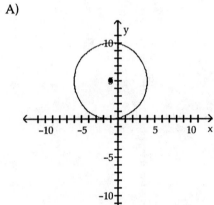

B)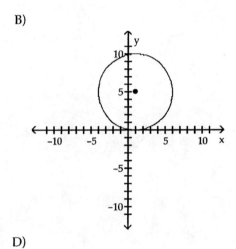

C)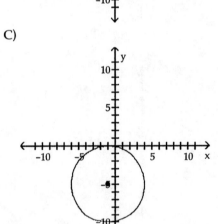

D)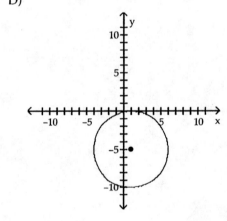

10) $x^2 + y^2 + 4x + 8y + 4 = 0$

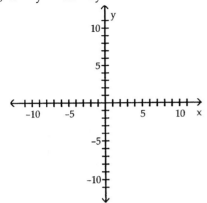

A)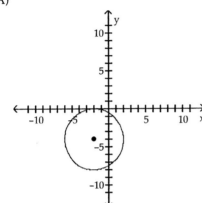

B)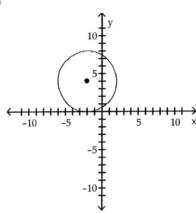

C)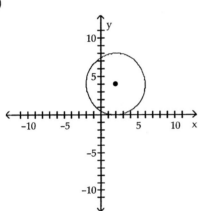

D)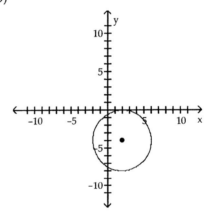

3 Work with the General Form of the Equation of a Circle

Write the general form of the equation of the circle with radius r and center (h, k).

1) $r = 2$; $(h, k) = (4, 2)$
A) $x^2 + y^2 - 8x + 4y + 16 = 0$
B) $x^2 + y^2 - 8x - 4y + 16 = 0$
C) $x^2 + y^2 + 8x - 4y + 16 = 0$
D) $x^2 + y^2 + 8x + 4y + 16 = 0$

2) $r = 2$; $(h, k) = (5, -5)$
A) $x^2 + y^2 - 10x - 10y + 46 = 0$
B) $x^2 + y^2 + 10x - 10y + 46 = 0$
C) $x^2 + y^2 + 10x + 10y + 46 = 0$
D) $x^2 + y^2 - 10x + 10y + 46 = 0$

Find the general form of the equation of the the circle.

3) Center at the point (−4, −3); containing the point (−3, 3)
   A) $x^2 + y^2 + 8x + 6y - 12 = 0$
   B) $x^2 + y^2 - 6x + 6y - 12 = 0$
   C) $x^2 + y^2 + 6x + 8y - 17 = 0$
   D) $x^2 + y^2 + 6x - 6y - 17 = 0$

4) Center at the point (2, −3); containing the point (5, −3)
   A) $x^2 + y^2 - 4x + 6y + 4 = 0$
   B) $x^2 + y^2 + 4x - 6y + 22 = 0$
   C) $x^2 + y^2 - 4x + 6y + 22 = 0$
   D) $x^2 + y^2 + 4x - 6y + 4 = 0$

5) Center at the point (3, 5); tangent to y–axis
   A) $x^2 + y^2 - 6x - 10y + 9 = 0$
   B) $x^2 + y^2 + 6x + 10y + 25 = 0$
   C) $x^2 + y^2 - 6x - 10y + 25 = 0$
   D) $x^2 + y^2 - 6x - 10y + 43 = 0$

6) With endpoints of a diameter at (6, −2) and (−4, 4)
   A) $x^2 + y^2 + 2x + 2y - 32 = 0$
   B) $x^2 + y^2 - 2x - 2y - 32 = 0$
   C) $x^2 + y^2 - 2x - 2y + 36 = 0$
   D) $x^2 + y^2 - 2x - 2y + 32 = 0$

7) With endpoints of a diameter at (5, 9) and (−1, 3)
   A) $x^2 + y^2 - 4x - 12y + 22 = 0$
   B) $x^2 + y^2 - 4x + 12y - 32 = 0$
   C) $x^2 + y^2 - 6x - 6y - 54 = 0$
   D) $x^2 + y^2 - 6x - 6y = 0$

4 Demonstrate Additional Skills and Understanding

Find the center (h, k) and radius r of the circle.

1) $x^2 + y^2 = 4$
   A) (h, k) = (2, 2); r = 4
   B) (h, k) = (0, 0); r = 4
   C) (h, k) = (2, 2); r = 2
   D) (h, k) = (0, 0); r = 2

2) $(x - 2)^2 + (y + 3)^2 = 16$
   A) (h, k) = (2, −3); r = 16
   B) (h, k) = (−3, 2); r = 4
   C) (h, k) = (2, −3); r = 4
   D) (h, k) = (−3, 2); r = 16

3) $(x - 5)^2 + y^2 = 25$
   A) (h, k) = (5, 0); r = 5
   B) (h, k) = (0, 5); r = 25
   C) (h, k) = (5, 0); r = 25
   D) (h, k) = (0, 5); r = 5

4) $x^2 + (y + 10)^2 = 36$
   A) (h, k) = (0, −10); r = 36
   B) (h, k) = (0, −10); r = 6
   C) (h, k) = (−10, 0); r = 36
   D) (h, k) = (−10, 0); r = 6

5) $2(x + 3)^2 + 2(y + 6)^2 = 30$
   A) (h, k) = (3, 6); r = $2\sqrt{15}$
   B) (h, k) = (3, 6); r = $\sqrt{15}$
   C) (h, k) = (−3, −6); r = $\sqrt{15}$
   D) (h, k) = (−3, −6); r = $2\sqrt{15}$

6) $x^2 + y^2 + 8x - 4y - 61 = 0$
   A) (h, k) = (2, −4); r = 9
   B) (h, k) = (−2, 4); r = 81
   C) (h, k) = (−4, 2); r = 9
   D) (h, k) = (4, −2); r = 81

7) $x^2 + y^2 + 4x - 2y + 1 = 0$
   A) $(h, k) = (-2, 1);\ r = 4$
   B) $(h, k) = (-2, 1);\ r = 2$
   C) $(h, k) = (2, -1);\ r = 4$
   D) $(h, k) = (2, -1);\ r = 2$

8) $13x^2 + 13y^2 = 169$
   A) $(h, k) = (0, 0);\ r = 169$
   B) $(h, k) = (13, 13);\ r = \sqrt{13}$
   C) $(h, k) = (0, 0);\ r = \sqrt{13}$
   D) $(h, k) = (0, 0);\ r = 13$

9) $4x^2 + 4y^2 - 12x + 16y - 5 = 0$
   A) $(h, k) = (-\frac{3}{2}, 2);\ r = \frac{3\sqrt{5}}{2}$
   B) $(h, k) = (-\frac{3}{2}, 2);\ r = \frac{\sqrt{30}}{2}$
   C) $(h, k) = (\frac{3}{2}, -2);\ r = \frac{\sqrt{30}}{2}$
   D) $(h, k) = (\frac{3}{2}, -2);\ r = \frac{3\sqrt{5}}{2}$

Solve the problem.

10) If a circle of radius 5 is made to roll along the x-axis, what is the equation for the path of the center of the circle?
    A) $y = 10$
    B) $y = 0$
    C) $y = 5$
    D) $x = 5$

11) Earth is represented on a map of the solar system so that its surface is a circle with the equation $x^2 + y^2 + 4x + 6y - 4083 = 0$. A weather satellite circles 0.6 units above the Earth with the center of its circular orbit at the center of the Earth. Find the general form of the equation for the orbit of the satellite on this map.
    A) $x^2 + y^2 + 4x + 6y - 4160.16 = 0$
    B) $x^2 + y^2 + 4x + 6y + 12.64 = 0$
    C) $x^2 + y^2 - 4x - 6y - 4160.16 = 0$
    D) $x^2 + y^2 + 4x + 6y - 50.64 = 0$

12) Find an equation of the line containing the centers of the two circles
    $x^2 + y^2 + 10x + 2y + 25 = 0$ and
    $x^2 + y^2 - 2x - 12y + 33 = 0$
    A) $-7x + 6y - 29 = 0$
    B) $5x + 4y - 29 = 0$
    C) $7x + 6y - 29 = 0$
    D) $-7x - 6y - 29 = 0$

Precalculus Enhanced with Graphing Utilities

Ch. 1 Graphs
Answer Key

1.1 Rectangular Coordinates; Graphing Utilities
1 Use the Distance Formula
1) A
2) C
3) C
4) B
5) A
6) B
7) D
8) B
9) C
10) B
11) A
12) A
13) A
14) C
15) D
16) B
17) D
18) B
19) C
20) C

2 Use the Midpoint Formula
1) B
2) C
3) B
4) D
5) C
6) A
7) C
8) D
9) D
10) C
11) D
12) C

3 Demonstrate Additional Skills and Understanding
1) C
2) C
3) D
4) D
5) C
6) A
7) B
8) B
9) B
10) C

1.2 Graphs of Equations in Two Variables
1 Graph Equations by Hand by Plotting Points
1) B
2) B
3) A
4) B

5) B
   6) B
   7) B
   8) A
   9) B
2 Graph Equations Using a Graphing Utility
   1) No Correct Answer Was Provided.
3 Use a Graphing Utility to Create Tables
   1) No Correct Answer Was Provided.
4 Find Intercepts from a Graph
   1) A
   2) B
   3) B
   4) D
   5) D
   6) B
   7) D
   8) B
   9) A
5 Find Intercepts from an Equation
   1) D
   2) A
   3) A
   4) B
   5) A
   6) D
   7) A
   8) C
   9) A
   10) B
   11) A
   12) A
   13) D
   14) A
6 Use a Graphing Utility to Approximate Intercepts
   1) B
   2) D
   3) C
   4) A
   5) C
   6) B
   7) C
   8) A
7 Demonstrate Additional Skills and Understanding
   1) C
   2) C
   3) B
   4) B
   5) D
   6) C
   7) B
   8) B
   9) C
   10) A

8 Know How to Graph Key Equations
   1) B
   2) B
   3) B
   4) D

1.3 Solving Equations in One Variable Using a Graphing Utility

1 Solve Equations in One Variable Using a Graphing Utility
   1) B
   2) D
   3) C
   4) B
   5) C
   6) C
   7) D
   8) D
   9) D
   10) D
   11) C
   12) B
   13) C
   14) A
   15) D
   16) D
   17) A
   18) C
   19) D
   20) A

1.4 Lines

1 Calculate and Interpret the Slope of a Line
   1) C
   2) B
   3) D
   4) B
   5) D
   6) B
   7) A
   8) A
   9) C
   10) C
   11) D
   12) B

2 Graph Lines Given a Point and the Slope
   1) A
   2) A
   3) B
   4) A
   5) B
   6) D
   7) D
   8) A
   9) A

3 Find the Equation of a Vertical Line
   1) D
   2) B
   3) C

4) B
5) C
6) A

4 Use the Point-Slope Form of a Line; Identify Horizontal Lines

1) C
2) C
3) A
4) D
5) A

5 Find the Equation of a Line Given Two Points

1) D

6 Write the Equation of a Line in Slope-Intercept Form

1) D
2) C
3) D
4) A
5) A
6) D
7) A
8) D
9) D

7 Identify the Slope and y-Intercept of a Line from Its Equation

1) A
2) A
3) C
4) C
5) C
6) A
7) D
8) C
9) A
10) C
11) B
12) B
13) B
14) B
15) C

8 Graph Lines Written in General Form Using Intercepts

1) A
2) D
3) C
4) B
5) D
6) B
7) B

9 Find Equations of Parallel Lines

1) B
2) A
3) C
4) B
5) C
6) A
7) A
8) B

10 Find Equations of Perpendicular Lines
1) B
2) C
3) D
4) C
5) D
6) C
7) C
8) A
9) B
10) C

11 Demonstrate Additional Skills and Understanding
1) C
2) C
3) B
4) B
5) C
6) C
7) C
8) D
9) B
10) A
11) B
12) C
13) A
14) C
15) B
16) B
17) C
18) D
19) B
20) A

1.5 Circles
1 Write the Standard Form of the Equation of a Circle
1) B
2) C
3) D
4) C
5) B
6) D
7) C
8) B
9) B
10) D

2 Graph a Circle by Hand and by Using a Graphing Utility
1) C
2) C
3) B
4) A
5) D
6) A
7) B
8) B
9) B
10) A

3 Work with the General Form of the Equation of a Circle
- 1) B
- 2) D
- 3) A
- 4) A
- 5) C
- 6) B
- 7) A

4 Demonstrate Additional Skills and Understanding
- 1) D
- 2) C
- 3) A
- 4) B
- 5) C
- 6) C
- 7) B
- 8) C
- 9) C
- 10) C
- 11) A
- 12) A

Ch. 2 Functions and Their Graphs

2.1 Functions

1 Determine Whether a Relation Represents a Function

Determine whether the relation represents a function. If it is a function, state the domain and range.

1)

```
4  → 12
6  → 18
8  → 24
10 → 30
```

A) function
   domain: {4, 6, 8, 10}
   range: {12, 18, 24, 30}

B) function
   domain: {12, 18, 24, 30}
   range: {4, 6, 8, 10}

C) not a function

2)

```
Alice ──→ snake
Brad  ──→ cat
Carl  ──→ dog
```

A) function
   domain: {Alice, Brad, Carl}
   range: {snake, cat, dog}

B) function
   domain: {snake, cat, dog}
   range: {Alice, Brad, Carl}

C) not a function

3)

```
Alice ──→ cat
Brad  ──→ dog
Carl  ──→
```

A) function
   domain: {cat, dog}
   range: {Alice, Brad, Carl}

B) function
   domain: {Alice, Brad, Carl}
   range: {cat, dog}

C) not a function

4) {(−3, −8), (0, 4), (4, −4), (8, −1)}

A) function
   domain: {−3, 0, 4, 8}
   range: {−8, 4, −4, −1}

B) function
   domain: {−8, 4, −4, −1}
   range: {−3, 0, 4, 8}

C) not a function

5) {(5, −2), (−4, −1), (−4, 0), (−3, 1), (5, 3)}

A) function
   domain: {5, −3, −4, 5}
   range: {−2, −1, 0, 1, 3}

B) function
   domain: {−2, −1, 0, 1, 3}
   range: {5, −3, −4, 5}

C) not a function

6) {(−2, 7), (−1, 4), (0, 3), (1, 4), (3, 12)}

A) function
   domain: {−2, −1, 0, 1, 3}
   range: {7, 4, 3, 12}

B) function
   domain: {7, 4, 3, 12}
   range: {−2, −1, 0, 1, 3}

C) not a function

7) $\{(2.11, 10.21), (2.111, -10.2), (\frac{9}{7}, 0), (1.29, -2)\}$

    A) function
       domain: $\{2.11, 2.111, \frac{9}{7}, 1.29\}$
       range: $\{10.21, -10.2, 0, -2\}$

    B) function
       domain: $\{10.21, -10.2, 0, -2\}$
       range: $\{2.11, 2.111, \frac{9}{7}, 1.29\}$

    C) not a function

Determine whether the equation is a function.

8) $y = x^2$
    A) function   B) not a function

9) $y = \frac{1}{x}$
    A) function   B) not a function

10) $y = |x|$
    A) function   B) not a function

11) $y^2 = 8 - x^2$
    A) function   B) not a function

12) $y = \pm\sqrt{1 - 2x}$
    A) function   B) not a function

13) $x = y^2$
    A) function   B) not a function

14) $y^2 + x = 4$
    A) function   B) not a function

15) $y = 2x^2 - 6x + 7$
    A) function   B) not a function

16) $y = \frac{5x - 1}{x + 2}$
    A) function   B) not a function

17) $x^2 + 2y^2 = 1$
    A) function   B) not a function

18) $x - 5y = 5$
    A) function   B) not a function

19) $7x + x^2 + 75 = y$
    A) function   B) not a function

2 Find the Value of a Function

Find the value for the function.

1) Find f(3) when $f(x) = x^2 - 5x + 2$.
   A) 22   B) 26   C) -8   D) -4

2) Find f(-1) when $f(x) = \dfrac{x^2 - 9}{x + 2}$.
   A) 10   B) 1   C) -8   D) $\dfrac{10}{3}$

3) Find f(-9) when $f(x) = |x| - 6$.
   A) -3   B) 3   C) 15   D) -15

4) Find f(2) when $f(x) = \sqrt{x^2 + 7x}$.
   A) $2\sqrt{14}$ B) $\sqrt{53}$ C) $\sqrt{11}$ D) $3\sqrt{2}$

5) Find f(-x) when $f(x) = -3x^2 - 5x + 4$.
   A) $-3x^2 + 5x - 4$ B) $-3x^2 + 5x + 4$ C) $3x^2 + 5x + 4$ D) $3x^2 + 5x - 4$

6) Find f(-x) when $f(x) = \dfrac{x}{x^2 + 6}$.
   A) $\dfrac{-x}{-x^2 + 6}$ B) $\dfrac{-x}{x^2 - 6}$ C) $\dfrac{-x}{x^2 + 6}$ D) $\dfrac{x}{-x^2 + 6}$

7) Find -f(x) when $f(x) = -2x^2 + 4x - 2$.
   A) $-2x^2 - 4x + 2$ B) $2x^2 - 4x + 2$ C) $2x^2 - 4x - 2$ D) $-2x^2 - 4x - 2$

8) Find -f(x) when $f(x) = |x| - 4$.
   A) $|-x| - 4$ B) $-|x| + 4$ C) $|-x| + 4$ D) $-|x| - 4$

9) Find f(x - 1) when $f(x) = 5x^2 + 2x + 1$.
   A) $5x^2 - 8x + 4$ B) $5x^2 + 7x + 8$ C) $-8x^2 + 5x + 4$ D) $5x^2 - 8x + 8$

10) Find f(x + 1) when $f(x) = \dfrac{x^2 - 6}{x - 2}$.
    A) $\dfrac{x^2 + 2x + 7}{x - 1}$ B) $\dfrac{x^2 + 2x - 5}{x + 3}$ C) $\dfrac{x^2 + 2x - 5}{x - 1}$ D) $\dfrac{x^2 - 5}{x - 1}$

11) Find f(2x) when $f(x) = 3x^2 - 4x + 4$.
    A) $6x^2 - 8x + 8$ B) $12x^2 - 8x + 4$ C) $6x^2 - 8x + 4$ D) $12x^2 - 8x + 8$

12) Find f(2x) when $f(x) = \sqrt{4x^2 + 3x}$.
    A) $\sqrt{8x^2 + 12x}$ B) $\sqrt{16x^2 + 6x}$ C) $\sqrt{8x^2 + 6x}$ D) $2\sqrt{4x^2 + 3x}$

13) Find $f(x + h)$ when $f(x) = 2x^2 + 3x - 4$.
  A) $2x^2 + 2xh + 2h^2 + 3x + 3h - 4$
  B) $2x^2 + 2h^2 + 3x + 3h - 4$
  C) $2x^2 + 2h^2 + 7x + 7h - 4$
  D) $2x^2 + 4xh + 2h^2 + 3x + 3h - 4$

14) Find $f(x + h)$ when $f(x) = \dfrac{-4x + 7}{3x - 8}$.
  A) $\dfrac{-4x + 7h}{3x - 8h}$
  B) $\dfrac{-4x + 3h}{3x - 5h}$
  C) $\dfrac{-4x - 4h + 7}{3x - 8}$
  D) $\dfrac{-4x - 4h + 7}{3x + 3h - 8}$

Solve the problem.

15) If $f(x) = 5x^3 + 2x^2 - x + C$ and $f(3) = 1$, what is the value of C?
  A) $C = -149$
  B) $C = 157$
  C) $C = 121$
  D) $C = -29$

16) If $f(x) = \dfrac{x - B}{x - A}$, $f(-9) = 0$, and $f(4)$ is undefined, what are the values of A and B?
  A) $A = -4, B = 9$
  B) $A = -9, B = 4$
  C) $A = 4, B = -9$
  D) $A = 9, B = -4$

17) If $f(x) = \dfrac{x - 4A}{-4x + 4}$ and $f(-4) = 8$, what is the value of A?
  A) $A = 41$
  B) $A = -21$
  C) $A = -41$
  D) $A = 21$

18) If a rock falls from a height of 80 meters on Earth, the height H (in meters) after x seconds is approximately
  $H(x) = 80 - 4.9x^2$.
  What is the height of the rock when $x = 1.3$ seconds? Round to the nearest hundredth, if necessary.
  A) 71.89 m
  B) 71.72 m
  C) 73.63 m
  D) 88.28 m

19) If a rock falls from a height of 100 meters on Earth, the height H (in meters) after x seconds is approximately
  $H(x) = 100 - 4.9x^2$.
  When does the rock strike the ground? Round to the nearest hundredth, if necessary.
  A) 4.16 sec
  B) 4.52 sec
  C) 2.04 sec
  D) 20.41 sec

3 Find the Domain of a Function

Find the domain of the function.

1) $f(x) = 5x + 4$
  A) $\{x \mid x > 0\}$
  B) $\{x \mid x \geq -4\}$
  C) $\{x \mid x \neq 0\}$
  D) all real numbers

2) $f(x) = x^2 + 2$
  A) $\{x \mid x \geq -2\}$
  B) $\{x \mid x \neq -2\}$
  C) all real numbers
  D) $\{x \mid x > -2\}$

3) $f(x) = \dfrac{x^2}{x^2 + 10}$
  A) all real numbers
  B) $\{x \mid x > -10\}$
  C) $\{x \mid x \neq 0\}$
  D) $\{x \mid x \neq -10\}$

4) $g(x) = \dfrac{2x}{x^2 - 16}$
  A) all real numbers
  B) $\{x \mid x \neq -4, 4\}$
  C) $\{x \mid x > 16\}$
  D) $\{x \mid x \neq 0\}$

5) $h(x) = \dfrac{x-4}{x^3 - 4x}$

    A) $\{x \mid x \neq -2, 0, 2\}$ B) all real numbers   C) $\{x \mid x \neq 0\}$ D) $\{x \mid x \neq 4\}$

6) $f(x) = \sqrt{24 - x}$

    A) $\{x \mid x \neq 2\sqrt{6}\}$ B) $\{x \mid x \leq 24\}$ C) $\{x \mid x \neq 24\}$ D) $\{x \mid x \leq 2\sqrt{6}\}$

7) $\dfrac{x}{\sqrt{x-1}}$

    A) $\{x \mid x \neq 1\}$ B) $\{x \mid x \geq 1\}$ C) $\{x \mid x > 1\}$ D) all real numbers

4   Form the Sum, Difference, Product, and Quotient of Two Functions

For the given functions f and g, find the requested function and state its domain.

1) $f(x) = 5 - 7x$; $g(x) = -9x + 7$
Find $f + g$.
  A) $(f + g)(x) = -9x + 5$; $\{x \mid x \neq \dfrac{5}{9}\}$ B) $(f + g)(x) = 2x + 12$; $\{x \mid x \neq 6\}$
  C) $(f + g)(x) = -4x$; all real numbers   D) $(f + g)(x) = -16x + 12$; all real numbers

2) $f(x) = 2x - 5$; $g(x) = 7x - 8$
Find $f - g$.
  A) $(f - g)(x) = -5x - 13$; $\{x \mid x \neq -\dfrac{13}{5}\}$ B) $(f - g)(x) = 5x - 3$; all real numbers
  C) $(f - g)(x) = -5x + 3$; all real numbers   D) $(f - g)(x) = 9x - 13$; $\{x \mid x \neq 1\}$

3) $f(x) = 8x + 6$; $g(x) = 2x - 5$
Find $f \cdot g$.
  A) $(f \cdot g)(x) = 10x^2 - 28x + 1$; all real numbers   B) $(f \cdot g)(x) = 16x^2 - 28x - 30$; all real numbers
  C) $(f \cdot g)(x) = 16x^2 - 30$; $\{x \mid x \neq -30\}$ D) $(f \cdot g)(x) = 16x^2 + 7x - 30$; $\{x \mid x \neq -30\}$

4) $f(x) = 4x + 1$; $g(x) = 2x - 3$
Find $\dfrac{f}{g}$.
  A) $\left(\dfrac{f}{g}\right)(x) = \dfrac{4x + 1}{2x - 3}$; $\{x \mid x \neq -\dfrac{1}{4}\}$ B) $\left(\dfrac{f}{g}\right)(x) = \dfrac{4x + 1}{2x - 3}$; $\{x \mid x \neq \dfrac{3}{2}\}$
  C) $\left(\dfrac{f}{g}\right)(x) = \dfrac{2x - 3}{4x + 1}$; $\{x \mid x \neq \dfrac{3}{2}\}$ D) $\left(\dfrac{f}{g}\right)(x) = \dfrac{2x - 3}{4x + 1}$; $\{x \mid x \neq -\dfrac{1}{4}\}$

5) $f(x) = 16 - x^2$; $g(x) = 4 - x$
Find $f + g$.
  A) $(f + g)(x) = -x^2 - x + 20$; $\{x \mid x \neq 4, x \neq -5\}$ B) $(f + g)(x) = x^3 - 4x^2 - 16x + 64$; all real numbers
  C) $(f + g)(x) = 4 + x$; $\{x \mid x \neq -4\}$ D) $(f + g)(x) = -x^2 + x + 12$; all real numbers

6) $f(x) = x + 1$; $g(x) = 5x^2$
   Find $f + g$.
   A) $(f + g)(x) = 5x^2 + x + 1$; $\{x \mid x \neq -1\}$
   B) $(f + g)(x) = 5x^2 + x + 1$; all real numbers
   C) $(f + g)(x) = 5x^2 - x - 1$; all real numbers
   D) $(f + g)(x) = -5x^2 + x + 1$; all real numbers

7) $f(x) = 4x^3 - 3$; $g(x) = 4x^2 - 1$
   Find $f \cdot g$.
   A) $(f \cdot g)(x) = 4x^3 + 4x^2 + 3$; all real numbers
   B) $(f \cdot g)(x) = 16x^5 - 4x^3 - 12x^2 + 3$; $\{x \mid x \neq 0\}$
   C) $(f \cdot g)(x) = 16x^5 - 4x^3 - 12x^2 + 3$; all real numbers
   D) $(f \cdot g)(x) = 16x^6 - 4x^3 - 12x^2 + 3$; all real numbers

8) $f(x) = \sqrt{x}$; $g(x) = 2x - 1$
   Find $\dfrac{f}{g}$.
   A) $\left(\dfrac{f}{g}\right)(x) = \dfrac{\sqrt{x}}{2x - 1}$; $\{x \mid x \neq \dfrac{1}{2}\}$
   B) $\left(\dfrac{f}{g}\right)(x) = \dfrac{2x - 1}{\sqrt{x}}$; $\{x \mid x \geq 0\}$
   C) $\left(\dfrac{f}{g}\right)(x) = \dfrac{\sqrt{x}}{2x - 1}$; $\{x \mid x \neq 0\}$
   D) $\left(\dfrac{f}{g}\right)(x) = \dfrac{\sqrt{x}}{2x - 1}$; $\{x \mid x \geq 0, x \neq \dfrac{1}{2}\}$

9) $f(x) = \sqrt{6 - x}$; $g(x) = \sqrt{x - 2}$
   Find $f \cdot g$.
   A) $(f \cdot g)(x) = \sqrt{-x^2 - 12}$; $\{x \mid x \neq 12\}$
   B) $(f \cdot g)(x) = \sqrt{(6 - x)(x - 2)}$; $\{x \mid x \neq 2, x \neq 6\}$
   C) $(f \cdot g)(x) = \sqrt{(6 - x)(x - 2)}$; $\{x \mid 2 \leq x \leq 6\}$
   D) $(f \cdot g)(x) = \sqrt{(6 - x)(x - 2)}$; $\{x \mid x \geq 0\}$

10) $f(x) = \dfrac{4x - 5}{5x - 8}$; $g(x) = \dfrac{6x}{5x - 8}$
    Find $f - g$.
    A) $(f - g)(x) = \dfrac{-2x - 5}{5x - 8}$; $\{x \mid x \neq 0\}$
    B) $(f - g)(x) = \dfrac{-2x - 5}{5x - 8}$; $\{x \mid x \neq \dfrac{8}{5}\}$
    C) $(f - g)(x) = \dfrac{-2x - 5}{5x - 8}$; $\{x \mid x \neq \dfrac{8}{5}, x \neq -\dfrac{5}{2}\}$
    D) $(f - g)(x) = \dfrac{10x + 5}{5x - 8}$; $\{x \mid x \neq \dfrac{8}{5}\}$

11) $f(x) = \sqrt{x + 9}$; $g(x) = \dfrac{3}{x}$
    Find $f \cdot g$.
    A) $(f \cdot g)(x) = \sqrt{\dfrac{12}{x}}$; $\{x \mid x \neq 0\}$
    B) $(f \cdot g)(x) = \sqrt{\dfrac{3x + 27}{x}}$; $\{x \mid x \geq -9, x \neq 0\}$
    C) $(f \cdot g)(x) = \dfrac{\sqrt{3x + 27}}{x}$; $\{x \mid x \geq -9, x \neq 0\}$
    D) $(f \cdot g)(x) = \dfrac{3\sqrt{x + 9}}{x}$; $\{x \mid x \geq -9, x \neq 0\}$

Solve the problem.

12) Given $f(x) = \dfrac{1}{x}$ and $\left(\dfrac{f}{g}\right)(x) = \dfrac{x + 7}{x^2 - 2x}$, find the function g.
    A) $g(x) = \dfrac{x + 2}{x - 7}$
    B) $g(x) = \dfrac{x - 7}{x + 2}$
    C) $g(x) = \dfrac{x + 7}{x - 2}$
    D) $g(x) = \dfrac{x - 2}{x + 7}$

13) Express the gross salary G of a person who earns $32 per hour as a function of the number x of hours worked.

   A) $G(x) = 32x^2$ B) $G(x) = \dfrac{32}{x}$ C) $G(x) = 32x$ D) $G(x) = 32 + x$

14) Jacey, a commissioned salesperson, earns $380 base pay plus $33 per item sold. Express Jacey's gross salary G as a function of the number x of items sold.

   A) $G(x) = 33(x + 380)$ B) $G(x) = 33x + 380$ C) $G(x) = 380x + 33$ D) $G(x) = 380(x + 33)$

Find and simplify the difference quotient of f, $\dfrac{f(x + h) - f(x)}{h}$, $h \neq 0$, for the function.

15) $f(x) = 5x - 7$

   A) $5 + \dfrac{10(x - 7)}{h}$ B) 0   C) $5 + \dfrac{-14}{h}$ D) 5

16) $f(x) = x^2 + 3x - 5$

   A) 1   B) $\dfrac{2x^2 + 2x + 2xh + h^2 + h - 10}{h}$

   C) $2x + h + 3$ D) $2x + h - 5$

17) $f(x) = \dfrac{1}{5x}$

   A) $\dfrac{1}{5x}$ B) $\dfrac{-1}{5x(x + h)}$ C) $\dfrac{-1}{x(x + h)}$ D) 0

Solve the problem.

18) Suppose that P(x) represents the percentage of income spent on food in year x and I(x) represents income in year x. Determine a function F that represents total food expenditures in year x.

   A) $F(x) = (P \cdot I)(x)$ B) $F(x) = (I - P)(x)$ C) $F(x) = \left(\dfrac{I}{P}\right)(x)$ D) $F(x) = (P + I)(x)$

2.2 The Graph of a Function

1 Identify the Graph of a Function

Determine whether the graph is that of a function. If it is, use the graph to find its domain and range, the intercepts, if any, and any symmetry with respect to the x-axis, the y-axis, or the origin.

1)

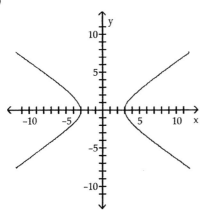

A) function
   domain: $\{x \mid -3 \leq x \leq 3\}$
   range: all real numbers
   intercepts: (-3, 0), (3, 0)
   symmetry: x-axis, y-axis

B) function
   domain: $\{x \mid x \leq -3 \text{ or } x \geq 3\}$
   range: all real numbers
   intercepts: (-3, 0), (3, 0)
   symmetry: x-axis, y-axis, origin

C) function
   domain: all real numbers
   range: $\{y \mid y \leq -3 \text{ or } y \geq 3\}$
   intercepts: (-3, 0), (3, 0)
   symmetry: y-axis

D) not a function

2)

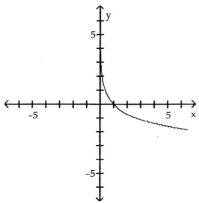

A) function
   domain: all real numbers
   range: $\{y \mid y > 0\}$
   intercept: (1, 0)
   symmetry: none

B) function
   domain: $\{x \mid x > 0\}$
   range: all real numbers
   intercept: (0, 1)
   symmetry: origin

C) function
   domain: $\{x \mid x > 0\}$
   range: all real numbers
   intercept: (1, 0)
   symmetry: none

D) not a function

3)

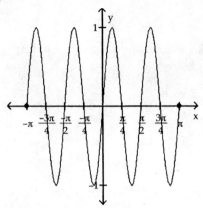

A) function
   domain: $\{x \mid -\pi \leq x \leq \pi\}$
   range: $\{y \mid -1 \leq y \leq 1\}$
   intercepts: $(-\pi, 0), (-\frac{3\pi}{4}, 0), (-\frac{\pi}{2}, 0), (-\frac{\pi}{4}, 0), (0, 0), (\frac{\pi}{4}, 0), (\frac{\pi}{2}, 0), (\frac{3\pi}{4}, 0), (\frac{3\pi}{4}, 0), (\pi, 0)$
   symmetry: origin

B) function
   domain: all real numbers
   range: $\{y \mid -1 \leq y \leq 1\}$
   intercepts: $(-\pi, 0), (-\frac{3\pi}{4}, 0), (-\frac{\pi}{2}, 0), (-\frac{\pi}{4}, 0), (0, 0), (\frac{\pi}{4}, 0), (\frac{\pi}{2}, 0), (\frac{3\pi}{4}, 0), (\frac{3\pi}{4}, 0), (\pi, 0)$
   symmetry: origin

C) function
   domain: $\{x \mid -1 \leq x \leq 1\}$
   range: $\{y \mid -\pi \leq y \leq \pi\}$
   intercepts: $(-\pi, 0), (-\frac{3\pi}{4}, 0), (-\frac{\pi}{2}, 0), (-\frac{\pi}{4}, 0), (0, 0), (\frac{\pi}{4}, 0), (\frac{\pi}{2}, 0), (\frac{3\pi}{4}, 0), (\frac{3\pi}{4}, 0), (\pi, 0)$
   symmetry: none

D) not a function

4)

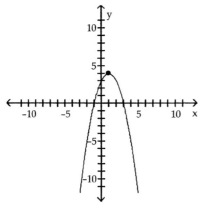

A) function
   domain: all real numbers
   range: {y | y ≤ 4}
   intercepts: (0, -1), (3, 0), (0, 3)
   symmetry: none

B) function
   domain: all real numbers
   range: {y | y ≤ 4}
   intercepts: (-1, 0), (0, 3), (3, 0)
   symmetry: none

C) function
   domain: {x | x ≤ 4}
   range: all real numbers
   intercepts: (-1, 0), (0, 3), (3, 0)
   symmetry: y-axis

D) not a function

5)

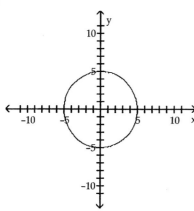

A) function
   domain: {x | -5 ≤ x ≤ 5}
   range: {y | -5 ≤ y ≤ 5}
   intercepts: (-5, 0), (0, -5), (0, 5), (5, 0)
   symmetry: x-axis, y-axis, origin

B) function
   domain: {x | -5 ≤ x ≤ 5}
   range: {y | -5 ≤ y ≤ 5}
   intercepts: (-5, 0), (0, -5), (0, 0), (0, 5), (5, 0)
   symmetry: origin

C) function
   domain: {x | -5 ≤ x ≤ 5}
   range: {y | -5 ≤ y ≤ 5}
   intercepts: (-5, 0), (0, -5), (0, 5), (5, 0)
   symmetry: x-axis, y-axis

D) not a function

6)

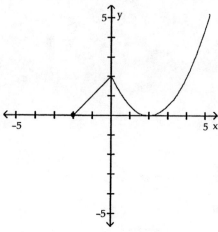

A) function
   domain: all real numbers
   range: all real numbers
   intercepts: (-2, 0), (0, 2), (2, 0)
   symmetry: none

B) function
   domain: {x | x ≥ -2}
   range: {y | y ≥ 0}
   intercepts: (-2, 0), (0, 2), (2, 0)
   symmetry: none

C) function
   domain: {x | x ≥ 0}
   range: {y | y ≥ -2}
   intercepts: (-2, 0), (0, 2), (2, 0)
   symmetry: y-axis

D) not a function

7)

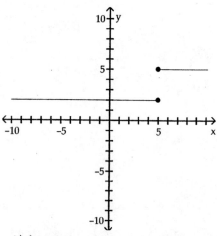

A) function
   domain: all real numbers
   range: {y | y = 5 or y = 2}
   intercept: (0, 2)
   symmetry: none

B) function
   domain: all real numbers
   range: all real numbers
   intercept: (0, 2)
   symmetry: none

C) function
   domain: {x | x = 5 or x = 2}
   range: all real numbers
   intercept: (2, 0)
   symmetry: x-axis

D) not a function

2 Obtain Information from or about the Graph of a Function

The graph of a function f is given. Use the graph to answer the question.

1) Use the graph of f given below to find f(20).

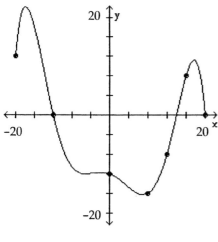

A) 20   B) 40   C) 0   D) 24

2) Is f(8) positive or negative?

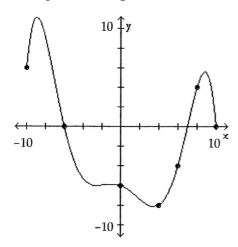

A) positive   B) negative

3) Is f(40) positive or negative?

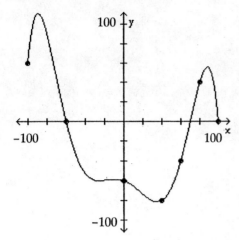

A) positive   B) negative

4) For what numbers x is f(x) = 0?

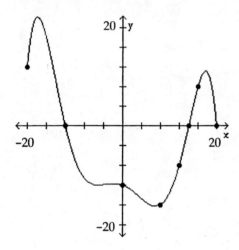

A) –12, 14, 20   B) (–12, 14)   C) (–20, –12), (14, 20)   D) –12

5) For what numbers x is f(x) > 0?

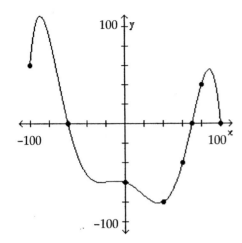

A) (-∞, -60)   B) (-60, ∞)   C) [-100, -60), (70, 100)   D) (-60, 70)

6) For what numbers x is f(x) < 0?

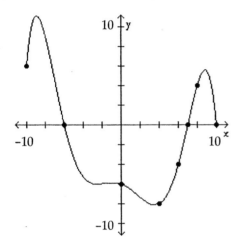

A) (-∞, -6)   B) [-10, -6), (7, 10)   C) (-6, 7)   D) (-6, ∞)

7) What is the domain of f?

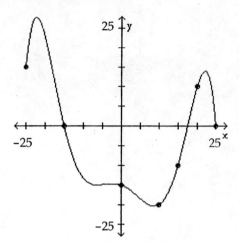

A) {x | x ≥ 0} B) {x | −25 ≤ x ≤ 25} C) {x | −20 ≤ x ≤ 27.5} D) all real numbers

8) What are the x-intercepts?

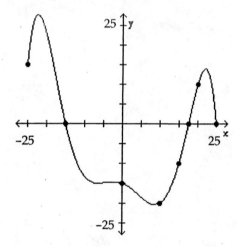

A) −25, −15, 17.5, 25 B) −15, 17.5, 25 C) −15, 17.5 D) −15

9) What is the y-intercept?

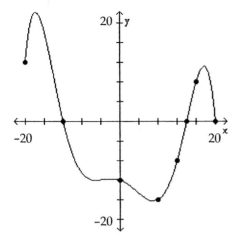

A) −16 B) 20 C) 14 D) −12

10) How often does the line y = −50 intersect the graph?

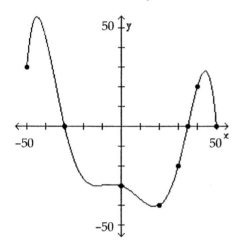

A) once B) twice C) three times D) does not intersect

11) How often does the line y = 2 intersect the graph?

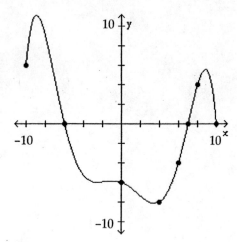

A) once   B) twice   C) three times   D) does not intersect

12) For which of the following values of x does f(x) = −20?

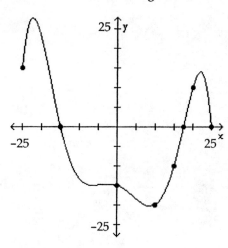

A) 0   B) 10   C) 15   D) −20

Answer the question about the given function.

13) Given the function $f(x) = -2x^2 - 4x - 8$, is the point (−1, −6) on the graph of f?
    A) Yes   B) No

14) Given the function $f(x) = -7x^2 - 14x + 4$, is the point (−2, −10) on the graph of f?
    A) Yes   B) No

15) Given the function $f(x) = 7x^2 - 14x + 7$, if x = 1, what is f(x)? What point is on the graph of f?
    A) 0; (1, 0)   B) 0; (0, 1)   C) 28; (28, 1)   D) 28; (1, 28)

16) Given the function $f(x) = -6x^2 - 12x - 9$, what is the domain of f?
    A) all real numbers   B) {x | x ≥ −1}   C) {x | x ≤ −1}   D) {x | x ≥ 1}

17) Given the function $f(x) = x^2 + 5x - 14$, list the x-intercepts, if any, of the graph of f.
    A) (7, 0), (−2, 0)   B) (7, 0), (2, 0)   C) (−7, 0), (1, 0)   D) (−7, 0), (2, 0)

18) Given the function $f(x) = -7x^2 - 14x - 7$, list the y-intercept, if there is one, of the graph of f.
   A) −7 B) −21 C) 0 D) −28

19) Given the function $f(x) = \dfrac{x^2 - 6}{x + 3}$, is the point $(1, -\dfrac{5}{4})$ on the graph of f?
   A) Yes B) No

20) Given the function $f(x) = \dfrac{x^2 - 6}{x + 3}$, is the point $(-1, \dfrac{7}{2})$ on the graph of f?
   A) Yes B) No

21) Given the function $f(x) = \dfrac{x^2 - 2}{x - 3}$, if $x = -1$, what is $f(x)$? What point is on the graph of f?
   A) $-\dfrac{3}{4}; (-1, -\dfrac{3}{4})$
   B) $\dfrac{1}{4}; (-1, \dfrac{1}{4})$
   C) $\dfrac{1}{4}; (\dfrac{1}{4}, -1)$
   D) $-\dfrac{3}{4}; (-\dfrac{3}{4}, -1)$

22) Given the function $f(x) = \dfrac{x^2 + 6}{x - 7}$, what is the domain of f?
   A) $\{x \mid x \ne \dfrac{6}{7}\}$
   B) $\{x \mid x \ne 7\}$
   C) $\{x \mid x \ne 6\}$
   D) $\{x \mid x \ne -7\}$

23) Given the function $f(x) = \dfrac{x^2 + 8}{x - 2}$, list the x-intercepts, if any, of the graph of f.
   A) (8, 0), (−8, 0) B) $(-2\sqrt{2}, 0)$ C) (2, 0) D) none

24) Given the function $f(x) = \dfrac{x^2 + 4}{x - 3}$, list the y-intercept, if there is one, of the graph of f.
   A) $(-\dfrac{4}{3}, 0)$
   B) $(0, -\dfrac{4}{3})$
   C) (0, −4)
   D) (0, 3)

Solve the problem.

25) If an object weighs m pounds at sea level, then its weight W (in pounds) at a height of h miles above sea level is given approximately by $W(h) = m\left(\dfrac{4000}{4000 + h}\right)^2$. How much will a man who weighs 165 pounds at sea level weigh on the top of a mountain which is 14,494 feet above sea level? Round to the nearest hundredth of a pound, if necessary.
   A) 7.72 pounds B) 165 pounds C) 165.23 pounds D) 164.77 pounds

Match the function with the graph that best describes the situation.

26) The amount of rainfall as a function of time, if the rain fell more and more softly.

A)

B)

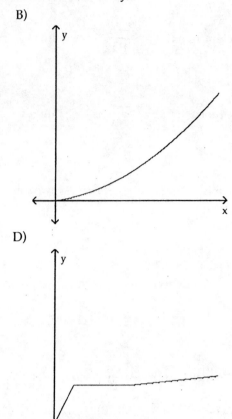

C)

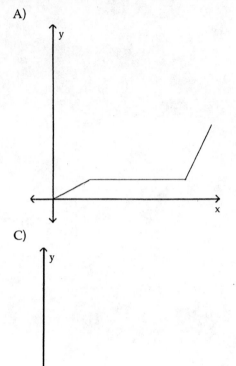

D)

27) The height of an animal as a function of time.

A)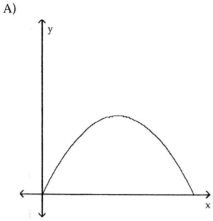

B)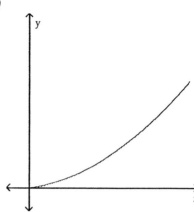

C)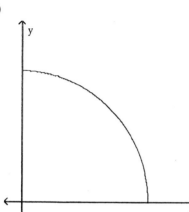

D)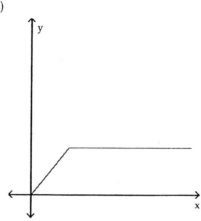

Solve the problem.

28) Michael decides to walk to the mall to do some errands. He leaves home, walks 3 blocks in 10 minutes at a constant speed, and realizes that he forgot his wallet at home. So Michael runs back in 6 minutes. At home, it takes him 4 minutes to find his wallet and close the door. Michael walks 2 blocks in 8 minutes and then decides to jog to the mall. It takes him 10 minutes to get to the mall which is 5 blocks away. Draw a graph of Michael's distance from home (in blocks) as a function of time.

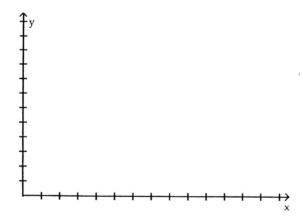

29) A steel can in the shape of a right circular cylinder must be designed to hold 650 cubic centimeters of juice (see figure). It can be shown that the total surface area of the can (including the ends) is given by $S(r) = 2\pi r^2 + \dfrac{1300}{r}$, where r is the radius of the can in centimeters. Using the TABLE feature of a graphing utility, find the radius that minimizes the surface area (and thus the cost) of the can. Round to the nearest tenth of a centimeter.

A) 4.7 cm   B) 3.9 cm   C) 5.9 cm   D) 0 cm

30) The concentration C (arbitrary units) of a certain drug in a patient's bloodstream can be modeled using $C(t) = \dfrac{t}{(0.412t + 2.429)^2}$, where t is the number of hours since a 500 milligram oral dose was administered.
Using the TABLE feature of a graphing utility, find the time at which the concentration of the drug is greatest. Round to the nearest tenth of an hour.

A) 6.7 hours   B) 7.4 hours   C) 8.2 hours   D) 5.9 hours

2.3 Properties of Functions

1 Determine Even and Odd Functions from a Graph

The graph of a function is given. Decide whether it is even, odd, or neither.

1)

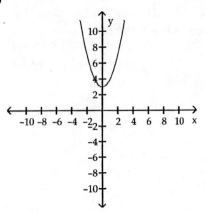

A) even   B) odd   C) neither

2)

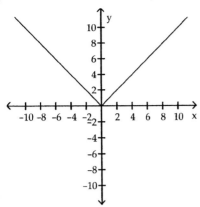

A) even   B) odd   C) neither

3)

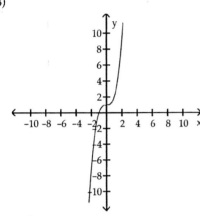

A) even   B) odd   C) neither

4)

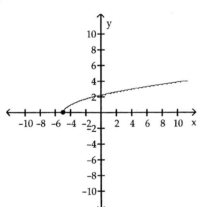

A) even   B) odd   C) neither

5)

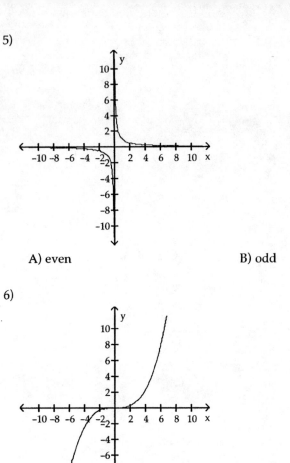

A) even   B) odd   C) neither

6)

A) even   B) odd   C) neither

7)

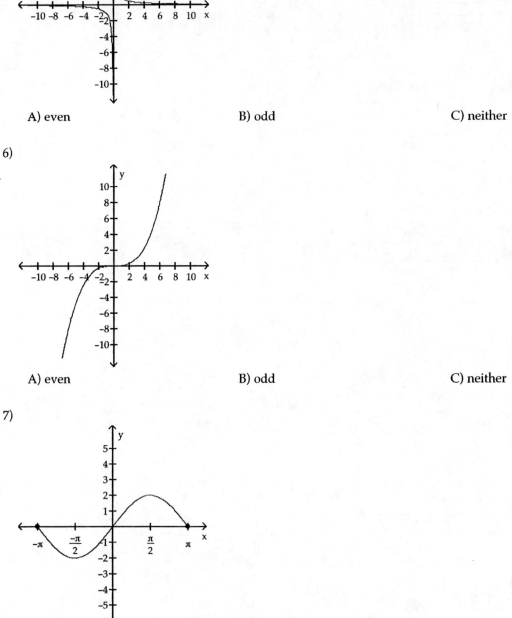

A) even   B) odd   C) neither

8)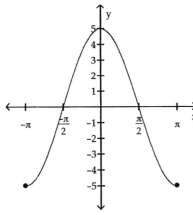

A) even B) odd C) neither

2 Identify Even and Odd Functions from the Equation

Determine algebraically whether the function is even, odd, or neither.

1) $f(x) = 2x^3$

A) even B) odd C) neither

2) $f(x) = -4x^4 - x^2$

A) even B) odd C) neither

3) $f(x) = -8x^2 - 3$

A) even B) odd C) neither

4) $f(x) = 2x^3 + 5$

A) even B) odd C) neither

5) $f(x) = \sqrt[3]{x}$

A) even B) odd C) neither

6) $f(x) = \sqrt{x}$

A) even B) odd C) neither

7) $\sqrt[3]{3x^2 + 8}$

A) even B) odd C) neither

8) $f(x) = \dfrac{1}{x^2}$

A) even B) odd C) neither

9) $f(x) = \dfrac{x}{x^2 - 3}$

A) even B) odd C) neither

10) $f(x) = \dfrac{-x^3}{4x^2 + 9}$

A) even B) odd C) neither

11) $f(x) = \dfrac{3x}{|x|}$

A) even B) odd C) neither

3 Use a Graph to Determine Where a Function Is Increasing, Decreasing, or Constant

The graph of a function is given. Determine whether the function is increasing, decreasing, or constant on the given interval.

1) $(-4, -2)$

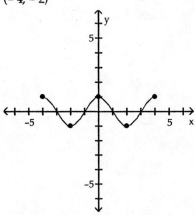

A) constant B) increasing C) decreasing

2) $(-1, 0)$

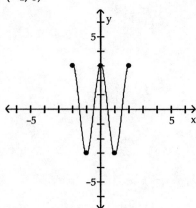

A) constant B) increasing C) decreasing

3) (0, 1)

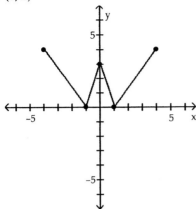

A) constant   B) increasing   C) decreasing

4) (1, 2)

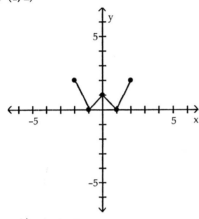

A) constant   B) decreasing   C) increasing

5) (0, 1)

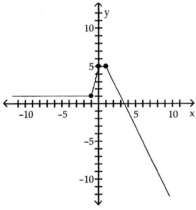

A) decreasing   B) constant   C) increasing

6) (−1, 0)

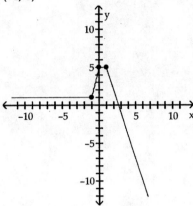

A) decreasing   B) constant   C) increasing

7) (3, ∞)

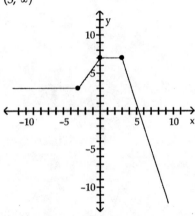

A) increasing   B) decreasing   C) constant

8) (0, ∞)

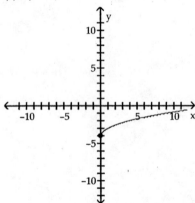

A) constant   B) increasing   C) decreasing

9) (−1, 1)

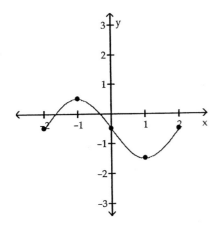

A) decreasing   B) constant   C) increasing

10) (−4, 0.5)

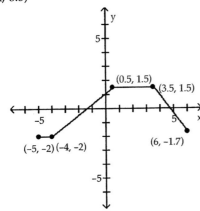

A) constant   B) decreasing   C) increasing

11) (2, 5)

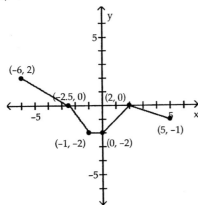

A) decreasing   B) constant   C) increasing

12) (-8, -2.5)

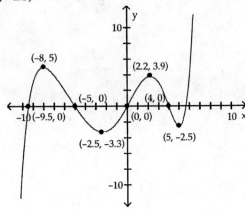

A) increasing   B) decreasing   C) constant

Use the graph to find the intervals on which it is increasing, decreasing, or constant.

13)

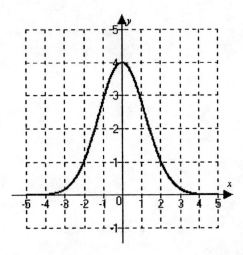

A) Decreasing on $(-\infty, 0)$; increasing on $(0, \infty)$
B) Decreasing on $(-\infty, \infty)$
C) Increasing on $(-\infty, \infty)$
D) Increasing on $(-\infty, 0)$; decreasing on $(0, \infty)$

14)

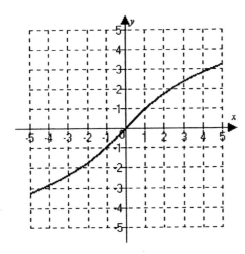

A) Decreasing on (-∞, 0); increasing on (0, ∞)
B) Decreasing on (-∞, ∞)
C) Increasing on (-∞, ∞)
D) Increasing on (-∞, 0); decreasing on (0, ∞)

15)

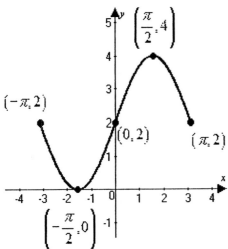

A) Increasing on $\left[-\pi, -\frac{\pi}{2}\right]$ and $\left[\frac{\pi}{2}, \pi\right]$; decreasing on $\left(-\frac{\pi}{2}, \frac{\pi}{2}\right)$

B) Increasing on (-∞, ∞)

C) Decreasing on $\left[-\pi, -\frac{\pi}{2}\right]$ and $\left[\frac{\pi}{2}, \pi\right]$; increasing on $\left(-\frac{\pi}{2}, \frac{\pi}{2}\right)$

D) Decreasing on $(-\pi, 0)$; increasing on $(0, \pi)$

16)

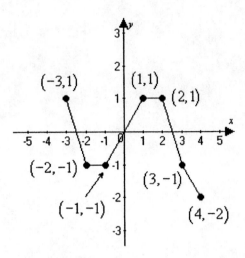

A) Decreasing on (-3, -2) and (2, 4); increasing on (-1, 1); constant on (-2, -1) and (1, 2)
B) Increasing on (-3, -2) and (2, 4); decreasing on (-1, 1); constant on (-2, -1) and (1, 2)
C) Decreasing on (-3, -1) and (1, 4); increasing on (-2, 1)
D) Decreasing on (-3, -2) and (2, 4); increasing on (-1, 1)

4 Use a Graph to Locate Local Maxima and Minima

The graph of a function f is given. Use the graph to answer the question.

1) Find the numbers, if any, at which f has a local maximum. What are the local maxima?

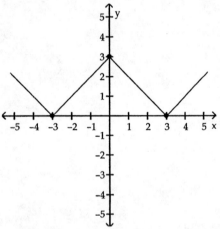

A) f has a local maximum at x = 0; the local maximum is 3
B) f has a local maximum at x = 3; the local maximum is 3
C) f has a local maximum at x = -3 and 3; the local maximum is 0
D) f has no local maximum

2) Find the numbers, if any, at which f has a local minimum. What are the local minima?

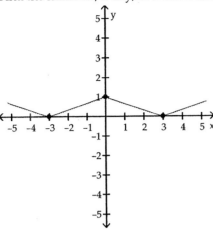

A) f has a local minimum at x = 0; the local minimum is 1
B) f has a local minimum at x = -3; the local minimum is 0
C) f has a local minimum at x = -3 and 3; the local minimum is 0
D) f has no local minimum

3) Find the numbers, if any, at which f has a local maximum. What are the local maxima?

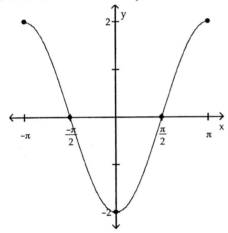

A) f has no local maximum
B) f has a local maximum at -π; the local maximum is 2
C) f has a local maximum at x = 0; the local maximum is -2
D) f has a local maximum at x = -π and π; the local maximum is 2

4) Find the numbers, if any, at which f has a local minimum. What are the local minima?

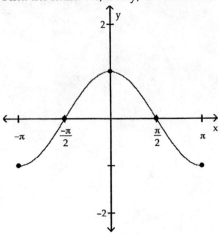

A) f has a local minimum at x = 0; the local minimum is 1
B) f has a local minimum at x = -π; the local minimum is -1
C) f has a local minimum at x = -π and π; the local minimum is -1
D) f has no local minimum

5)

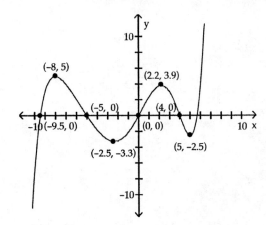

Find the numbers, if any, at which f has a local maximum. What are the local maxima?
A) f has a local minimum at x = -8 and 2.2; the local minimum at -8 is 5; the local minimum at 2.2 is 3.9
B) f has a local minimum at x = 5 and 3.9; the local minimum at 5 is -8; the local minimum at 3.9 is 2.2
C) f has a local maximum at x = -8 and 2.2; the local maximum at -8 is 5; the local maximum at 2.2 is 3.9
D) f has a local maximum at x = 5 and 3.9; the local maximum at 5 is -8; the local maximum at 3.9 is 2.2

Solve the problem.

6) The height s of a ball (in feet) thrown with an initial velocity of 70 feet per second from an initial height of 3 feet is given as a function of time t (in seconds) by $s(t) = -16t^2 + 70t + 3$. What is the maximum height? Round to the nearest hundredth, if necessary.

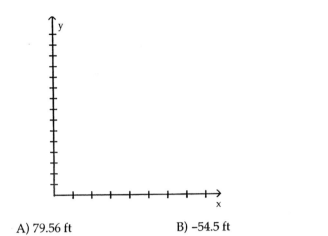

A) 79.56 ft   B) -54.5 ft   C) 90.5 ft   D) 76.75 ft

5 Use a Graphing Utility to Approximate Local Maxima and Minima and to Determine Where a Function Is Increasing or Decreasing

Use a graphing utility to graph the function over the indicated interval and approximate any local maxima and local minima. Determine where the function is increasing and where it is decreasing. If necessary, round answers to two decimal places.

1) $f(x) = x^3 - 3x^2 + 1$, $(-1, 3)$

    A) local maximum at (0, 1)
       local minimum at (2, -3)
       increasing on (0, 2)
       decreasing on (-1, 0) and (2, 3)

    B) local maximum at (0, 1)
       local minimum at (2, -3)
       increasing on (-1, 0) and (2, 3)
       decreasing on (0, 2)

    C) local maximum at (2, -3)
       local minimum at (0, 1)
       increasing on (-1, 0) and (2, 3)
       decreasing on (0, 2)

    D) local maximum at (2, -3)
       local minimum at (0, 1)
       increasing on (-1, 0)
       decreasing on (0, 2)

2) $f(x) = x^3 - 4x^2 + 6$; $(-1, 4)$

3) $f(x) = x^5 - x^2$; $(-2, 2)$

4) $f(x) = -0.3x^3 + 0.2x^2 + 4x - 5$; $(-4, 5)$

5) $f(x) = 0.15x^4 + 0.3x^3 - 0.8x^2 + 5$; $(-4, 2)$

Use a graphing utility to graph the function over the indicated interval and approximate any local maxima and local minima. If necessary, round answers to two decimal places.

6) $f(x) = x^2 + 2x - 3$; $(-5, 5)$

    A) local minimum at (-1, -4)
    C) local maximum at (-1, 4)

    B) local maximum at (1, -4)
    D) local minimum at (1, 4)

7) $f(x) = 2 + 8x - x^2$; $(-5, 5)$
   A) local maximum at $(4, 18)$
   B) local maximum at $(-4, 50)$
   C) local minimum at $(4, 50)$
   D) local minimum at $(-4, 18)$

8) $f(x) = x^3 - 3x^2 + 1$; $(-5, 5)$
   A) local minimum at $(2, -3)$
   B) local maximum at $(0, 1)$
      local minimum at $(2, -3)$
   C) local minimum at $(0, 1)$
      local maximum at $(2, -3)$
   D) none

9) $f(x) = x^3 - 12x + 2$; $(-5, 5)$
   A) local minimum at $(0, 0)$
   B) local maximum at $(-2, 18)$
      local minimum at $(0, 0)$
      local minimum at $(2, -14)$
   C) local maximum at $(-2, 18)$
      local minimum at $(2, -14)$
   D) none

10) $f(x) = x^4 - 5x^3 + 3x^2 + 9x - 3$; $(-5, 5)$
    A) local minimum at $(-3, -3)$
       local maximum at $(-1.32, 5.64)$
       local minimum at $(0.57, -6.12)$
    B) local minimum at $(-0.57, -6.12)$
       local maximum at $(1.32, 5.64)$
       local minimum at $(3, -3)$
    C) local minimum at $(-1, -6)$
       local maximum at $(1, 6)$
       local minimum at $(3, -3)$
    D) local minimum at $(-0.61, -5.64)$
       local maximum at $(1.41, 6.12)$
       local minimum at $(3, -3)$

6 Find the Average Rate of Change of a Function

For the function, find the average rate of change of f from 1 to x:
$$\frac{f(x) - f(1)}{x - 1}, x \neq 1$$

1) $f(x) = 9x$
   A) 8
   B) $\frac{9}{x-1}$
   C) 9
   D) 0

2) $f(x) = x^3 - x$
   A) $x^2 - x$
   B) 1
   C) $\frac{x^3 - x - 1}{x - 1}$
   D) $x^2 + x$

3) $f(x) = \frac{3}{x+2}$
   A) $-\frac{1}{x+2}$
   B) $\frac{1}{x+2}$
   C) $\frac{3}{x(x+2)}$
   D) $\frac{3}{(x-1)(x+2)}$

4) $f(x) = \sqrt{x + 80}$
   A) $\frac{\sqrt{x+80} - 9}{x - 1}$
   B) $\frac{\sqrt{x+80} + 9}{x - 1}$
   C) $\frac{\sqrt{x+80} - 9}{x + 1}$
   D) $\frac{\sqrt{x+80} + 9}{x + 1}$

Precalculus Enhanced with Graphing Utilities

Find the average rate of change for the function between the given values.

5) $f(x) = -2x + 6$; from 1 to 3
   A) -2
   B) -6
   C) 6
   D) 2

6) $f(x) = x^2 + 3x$; from 2 to 5
   A) $\dfrac{40}{3}$
   B) 8
   C) 10
   D) 6

7) $f(x) = 2x^3 + 3x^2 - 7$; from -6 to 8
   A) $\dfrac{1209}{14}$
   B) 110
   C) $\dfrac{1209}{8}$
   D) $\dfrac{385}{2}$

8) $f(x) = \sqrt{2x}$; from 2 to 8
   A) $\dfrac{1}{3}$
   B) $-\dfrac{3}{10}$
   C) 7
   D) 2

9) $f(x) = \dfrac{3}{x-2}$; from 4 to 7
   A) $\dfrac{1}{3}$
   B) $-\dfrac{3}{10}$
   C) 7
   D) 2

10) $f(x) = 4x^2$; from 0 to $\dfrac{7}{4}$
    A) 7
    B) $-\dfrac{3}{10}$
    C) $\dfrac{1}{3}$
    D) 2

11) $f(x) = -3x^2 - x$; from 5 to 6
    A) -34
    B) $-\dfrac{1}{6}$
    C) -2
    D) $\dfrac{1}{2}$

12) $f(x) = x^3 + x^2 - 8x - 7$; from 0 to 2
    A) $\dfrac{1}{2}$
    B) $-\dfrac{1}{6}$
    C) -28
    D) -2

13) $f(x) = \sqrt{2x - 1}$; from 1 to 5
    A) $-\dfrac{1}{6}$
    B) $\dfrac{1}{2}$
    C) -2
    D) -28

14) $f(x) = \dfrac{3}{x+2}$; from 1 to 4
    A) $\dfrac{1}{2}$
    B) -2
    C) $-\dfrac{1}{6}$
    D) -28

Find an equation of the secant line containing $(1, f(1))$ and $(2, f(2))$.

15) $f(x) = x^2 - 2x$
    A) $y = x - 2$
    B) $y = -x + 2$
    C) $y = -x - 2$
    D) $y = x + 2$

16) $f(x) = \dfrac{8}{x+7}$

A) $y = \dfrac{1}{9}x + \dfrac{5}{4}$ B) $y = \dfrac{1}{9}x + \dfrac{8}{9}$ C) $y = -\dfrac{1}{9}x + \dfrac{10}{9}$ D) $y = \dfrac{8}{9}x + \dfrac{1}{9}$

17) $f(x) = \sqrt{x+48}$

A) $y = (5\sqrt{2} - 7)x + 5\sqrt{2} - 14$ B) $y = (-5\sqrt{2} + 7)x + 5\sqrt{2} - 14$
C) $y = (5\sqrt{2} - 7)x - 5\sqrt{2} + 14$ D) $y = (-5\sqrt{2} - 7)x - 5\sqrt{2} + 14$

2.4 Linear Functions and Models

1 Graph Linear Functions

Graph the linear function by hand.

1) $f(x) = 3x - 5$

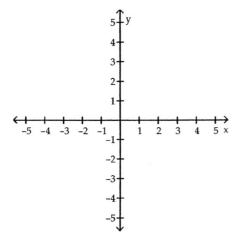

A)
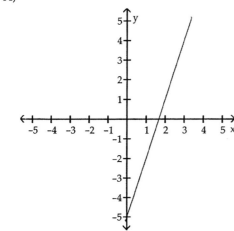

B)
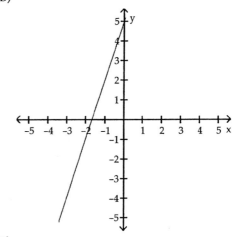

C)
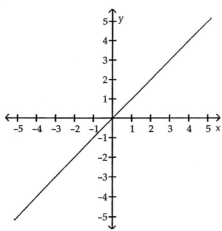

D)
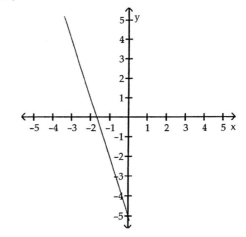

2) $g(x) = -\frac{5}{2}x + 1$

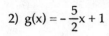

A)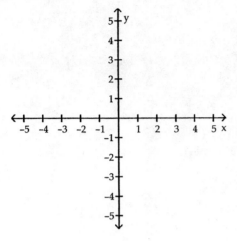

B)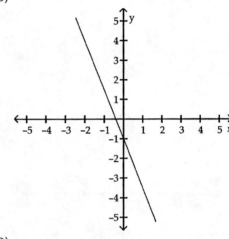

C)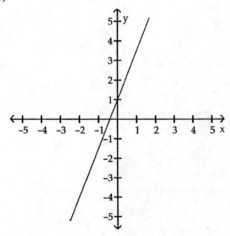

D)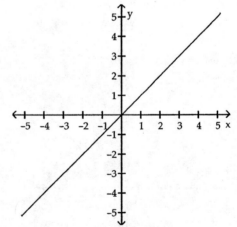

3) $G(x) = 9$

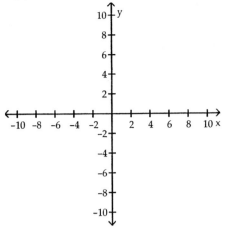

A)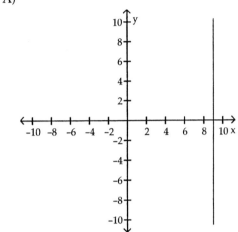

B)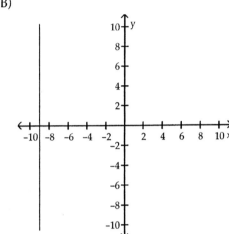

C)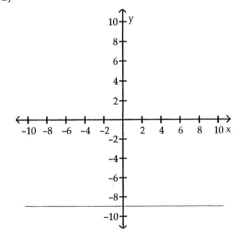

D)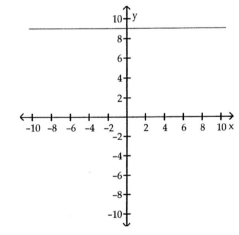

Solve the problem.

4) Suppose that a school has just purchased new computer equipment for $16,000.00. The school chooses to depreciate the equipment using the straight line method over 5 years. (a) Write a linear function that expresses the book value of the equipment as a function of its age. (b) Graph the linear function. (c) What is the value of the machine after 3 years?

A) $f(x) = -16{,}000x + 16{,}000$;
value after 3 years is -$32,000.00

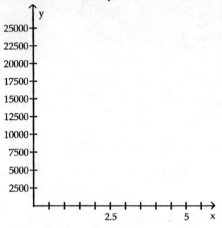

B) $f(x) = 16{,}000x + 5$;
value after 3 years is $6400.00

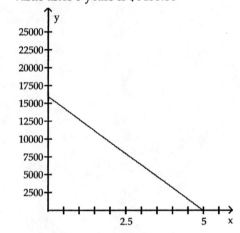

C) $f(x) = 3200x - 16{,}000$;
value after 3 years is $6400.00

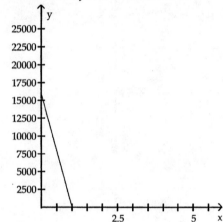

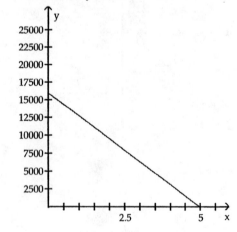

D) $f(x) = -3200x + 16{,}000$;
value after 3 years is $6400.00

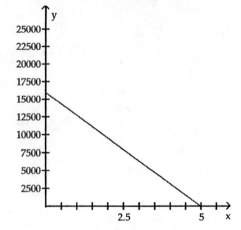

2 Work with Applications of Linear Functions

Solve the problem.

1) The cost C, in dollars, to produce graphing calculators is given by the function $C(x) = 55x + 4000$, where x is the number of calculators produced. What is the cost to produce 2300 calculators?

   A) $130,500   B) $122,500   C) $130,850   D) $126,500

2) The cost C, in dollars, to produce graphing calculators is given by the function $C(x) = 57x + 3000$, where x is the number of calculators produced. How many calculators can be produced if the cost is limited to $117,000?

   A) 1800 calculators   B) 2280 calculators   C) 2105 calculators   D) 2000 calculators

3 Draw and Interpret Scatter Diagrams

Plot a scatter diagram.

1) | x: | 9.6 | 15.9 | 16.3 | 19.4 | 1.1 | 2.7 |
   |----|-----|------|------|------|-----|-----|
   | y: | 4.2 | 2.4 | 0.9 | 1.1 | 3.3 | 2.4 |

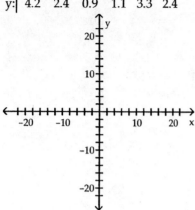

A)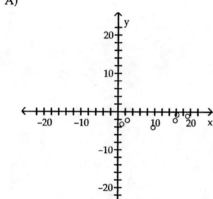

B)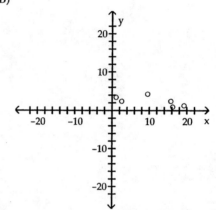

C)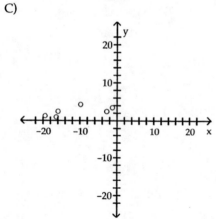

D)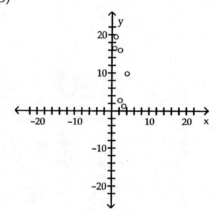

2)
x:	19	-13	16	-12	-2	11	2	15	8	-3
y:	57	30	42	-10	1	36	9	60	-3	5

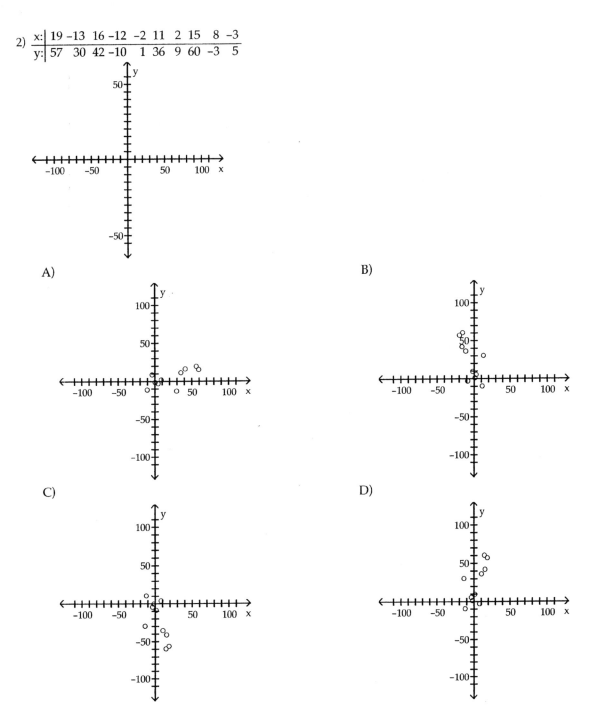

A)

B)

C)

D)

3)
x:	-3.30	0.41	1.99	-0.09	-0.21	-0.33	-15.05
y:	17.18	37.59	30.39	31.97	37.49	42.58	-16.44

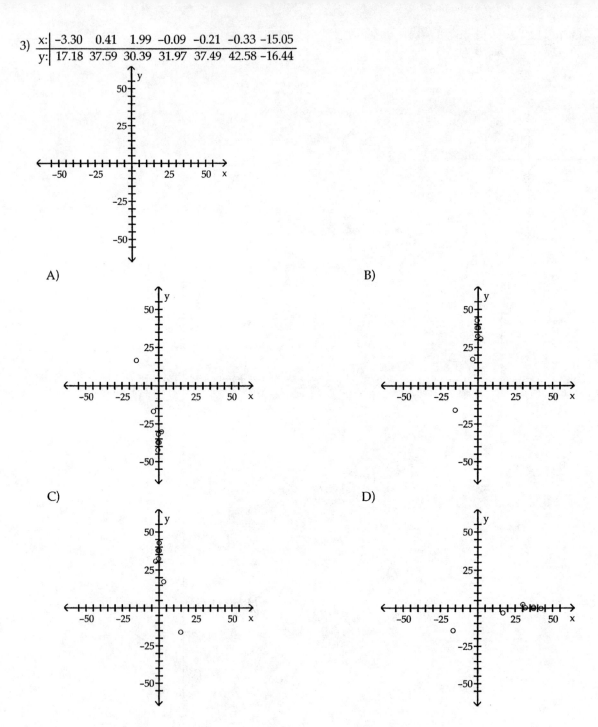

A)

B)

C)

D)

4) A study compared the number of newly planted trees in a series of 1-acre lots with the change in average temperature from a lot with no trees, during August. In order to compensate for differences between the lots, three lots with 5 trees, three with 10 trees, and three with 15 trees are randomly chosen. The results are given below. Plot a scatter diagram of the data.

x: Number of trees	5	10	15
y: Change in average temperature (°F)	-3	-7	-9
	-5	-5	-12
	-2	-10	-11

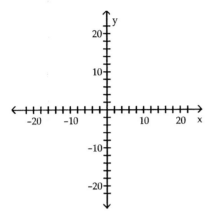

A)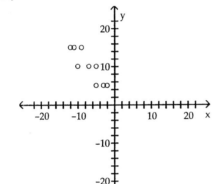

B)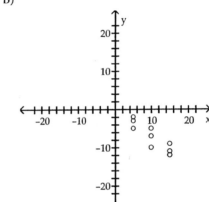

C)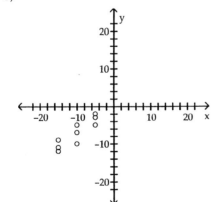

D)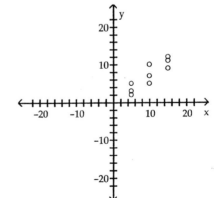

4   Distinguish Between Linear and Nonlinear Relations

Determine if the type of relation is linear, nonlinear, or none.

1)
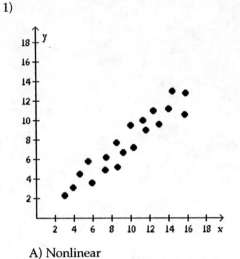

A) Nonlinear   B) None   C) Linear

2)
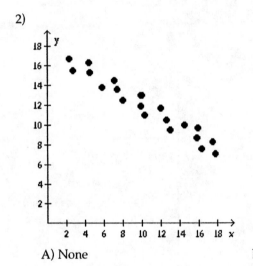

A) None   B) Linear   C) Nonlinear

3)
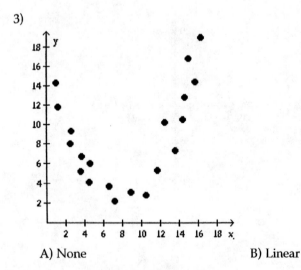

A) None   B) Linear   C) Nonlinear

4)

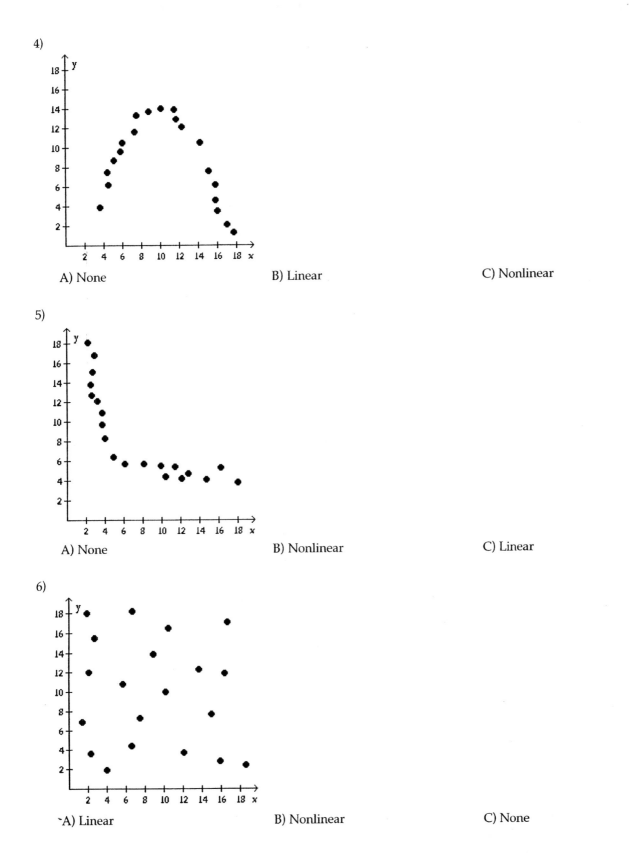

A) None B) Linear C) Nonlinear

5)

A) None B) Nonlinear C) Linear

6)

A) Linear B) Nonlinear C) None

7) Examine the scatter diagram and determine whether the relation is linear, non-linear, or none.

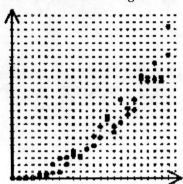

Solve the problem.

8) Identify the scatter diagram of the relation that appears linear.

A)

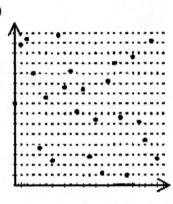

B)

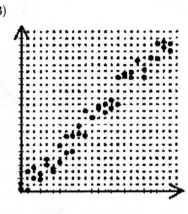

C)

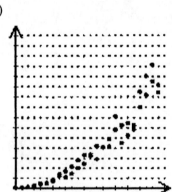

D)

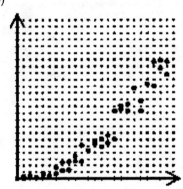

5 Use a Graphing Utility to Find the Line of Best Fit

Use a graphing utility to find the equation of the line of best fit.

1)
x	2	4	5	6
y	7	11	13	20

A) $y = 2.8x + 0.15$ B) $y = 2.8x$ C) $y = 3.0x + 0.15$ D) $y = 3.0x$

2)
x	6	8	20	28	36
y	2	4	13	20	30

A) $y = 0.95x - 2.79$ B) $y = 0.90x - 3.79$ C) $y = 0.80x - 3.79$ D) $y = 0.85x - 2.79$

3)
x	1	3	5	7	9
y	143	116	100	98	90

A) y = 6.8x − 150.7   B) y = −6.2x + 140.4   C) y = 6.2x − 140.4   D) y = −6.8x + 150.7

4)
x	1	2	3	4	5	6
y	17	20	19	22	21	24

A) y = 1.03x + 18.9   B) y = 1.17x + 16.4   C) y = 1.03x + 16.4   D) y = 1.17x + 18.9

Solve the problem. Use a graphing calculator to graph the data.

5) A drug company establishes that the most effective dose of a new drug relates to body weight as shown below. Let body weight be the independent variable and drug dosage be the dependent variable. Use a graphing utility to draw a scatter diagram and to find the line of best fit. What is the most effective dosage for a person weighing 120 lbs?

Body Weight (lbs)	Drug Dosage (mg)
50	10
100	10
150	15
200	17
250	20

A) 12.78 mg   B) 10.07 mg   C) 24.44 mg   D) 12.5 mg

6) A marina owner wishes to estimate a linear function that relates boat length in feet and its draft (depth of boat below water line) in feet. He collects the following data. Let boat length represent the independent variable and draft represent the dependent variable. Use a graphing utility to draw a scatter diagram and to find the line of best fit. What is the draft for a boat 60 ft in length (to the nearest tenth)?

Boat Length (ft)	Draft (ft)
25	2.5
25	2
30	3
30	3.5
45	6
45	7
50	7
50	8

A) 10.5   B) 15.7   C) 9.7   D) 10.3

Solve the problem.

7) Super Sally, a truly amazing individual, picks up a rock and throws it as hard as she can. The table below displays the relationship between the rock's horizontal distance, d (in feet) from Sally and the initial speed with which she throws.

Initial speed (in ft/sec), v	10	15	20	25	30
Horizontal distance of the rock (in feet), d	9.9	14.8	19.1	24.5	28.2

Assume that the horizontal distance travelled varies linearly with the speed with which the rock is thrown. Using a graphing utility, find the line of best fit, and estimate, rounded to two decimal places, the horizontal distance of the rock if the initial speed is 33 ft/sec.

A) 31.34 feet   B) 34.76 feet   C) 31.33 feet   D) 26.67 feet

6 Construct a Linear Model Using Direct Variation

Solve the problem.

1) On planet X, an object falls 18 feet in 3 seconds. Knowing the distance it falls varies directly with the square of the time of fall, how long does it take an object to fall 100 feet? Round your answer to three decimal places.

2) The monthly payment p on a mortgage varies directly with the amount borrowed B. If the monthly payment on a 30-year mortgage is $5.45 for every $1000 borrowed, find a linear function that relates the monthly payment p to the amount borrowed B for a mortgage with the same terms. Then find the monthly payment p when the amount borrowed is $202,000.

   A) $p = \dfrac{B}{163.5}$; $1235.47

   B) $p = \dfrac{B}{1000}$; $0.02

   C) $p = 0.00545B$; $1100.90

   D) $p = \dfrac{B}{30}$; $6733.33

3) The cost C of double-dipped chocolate pretzel O's varies directly with the number of pounds of pretzels purchased, P. If the cost is $43.88 when 3.2 pounds are purchased, find a linear function that relates the cost C to the number of pounds of pretzels purchased P. Then find the cost C when 9.5 pounds are purchased.

   A) $C = 4.619P$; $14.78

   B) $C = 13.713P$; $130.27

   C) $C = 0.073P$; $0.69

   D) $C = \dfrac{140.416}{P}$; $14.78

2.5 Library of Functions; Piecewise-Defined Functions

1 Graph the Functions Listed in the Library of Functions

Match the graph to the function listed whose graph most resembles the one given.

1)

A) square function
B) absolute value function
C) reciprocal function
D) cube function

2)

_____

A) constant function
B) reciprocal function
C) linear function
D) absolute value function

3)

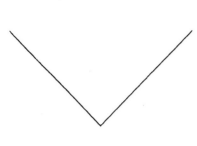

A) square function
C) cube root function

B) square root function
D) cube function

4)

A) linear function
C) reciprocal function

B) square function
D) absolute value function

5)

A) reciprocal function
C) constant function

B) linear function
D) absolute value function

6)

A) cube root function
C) square function

B) square root function
D) cube function

7)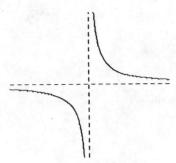

A) square root function
C) reciprocal function
B) absolute value function
D) square function

8)

A) square function
C) square root function
B) cube root function
D) cube function

Graph the function.

9) $f(x) = x$

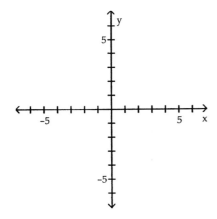

A)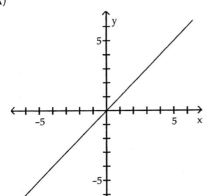

B)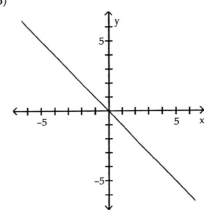

C)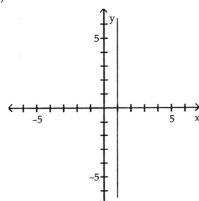

D)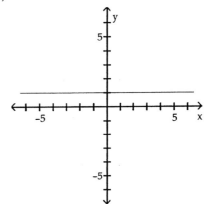

10) $f(x) = x^2$

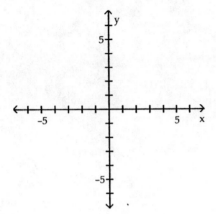

A)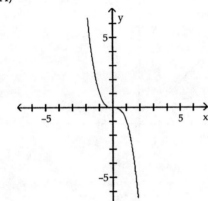

B)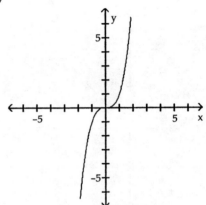

C)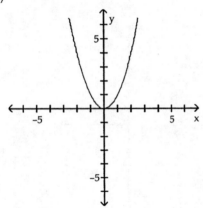

D)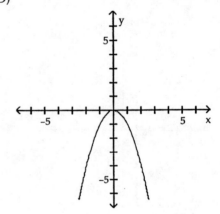

11) $f(x) = x^3$

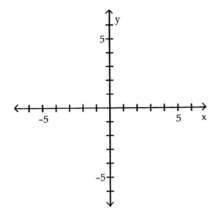

A)
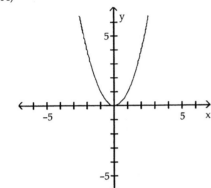

B)
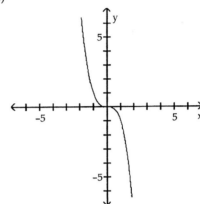

C)
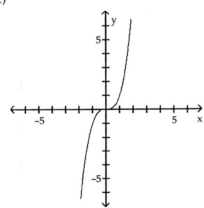

D)
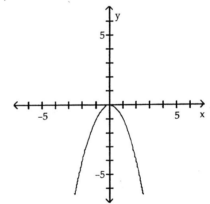

12) $f(x) = \sqrt{x}$

A)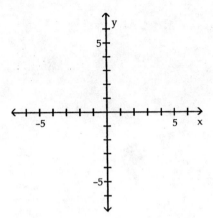

B)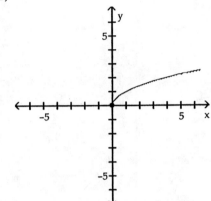

C)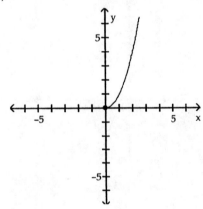

D)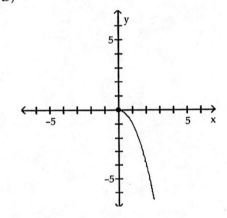

13) $f(x) = \dfrac{1}{x}$

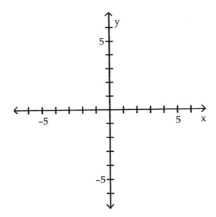

A)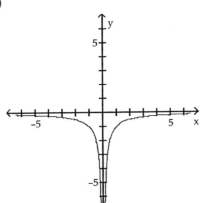

B)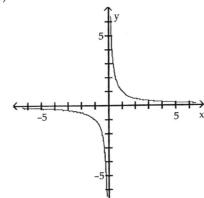

C)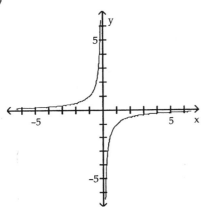

D)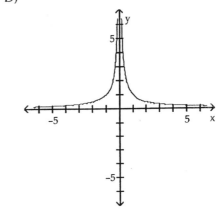

14) $f(x) = |x|$

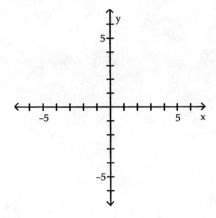

A)
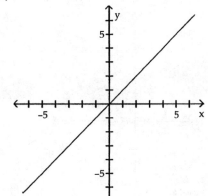

B)
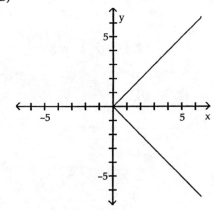

C)
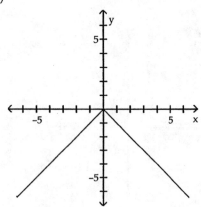

D)
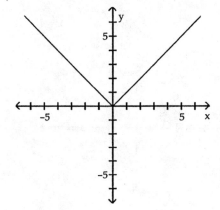

15) $f(x) = \sqrt[3]{x}$

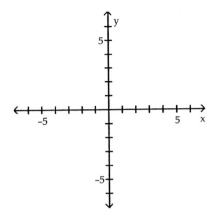

A)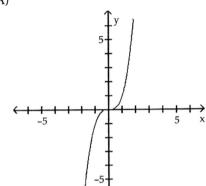

B)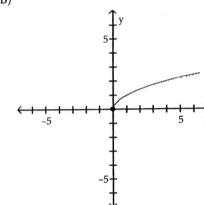

C)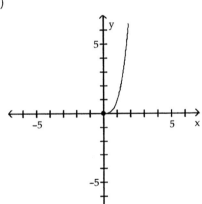

D)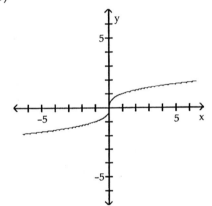

16) $f(x) = -1$

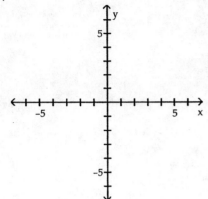

A)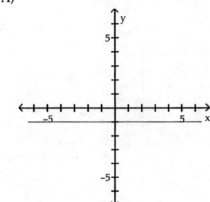

B)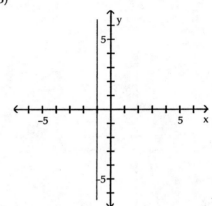

C)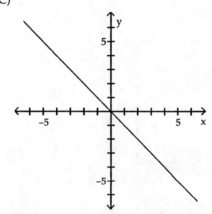

D)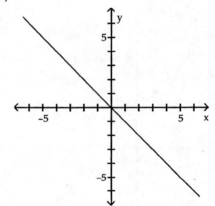

2 Graph Piecewise-Defined Functions

Graph the function.

1)
$$f(x) = \begin{cases} x+1 & \text{if } x < 1 \\ -5 & \text{if } x \geq 1 \end{cases}$$

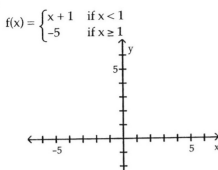

A)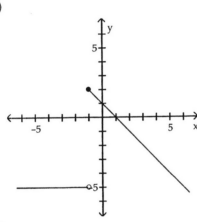

B)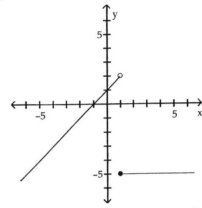

C)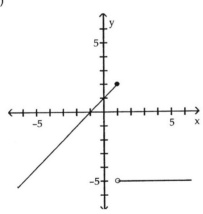

D)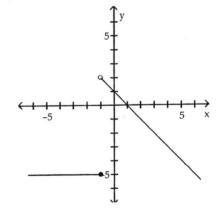

2) $f(x) = \begin{cases} -x + 3 & \text{if } x < 2 \\ 2x - 3 & \text{if } x \geq 2 \end{cases}$

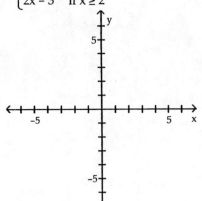

A)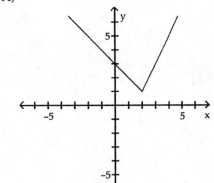

B)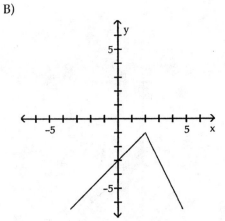

C)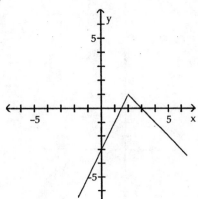

D)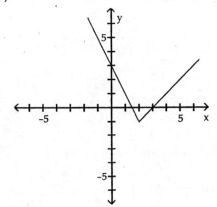

3) $f(x) = \begin{cases} -x + 2 & x < 0 \\ \sqrt{x} + 3 & x \geq 0 \end{cases}$

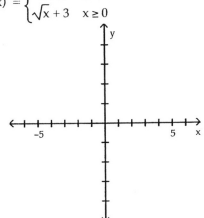

A)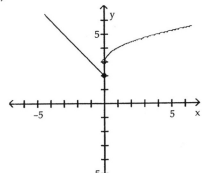

B)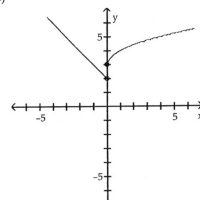

C)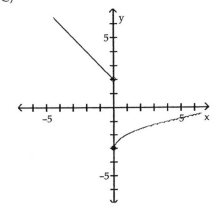

D)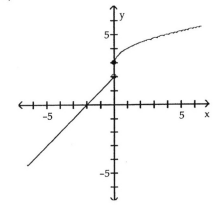

4) $f(x) = \begin{cases} x+1 & \text{if } -9 \le x < 6 \\ -4 & \text{if } x = 6 \\ -x+9 & \text{if } x > 6 \end{cases}$

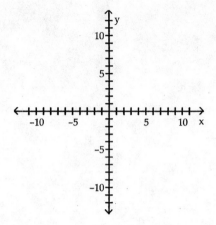

A)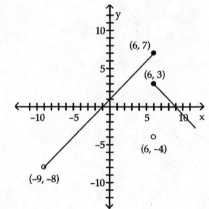

B)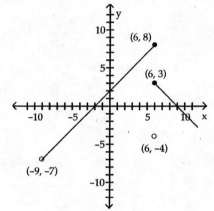

C)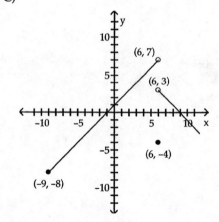

D)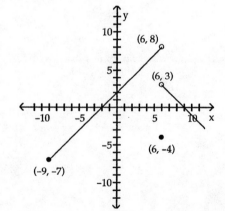

5)
$$f(x) = \begin{cases} 1 & \text{if } 0 \le x < 4 \\ |x| & \text{if } 4 \le x < 8 \\ \sqrt[3]{x} & \text{if } 8 \le x \le 14 \end{cases}$$

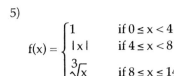

A)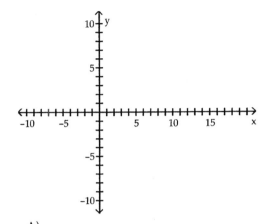

B)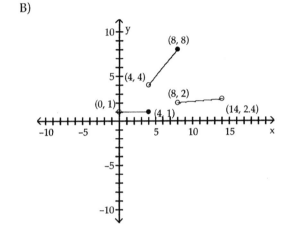

C)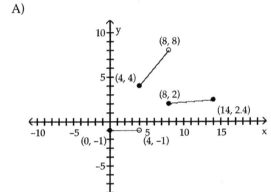

D)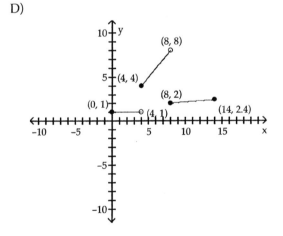

Find the domain of the function.

6) $f(x) = \begin{cases} 4x & \text{if } x \ne 0 \\ 5 & \text{if } x = 0 \end{cases}$

A) all real numbers B) $\{x \mid x \ne 0\}$ C) $\{x \mid x \le 0\}$ D) $\{0\}$

7)
$$f(x) = \begin{cases} 1 & \text{if } -8 \leq x < -6 \\ |x| & \text{if } -6 \leq x < 8 \\ \sqrt[3]{x} & \text{if } 8 \leq x \leq 32 \end{cases}$$

A) $\{x \mid 8 \leq x \leq 32\}$
B) $\{x \mid -8 \leq x < 8 \text{ or } 8 < x \leq 32\}$
C) $\{x \mid x \geq -8\}$
D) $\{x \mid -8 \leq x \leq 32\}$

Locate any intercepts of the function.

8)
$$f(x) = \begin{cases} -7x + 8 & \text{if } x < 1 \\ 8x - 7 & \text{if } x \geq 1 \end{cases}$$

A) $(0, -7)$
B) $(0, 8), (\frac{8}{7}, 0), (\frac{7}{8}, 0)$
C) $(0, 8)$
D) $(0, -7), (\frac{8}{7}, 0), (\frac{7}{8}, 0)$

9)
$$f(x) = \begin{cases} 1 & \text{if } -6 \leq x < -1 \\ |x| & \text{if } -1 \leq x < 6 \\ \sqrt{x} & \text{if } 6 \leq x \leq 21 \end{cases}$$

A) $(0, 0), (0, 1)$
B) $(0, 0), (1, 0)$
C) $(0, 0)$
D) none

Based on the graph, find the range of y = f(x).

10)
$$f(x) = \begin{cases} -\frac{1}{3}x & \text{if } x \neq 0 \\ -8 & \text{if } x = 0 \end{cases}$$

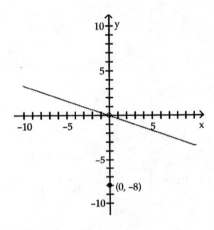

A) $(-\infty, 0)$ or $\{0\}$ or $(0, \infty)$
B) $(-\infty, \infty)$
C) $(-10, 10)$
D) $(-\infty, 0)$ or $(0, \infty)$

11)
$$f(x) = \begin{cases} 4 & \text{if } -4 \leq x < -3 \\ |x| & \text{if } -3 \leq x < 7 \\ \sqrt[3]{x} & \text{if } 7 \leq x \leq 14 \end{cases}$$

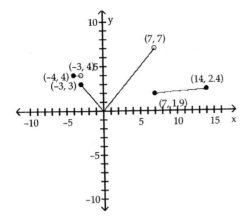

A) [0, 7] B) $[0, \sqrt[3]{14}]$ C) [0, ∞) D) [0, 7)

The graph of a piecewise-defined function is given. Write a definition for the function.

12)

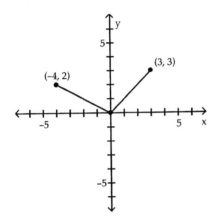

A) $f(x) = \begin{cases} \frac{1}{2}x & \text{if } -4 < x < 0 \\ x & \text{if } 0 < x < 3 \end{cases}$

B) $f(x) = \begin{cases} -\frac{1}{2}x & \text{if } -4 < x < 0 \\ x & \text{if } 0 < x < 3 \end{cases}$

C) $f(x) = \begin{cases} -\frac{1}{2}x & \text{if } -4 \leq x \leq 0 \\ x & \text{if } 0 < x \leq 3 \end{cases}$

D) $f(x) = \begin{cases} -2x & \text{if } -4 \leq x \leq 0 \\ x & \text{if } 0 < x \leq 3 \end{cases}$

13)

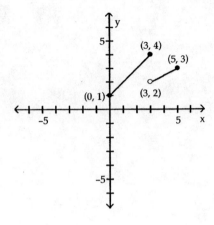

A) $f(x) = \begin{cases} x+1 & \text{if } 0 \le x \le 3 \\ \frac{1}{2}x+2 & \text{if } 3 < x \le 5 \end{cases}$

B) $f(x) = \begin{cases} x+1 & \text{if } 0 \le x \le 3 \\ \frac{1}{2}x+\frac{1}{2} & \text{if } 3 < x \le 5 \end{cases}$

C) $f(x) = \begin{cases} x+1 & \text{if } 0 \le x \le 3 \\ \frac{1}{2}x-\frac{1}{2} & \text{if } 3 < x \le 5 \end{cases}$

D) $f(x) = \begin{cases} x+1 & \text{if } 0 \le x \le 3 \\ \frac{1}{2}x & \text{if } 3 < x \le 5 \end{cases}$

14)

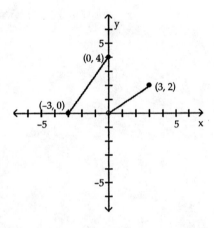

A) $f(x) = \begin{cases} \frac{4}{3}x+4 & \text{if } -3 \le x \le 0 \\ \frac{2}{3}x & \text{if } 0 < x \le 3 \end{cases}$

B) $f(x) = \begin{cases} \frac{3}{4}x+4 & \text{if } -3 \le x \le 0 \\ \frac{3}{2}x & \text{if } 0 < x \le 3 \end{cases}$

C) $f(x) = \begin{cases} \frac{4}{3}x+4 & \text{if } -3 \le x \le 0 \\ \frac{2}{3}x+2 & \text{if } 0 < x \le 3 \end{cases}$

D) $f(x) = \begin{cases} \frac{4}{3}x-4 & \text{if } -3 \le x \le 0 \\ \frac{2}{3}x & \text{if } 0 \le x \le 3 \end{cases}$

15)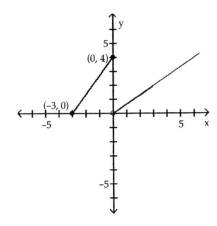

A)
$$f(x) = \begin{cases} \frac{4}{3}x + 4 & \text{if } -3 \le x \le 0 \\ \frac{2}{3}x & \text{if } 0 < x \le 3 \end{cases}$$

B)
$$f(x) = \begin{cases} \frac{3}{4}x + 4 & \text{if } -3 \le x \le 0 \\ \frac{3}{2}x & \text{if } x \ge 0 \end{cases}$$

C)
$$f(x) = \begin{cases} \frac{3}{4}x + 4 & \text{if } -3 \le x \le 0 \\ \frac{3}{2}x & \text{if } x > 0 \end{cases}$$

D)
$$f(x) = \begin{cases} \frac{4}{3}x + 4 & \text{if } -3 \le x \le 0 \\ \frac{2}{3}x & \text{if } x > 0 \end{cases}$$

Solve the problem.

16) If $f(x) = \text{int}(3x)$, find $f(1.6)$.

 A) 1 B) 4 C) 2 D) 5

17) A gas company has the following rate schedule for natural gas usage in single-family residences:

Monthly service charge	$8.80
Per therm service charge	
1st 25 therms	$0.6686/therm
Over 25 therms	$0.85870/therm

What is the charge for using 25 therms in one month?
What is the charge for using 45 therms in one month?
Construct a function that gives the monthly charge C for x therms of gas.

18) An electric company has the following rate schedule for electricity usage in single-family residences:

    Monthly service charge $4.93

    Per kilowatt service charge
        1st 300 kilowatts $0.11589/kW
        Over 300 kilowatts $0.13321/kW

What is the charge for using 300 kilowatts in one month?
What is the charge for using 375 kilowatts in one month?
Construct a function that gives the monthly charge C for x kilowatts of electricity.

19) One Internet service provider has the following rate schedule for high-speed Internet service:

    Monthly service charge $18.00

    1st 50 hours of use   free
    Next 50 hours of use $0.25/hour
    Over 100 hours of use $1.00/hour

What is the charge for 50 hours of high-speed Internet use in one month?
What is the charge for 75 hours of high-speed Internet use in one month?
What is the charge for 135 hours of high-speed Internet use in one month?

20) The wind chill factor represents the equivalent air temperature at a standard wind speed that would produce the same heat loss as the given temperature and wind speed. One formula for computing the equivalent temperature is

$$W(t) = \begin{cases} t & \text{if } 0 \le v < 1.79 \\ 33 - \dfrac{(10.45 + 10\sqrt{v} - v)(33 - t)}{22.04} & \text{if } 1.79 \le v < 20 \\ 33 - 1.5958(33 - t) & \text{if } v \ge 20 \end{cases}$$

where v represents the wind speed (in meters per second) and t represents the air temperature (°C). Compute the wind chill for an air temperature of 15°C and a wind speed of 12 meters per second. (Round the answer to one decimal place.)

21) A cellular phone plan had the following schedule of charges:

    Basic service, including 100 minutes of calls $20.00 per month
    2nd 100 minutes of calls $0.075 per minute
    Additional minutes of calls $0.10 per minute

What is the charge for 200 minutes of calls in one month?
What is the charge for 250 minutes of calls in one month?
Construct a function that relates the monthly charge C for x minutes of calls.

2.6 Graphing Techniques: Transformations

1 Graph Functions Using Vertical and Horizontal Shifts

Graph the function by starting with the graph of the basic function and then using the techniques of shifting, compressing, stretching, and/or reflecting.

1) $f(x) = x^2 - 3$

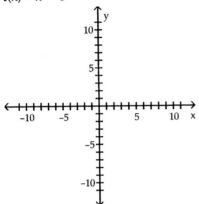

A)
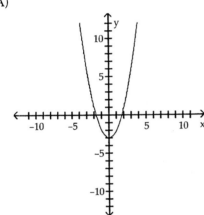

B)
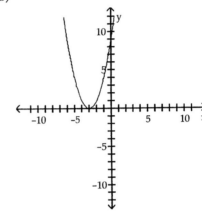

C)
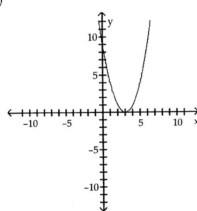

D)
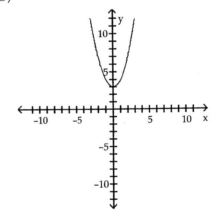

2) $f(x) = (x+2)^2$

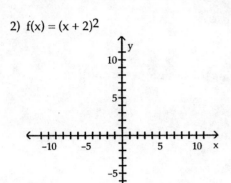

A)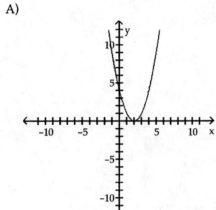

B)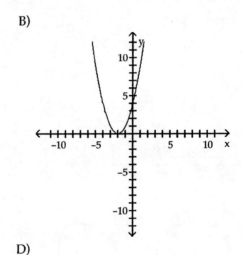

C)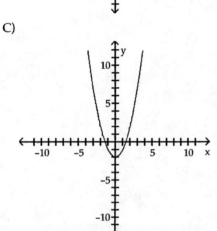

D)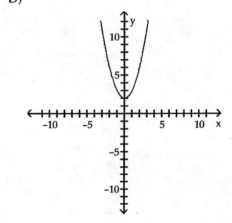

3) $f(x) = (x+1)^2 - 7$

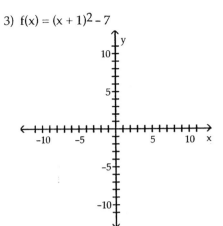

A)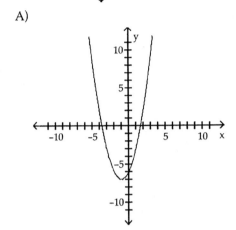

B)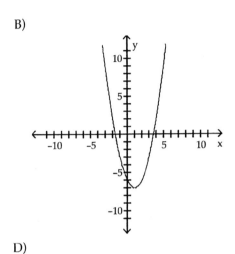

C)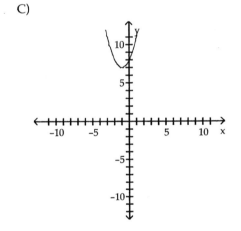

D)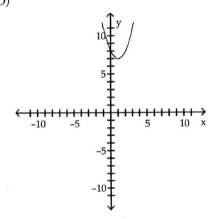

4) $f(x) = x^3 - 1$

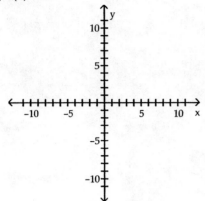

A)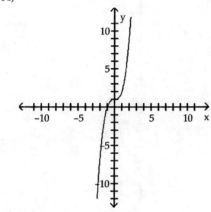

B)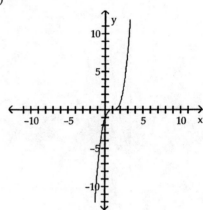

C)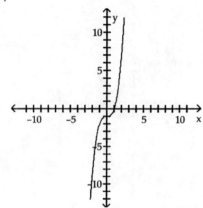

D)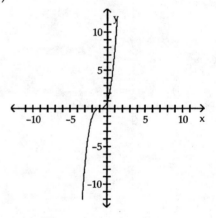

5) $f(x) = (x-4)^3$

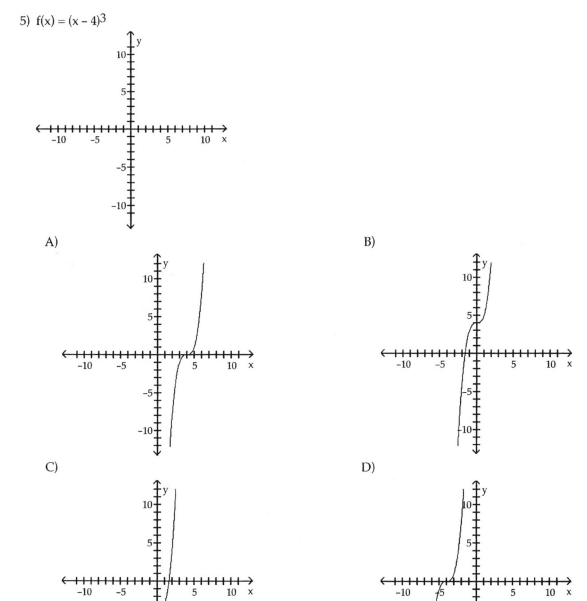

6) $f(x) = (x-3)^3 - 2$

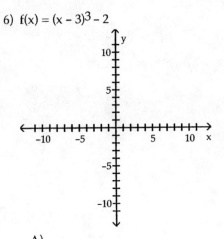

A)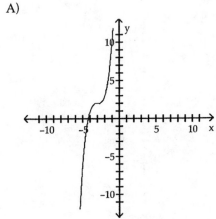

B)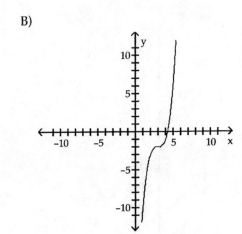

C)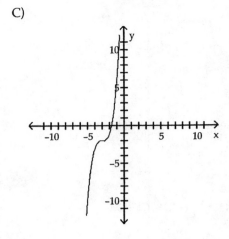

D)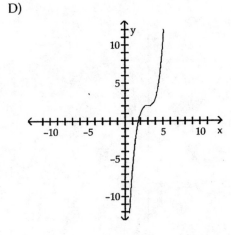

7) $f(x) = \sqrt{x} - 3$

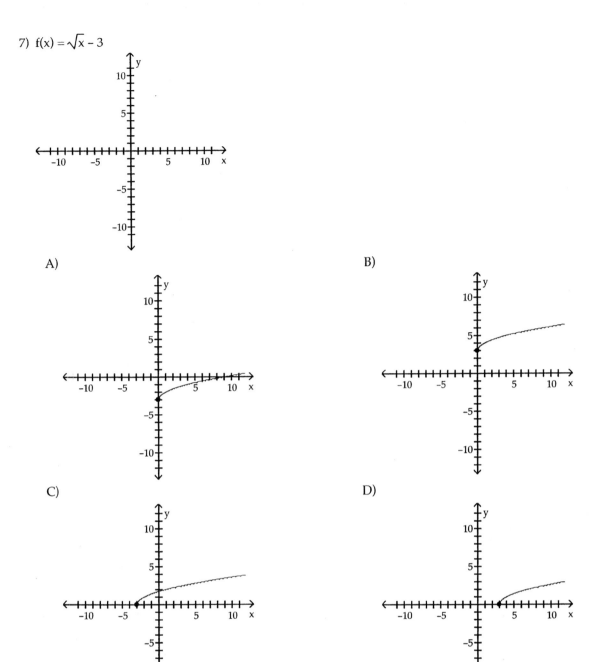

8) $f(x) = \sqrt{x+1}$

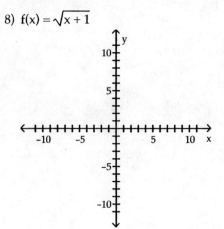

A)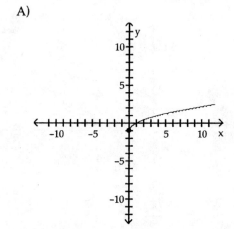

B)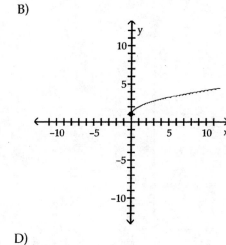

C)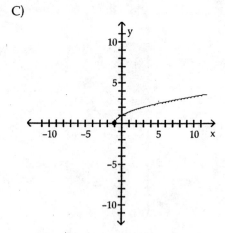

D)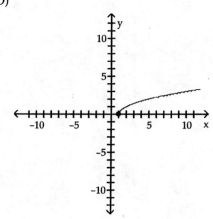

9) $f(x) = \sqrt{x-4} - 5$

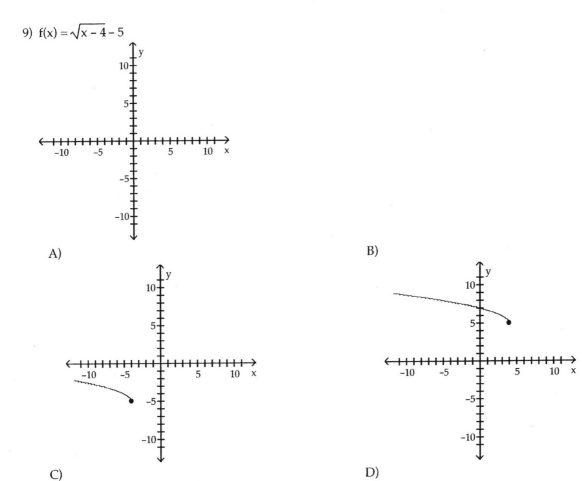

10) $f(x) = |x| + 5$

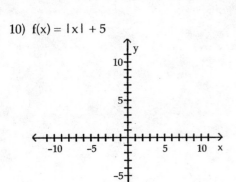

A)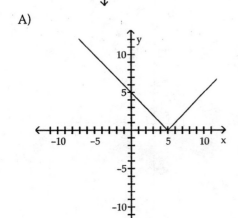

B)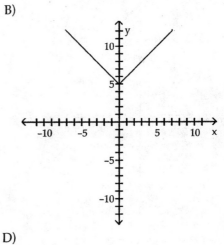

C)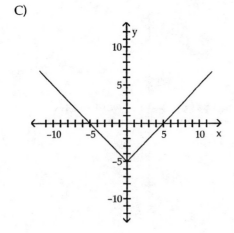

D)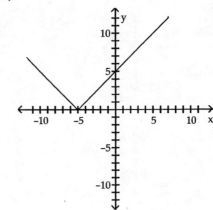

11) $f(x) = |x + 4|$

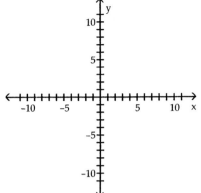

A)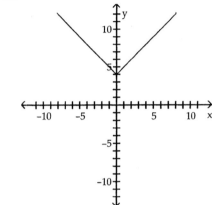

B)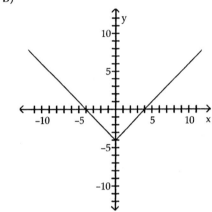

C)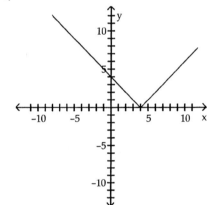

D)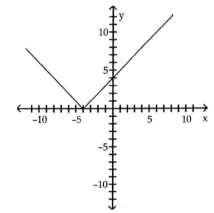

12) $f(x) = |x+5| + 7$

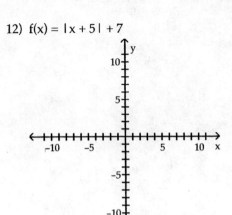

A)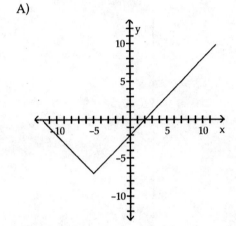

B)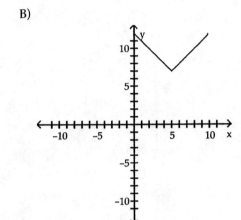

C)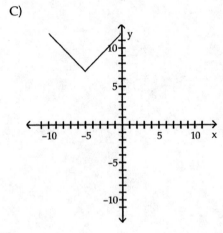

D)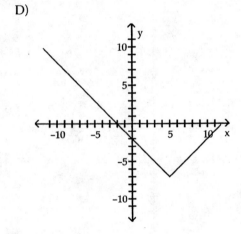

13) $f(x) = \dfrac{1}{x} + 3$

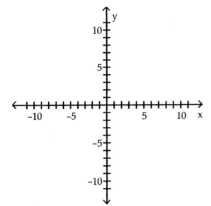

A)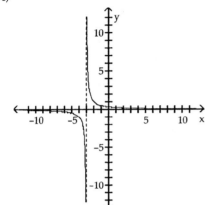

B)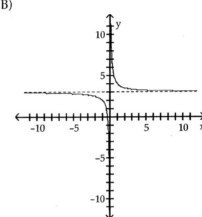

C)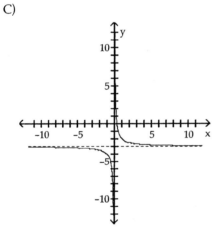

D)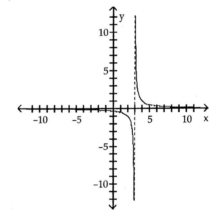

14) $f(x) = \dfrac{1}{x+2}$

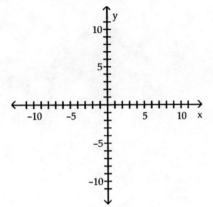

A)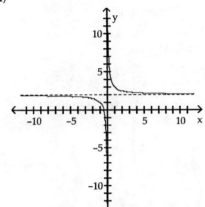

B)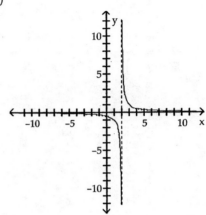

C)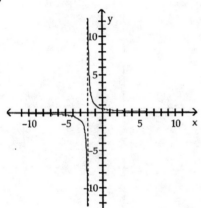

D)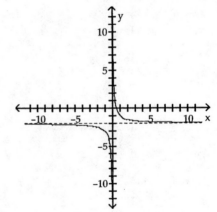

15) $f(x) = \dfrac{1}{x+4} - 7$

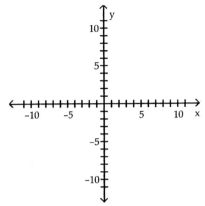

A)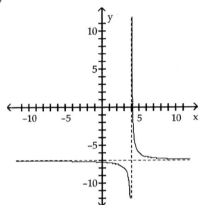

B)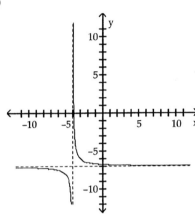

C)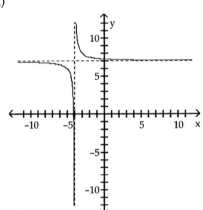

D)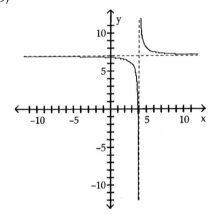

Match the correct function to the graph.

16)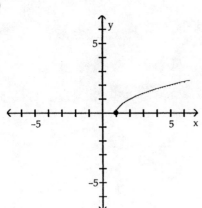

A) $y = \sqrt{x}$ B) $y = \sqrt{x+1}$ C) $y = \sqrt{x-1}$ D) $y = x - 1$

17)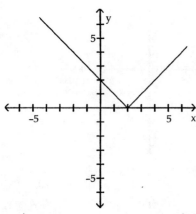

A) $y = x - 2$ B) $y = |x + 2|$ C) $y = |1 - x|$ D) $y = |2 - x|$

Find the function.

18) Find the function that is finally graphed after the following transformations are applied to the graph of $y = |x|$. The graph is shifted right 3 units, stretched by a factor of 3, shifted vertically down 2 units, and finally reflected across the x-axis.

A) $y = 3|-x - 3| - 2$ B) $y = -(3|x - 3| - 2)$ C) $y = -3|x - 3| - 2$ D) $y = -(3|x + 3| - 2)$

Find the function that is finally graphed after the following transformations are applied to the graph of $y = \sqrt{x}$.

19) i) Shift up 5 units
    ii) Reflect about the y-axis
    iii) Shift right 2 units

A) $y = -\sqrt{x - 2} + 5$ B) $y = \sqrt{-x - 2} - 5$ C) $y = \sqrt{-x + 2} - 5$ D) $y = \sqrt{-x + 2} + 5$

2 Graph Functions Using Compressions and Stretches

Graph the function by starting with the graph of the basic function and then using the techniques of shifting, compressing, stretching, and/or reflecting.

1) $f(x) = 3x^2$

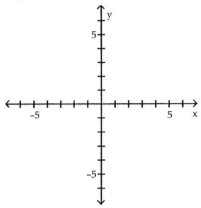

A)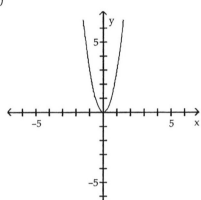

B)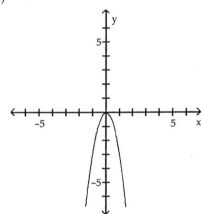

C)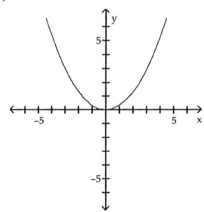

D)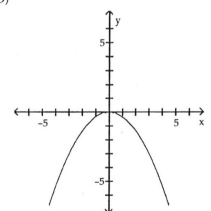

2) $f(x) = \frac{1}{3}x^2$

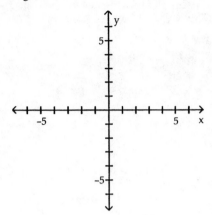

A)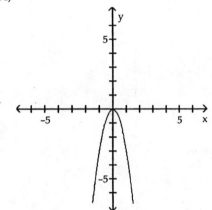

B)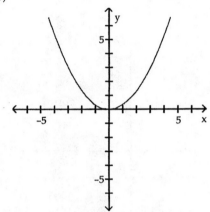

C)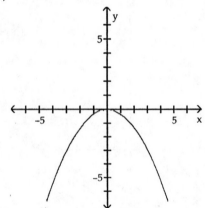

D)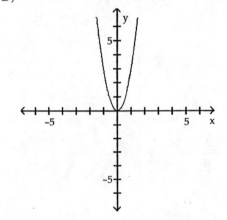

3) $f(x) = 2x^3$

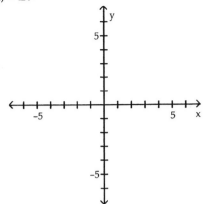

A)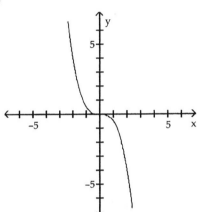

B)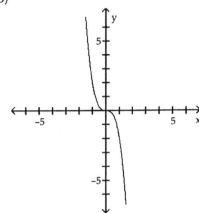

C)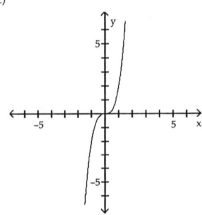

D)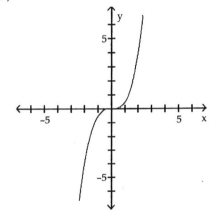

4) $f(x) = \frac{1}{5}x^3$

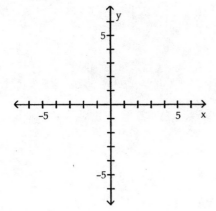

A)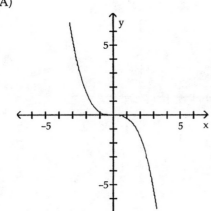

B)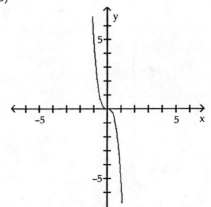

C)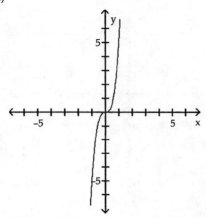

D)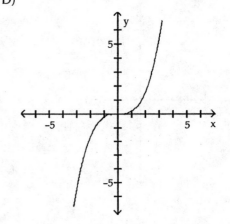

5) $f(x) = 2\sqrt{x}$

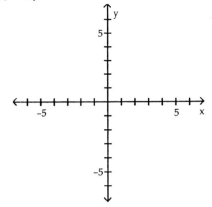

A)
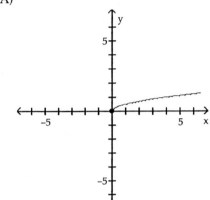

B)
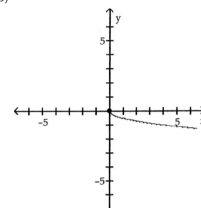

C)
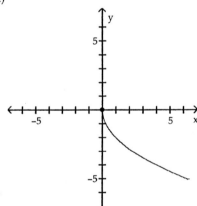

D)
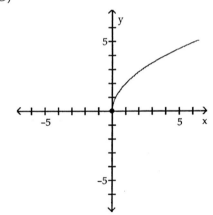

6) $f(x) = \frac{1}{7}\sqrt{x}$

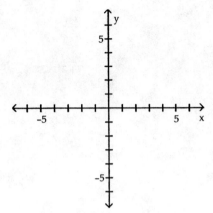

A)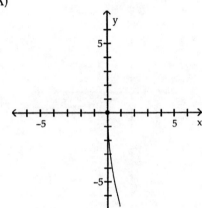

B)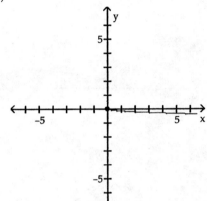

C)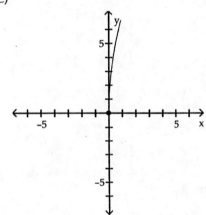

D)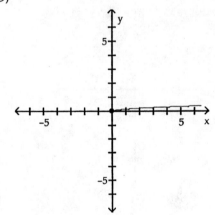

7) $f(x) = 4|x|$

A)
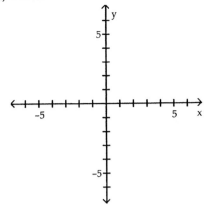

B)
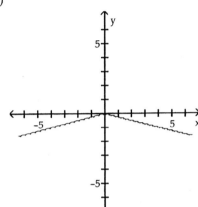

C)
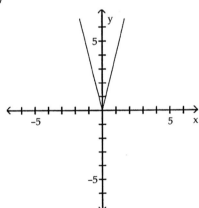

D)
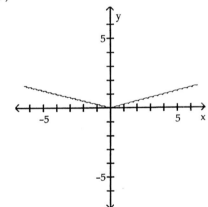

8) $f(x) = \dfrac{1}{4}|x|$

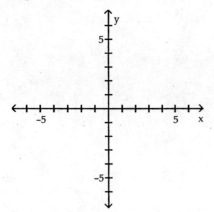

A)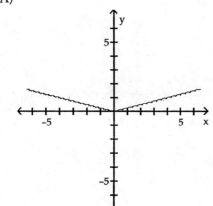

B)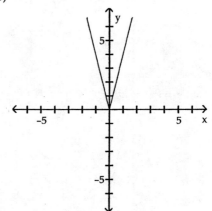

C)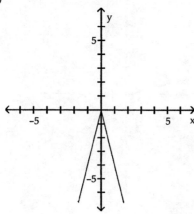

D)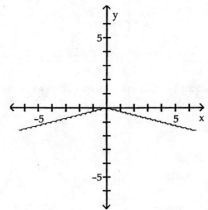

9) $f(x) = \dfrac{6}{x}$

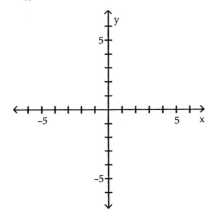

A)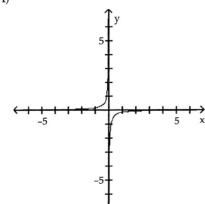

B)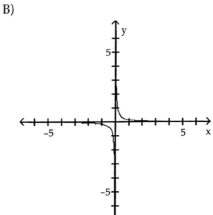

C)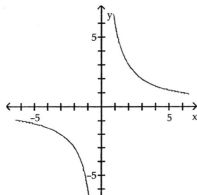

D)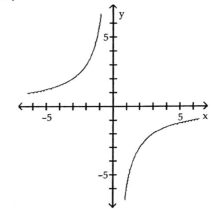

10) $f(x) = \dfrac{1}{2x}$

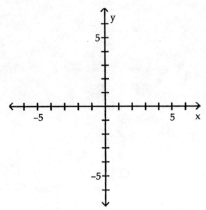

A)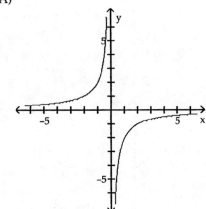

B)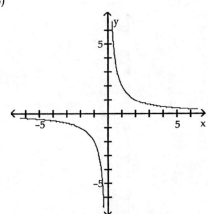

C)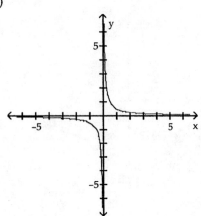

D)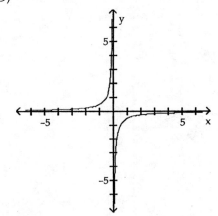

3 Graph Functions Using Reflections about the x-Axis or y-Axis

Graph the function by starting with the graph of the basic function and then using the techniques of shifting, compressing, stretching, and/or reflecting.

1) $f(x) = -x^2$

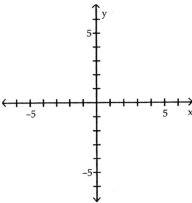

A)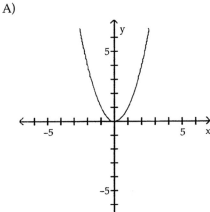

B)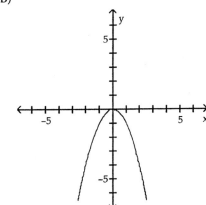

C)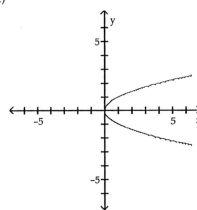

D)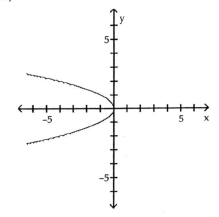

2) $f(x) = (-x)^2$

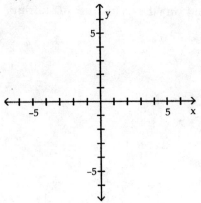

A)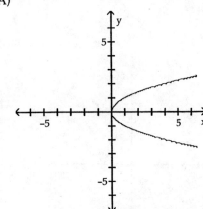

B)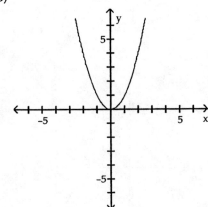

C)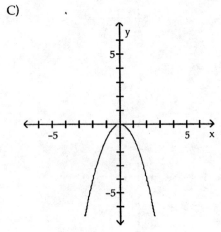

D)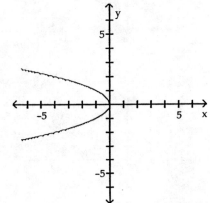

3) $f(x) = -x^3$

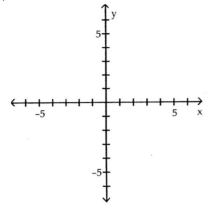

A)
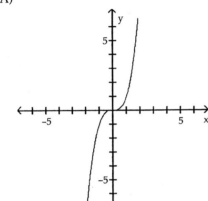

B)
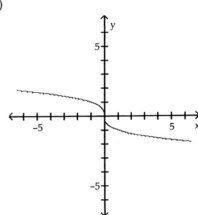

C)
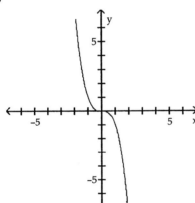

D)
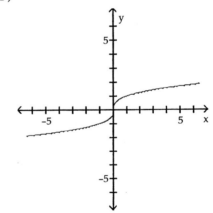

4) $f(x) = (-x)^3$

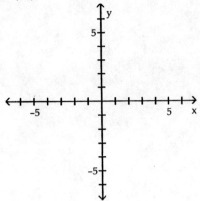

A)
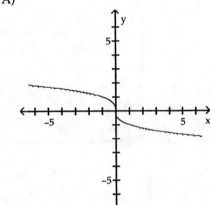

B)
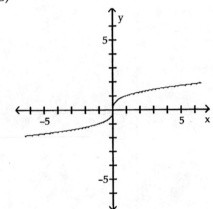

C)
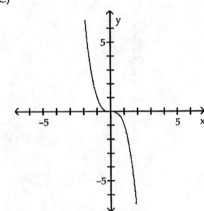

D)
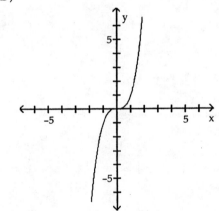

5) $f(x) = -\sqrt{x}$

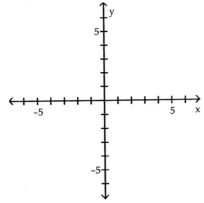

A)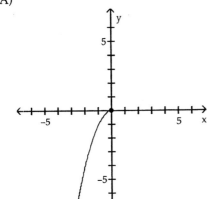

B)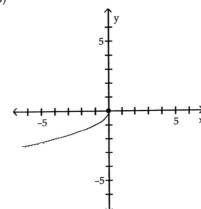

C)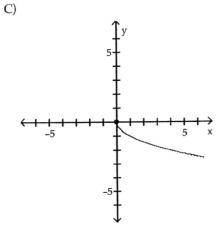

D)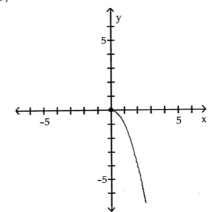

6) $f(x) = \sqrt{-x}$

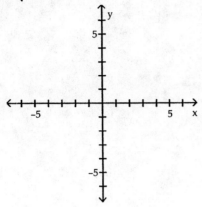

A)
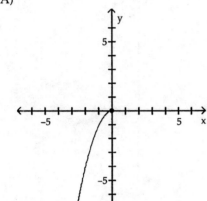

B)
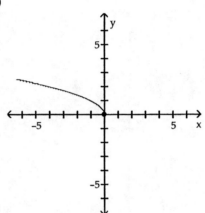

C)
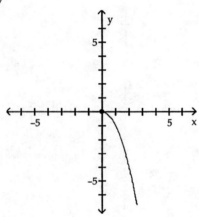

D)
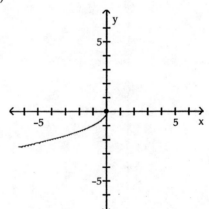

7) $f(x) = -|x|$

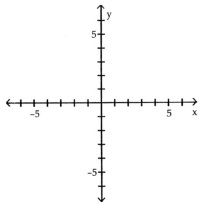

A)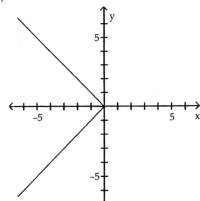

B)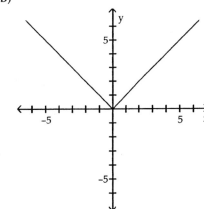

C)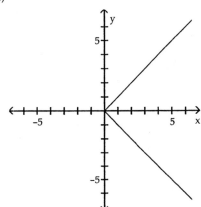

D)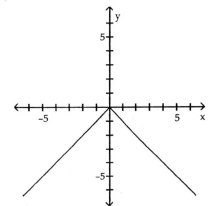

8) $f(x) = |-x|$

A)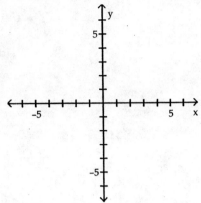

B)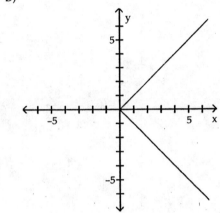

C)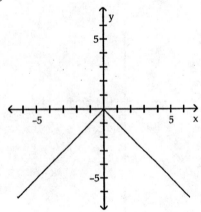

D)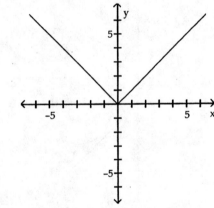

9) $f(x) = -\dfrac{1}{x}$

A)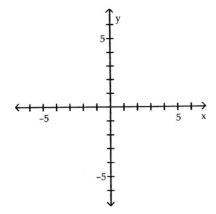

B)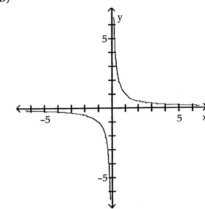

C)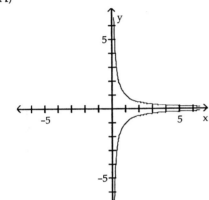

D)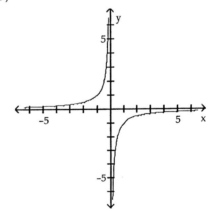

4 Demonstrate Additional Understanding and Skills: Function Shifts, Compressions, and Reflections

Graph the function by starting with the graph of the basic function and then using the techniques of shifting, compressing, stretching, and/or reflecting.

1) $f(x) = x^2 - 3$

A)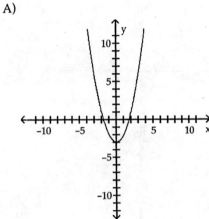

B)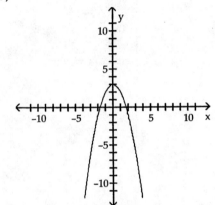

C)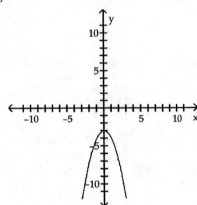

D)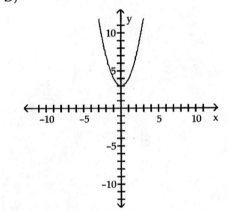

2) $f(x) = 2(x+1)^2 + 3$

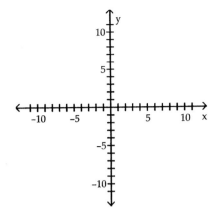

A)
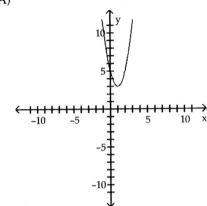

B)
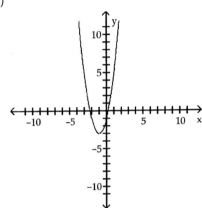

C)
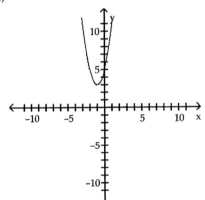

D)
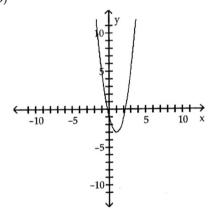

3) $f(x) = -(x-1)^2 - 2$

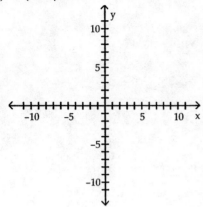

A)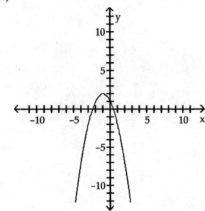

B)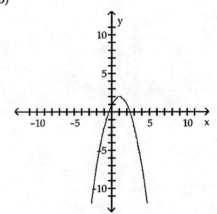

C)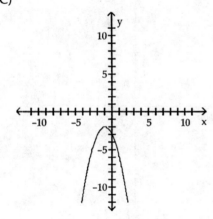

D)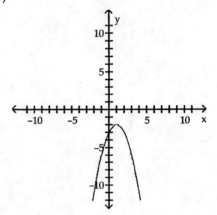

4) $f(x) = -2(x+1)^2 - 2$

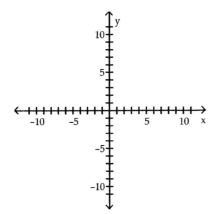

A)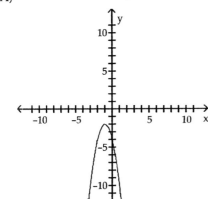

B)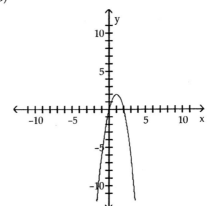

C)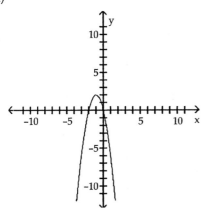

D)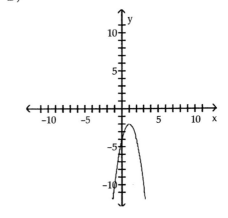

5) $f(x) = \sqrt{x-4} - 6$

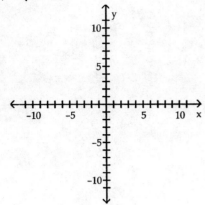

A)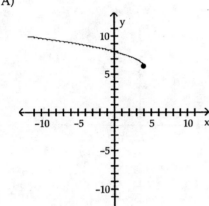

B)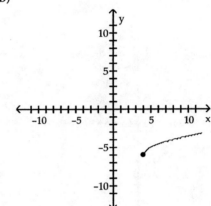

C)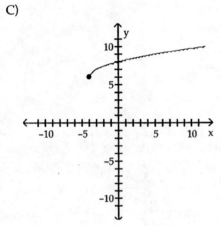

D)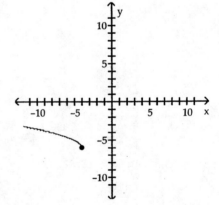

Using transformations, sketch the graph of the requested function.

6) The graph of a function f is illustrated. Use the graph of f as the first step toward graphing the function F(x), where F(x) = f(x + 2) − 1.

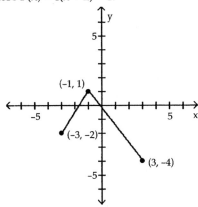

A)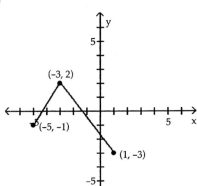

B)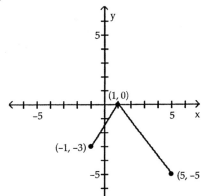

C)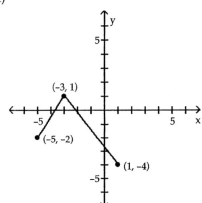

D)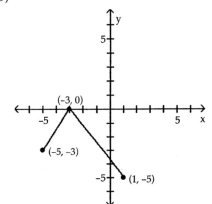

Match the correct function to the graph.

7)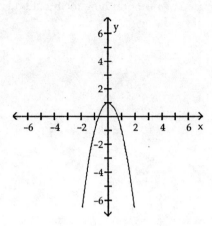

A) $y = -2x^2 + 1$ B) $y = -2x^2$ C) $y = 1 - x^2$ D) $y = -2x^2 - 1$

Find the function.

8) Find the function that is finally graphed after the following transformations are applied to the graph of $y = |x|$. The graph is shifted right 3 units, stretched by a factor of 3, shifted vertically down 2 units, and finally reflected across the x-axis.

   A) $y = -(3|x + 3| - 2)$ B) $y = -(3|x - 3| - 2)$ C) $y = 3|-x - 3| - 2$ D) $y = -3|x - 3| - 2$

Find the function that is finally graphed after the following transformations are applied to the graph of $y = \sqrt{x}$.

9) i) Shift up 8 units
   ii) Reflect about the y-axis
   iii) Shift right 4 units

   A) $y = \sqrt{-x + 4} + 8$ B) $y = \sqrt{-x - 4} - 8$ C) $y = \sqrt{-x + 4} - 8$ D) $y = -\sqrt{x - 4} + 8$

Provide an appropriate answer.

10) Complete the square of the given quadratic function. Then, graph the function by hand using the technique of shifting.

$f(x) = x^2 + 2x + 3$

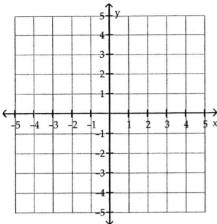

A)
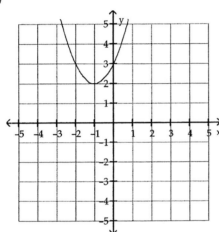

B)
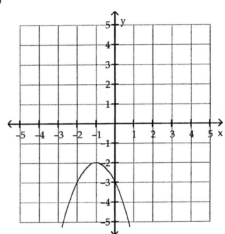

C)
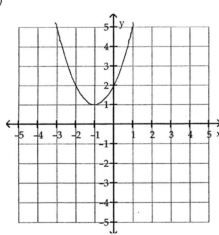

D)
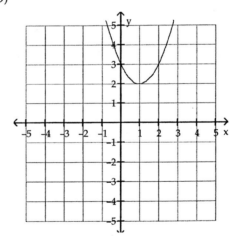

2.7 Mathematical Models: Constructing Functions

1 Construct and Analyze Functions

Solve the problem.

1) The volume V of a square-based pyramid with base sides s and height h is $V = \frac{1}{3}s^2h$. If the height is half of the length of a base side, express the volume V as a function of s.

2) The price p and the quantity x sold of a certain product obey the demand equation:
$$p = -\frac{1}{9}x + 200, \{x \mid 0 \le x \le 900\}$$

   What is the revenue to the nearest dollar when 600 units are sold?

   A) $210,000 B) $30,000 C) $80,000 D) $160,000

3) A farmer has 400 yards of fencing to enclose a rectangular garden. Express the area A of the rectangle as a function of the width x of the rectangle. What is the domain of A?

   A) $A(x) = -x^2 + 200x$; $\{x \mid 0 < x < 400\}$
   B) $A(x) = x^2 + 200x$; $\{x \mid 0 < x < 200\}$
   C) $A(x) = -x^2 + 400x$; $\{x \mid 0 < x < 400\}$
   D) $A(x) = -x^2 + 200x$; $\{x \mid 0 < x < 200\}$

4) A wire of length 5x is bent into the shape of a square. Express the area A of the square as a function of x.

   A) $A(x) = \frac{5}{4}x^2$ B) $A(x) = \frac{25}{16}x^2$ C) $A(x) = \frac{25}{8}x^2$ D) $A(x) = \frac{1}{16}x^2$

5) Let P = (x, y) be a point on the graph of $y = \sqrt{x}$. Express the distance d from P to the point (1, 0) as a function of x.

   A) $d(x) = \sqrt{x^2 - x + 1}$ B) $d(x) = x^2 - x + 1$ C) $d(x) = x^2 + 2x + 2$ D) $d(x) = \sqrt{x^2 + 2x + 2}$

6) The price p and x, the quantity of a certain product sold, obey the demand equation
$$p = -\frac{1}{10}x + 100, \{x \mid 0 \le x \le 1000\}$$

   a) Express the revenue R as a function of x.
   b) What is the revenue if 450 units are sold?
   c) Graph the revenue function using a graphing utility.
   d) What quantity x maximizes revenue? What is the maximum revenue?
   e) What price should the company charge to maximize revenue?

7) A right triangle has one vertex on the graph of $y = x^2$ at (x, y), another at the origin, and the third on the (positive) y-axis at (0, y). Express the area A of the triangle as a function of x.

8) A wire 20 feet long is to be cut into two pieces. One piece will be shaped as a square and the other piece will be shaped as an equilateral triangle. Express the total area A enclosed by the pieces of wire as a function of the length x of a side of the equilateral triangle. What is the domain of A?

9) Two boats leave a dock at the same time. One boat is headed directly east at a constant speed of 35 knots (nautical miles per hour), and the other is headed directly south at a constant speed of 22 knots. Express the distance d between the boats as a function of the time t.

Ch. 2 Functions and Their Graphs
Answer Key

2.1 Functions
1 Determine Whether a Relation Represents a Function
1) A
2) C
3) B
4) A
5) C
6) A
7) A
8) A
9) A
10) A
11) B
12) B
13) B
14) B
15) A
16) A
17) B
18) A
19) A

2 Find the Value of a Function
1) D
2) C
3) B
4) D
5) B
6) C
7) B
8) B
9) A
10) C
11) B
12) B
13) D
14) D
15) A
16) C
17) C
18) B
19) B

3 Find the Domain of a Function
1) D
2) C
3) A
4) B
5) A
6) B
7) C

4 Form the Sum, Difference, Product, and Quotient of Two Functions
1) D
2) C

3) B
4) B
5) A
6) B
7) C
8) D
9) C
10) B
11) D
12) D
13) C
14) B
15) D
16) C
17) B
18) A

2.2 The Graph of a Function
1 Identify the Graph of a Function
1) D
2) C
3) A
4) B
5) D
6) B
7) D

2 Obtain Information from or about the Graph of a Function
1) C
2) A
3) B
4) A
5) C
6) C
7) B
8) B
9) D
10) D
11) C
12) B
13) A
14) B
15) A
16) A
17) D
18) B
19) A
20) B
21) B
22) B
23) D
24) B
25) D
26) C
27) D

28)

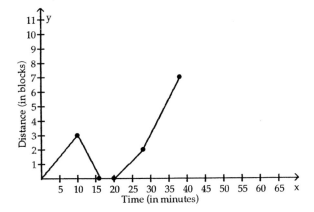

29) A
30) D

2.3 Properties of Functions
1 Determine Even and Odd Functions from a Graph

1) A
2) A
3) C
4) C
5) B
6) B
7) B
8) A

2 Identify Even and Odd Functions from the Equation

1) B
2) A
3) A
4) C
5) B
6) C
7) A
8) A
9) B
10) B
11) B

3 Use a Graph to Determine Where a Function Is Increasing, Decreasing, or Constant

1) C
2) B
3) C
4) C
5) B
6) C
7) B
8) B
9) A
10) C
11) A
12) B
13) D
14) C

15) C
16) A

4 Use a Graph to Locate Local Maxima and Minima

1) A
2) C
3) D
4) C
5) C
6) A

5 Use a Graphing Utility to Approximate Local Maxima and Minima and to Determine Where a Function Is Increasing or Decreasing

1) B
2) local maximum at (0, 6)
   local minimum at (2.67, –3.48)
   increasing on (–1, 0) and (2.67, 4)
   decreasing on (0, 2.67)
3) local maximum at (0, 0)
   local minimum at (0.74, –0.33)
   increasing on (–2, 0) and (0.74, 2)
   decreasing on (0, 0.74)
4) local maximum at (2.34, 1.61)
   local minimum at (–1.9, –9.82)
   increasing on (–1.9, 2.34)
   decreasing on (–4, –1.9) and (2.34, 5)
5) local maximum at (0, 5)
   local minima at (–2.55, 1.17) and (1.05, 4.65)
   increasing on (–2.55, 0) and (1.05, 2)
   decreasing on (–4, –2.55) and (0, 1.05)
6) A
7) A
8) B
9) C
10) B

6 Find the Average Rate of Change of a Function

1) C
2) D
3) A
4) A
5) A
6) C
7) B
8) A
9) B
10) A
11) A
12) D
13) B
14) C
15) A
16) C
17) C

2.4 Linear Functions and Models

1 Graph Linear Functions

1) A
2) A

        3) D
        4) D
2   Work with Applications of Linear Functions
        1) A
        2) D
3   Draw and Interpret Scatter Diagrams
        1) B
        2) D
        3) B
        4) B
4   Distinguish Between Linear and Nonlinear Relations
        1) C
        2) B
        3) C
        4) C
        5) B
        6) C
        7) non-linear
        8) B
5   Use a Graphing Utility to Find the Line of Best Fit
        1) D
        2) B
        3) B
        4) B
        5) A
        6) C
        7) A
6   Construct a Linear Model Using Direct Variation
        1) 7.071 sec
        2) C
        3) B

2.5   Library of Functions; Piecewise-Defined Functions
1   Graph the Functions Listed in the Library of Functions
        1) A
        2) A
        3) B
        4) D
        5) B
        6) D
        7) C
        8) B
        9) A
        10) C
        11) C
        12) B
        13) B
        14) D
        15) D
        16) A
2   Graph Piecewise-Defined Functions
        1) B
        2) A
        3) B
        4) C
        5) D

6) A
7) D
8) C
9) C
10) D
11) D
12) C
13) B
14) A
15) D
16) B
17) $25.52
$42.69
$$C(x) = \begin{cases} 8.8 + 0.6686x & \text{if } 0 \leq x \leq 25 \\ 4.0475 + 0.8587x & \text{if } x > 25 \end{cases}$$
18) $39.70
$49.69
$$C(x) = \begin{cases} 4.93 + 0.11589x & \text{if } 0 \leq x \leq 300 \\ -0.266 + 0.13321x & \text{if } x > 300 \end{cases}$$
19) $18.00
$24.25
$65.50
20) 6.0°C
21) $27.50
$32.50;
$$C(x) = \begin{cases} 20 & \text{if } 0 \leq x \leq 100 \\ 12.5 + 0.075x & \text{if } 100 < x \leq 200 \\ 7.5 + 0.1x & \text{if } x > 200 \end{cases}$$

2.6 Graphing Techniques: Transformations
1 Graph Functions Using Vertical and Horizontal Shifts
1) A
2) B
3) A
4) C
5) A
6) B
7) A
8) C
9) D
10) B
11) D
12) C
13) B
14) C
15) B
16) C
17) D
18) B
19) D

2 Graph Functions Using Compressions and Stretches
1) A
2) B
3) C
4) D

Precalculus Enhanced with Graphing Utilities

- 5) D
- 6) D
- 7) C
- 8) A
- 9) C
- 10) C

3 Graph Functions Using Reflections about the x-Axis or y-Axis
- 1) B
- 2) B
- 3) C
- 4) C
- 5) C
- 6) B
- 7) D
- 8) D
- 9) D

4 Demonstrate Additional Understanding and Skills: Function Shifts, Compressions, and Reflections
- 1) A
- 2) C
- 3) D
- 4) A
- 5) B
- 6) D
- 7) A
- 8) B
- 9) A
- 10) A

2.7 Mathematical Models: Constructing Functions

1 Construct and Analyze Functions

1) $V(s) = \dfrac{1}{6}s^3$

2) C

3) D

4) B

5) A

6) a. $R(x) = -\dfrac{1}{10}x^2 + 100x$

   b. $R(450) = \$24{,}750.00$

   c.
   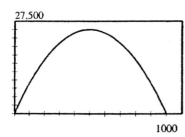

   d. 500; $25,000.00

   e. $50.00

7) $A(x) = \dfrac{1}{2}x^3$

8) $A(x) = \dfrac{4\sqrt{3}+9}{16}x^2 - \dfrac{15}{2}x + 25;\ \{x \mid 0 \le x \le \dfrac{20}{3}\}$

9) $d(t) = \sqrt{1709t}$

Ch. 3 Polynomial and Rational Functions

3.1 Quadratic Functions and Models

1 Graph a Quadratic Function Using Transformations

Match the graph to one of the listed functions without using a graphing utility.

1)

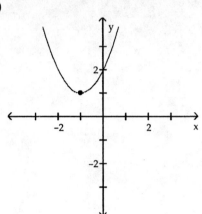

A) $f(x) = x^2 - 2x + 1$ B) $f(x) = x^2 + 2x + 1$ C) $f(x) = x^2 + 2x + 2$ D) $f(x) = x^2 - 2x + 2$

2)

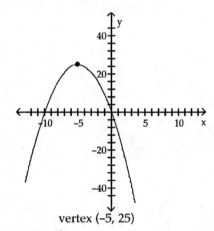

vertex (−5, 25)

A) $f(x) = x^2 - 10$ B) $f(x) = -x^2 - 10$ C) $f(x) = x^2 - 10x$ D) $f(x) = -x^2 - 10x$

3)

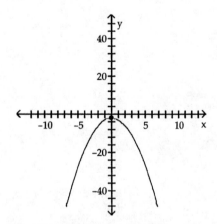

A) $f(x) = -x^2 - 2x$ B) $f(x) = x^2 - 2x$ C) $f(x) = -x^2 - 2$ D) $f(x) = x^2 - 2$

4)

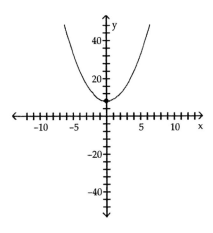

A) $f(x) = x^2 + 8$ B) $f(x) = -x^2 + 8x$ C) $f(x) = x^2 + 8x$ D) $f(x) = -x^2 + 8$

Graph the function f by starting with the graph of $y = x^2$ and using transformations (shifting, compressing, stretching, and/or reflection).

5) $f(x) = x^2 - 2$

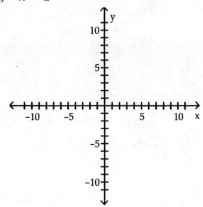

A)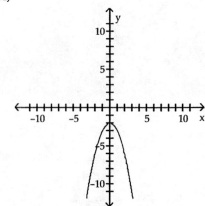

B)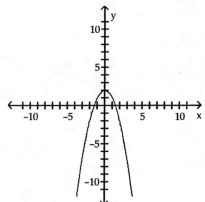

C)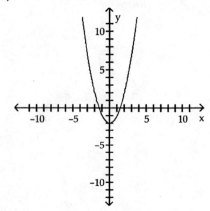

D)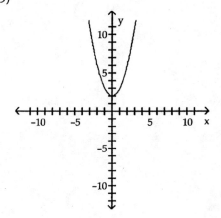

6) $f(x) = -3x^2$

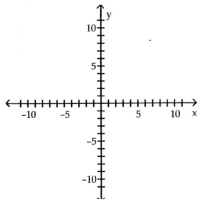

A)
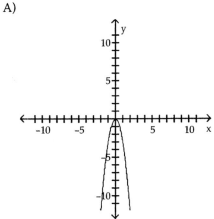

B)
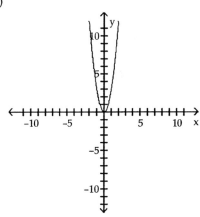

C)
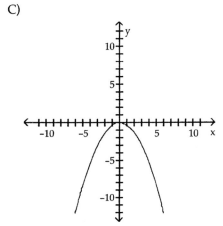

D)
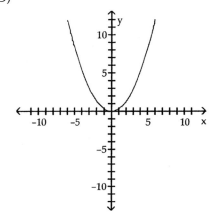

7) $f(x) = -\dfrac{1}{2}x^2$

A)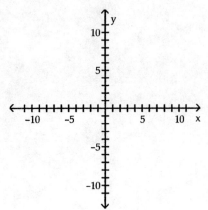

B)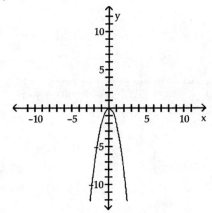

C)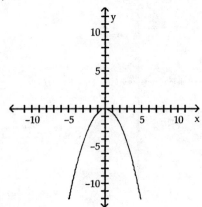

D)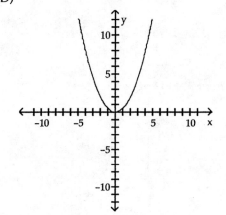

8) $f(x) = \frac{1}{4}x^2 + 2$

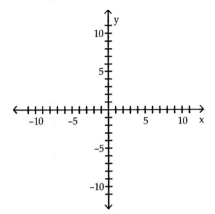

A)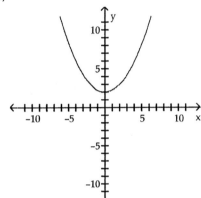

B)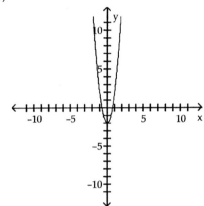

C)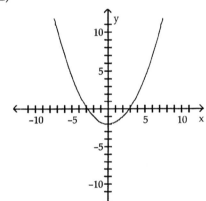

D)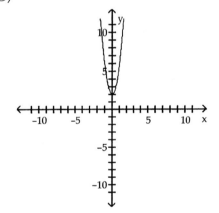

9) $f(x) = -4x^2 - 1$

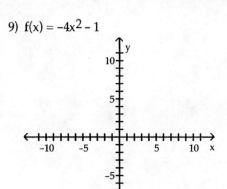

A)

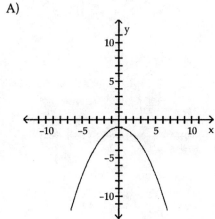

B)

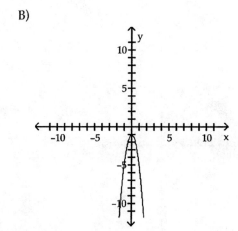

C)

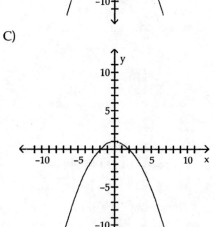

D)

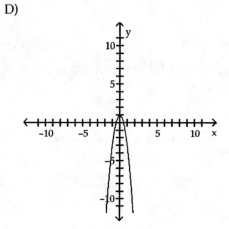

Precalculus Enhanced with Graphing Utilities 204

10) $f(x) = 3x^2 - 4$

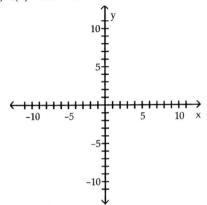

A)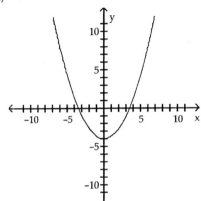

B)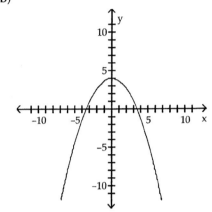

C)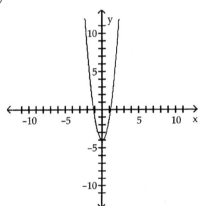

D)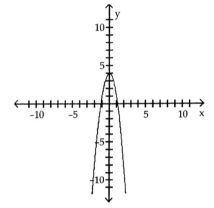

11) $f(x) = -x^2 - 4x$

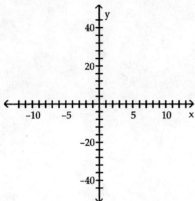

A)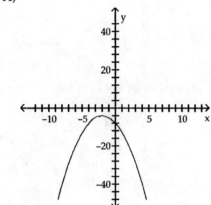

B)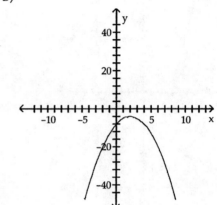

C)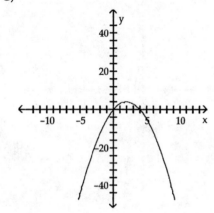

D)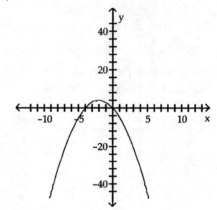

12) $f(x) = x^2 + 4x - 5$

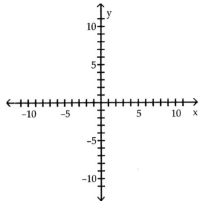

A)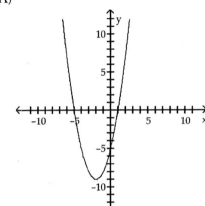

B)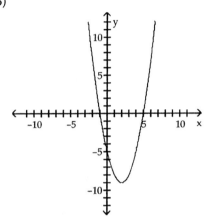

C)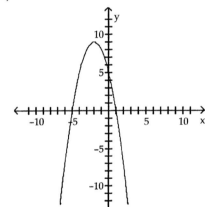

D)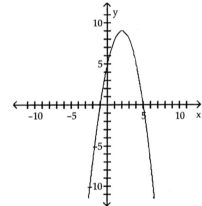

13) $f(x) = -4x^2 - 8x + 5$

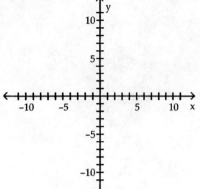

A)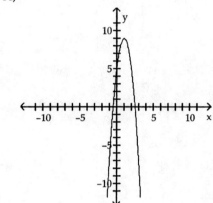

B)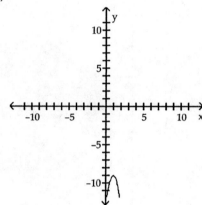

C)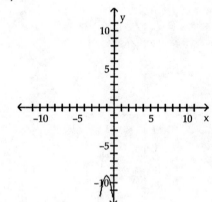

D)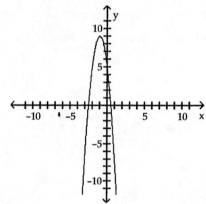

14) $f(x) = \frac{2}{3}x^2 + \frac{4}{3}x - 1$

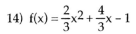

A)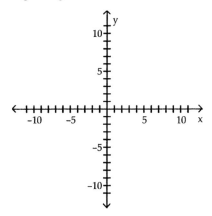

B)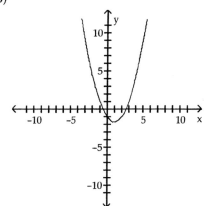

C)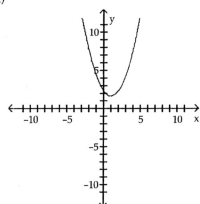

D)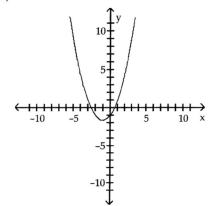

15) $f(x) = -x^2 + 6x - 3$

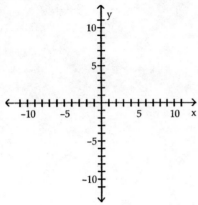

A)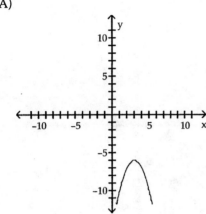

B)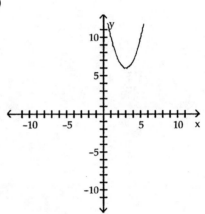

C)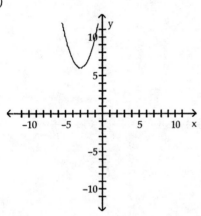

D)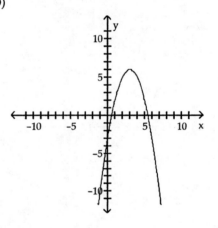

2 Identify the Vertex and Axis of Symmetry of a Quadratic Function

Find the vertex and axis of symmetry of the graph of the function.

1) $f(x) = x^2 + 10x$
   A) $(25, -5); x = 25$ B) $(-5, -25); x = -5$ C) $(-25, 5); x = -25$ D) $(5, -25); x = 5$

2) $f(x) = x^2 - 6x$
   A) $(3, -9); x = 3$ B) $(9, -3); x = 9$ C) $(-3, 9); x = -3$ D) $(-9, 3); x = -9$

3) $f(x) = -x^2 + 8x$
   A) $(-4, -16); x = -4$ B) $(-16, 4); x = -16$ C) $(4, 16); x = 4$ D) $(16, -4); x = 16$

Precalculus Enhanced with Graphing Utilities 210

4) $f(x) = -x^2 - 4x$
   A) $(2, -4)$; $x = 2$
   B) $(-4, 2)$; $x = -4$
   C) $(4, -2)$; $x = 4$
   D) $(-2, 4)$; $x = -2$

5) $f(x) = 3x^2 - 6x$
   A) $(1, -3)$; $x = 1$
   B) $(1, 0)$; $x = 1$
   C) $(-1, -3)$; $x = -1$
   D) $(-1, 0)$; $x = -1$

6) $f(x) = x^2 + 4x - 5$
   A) $(2, -9)$; $x = 2$
   B) $(-2, -9)$; $x = -2$
   C) $(-2, 9)$; $x = -2$
   D) $(2, 9)$; $x = 2$

7) $f(x) = -x^2 - 14x - 3$
   A) $(-14, -3)$; $x = -14$
   B) $(7, -52)$; $x = 7$
   C) $(-7, 46)$; $x = -7$
   D) $(7, -150)$; $x = 7$

8) $f(x) = -3x^2 - 6x + 1$
   A) $(-2, -5)$; $x = -2$
   B) $(1, -8)$; $x = 1$
   C) $(-1, 4)$; $x = -1$
   D) $(2, -23)$; $x = 2$

9) $f(x) = x^2 + 15x - 8$
   A) $(15, 442)$; $x = 15$
   B) $(-13, -8)$; $x = -13$
   C) $\left(\dfrac{15}{2}, \dfrac{643}{4}\right)$; $x = \dfrac{15}{2}$
   D) $\left(-\dfrac{15}{2}, -\dfrac{257}{4}\right)$; $x = -\dfrac{15}{2}$

10) $f(x) = -11x^2 - 2x - 2$
   A) $\left(\dfrac{1}{11}, \dfrac{21}{11}\right)$; $x = \dfrac{1}{11}$
   B) $(11, -2)$; $x = 11$
   C) $\left(-11, -\dfrac{21}{11}\right)$; $x = -11$
   D) $\left(-\dfrac{1}{11}, -\dfrac{21}{11}\right)$; $x = -\dfrac{1}{11}$

3 Graph a Quadratic Function Using Its Vertex, Axis and Intercepts

Graph the function using its vertex, axis of symmetry, and intercepts.

1) $f(x) = x^2 + 4x$

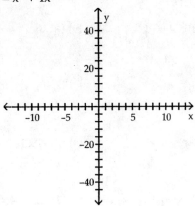

A) vertex (-2, 4)
intercept (0, 8)

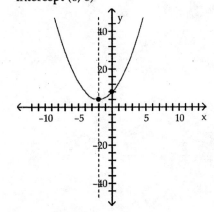

B) vertex (2, -4)
intercepts (0, 0), (4, 0)

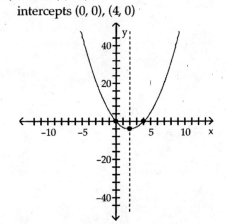

C) vertex (2, 4)
intercept (0, 8)

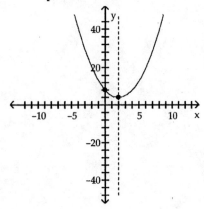

D) vertex (-2, -4)
intercepts (0, 0), (-4, 0)

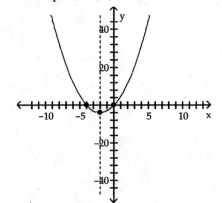

2) $f(x) = -x^2 - 12x$

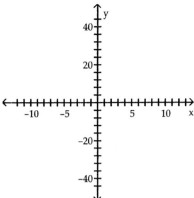

A) vertex (−6, −36)
   intercept (0, −72)

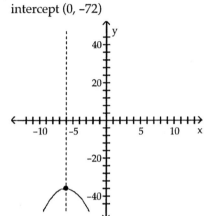

B) vertex (−6, 36)
   intercepts (0, 0), (−12, 0)

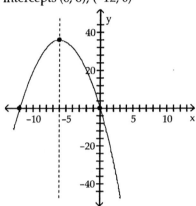

C) vertex (6, −36)
   intercept (0, −72)

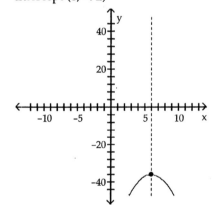

D) vertex (6, 36)
   intercepts (0, 0), (12, 0)

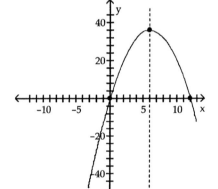

3) $f(x) = x^2 + 10x + 25$

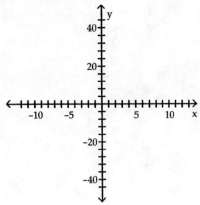

A) vertex (-5, 25)
   intercept (0, 50)

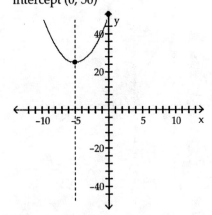

B) vertex (5, 0)
   intercepts (0, 25), (5, 0)

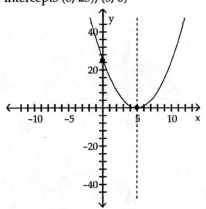

C) vertex (-5, 0)
   intercepts (0, 25), (-5, 0)

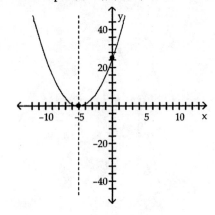

D) vertex (5, 25)
   intercept (0, 50)

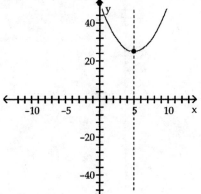

4) $f(x) = x^2 + 8x + 7$

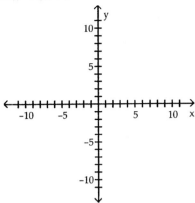

A) vertex $(4, -9)$
intercepts $(1, 0), (7, 0), (0, 7)$

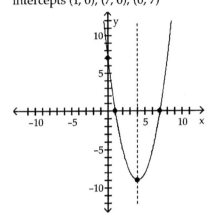

B) vertex $(-4, -9)$
intercepts $(-1, 0), (-7, 0), (0, 7)$

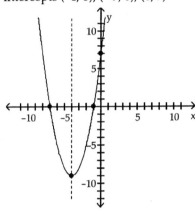

C) vertex $(4, 9)$
intercepts $(1, 0), (7, 0), (0, -7)$

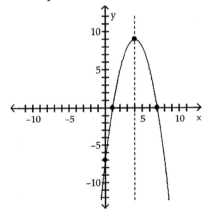

D) vertex $(-4, 9)$
intercepts $(-1, 0), (-7, 0), (0, -7)$

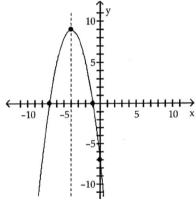

5) $f(x) = -x^2 - 6x - 8$

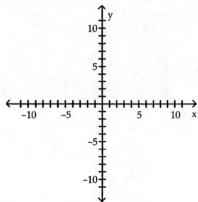

A) vertex $(3, -1)$
   intercepts $(2, 0), (4, 0), (0, 8)$

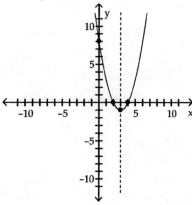

B) vertex $(-3, -1)$
   intercepts $(-2, 0), (-4, 0), (0, 8)$

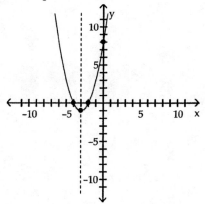

C) vertex $(3, 1)$
   intercepts $(2, 0), (4, 0), (0, -8)$

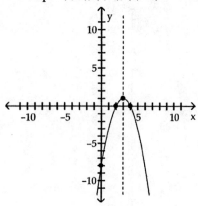

D) vertex $(-3, 1)$
   intercepts $(-2, 0), (-4, 0), (0, -8)$

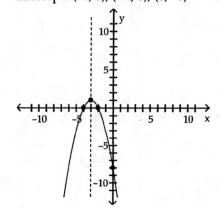

6) $f(x) = x^2 - 6x + 5$

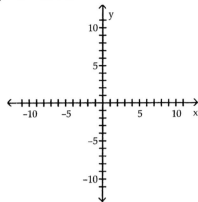

A) vertex (3, −4)
   intercepts (5, 0), (1, 0), (0, 5)

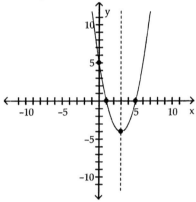

B) vertex (−3, −4)
   intercepts (−5, 0), (−1, 0), (0, 5)

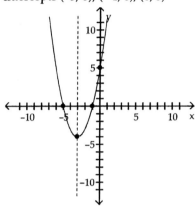

C) vertex (3, 4)
   intercepts (5, 0), (1, 0), (0, −5)

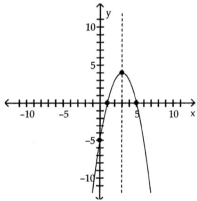

D) vertex (−3, 4)
   intercepts (−5, 0), (−1, 0), (0, −5)

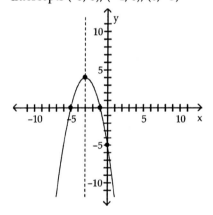

7) $f(x) = -x^2 + 6x - 5$

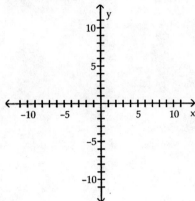

A) vertex (3, 4)
   intercepts (5, 0), (1, 0), (0, -5)

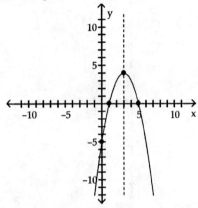

B) vertex (-3, 4)
   intercepts (-5, 0), (-1, 0), (0, -5)

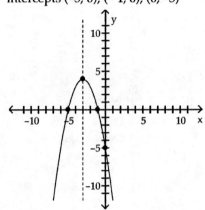

C) vertex (3, -4)
   intercepts (5, 0), (1, 0), (0, 5)

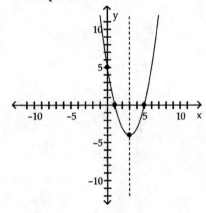

D) vertex (-3, -4)
   intercepts (-5, 0), (-1, 0), (0, 5)

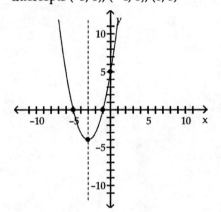

8) $f(x) = 2x^2 + 20x + 54$

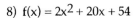

A) vertex (5, 4) intercept $\left(0, \dfrac{33}{2}\right)$

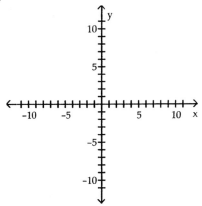

B) vertex (−5, 4) intercept (0, 54)

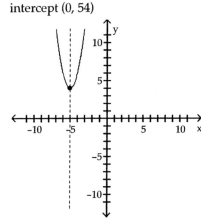

C) vertex (5, 4) intercept (0, 54)

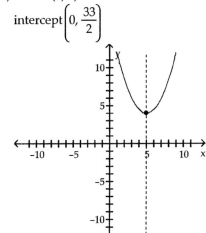

D) vertex (−5, 4) intercept $\left(0, \dfrac{33}{2}\right)$

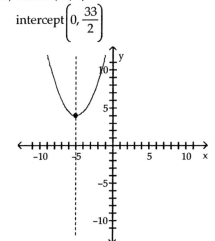

9) $f(x) = -8x^2 - 2x - 2$

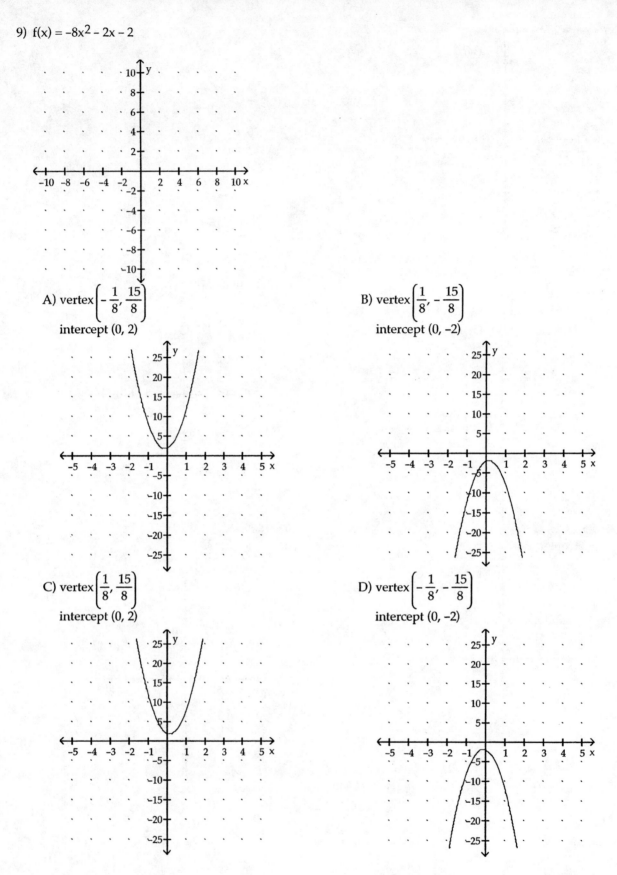

A) vertex $\left(-\frac{1}{8}, \frac{15}{8}\right)$
intercept $(0, 2)$

B) vertex $\left(\frac{1}{8}, -\frac{15}{8}\right)$
intercept $(0, -2)$

C) vertex $\left(\frac{1}{8}, \frac{15}{8}\right)$
intercept $(0, 2)$

D) vertex $\left(-\frac{1}{8}, -\frac{15}{8}\right)$
intercept $(0, -2)$

Determine the domain and the range of the function.

10) $f(x) = x^2 + 4x$

    A) domain: $\{x \mid x \geq -2\}$
       range: $\{y \mid y \geq -4\}$

    B) domain: all real numbers
       range: $\{y \mid y \geq -4\}$

    C) domain: $\{x \mid x \geq 2\}$
       range: $\{y \mid y \geq 4\}$

    D) domain: all real numbers
       range: $\{y \mid y \geq 4\}$

11) $f(x) = -x^2 + 4x$

    A) domain: all real numbers
       range: $\{y \mid y \leq -4\}$

    B) domain: $\{x \mid x \leq -2\}$
       range: $\{y \mid y \leq 4\}$

    C) domain: all real numbers
       range: $\{y \mid y \leq 4\}$

    D) domain: $\{x \mid x \leq 2\}$
       range: $\{y \mid y \leq 4\}$

12) $f(x) = x^2 - 8x + 16$

    A) domain: $\{x \mid x \geq 4\}$
       range: $\{y \mid y \geq 0\}$

    B) domain: all real numbers
       range: $\{y \mid y \geq 0\}$

    C) domain: all real numbers
       range: $\{y \mid y \geq 16\}$

    D) domain: $\{x \mid x \geq -4\}$
       range: $\{y \mid y \geq 0\}$

13) $f(x) = x^2 + 6x + 5$

    A) domain: all real numbers
       range: $\{y \mid y \geq -4\}$

    B) domain: range: $\{x \mid x \geq 3\}$
       range: $\{y \mid y \geq 4\}$

    C) domain: all real numbers
       range: $\{y \mid y \geq 4\}$

    D) domain: range: $\{x \mid x \geq 3\}$
       range: $\{y \mid y \geq -4\}$

14) $f(x) = -x^2 - 6x - 5$

    A) domain: $\{x \mid x \leq -3\}$
       range: $\{y \mid y \leq -4\}$

    B) domain: all real numbers
       range: $\{y \mid y \leq 4\}$

    C) domain: $\{x \mid x \leq -3\}$
       range: $\{y \mid y \leq 4\}$

    D) domain: all real numbers
       range: $\{y \mid y \leq -4\}$

15) $f(x) = x^2 - 2x - 8$

    A) domain: all real numbers
       range: $\{y \mid y \leq 9\}$

    B) domain: $\{x \mid x \geq -1\}$
       range: $\{y \mid y \geq -9\}$

    C) domain: all real numbers
       range: $\{y \mid y \geq -9\}$

    D) domain: all real numbers
       range: all real numbers

16) $f(x) = -x^2 + 2x + 8$

    A) domain: all real numbers
       range: $\{y \mid y \leq -9\}$

    B) domain: $\{x \mid x \leq -1\}$
       range: $\{y \mid y \leq 9\}$

    C) domain: all real numbers
       range: $\{y \mid y \leq 9\}$

    D) domain: all real numbers
       range: all real numbers

17) $g(x) = 8x^2 + 32x - 32$

    A) domain: all real numbers
       range: $\{y \mid y \leq -2\}$

    B) domain: all real numbers
       range: $\{y \mid y \leq -64\}$

    C) domain: all real numbers
       range: $\{y \mid y \geq -2\}$

    D) domain: all real numbers
       range: $\{y \mid y \geq -64\}$

18) $f(x) = -11x^2 - 2x - 7$

A) domain: all real numbers
range: $\left\{y \mid y \leq -\dfrac{76}{11}\right\}$

B) domain: all real numbers
range: $\left\{y \mid y \geq \dfrac{76}{11}\right\}$

C) domain: all real numbers
range: $\left\{y \mid y \leq \dfrac{76}{11}\right\}$

D) domain: all real numbers
range: $\left\{y \mid y \geq -\dfrac{76}{11}\right\}$

Determine where the function is increasing and where it is decreasing.

19) $f(x) = x^2 - 6x$

A) increasing on $(-\infty, 3)$
decreasing on $(3, \infty)$

B) increasing on $(-\infty, -3)$
decreasing on $(-3, \infty)$

C) increasing on $(-3, \infty)$
decreasing on $(-\infty, -3)$

D) increasing on $(3, \infty)$
decreasing on $(-\infty, 3)$

20) $f(x) = -x^2 + 2x$

A) increasing on $(-\infty, 1)$
decreasing on $(1, \infty)$

B) increasing on $(-1, \infty)$
decreasing on $(-\infty, -1)$

C) increasing on $(1, \infty)$
decreasing on $(-\infty, 1)$

D) increasing on $(-\infty, -1)$
decreasing on $(-1, \infty)$

21) $f(x) = x^2 + 8x + 16$

A) increasing on $(-4, \infty)$
decreasing on $(-\infty, -4)$

B) increasing on $(-\infty, -4)$
decreasing on $(-4, \infty)$

C) increasing on $(4, \infty)$
decreasing on $(-\infty, 4)$

D) increasing on $(-\infty, 4)$
decreasing on $(4, \infty)$

22) $f(x) = x^2 + 4x + 3$

A) increasing on $(-2, \infty)$
decreasing on $(-\infty, -2)$

B) increasing on $(-1, \infty)$
decreasing on $(-\infty, -1)$

C) increasing on $(-\infty, -1)$
decreasing on $(-1, \infty)$

D) increasing on $(-\infty, -2)$
decreasing on $(-2, \infty)$

23) $f(x) = -x^2 - 6x - 5$

A) increasing on $(4, \infty)$
decreasing on $(-\infty, 4)$

B) increasing on $(-3, \infty)$
decreasing on $(-\infty, -3)$

C) increasing on $(-\infty, -3)$
decreasing on $(-3, \infty)$

D) increasing on $(-\infty, 4)$
decreasing on $(4, \infty)$

24) $f(x) = x^2 - 4x - 5$

A) increasing on $(2, \infty)$
decreasing on $(-\infty, 2)$

B) increasing on $(-\infty, -9)$
decreasing on $(-9, \infty)$

C) increasing on $(-\infty, 2)$
decreasing on $(2, \infty)$

D) increasing on $(-9, \infty)$
decreasing on $(-\infty, -9)$

25) $f(x) = -x^2 + 2x + 3$

   A) increasing on $(-\infty, 4)$
      decreasing on $(4, \infty)$
   B) increasing on $(1, \infty)$
      decreasing on $(-\infty, 1)$
   C) increasing on $(4, \infty)$
      decreasing on $(-\infty, 4)$
   D) increasing on $(-\infty, 1)$
      decreasing on $(1, \infty)$

26) $g(x) = 10x^2 + 120x + 260$

   A) decreasing on $(-\infty, -6)$
      increasing on $(-6, \infty)$
   B) decreasing on $(-\infty, 6)$
      increasing on $(6, \infty)$
   C) increasing on $(-\infty, -60)$
      decreasing on $(-60, \infty)$
   D) increasing on $(-\infty, -6)$
      decreasing on $(-6, \infty)$

27) $f(x) = -3x^2 - 2x - 4$

   A) increasing on $\left(-\infty, -\frac{1}{3}\right]$
      decreasing on $\left[-\frac{1}{3}, \infty\right)$
   B) increasing on $\left(-\infty, -\frac{11}{3}\right]$
      decreasing on $\left[-\frac{11}{3}, \infty\right)$
   C) decreasing on $\left(-\infty, -\frac{1}{3}\right]$
      increasing on $\left[-\frac{1}{3}, \infty\right)$
   D) increasing on $\left(-\infty, \frac{1}{3}\right]$
      decreasing on $\left[\frac{1}{3}, \infty\right)$

Determine the quadratic function whose graph is given.

28)

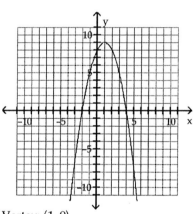

Vertex: (1, 9)
y-intercept: (0, 8)

A) $f(x) = -x^2 - 4x + 8$ B) $f(x) = -x^2 + 2x - 8$ C) $f(x) = x^2 - 4x + 8$ D) $f(x) = -x^2 + 2x + 8$

29)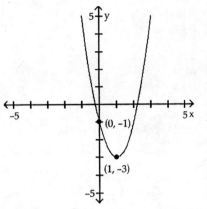

A) $f(x) = 2x^2 - 4x - 1$ B) $f(x) = 2x^2 + 8x - 1$ C) $f(x) = -2x^2 - 4x - 1$ D) $f(x) = -2x^2 + 4x + 1$

4 Use the Maximum or Minimum of a Quadratic Function to Solve Applied Problems

Determine, without graphing, whether the given quadratic function has a maximum value or a minimum value and then find that value.

1) $f(x) = x^2 + 9$
   A) maximum; 0   B) maximum; 9   C) minimum; 0   D) minimum; 9

2) $f(x) = x^2 - 3$
   A) maximum; 0   B) maximum; -3   C) minimum; 0   D) minimum; -3

3) $f(x) = x^2 - 2x - 3$
   A) maximum; 1   B) minimum; -4   C) maximum; -4   D) minimum; 1

4) $f(x) = -x^2 + 2x - 9$
   A) minimum; 1   B) minimum; -8   C) maximum; -8   D) maximum; 1

5) $f(x) = 2x^2 + 2x - 9$
   A) maximum; $-\frac{1}{2}$ B) minimum; $-\frac{1}{2}$ C) maximum; $-\frac{19}{2}$ D) minimum; $-\frac{19}{2}$

6) $f(x) = 3x^2 + 3x$
   A) minimum; $-\frac{3}{4}$ B) maximum; $-\frac{3}{4}$ C) maximum; $-\frac{1}{2}$ D) minimum; $-\frac{1}{2}$

7) $f(x) = -4x^2 - 8x$
   A) maximum; -4   B) maximum; 4   C) minimum; -4   D) minimum; 4

8) $f(x) = -4x^2 - 2x - 5$
   A) maximum; $\frac{19}{4}$ B) minimum; $\frac{19}{4}$ C) minimum; $-\frac{19}{4}$ D) maximum; $-\frac{19}{4}$

Solve the problem.

9) The manufacturer of a CD player has found that the revenue R (in dollars) is
$R(p) = -5p^2 + 1120p$, when the unit price is p dollars. If the manufacturer sets the price p to maximize revenue, what is the maximum revenue to the nearest whole dollar?
   A) $501,760   B) $125,440   C) $62,720   D) $250,880

10) The owner of a video store has determined that the cost C, in dollars, of operating the store is approximately given by $C(x) = 2x^2 - 18x + 780$, where x is the number of videos rented daily. Find the lowest cost to the nearest dollar.
   A) $821   B) $740   C) $699   D) $618

11) The price p and the quantity x sold of a certain product obey the demand equation
$$p = -\frac{1}{7}x + 140, \quad 0 \le x \le 980.$$
What quantity x maximizes revenue? What is the maximum revenue?
   A) 980; $34,300   B) 245; $25,725   C) 735; $25,725   D) 490; $34,300

12) The price p and the quantity x sold of a certain product obey the demand equation
$$p = -\frac{1}{4}x + 140, \quad 0 \le x \le 560.$$
What price should the company charge to maximize revenue?
   A) $105   B) $70   C) $84   D) $35

13) The price p (in dollars) and the quantity x sold of a certain product obey the demand equation
$x = -9p + 180, \quad 0 \le p \le 20.$
What quantity x maximizes revenue? What is the maximum revenue?
   A) 90; $900   B) 45; $675   C) 180; $900   D) 135; $675

14) The price p (in dollars) and the quantity x sold of a certain product obey the demand equation
$p = -10x + 280, \quad 0 \le x \le 28.$
What price should the company charge to maximize revenue?
   A) $7   B) $14   C) $21   D) $16.8

15) The profit that the vendor makes per day by selling x pretzels is given by the function
$P(x) = -0.004x^2 + 2.4x - 200$. Find the number of pretzels that must be sold to maximize profit.
   A) 1.2 pretzels   B) 160 pretzels   C) 300 pretzels   D) 600 pretzels

16) The owner of a video store has determined that the profits P of the store are approximately given by
$P(x) = -x^2 + 150x + 69$, where x is the number of videos rented daily. Find the maximum profit to the nearest dollar.
   A) $5625   B) $5694   C) $11,250   D) $11,319

17) You have 252 feet of fencing to enclose a rectangular region. Find the dimensions of the rectangle that maximize the enclosed area.
   A) 65 ft by 61 ft   B) 126 ft by 31.5 ft   C) 63 ft by 63 ft   D) 126 ft by 126 ft

18) A developer wants to enclose a rectangular grassy lot that borders a city street for parking. If the developer has 328 feet of fencing and does not fence the side along the street, what is the largest area that can be enclosed?

   A) 13,448 ft^2 B) 20,172 ft^2 C) 26,896 ft^2 D) 6724 ft^2

19) You have 136 feet of fencing to enclose a rectangular region. What is the maximum area?

   A) 1152 square feet   B) 1156 square feet   C) 4624 square feet   D) 18,496 square feet

20) You have 92 feet of fencing to enclose a rectangular plot that borders on a river. If you do not fence the side along the river, find the length and width of the plot that will maximize the area.

   A) length: 46 feet, width: 46 feet   B) length: 46 feet, width: 23 feet
   C) length: 23 feet, width: 23 feet   D) length: 69 feet, width: 23 feet

21) A projectile is fired from a cliff 300 feet above the water at an inclination of 45° to the horizontal, with a muzzle velocity of 350 feet per second. The height h of the projectile above the water is given by $h(x) = \dfrac{-32x^2}{(350)^2} + x + 300$, where x is the horizontal distance of the projectile from the base of the cliff. Find the maximum height of the projectile.

   A) 1914.06 ft   B) 3171.09 ft   C) 1257.03 ft   D) 957.03 ft

22) A projectile is fired from a cliff 300 feet above the water at an inclination of 45° to the horizontal, with a muzzle velocity of 250 feet per second. The height h of the projectile above the water is given by $h(x) = \dfrac{-32x^2}{(250)^2} + x + 300$, where x is the horizontal distance of the projectile from the base of the cliff. How far from the base of the cliff is the height of the projectile a maximum?

   A) 1764.84 ft   B) 976.56 ft   C) 488.28 ft   D) 788.28 ft

23) Consider the quadratic model $h(t) = -16t^2 + 40t + 50$ for the height (in feet), h, of an object t seconds after the object has been projected straight up into the air. Find the maximum height attained by the object. How much time does it take to fall back to the ground? Assume that it takes the same time for going up and coming down.

   A) maximum height = 50 ft; time to reach ground = 2.5 seconds
   B) maximum height = 50 ft; time to reach ground = 1.25 seconds
   C) maximum height = 75 ft; time to reach ground = 1.25 seconds
   D) maximum height = 75 ft; time to reach ground = 2.5 seconds

Solve.

24) An object is propelled vertically upward from the top of a 80-foot building. The quadratic function $s(t) = -16t^2 + 112t + 80$ models the ball's height above the ground, s(t), in feet, t seconds after it was thrown. How many seconds does it take until the object finally hits the ground? Round to the nearest tenth of a second if necessary.

   A) 3.5 seconds   B) 7.7 seconds   C) 0.7 seconds   D) 2 seconds

Solve the problem.

25) A suspension bridge has twin towers that are 1300 feet apart. Each tower extends 180 feet above the road surface. The cables are parabolic in shape and are suspended from the tops of the towers. The cables touch the road surface at the center of the bridge. Find the height of the cable at a point 200 feet from the center of the bridge.

26) Alan is building a garden shaped like a rectangle with a semicircle attached to one short side. If he has 50 feet of fencing to go around it, what dimensions will give him the maximum area in the garden?

A) width = $\dfrac{100}{\pi + 4} \approx 14$, length = 18

B) width = $\dfrac{50}{\pi + 4} \approx 7$, length = 14

C) width = $\dfrac{100}{\pi + 8} \approx 9$, length = 13.5

D) width = $\dfrac{100}{\pi + 4} \approx 14$, length = 7

27) The quadratic function $f(x) = 0.0037x^2 - 0.41x + 36.14$ models the median, or average, age, y, at which U.S. men were first married x years after 1900. In which year was this average age at a minimum? (Round to the nearest year.) What was the average age at first marriage for that year? (Round to the nearest tenth.)

A) 1951, 36 years old
B) 1955, 47.5 years old
C) 1955, 24.8 years old
D) 1936, 47.5 years old

5 Use a Graphing Utility to Find the Quadratic Function of Best Fit to Data

Use a graphing calculator to plot the data and find the quadratic function of best fit.

1) Southern Granite and Marble sells granite and marble by the square yard. One of its granite patterns is price sensitive. If the price is too low, customers perceive that it has less quality. If the price is too high, customers perceive that it is overpriced. The company conducted a pricing test with potential customers. The following data was collected. Use a graphing calculator to plot the data. What is the quadratic function of best fit?

Price, x	Buyers, B
$20	30
$30	50
$40	65
$60	75
$80	72
$100	50
$110	25

A) $B(x) = -0.0243x^2 + 3.115x + 22.13$

B) $B(x) = -0.243x^2 + 3.115x - 22.13$

C) $B(x) = -0.0243x^2 + 3.115x - 22.13$

D) $B(x) = 0.0243x^2 - 3.115x - 22.13$

2) An engineer collects data showing the speed s of a given car model and its average miles per gallon M. Use a graphing calculator to plot the scatter diagram. What is the quadratic function of best fit?

Speed, s	mph, M
20	18
30	20
40	23
50	25
60	28
70	24
80	22

A) $M(s) = -0.631x^2 + 0.720x + 5.142$

B) $M(s) = -6.309x^2 + 0.720x + 5.142$

C) $M(s) = 0.063x^2 + 0.720x + 5.142$

D) $M(s) = -0.0063x^2 + 0.720x + 5.142$

3) The number of housing starts in one beachside community remained fairly level until 1992 and then began to increase. The following data shows the number of housing starts since 1992 (x = 1). Use a graphing calculator to plot a scatter diagram. What is the quadratic function of best fit?

Year, x	Housing Starts, H
1	200
2	205
3	210
4	240
5	245
6	230
7	220
8	210

A) $H(x) = -2.679x^2 + 26.607x + 168.571$

B) $H(x) = -2.679x^2 - 26.607x + 168.571$

C) $H(x) = 2.679x^2 + 26.607x + 168.571$

D) $H(x) = -2.679x^2 + 26.607x - 168.571$

4) The number of housing starts in one beachside community remained fairly level until 1992 and then began to increase. The following data shows the number of housing starts since 1992 (x = 1). Use a graphing calculator to plot a scatter diagram. What is the quadratic function of best fit?

Year, x	Housing Starts, H
1	200
2	210
3	230
4	240
5	250
6	230
7	215
8	208

A) $H(x) = -3.268x^2 + 30.494x - 168.982$

B) $H(x) = -3.268x^2 + 30.494x + 168.982$

C) $H(x) = 3.268x^2 + 30.494x + 168.982$

D) $H(x) = -3.268x^2 - 30.494x + 168.982$

5) A small manufacturing firm collected the following data on advertising expenditures (in thousands of dollars) and total revenue (in thousands of dollars).

Advertising, x	Total Revenue, R
25	6430
28	6432
31	6434
32	6434
34	6434
39	6431
40	6432
45	6420

Find the quadratic function of best fit.

A) $R(x) = -0.015x^2 + 4.53x + 6123$

B) $R(x) = -0.31x^2 + 2.63x + 6128$

C) $R(x) = -0.024x^2 + 7.13x + 6209$

D) $R(x) = -0.091x^2 + 5.95x + 6337$

Precalculus Enhanced with Graphing Utilities

6) The following data represents the total revenue, R (in dollars), received from selling x bicycles at Tunney's Bicycle Shop. Using a graphing utility, find the quadratic function of best fit using coefficients rounded to the nearest hundredth.

Number of Bicycles, x	Total Revenue, R (in dollars)
0	0
22	27,000
70	46,000
96	55,200
149	61,300
200	64,000
230	64,500
250	67,000

7) The following table shows the median number of hours of leisure time that Americans had each week in various years.

Year	1973	1980	1987	1993	1997
Median # of Leisure hrs per Week	26.2	19.2	16.6	18.8	19.5

Use x = 0 to represent the year 1973. Using a graphing utility, determine the quadratic regression equation for the data given. What year corresponds to the time when Americans had the least time to spend on leisure?

3.2 Polynomial Functions and Models

1 Identify Polynomial Functions and Their Degree

State whether the function is a polynomial function or not. If it is, give its degree. If it is not, tell why not.

1) $f(x) = 3x + 6x^2$

   A) Yes; degree 3   B) Yes; degree 2   C) Yes; degree 1   D) Yes; degree 6

2) $f(x) = 13x^4 + 9x^3 + 5$

   A) Yes; degree 8
   B) Yes; degree 7
   C) No; the last term has no variable
   D) Yes; degree 4

3) $f(x) = \dfrac{7 - x^5}{2}$

   A) Yes; degree 5
   B) No; it is a ratio
   C) Yes; degree 1
   D) No; x is a negative term

4) $f(x) = \dfrac{9}{2} - \dfrac{1}{4}x$

   A) Yes; degree 0
   B) Yes; degree 4
   C) Yes; degree 1
   D) No; x has a fractional coefficient

5) $f(x) = 14$

   A) Yes; degree 0
   B) No; it contains no variables
   C) Yes; degree 1
   D) No; it is a constant

6) $f(x) = 1 + \dfrac{15}{x}$

    A) Yes; degree 15   B) Yes; degree 0

    C) Yes; degree 1   D) No; x is raised to a negative power

7) $f(x) = x(x - 7)$

    A) Yes; degree 2   B) Yes; degree 1   C) Yes; degree 0   D) No; it is a product

8) $f(x) = 1 - \dfrac{3}{x^2}$

    A) No; x is raised to the negative 2 power   B) Yes; degree 2

    C) Yes; degree $\dfrac{1}{2}$ D) Yes; degree –2

9) $f(x) = \dfrac{x^5 - 7}{x^6}$

    A) No; it is a ratio of polynomials   B) Yes; degree 5

    C) Yes; degree –6   D) Yes; degree 6

10) $f(x) = x^{3/2} - x^6 + 3$

    A) Yes; degree 3   B) Yes; degree 6

    C) No; x is raised to non-integer 3/2 power   D) Yes; degree 3/2

11) $2(x - 1)^{11}(x + 1)^8$

    A) Yes; degree 22   B) Yes; degree 11   C) Yes; degree 19   D) Yes; degree 2

12) $f(x) = \sqrt{x}(\sqrt{x} - 7)$

    A) No; it is a product   B) No; x is raised to non-integer power

    C) Yes; degree 1   D) Yes; degree 2

13) $f(x) = -9x^5 + \pi x^4 - 1$

    A) Yes; degree 9   B) No; x^4 has a non-integer coefficient

    C) Yes; degree 10   D) Yes; degree 5

2 Graph Polynomial Functions Using Transformations

Use transformations of the graph of $y = x^4$ or $y = x^5$ to graph the function.

1) $f(x) = (x - 5)^4$

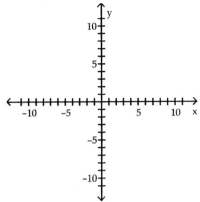

A)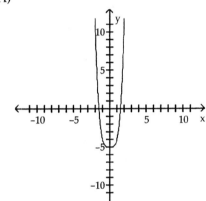

B)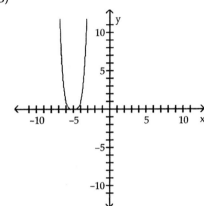

C)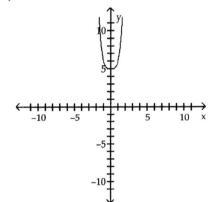

D)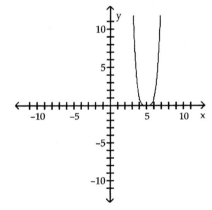

2) $f(x) = x^4 + 4$

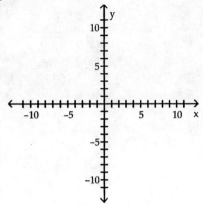

A)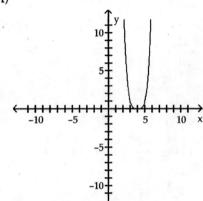

B)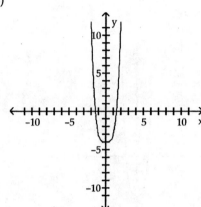

C)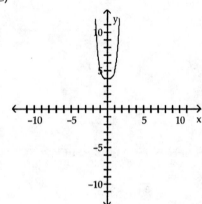

D)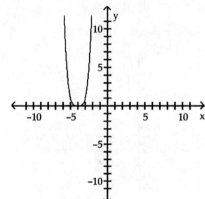

3) $f(x) = \frac{1}{2}x^4$

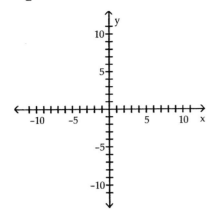

A)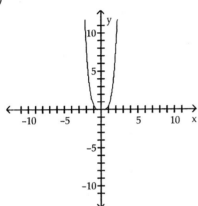

B)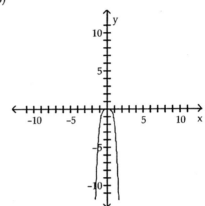

C)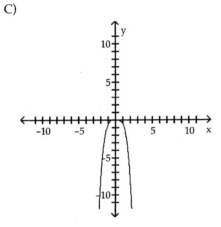

D)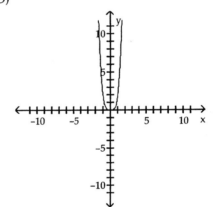

4) $f(x) = 5x^4$

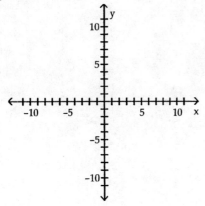

A)
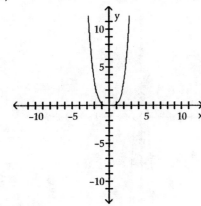

B)
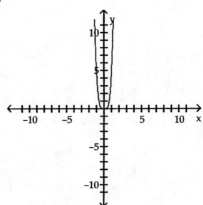

C)
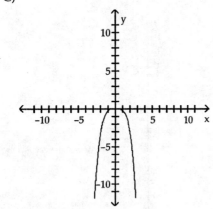

D)
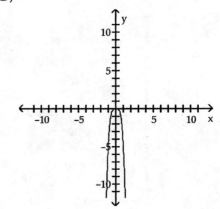

5) $f(x) = (x+3)^4 + 4$

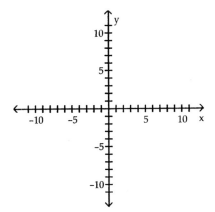

A)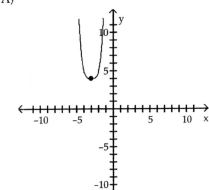

B)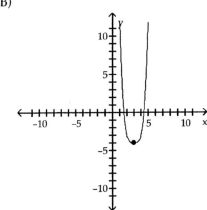

C)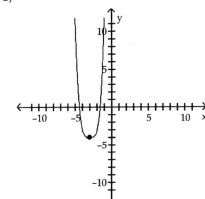

D)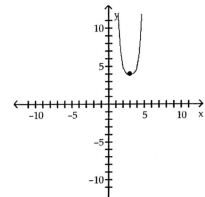

6) $f(x) = \frac{1}{2}(x-5)^4 + 4$

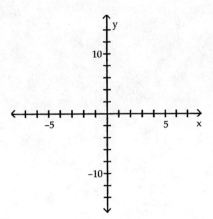

A)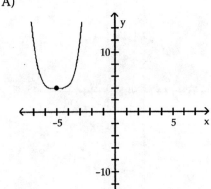

B)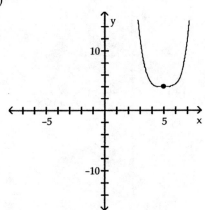

C)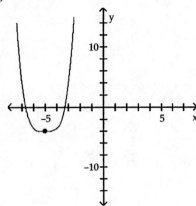

D)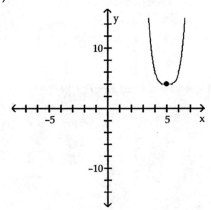

7) $f(x) = -2(x+5)^4 + 4$

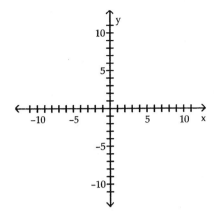

A)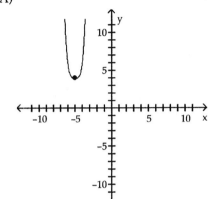

B)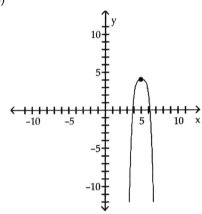

C)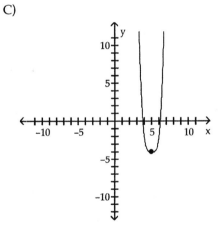

D)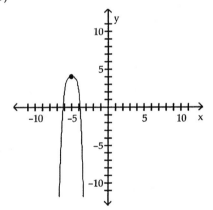

8) $f(x) = 3 - (x-5)^4$

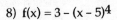

A)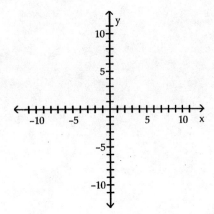

B)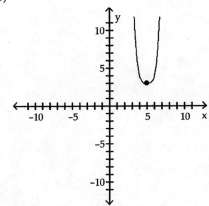

C)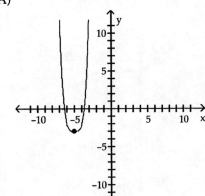

D)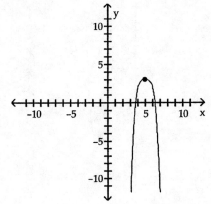

9) $f(x) = (x-4)^5$

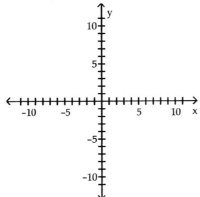

A)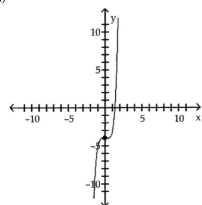

B)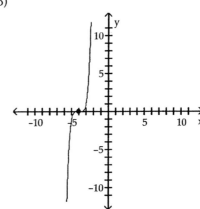

C)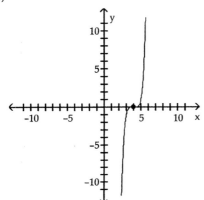

D)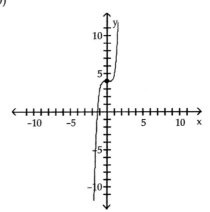

10) $f(x) = x^5 - 3$

A)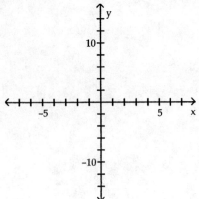

B)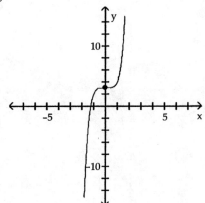

C)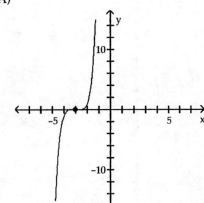

D)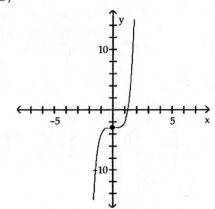

11) $f(x) = -\dfrac{1}{3}x^5$

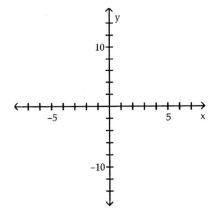

A)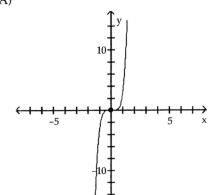

B)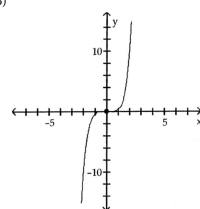

C)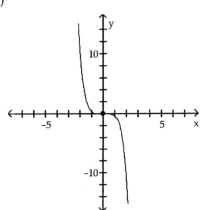

D)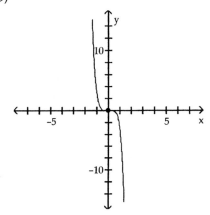

12) $f(x) = 3x^5$

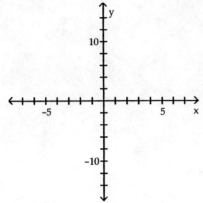

A)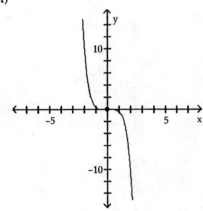

B)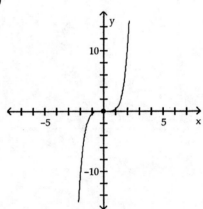

C)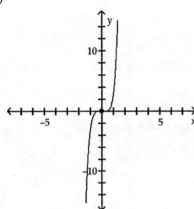

D)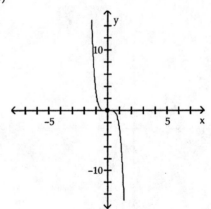

13) $f(x) = (x+5)^5 + 2$

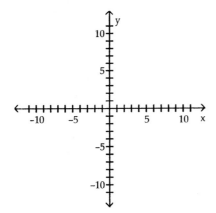

A)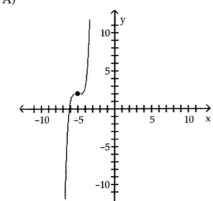

B)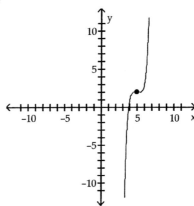

C)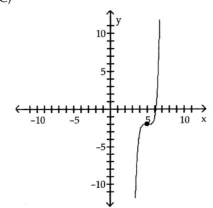

D)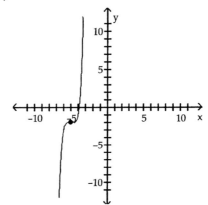

14) $f(x) = \frac{1}{2}(x+5)^5 + 2$

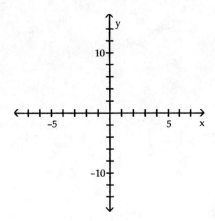

A)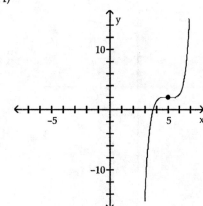

B)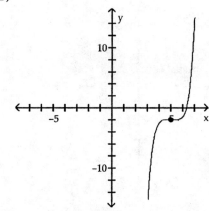

C)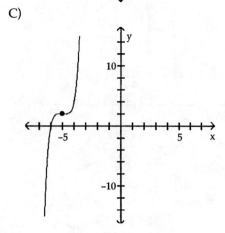

D)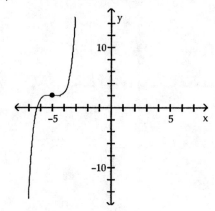

15) $f(x) = -2(x+4)^5 + 2$

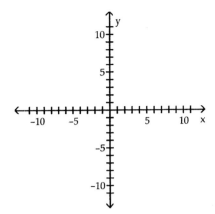

A)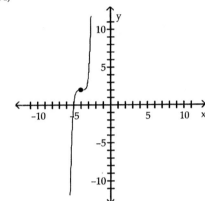

B)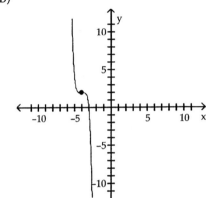

C)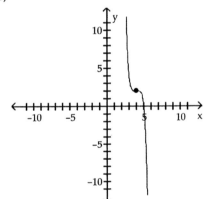

D)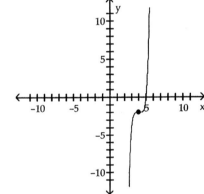

16) $f(x) = 3 - (x + 5)^5$

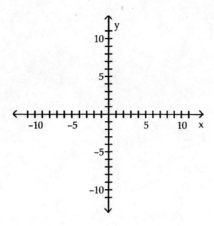

A)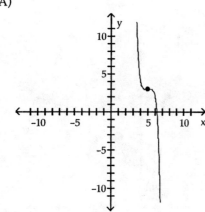

B)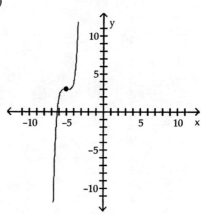

C)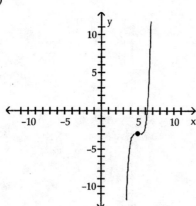

D)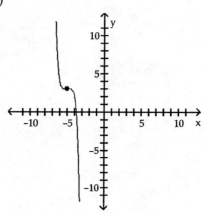

3 Identify the Zeros of a Polynomial Function and Their Multiplicity

Form a polynomial whose zeros and degree are given.

1) Zeros: -3, -2, 3; degree 3
   A) $f(x) = x^3 + 2x^2 + 9x + 18$ for $a = 1$
   B) $f(x) = x^3 + 2x^2 - 9x - 18$ for $a = 1$
   C) $f(x) = x^3 - 2x^2 - 9x + 18$ for $a = 1$
   D) $f(x) = x^3 - 2x^2 + 9x - 18$ for $a = 1$

2) Zeros: 0, -5, 4; degree 3
   A) $f(x) = x^3 + x^2 + x + 20$ for $a = 1$
   B) $f(x) = x^3 + x^2 - 20x$ for $a = 1$
   C) $f(x) = x^3 + x^2 + 20x$ for $a = 1$
   D) $f(x) = x^3 + x^2 + x - 20$ for $a = 1$

Precalculus Enhanced with Graphing Utilities

3) Zeros: −1, 1, −5; degree 3
   A) $f(x) = x^3 - 5x^2 - x + 5$ for $a = 1$
   B) $f(x) = x^3 + 5x^2 - x - 5$ for $a = 1$
   C) $f(x) = x^3 - 5x^2 + x - 5$ for $a = 1$
   D) $f(x) = x^3 + 5x^2 + x + 5$ for $a = 1$

4) Zeros: −4, −5, 5; degree 3
   A) $f(x) = x^3 + 25x + 4x^2 + 100$ for $a = 1$
   B) $f(x) = x^3 + 25x - 4x^2 - 100$ for $a = 1$
   C) $f(x) = x^3 - 25x - 4x^2 + 100$ for $a = 1$
   D) $f(x) = x^3 - 25x + 4x^2 - 100$ for $a = 1$

5) Zeros: 3, multiplicity 2; −3, multiplicity 2; degree 4
   A) $f(x) = x^4 + 18x^2 + 81$
   B) $f(x) = x^4 + 6x^3 - 18x^2 + 27x - 81$
   C) $f(x) = x^4 - 18x^2 + 81$
   D) $f(x) = x^4 - 6x^3 + 18x^2 - 27x + 81$

6) Zeros: −3, multiplicity 2; 3, multiplicity 1; degree 3
   A) $x^3 + 3x^2 - 9x - 27$
   B) $x^3 - 3x^2 - 18x + 27$
   C) $x^3 - 3x^2 - 9x + 27$
   D) $x^3 + 6x^2 - 9x - 27$

7) Zeros: −5, 2, 4, 5; degree 4
   A) $x^4 + 10x^2 - 200$
   B) $x^4 + 6x^3 - 17x^2 - 150x - 200$
   C) $x^4 - 6x^3 - 17x^2 - 200x - 200$
   D) $x^4 - 6x^3 - 17x^2 + 150x - 200$

For the polynomial, list each real zero and its multiplicity. Determine whether the graph crosses or touches the x-axis at each x-intercept.

8) $f(x) = 4(x + 1)(x + 7)^4$
   A) −1, multiplicity 1, crosses x-axis; −7, multiplicity 4, touches x-axis
   B) −1, multiplicity 1, touches x-axis; −7, multiplicity 4, crosses x-axis
   C) 1, multiplicity 1, crosses x-axis; 7, multiplicity 4, touches x-axis
   D) 1, multiplicity 1, touches x-axis; 7, multiplicity 4, crosses x-axis

9) $f(x) = 5(x + 1)(x - 3)^3$
   A) 1, multiplicity 1, crosses x-axis; −3, multiplicity 3, crosses x-axis
   B) −1, multiplicity 1, touches x-axis; 3, multiplicity 3
   C) 1, multiplicity 1, touches x-axis; −3, multiplicity 3
   D) −1, multiplicity 1, crosses x-axis; 3, multiplicity 3, crosses x-axis

10) $f(x) = 4(x^2 + 5)(x + 4)^2$
    A) −5, multiplicity 1, crosses x-axis; −4, multiplicity 2, touches x-axis
    B) −4, multiplicity 2, crosses x-axis
    C) −5, multiplicity 1, touches x-axis; −4, multiplicity 2, crosses x-axis
    D) −4, multiplicity 2, touches x-axis

11) $f(x) = \left(x + \dfrac{1}{3}\right)^2 (x - 5)^5$

    A) $\dfrac{1}{3}$, multiplicity 2, crosses x-axis; -5, multiplicity 5, touches x-axis

    B) $-\dfrac{1}{3}$, multiplicity 2, touches x-axis; 5, multiplicity 5, crosses x-axis

    C) $\dfrac{1}{3}$, multiplicity 2, touches x-axis; -5, multiplicity 5, crosses x-axis

    D) $-\dfrac{1}{3}$, multiplicity 2, crosses x-axis; 5, multiplicity 5, touches x-axis

12) $f(x) = \left(x + \dfrac{1}{5}\right)^4 (x^2 + 6)^5$

    A) $-\dfrac{1}{5}$, multiplicity 4, crosses x-axis

    B) $-\dfrac{1}{5}$, multiplicity 4, touches x-axis

    C) $\dfrac{1}{5}$, multiplicity 4, touches x-axis; 6, multiplicity 5, crosses x-axis

    D) $-\dfrac{1}{5}$, multiplicity 4, touches x-axis; -6, multiplicity 5, crosses x-axis

13) $f(x) = \dfrac{1}{5}x(x^2 - 3)$

    A) 0, multiplicity 1, touches x-axis;
        $\sqrt{3}$, multiplicity 1, touches x-axis;
        $-\sqrt{3}$, multiplicity 1, touches x-axis

    B) $\sqrt{3}$, multiplicity 1, touches x-axis;
        $-\sqrt{3}$, multiplicity 1, touches x-axis

    C) 0, multiplicity 1, crosses x-axis;
        $\sqrt{3}$, multiplicity 1, crosses x-axis;
        $-\sqrt{3}$, multiplicity 1, crosses x-axis

    D) 0, multiplicity 1

14) $f(x) = \dfrac{1}{5}x^2(x^2 - 5)$

    A) 0, multiplicity 2, crosses x-axis

    B) 0, multiplicity 2, touches x-axis;
        $\sqrt{5}$, multiplicity 1, crosses x-axis;
        $-\sqrt{5}$, multiplicity 1, crosses x-axis

    C) 0, multiplicity 2, touches x-axis

    D) 0, multiplicity 2, crosses x-axis;
        $\sqrt{5}$, multiplicity 1, touches x-axis;
        $-\sqrt{5}$, multiplicity 1, touches x-axis

15) $f(x) = 3(x^2 + 7)(x^2 + 6)^2$

    A) -7, multiplicity 1, crosses x-axis; -6, multiplicity 2, touches x-axis

    B) $\sqrt{7}$, multiplicity 1, crosses x-axis; $-\sqrt{7}$, multiplicity 1, crosses x-axis;
       $\sqrt{6}$, multiplicity 2, touches x-axis; $-\sqrt{6}$, multiplicity 2, touches x-axis

    C) -7, multiplicity 1, touches x-axis; -6, multiplicity 2, crosses x-axis

    D) No real zeros

16) $f(x) = \frac{1}{2}x^2(x^2 - 5)(x - 9)$

   A) 0, multiplicity 2, touches x-axis;
      9, multiplicity 1, crosses x-axis

   B) 0, multiplicity 2, crosses x-axis;
      9, multiplicity 1, touches x-axis;
      $\sqrt{5}$, multiplicity 1, touches x-axis;
      $-\sqrt{5}$, multiplicity 1, touches x-axis

   C) 0, multiplicity 2, crosses x-axis;
      9, multiplicity 1, touches x-axis

   D) 0, multiplicity 2, touches x-axis;
      9, multiplicity 1, crosses x-axis;
      $\sqrt{5}$, multiplicity 1, crosses x-axis;
      $-\sqrt{5}$, multiplicity 1, crosses x-axis

4 Analyze the Graph of a Polynomial Function

Find the x- and y-intercepts of f.

1) $f(x) = (x + 2)^2$
   A) x-intercept: 2; y-intercept: 4
   B) x-intercept: -2; y-intercept: 0
   C) x-intercept: 2; y-intercept: 0
   D) x-intercept: -2; y-intercept: 4

2) $f(x) = 2x^5(x + 8)^5$
   A) x-intercepts: 0, 8; y-intercept: 0
   B) x-intercepts: 0, -8; y-intercept: 0
   C) x-intercepts: 0, -8; y-intercept: 2
   D) x-intercepts: 0, 8; y-intercept: 2

3) $f(x) = (x + 6)(x - 2)(x + 2)$
   A) x-intercepts: -6, -2, 2; y-intercept: 24
   B) x-intercepts: -2, 2, 6; y-intercept: 24
   C) x-intercepts: -2, 2, 6; y-intercept: -24
   D) x-intercepts: -6, -2, 2; y-intercept: -24

4) $f(x) = 7x - x^3$
   A) x-intercepts: 0, -7; y-intercept: 0
   B) x-intercepts: 0, $\sqrt{7}$, $-\sqrt{7}$; y-intercept: 0
   C) x-intercepts: 0, $\sqrt{7}$, $-\sqrt{7}$; y-intercept: 7
   D) x-intercepts: 0, -7; y-intercept: 7

5) $f(x) = (x + 1)(x - 7)(x - 1)^2$
   A) x-intercepts: -1, 1, 7; y-intercept: 7
   B) x-intercepts: -1, 1, 7; y-intercept: -7
   C) x-intercepts: -1, 1, -7; y-intercept: -7
   D) x-intercepts: -1, 1, -7; y-intercept: 7

6) $f(x) = -x^2(x + 6)(x^2 - 1)$
   A) x-intercepts: -6, -1, 0, 1; y-intercept: -6
   B) x-intercepts: -6, 0, 1; y-intercept: -6
   C) x-intercepts: -1, 0, 1, 6; y-intercept: 0
   D) x-intercepts: -6, -1, 0, 1; y-intercept: 0

7) $f(x) = -x^2(x + 7)(x^2 + 1)$
   A) x-intercepts: -7, -1, 0; y-intercept: 7
   B) x-intercepts: -7, -1, 0; y-intercept: -7
   C) x-intercepts: -7, 0; y-intercept: 0
   D) x-intercepts: -7, -1, 0, 1; y-intercept: 0

8) $f(x) = (x - 3)(x - 4)$
   A) x-intercepts: 3, 4; y-intercept: -7
   B) x-intercepts: -3, -4; y-intercept: -7
   C) x-intercepts: -3, -4; y-intercept: 12
   D) x-intercepts: 3, 4; y-intercept: 12

9) $f(x) = x^2(x - 6)(x - 2)$

   A) x-intercepts: 0, 6, 2; y-intercept: 12
   B) x-intercepts: 0, -6, -2; y-intercept: 12
   C) x-intercepts: 0, -6, -2; y-intercept: 0
   D) x-intercepts: 0, 6, 2; y-intercept: 0

10) $f(x) = (x - 4)^2(x^2 - 25)$

    A) x-intercepts: -5, 4, 5; y-intercept: 400
    B) x-intercepts: 4, 25; y-intercept: 100
    C) x-intercepts: -5, 4, 5; y-intercept: -400
    D) x-intercepts: -4, -25; y-intercept: 100

Find the power function that the graph of f resembles for large values of |x|.

11) $f(x) = (x + 5)^2$

    A) $y = x^{25}$
    B) $y = x^{10}$
    C) $y = x^5$
    D) $y = x^2$

12) $f(x) = (x - 5)^3$

    A) $y = x^3$
    B) $y = x^{-125}$
    C) $y = x^{15}$
    D) $y = x^{-5}$

13) $f(x) = (x + 3)^2(x - 11)^4$

    A) $y = x^6$
    B) $y = x^2$
    C) $y = x^8$
    D) $y = x^4$

14) $f(x) = -x^2(x + 2)^3(x^2 - 1)$

    A) $y = x^7$
    B) $y = x^2$
    C) $y = x^3$
    D) $y = -x^7$

15) $f(x) = 6x - x^3$

    A) $y = x^3$
    B) $y = x^4$
    C) $y = -x^3$
    D) $y = x^2$

Determine the maximum number of turning points of f.

16) $f(x) = -x^2(x + 3)^3(x^2 - 1)$

    A) 5
    B) 2
    C) 7
    D) 6

17) $f(x) = 9x - x^3$

    A) 2
    B) 1
    C) 3
    D) 4

18) $f(x) = (x - 4)^2(x + 5)^2$

    A) 3
    B) 2
    C) 1
    D) 4

Use the x-intercepts to find the intervals on which the graph of f is above and below the x-axis.

19) $f(x) = (x + 9)^2$

    A) above the x-axis: no intervals
       below the x-axis: $(-\infty, -9), (-9, \infty)$
    B) above the x-axis: $(-9, \infty)$
       below the x-axis: $(-\infty, -9)$
    C) above the x-axis: $(-\infty, -9)$
       below the x-axis: $(-9, \infty)$
    D) above the x-axis: $(-\infty, -9), (-9, \infty)$
       below the x-axis: no intervals

20) $f(x) = (x - 3)^3$

    A) above the x-axis: $(-\infty, 3)$
       below the x-axis: $(3, \infty)$
    B) above the x-axis: no intervals
       below the x-axis: $(-\infty, 3), (3, \infty)$
    C) above the x-axis: $(-\infty, 3), (3, \infty)$
       below the x-axis: no intervals
    D) above the x-axis: $(3, \infty)$
       below the x-axis: $(-\infty, 3)$

21) $f(x) = (x-2)^2(x+3)^2$

A) above the x-axis: $(-\infty, -3), (2, \infty)$
below the x-axis: $(-3, 2)$

B) above the x-axis: $(-\infty, -3), (-3, 2), (2, \infty)$
below the x-axis: no intervals

C) above the x-axis: no intervals
below the x-axis: $(-\infty, -3), (-3, 2), (2, \infty)$

D) above the x-axis: $(-3, 2)$
below the x-axis: $(-\infty, -3), (2, \infty)$

22) $f(x) = \left(x - \dfrac{1}{7}\right)^2 (x-2)^3$

A) above the x-axis: $\left(-\infty, \dfrac{1}{7}\right), \left(\dfrac{1}{7}, 2\right)$
below the x-axis: $(2, \infty)$

B) above the x-axis: $\left(-\infty, \dfrac{1}{7}\right), (2, \infty)$
below the x-axis: $\left(\dfrac{1}{7}, 2\right)$

C) above the x-axis: $(2, \infty)$
below the x-axis: $\left(-\infty, \dfrac{1}{7}\right), \left(\dfrac{1}{7}, 2\right)$

D) above the x-axis: $\left(\dfrac{1}{7}, 2\right)$
below the x-axis: $\left(-\infty, \dfrac{1}{7}\right), (2, \infty)$

Analyze the graph of the given function f as follows:
(a) Determine the end behavior: find the power function that the graph of f resembles for large values of $|x|$.
(b) Find the x- and y-intercepts of the graph.
(c) Determine whether the graph crosses or touches the x-axis at each x-intercept.
(d) Graph f using a graphing utility.
(e) Use the graph to determine the local maxima and local minima, if any exist. Round turning points to two decimal places.
(f) Use the information obtained in (a) – (e) to draw a complete graph of f by hand. Label all intercepts and turning points.
(g) Find the domain of f. Use the graph to find the range of f.
(h) Use the graph to determine where f is increasing and where f is decreasing.

23) $f(x) = x^2(x+3)$

24) $f(x) = (x+3)(x-2)^2$

25) $f(x) = -2(x-2)(x+2)^3$

26) $f(x) = (x-3)(x-1)(x+2)$

27) $f(x) = -x^2(x-1)(x+3)$

28) $f(x) = x^2(x^2 - 4)(x+4)$

Analyze the graph of the given function f as follows:
(a) Determine the end behavior: find the power function that the graph of f resembles for large values of $|x|$.
(b) Graph f using a graphing utility.
(c) Find the x- and y-intercepts of the graph.
(d) Use the graph to determine the local maxima and local minima, if any exist. Round turning points to two decimal places.
(e) Use the information obtained in (a) – (d) to draw a complete graph of f by hand. Label all intercepts and turning points.
(f) Find the domain of f. Use the graph to find the range of f.
(g) Use the graph to determine where f is increasing and where f is decreasing.

29) $f(x) = x^3 - 0.4x^2 - 2.5861x + 3.0912$

Solve the problem.

30) For the polynomial function $f(x) = 2x^4 - 7x^3 + 11x - 4$
   a) Find the x- and y-intercepts of the graph of f. Round to two decimal places, if necessary.
   b) Determine whether the graph crosses or touches the x-axis at each x-intercept.
   c) End behavior: find the power function that the graph of f resembles for large values of $|x|$.
   d) Use a graphing utility to graph the function. Approximate the local maxima rounded to two decimal places, if necessary. Approximate the local minima rounded to two decimal places, if necessary.
   e) Determine the number of turning points on the graph.
   f) Put all the information together, and connect the points with a smooth, continuous curve to obtain the graph of f.

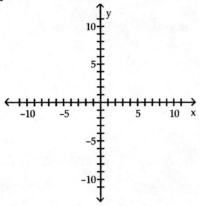

31) Which of the following polynomial functions might have the graph shown in the illustration below?

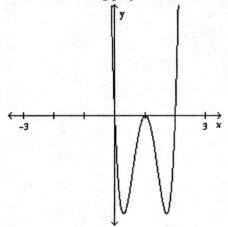

A) $f(x) = x(x - 2)^2(x - 1)$
C) $f(x) = x^2(x - 2)(x - 1)$

B) $f(x) = x(x - 2)(x - 1)^2$
D) $f(x) = x^2(x - 2)^2(x - 1)^2$

5 Find the Cubic Function of Best Fit to Data

Solve the problem.

1) The profits (in millions) for a company for 8 years was as follows:

Year, x	Profits
1993, 1	1.1
1994, 2	1.7
1995, 3	2.0
1996, 4	1.4
1997, 5	1.3
1998, 6	1.5
1999, 7	1.8
2000, 8	2.1

Find the cubic function of best fit to the data.

2) The amount of water (in gallons) in a leaky bathtub is given in the table below. Using a graphing utility, fit the data to a third degree polynomial (or a cubic). Then approximate the time at which there is maximum amount of water in the tub, and estimate the time when the water runs out of the tub. Express all your answers rounded to two decimal places.

t (in minutes)	0	1	2	3	4	5	6	7
V (in gallons)	20	26	45	63	86	94	90	67

A) maximum amount of water after 8.23 minutes; water runs out after 19.73 minutes
B) maximum amount of water after 5.31 minutes; water runs out after 8.23 minutes
C) maximum amount of water after 5.31 minutes; water never runs out
D) maximum amount of water after 5.37 minutes; water runs out after 11.06 minutes

3.3 Properties of Rational Functions

1 Find the Domain of a Rational Function

Find the domain of the rational function.

1) $f(x) = \dfrac{3x}{x+2}$

　A) $\{x \mid x \neq -2\}$　　B) all real numbers　　C) $\{x \mid x \neq 2\}$　　D) $\{x \mid x \neq 0\}$

2) $h(x) = \dfrac{7x}{(x-4)(x-8)}$

　A) $\{x \mid x \neq 4, x \neq 8\}$
　B) all real numbers
　C) $\{x \mid x \neq -4, x \neq -8\}$
　D) $\{x \mid x \neq 4, x \neq 8, x \neq -7\}$

3) $g(x) = \dfrac{x+5}{x^2 - 9}$

　A) all real numbers
　B) $\{x \mid x \neq -3, x \neq 3\}$
　C) $\{x \mid x \neq 0, x \neq 9\}$
　D) $\{x \mid x \neq -3, x \neq 3, x \neq -5\}$

4) $h(x) = \dfrac{x+6}{x^2+25}$

   A) $\{x \mid x \neq -5, x \neq 5\}$
   B) $\{x \mid x \neq -5, x \neq 5, x \neq -6\}$
   C) all real numbers
   D) $\{x \mid x \neq 0, x \neq -25\}$

5) $g(x) = \dfrac{x+5}{x^2-36x}$

   A) $\{x \mid x \neq -6, x \neq 6\}$
   B) all real numbers
   C) $\{x \mid x \neq 0, x \neq 36\}$
   D) $\{x \mid x \neq -6, x \neq 6, x \neq -5\}$

6) $R(x) = \dfrac{-3x^2}{x^2+3x-108}$

   A) $\{x \mid x \neq -108, 1\}$
   B) $\{x \mid x \neq 12, 9\}$
   C) $\{x \mid x \neq 12, -9\}$
   D) $\{x \mid x \neq -12, 9\}$

7) $f(x) = \dfrac{2x^2-4}{3x^2+6x-45}$.

   A) all real numbers
   B) $\{x \mid x \neq 3, x \neq -5\}$
   C) $\{x \mid x \neq 3, x \neq -3, x \neq -5\}$
   D) $\{x \mid x \neq -3, x \neq 5\}$

8) $f(x) = \dfrac{-2x(x+2)}{3x^2-4x-7}$

   A) $\left\{x \mid x \neq -\dfrac{3}{7}, 1\right\}$
   B) $\left\{x \mid x \neq \dfrac{7}{3}, -1\right\}$
   C) $\left\{x \mid x \neq \dfrac{3}{7}, -1\right\}$
   D) $\left\{x \mid x \neq -\dfrac{7}{3}, 1\right\}$

9) $f(x) = \dfrac{x(x-1)}{9x^2+36x+20}$

   A) $\left\{x \mid x \neq -\dfrac{2}{9}, -\dfrac{10}{9}\right\}$
   B) $\left\{x \mid x \neq -\dfrac{2}{3}, -\dfrac{10}{3}\right\}$
   C) $\left\{x \mid x \neq \dfrac{2}{3}, \dfrac{10}{3}\right\}$
   D) $\left\{x \mid x \neq -\dfrac{10}{3}, \dfrac{10}{3}\right\}$

10) $g(x) = \dfrac{x}{x^3-64}$

    A) $\{x \mid x \neq -4, 4\}$
    B) $\{x \mid x \neq 16\}$
    C) $\{x \mid x \neq 4\}$
    D) $\{x \mid x \neq -4\}$

Use the graph to determine the domain and range of the function.

11)

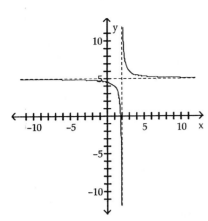

A) domain: $\{x \mid x \neq 2\}$
range: $\{y \mid y \neq 5\}$

B) domain: $\{x \mid x \neq 5\}$
range: $\{y \mid y \neq -2\}$

C) domain: $\{x \mid x \neq -2\}$
range: $\{y \mid y \neq 5\}$

D) domain: $\{x \mid x \neq 5\}$
range: $\{y \mid y \neq 2\}$

12)

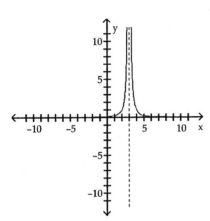

A) domain: $\{x \mid x \neq 3\}$
range: $\{y \mid y \geq 0\}$

B) domain: $\{x \mid x \neq 3\}$
range: $\{y \mid y > 0\}$

C) domain: $\{x \mid x > 0\}$
range: $\{y \mid y \neq 3\}$

D) domain: $\{x \mid x \geq 0\}$
range: $\{y \mid y \neq 3\}$

13)

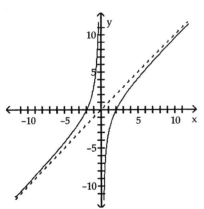

A) domain: $\{x \mid x \neq 0\}$
range: $\{y \mid y \neq 0\}$

B) domain: all real numbers
range: $\{y \mid y \neq 0\}$

C) domain: $\{x \mid x \neq 0\}$
range: all real numbers

D) domain: all real numbers
range: all real numbers

14)

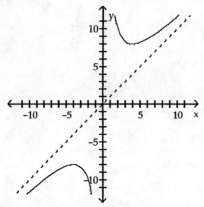

A) domain: {x | x ≠ 0}
   range: {y | y ≤ -8 or y ≥ 8}

B) domain: {x | x ≠ 0}
   range: all real numbers

C) domain: {x | x ≤ -8 or x ≥ 8}
   range: {y | y ≠ 0}

D) domain: all real numbers
   range: {y | y ≤ -8 or y ≥ 8}

15)

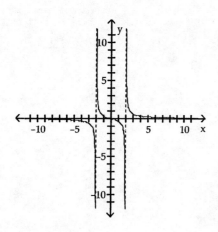

A) domain: {x | x ≠ -2, x ≠ 2}
   range: {y | y ≠ 0}

B) domain: {x | x ≠ -2, x ≠ 2}
   range: all real numbers

C) domain: all real numbers
   range: {y | y ≠ -2, y ≠ 2}

D) domain: all real numbers
   range: all real numbers

16)

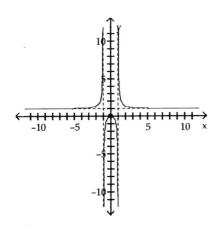

A) domain: all real numbers
   range: all real numbers

B) domain: {x | x ≠ −1, x ≠ 1}
   range: {y | y ≤ 0 or y > 1}

C) domain: {x | x ≤ 0 or x > 1}
   range: {y | y ≠ −1, y ≠ 1}

D) domain: {x | x ≠ −1, x ≠ 1}
   range: {y | y ≤ 0 or y ≥ 1}

2 Find the Vertical Asymptotes of a Rational Function

Find the vertical asymptotes of the rational function.

1) $f(x) = \dfrac{8x}{x+1}$

   A) $x = 8$ B) none   C) $x = 1$ D) $x = -1$

2) $h(x) = \dfrac{3x^2}{(x+8)(x+2)}$

   A) $x = -3$
   C) $x = 8, x = 2$
   B) $x = -8, x = -2$
   D) $x = -8, x = -2, x = -3$

3) $f(x) = \dfrac{x+6}{x^2-9}$

   A) $x = 0, x = 9$ B) $x = -3, x = 3, x = -6$ C) $x = 9, x = -6$ D) $x = -3, x = 3$

4) $g(x) = \dfrac{x+8}{x^2+49}$

   A) $x = -7, x = -8$ B) none   C) $x = -7, x = 7, x = -8$ D) $x = -7, x = 7$

5) $h(x) = \dfrac{x+11}{x^2-36x}$

   A) $x = 0, x = 36$ B) $x = -6, x = 6$ C) $x = 36, x = -11$ D) $x = 0, x = -6, x = 6$

6) $f(x) = \dfrac{x(x-1)}{x^3+25x}$

   A) $x = 0, x = -25$ B) $x = 0$ C) $x = 0, x = -5, x = 5$ D) $x = -5, x = 5$

7) $R(x) = \dfrac{-3x^2}{x^2 + 9x - 22}$

   A) $x = 11, x = -2$
   B) $x = -22$
   C) $x = -11, x = 2$
   D) $x = -11, x = 2, x = -3$

8) $f(x) = \dfrac{-2x(x + 2)}{5x^2 - 3x - 8}$

   A) $x = -\dfrac{8}{5}, x = 1$
   B) $x = -\dfrac{5}{8}, x = 1$
   C) $x = \dfrac{8}{5}, x = -1$
   D) $x = \dfrac{5}{8}, x = -1$

9) $f(x) = \dfrac{x(x - 1)}{4x^2 + 28x + 13}$

   A) $x = -\dfrac{1}{2}, x = -\dfrac{13}{2}$
   B) $x = -\dfrac{1}{4}, x = -\dfrac{13}{4}$
   C) $x = -\dfrac{13}{2}, x = \dfrac{13}{2}$
   D) $x = \dfrac{1}{2}, x = \dfrac{13}{2}$

10) $g(x) = \dfrac{x}{x^3 - 125}$

    A) $x = 5$
    B) $x = -5$
    C) $x = 25$
    D) $x = -5, x = 5$

11) $f(x) = \dfrac{x - 7}{49x - x^3}$

    A) $x = 0, x = -7$
    B) $x = 0, x = -7, x = 7$
    C) $x = -7, x = 7$
    D) $x = 0, x = 7$

12) $f(x) = \dfrac{-x^2 + 16}{x^2 + 5x + 4}$

    A) $x = 1, x = -4$
    B) $x = -1$
    C) $x = -1, x = -4$
    D) $x = -1, x = 4$

Use the graph to find the vertical asymptotes, if any, of the function.

13)

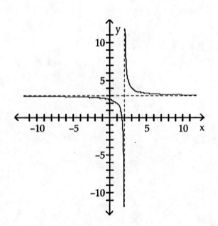

   A) $x = 2, y = 3$
   B) $x = 2$
   C) $x = 2, x = 0$
   D) $y = 3$

14)

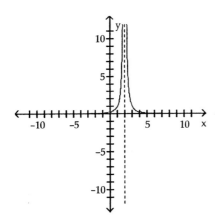

A) y = 2 B) x = 2, x = 0 C) none D) x = 2

15)

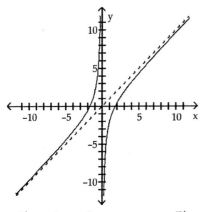

A) x = 0, y = 0 B) x = 0 C) none D) y = 0

16)

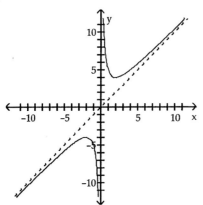

A) x = 0 B) none C) x = 0, y = 0 D) y = −4, y = 4

17)

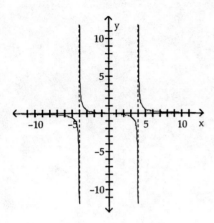

A) $x = -4, x = 4, y = 0$ B) $x = -4, x = 4$ C) none   D) $x = -4, x = 4, x = 0$

18)

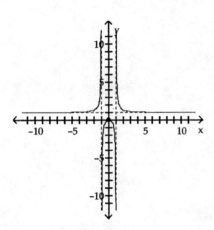

A) $x = -1, x = 1, y = 1$ B) $x = -1, x = 1, x = 0, y = 1$
C) $x = -1, x = 1, x = 0$ D) $x = -1, x = 1$

3 Find the Horizontal or Oblique Asymptotes of a Rational Function

Give the equation of the horizontal asymptote, if any, of the function.

1) $h(x) = \dfrac{6x - 4}{x - 5}$

A) $y = 0$ B) $y = 5$
C) $y = 6$ D) no horizontal asymptotes

2) $g(x) = \dfrac{x^2 + 9x - 6}{x - 6}$

A) $y = 6$ B) $y = 1$
C) $y = 0$ D) no horizontal asymptotes

3) $f(x) = \dfrac{5x^2 + 8}{5x^2 - 8}$

A) $y = 1$ B) $y = 5$
C) $y = 8$ D) no horizontal asymptotes

4) $h(x) = \dfrac{5x^2 - 5x - 6}{6x^2 - 2x + 9}$

   A) $y = 0$ B) $y = \dfrac{5}{2}$

   C) $y = \dfrac{5}{6}$ D) no horizontal asymptotes

5) $h(x) = \dfrac{8x^3 - 2x - 4}{7x + 2}$

   A) $y = \dfrac{8}{7}$ B) $y = 8$

   C) $y = 0$ D) no horizontal asymptotes

6) $h(x) = \dfrac{x + 3}{x^2 - 16}$

   A) $y = 1$ B) $y = 0$

   C) no horizontal asymptotes   D) $y = -4$, $y = 4$

7) $f(x) = \dfrac{x(x - 1)}{x^3 + 25x}$

   A) $x = 0$, $x = -25$ B) $y = 1$

   C) no horizontal asymptotes   D) $y = 0$

8) $R(x) = \dfrac{-3x^2}{x^2 + 4x - 32}$

   A) $y = -8$, $y = 4$ B) no horizontal asymptotes

   C) $y = -3$ D) $y = 0$

9) $f(x) = \dfrac{x^2 - 5}{25x - x^4}$

   A) $y = -1$ B) $y = 0$

   C) $y = -5$, $y = 5$ D) no horizontal asymptotes

10) $f(x) = \dfrac{36x^5 - 6}{x - x^3}$

    A) $y = -1$, $y = 1$ B) $y = -36$

    C) $y = 0$ D) no horizontal asymptotes

11) $f(x) = \dfrac{-x^2 + 16}{x^2 + 5x + 4}$

    A) $y = -16$ B) no horizontal asymptotes

    C) $y = 0$ D) $y = -1$

Use the graph to find the horizontal asymptote, if any, of the function.

12)

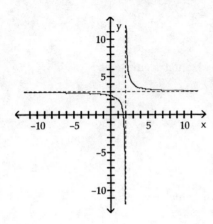

A) y = 3   B) x = 2   C) y = 0, y = 3   D) y = 0

13)

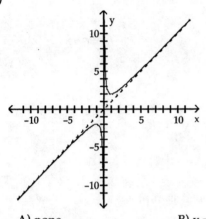

A) none   B) y = 0   C) y = -2   D) y = 2

14)

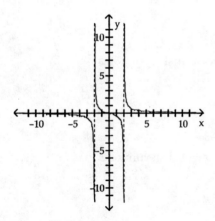

A) no horizontal asymptotes   B) x = -2, x = 2, y = 0
C) y = 0   D) y = -2, y = 2

15)

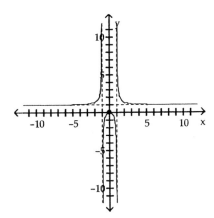

A) $y = -1, y = 1$ B) $x = -1, x = 1, y = 1$ C) $y = 0, y = 1$ D) $y = 1$

Give the equation of the oblique asymptote, if any, of the function.

16) $f(x) = \dfrac{x^2 + 9x - 4}{x - 6}$

A) $x = y + 15$ B) $y = x + 3$
C) no oblique asymptotes   D) $y = x + 15$

17) $h(x) = \dfrac{3x^2 - 5x - 2}{2x^2 - 8x + 4}$

A) $y = \dfrac{3}{2}$ B) $y = x + \dfrac{3}{2}$
C) $y = \dfrac{3}{2}x$ D) no oblique asymptote

18) $f(x) = \dfrac{x^2 - 3x + 2}{x + 4}$

A) $x = y + 3$ B) $y = x + 5$
C) $y = x - 7$ D) no oblique asymptote

19) $f(x) = \dfrac{x^2 + 9x + 3}{x + 6}$

A) $y = x - 15$ B) $y = x + 3$
C) $x = y + 3$ D) no oblique asymptotes

20) $f(x) = \dfrac{2x^3 + 11x^2 + 5x - 1}{x^2 + 6x + 5}$.

A) $y = 2x$ B) $y = 2x - 1$ C) $y = 0$ D) $y = 2x + 1$

21) $g(x) = \dfrac{x + 7}{x^2 - 4}$

A) no oblique asymptote   B) $y = 0$
C) $y = x + 7$ D) $y = 7x$

22) $f(x) = \dfrac{x^2 - 2}{4x - x^4}$

   A) $y = 0$
   B) $y = x - 2$
   C) no oblique asymptote
   D) $y = 4x$

23) $f(x) = \dfrac{6x^3 - 7x^2 - 15x + 15}{-2x - 1}$

   A) $y = -3x + 5$
   B) $y = 0$
   C) no oblique asymptote
   D) $y = -3x^2 + 5x + 5$

Use the graph to find the oblique asymptote, if any, of the function.

24)

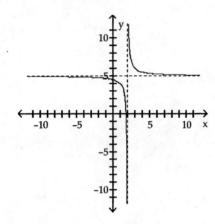

   A) $y = 5$
   B) $y = x + 2$
   C) $y = 5x + 2$
   D) no oblique asymptote

25)

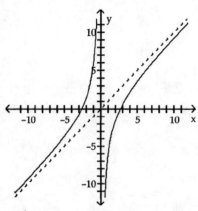

   A) $y = -x$
   B) $y = x$
   C) $y = x + 1$
   D) no oblique asymptote

26)

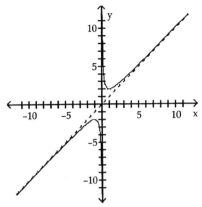

A) no oblique asymptote
C) y = x
B) y = 2x
D) y = -x

27)

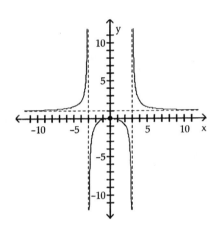

A) no oblique asymptote
C) y = 1
B) y = x
D) y = x + 1

4 Demonstrate Additional Understanding and Skills

Graph the function using transformations.

1) $f(x) = \dfrac{4}{(6+x)^2}$

A)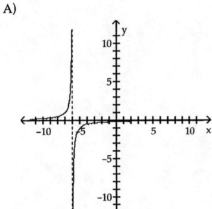

B)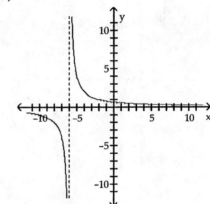

C)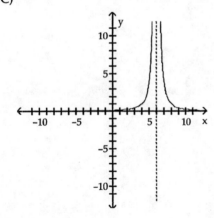

D)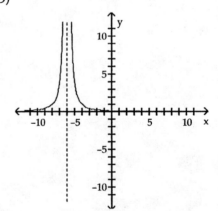

2) $f(x) = \dfrac{1}{x} - 1$

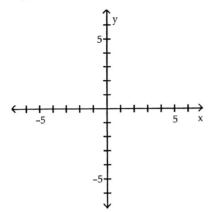

A)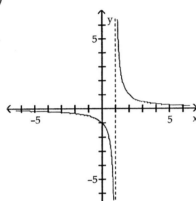

B)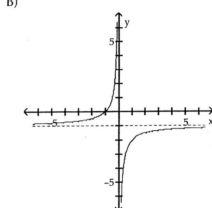

C)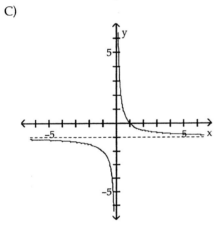

D)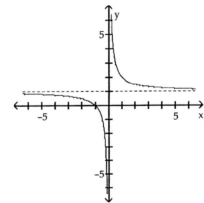

3) $f(x) = \dfrac{-2}{x-4}$

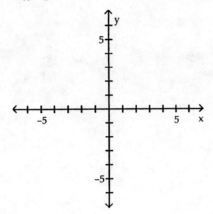

A)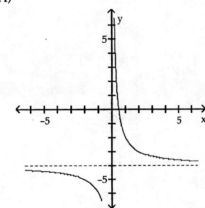

B)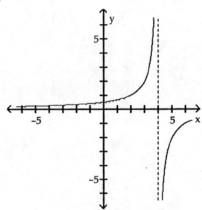

C)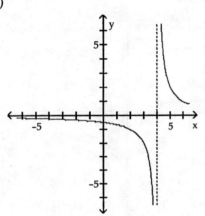

D)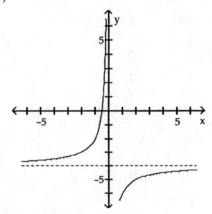

4) $f(x) = \dfrac{1}{x-4} + 1$

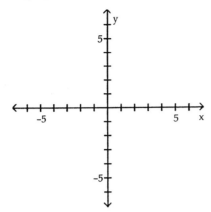

A)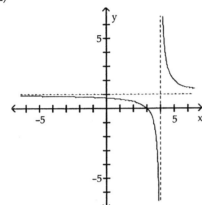

B)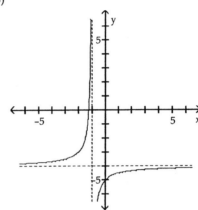

C)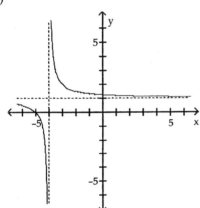

D)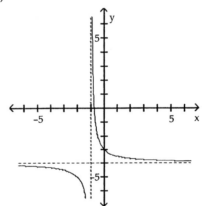

5) $f(x) = \dfrac{1}{x^2} + 4$

A)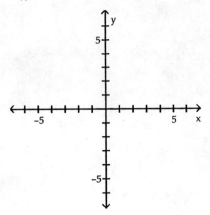

B)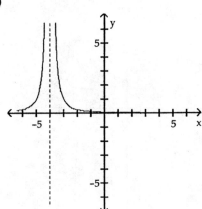

C)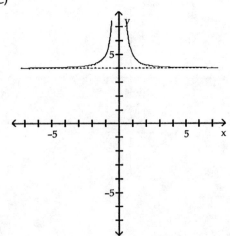

D)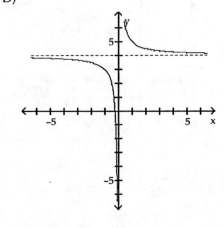

6) $f(x) = 4 - \dfrac{1}{(x+3)^2}$

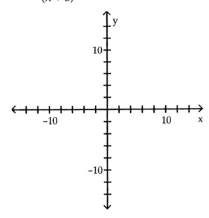

A)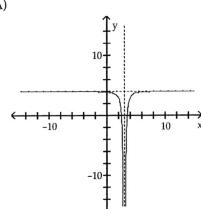

B)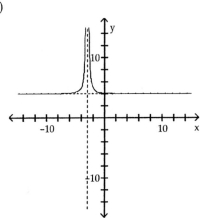

C)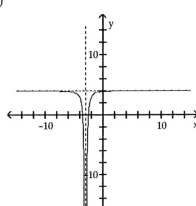

D)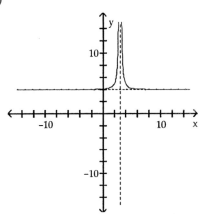

Solve the problem.

7) The acceleration due to gravity g (in meters per second per second) at a height h meters above sea level is given by $g(h) = \dfrac{3.99 \times 10^{14}}{(6.374 \times 10^6 + h)^2}$ where 6.374×10^6 is the radius of Earth in meters. Death Valley in California is 86 m below sea level.
   a) Find the value of g(h) at Death Valley to four decimal places.
   b) Compare the value in (a) to the value of g(h) at sea level.

8) The distance formula states that d = rt. If a car drives 50 miles, the function r = 50/t is a rational function. Find the asymptotes of this function.

9) A lens can be used to create an image of an object on the opposite side of the lens, such as the image created on a movie screen. Every lens has a measurement called its focal length, f. The distance s_1 of the object to the lens is related to the distance s_2 of the lens to the image by the function

$$s_1 = \frac{fs_2}{s_2 - f}.$$

For a lens with f = 0.3 m, what are the asymptotes of this function?

10) When two lenses are placed next to each other, their combined focal length (a measurement that can be negative or positive) is described by the equation

$$f = \frac{f_1 f_2}{f_1 + f_2}.$$

If $f_1 = 0.001$, what are the asymptotes of this function?

3.4 The Graph of a Rational Function; Inverse and Joint Variation

1 Analyze the Graph of a Rational Function

Find the indicated intercept(s) of the graph of the function.

1) x-intercepts of $f(x) = \dfrac{2x + 7}{x - 9}$

   A) $\left(\dfrac{7}{2}, 0\right)$ B) $\left(-\dfrac{7}{2}, 0\right)$ C) (9, 0)   D) (-9, 0)

2) x-intercepts of $f(x) = \dfrac{x - 4}{x^2 + 2x - 2}$

   A) (-4, 0)   B) (4, 0)   C) (2, 0)   D) none

3) x-intercepts of $f(x) = \dfrac{x^2 + 5}{x^2 + 2x + 3}$

   A) $(\sqrt{5}, 0), (-\sqrt{5}, 0)$ B) (-5, 0)   C) (3, 0)   D) none

4) x-intercepts of $f(x) = \dfrac{5}{x^2 - x - 6}$.

   A) (2, 0), (-3, 0)   B) (5, 0)   C) none   D) (3, 0), (-2, 0)

5) x-intercepts of $f(x) = \dfrac{8x}{x^2 - 1}$

   A) (1, 0)   B) (-1, 0), (1, 0)   C) (0, 0)   D) (8, 0)

6) x-intercepts of $f(x) = \dfrac{x^2 - 64}{4 + x^4}$

   A) (64, 0)   B) (-8, 0), (8, 0)   C) (4, 0)   D) none

7) x-intercepts of $f(x) = \dfrac{x^2 + 7x}{x^2 + 9x - 5}$

   A) (0, 0) and (7, 0)　　　B) (-7, 0)　　　C) (7, 0)　　　D) (0, 0) and (-7, 0)

8) x-intercepts of $f(x) = \dfrac{(x - 9)(2x + 5)}{x^2 + 9x - 7}$

   A) (9, 0) and (-5, 0)　　　B) (9, 0) and $\left(-\dfrac{5}{2}, 0\right)$　　　C) (-9, 0) and $\left(\dfrac{5}{2}, 0\right)$　　　D) none

9) x-intercepts of $f(x) = \dfrac{x^2 - x - 56}{x^2 + 5}$.

   A) (-56, 0)　　　B) (-8, 0), (7, 0)　　　C) (-7, 0), (8, 0)　　　D) (-8, 0), (0, 0)

10) x-intercepts of $f(x) = \dfrac{x^3 - 8}{x^2 - 25}$

    A) (5, 0)　　　B) (-8, 0)　　　C) (-2, 0), (2, 0)　　　D) (2, 0)

11) x-intercepts of $f(x) = x + \dfrac{1}{x}$

    A) (-1, 0), (1, 0)　　　B) none　　　C) (1, 0)　　　D) (-1, 0)

12) y-intercept of $f(x) = \dfrac{x - 5}{3x - 13}$

    A) $\left(0, \dfrac{5}{13}\right)$　　　B) $\left(0, -\dfrac{13}{5}\right)$　　　C) (0, 5)　　　D) none

13) y-intercept of $f(x) = \dfrac{15}{x^2 - 3x - 23}$

    A) none　　　B) (0, 15)　　　C) $\left(0, -\dfrac{15}{23}\right)$　　　D) $\left(0, \dfrac{15}{23}\right)$

14) y-intercept of $f(x) = \dfrac{12x}{x^2 - 19}$

    A) (0, 12)　　　B) (0, 0)　　　C) none　　　D) $\left(0, -\dfrac{12}{19}\right)$

15) y-intercept of $f(x) = \dfrac{x - 15}{x^2 + 12x - 3}$

    A) (0, 15)　　　B) $\left(0, -\dfrac{1}{5}\right)$　　　C) (0, 5)　　　D) none

16) y-intercept of $f(x) = \dfrac{(5x - 10)(x - 3)}{x^2 + 6x - 19}$

    A) $\left(0, \dfrac{30}{19}\right)$　　　B) (0, 2)　　　C) (0, 3)　　　D) $\left(0, -\dfrac{30}{19}\right)$

17) y-intercept of $f(x) = \dfrac{(x-5)^2}{(x+11)^3}$

A) $(0, 5)$ B) $\left(0, -\dfrac{25}{1331}\right)$ C) $\left(0, -\dfrac{5}{11}\right)$ D) $\left(0, \dfrac{25}{1331}\right)$

18) y-intercept of $f(x) = \dfrac{23}{(x+10)(x^2-6)}$

A) $\left(0, -\dfrac{23}{60}\right)$ B) $\left(0, \dfrac{23}{60}\right)$ C) none D) $(0, 23)$

19) y-intercept of $f(x) = \dfrac{x}{(x+10)(x-6)}$

A) $\left(0, -\dfrac{1}{60}\right)$ B) $(0, 6)$ C) none D) $(0, 0)$

20) y-intercept of $f(x) = \dfrac{x^2 - 3x}{x^2 + 6x - 3}$

A) $(0, -1)$ B) $(0, 1)$ C) $(0, 0)$ D) $(0, 3)$

21) y-intercept of $f(x) = \dfrac{x^2 - 5}{x^2 + 6x - 4}$

A) $\left(0, -\dfrac{4}{5}\right)$ B) $\left(0, \dfrac{5}{4}\right)$ C) $(0, 5)$ D) none

22) y-intercept of $f(x) = \dfrac{x^2 - 14x + 3}{12x}$

A) $(0, -4)$ B) $\left(0, \dfrac{1}{4}\right)$ C) $(0, 3)$ D) none

23) y-intercept of $f(x) = \dfrac{x^2 - 9x + 9}{x^2 + 10x - 3}$

A) $(0, 10)$ B) $(0, -3)$ C) $(0, 9)$ D) none

24) y-intercept of $f(x) = \dfrac{x^3 - 2}{x^2 + 2}$

A) $(0, 13)$ B) $(0, -2)$ C) $(0, -1)$ D) none

25) y-intercept of $f(x) = x + \dfrac{25}{x}$

A) $(0, 25)$ B) $(0, 5)$ C) $(0, 0)$ D) none

Determine whether the rational function has symmetry with respect to the origin, symmetry with respect to the y-axis, or neither.

26) $f(x) = \dfrac{-7x^2}{-7x^4 - 12}$

   A) symmetry with respect to the origin
   B) symmetry with respect to the y-axis
   C) neither

27) $f(x) = \dfrac{-7x^2}{-8x^3 - 6}$

   A) symmetry with respect to the y-axis
   B) symmetry with respect to the origin
   C) neither

28) $f(x) = \dfrac{4x^2 + 4x - 10}{-5x + 15}$

   A) symmetry with respect to the y-axis
   B) symmetry with respect to the origin
   C) neither

29) $f(x) = \dfrac{-2x^2 + 3}{-8x}$

   A) symmetry with respect to the origin
   B) symmetry with respect to the y-axis
   C) neither

30) $f(x) = \dfrac{x - 2}{3x - 15}$

   A) symmetry with respect to the y-axis
   B) symmetry with respect to the origin
   C) neither

31) $f(x) = \dfrac{9}{x^2 - 3x - 23}$

   A) symmetry with respect to the origin
   B) symmetry with respect to the y-axis
   C) neither

32) $f(x) = \dfrac{11}{x^2 - 19}$

   A) symmetry with respect to the y-axis
   B) symmetry with respect to the origin
   C) neither

33) $f(x) = \dfrac{(5x - 10)(x - 2)}{x^2 + 4x - 19}$

   A) symmetry with respect to the origin
   B) symmetry with respect to the y-axis
   C) neither

34) $f(x) = \dfrac{23}{x(x^2 - 3)}$

   A) symmetry with respect to the origin
   B) symmetry with respect to the y-axis
   C) neither

35) $f(x) = \dfrac{x}{(x + 15)(x - 6)}$

   A) symmetry with respect to the y-axis
   B) symmetry with respect to the origin
   C) neither

36) $f(x) = x + \dfrac{64}{x}$

   A) symmetry with respect to the origin
   B) symmetry with respect to the y-axis
   C) neither

Graph the function.

37) $f(x) = \dfrac{2x+5}{x-2}$

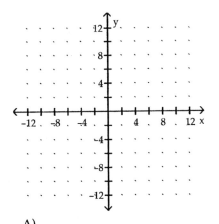

A)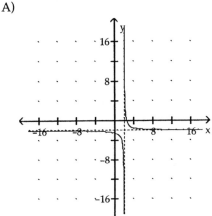

B)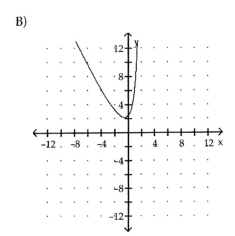

C)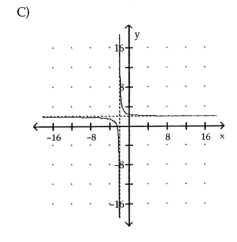

D)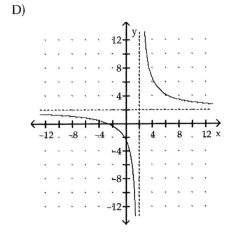

38) $f(x) = \dfrac{2x}{(x-4)(x+4)}$

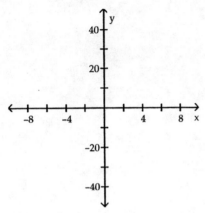

A)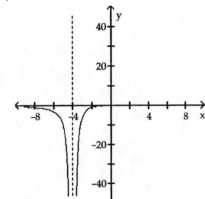

B)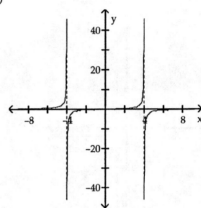

C)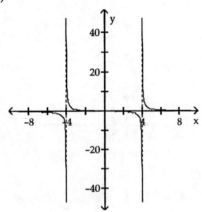

D)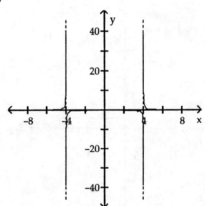

39) $f(x) = x + \dfrac{9}{x}$

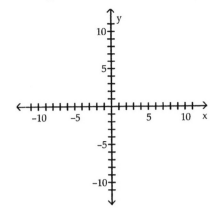

A)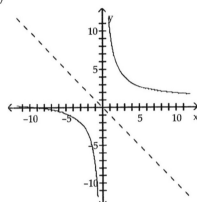

B)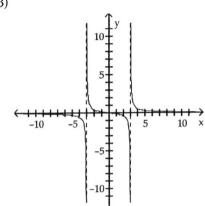

C)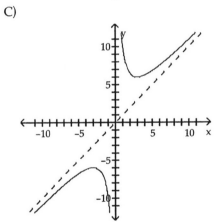

D)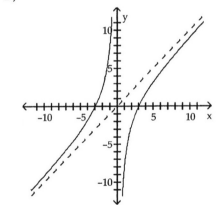

40) $f(x) = x^2 + \dfrac{16}{x}$

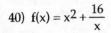

A)
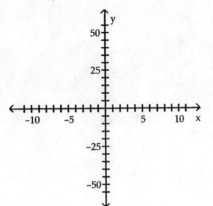

B)
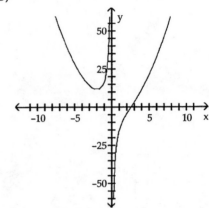

C)
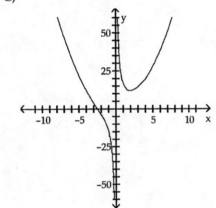

D)
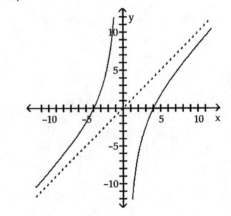

41) $f(x) = \dfrac{x}{x^2 - 36}$

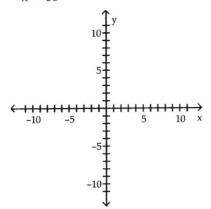

A)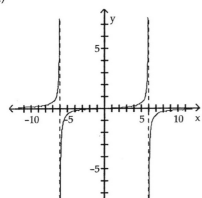

B)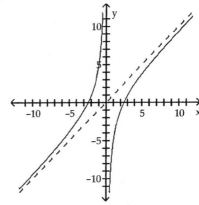

C)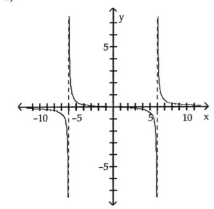

D)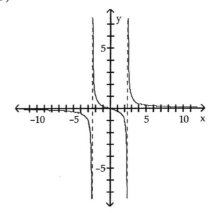

42) $f(x) = \dfrac{x^4 - 1}{x^2 - 9}$

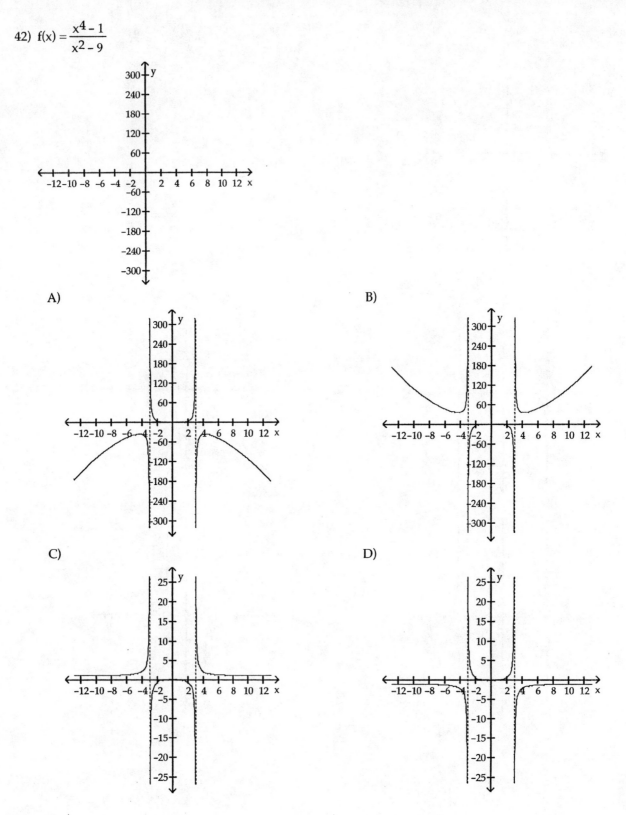

43) $f(x) = \dfrac{x^2 + x - 6}{x^2 - x - 12}$

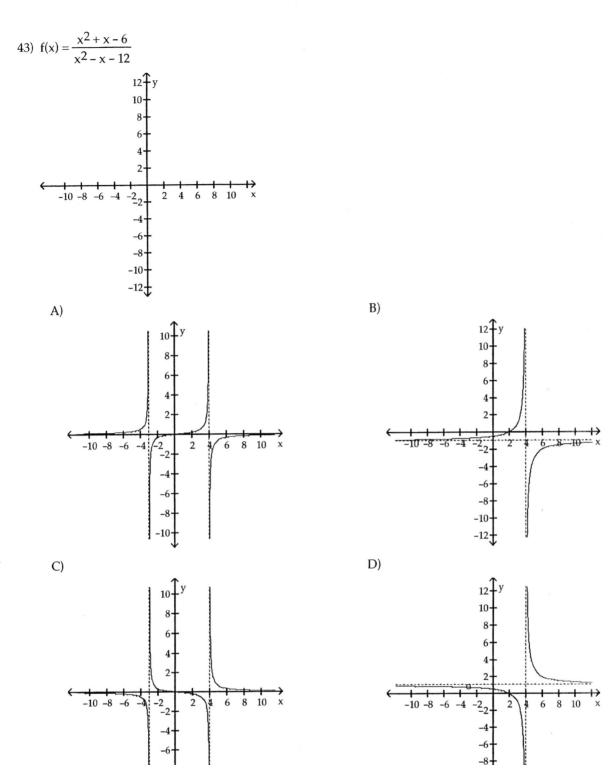

44) $f(x) = \dfrac{x^2 + 3x + 2}{(x-1)^2}$

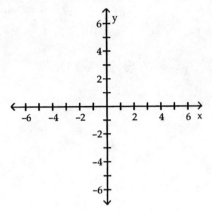

A)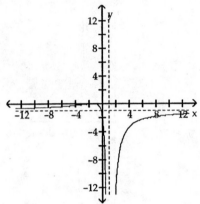

B)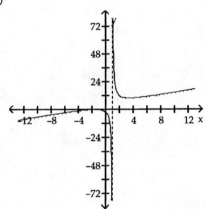

C)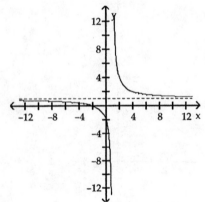

D)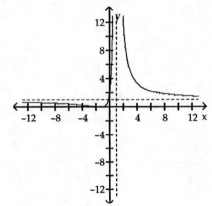

45) $f(x) = \dfrac{x^2 + 7x + 12}{x - 4}$

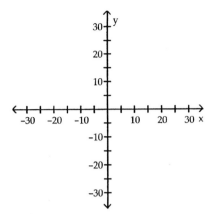

A)
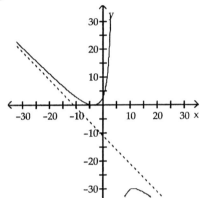

B)
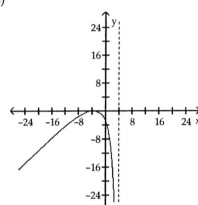

C)
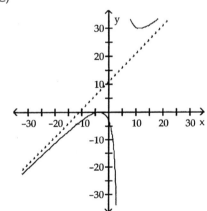

D)
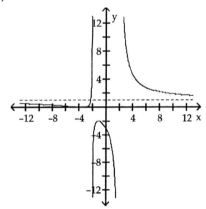

46) $f(x) = \dfrac{3}{(x^2-1)(x+7)}$.

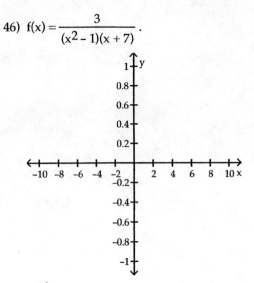

A)
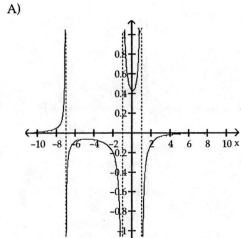

B)
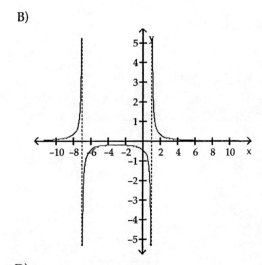

C)
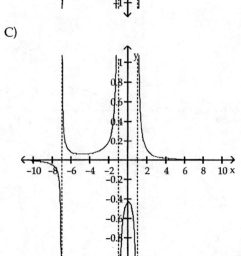

D)
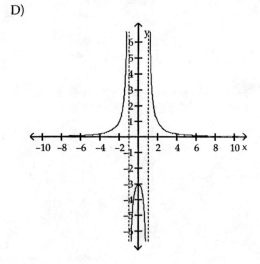

47) $f(x) = \dfrac{(x-4)(x+4)}{x^2 - 36}$

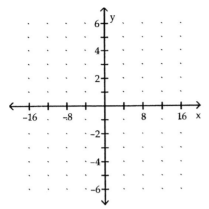

A)
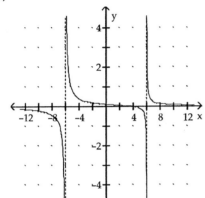

B)
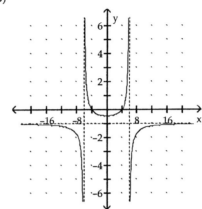

C)
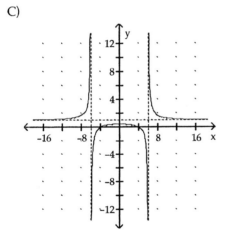

D)
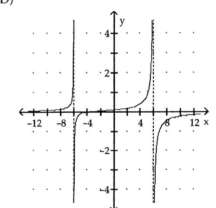

48) $f(x) = \dfrac{x^2 + 10x + 25}{x^2 - 2}$

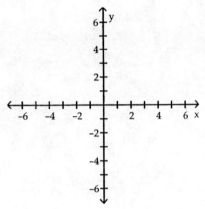

A)
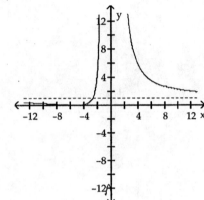

B)
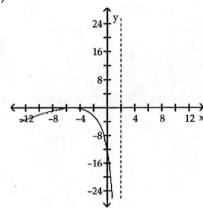

C)
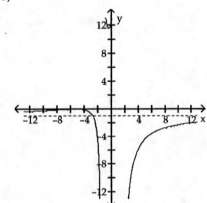

D)
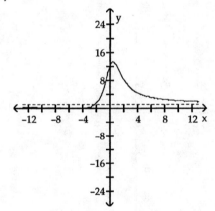

49) $f(x) = \dfrac{x^2 - 2x}{(x-3)^2}$

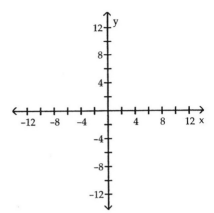

A)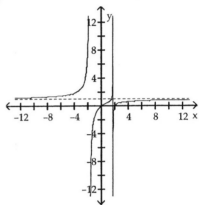

B)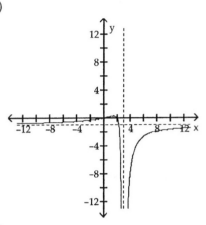

C)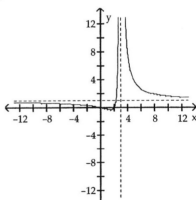

D)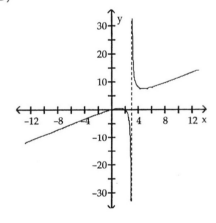

Analyze the graph of the rational function for the given step.

50) Find the vertical asymptote(s) and/or hole(s) for $R(x) = \dfrac{x^2 + x - 72}{x^2 - x - 56}$.

A) vertical asymptote: $x = -7$; hole at $\left(8, \dfrac{17}{15}\right)$

B) vertical asymptotes: $x = -8$, $x = 7$

C) vertical asymptotes: $x = 8$, $x = -7$

D) vertical asymptote: $x = -7$

51) Find the vertical asymptote(s) and/or hole(s) for $R(x) = \dfrac{3}{(x^2 - 81)(x + 7)}$.

   A) vertical asymptote: $x = -7$
   B) vertical asymptotes: $x = -9$, $x = 9$, $x = -7$
   C) vertical asymptotes: $x = 81$, $x = -7$
   D) vertical asymptotes: $x = -9$, $x = 9$; hole at $\left(-7, -\dfrac{3}{32}\right)$

Solve the problem.

52) Decide which of the rational functions might have the given graph.

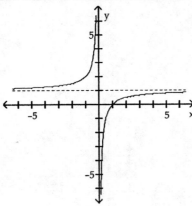

   A) $f(x) = 1 - x$
   B) $f(x) = \dfrac{1}{x} - 1$
   C) $f(x) = 1 + \dfrac{1}{x}$
   D) $f(x) = 1 - \dfrac{1}{x}$

53) Decide which of the rational functions might have the given graph.

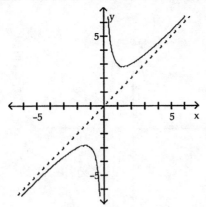

   A) $f(x) = x + \dfrac{2}{x}$
   B) $f(x) = x + 2$
   C) $f(x) = 2x + \dfrac{1}{x}$
   D) $f(x) = x + \dfrac{1}{x}$

54) Decide which of the rational functions might have the given graph.

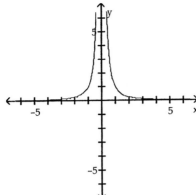

A) $f(x) = \dfrac{1}{x}$ 	B) $f(x) = \dfrac{1}{2x}$ 	C) $f(x) = \dfrac{1}{x^2}$ 	D) $f(x) = x^2$

55) Decide which of the rational functions might have the given graph.

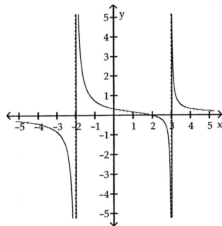

A) $R(x) = \dfrac{x - 2}{(x + 2)(x - 3)}$ 	B) $R(x) = \dfrac{2 - x}{(x + 2)(x - 3)}$

C) $R(x) = \dfrac{x + 2}{(x - 2)(x + 3)}$ 	D) $R(x) = \dfrac{x - 2}{(x + 2)^2(x - 3)^2}$

56) Determine which rational function $R(x)$ has a graph that crosses the x-axis at −1, touches the x-axis at −4, has vertical asymptotes at $x = -2$ and $x = 3$, and has one horizontal asymptote at $y = -2$.

A) $R(x) = \dfrac{-2(x + 1)(x + 4)}{(x + 2)(x - 3)}$, $x \neq -2, 3$ 	B) $R(x) = \dfrac{-2(x + 1)(x + 4)^2}{(x + 2)^2(x - 3)}$, $x \neq -2, 3$

C) $R(x) = \dfrac{-(x + 1)(x + 4)^2}{2(x - 2)^2(x + 3)}$, $x \neq 2, -3$ 	D) $R(x) = \dfrac{-2(x - 3)(x + 2)^2}{(x + 4)^2(x + 1)}$, $x \neq -4, -1$

2 Solve Applied Problems Involving Rational Functions

Solve the problem.

1) A rare species of insect was discovered in the rain forest of Costa Rica. Environmentalists transplant the insect into a protected area. The population of the insect t months after being transplanted is
$$P(t) = \frac{45(1 + 0.6t)}{(3 + 0.02t)}.$$
   a) What was the population when $t = 0$?
   b) What will the population be after 10 years?
   c) What is the largest value the population could reach?

2) The concentration C of a certain drug in a patient's bloodstream is given by
$$\frac{30t}{t^2 + 49}.$$
   a) Find the horizontal asymptote of C(t).
   b) Using a graphing utility, determine the time at which the concentration is highest.

3) A can in the shape of a right circular cylinder is required to have a volume of 700 cubic centimeters. The top and bottom are made up of a material that costs 8¢ per square centimeter, while the sides are made of material that costs 5¢ per square centimeter. Which function below describes the total cost of the material as a function of the radius r of the cylinder?

   A) $C(r) = 0.16\pi r^2 + \frac{140}{r}$
   B) $C(r) = 0.08\pi r^2 + \frac{70}{r}$
   C) $C(r) = 0.16\pi r^2 + \frac{70}{r}$
   D) $C(r) = 0.08\pi r^2 + \frac{140}{r}$

4) The concentration of a drug in the bloodstream, measured in milligrams per liter, can be modeled by the function, $C(t) = \frac{12t + 4}{3t^2 + 2}$, where t is the number of minutes after injection of the drug. When will the drug be at its highest concentration? Approximate your answer rounded to two decimal places.

   A) at the time of injection
   B) $t = 3.65$ minutes after the injection is given
   C) $t = 0.55$ minutes after the injection is given
   D) $t = 4$ minutes after the injection is given

5) A closed box with a square base has to have a volume of 11,000 cubic inches. Find a function for the surface area of the box.

   A) $S(x) = x^2 + \frac{44{,}000}{x}$
   B) $S(x) = 2x^2 + \frac{11{,}000}{x}$
   C) $S(x) = 2x^2 + \frac{44{,}000}{x}$
   D) $S(x) = 2x^2 + \frac{66{,}000}{x}$

6) Economists use what is called a Leffer curve to predict the government revenue for tax rates from 0% to 100%. Economists agree that the end points of the curve generate 0 revenue, but disagree on the tax rate that produces the maximum revenue. Suppose an economist produces this rational function
$$R(x) = \frac{10x(100 - x)}{50 + x},$$
where R is revenue in millions at a tax rate of x percent. Use a graphing calculator to graph the function. What tax rate produces the maximum revenue? What is the maximum revenue?

   A) 41.2%; $264 million
   B) 36.6%; $268 million
   C) 35.8%; $276 million
   D) 34.0%; $271 million

7) Economists use what is called a Leffer curve to predict the government revenue for tax rates from 0% to 100%. Economists agree that the end points of the curve generate 0 revenue, but disagree on the tax rate that produces the maximum revenue. Suppose an economist produces this rational function
$$R(x) = \frac{10x(100-x)}{25+x},$$ where R is revenue in millions at a tax rate of x percent. Use a graphing calculator to graph the function. What tax rate produces the maximum revenue? What is the maximum revenue?

A) 30.9%; $382 million   B) 28.8%; $272 million   C) 38.4%; $383 million   D) 27.0%; $379 million

8) Economists use what is called a Leffer curve to predict the government revenue for tax rates from 0% to 100%. Economists agree that the end points of the curve generate 0 revenue, but disagree on the tax rate that produces the maximum revenue. Suppose an economist produces this rational function
$$R(x) = \frac{10x(100-x)}{75+x},$$ where R is revenue in millions at a tax rate of x percent. Use a graphing calculator to graph the function. What tax rate produces the maximum revenue? What is the maximum revenue?

A) 37.5%; $210 million   B) 39.6%; $209 million   C) 35.8%; $209 million   D) 34.9%; $207 million

9) Economists use what is called a Leffer curve to predict the government revenue for tax rates from 0% to 100%. Economists agree that the end points of the curve generate 0 revenue, but disagree on the tax rate that produces the maximum revenue. Suppose an economist produces this rational function
$$R(x) = \frac{10x(100-x)}{15+x},$$ where R is revenue in millions at a tax rate of x percent. Use a graphing calculator to graph the function. What tax rate produces the maximum revenue? What is the maximum revenue?

A) 28.1%; $470 million   B) 31.4%; $464 million   C) 29.7%; $467 million   D) 26.5%; $469 million

10) A box has a base whose length is twice its width. The volume of the box is 7000 cubic inches.
   a) Find a function for the surface area of the box.
   b) What are the dimensions of the box that minimizes surface area?

11) The formula $y = \dfrac{fx}{x-f}$ models the relationships needed to focus an image where y is the distance between the film and projector lens, x is the distance between the move screen and the projector lens, and f is the focal length.
   a) Sketch the graph of this rational function for f = 5 centimeters.

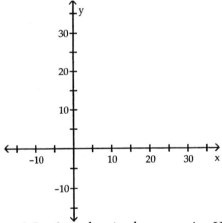

   b) Bob and Carol are showing home movies. Use the graph to describe the desired distance between the film and the projector lens as Carol moves the projector further from the screen.

12) Which of the following functions could have this graph?

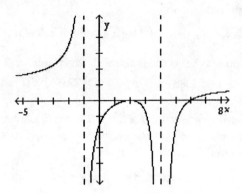

A) $y = \dfrac{(x-2)(x-6)^2}{(x+1)2(x-4)}$ B) $y = \dfrac{(x+1)(x-4)^2}{(x-2)2(x-6)}$ C) $y = \dfrac{(x-2)^2(x-6)}{(x+1)(x-4)^2}$ D) $y = \dfrac{2(x-2)^2(x-6)}{(x+1)(x-4)^2}$

3 Construct a Model Using Inverse Variation

Solve the problem.

1) When the temperature stays the same, the volume of a gas is inversely proportional to the pressure of the gas. If a balloon is filled with 246 cubic inches of a gas at a pressure of 14 pounds per square inch, find the new pressure of the gas if the volume is decreased to 41 cubic inches.

   A) 78 pounds per square inch
   B) $\dfrac{41}{14}$ pounds per square inch
   C) 84 pounds per square inch
   D) 70 pounds per square inch

2) If the force acting on an object stays the same, then the acceleration of the object is inversely proportional to its mass. If an object with a mass of 42 kilograms accelerates at a rate of 3 meters per second per second by a force, find the rate of acceleration of an object with a mass of 7 kilograms that is pulled by the same force.

   A) $\dfrac{1}{2}$ meters per second per second
   B) 18 meters per second per second
   C) 12 meters per second per second
   D) 15 meters per second per second

3) While traveling at a constant speed in a car, the centrifugal acceleration passengers feel while the car is turning is inversely proportional to the radius of the turn. If the passengers feel an acceleration of 12 feet per second per second when the radius of the turn is 40 feet, find the acceleration the passengers feel when the radius of the turn is 120 feet.

   A) 5 feet per second per second
   B) 4 feet per second per second
   C) 7 feet per second per second
   D) 6 feet per second per second

4) The amount of time it takes a swimmer to swim a race is inversely proportional to the average speed of the swimmer. A swimmer finishes a race in 200 seconds with an average speed of 3 feet per second. Find the average speed of the swimmer if it takes 150 seconds to finish the race.

   A) 6 feet per second   B) 3 feet per second   C) 5 feet per second   D) 4 feet per second

5) If the voltage, V, in an electric circuit is held constant, the current, I, is inversely proportional to the resistance, R. If the current is 200 milliamperes when the resistance is 2 ohms, find the current when the resistance is 10 ohms.

   A) 995 milliamperes   B) 40 milliamperes   C) 80 milliamperes   D) 1000 milliamperes

6) The gravitational attraction A between two masses varies inversely as the square of the distance between them. The force of attraction is 4 lb when the masses are 3 ft apart, what is the attraction when the masses are 6 ft apart?

   A) 4 lb   B) 1 lb   C) 3 lb   D) 2 lb

4 Construct a Model Using Joint or Combined Variation

Solve.

1) The amount of paint needed to cover the walls of a room varies jointly as the perimeter of the room and the height of the wall. If a room with a perimeter of 80 feet and 6-foot walls requires 4.8 quarts of paint, find the amount of paint needed to cover the walls of a room with a perimeter of 40 feet and 8-foot walls.

   A) 3.2 quarts   B) 32 quarts   C) 320 quarts   D) 6.4 quarts

2) The amount of simple interest earned on an investment over a fixed amount of time is jointly proportional to the principle invested and the interest rate. A principle investment of $4300.00 with an interest rate of 4% earned $516.00 in simple interest. Find the amount of simple interest earned if the principle is $3900.00 and the interest rate is 7%.

   A) $468.00   B) $903.00   C) $81,900.00   D) $819.00

3) The voltage across a resistor is jointly proportional to the resistance of the resistor and the current flowing through the resistor. If the voltage across a resistor is 18 volts for a resistor whose resistance is 2 ohms and when the current flowing through the resistor is 9 amperes, find the voltage across a resistor whose resistance is 5 ohms and when the current flowing through the resistor is 8 amperes.

   A) 45 volts   B) 72 volts   C) 16 volts   D) 40 volts

4) The power that a resistor must dissipate is jointly proportional to the square of the current flowing through the resistor and the resistance of the resistor. If a resistor needs to dissipate 256 watts of power when 8 amperes of current is flowing through the resistor whose resistance is 4 ohms, find the power that a resistor needs to dissipate when 4 amperes of current are flowing through a resistor whose resistance is 2 ohms.

   A) 16 watts   B) 32 watts   C) 64 watts   D) 8 watts

5) While traveling in a car, the centrifugal force a passenger experiences as the car drives in a circle varies jointly as the mass of the passenger and the square of the speed of the car. If the a passenger experiences a force of 144 newtons when the car is moving at a speed of 40 kilometers per hour and the passenger has a mass of 100 kilograms, find the force a passenger experiences when the car is moving at 70 kilometers per hour and the passenger has a mass of 70 kilograms.

   A) 343 newtons   B) 274.4 newtons   C) 308.7 newtons   D) 392 newtons

6) The volume V of a given mass of gas varies directly as the temperature T and inversely as the pressure P. A measuring device is calibrated to give $V = 176 \text{ in}^3$ when $T = 320°$ and $P = 20 \text{ lb/in}^2$. What is the volume on this device when the temperature is 150° and the pressure is 10 lb/in²?

   A) $V = 15 \text{ in}^3$ B) $V = 125 \text{ in}^3$ C) $V = 205 \text{ in}^3$ D) $V = 165 \text{ in}^3$

7) The time in hours it takes a satellite to complete an orbit around the earth varies directly as the radius of the orbit (from the center of the earth) and inversely as the orbital velocity. If a satellite completes an orbit 730 miles above the earth in 10 hours at a velocity of 29,000 mph, how long would it take a satellite to complete an orbit if it is at 1200 miles above the earth at a velocity of 30,000 mph? (Use 3960 miles as the radius of the earth.)

    A) 106.35 hours   B) 10.64 hours   C) 15.89 hours   D) 2.47 hours

3.5 Polynomial and Rational Inequalities

1 Solve Polynomial Inequalities Algebraically and Graphically

Solve the inequality.

1) $(x - 1)(x - 5) > 0$
    A) $(5, \infty)$ B) $(-\infty, 1)$ or $(5, \infty)$ C) $(-\infty, 1)$ D) $(1, 5)$

2) $(x + 2)(x - 6) \leq 0$
    A) $(-\infty, -2]$ B) $(-\infty, -2]$ or $[6, \infty)$ C) $[-2, 6]$ D) $[6, \infty)$

3) $x^2 - 4x \geq 0$
    A) $[0, 4]$ B) $(-\infty, -4]$ or $[0, \infty)$ C) $[-4, 0]$ D) $(-\infty, 0]$ or $[4, \infty)$

4) $x^2 + 2x \geq 0$
    A) $[0, 2]$ B) $(-\infty, -2]$ or $[0, \infty)$ C) $[-2, 0]$ D) $(-\infty, 0]$ or $[2, \infty)$

5) $x^2 + 4x \leq 0$
    A) $(-\infty, -4]$ or $[0, \infty)$ B) $[0, 4]$ C) $(-\infty, 0]$ or $[4, \infty)$ D) $[-4, 0]$

6) $x^2 - 4x \leq 0$
    A) $[0, 4]$ B) $(-\infty, -4]$ or $[0, \infty)$ C) $(-\infty, 0]$ or $[4, \infty)$ D) $[-4, 0]$

7) $x^2 - 36 > 0$
    A) $(-\infty, -36)$ or $(36, \infty)$ B) $(-\infty, -6)$ or $(6, \infty)$ C) $(-6, 6)$ D) $(-36, 36)$

8) $x^2 - 64 \leq 0$
    A) $(-\infty, -8]$ or $[8, \infty)$ B) $[-64, 64]$ C) $[-8, 8]$ D) $(-\infty, -64]$ or $[64, \infty)$

9) $56(x^2 - 1) > 15x$
    A) $\left(-\dfrac{7}{8}, \dfrac{8}{7}\right)$ B) $\left(-\infty, -\dfrac{7}{8}\right)$ or $\left(\dfrac{8}{7}, \infty\right)$ C) $\left(-\infty, -\dfrac{8}{7}\right)$ or $\left(\dfrac{7}{8}, \infty\right)$ D) $\left(-\dfrac{8}{7}, \dfrac{7}{8}\right)$

10) $x^2 + 5x \geq -6$
    A) $(-\infty, -3]$ or $[-2, \infty)$ B) $[-2, \infty)$ C) $(-\infty, -3]$ D) $[-3, -2]$

11) $x(x - 3) \geq -2$
    A) $(-\infty, 1]$ or $[2, \infty)$ B) $[2, \infty)$ C) $[1, 2]$ D) $(-\infty, 1]$

12) $3x^2 + 2x < 1$
    A) $(-1, \infty)$ B) $\left(-1, \dfrac{1}{3}\right)$ C) $\left(-\infty, \dfrac{1}{3}\right)$ D) $(-\infty, -1)$ or $\left(\dfrac{1}{3}, \infty\right)$

13) $(a + 6)(a + 5)(a - 4) > 0$
   A) $(-6, -5)$ or $(4, \infty)$
   B) $(4, \infty)$
   C) $(-\infty, -5)$
   D) $(-\infty, -6)$ or $(-5, 4)$

14) $(b + 7)(b + 5)(b - 5) < 0$
   A) $(5, \infty)$
   B) $(-\infty, -5)$
   C) $(-7, -5)$ or $(5, \infty)$
   D) $(-\infty, -7)$ or $(-5, 5)$

15) $(x + 1)(x^2 + x + 1) >$
   A) $(-\infty, -1)$
   B) $(-1, \infty)$
   C) $(-1, 1)$
   D) $(-\infty, -1)$ or $(1, \infty)$

16) $x^2 - 9x + 18 > 0$
   A) $(-\infty, 3)$
   B) $(-\infty, 3)$ or $(6, \infty)$
   C) $(3, 6)$
   D) $(6, \infty)$

17) $x^2 - 2x - 15 \leq 0$
   A) $(-\infty, -3]$
   B) $[-3, 5]$
   C) $(-\infty, -3]$ or $[5, \infty)$
   D) $[5, \infty)$

18) $x^3 - 7x^2 - 18x > 0$
   A) $(-9, 0)$ or $(2, \infty)$
   B) $(-\infty, -2)$ or $(0, 9)$
   C) $(-2, \infty)$
   D) $(-2, 0)$ or $(9, \infty)$

19) $x(x + 3)(5 - x) \geq 0$
   A) $(-\infty, -3]$ or $[0, 5]$
   B) $[0, 5]$
   C) $[-3, 0]$ or $[5, \infty)$
   D) $[-3, 5]$

20) $x^4 < 36x^2$
   A) $(-6, 0)$ or $(6, \infty)$
   B) $(-6, 0)$ or $(0, 6)$
   C) $(-\infty, -6)$ or $(6, \infty)$
   D) $(-\infty, -6)$ or $(0, 6)$

21) $x^3 > 6x^2$
   A) $(6, \infty)$
   B) $(-\infty, 0)$ or $(6, \infty)$
   C) $(0, 6)$
   D) $(-\infty, 6)$

22) $x^4 - 5x^2 - 36 > 0$
   A) $(-3, 3)$
   B) $(-3, -2)$ or $(2, 3)$
   C) $(-\infty, -3)$ or $(-2, 2)$ or $(3, \infty)$
   D) $(-\infty, -3)$ or $(3, \infty)$

23) $x^3 \geq 8$
   A) $(-\infty, -2]$ or $[2, \infty)$
   B) $[-2, 2]$
   C) $[2, \infty)$
   D) $(-\infty, 2]$

Solve the problem.

24) For what positive numbers will the cube of a number exceed 8 times its square?
   A) $\{x | x > 64\}$
   B) $\{x | 0 < x < 8\}$
   C) $\{x | x > 8\}$
   D) $\{x | 0 < x < 64\}$

25) A ball is thrown vertically upward with an initial velocity of 128 feet per second. The distance in feet of the ball from the ground after t seconds is $s = 128t - 16t^2$. For what interval of time is the ball more than 112 above the ground?
   A) $\{x | 0.5 \text{ sec} < x < 7.5 \text{ sec}\}$
   B) $\{x | 1 \text{ sec} < x < 7 \text{ sec}\}$
   C) $\{x | 3.5 \text{ sec} < x < 4.5 \text{ sec}\}$
   D) $\{x | 5 \text{ sec} < x < 11 \text{ sec}\}$

26) A ball is thrown vertically upward with an initial velocity of 192 feet per second. The distance in feet of the ball from the ground after t seconds is s = 192t − 16t². For what intervals of time is the ball less than 512 above the ground (after it is tossed until it returns to the ground)?

   A) {x | 0 sec < x < 4 sec and 8 sec > x > 12 sec}
   B) {x | 0 sec < x < 3.5 sec and 8.5 sec > x > 12 sec}
   C) {x | 0 sec < x < 5.5 sec and 6.5 sec > x > 12 sec}
   D) {x | 4 sec < x < 8 sec}

27) The revenue achieved by selling x graphing calculators is figured to be x(42 − 0.2x) dollars. The cost of each calculator is $22. How many graphing calculators must be sold to make a profit (revenue − cost) of at least $483.80?

   A) {x | 41 < x < 59}   B) {x | 42 < x < 40}   C) {x | 43 < x < 57}   D) {x | 16 < x < 34}

28) The revenue achieved by selling x graphing calculators is figured to be x(49 − 0.5x) dollars. The cost of each calculator is $13. How many graphing calculators must be sold to make a profit (revenue − cost) of at least $635.50?

   A) {x | 32 < x < 40}   B) {x | 33 < x < 39}   C) {x | 31 < x < 41}   D) {x | 40 < x < 50}

2 Solve Rational Inequalities Algebraically and Graphically

Solve the inequality.

1) $\dfrac{x-4}{x+3} < 0$

   A) (−∞, −3) or (4, ∞)   B) (4, ∞)   C) (−3, 4)   D) (−∞, −3)

2) $\dfrac{x-7}{x+3} > 0$

   A) (−∞, −3)   B) (7, ∞)   C) (−3, 7)   D) (−∞, −3) or (7, ∞)

3) $\dfrac{x-9}{x+8} < 1$

   A) (−∞, −8) or (9, ∞)   B) (−∞, −8)   C) (−8, ∞)   D) (−8, 9)

4) $\dfrac{x+7}{x+2} < 2$

   A) (−∞, −2) or (2, ∞)   B) (−∞, 3) or (2, ∞)   C) (−2, 3)   D) (−∞, −2) or (3, ∞)

5) $x + \dfrac{48}{x} < 14$

   A) (0, 6) or (8, ∞)   B) (−∞, 0) or (8, ∞)   C) (0, 6) or (6, 8)   D) (−∞, 0) or (6, 8)

6) $\dfrac{(x-5)(x+5)}{x} \le 0$

   A) [−5, 0) or (0, 5]   B) [−5, 0) or [5, ∞)   C) (−∞, −5] or [5, ∞)   D) (−∞, −5] or (0, 5]

7) $\dfrac{(x+11)(x-6)}{x-1} \ge 0$

   A) [−11, 1) or [6, ∞)   B) (−∞, −11] or [6, ∞)   C) [−11, 1) or (1, 6]   D) (−∞, −11] or (1, 6]

8) $\dfrac{(x-4)^2}{x^2-36} > 0$

   A) $(-\infty, -6)$ or $(4, 6)$ B) $(-6, 4)$ or $(4, 6)$ C) $(-\infty, -6)$ or $(6, \infty)$ D) $(-6, 4)$ or $(6, \infty)$

9) $\dfrac{(x-1)(3-x)}{(x-2)^2} \le 0$

   A) $(-\infty, 1]$ or $[3, \infty)$ B) $(-\infty, -3]$ or $(-2, -1)$ or $[1, \infty)$
   C) $(-\infty, 1)$ or $(3, \infty)$ D) $(-\infty, -3)$ or $(-1, \infty)$

10) $\dfrac{2x}{4-x} < x$

    A) $(2, 4)$ B) $(-\infty, 2)$ or $(4, \infty)$ C) $(4, \infty)$ D) $(0, 2)$ or $(4, \infty)$

11) $\dfrac{12x}{4-x} \ge 6x$

    A) $[0, 2]$ or $[4, \infty)$ B) $[4, \infty)$ C) $(-\infty, 2]$ or $[4, \infty)$ D) $(-\infty, 0]$ or $[2, 4)$

12) $\dfrac{12}{x-3} > \dfrac{9}{x+1}$

    A) $(-\infty, -13)$ or $(3, \infty)$ B) $(-\infty, -13)$ or $(-1, 3)$ C) $(-13, -1)$ or $(3, \infty)$ D) $(-13, -1)$ or $(-1, 3)$

13) $\dfrac{x^2(x-10)(x+2)}{(x-4)(x+8)} \ge 0$

    A) $(-8, -2]$ or $(4, 10]$ B) $(-\infty, -8)$ or $[-2, 0)$ or $(0, 4)$ or $[10, \infty)$
    C) $(-\infty, -8)$ or $[-2, 4)$ or $[10, \infty)$ D) $(-\infty, -8)$ or $[10, \infty)$

14) $\dfrac{3x^2 - 4x - 7}{x + 3} \le 0$

    A) $(-\infty, -3]$ or $\left[-1, \dfrac{7}{3}\right]$ B) $(-\infty, -1]$ or $\left(\dfrac{7}{3}, \infty\right)$ C) $(-3, -1]$ or $\left(\dfrac{7}{3}, \infty\right)$ D) $(-\infty, -3)$ or $\left[-1, \dfrac{7}{3}\right]$

3.6 The Real Zeros of a Polynomial Function

1 Use the Remainder and Factor Theorems

Use the Factor Theorem to determine whether x − c is a factor of f(x).

1) $f(x) = x^3 + 5x^2 - 12x + 14$; $x + 7$

   A) Yes   B) No

2) $f(x) = x^3 + 7x^2 - 16x + 18$; $x - 9$

   A) Yes   B) No

3) $f(x) = x^4 - 32x^2 - 144$; $x - 6$

   A) Yes   B) No

4) $f(x) = x^4 - 5x^2 - 36$; $x - 6$

   A) Yes   B) No

5) $f(x) = x^4 + 10x^3 + 7x^2 + 62x - 80$; $x + 10$

   A) Yes   B) No

6) $f(x) = x^4 + 10x^3 + 3x^2 + 26x - 40$; $x - 10$

   A) Yes   B) No

7) $f(x) = 24x^3 + 58x^2 - 42x - 55$; $x + \dfrac{11}{4}$

   A) Yes   B) No

8) $f(x) = 42x^3 + 52x^2 - 36x - 22$; $x - \dfrac{11}{7}$

   A) Yes   B) No

9) $9x^4 + 35x^3 - 4x^2 + x - 4$; $x + 4$

   A) Yes   B) No

10) $8x^4 + 31x^3 - 4x^2 + x + 4$; $x + 4$

   A) Yes   B) No

11) $3x^3 + 11x^2 - 19x + 5$; $x + 5$

   A) Yes   B) No

12) $6x^3 + 27x^2 - 14x - 5$; $x + 5$

   A) Yes   B) No

Use the Factor Theorem to determine whether $x - c$ is a factor of f. If it is, write f in factored form, that is, write f in the form $f(x) = (x - c)(\text{quotient})$.

13) $f(x) = 4x^3 + 27x^2 + 26x + 48$; $c = -6$

   A) Yes; $f(x) = (x + 6)(4x^2 + x + 8)$ B) No

   C) Yes; $f(x) = (x + 6)(4x^2 + 3x + 8)$ D) Yes; $f(x) = (x - 6)(4x^2 + 3x - 8)$

14) $f(x) = x^4 + 10x^3 + 9x^2 + 82x - 80$; $c = -10$

   A) Yes; $f(x) = (x + 10)(x^3 + x - 8)$ B) Yes; $f(x) = (x - 10)(x^3 + 9x + 8)$

   C) Yes; $f(x) = (x + 10)(x^3 + 9x - 8)$ D) No

15) $f(x) = 2x^3 + 11x^2 + 12x - 9$; $c = \dfrac{1}{2}$

   A) Yes; $f(x) = \left(x + \dfrac{1}{2}\right)(2x^2 + 12x - 18)$ B) Yes; $f(x) = \left(x - \dfrac{1}{2}\right)(2x^2 + 13x + 18)$

   C) Yes; $f(x) = \left(x - \dfrac{1}{2}\right)(2x^2 + 12x + 18)$ D) No

16) $f(x) = 3x^4 - 7x^3 + 7x^2 - 7x + 4$; $c = 1$

   A) Yes; $f(x) = (x - 1)(3x^3 - 4x^2 - 4x - 4)$ B) Yes; $f(x) = (x - 1)(3x^3 - 4x^2 + 3x - 4)$

   C) No   D) Yes; $f(x) = (x - 1)(3x^3 + 5x^2 + 3x - 4)$

17) $f(x) = x^6 - 12x^4 - 69x^2 + 80; c = 4$
   A) Yes; $f(x) = (x - 4)(x^5 - 5x^4 + 5x^3 + 16x^2 - 5x - 20)$
   B) Yes; $f(x) = (x - 4)(x^5 + 4x^4 - 4x^3 + 20x^2 - 5x - 20)$
   C) No
   D) Yes; $f(x) = (x - 4)(x^5 + 4x^4 + 4x^3 + 16x^2 - 5x - 20)$

2 Use the Rational Zeros Theorem

List the potential rational zeros of the polynomial function. Do not find the zeros.

1) $f(x) = 11x^4 - x^2 + 5$
   A) $\pm \frac{1}{11}, \pm \frac{5}{11}, \pm 1, \pm 5$
   B) $\pm \frac{1}{11}, \pm \frac{5}{11}, \pm 1, \pm 5, \pm 11$
   C) $\pm \frac{1}{11}, \pm \frac{1}{5}, \pm 1, \pm 5, \pm 11$
   D) $\pm \frac{1}{5}, \pm \frac{11}{5}, \pm 1, \pm 11$

2) $f(x) = 6x^4 + 4x^3 - 3x^2 + 2$
   A) $\pm \frac{1}{6}, \pm \frac{1}{3}, \pm \frac{1}{2}, \pm \frac{2}{3}, \pm 1, \pm 2$
   B) $\pm \frac{1}{2}, \pm \frac{3}{2}, \pm 1, \pm 2, \pm 3, \pm 6$
   C) $\pm \frac{1}{6}, \pm \frac{1}{3}, \pm \frac{1}{2}, \pm 1, \pm 2$
   D) $\pm \frac{1}{6}, \pm \frac{1}{3}, \pm \frac{1}{2}, \pm \frac{2}{3}, \pm 1, \pm 2, \pm 3$

3) $f(x) = -2x^3 + 2x^2 - 4x + 8$
   A) $\pm \frac{1}{2}, \pm 1, \pm 2, \pm 4$
   B) $\pm \frac{1}{8}, \pm \frac{1}{4}, \pm \frac{1}{2}, \pm 1, \pm 2, \pm 4, \pm 8$
   C) $\pm \frac{1}{2}, \pm 1, \pm 2, \pm 4, \pm 8$
   D) $\pm \frac{1}{4}, \pm \frac{1}{2}, \pm 1, \pm 2, \pm 4, \pm 8$

4) $f(x) = -4x^4 + 3x^2 - 4x + 6$
   A) $\pm \frac{1}{6}, \pm \frac{1}{2}, \pm \frac{1}{3}, \pm \frac{2}{3}, \pm \frac{4}{3}, \pm 1, \pm 2, \pm 4$
   B) $\pm \frac{1}{4}, \pm \frac{1}{2}, \pm \frac{2}{3}, \pm \frac{3}{4}, \pm \frac{3}{2}, \pm 1, \pm 2, \pm 3, \pm 6$
   C) $\pm \frac{1}{4}, \pm \frac{1}{2}, \pm \frac{3}{4}, \pm \frac{3}{2}, \pm 1, \pm 2, \pm 3, \pm 4, \pm 6$
   D) $\pm \frac{1}{4}, \pm \frac{1}{2}, \pm \frac{3}{4}, \pm \frac{3}{2}, \pm 1, \pm 2, \pm 3, \pm 6$

5) $f(x) = 5x^5 - 3x^2 + 2x - 1$
   A) $\pm 1, \pm 5, \pm \frac{1}{5}$
   B) $\pm 1, \pm \frac{1}{5}$
   C) $\pm 5, \pm \frac{1}{5}$
   D) $\pm 1, \pm 5$

6) $f(x) = x^5 - 6x^2 + 4x + 2$
   A) $\pm 1, \pm \frac{1}{2}$
   B) $\pm \frac{1}{6}, \pm \frac{1}{3}, \pm 2$
   C) $\pm 2, \pm \frac{1}{2}$
   D) $\pm 1, \pm 2$

7) $f(x) = x^5 - 2x^2 + 6x + 14$
   A) $\pm 1, \pm 7, \pm 2, \pm 14$
   B) $\pm 1, \pm 7, \pm 2$
   C) $\pm 1, \pm \frac{1}{7}, \pm \frac{1}{2}, \pm \frac{1}{14}$
   D) $\pm 1, \pm \frac{1}{7}, \pm \frac{1}{2}, \pm \frac{1}{14}, \pm 7, \pm 2, \pm 14$

3 Find the Real Zeros of a Polynomial Function

Find all of the real zeros of the polynomial function, then use the real zeros to factor f over the real numbers.

1) $f(x) = x^3 + 2x^2 - 9x - 18$
   A) -3; $f(x) = (x+3)(x^2-x-6)$
   B) $-3, -2, 3$; $f(x) = (x+3)(x+2)(x-3)$
   C) -2; $f(x) = (x+2)(x^2+x-9)$
   D) $-3, 2, 3$; $f(x) = (x+3)(x-2)(x-3)$

2) $f(x) = 2x^3 + 11x^2 + 2x + 15$
   A) $1, \frac{2}{3}, -5$; $f(x) = (2x-3)(x-1)(x+5)$
   B) $-1, \frac{3}{2}, 5$; $f(x) = (2x-3)(x-5)(x+1)$
   C) $-5, \frac{3}{2}, 1$; $f(x) = (2x-3)(x-1)(x+5)$
   D) $-1, \frac{2}{3}, -5$; $f(x) = (2x-3)(x-5)(x+1)$

3) $f(x) = 3x^3 - 2x^2 + 6x - 4$
   A) $-2, -1, \frac{2}{3}$; $f(x) = (3x-2)(x+1)(x+2)$
   B) $2, \frac{2}{3}, 1$; $f(x) = (3x-2)(x-1)(x-2)$
   C) 4; $f(x) = (x-4)(3x^2+1)$
   D) $\frac{2}{3}$; $f(x) = (3x-2)(x^2+2)$

4) $f(x) = 4x^4 - 9x^3 + 21x^2 - 36x + 20$
   A) $4, \frac{5}{4}$; $f(x) = (x-4)(4x-5)(x^2+1)$
   B) $-4, -1, 1, \frac{5}{4}$; $f(x) = (x-1)(4x-5)(x+1)(x+4)$
   C) $-4, -1, 1, -\frac{5}{4}$; $f(x) = (x-1)(4x+5)(x+1)(x+4)$
   D) $1, \frac{5}{4}$; $f(x) = (x-1)(4x-5)(x^2+4)$

5) $f(x) = x^4 - 9x^2 - 400$
   A) $-5, 5$; $f(x) = (x-5)(x+5)(x^2+16)$
   B) $-4, 4$; $f(x) = (x-4)(x+4)(x^2+25)$
   C) $-5, -4, 5, 4$; $f(x) = (x-5)(x+5)(x-4)(x+4)$
   D) 5; $f(x) = (x-5)^2(x^2+16)$

6) $f(x) = 2x^4 - 20x^3 + 51x^2 - 10x + 25$
   A) no real roots; $f(x) = (x^2+25)(2x^2+1)$
   B) -5, multiplicity 2; $f(x) = (x+5)^2(2x^2+1)$
   C) $-5, 5$; $f(x) = (x-5)(x+5)(2x^2+1)$
   D) 5, multiplicity 2; $f(x) = (x-5)^2(2x^2+1)$

The equation has a solution r in the interval indicated. Approximate this solution correct to two decimal places.

7) $x^3 - 8x - 3 = 0$; $-3 \le r \le -2$

8) $x^3 - 8x - 3 = 0$; $-1 \le r \le 0$

9) $x^4 - x^3 - 7x^2 + 5x + 10 = 0$; $-3 \le r \le -2$

10) $x^4 - x^3 - 7x^2 + 5x + 10 = 0$; $2 < r \le 3$

Find the rational zeros of the polynomial. List any irrational zeros correct to two decimal places.

11) $f(x) = x^3 - 8x - 3$
    A) rational zero: -3; irrational zeros $\approx 5.23, 0.76$
    B) rational zero: 3; irrational zeros $\approx -0.76, -5.23$
    C) rational zero: 3; irrational zeros $\approx -0.38, -2.61$
    D) rational zero: -3; irrational zeros $\approx 2.61, 0.38$

12) $f(x) = x^4 - x^3 - 7x^2 + 5x + 10$

    A) rational zeros: 2, −1; irrational zeros ≈ −2.47, 2.47
    B) rational zeros: −2, 1; irrational zeros ≈ −2.47, 2.47
    C) rational zeros: −2, 1; irrational zeros ≈ 2.24, −2.24
    D) rational zeros: 2, −1; irrational zeros ≈ 2.24, −2.24

4 Solve Polynomial Equations

Solve the equation in the real number system.

1) $x^3 + 4x^2 - 9x - 36 = 0$
    A) {3, 4}   B) {−3, 3, 4}   C) {−4, −3}   D) {−4, −3, 3}

2) $x^3 + 8x^2 - 18x + 20 = 0$
    A) {1}   B) {−10, 10}   C) {10}   D) {−10}

3) $3x^3 - x^2 - 15x + 5 = 0$
    A) $\{3, \sqrt{5}, -\sqrt{5}\}$ B) $\left\{\frac{1}{3}, \sqrt{5}, -\sqrt{5}\right\}$ C) $\{-3, \sqrt{5}, -\sqrt{5}\}$ D) $\left\{-\frac{1}{3}, \sqrt{5}, -\sqrt{5}\right\}$

4) $2x^3 - 13x^2 + 22x - 8 = 0$
    A) $\left\{-\frac{1}{2}, 2, -4\right\}$ B) {2, 1, 2}   C) {−2, 1, −2}   D) $\left\{\frac{1}{2}, 2, 4\right\}$

5) $3x^3 - x^2 + 3x - 1 = 0$
    A) $\left\{-3, \frac{1}{3}, -1\right\}$ B) $\left\{\frac{1}{3}, -1\right\}$ C) $\left\{\frac{1}{3}\right\}$ D) $\left\{-3, -\frac{1}{3}, -1\right\}$

6) $x^4 - 3x^3 + 5x^2 - x - 10 = 0$
    A) {−1, 2}   B) {1, 2}   C) {−2, 1}   D) {−1, −2}

7) $x^4 - 12x^2 - 64 = 0$
    A) {−4, 4}   B) {−2, 2}   C) {−4, −2, 2, 4}   D) {−8, 8}

8) $x^4 - 8x^3 + 16x^2 + 8x - 17 = 0$
    A) {−4, 4}   B) {−1, 1}   C) {−4, 1}   D) {−1, 4}

9) $2x^4 - 2x^3 + x^2 - 5x - 10 = 0$
    A) {1, −2}   B) $\left\{-\frac{\sqrt{10}}{2}, \frac{\sqrt{10}}{2}\right\}$ C) {−1, 2}   D) $\left\{-\frac{5}{2}, \frac{5}{2}\right\}$

10) $3x^4 - 29x^3 + 111x^2 - 179x + 78 = 0$
    A) $\left\{3, -\frac{2}{3}\right\}$ B) $\left\{-3, -\frac{2}{3}\right\}$ C) $\left\{3, \frac{2}{3}\right\}$ D) $\left\{-3, \frac{2}{3}\right\}$

Solve the problem.

11) One solution of $x^3 - 5x^2 + 5x - 1 = 0$ is 1. Find the other two solutions.
    A) $\{2 + \sqrt{3}, 2 - \sqrt{3}\}$ B) $\{4 + \sqrt{3}, 4 - \sqrt{3}\}$ C) $\{2 + 2\sqrt{3}, 2 - 2\sqrt{3}\}$ D) $\{4 + 2\sqrt{3}, 4 - 2\sqrt{3}\}$

12) Find k such that $f(x) = x^4 + kx^3 + 2$ has the factor $x + 1$.
    A) −3   B) 3   C) 2   D) −2

5 Use the Theorem for Bounds on Zeros

Use the Theorem for bounds on zeros to find a bound on the real zeros of the polynomial function.

1) $f(x) = x^4 - 15x^2 - 16$

   A) -31 and 31   B) -17 and 17   C) -16 and 16   D) -32 and 32

2) $f(x) = 4x^4 + 3x^3 + 2x^2 - 3x + 4$

   A) -11 and 11   B) -16 and 16   C) -4 and 4   D) -5 and 5

3) $f(x) = 6x^3 - x^2 + 0.3x - 0.06$

   A) -2 and 2   B) -1.36 and 1.36   C) -6 and 6   D) -1 and 1

6 Use the Intermediate Value Theorem

Use the intermediate value theorem to determine whether the polynomial function has a zero in the given interval.

1) $f(x) = -2x^3 - 9x^2 - 2x + 10$; $[-2, -1]$

   A) $f(-2) = -6$ and $f(-1) = 5$; yes
   B) $f(-2) = 6$ and $f(-1) = 5$; no
   C) $f(-2) = 6$ and $f(-1) = -5$; yes
   D) $f(-2) = -6$ and $f(-1) = -5$; no

2) $f(x) = 6x^5 + 4x^3 + 4x^2 + 2$; $[-1, 0]$

   A) $f(-1) = -4$ and $f(0) = -2$; no
   B) $f(-1) = -4$ and $f(0) = 2$; yes
   C) $f(-1) = 4$ and $f(0) = 2$; no
   D) $f(-1) = 4$ and $f(0) = -2$; yes

3) $f(x) = -4x^4 + 9x^2 - 4$; $[-2, -1]$

   A) $f(-2) = -32$ and $f(-1) = -1$; no
   B) $f(-2) = -32$ and $f(-1) = 1$; yes
   C) $f(-2) = 32$ and $f(-1) = 2$; no
   D) $f(-2) = 32$ and $f(-1) = -1$; yes

4) $f(x) = 7x^4 + 2x^3 - 6x - 2$; $[-1, 0]$

   A) $f(-1) = -9$ and $f(0) = 2$; yes
   B) $f(-1) = 9$ and $f(0) = 2$; no
   C) $f(-1) = 9$ and $f(0) = -2$; yes
   D) $f(-1) = -9$ and $f(0) = -2$; no

5) $f(x) = 6x^3 - 5x + 9$; $[-2, -1]$

   A) $f(-2) = -29$ and $f(-1) = 8$; yes
   B) $f(-2) = -29$ and $f(-1) = -8$; no
   C) $f(-2) = 29$ and $f(-1) = -8$; yes
   D) $f(-2) = 29$ and $f(-1) = 8$; no

3.7 Complex Zeros; Fundamental Theorem of Algebra

1 Use the Conjugate Pairs Theorem

Information is given about a polynomial f(x) whose coefficients are real numbers. Find the remaining zeros of f.

1) Degree 3; zeros: 2, 4 - i

   A) -4 + i   B) -2   C) 4 + i   D) no other zeros

2) Degree 4; zeros: i, 3 + i

   A) 3 - i   B) -3 + i, 3 - i   C) -i, -3 + i   D) -i, 3 - i

3) Degree 4; zeros: 7 - 5i, 8i

   A) 7 + 5i, -8i   B) 7 + 5i, 8 - i   C) -7 - 5i, -8i   D) -7 + 5i, -8i

4) Degree 3; zeros: −5, 5 − 5i
   A) 5, 5 + 5i
   B) 5, −5 + 5i
   C) −5 + 5i
   D) 5 + 5i

5) Degree 5; zeros: 4, 8 + 5i, −4i
   A) −8 − 5i, 4i
   B) −8 + 5i, 4i
   C) −4, 8 − 5i, 4i
   D) 8 − 5i, 4i

6) Degree 5; zeros: 1, i, 2i
   A) −1, −i, −2i
   B) −1, −i
   C) −1, −2i
   D) −i, −2i

7) Degree 6; zeros: 1, 2 + i, −3 − i, 0
   A) −2 + i, 3 − i
   B) 2 − i, −3 + i
   C) −2 − i, 3 + i
   D) −1, 2 − i, −3 + i

8) Degree 6; zeros: −2, 9, 2 − 5i, −9 + i
   A) 2 + 5i, −9 − i
   B) 2, 2 + 5i, −9 − i
   C) 2, 2 + 5i
   D) −2 + 5i, 9 − i

2 Find a Polynomial Function with Specified Zeros

Form a polynomial f(x) with real coefficients having the given degree and zeros.

1) Degree 3: zeros: 1 + i and −6
   A) $f(x) = x^3 + 4x^2 + 12x − 10$
   B) $f(x) = x^3 + 4x^2 − 10x + 12$
   C) $f(x) = x^3 − 6x^2 − 10x − 12$
   D) $f(x) = x^3 + x^2 − 10x + 12$

2) Degree: 3; zeros: −4 and 3 − 2i
   A) $f(x) = x^3 − 2x^2 + 5x − 52$
   B) $f(x) = x^3 − 2x^2 − 11x + 52$
   C) $f(x) = x^3 − x^2 + 11x + 52$
   D) $f(x) = x^3 − x^2 − 11x + 52$

3) Degree: 3; zeros: −2 and 3 + i.
   A) $f(x) = x^3 − 6x^2 − 10x + 20$
   B) $f(x) = x^3 − 4x^2 − 10x + 20$
   C) $f(x) = x^3 − 4x^2 − 2x + 20$
   D) $f(x) = x^3 − 8x^2 + 2x + 20$

4) Degree: 4; zeros: −1, 2, and 1 − 2i.
   A) $f(x) = x^4 − x^3 + 3x^2 − 5x − 10$
   B) $f(x) = x^4 − 3x^3 + 5x^2 − x − 10$
   C) $f(x) = x^4 − x^3 + x^2 + 9x − 10$
   D) $f(x) = x^4 − 3x^3 − 3x^2 + 7x + 6$

5) Degree: 4; zeros: 3i and −5i
   A) $f(x) = x^4 + 34x^2 + 225$
   B) $f(x) = x^4 − 3x^3 + 34x^2 + 225$
   C) $f(x) = x^4 − 5x^2 + 225$
   D) $f(x) = x^4 + 34x^2 − 5x + 225$

6) Degree: 4; zeros: 1, −1, and 4 − 2i
   A) $f(x) = x^4 − 8x^3 + 16x^2 + 8x + 17$
   B) $f(x) = x^4 + 8x^3 + 16x^2 − 8x + 17$
   C) $f(x) = x^4 + 8x^3 + 16x^2 − 8x − 17$
   D) $f(x) = x^4 − 8x^3 + 16x^2 + 8x − 17$

7) Degree: 5; zeros: 2, −3i, and 4 − i
   A) $f(x) = x^5 − 10x^4 + 42x^3 − 124x^2 + 297x − 306$
   B) $f(x) = x^5 − 10x^4 + 26x^3 − 124x^2 − 72x − 306$
   C) $f(x) = x^5 − 10x^4 + 26x^3 − 124x^2 + 72x + 306$
   D) $f(x) = x^5 − 10x^4 − 42x^3 − 124x^2 + 297x + 306$

3 Find the Complex Zeros of a Polynomial

Use the given zero to find the remaining zeros of the function.

1) $f(x) = x^4 - 32x^2 - 144$; zero: $-2i$
   A) $2i, 12, -12$
   B) $2i, 6i, -6i$
   C) $2i, 12i, -12i$
   D) $2i, 6, -6$

2) $f(x) = x^3 + 2x^2 - 6x + 8$; zero: $1 + i$
   A) $1 - i, 4$
   B) $1 - i, -4$
   C) $-4, 4$
   D) $1 - i, 4i$

3) $f(x) = x^3 - 2x^2 - 11x + 52$; zero: -4
   A) $1 + 2\sqrt{13}i, 1 - 2\sqrt{13}i$
   B) $6 + 4i, 6 - 4i$
   C) $1 + 2i, 1 - 2i$
   D) $3 + 2i, 3 - 2i$

4) $f(x) = x^3 - 6x^2 + 21x - 26$; zero: $2 + 3i$
   A) $3 - 2i, -2$
   B) $2 - 3i, 2$
   C) $2 - 3i, -2$
   D) $3 - 2i, 2$

5) $f(x) = 3x^4 - 19x^3 + 69x^2 - 99x + 26$; zero: $2 + 3i$
   A) $3 - 2i, -2, -\frac{1}{3}$
   B) $2 - 3i, -2, \frac{1}{3}$
   C) $3 - 2i, 2, -\frac{1}{3}$
   D) $2 - 3i, 2, \frac{1}{3}$

6) $f(x) = x^5 - 10x^4 + 42x^3 - 124x^2 + 297x - 306$; zero: $3i$
   A) $-2, -3i, 4 - i, 4 + i$
   B) $-2, -3i, -4 - i, -4 + i$
   C) $2, -3i, 4 - i, 4 + i$
   D) $2, -3i, -4 - i, -4 + i$

Find all zeros of the function and write the polynomial as a product of linear factors.

7) $f(x) = x^3 - x^2 + 36x - 36$
   A) $f(x) = (x - 25)(x + i)(x - i)$
   B) $f(x) = (x - 1)(x + 6i)(x - 6i)$
   C) $f(x) = (x - 1)(x + 1)(x + 36)$
   D) $f(x) = (x - 1)(x + 6)(x - 6)$

8) $f(x) = x^3 + 12x^2 + 46x + 52$
   A) $f(x) = (x + 2)(x + 5 + i)(x - 5 - i)$
   B) $f(x) = (x + 1)(x + 5 + i\sqrt{3})(x - 2 - i\sqrt{3})$
   C) $f(x) = (x - 1)(x + 5 + i\sqrt{3})(x + 5 - i\sqrt{3})$
   D) $f(x) = (x + 2)(x + 5 + i)(x + 5 - i)$

9) $f(x) = x^3 + 3x^2 - 4x - 42$
   A) $f(x) = (x - 3)(x + 3 + i\sqrt{5})(x - 3 - i\sqrt{5})$
   B) $f(x) = (x - 3)(x + 3 + i\sqrt{5})(x + 3 - i\sqrt{5})$
   C) $f(x) = (x + 3)(x + 5 + 3i)(x - 5 - 3i)$
   D) $f(x) = (x + 3)(x + 5 + 3i)(x + 5 - 3i)$

10) $f(x) = x^4 + 34x^2 + 225$
    A) $f(x) = (x + 3i)(x - 3i)(x + 5i)(x - 5i)$
    B) $f(x) = (x + 3 + 5i)^2(x + 3 - 5i)^2$
    C) $f(x) = (x + 3i)^2(x + 5i)^2$
    D) $f(x) = (x + i)(x - i)(x + 15i)(x - 15i)$

11) $f(x) = x^4 + 5x^3 + 15x^2 + 45x + 54$
    A) $f(x) = (x - 2)(x + 3)(x - 3)(x + 3)$
    B) $f(x) = (x - 1)(x - 6)(x - 3i)(x + 3i)$
    C) $f(x) = (x + 2)(x + 3)(x - 3i)(x + 3i)$
    D) $f(x) = (x - i\sqrt{6})(x + i\sqrt{6})(x - 3)(x + 3)$

12) $f(x) = 3x^4 - 7x^3 + 29x^2 - 63x + 18$
    A) $f(x) = (3x - 1)(x - 2)(x + 3i)(x - 3i)$
    B) $f(x) = (3x + 1)(x + 2)(x + 3)(x - 3)$
    C) $f(x) = (3x - 1)(x - 2)(x + 3)(x - 3)$
    D) $f(x) = (3x + 1)(x + 2)(x + 3i)(x - 3i)$

Ch. 3 Polynomial and Rational Functions
Answer Key

3.1 Quadratic Functions and Models
1 Graph a Quadratic Function Using Transformations
1) C
2) D
3) C
4) A
5) C
6) B
7) A
8) A
9) B
10) C
11) D
12) A
13) D
14) D
15) D

2 Identify the Vertex and Axis of Symmetry of a Quadratic Function
1) B
2) A
3) C
4) D
5) A
6) B
7) C
8) C
9) D
10) D

3 Graph a Quadratic Function Using Its Vertex, Axis and Intercepts
1) D
2) B
3) C
4) B
5) D
6) A
7) A
8) B
9) D
10) B
11) C
12) B
13) A
14) B
15) C
16) C
17) D
18) A
19) D
20) A
21) A
22) A
23) C

24) A
25) D
26) A
27) A
28) D
29) A

4 Use the Maximum or Minimum of a Quadratic Function to Solve Applied Problems

1) D
2) D
3) B
4) C
5) D
6) A
7) B
8) D
9) C
10) B
11) D
12) B
13) A
14) B
15) C
16) B
17) C
18) A
19) B
20) B
21) C
22) B
23) D
24) B
25) approximately 17 ft.
26) D
27) C

5 Use a Graphing Utility to Find the Quadratic Function of Best Fit to Data

1) C
2) D
3) A
4) B
5) D
6) $R(x) = -1.65x^2 + 634.42x + 7089.93$
7) $M(x) = 0.04x^2 - 1.21x + 26.03$; 1988

3.2 Polynomial Functions and Models

1 Identify Polynomial Functions and Their Degree

1) B
2) D
3) A
4) C
5) A
6) D
7) A
8) A
9) A
10) C

11) C
12) B
13) D

2 Graph Polynomial Functions Using Transformations

1) D
2) C
3) A
4) B
5) A
6) B
7) D
8) D
9) C
10) D
11) C
12) C
13) A
14) D
15) B
16) D

3 Identify the Zeros of a Polynomial Function and Their Multiplicity

1) B
2) B
3) B
4) D
5) C
6) A
7) D
8) A
9) D
10) D
11) B
12) B
13) C
14) B
15) D
16) D

4 Analyze the Graph of a Polynomial Function

1) D
2) B
3) D
4) B
5) B
6) D
7) C
8) D
9) D
10) C
11) D
12) A
13) A
14) D
15) C
16) D
17) A

18) A
19) D
20) D
21) B
22) C
23) (a) For large values of |x|, the graph of f(x) will resemble the graph of $y = x^3$.
  (b) y-intercept: (0, 0), x-intercepts: (0, 0) and (−3, 0)
  (c) The graph of f crosses the x-axis at (−3, 0) and touches the x-axis at (0, 0).
  (e) Local minimum at (0, 0), Local maximum at (−2.00, 4.00)
  (f)

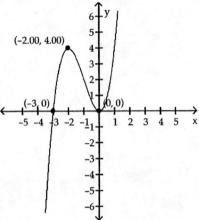

  (g) Domain of f: all real numbers; range of f: all real numbers
  (h) f is increasing on (−∞, −2.00) and (0, ∞); f is decreasing on (−2.00, 0)

24) (a) For large values of |x|, the graph of f(x) will resemble the graph of $y = x^3$.
  (b) y-intercept: (0, 12), x-intercepts: (2, 0) and (−3, 0)
  (c) The graph of f crosses the x-axis at (−3, 0) and touches the x-axis at (2, 0).
  (e) Local minimum at (2, 0); Local maximum at (−1.33, 18.52)
  (f)

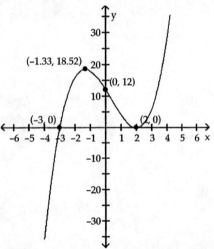

  (g) Domain of f: all real numbers; range of f: all real numbers
  (h) f is increasing on (−∞, −1.33) and (2, ∞); f is decreasing on (18.52, 2)

25) (a) For large values of $|x|$, the graph of f(x) will resemble the graph of $y = -2x^4$.
   (b) y-intercept: (0, 32), x-intercepts: (-2, 0) and (2, 0)
   (c) The graph of f crosses the x-axis at (2, 0) and crosses the x-axis at (-2, 0).
   (e) Local maximum at (1.00, 54.00)
   (f)

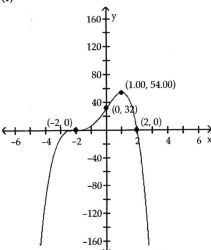

   (g) Domain of f: all real numbers; range of f: $(-\infty, 54.00]$
   (h) f is increasing on $(-\infty, -2)$ and $(-2, 1.00)$; f is decreasing on $(1.00, \infty)$

26) (a) For large values of $|x|$, the graph of f(x) will resemble the graph of $y = x^3$.
   (b) y-intercept: (0, 6), x-intercepts: (-2, 0), (1, 0), and (3, 0)
   (c) The graph of f crosses the x-axis at each of the intercepts (-2, 0), (1, 0), and (3, 0)
   (e) Local maximum at (-0.79, 8.21); Local minimum at (2.12, -4.06)
   (f)

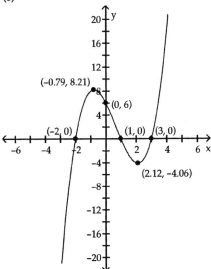

   (g) Domain of f: all real numbers; range of f: all real numbers
   (h) f is increasing on $(-\infty, -0.79)$ and $(2.12, \infty)$; f is decreasing on $(-0.79, 2.12)$

27) (a) For large values of $|x|$, the graph of $f(x)$ will resemble the graph of $y = -x^4$.
 (b) y-intercept: (0, 0), x-intercepts: (-3, 0), (0, 0), and (1, 0)
 (c) The graph of f crosses the x-axis at (1, 0) and (-3, 0) and touches the x-axis at (0, 0).
 (e) Local maxima at (-2.19, 12.39) and (0.69, 0.55); Local minimum at (0, 0)
 (f)

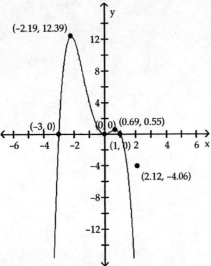

 (g) Domain of f: all real numbers; range of f: $(-\infty, 12.39]$
 (h) f is increasing on $(-\infty, -2.19)$ and $(0, 0.69)$; f is decreasing on $(-2.19, 0)$ and $(0.69, \infty)$

28) (a) For large values of $|x|$, the graph of $f(x)$ will resemble the graph of $y = x^5$.
 (b) y-intercept: (0, 0), x-intercepts: (-4, 0), (-2, 0), (0, 0), and (2, 0)
 (c) The graph of f crosses the x-axis at (-4, 0), (-2, 0), and (2, 0) and touches the x-axis at (0, 0).
 (e) Local maxima at (-3.35, 52.69) and (0,0); Local minima at (-1.31, -10.54) and (1.46, -21.75)
 (f)

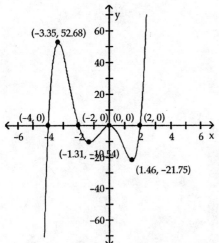

 (g) Domain of f: all real numbers; range of f: all real numbers
 (h) f is increasing on $(-\infty, -3.35)$, $(-1.31, 0)$, and $(1.46, \infty)$; f is decreasing on $(-3.35, -1.31)$ and $(0, 1.46)$

29) (a) For large values of $|x|$, the graph of f(x) will resemble the graph of $y = x^3$.
(c) y-intercept: (0, 3.0912), x-intercept: (-1.87, 0)
(d) Local maximum at (-0.80, 4.39); Local minimum at (1.07, 1.09)
(e)

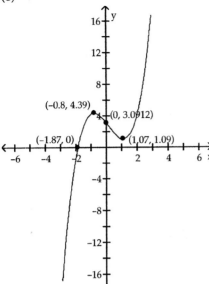

(f) domain of f: all real numbers; range of f: all real numbers
(g) f is increasing on $(-\infty, -0.80)$ and $(1.07, \infty)$; f is decreasing on $(-0.80, 1.07)$

30) a) The x-intercepts are -1.23, 0.40, 1.38, and 2.94. The y-intercept is -4.
b) The graph crosses the x-axis at each x-intercept.
c) The graph resembles $f(x) = 2x^4$ for large values of $|x|$.
d) Maximum at (0.89, 2.11); minima at (-0.65, -8.87) and (2.38, -8.02)
e) The graph has 3 turning points.
f)

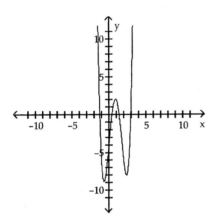

31) B

5 Find the Cubic Function of Best Fit to Data

1) $y = 0.03x^3 - 0.34x^2 + 1.31x + 0.17$
2) B

3.3 Properties of Rational Functions
1 Find the Domain of a Rational Function

1) A
2) A
3) B
4) C
5) C
6) D

7) B
8) B
9) B
10) C
11) A
12) B
13) C
14) A
15) B
16) B

2 Find the Vertical Asymptotes of a Rational Function

1) D
2) B
3) D
4) B
5) A
6) B
7) C
8) C
9) A
10) A
11) A
12) B
13) B
14) D
15) B
16) A
17) B
18) D

3 Find the Horizontal or Oblique Asymptotes of a Rational Function

1) C
2) D
3) A
4) C
5) D
6) B
7) D
8) C
9) B
10) D
11) D
12) A
13) A
14) C
15) D
16) D
17) D
18) C
19) B
20) B
21) A
22) C
23) C
24) D
25) B

26) C
27) A

4 Demonstrate Additional Understanding and Skills

1) D
2) C
3) B
4) A
5) C
6) C
7) a) $g(-86) \approx 9.8211$ m/sec^2
   b) $g(0) \approx 9.8208$ m/sec^2
8) horizontal asymptote: $r = 0$; vertical asymptote: $t = 0$
9) vertical asymptote: $s_2 = 0.3$; horizontal asymptote: $s_1 = 0.3$
10) vertical asymptote: $f_2 = -0.001$; horizontal asymptote: $f = 0.001$

3.4 The Graph of a Rational Function; Inverse and Joint Variation

1 Analyze the Graph of a Rational Function

1) B
2) B
3) D
4) C
5) C
6) B
7) D
8) B
9) C
10) D
11) B
12) A
13) C
14) B
15) C
16) D
17) D
18) A
19) D
20) C
21) B
22) D
23) B
24) C
25) D
26) B
27) C
28) C
29) A
30) C
31) C
32) A
33) C
34) A
35) C
36) A
37) D

38) C
39) C
40) C
41) C
42) B
43) D
44) D
45) C
46) C
47) C
48) A
49) C
50) A
51) B
52) D
53) A
54) C
55) A
56) B

2 Solve Applied Problems Involving Rational Functions
1) a) P(0) = 15 insects
   b) P(120) ≈ 608 insects
   c) 1,350
2) a) y = 0
   b) t = 7
3) C
4) C
5) C
6) B
7) A
8) B
9) D
10) a) Surface area equals $A(x) = 4x^2 + \dfrac{21{,}000}{x}$, x > 0, where x equals the width of the box.
    b) width ≈ 13.8 in, length ≈ 27.6 in, height ≈ 18.4 in
11) a)

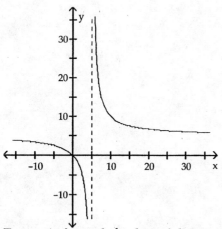

    b) To remain focused, the desired distance between the film and the projector lens decreases and approaches 5 centimeters as the distance between the movie screen and the projector lens increases.
12) C

3 Construct a Model Using Inverse Variation
1) C

2) B
3) B
4) D
5) B
6) B

4 Construct a Model Using Joint or Combined Variation

1) A
2) D
3) D
4) B
5) C
6) D
7) B

3.5 Polynomial and Rational Inequalities

1 Solve Polynomial Inequalities Algebraically and Graphically

1) B
2) C
3) D
4) B
5) D
6) A
7) B
8) C
9) B
10) A
11) A
12) B
13) A
14) D
15) B
16) B
17) B
18) D
19) A
20) B
21) A
22) D
23) C
24) C
25) B
26) A
27) A
28) C

2 Solve Rational Inequalities Algebraically and Graphically

1) C
2) D
3) C
4) D
5) D
6) D
7) A
8) C
9) A
10) D
11) D

12) C
13) C
14) D

3.6 The Real Zeros of a Polynomial Function

1 Use the Remainder and Factor Theorems

1) A
2) B
3) A
4) B
5) A
6) B
7) A
8) B
9) B
10) A
11) A
12) B
13) C
14) C
15) C
16) B
17) D

2 Use the Rational Zeros Theorem

1) A
2) A
3) C
4) D
5) B
6) D
7) A

3 Find the Real Zeros of a Polynomial Function

1) B
2) B
3) D
4) D
5) A
6) D
7) −2.62
8) −0.38
9) −2.24
10) 2.24
11) C
12) D

4 Solve Polynomial Equations

1) D
2) D
3) B
4) D
5) C
6) A
7) A
8) B
9) C
10) C
11) A

12) B
5 Use the Theorem for Bounds on Zeros
- 1) B
- 2) D
- 3) B

6 Use the Intermediate Value Theorem
- 1) A
- 2) B
- 3) B
- 4) C
- 5) A

3.7 Complex Zeros; Fundamental Theorem of Algebra
1 Use the Conjugate Pairs Theorem
- 1) C
- 2) D
- 3) A
- 4) D
- 5) D
- 6) D
- 7) B
- 8) A

2 Find a Polynomial Function with Specified Zeros
- 1) B
- 2) B
- 3) C
- 4) B
- 5) A
- 6) B
- 7) C

3 Find the Complex Zeros of a Polynomial
- 1) D
- 2) B
- 3) D
- 4) B
- 5) D
- 6) B
- 7) B
- 8) D
- 9) B
- 10) A
- 11) C
- 12) A

Ch. 4 Exponential and Logarithmic Functions

4.1 Composite Functions

1 Form a Composite Function

For the given functions f and g, find the requested composite function value.

1) $f(x) = \sqrt{x+5}$; $g(x) = 2x$; Find $(f \circ g)(0)$.
   A) $\sqrt{5}$
   B) $\sqrt{10}$
   C) $2\sqrt{5}$
   D) $2\sqrt{10}$

2) $f(x) = 2x + 4$; $g(x) = 2x^2 + 5$; Find $(g \circ g)(3)$.
   A) 1063
   B) 50
   C) 24
   D) 205

3) Given $f(x) = \dfrac{x-6}{x}$ and $g(x) = x^2 + 9$, find $(g \circ f)(-2)$.
   A) 13
   B) $\dfrac{7}{13}$
   C) $\dfrac{145}{16}$
   D) 25

For the functions f and g and the number c, compute $(f \circ g)(c)$.

4) $f(x) = 14x^2 - 5x$
   $g(x) = 20x - 10$
   $c = 3$
   A) 5550
   B) 34,750
   C) 2210
   D) 32,540

5) $f(t) = \sqrt{t^4 + 6t^2 + 9}$
   $g(t) = \dfrac{t+3}{3}$
   $c = 3$
   A) 49
   B) 7
   C) 24
   D) 5

6) $f(x) = x^2 + 2x - 3$
   $g(x) = x^2 - 2x + 5$
   $c = -4$
   A) 12
   B) 828
   C) 80
   D) 896

Find the indicated composite for the pair of functions.

7) $(f \circ g)(x)$: $f(x) = 4x + 12$, $g(x) = 5x - 1$
   A) $20x + 11$
   B) $20x + 16$
   C) $20x + 8$
   D) $20x + 59$

8) $(g \circ f)(x)$: $f(x) = -2x + 3$, $g(x) = 6x + 4$
   A) $12x + 22$
   B) $-12x - 14$
   C) $-12x + 22$
   D) $-12x + 11$

9) $(f \circ g)(x)$: $f(x) = \dfrac{1}{x+5}$, $g(x) = \dfrac{7}{8x}$
   A) $\dfrac{8x}{7 + 40x}$
   B) $\dfrac{8x}{7 - 40x}$
   C) $\dfrac{1x}{7 + 40x}$
   D) $\dfrac{7x + 35}{8x}$

10) $(g \circ f)(x)$: $f(x) = \dfrac{x-7}{3}$, $g(x) = 3x + 7$

　　A) $3x + 14$　　　　　B) $x - \dfrac{7}{3}$　　　　　C) $x + 14$　　　　　D) x

11) $(f \circ g)(x)$: $f(x) = \sqrt{x+4}$, $g(x) = 8x - 8$
　　A) $2\sqrt{2x-1}$　　　B) $8\sqrt{x+4} - 8$　　　C) $2\sqrt{2x+1}$　　　D) $8\sqrt{x-4}$

12) $(g \circ f)(x)$: $f(x) = 4x^2 + 3x + 5$, $g(x) = 3x - 8$
　　A) $12x^2 + 9x + 23$　　B) $12x^2 + 9x + 7$　　C) $4x^2 + 9x + 7$　　D) $4x^2 + 3x - 3$

Decide whether the composite functions, $f \circ g$ and $g \circ f$, are equal to x.

13) $f(x) = \sqrt[5]{x-8}$, $g(x) = x^5 + 8$
　　A) No, yes　　　　B) Yes, no　　　　C) No, no　　　　D) Yes, yes

14) $f(x) = x^2 + 5$, $g(x) = \sqrt{x} - 5$
　　A) No, no　　　　B) No, yes　　　　C) Yes, yes　　　　D) Yes, no

15) $f(x) = \sqrt{x}$, $g(x) = x^2$
　　A) Yes, no　　　　B) No, yes　　　　C) Yes, yes　　　　D) No, no

16) $f(x) = \dfrac{1}{x}$, $g(x) = x$
　　A) Yes, no　　　　B) No, no　　　　C) Yes, yes　　　　D) No, yes

17) $f(x) = \sqrt{x+1}$, $g(x) = x^2$
　　A) Yes, no　　　　B) Yes, yes　　　　C) No, yes　　　　D) No, no

18) $f(x) = x^3 + 3$, $g(x) = \sqrt[3]{x-3}$
　　A) Yes, yes　　　　B) Yes, no　　　　C) No, no　　　　D) No, yes

Find functions f and g so that the composition of f and g is H.

19) $H(x) = \sqrt[3]{x+1}$
　　A) $f(x) = x + 1$; $g(x) = \sqrt[3]{x}$　　　　　　B) $f(x) = \sqrt{x}$; $g(x) = x + 1$
　　C) $f(x) = \sqrt[3]{x}$; $g(x) = 1$　　　　　　　D) $f(x) = \sqrt[3]{x}$; $g(x) = x + 1$

20) $H(x) = (5 - 2x^3)^2$
　　A) $f(x) = x^2$; $g(x) = 5 - 2x^3$　　　　　　B) $f(x) = x^3$; $g(x) = (5 - 2x)^2$
　　C) $f(x) = 5 - 2x^3$; $g(x) = x^2$　　　　　　D) $f(x) = (5 - 2x)^2$; $g(x) = x^3$

21) $H(x) = \dfrac{1}{x^2 - 7}$

   A) $f(x) = x^2 - 7;\quad g(x) = \dfrac{1}{x}$
   B) $f(x) = \dfrac{1}{x^2};\quad g(x) = -1/7$
   C) $f(x) = \dfrac{1}{x};\quad g(x) = x^2 - 7$
   D) $f(x) = \dfrac{1}{x^2};\quad g(x) = x - 7$

22) $H(x) = \dfrac{8}{\sqrt{9x + 4}}$

   A) $f(x) = \sqrt{9x + 4};\quad g(x) = 8$
   B) $f(x) = \dfrac{8}{\sqrt{x}};\quad g(x) = 9x + 4$
   C) $f(x) = \dfrac{8}{x};\quad g(x) = 9x + 4$
   D) $f(x) = 8;\quad g(x) = \sqrt{9 + 4}$

23) $H(x) = |4 - 3x^2|$

   A) $f(x) = |x|;\ g(x) = 4 - 3x^2$
   B) $f(x) = 4 - 3|x|;\ g(x) = x^2$
   C) $f(x) = 4 - 3x^2;\ g(x) = |x|$
   D) $f(x) = x^2;\ g(x) = 4 - 3|x|$

24) $H(x) = |5x + 1|$

   A) $f(x) = |x|;\quad g(x) = 5x + 1$
   B) $f(x) = |-x|;\quad g(x) = 5x - 1$
   C) $f(x) = x;\quad g(x) = 5x + 1$
   D) $f(x) = -|x|;\quad g(x) = 5x + 1$

Solve the problem.

25) The population P of a predator mammal depends upon the number x of a smaller animal that is its primary food source. The population s of the smaller animal depends upon the amount a of a certain plant that is its primary food source. If $P(x) = 2x^2 + 4$ and $s(a) = 3a + 2$, what is the relationship between the predator mammal and the plant food source?

   A) $P(s(a)) = 9a^2 + 12a + 8$
   B) $P(s(a)) = 6a + 6$
   C) $P(s(a)) = 18a^2 + 24a + 12$
   D) $P(s(a)) = 18a^2 + 12a + 12$

26) An oil well off the Gulf Coast is leaking, with the leak spreading oil over the surface of the gulf as a circle. At any time t, in minutes, after the beginning of the leak, the radius of the oil slick on the surface is $r(t) = 2t$ ft. Find the area A of the oil slick as a function of time.

   A) $A(r(t)) = 4\pi t$
   B) $A(r(t)) = 4\pi t^2$
   C) $A(r(t)) = 2\pi t^2$
   D) $A(r(t)) = 4t^2$

27) An airline charter service charges a fare per person of $500 plus $20 for each unsold seat. The airplane holds 75 passengers. Let x represent the number of unsold seats and write an expression for the total revenue R for a charter flight.

   A) $R(x) = (75 - x)(500 + 20x)$ or $37,500 + 1000x - 20x^2$
   B) $R(x) = x(500 + 20x)$ or $500x + 20x^2$
   C) $R(x) = 75(500 + 20x)$ or $37,500 + 1500x$
   D) $R(x) = (75 - x)(500 + 20x)$ or $37,500 + 1500x - 20x^2$

28) The surface area of a balloon is given by $S(r) = 4\pi r^2$, where r is the radius of the balloon. If the radius is increasing with time t, as the balloon is being blown up, according to the formula $r(t) = \frac{3}{4}t^3$, $t \geq 0$, find the surface area S as a function of the time t.

   A) $S(r(t)) = \frac{9}{4}\pi t^6$
   B) $S(r(t)) = \frac{9}{4}\pi t^9$
   C) $S(r(t)) = \frac{9}{4}\pi t^3$
   D) $S(r(t)) = \frac{9}{16}\pi t^6$

29) The surface area S (in square inches) of a cylindrical pipe with length 12 inches is given by $S(r) = 2\pi r^2 + 24\pi r$, where r is the radius of the piston (in inches). If the radius is increasing with time t (in minutes) according to the formula $r(t) = \frac{1}{6}t^2$, $t \geq 0$, find the surface area S of the pipe as a function of the time t.

30) The volume V (in cubic inches) of a cylindrical pipe with length 12 inches is given by $V(r) = 12\pi r^2$, where r is the radius of the piston (in inches). If the radius is increasing with time t (in minutes) according to the formula $r(t) = \frac{1}{6}t^2$, $t \geq 0$, find the volume V of the pipe as a function of the time t.

31) The price p of a certain product and the quantity sold x obey the demand equation $p = -\frac{2}{3}x + 200$, $0 \leq x \leq 300$. Suppose that the cost C of producing x units is $C = \frac{\sqrt{x}}{20} + 800$. Assuming that all items produced are sold, find the cost C as a function of the price p.

32) If $f(x) = \frac{1}{2}x^2 + 4$ and $g(x) = 2x - a$, find a so that the graph of $f \circ g$ crosses the y-axis at 36.

2 Find the Domain of a Composite Function

Find the domain of the composite function f ∘ g.

1) $f(x) = 8x + 32$; $g(x) = x + 8$
   A) $\{x \mid x \neq -12\}$
   B) $\{x \mid x \neq 12\}$
   C) $\{x \mid x \neq -8, x \neq -4\}$
   D) $\{x \mid x \text{ is any real number}\}$

2) $f(x) = \frac{4}{x+1}$; $g(x) = x + 4$
   A) $\{x \mid x \neq -1, x \neq -4\}$
   B) $\{x \mid x \neq -5\}$
   C) $\{x \mid x \neq -1\}$
   D) $\{x \mid x \text{ is any real number}\}$

3) $f(x) = x + 10$; $g(x) = \frac{7}{x+4}$
   A) $\{x \mid x \text{ is any real number}\}$
   B) $\{x \mid x \neq -4\}$
   C) $\{x \mid x \neq -4, x \neq -10\}$
   D) $\{x \mid x \neq -14\}$

4) $f(x) = \frac{-9}{x+1}$; $g(x) = \frac{-2}{x}$
   A) $\{x \mid x \neq 0, x \neq -1\}$
   B) $\{x \mid x \text{ is any real number}\}$
   C) $\{x \mid x \neq 0, x \neq -1, x \neq 2\}$
   D) $\{x \mid x \neq 0, x \neq 2\}$

Precalculus Enhanced with Graphing Utilities 323

5) $f(x) = \dfrac{40}{x}$; $g(x) = \dfrac{9}{x-5}$

   A) $\{x \mid x \neq 5\}$ B) $\{x \mid x \neq 0, x \neq 5, x \neq 8\}$
   C) $\{x \mid x \neq 5, x \neq 0\}$ D) $\{x \mid x$ is any real number$\}$

6) $f(x) = \dfrac{x}{x+5}$; $g(x) = \dfrac{20}{x+3}$

   A) $\{x \mid x \neq -3, x \neq -7\}$ B) $\{x \mid x \neq 0, x \neq -3, x \neq -7\}$
   C) $\{x \mid x \neq -3, x \neq -5\}$ D) $\{x \mid x$ is any real number$\}$

7) $f(x) = \sqrt{x}$; $g(x) = 5x + 10$
   A) $\{x \mid x \leq -2 \text{ or } x \geq 0\}$ B) $\{x \mid x$ is any real number$\}$
   C) $\{x \mid x \geq 0\}$ D) $\{x \mid x \geq -2\}$

8) $f(x) = 4x + 24$; $g(x) = \sqrt{x}$
   A) $\{x \mid x \leq -6 \text{ or } x \geq 0\}$ B) $\{x \mid x \geq -6\}$
   C) $\{x \mid x \geq 0\}$ D) $\{x \mid x$ is any real number$\}$

9) $f(x) = \sqrt{x-3}$; $g(x) = \dfrac{3}{x-10}$

   A) $\{x \mid x \neq 10, x \neq 3\}$ B) $\{x \mid x$ is any real number$\}$
   C) $\{x \mid 10 < x \leq 11\}$ D) $\{x \mid x \geq 3, x \neq 10\}$

10) $f(x) = \dfrac{5}{x-6}$; $g(x) = \sqrt{x-5}$

   A) $\{x \mid x \geq 5, x \neq 41\}$ B) $\{x \mid x$ is any real number$\}$
   C) $\{x \mid x \geq 5, x \neq 6\}$ D) $\{x \mid x \geq 5, x \neq 6, x \neq 41\}$

11) $f(x) = \sqrt{2-x}$; $g(x) = |2x - 1|$
   A) $\left\{x \mid -\dfrac{1}{2} \leq x \leq \dfrac{3}{2}\right\}$ B) $\{x \mid x \leq 2\}$ C) $\{x \mid x \geq 2\}$ D) all real numbers

Solve the problem.

12) If $f(x) = x^2$ and $g(x) = -1 + 5x$, find $(f \circ g)(x)$ and find the domain of $(f \circ g)(x)$.

4.2 One-to-One Functions; Inverse Functions

1 Determine Whether a Function Is One-to-One

Indicate whether the function is one-to-one.

1) $\{(-13, 18), (-20, 7), (11, 13)\}$
   A) No   B) Yes

2) $\{(9, 5), (17, 5), (-1, -16)\}$
   A) No   B) Yes

3) $\{(-2, -2), (-1, -2), (0, 1), (1, 2)\}$
   A) No   B) Yes

4) {(6, -9), (-2, -8), (-4, -7), (-6, -6)}

   A) No   B) Yes

5) {(-5, -6), (6, 5), (-3, -5), (3, 5)}

   A) Yes   B) No

Use the horizontal line test to determine whether the function is one-to-one.

6)

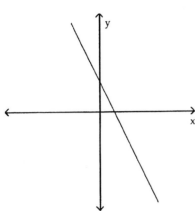

   A) Yes   B) No

7)

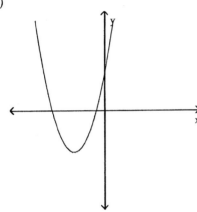

   A) No   B) Yes

8)

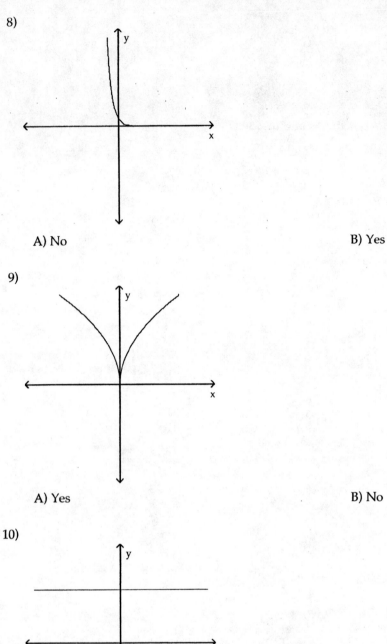

A) No

B) Yes

9)

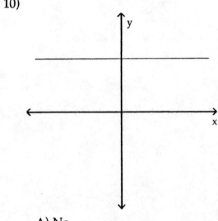

A) Yes

B) No

10)

A) No

B) Yes

11)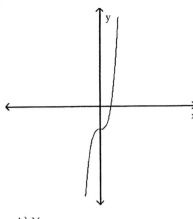

A) Yes   B) No

2 Determine the Inverse of a Function Defined by a Map or an Ordered Pair

Find the inverse of the function.

1) {(12, −7), (10, −6), (8, −5), (6, −4)}
   A) {(−7, 12), (−6, 10), (−5, 8), (−4, 6)}
   B) {(−6, −7), (−4, 8), (12, 8), (−6, −5)}
   C) {(−6, −7), (−7, 8), (12, 10), (−6, −5)}
   D) $\left\{\left(12, -\frac{1}{7}\right), \left(10, -\frac{1}{6}\right), \left(8, -\frac{1}{5}\right), \left(6, -\frac{1}{4}\right)\right\}$

2) {(−2, −1), (1, 2), (−6, 3), (6, −3)}
   A) $\left\{(-2, -1), \left(1, \frac{1}{2}\right), \left(-6, \frac{1}{3}\right), \left(6, -\frac{1}{3}\right)\right\}$
   B) {(−1, −2), (2, 1), (3, −6), (−3, 6)}
   C) {(−3, −6), (2, 1), (−1, 1), (3, 6)}
   D) {(−3, −6), (−6, 1), (−1, −2), (3, 6)}

3) {(−3, 4), (−1, 5), (0, 2), (2, 4), (5, 7)}
   A) {(3, 4), (1, 5), (0, 2), (−2, 4), (−5, 7)}
   B) {(−3, −4), (−1, −5), (0, −2), (2, −4), (5, −7)}
   C) {(3, −4), (1, −5), (0, −2), (−2, −4), (−5, −7)}
   D) {(4, −3), (5, −1), (2, 0), (4, 2), (7, 5)}

Find the inverse. Determine whether the inverse represents a function.

4) {(6, 4), (2, 5), (0, 6), (−2, 7)}
   A) {(4, 6), (5, 2), (6, 0), (7, −2)}; a function
   B) {(4, 6), (5, 2), (6, 0), (7, −2)}; not a function
   C) {(5, 4), (7, 0), (6, 0), (5, 6)}; a function
   D) {(5, 4), (4, 0), (6, 2), (5, 6)}; not a function

If the following defines a one-to-one function, find the inverse.

5) {(6, 3), (−1, 4), (−3, 5), (−5, 6)}
   A) {(4, 3), (3, −3), (6, −1), (4, 5)}
   B) {(3, 6), (4, −1), (5, −3), (6, −5)}
   C) {(4, 3), (6, −3), (6, −3), (4, 5)}
   D) Not a one-to-one function

3 Obtain the Graph of the Inverse Function from the Graph of the Function

The graph of a one-to-one function f is given. Draw the graph of the inverse function f^{-1} as a dashed line or curve.

1) $f(x) = 5x$

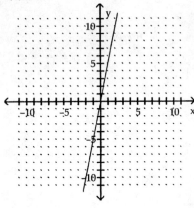

A)

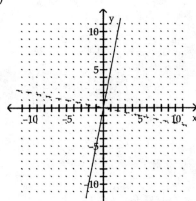

B)

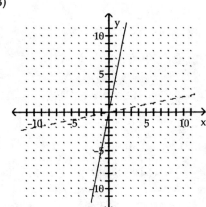

2) $f(x) = \sqrt{x} + 2$

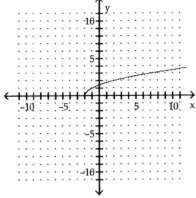

A)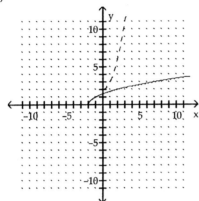

B)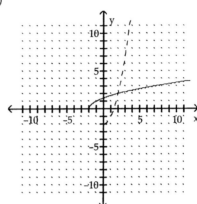

3) $f(x) = x^3 + 3$

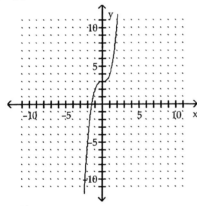

A)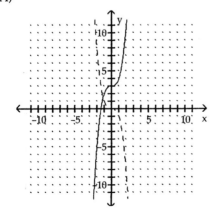

B)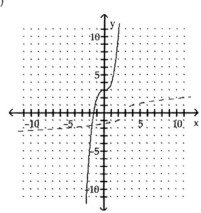

Precalculus Enhanced with Graphing Utilities 329

4) $f(x) = \dfrac{6}{x}$

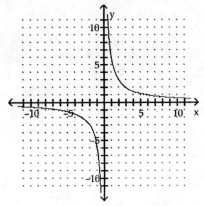

A)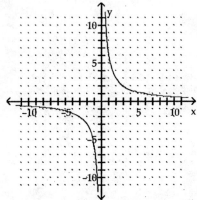

Function is its own inverse

B)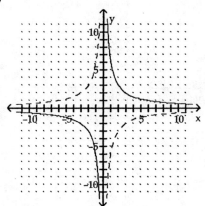

Use the graph of the given one-to-one function to sketch the graph of the inverse function. For convenience, the graph of $y = x$ is also given.

5)
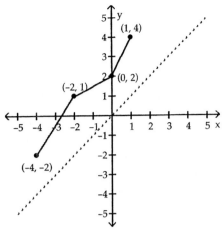

A)
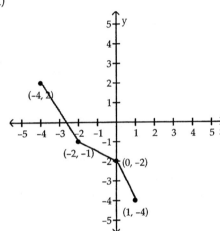

B)
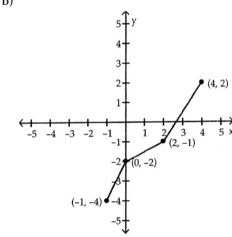

C)
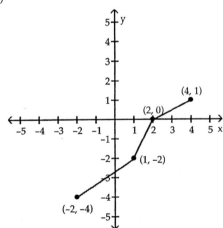

D)
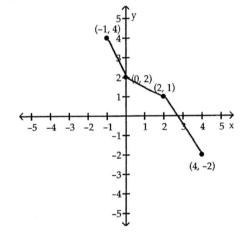

4 Find the Inverse of a Function Defined by an Equation

The function f is one-to-one. Find its inverse.

1) $f(x) = 5x + 8$

A) $f(x) = \dfrac{x-8}{5}$

B) $f^{-1}(x) = \dfrac{x-8}{5}$

C) $f^{-1}(x) = -\dfrac{x+5}{8}$

D) $f^{-1}(x) = \dfrac{x+8}{5}$

2) $f(x) = \dfrac{7x - 2}{3}$

A) $f^{-1}(x) = \dfrac{3}{7x - 2}$ B) $f^{-1}(x) = \dfrac{3x + 2}{7}$ C) $f^{-1}(x) = \dfrac{3x - 2}{7}$ D) $f^{-1}(x) = \dfrac{3}{7x + 2}$

3) $f(x) = \dfrac{3}{2x - 5}$

A) $f^{-1}(x) = \dfrac{2x - 5}{3}$ B) $f^{-1}(x) = \dfrac{3}{2x} + \dfrac{5}{2}$ C) $f^{-1}(x) = -\dfrac{5}{2} - \dfrac{3}{2x}$ D) $f^{-1}(x) = \dfrac{3}{2y} + \dfrac{5}{2}$

4) $f(x) = (x + 6)^3$

A) $f^{-1}(x) = \sqrt[3]{x} - 6$ B) $f^{-1}(x) = \sqrt[3]{x} + 6$ C) $f^{-1}(x) = \sqrt[3]{x} - 216$ D) $f^{-1}(x) = \sqrt{x} - 6$

5) $f(x) = \sqrt[3]{x + 8}$

A) $f^{-1}(x) = x^3 + 64$ B) $f^{-1}(x) = x^3 - 8$ C) $f^{-1}(x) = x - 8$ D) $f^{-1}(x) = \dfrac{1}{x^3 - 8}$

6) $f(x) = (x + 2)^3 - 8$.

A) $f^{-1}(x) = \sqrt[3]{x - 2} + 8$ B) $f^{-1}(x) = \sqrt[3]{x + 10}$ C) $f^{-1}(x) = \sqrt[3]{x + 8} - 2$ D) $f^{-1}(x) = \sqrt[3]{x + 6}$

Find the inverse function of f. State the domain and range of f.

7) $f(x) = \dfrac{3x - 2}{x + 5}$

A) $f^{-1}(x) = \dfrac{5x + 2}{3 + x}$; domain of f: $\{x \mid x \ne -5\}$; range of f: $\{y \mid y \ne -3\}$

B) $f^{-1}(x) = \dfrac{3x + 2}{x - 5}$; domain of f: $\{x \mid x \ne -5\}$; range of f: $\{y \mid y \ne 5\}$

C) $f^{-1}(x) = \dfrac{x + 5}{3x - 2}$; domain of f: $\{x \mid x \ne -5\}$; range of f: $\{y \mid y \ne \dfrac{2}{3}\}$

D) $f^{-1}(x) = \dfrac{5x + 2}{3 - x}$; domain of f: $\{x \mid x \ne -5\}$; range of f: $\{y \mid y \ne 3\}$

Solve the problem.

8) The profit P for selling x items is given by the equation $P(x) = 2x - 500$. Express the sales amount x as a function of the profit P.

9) The function $f(x) = |x| - 5$ is not one-to-one.
   (a) Find a suitable restriction on the domain of f so that the new function that results is one-to-one.
   (b) Find the inverse of f.

10) The weight W of a bird's brain (in ounces) is related to the volume V of the bird's skull (in cubic ounces) through the function $W(V) = 3.51\sqrt[3]{V} + 1.2$.
   (a) Express the skull volume V as a function of brain weight W.
   (b) Predict the skull volume of a bird whose brain weighs 3 oz.

5 Demonstrate Additional Understanding and Skills

Decide whether or not the functions are inverses of each other.

1) $f(x) = 4x + 7$; $g(x) = \frac{1}{4}(x - 7)$

   A) Yes   B) No

2) $f(x) = 2x + 6$; $g(x) = \frac{x}{2} - 6$

   A) No   B) Yes

3) $f(x) = 4 - 8x$; $g(x) = -\frac{x}{8}(x - 4)$

   A) Yes   B) No

4) $f(x) = 9 - 4x$; $g(x) = \frac{x}{4}(x - 9)$

   A) Yes   B) No

5) $f(x) = (x - 5)^2$, $x \geq 5$; $g(x) = \sqrt{x} + 5$

   A) No   B) Yes

6) $f(x) = (x - 5)^2$, $x \geq 5$; $g(x) = \sqrt{x + 5}$

   A) No   B) Yes

Solve the problem.

7) Show that f and g are inverse functions or state that they are not.

$$f(x) = \sqrt[3]{-8x - 6}\,;\, g(x) = -\frac{x^3 + 6}{8}$$

4.3 Exponential Functions

1 Evaluate Exponential Functions

Approximate the value using a calculator. Express answer rounded to three decimal places.

1) $6^{1.2}$

   A) 7.200   B) 2.986   C) 8.586   D) 46,656.000

2) $3.7^{4.28}$

   A) 270.336   B) 216.941   C) 126.575   D) 15.836

3) $4.548^{2.081}$

   A) 23.384   B) 981.222   C) 9.464   D) 28.022

4) 3.9^{π}

   A) 36.462   B) 71.926   C) 86.873   D) 12.252

5) $2^{\sqrt{5}}$

   A) 4.472   B) 16.000   C) 4.711   D) 5.000

6) $e^{2.98}$

A) 19.688   B) 8.100   C) 15.154   D) 19.456

Determine whether the given function is exponential or not. If it is exponential, identify the value of the base a.

7)

x	H(x)
-1	7
0	12
1	17
2	22
3	27

A) Exponential; a = 5   B) Exponential; a = 7   C) Exponential; a = 12   D) Not exponential

8)

x	H(x)
-1	$\frac{8}{3}$
0	1
1	$\frac{3}{8}$
2	$\frac{9}{64}$
3	$\frac{27}{512}$

A) Exponential; a = 3   B) Exponential; a = $\frac{3}{8}$ C) Exponential; a = $\frac{8}{3}$ D) Not exponential

Solve the problem.

9) The function $D(h) = 6e^{-0.4h}$ can be used to determine the milligrams D of a certain drug in a patient's bloodstream h hours after the drug has been given. How many milligrams (to two decimals) will be present after 7 hours?

A) 0.64 mg   B) 4.36 mg   C) 0.36 mg   D) 98.67 mg

10) The formula $P = 14.7e^{-0.21x}$ gives the average atmospheric pressure, P, in pounds per square inch, at an altitude x, in miles above sea level. Find the average atmospheric pressure for an altitude of 2.3 miles. Round your answer to the nearest tenth.

A) 11.0 lb/in^2 B) 9.1 lb/in^2 C) 7.8 lb/in^2 D) 8.4 lb/in^2

11) A rumor is spread at an elementary school with 1200 students according to the model $N = 1200(1 - e^{-0.16d})$ where N is the number of students who have heard the rumor and d is the number of days that have elapsed since the rumor began. How many students will have heard the rumor after 5 days?

12) The function $f(x) = 200(0.5)^{x/50}$ models the amount in pounds of a particular radioactive material stored in a concrete vault, where x is the number of years since the material was put into the vault. Find the amount of radioactive material in the vault after 60 years. Round to the nearest whole number.

A) 87 pounds   B) 83 pounds   C) 120 pounds   D) 112 pounds

13) Instruments on a satellite measure the amount of power generated by the satellite's power supply. The time t and the power P can be modeled by the function $P = 50e^{-t/300}$, where t is in days and P is in watts. How much power will be available after 378 days? Round to the nearest hundredth.

14) A cancer patient undergoing chemotherapy is injected with a particular drug. The function $D(h) = 4e^{-0.35h}$ gives the number of milligrams D of this drug that is in the patient's bloodstream h hours after the drug has been administered. How many milligrams of the drug were injected? To the nearest milligram, how much of the drug will be present after 2 hours?

15) A grocery store normally sells 7 jars of caviar per week. Use the Poisson Distribution $P(x) = \dfrac{7^x e^{-7}}{x!}$ to find the probability (to three decimals) of selling 5 jars in a week. $(x! = x \cdot (x-1) \cdot (x-2) \cdot \ldots \cdot (3)(2)(1))$.

A) 0.613   B) 0.639   C) 0.255   D) 0.128

16) If $5^x = 3$, what does 5^{-2x} equal?

A) -9   B) $\dfrac{1}{6}$ C) $\dfrac{1}{9}$ D) 9

17) If $6^{-x} = \dfrac{1}{6}$, what does 36^x equal?

A) 6   B) $\dfrac{1}{36}$ C) -36   D) 36

2 Graph Exponential Functions

The graph of an exponential function is given. Match the graph to one of the following functions.

1)

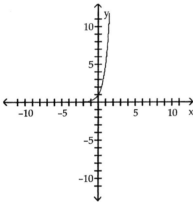

A) $f(x) = 5^x + 1$ B) $f(x) = 5^{x-1}$ C) $f(x) = 5^x - 1$ D) $f(x) = 5^x$

2)

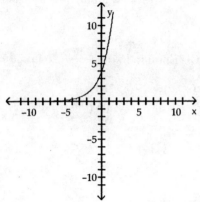

A) $f(x) = 2^x + 2$ B) $f(x) = 2^{x+2}$ C) $f(x) = 2^x$ D) $f(x) = 2^x - 2$

3)

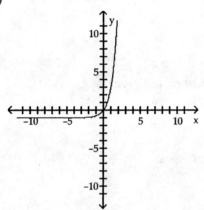

A) $f(x) = 4^x - 1$ B) $f(x) = 4^{x-1}$ C) $f(x) = 4^x + 1$ D) $f(x) = 4^x$

4)

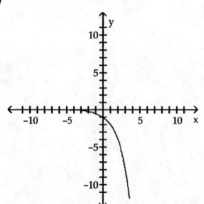

A) $f(x) = 2^{-x}$ B) $f(x) = -2^x$ C) $f(x) = -2^{-x}$ D) $f(x) = 2^x$

5)

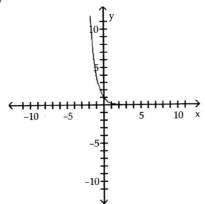

A) $f(x) = -4^{-x}$ B) $f(x) = -4^x$ C) $f(x) = 4^{-x}$ D) $f(x) = 4^x$

6)

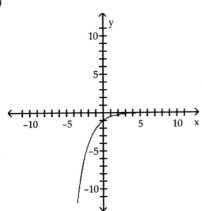

A) $f(x) = -2^{-x}$ B) $f(x) = -2^x$ C) $f(x) = 2^x$ D) $f(x) = 2^{-x}$

7)

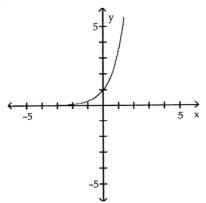

A) $y = 2.4^x$ B) $y = 0.32^x$ C) $y = 3.5^x$ D) $y = 0.65^x$

Use transformations to graph the function. Determine the domain, range, and horizontal asymptote of the function.

8) $f(x) = -1 + e^x$

A)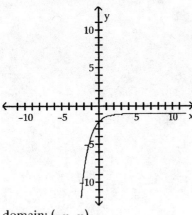
domain: $(-\infty, \infty)$
range: $(-1, \infty)$
horizontal asymptote: $y = -1$

B)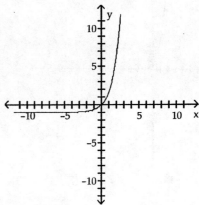
domain: $(-\infty, \infty)$
range: $(-1, \infty)$
horizontal asymptote: $y = -1$

C)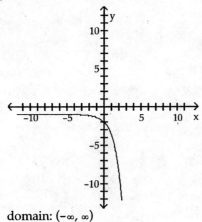
domain: $(-\infty, \infty)$
range: $(-1, \infty)$
horizontal asymptote: $y = -1$

D)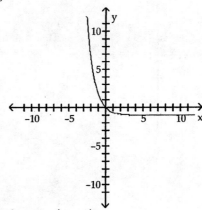
domain: $(-\infty, \infty)$
range: $(-1, \infty)$
horizontal asymptote: $y = -1$

9) $f(x) = -2^{x+3} + 4$

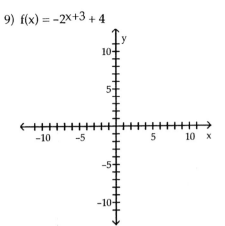

A)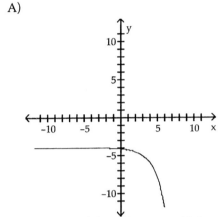

domain of f: $(-\infty, \infty)$; range of f: $(-\infty, -4)$;
horizontal asymptote: $y = -4$

B)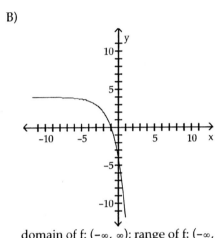

domain of f: $(-\infty, \infty)$; range of f: $(-\infty, 4)$;
horizontal asymptote: $y = 4$

C)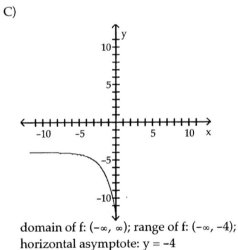

domain of f: $(-\infty, \infty)$; range of f: $(-\infty, -4)$;
horizontal asymptote: $y = -4$

D)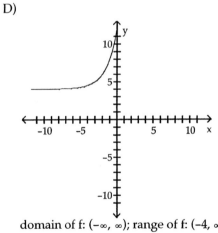

domain of f: $(-\infty, \infty)$; range of f: $(-4, \infty)$;
horizontal asymptote: $y = 4$

10) $f(x) = 3^{(x-2)}$

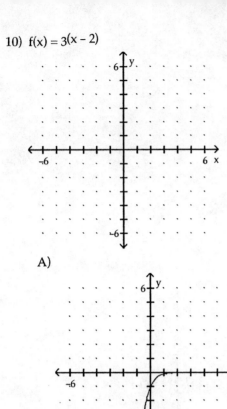

A)

domain of f: $(-\infty, \infty)$; range of f: $(-\infty, 0)$
horizontal asymptote: $y = 0$

B)

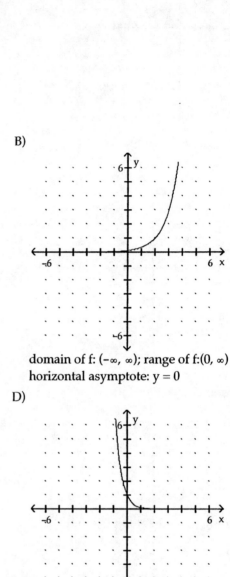

domain of f: $(-\infty, \infty)$; range of f: $(0, \infty)$
horizontal asymptote: $y = 0$

C)

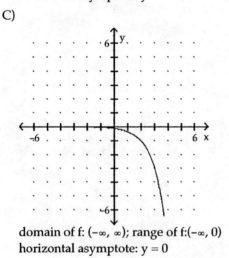

domain of f: $(-\infty, \infty)$; range of f: $(-\infty, 0)$
horizontal asymptote: $y = 0$

D)

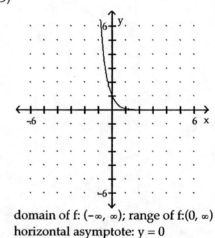

domain of f: $(-\infty, \infty)$; range of f: $(0, \infty)$
horizontal asymptote: $y = 0$

11) $f(x) = 4^{-x} + 3$

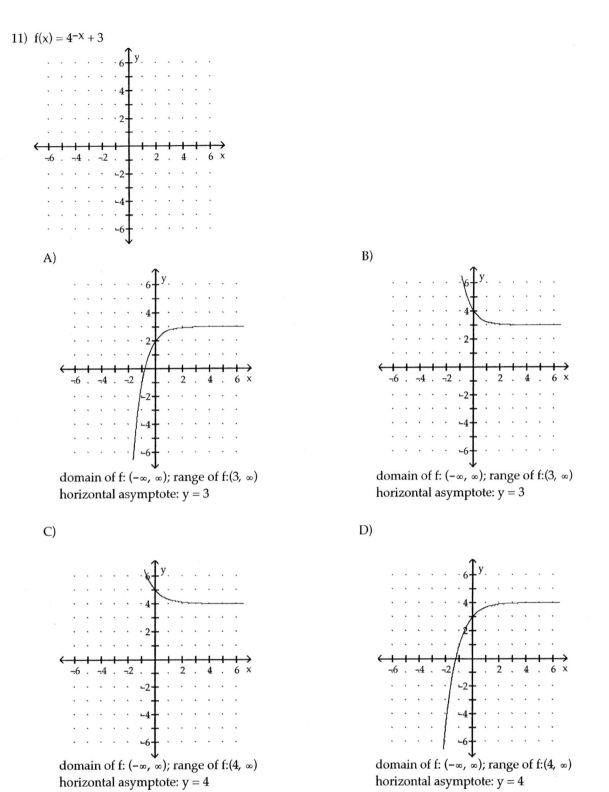

A) domain of f: $(-\infty, \infty)$; range of f: $(3, \infty)$
horizontal asymptote: $y = 3$

B) domain of f: $(-\infty, \infty)$; range of f: $(3, \infty)$
horizontal asymptote: $y = 3$

C) domain of f: $(-\infty, \infty)$; range of f: $(4, \infty)$
horizontal asymptote: $y = 4$

D) domain of f: $(-\infty, \infty)$; range of f: $(4, \infty)$
horizontal asymptote: $y = 4$

Graph the function.

12) $f(x) = 4^x$

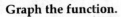

A)
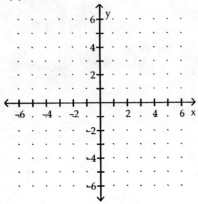

B)
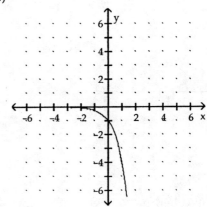

C)
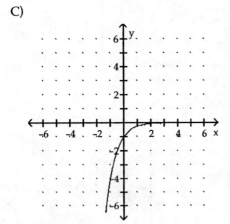

D)
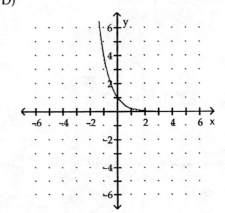

13) $f(x) = \left(\dfrac{1}{2}\right)^x$

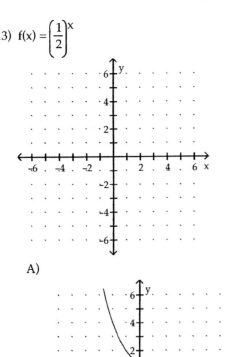

A)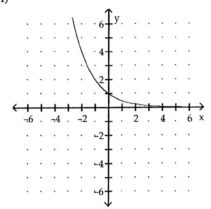

B)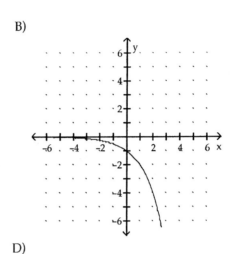

C)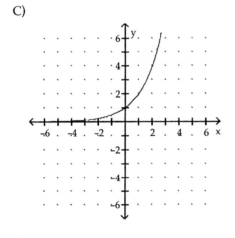

D)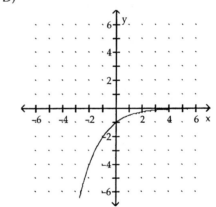

14) $f(x) = \left(\dfrac{2}{5}\right)^x$

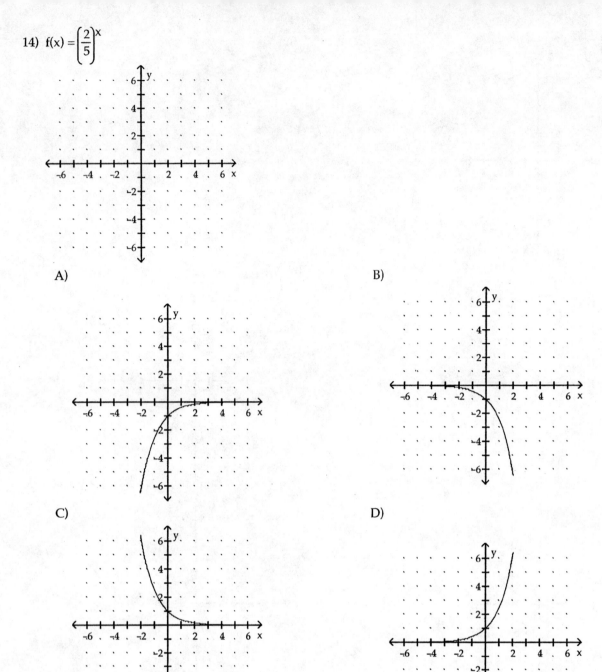

15) $f(x) = 0.7^x$

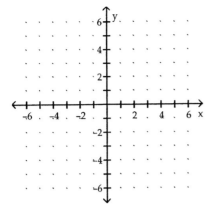

A)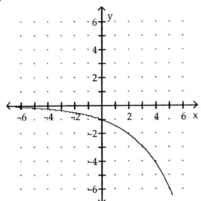

B)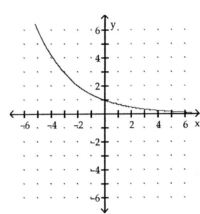

C)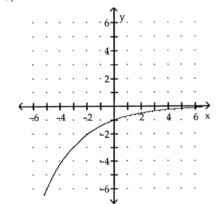

D)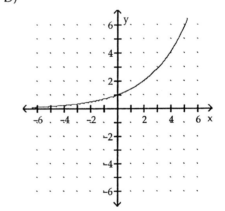

16) $f(x) = 3^x - 3$

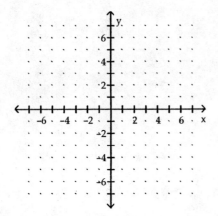

A)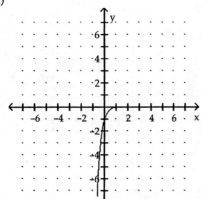

B)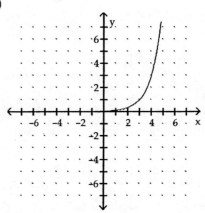

C)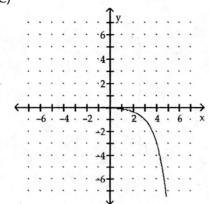

D)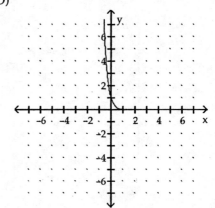

17) $f(x) = 5^x - 2$

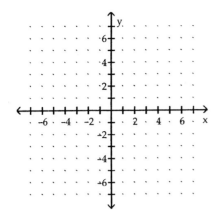

A)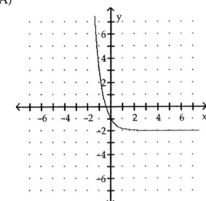

B)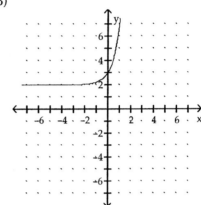

C)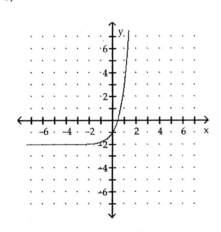

D)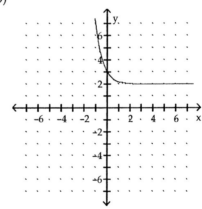

18) $f(x) = 5^{-x}$

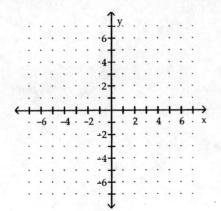

A)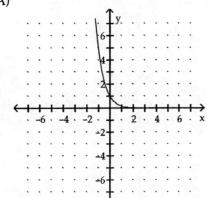

B)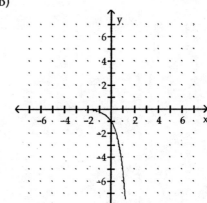

C)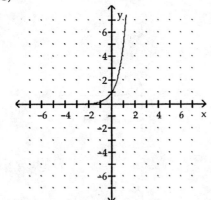

D)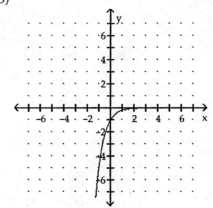

19) $f(x) = e^{5x}$

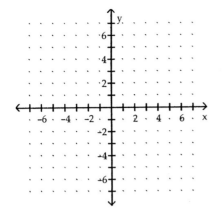

A)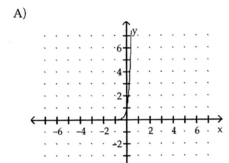

B)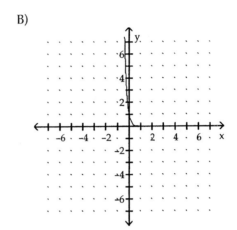

C)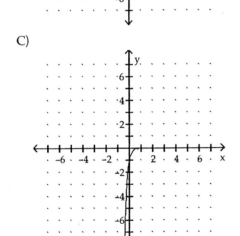

D)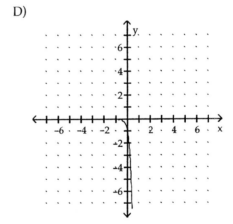

20) $f(x) = e^x + 2$

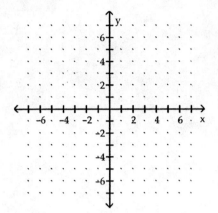

A)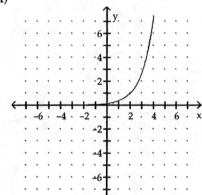

B)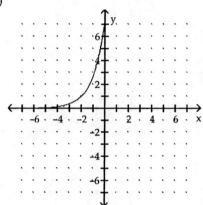

C)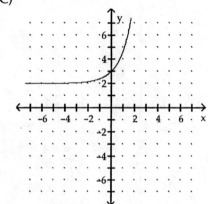

D)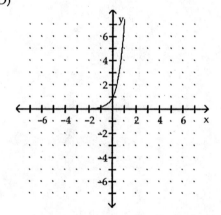

21) $f(x) = e^x$

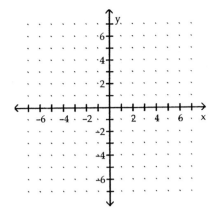

A)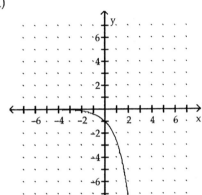

B)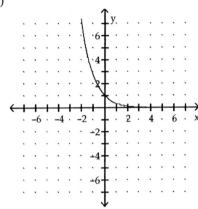

C)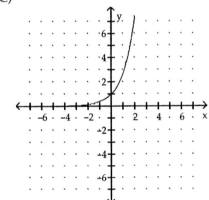

D)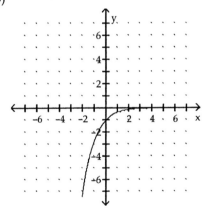

22) $f(x) = 5e^x$

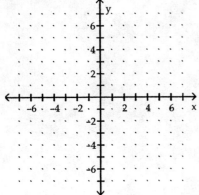

A)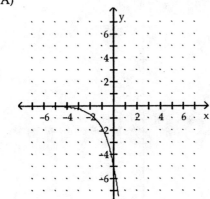

B)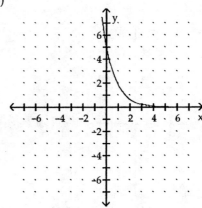

C)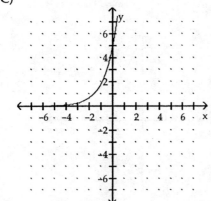

D)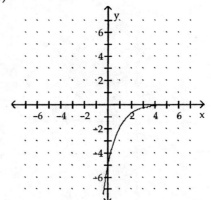

Use transformations to graph the function.

23) Use the graph of $f(x) = 4^x$ to obtain the graph of $g(x) = 4^x + 1$.

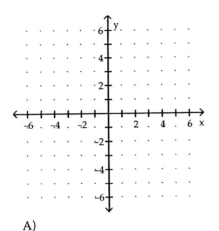

A)
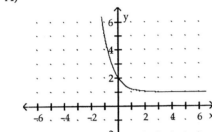

B)
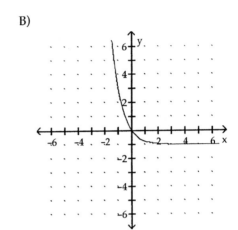

C)
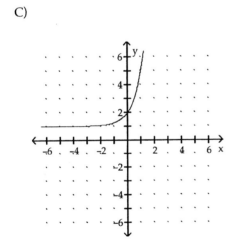

D)
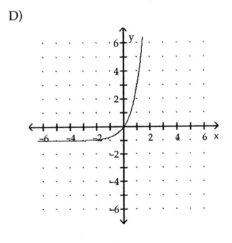

24) Use the graph of $f(x) = 3^x$ to obtain the graph of $g(x) = 3^{-x}$.

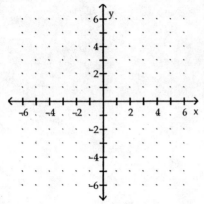

A)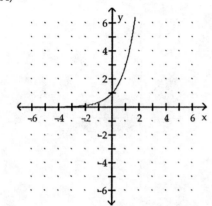

B)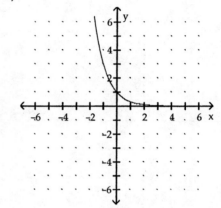

C)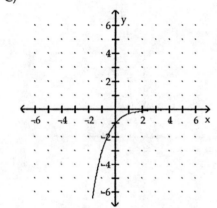

D)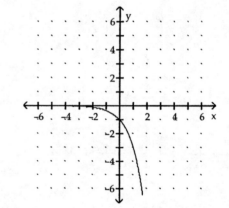

25) Use the graph of $f(x) = 3^x$ to obtain the graph of $g(x) = -3^x$.

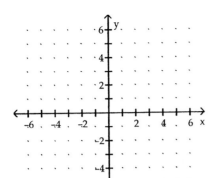

A)
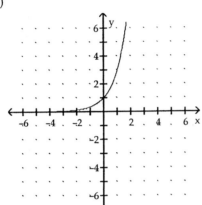

B)
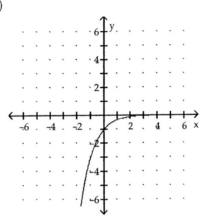

C)
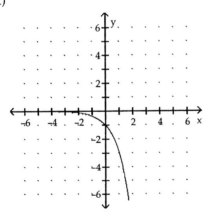

D)
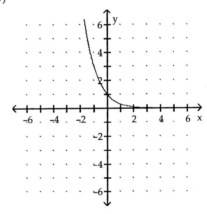

3 Define the Number e

Answer the question.

1) Define the number e.

A) The number that the expression, $\left(1 + \dfrac{1}{n}\right)^n$, approaches as $n \to \infty$.

B) The number approximately equal to 2.72.

C) The number defined by $e = \lim\limits_{n \to \infty} \left(1 + \dfrac{1}{n}\right)^n$ in Calculus.

D) All of the above.

Approximate the value using a calculator. Express answer rounded to three decimal places.

2) $e^{1.7}$

   A) 4.231   B) 5.474   C) 5.774   D) 4.621

3) $e^{4.11}$

   A) 60.947   B) 15.154   C) 11.172   D) 46.622

4) $e^{-1.2}$

   A) -0.301   B) -3.262   C) 0.601   D) 0.301

5) 2^e

   A) 4.718   B) 7.389   C) 5.437   D) 6.581

6) π^e

   A) 8.540   B) 22.459   C) 5.860   D) 23.141

7) e^π

   A) 5.860   B) 8.540   C) 22.459   D) 23.141

4 Solve Exponential Equations

Solve the equation.

1) $3^{1+2x} = 243$

   A) {2}   B) {6}   C) {-2}   D) {81}

2) $3^{-x} = \dfrac{1}{81}$

   A) $\left\{\dfrac{1}{27}\right\}$ B) {-4}   C) {4}   D) $\left\{\dfrac{1}{4}\right\}$

3) $4^{5-3x} = \dfrac{1}{256}$

   A) {3}   B) {128}   C) {-3}   D) $\left\{\dfrac{1}{64}\right\}$

4) $5^x = \dfrac{1}{25}$

   A) {2}   B) $\left\{\dfrac{1}{2}\right\}$ C) $\left\{\dfrac{1}{5}\right\}$ D) {-2}

5) $2^{x^2-3} = 64$

   A) {3, -3}   B) $\{\sqrt{35}, -\sqrt{35}\}$ C) {6}   D) {3}

6) $9^{2x} \cdot 27^{(3-x)} = \dfrac{1}{9}$

   A) {10} B) {-11}

   C) {-8} D) $\left\{\dfrac{9+\sqrt{87}}{6}, \dfrac{9-\sqrt{87}}{6}\right\}$

7) $5^x = 125$

   A) {2}   B) {25}   C) {4}   D) {3}

8) $125^x = 25$

   A) $\left\{\dfrac{3}{2}\right\}$ B) $\left\{\dfrac{1}{2}\right\}$ C) $\left\{\dfrac{2}{3}\right\}$ D) $\left\{\dfrac{1}{3}\right\}$

9) $3^{(3x-6)} = 27$

   A) {3}   B) {-3}   C) {9}   D) $\left\{\dfrac{1}{9}\right\}$

10) $4^{(11-3x)} = 16$

    A) {-3}   B) {2}   C) {3}   D) {4}

11) $243^{5x-1} = 27^{4x}$

    A) $\left\{-\dfrac{13}{5}\right\}$ B) $\left\{\dfrac{13}{5}\right\}$ C) $\left\{\dfrac{5}{13}\right\}$ D) $\left\{-\dfrac{5}{13}\right\}$

12) $(e^x)^x \cdot e^{18} = e^{9x}$

    A) {6}   B) {-3, -6}   C) {3, 6}   D) {3}

4.4 Logarithmic Functions

1 Change Exponential Expressions to Logarithmic Expressions and Logarithmic Expressions to Exponential Expressions

Change the exponential expression to an equivalent expression involving a logarithm.

1) $5^2 = 25$

   A) $\log_{25} 5 = 2$ B) $\log_5 2 = 25$ C) $\log_5 25 = 2$ D) $\log_2 25 = 5$

2) $5^{-2} = \dfrac{1}{25}$

   A) $\log_5 \dfrac{1}{25} = -2$ B) $\log_5 -2 = \dfrac{1}{25}$ C) $\log_{1/25} 5 = -2$ D) $\log_{-2} \dfrac{1}{25} = 5$

3) $6^3 = x$

   A) $\log_3 x = 6$ B) $\log_6 x = 3$ C) $\log_x 6 = 3$ D) $\log_6 3 = x$

4) $8^{1/3} = 2$

   A) $\log_2 8 = \dfrac{1}{3}$ B) $\dfrac{\log_3 2}{\log_1 8} = 8$ C) $\log_1 8 = \dfrac{1}{3}$ D) $\log_8 2 = \dfrac{1}{3}$

5) $10^x = 1000$
   A) $\log_{1000} x = 10$
   B) $\log_{1000} 10 = x$
   C) $\log_x 1000 = 10$
   D) $\log_{10} 1000 = x$

6) $2.3^{x+5} = 15$
   A) $\log_{(x+5)} 15 = 2.3$
   B) $\log_{2.3} x = 10$
   C) $\log_{15}(x+5) = 2.3$
   D) $\log_{2.3} 15 = x + 5$

7) $x^{\sqrt{5}} = \pi$
   A) $x = \log_{\sqrt{5}} \pi$
   B) $\sqrt{5} = \log_\pi x$
   C) $x = \log_\pi \sqrt{5}$
   D) $\sqrt{5} = \log_x \pi$

8) $6^5 = 7776$
   A) $\log_{7776} 6 = 5$
   B) $\log_5 7776 = 6$
   C) $\log_{1/5} 7776 = 6$
   D) $\log_6 7776 = 5$

9) $16^{3/2} = 64$

10) $e^x = 20$
    A) $\log_x e = 20$
    B) $\log_{20} x = e$
    C) $\ln x = 20$
    D) $\ln 20 = x$

Change the logarithmic expression to an equivalent expression involving an exponent.

11) $\log_{1/5} 125 = -3$
    A) $(-3)^{1/5} = 125$
    B) $\left(\frac{1}{5}\right)^3 = 125$
    C) $125^{1/5} = 3$
    D) $\left(\frac{1}{5}\right)^{-3} = 125$

12) $\log_5 \frac{1}{25} = -2$
    A) $5^{-2} = \frac{1}{25}$
    B) $2^5 = \frac{1}{25}$
    C) $\left(\frac{1}{25}\right)^2 = 5$
    D) $5^{25} = 2$

13) $\log_5 125 = 3$
    A) $125^3 = 5$
    B) $5^3 = 125$
    C) $3^5 = 125$
    D) $5^{125} = 3$

14) $\log_2 x = 3$
    A) $2^x = 3$
    B) $2^3 = x$
    C) $x^3 = 2$
    D) $3^2 = x$

15) $\log_b 36 = 2$
    A) $2^b = 36$
    B) $36^b = 2$
    C) $36^2 = b$
    D) $b^2 = 36$

16) $\log_5 125 = x$
    A) $125^5 = x$
    B) $125^x = 5$
    C) $5^x = 125$
    D) $x^5 = 125$

17) $\log_b 49 = \frac{2}{3}$
    A) $\left(\frac{2}{3}\right)^b = 49$
    B) $b^{2/3} = 49$
    C) $49^{2/3} = b$
    D) $b^{3/2} = 49$

18) $\log_{\sqrt{3}} 64 = \frac{7}{11}$

   A) $\sqrt{3}^{11/7} = 64$
   B) $64^{7/11} = \sqrt{3}$
   C) $\sqrt{3}^{7/11} = 64$
   D) $\left(\frac{7}{11}\right)^{\sqrt{3}} = 64$

19) $\log_\pi 37 = x$

   A) $\pi^x = \frac{1}{37}$
   B) $37^x = \pi$
   C) $\pi^x = 37$
   D) $x^\pi = 37$

20) $\log_{1024} 256 = \frac{4}{5}$

   A) $1024^{4/5} = 256$
   B) $256^{5/4} = 1024$
   C) $256^{4/5} = 1024$
   D) $\left(\frac{4}{5}\right)^{1024} = 256$

21) $y = \log_{61} x$

22) $\ln y = 8$

   A) $y^8 = e$
   B) $e^y = 8$
   C) $8^e = y$
   D) $e^8 = y$

23) $\ln \frac{1}{e^3} = -3$

   A) $\left(\frac{1}{e^3}\right)^e = -3$
   B) $\left(\frac{1}{e^3}\right)^{-3} = e$
   C) $e^{-3} = \frac{1}{e^3}$
   D) $-3^e = \frac{1}{e^3}$

2 Evaluate Logarithmic Expressions

Find the exact value of the logarithmic expression.

1) $\log_4 64$

   A) 12
   B) 64
   C) 3
   D) 4

2) $\log_6 \frac{1}{36}$

   A) 6
   B) 2
   C) -6
   D) -2

3) $\log_9 \frac{1}{729}$

   A) 3
   B) 81
   C) -3
   D) -81

4) $\log_4 \frac{1}{64}$

   A) 3
   B) $-\frac{1}{3}$
   C) $\frac{1}{3}$
   D) -3

5) $\log_6 1$

   A) 6
   B) 0
   C) $\frac{1}{6}$
   D) 1

6) $\log_3 \sqrt{3}$
   A) $\frac{1}{2}$ B) 1   C) $\frac{1}{3}$ D) 3

7) $\log_{10} 10{,}000$
   A) 40   B) -4   C) 4   D) $\frac{1}{10000}$

8) $\ln 1$
   A) 1   B) 0   C) e   D) -1

9) $\ln e$
   A) -1   B) e   C) 1   D) 0

10) $\ln e^8$
    A) $\frac{1}{8}$ B) 1   C) 8   D) e

Use a calculator to find the natural logarithm correct to four decimal places.

11) $\ln 138$
    A) 0.2023   B) 2.1399   C) 4.9273   D) 50.9225

12) $\ln 0.998$
    A) 0.0020   B) -0.0020   C) -0.0009   D) 0.0009

13) $\ln 0.00031$
    A) 8.0789   B) -8.0789   C) 3.5086   D) -3.5086

14) $\ln 12{,}200{,}000$
    A) 4.8040   B) 0.0611   C) 7.0864   D) 16.3169

15) $\ln (6.66 \cdot 10^{-17})$
    A) -37.2478   B) 4.7293   C) 1.8961   D) -32.2340

16) $\ln \sqrt{68}$
    A) 0.9163   B) 2.0541   C) 2.1098   D) 4.2195

17) $\ln(8.2 \cdot e^{-5})$
    A) 2.1050   B) 7.1041   C) 2.1041   D) -2.8959

Solve the problem.

18) The pH of a chemical solution is given by the formula
$$pH = -\log_{10}[H^+]$$
where $[H^+]$ is the concentration of hydrogen ions in moles per liter.
Find the pH if the $[H^+] = 4.7 \times 10^{-13}$.
    A) 12.67   B) 12.33   C) 13.67   D) 13.33

19) The long jump record, in feet, at a particular school can be modeled by f(x) = 19.6 + 2.2 ln (x + 1) where x is the number of years since records began to be kept at the school. What is the record for the long jump 7 years after record started being kept? Round your answer to the nearest tenth.

   A) 21.8 feet   B) 24.2 feet   C) 23.9 feet   D) 23.5 feet

20) The function f(x) = 1 + 1.4 ln (x+1) models the average number of free-throws a basketball player can make consecutively during practice as a function of time, where x is the number of consecutive days the basketball player has practiced for two hours. After 147 days of practice, what is the average number of consecutive free throws the basketball player makes?

   A) 11 consecutive free throws   B) 8 consecutive free throws
   C) 9 consecutive free throws   D) 12 consecutive free throws

21) Find a so that the graph of f(x) = log$_a$ x contains the point (14, 15).

   A) $\sqrt[14]{15}$ B) $\frac{15}{14}$ C) $\sqrt[15]{14}$ D) $\frac{14}{15}$

22) The number of men dying of AIDS (in thousands) since 1987 is modeled by y = 17.3 + 10.06(ln x), where x represents the number of years after 1987. Use this model to predict the number of AIDS deaths among men in 1994. Express answer rounded to the nearest hundred men.

   A) 36,900   B) 25,800   C) 37,000   D) 26,000

3 Determine the Domain of a Logarithmic Function

Find the domain of the function.

1) f(x) = log (x - 2)

   A) x > 2   B) x > -2   C) x > 1   D) x > 0

2) f(x) = ln (-1 - x)

   A) x > 1   B) x < -1   C) x > -1   D) x < 1

3) f(x) = log$_9$ (4 - x^2)

   A) x < -2 and x > 2   B) -4 < x < 4   C) -2 ≤ x ≤ 2   D) -2 < x < 2

4) f(x) = ln (10x - x^2)

   A) -10 < x < 10   B) 0 < x < 10   C) -10 ≤ x < 0   D) x ≤ 10

5) f(x) = log$_{10}$ (x^2 - 7x + 10)

   A) (5, ∞)   B) (-∞, -2)   C) (-∞, 2) ∪ (5, ∞)   D) (-2, 5)

6) f(x) = log$_{10}$ $\left(\frac{x+8}{x-6}\right)$

   A) (-∞, -8)   B) (-8, 6)   C) (6, ∞)   D) (-∞, -8) ∪ (6, ∞)

7) f(x) = ln $\left(\frac{1}{x-4}\right)$

   A) x > 0   B) x > 4   C) x > 1   D) x > -4

Precalculus Enhanced with Graphing Utilities 361

8) $f(x) = 6 - \ln(7x)$

   A) $(7, \infty)$ B) $(-\infty, 6) \cup (7, \infty)$ C) $(-6, 7)$ D) $(0, \infty)$

9) $f(x) = \log_2 (x-3)^2$

   A) $(3, \infty)$ B) $(-3, \infty)$ C) $(-\infty, 0)$ or $(0, \infty)$ D) $(-\infty, 3)$ or $(3, \infty)$

10) $f(x) = \log_7 (x+9)^2$

    A) $(-\infty, 0)$ or $(0, \infty)$ B) $(-9, \infty)$ C) $(-\infty, -9)$ or $(-9, \infty)$ D) $(9, \infty)$

11) $f(x) = \ln \sqrt{x}$

    A) $(0, \infty)$ B) $(-\infty, 1)$ C) $(1, \infty)$ D) $(-\infty, 0)$

12) $f(x) = \log_{1/2}(x+4)$

    A) $(4, \infty)$ B) $(-4, \infty)$ C) $(-\infty, 4)$ D) $(-\infty, -4)$

Solve the problem.

13) Determine the domain of the function $f(x) = \log_5(x+2)$.

4 Graph Logarithmic Functions

The graph of a logarithmic function is shown. Select the function which matches the graph.

1)

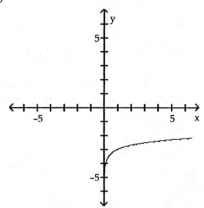

   A) $y = \log(x - 3)$ B) $y = \log(3 - x)$ C) $y = \log(x) - 3$ D) $y = 3 - \log(x)$

2)

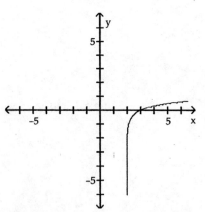

   A) $y = \log(x - 2)$ B) $y = \log(2 - x)$ C) $y = 2 - \log(x)$ D) $y = \log(x) - 2$

3)

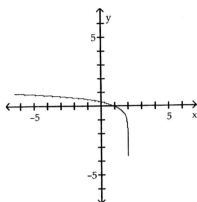

A) y = log (x) − 2   B) y = log (2 − x)   C) y = log (x − 2)   D) y = 2 − log (x)

4)

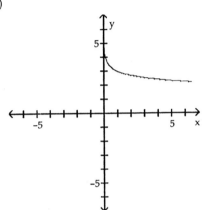

A) y = 3 − log (x)   B) y = log (3 − x)   C) y = log (x) − 3   D) y = log (x − 3)

5)

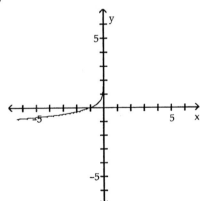

A) y = log (x)   B) y = − log (x)   C) y = log (−x)   D) y = − log (−x)

6)

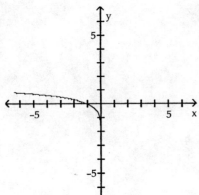

A) $y = -\log(-x)$ B) $y = \log(-x)$ C) $y = \log(x)$ D) $y = -\log(x)$

7)

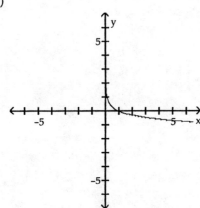

A) $y = -\log(x)$ B) $y = -\log(-x)$ C) $y = \log(x)$ D) $y = \log(-x)$

8)

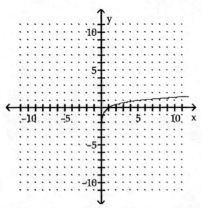

A) $f(x) = \log_5(x-2)$ B) $f(x) = \log_5(x+2)$ C) $f(x) = \log_5 x$ D) $f(x) = \log_5 x - 2$

9)

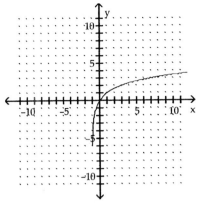

A) $f(x) = \log_2 (x - 1)$ B) $f(x) = \log_2 x + 1$ C) $f(x) = \log_2 (x + 1)$ D) $f(x) = \log_2 x$

10)

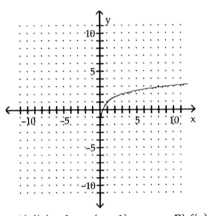

A) $f(x) = \log_3 (x + 1)$ B) $f(x) = \log_3 x + 1$ C) $f(x) = \log_3 x$ D) $f(x) = \log_3 (x - 1)$

11)

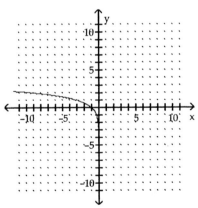

A) $f(x) = -\log_3 x$ B) $f(x) = \log_3 x$ C) $f(x) = 1 - \log_3 x$ D) $f(x) = \log_3 (-x)$

12)

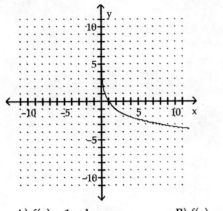

A) $f(x) = 1 - \log_2 x$ B) $f(x) = \log_2 x$ C) $f(x) = -\log_2 x$ D) $f(x) = \log_2(-x)$

13)

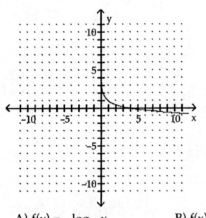

A) $f(x) = -\log_4 x$ B) $f(x) = \log_4(-x)$ C) $f(x) = \log_4 x$ D) $f(x) = 1 - \log_4 x$

Graph the function.

14) $y = \log_4 x$

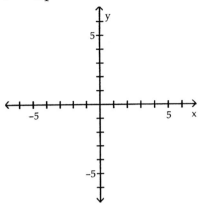

A)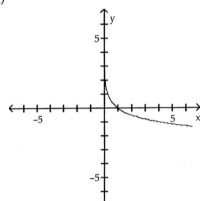

B)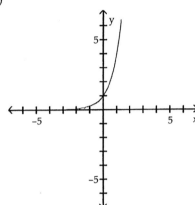

C)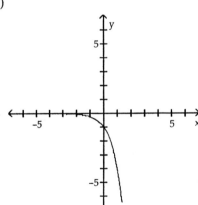

D)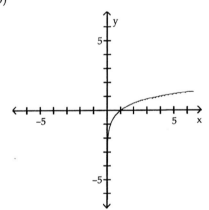

15) $y = \log_{1/3} x$

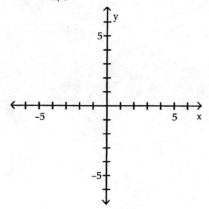

A)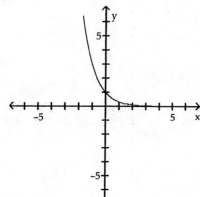

B)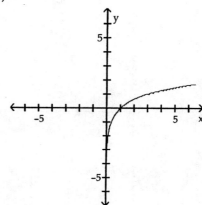

C)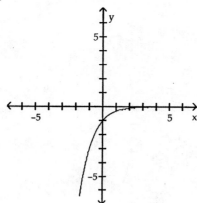

D)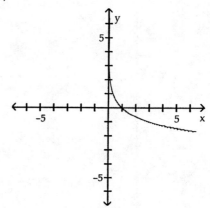

16) $f(x) = 3 \ln x$

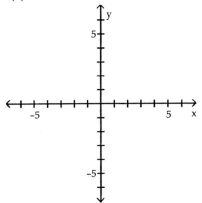

A)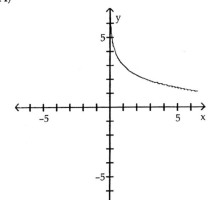

B)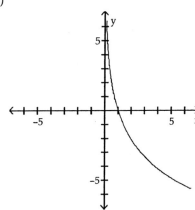

C)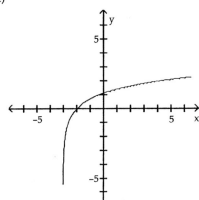

D)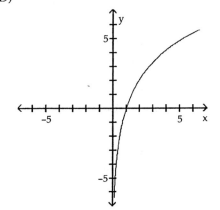

17) $f(x) = -1 - \ln x$

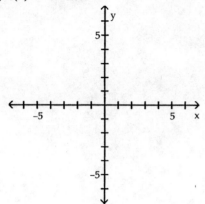

A)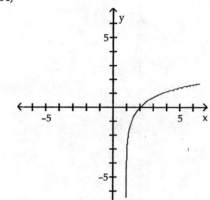

B)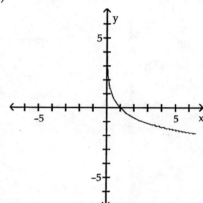

C)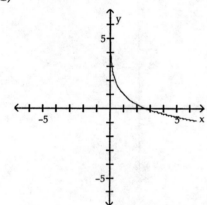

D)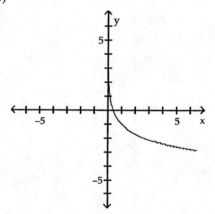

18) $f(x) = 2 - \ln(x + 4)$

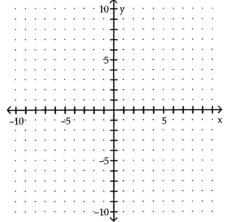

A)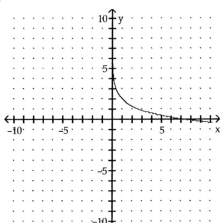

B)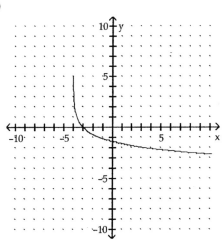

C)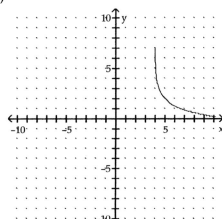

D)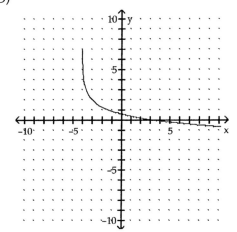

Use transformations to graph the function.

19) Use the graph of $\log_3 x$ to obtain the graph of $f(x) = \log_3(x-1)$.

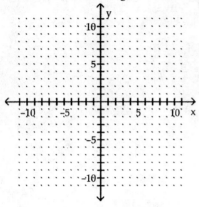

A)
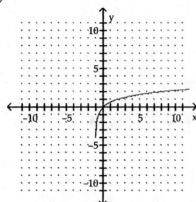

B)
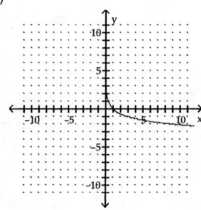

C)
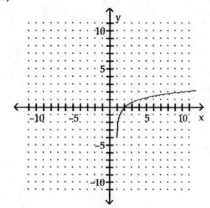

D)
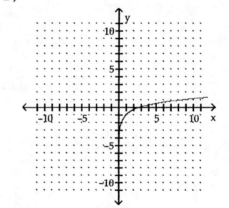

20) Use the graph of $\log_5 x$ to obtain the graph of $f(x) = 2 + \log_5 x$.

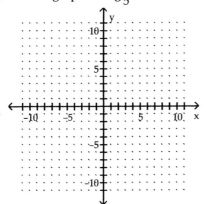

A)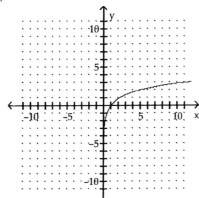

B)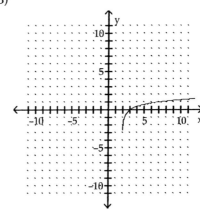

C)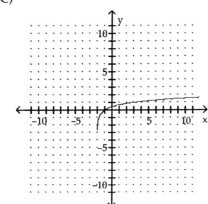

D)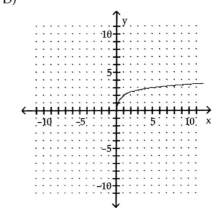

21) Use the graph of $\log_3 x$ to obtain the graph of $f(x) = 2\log_3 x$.

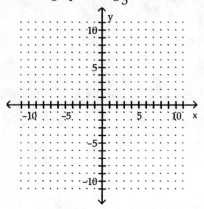

A)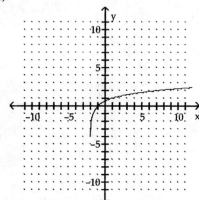

B)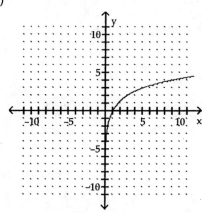

C)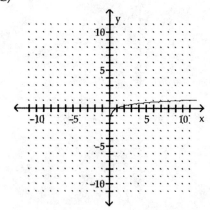

D)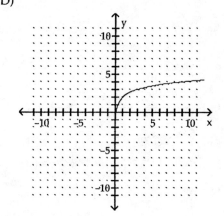

22) Use the graph of $\log_3 x$ to obtain the graph of $f(x) = \frac{1}{3}\log_3 x$.

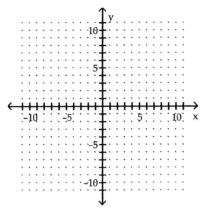

A)
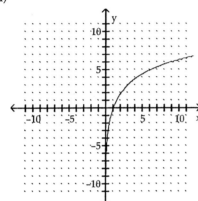

B)
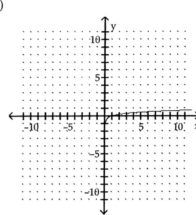

C)
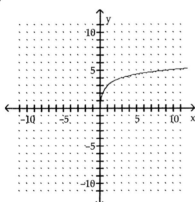

D)
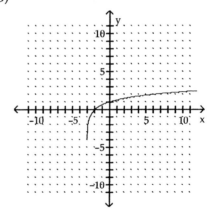

Use transformations to graph the function. Determine the domain, range, and vertical asymptote of the function.

23) $h(x) = 3 + \log(x + 2)$

5 Solve Logarithmic Equations

Solve the equation.

1) $\log_4 x = 3$

   A) {7}   B) {81}   C) {12}   D) {64}

2) $\log_2 8 = x$

   A) {10}   B) {3}   C) {16}   D) {4}

Precalculus Enhanced with Graphing Utilities 375

3) $\log_x \left(\frac{125}{8}\right) = 3$

A) 125 B) 3 C) $\left\{\frac{2}{5}\right\}$ D) $\left\{\frac{5}{2}\right\}$

4) $\log_6 x^2 = 4$

A) $\{2\sqrt{6}, -2\sqrt{6}\}$ B) $\{64\}$ C) $\{1296\}$ D) $\{36, -36\}$

5) $\log_6 (x - 4) = 3$

A) $\{220\}$ B) $\{733\}$ C) $\{725\}$ D) $\{212\}$

6) $\log_5 (x + 4) = -2$

A) $\left\{\frac{101}{32}\right\}$ B) $\left\{-\frac{99}{32}\right\}$ C) $\left\{-\frac{99}{25}\right\}$ D) $\left\{\frac{101}{25}\right\}$

7) $\log_4 (x^2 - 3x) = 1$

A) $\{-4, 1\}$ B) $\{1\}$ C) $\{4\}$ D) $\{4, -1\}$

8) $\log_{72} (x^2 - x) = 1$

A) $\{1, 72\}$ B) $\{8, 9\}$ C) $\{-8, 9\}$ D) $\{-8, -9\}$

9) $4 \ln 8x = 16$

A) $\left\{\frac{4}{\ln 8}\right\}$ B) $\{e^{1/2}\}$ C) $\left\{\frac{e^4}{8}\right\}$ D) $\{e^4\}$

10) $4 + 9 \ln x = 10$

A) $\left\{\ln\left(\frac{2}{3}\right)\right\}$ B) $\{e^{2/3}\}$ C) $\left\{\frac{6}{9 \ln 1}\right\}$ D) $\left\{\frac{e^6}{9}\right\}$

11) $\ln \sqrt{x + 8} = 9$

A) $\{e^{18} - 8\}$ B) $\{e^{18} + 8\}$ C) $\{e^9 - 8\}$ D) $\left\{\frac{e^9}{2} + 8\right\}$

12) $e^{5x} = 8$

A) $\left\{\frac{\ln 5}{8}\right\}$ B) $\left\{\frac{\ln 8}{5}\right\}$ C) $\{5 \ln 8\}$ D) $\left\{\frac{8}{5}e\right\}$

13) $e^{x+8} = 3$

A) $\{\ln 11\}$ B) $\{\ln 3 - 8\}$ C) $\{e^{24}\}$ D) $\{e^3 + 8\}$

The loudness of a sound of intensity x, measured in watts per square meter, is defined as $L(x) = \log\left(\frac{x}{x_0}\right)$, where $x_0 = 10^{-3}$.

14) A company with loud machinery needs to cut its sound intensity to 26% of its original level. By how many decibels should the loudness be reduced?

A) 5.760 decibels B) 6.237 decibels C) 6.880 decibels D) 5.850 decibels

Precalculus Enhanced with Graphing Utilities 376

15) A company with load machinery needs to cut its sound intensity to 72% of its original level. By how many decibels should the loudness be reduced?

16) A particular Boeing 747 jetliner produces noise at a loudness level of 113 decibels. Find the intensity level (round to the nearest hundredth) in watt per square meter for this noise.

17) At a recent Phish rock concert, sound intensity reached a level of 0.50 watt per square meter. To the nearest whole number, calculate the loudness of this sound in decibels.

18) At a rock concert by The Who, the music registered a loudness level of 120 decibels. The human threshold of pain due to sound averages 130 decibels. Compute the ratio of the intensities associated with these two loudness level to determine by how much the intensity of a sound that crosses the human threshold of pain exceeds that of this particular rock concert.

19) You have two friends, Jim and Amy. Jim always yells when he speaks, and Amy always whispers. The loudness of Jim's voice is 120 decibels, and the loudness of Amy's voice is 20 decibels. Determine how many times as intense Jim's voice is as compared at Amy's.

The Richter scale converts seismographic readings into numbers for measuring the magnitude of an earthquake according to this function

$$M(x) = \log\left(\frac{x}{x_0}\right), \text{ where } x_0 = 10^{-3}.$$

20) What is the magnitude of an earthquake whose seismographic reading is 7.9 millimeters at a distance of 100 kilometers from its epicenter? Round the answer to the nearest tenth.
   A) 2.7   B) 3.9   C) 897.6   D) 2.9

21) What is the magnitude of an earthquake whose seismographic reading is 7.6 millimeters at a distance of 100 kilometers from its epicenter?
   A) 0.38808   B) 0.20281   C) 2.0281   D) 3.8808

22) What is the magnitude of an earthquake whose seismographic reading is 0.94 millimeters at a distance of 100 kilometers from its epicenter?

23) Find the magnitude (to one decimal place) of an earthquake whose seismographic reading is 2000 millimeters at a distance of 100 kilometers from its epicenter.

24) Two earthquakes differ by 0.1 when measured on the Richter scale. How would the seismographic readings differ at a distance of 100 kilometers from the epicenter?

Solve the problem.

25) The Richter scale converts seismographic readings into numbers for measuring the magnitude of an earthquake according to this function

$$M(x) = \log\left(\frac{x}{x_0}\right), \text{ where } x_0 = 10^{-3}.$$

What would be the readings x (to the nearest tenth) for magnitudes of 4.9 and 6.9?
   A) 3.7 and 3.8   B) 690.2 and 838.8   C) 2.1 and 2.5   D) 2.7 and 2.8

26) The formula $D = 6e^{-0.04h}$ can be used to find the number of milligrams D of a certain drug in a patient's bloodstream h hours after the drug has been given. When the number of milligrams reaches 3, the drug is to be given again. What is the time between injections?

   A) 18.61 hrs   B) 44.79 hrs   C) 17.33 hrs   D) 27.47 hrs

27) Between 7:00 AM and 8:00 AM, trains arrive at a subway station at a rate of 12 trains per hour (0.2 trains per minute). The following formula from statistics can be used to determine the probability that a train will arrive within t minutes of 7:00 AM.

$$F(t) = 1 - e^{-0.2t}$$

   Determine how many minutes are needed for the probability to reach 50%.

   A) 4.58 min   B) 3.47 min   C) 6.26 min   D) 2.03 min

28) The concentration of alcohol in a person's blood is measurable. Suppose that the risk R (given as a percent) of having an accident while driving a car can be modeled by the equation

$$R = 6e^{kx}$$

   where x is the variable concentration of alcohol in the blood and k is a constant.

   (a) Suppose that a concentration of alcohol in the blood of 0.07 results in a 12% risk (R = 12) of an accident. Find the constant k in the equation.
   (b) Using this value of k, what is the risk if the concentration is 0.012?
   (c) Using the same value of k, what concentration of alcohol corresponds to a risk of 100%?
   (d) If the law asserts that anyone with a risk of having an accident of 12 or more should not have driving privileges, at what concentration of alcohol in the blood should a driver be arrested and charged with a DUI?

29) $pH = -\log_{10}[H^+]$ Find the $[H^+]$ if the pH = 5.4.

   A) 3.98×10^{-6} B) 2.51×10^{-6} C) 3.98×10^{-5} D) 2.51×10^{-5}

30) $pH = -\log_{10}[H^+]$ Find the pH if the $[H^+] = 6.4 \times 10^{-12}$.

   A) 12.81   B) 12.19   C) 11.19   D) 11.81

4.5 Properties of Logarithms

1 Work with the Properties of Logarithms

Use the properties of logarithms to find the exact value of the expression. Do not use a calculator.

1) $\log_4 4^{-16}$

   A) 1   B) 4   C) −16   D) −64

2) $\ln e\sqrt{6}$

   A) $\sqrt{6}$ B) 6   C) e   D) 36

3) $\log_{140} 10 + \log_{140} 14$

   A) 140   B) 14   C) 1   D) 10

4) $\log_6 30 - \log_6 5$

   A) 1   B) 30   C) 5   D) 6

5) $\log_4 24 - \log_4 6$

   A) 1   B) 6   C) 4   D) 24

6) $\log_4 11 \cdot \log_{11} 64$

   A) 3   B) 11   C) 4   D) 64

7) $2 \ln e^{4.2}$

   A) 2.1   B) 8.4   C) 4.2   D) $e^{8.4}$

8) $\log_5 5^{38.8}$

   A) 0.026   B) 393   C) 5   D) 38.8

9) $5^{\log_5 0.392}$

   A) 5   B) 0.392   C) 2.551   D) 397

Find the value of the expression.

10) Let $\log_b A = 5$ and $\log_b B = -4$. Find $\log_b AB$.

    A) 20   B) −20   C) 1   D) 9

11) Let $\log_b A = 3$ and $\log_b B = -15$. Find $\log_b \frac{A}{B}$.

    A) −12   B) $-\frac{1}{5}$ C) $\frac{1}{5}$ D) 18

12) Let $\log_b A = 2$ and $\log_b B = -5$. Find $\log_b B^2$.

    A) −10   B) 4   C) 25   D) −25

13) Let $\log_b A = 5$ and $\log_b B = -4$. Find $\log_b \sqrt[3]{AB}$.

    A) 0.333   B) $3\sqrt{-20}$ C) −2.714   D) 2.714

14) Let $\log_b A = 2.226$ and $\log_b B = 0.125$. Find $\log_b AB$.

    A) 0.279   B) 17.765   C) 2.351   D) 2.101

15) Let $\log_b A = 1.480$ and $\log_b B = 0.105$. Find $\log_b \frac{A}{B}$.

    A) 1.585   B) 1.375   C) 1.480   D) 0.156

Express y as a function of x. The constant C is a positive number.

16) $\ln y = \ln 4x + \ln C$

    A) $y = 4Cx$ B) $y = x + 4C$ C) $y = (4x)^C$ D) $y = 4x + C$

17) $3\ln y = \frac{1}{3}\ln x - \ln \frac{x^2-1}{x^{4/3}} + \ln C$

A) $y = \sqrt[3]{\dfrac{C}{x^{1/3}(x^2-1)}}$
B) $y = \sqrt[3]{\dfrac{Cx}{x^2-1}}$
C) $y = \sqrt[3]{\dfrac{Cx^{1/3}}{x^2-1}}$
D) $y = \sqrt[3]{\dfrac{Cx^{5/3}}{x^2-1}}$

2 Write a Logarithmic Expression as a Sum or Difference of Logarithms

Write as the sum and/or difference of logarithms. Express powers as factors.

1) $\log_{19} \dfrac{5}{2}$

A) $\log_{19} 2 - \log_{19} 5$
B) $\log_{19} 5 + \log_{19} 5$
C) $\log_{19} 5 - \log_{19} 2$
D) $\log_{19} 5 \div \log_{19} 2$

2) $\log_{11} \dfrac{5\sqrt{m}}{n}$

A) $\log_{11} n - \log_{11} 5 - \dfrac{1}{2}\log_{11} m$
B) $\log_{11} 5 + \dfrac{1}{2}\log_{11} m - \log_{11} n$

C) $\log_{11} 5 \cdot \dfrac{1}{2}\log_{11} m \div \log_{11} n$
D) $\log_{11}(5\sqrt{m}) - \log_{11} n$

3) $\log_4 \left(\dfrac{x^4}{y^6}\right)$

A) $4\log_4 x + 6\log_4 y$
B) $\dfrac{2}{3}\log_4\left(\dfrac{x}{y}\right)$
C) $4\log_4 x - 6\log_4 y$
D) $6\log_4 y - 4\log_4 x$

4) $\log_3 \left(\dfrac{x-5}{x^8}\right)$

A) $\log_3(x-5) - \log_3 x$
B) $8\log_3 x - \log_3(x-5)$

C) $\log_3(x-5) - 8\log_3 x$
D) $\log_3(x-5) + 8\log_3 x$

5) $\log_w \left(\dfrac{13x}{5}\right)$

A) $\log_w 8x$
B) $\log_w 13 + \log_w x + \log_w 5$

C) $\log_w 13x - \log_w 5$
D) $\log_w 13 + \log_w x - \log_w 5$

6) $\log_2 \sqrt{10x}$

A) $\dfrac{1}{2}\log_2 10 + \dfrac{1}{2}\log_2 x$
B) $\log_2 10 + \dfrac{1}{2}\log_2 x$

C) $\log_2 \sqrt{10} + \log_2 \sqrt{x}$
D) $\dfrac{1}{2}\log_2 10x$

7) $\log_2 \left(\dfrac{\sqrt{x}}{8}\right)$

A) $-3\log_2 x$
B) $6 - \dfrac{1}{2}\log_2 x$
C) $\dfrac{1}{2}\log_2 x - 3$
D) $\log_2 x - 3$

8) $\ln \sqrt[5]{ey}$

   A) $5 \ln y + 5$ B) $\frac{1}{5} \ln y + \frac{1}{5}$ C) $\frac{y}{5}$ D) $\frac{1}{5} \ln \sqrt[5]{ey} + \frac{1}{5}$

9) $\log_8 \frac{\sqrt[3]{7}}{s^2 r}$

   A) $\frac{1}{3} \log_8 7 - 2 \log_8 s - \log_8 r$ B) $\frac{1}{3} \log_8 7 - 2 \log_8 s - 2 \log_8 r$

   C) $3 \log_8 7 - 2 \log_8 s - \log_8 3$ D) $\log_8 7 - \log_8 s - \log_8 r$

10) $\log_7 \sqrt{\frac{mn}{6}}$

    A) $\frac{1}{2} \log_7 m + \frac{1}{2} \log_7 n - \log_7 6$ B) $\frac{1}{2} \log_7 m + \frac{1}{2} \log_7 n - \frac{1}{2} \log_7 6$

    C) $\frac{1}{2} \log_7 m \cdot \frac{1}{2} \log_7 n \div \frac{1}{2} \log_7 6$ D) $\frac{1}{2} \log_7 mn - \frac{1}{2} \log_7 6$

11) $\log_3 \frac{\sqrt[5]{r} \sqrt[7]{s}}{u^2}$

    A) $\frac{1}{5} \log_3 r \cdot \frac{1}{7} \log_3 s \div 2 \log_3 u$ B) $\frac{1}{5} \log_3 r + \frac{1}{7} \log_3 s - 2 \log_3 u$

    C) $5 \log_3 r + 7 \log_3 s - 2 \log_3 u$ D) $\frac{5}{3} \log_3 r + \frac{7}{3} \log_3 s - \frac{2}{3} \log_3 u$

12) $\log_b \sqrt[3]{\frac{x^5 y^8}{z^2}}$

13) $\log \left(1 - \frac{1}{x^3}\right)$

    A) $\log(x - 1) + \log(x^2 + x + 1) - 3 \log x$ B) $\log(x - 1) + \log(x^2 + 1) - 3 \log x$
    C) $\log 1 - 3 \log 1 - 3 \log x$ D) $\log x^3 - \log 1 - 3 \log x$

14) $\ln \left[\frac{(x+6)(x-7)}{(x-2)^3}\right]^{2/5}$, $x > 7$

    A) $\frac{2}{5}\ln(x + 6) + \frac{2}{5}\ln(x - 7) - \frac{6}{5}\ln(x - 2)$ B) $\frac{2}{5}\ln(x^2 + 13x - 42) - \frac{6}{5}\ln(x - 2)$

    C) $\ln(x + 6) + \ln(x - 7) + \ln 2 - 6\ln(x - 2) - \ln 5$ D) $2\ln(x + 6) - 5\ln(x - 7) - \frac{6}{5}\ln(x - 2)$

15) $\ln \dfrac{(7x)\sqrt[5]{1+4x}}{(x-8)^3}$, $\quad x > 8$

A) $7\ln x + \dfrac{4}{5}\ln(1+4x) - 3\ln(x-8)$

B) $\ln 7 + \ln x - 5\ln(1+4x) - 3\ln(x-8)$

C) $\ln 7 + \ln x + \dfrac{1}{5}\ln(1+4x) - \ln 3 - \ln(x-8)$

D) $\ln 7 + \ln x + \dfrac{1}{5}\ln(1+4x) - 3\ln(x-8)$

3 Write a Logarithmic Expression as a Single Logarithm

Express as a single logarithm.

1) $\log_c t + \log_c s$

A) $\log_c \dfrac{t}{s}$
B) $\log_c (ts)$
C) $\log_c ts$
D) $\log_c t \cdot \log_c s$

2) $3\log_b t - \log_b s$

A) $\log_b \dfrac{3t}{s}$
B) $\log_b \dfrac{t^3}{s}$
C) $\log_b (t^3 - s)$
D) $\log_b t^3 \div \log_b s$

3) $10\log_c 7 + 5\log_c 4$

A) $\log_c (70 + 20)$
B) $\log_c 7^{10} \cdot \log_c 4^5$
C) $\log_c 7^{10} 4^5$
D) $\log_c \dfrac{7^{10}}{4^5}$

4) $(\log_a q - \log_a r) + 6\log_a p$

A) $\log_a \dfrac{6qp}{r}$
B) $\log_a \dfrac{qp^6}{r}$
C) $\log_a \dfrac{q}{p^6 r}$
D) $\log_a qp^6 r$

5) $3\log_b m - \dfrac{4}{5}\log_b n + \dfrac{1}{2}\log_b j - 2\log_b k$

A) $\log_b \dfrac{m^3 k^2}{j^{1/2} n^{4/5}}$

B) $\log_b \dfrac{m^3 n^{4/5}}{j^{1/2} k^2}$

C) $\log_b \dfrac{m^3 j^{1/2}}{n^{4/5} k^2}$

D) $\log_b \left(3m - \dfrac{4}{5}n + \dfrac{1}{2}j - 2k\right)$

6) $3\log_6 x + 5\log_6 (x-6)$

A) $15 \log_6 x(x-6)$
B) $\log_6 x(x-6)^{15}$
C) $\log_6 x^3(x-6)^5$
D) $\log_6 x(x-6)$

7) $3 \log_a (2x+1) - 2\log_a (2x-1) + 2$

A) $\log_a (2x+3)$
B) $\log_a \dfrac{a^2(2x+1)^3}{(2x-1)^2}$
C) $\log_a (2x+1) + 2$
D) $\log_a 2(x+1)$

8) $\log_a 6x + 3(\log_a x - \log_a y)$

9) $\ln \dfrac{x^2+6x-27}{x-1} - \ln \dfrac{x^2+8x-9}{x+1} + \ln(x^2-6x+9)$, $x>0$

A) $\ln \dfrac{3(x-3)}{2(x-1)(x+1)}$ B) $\ln \dfrac{(x-3)^3(x+1)}{(x-1)^2}$ C) $\ln \dfrac{(x-3)^3}{(x-1)^2(x+1)}$ D) $\ln \dfrac{3(x-3)(x+1)}{2(x-1)}$

10) $30 \log_5 \sqrt[5]{x} + \log_5(30x^8) - \log_5 30$

A) $\log_5 x^{13/6}$ B) $\log_5 x^{11/8}$ C) $\log_5 x^{14/5}$ D) $\log_5 x^{14}$

4 Evaluate Logarithms Whose Base Is Neither 10 nor e

Use the Change-of-Base Formula and a calculator to evaluate the logarithm. Round your answer to three decimal places.

1) $\log_6 65.16$

A) 10.860 B) 0.429 C) 1.814 D) 2.331

2) $\log_5 0.745$

A) -5.467 B) -0.183 C) -0.128 D) 6.711

3) $\log_{7.4} 192$

A) 2.627 B) 0.381 C) 25.946 D) 2.283

4) $\log_{7.3} 3.4$

A) 1.624 B) 0.531 C) 0.616 D) 0.466

5) $\log_2 139.4$

A) 3.562 B) 0.140 C) 0.301 D) 7.123

Use the Change-of-Base Formula and a calculator to evaluate the logarithm. Round your answer to two decimal places.

6) $\log_{6.9} 198$

A) 28.70 B) 2.74 C) 2.30 D) 0.37

7) $\log_{5.4} 2.3$

A) 0.43 B) 0.36 C) 0.49 D) 2.02

8) $\log_{\sqrt{2}} 239.6$

A) 15.81 B) 0.15 C) 0.06 D) 7.90

9) $\log_3 25$

A) 2.93 B) 0.34 C) 1.10 D) 3.22

10) $\log_{(2/3)} 19$

Solve the problem.

11) Find the value of $\log_3 4 \cdot \log_4 5 \cdot \log_5 6 \cdot \log_6 7 \cdot \log_7 8 \cdot \log_8 9$

5 Graph Logarithmic Functions Whose Base Is Neither 10 nor e

Graph the function using a graphing utility and the Change-of-Base Formula.

1) $\log_{x-7}(x+7)$

A)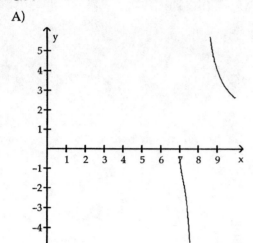

B)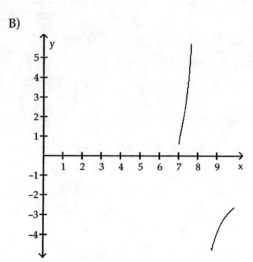

C)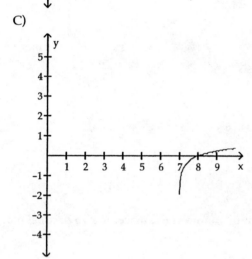

D)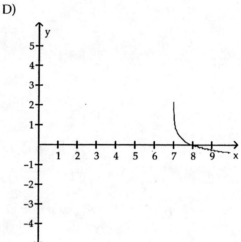

2) $y = \log_2 x$

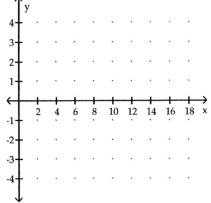

A)
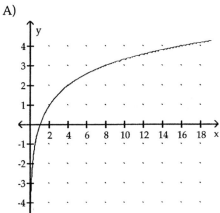

B)
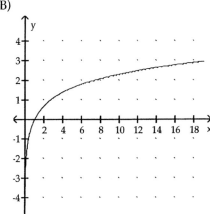

C)
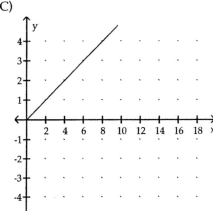

D)
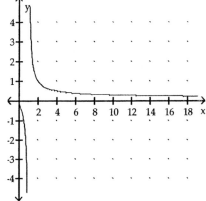

3) $y = \log_5(x - 3)$

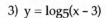

A)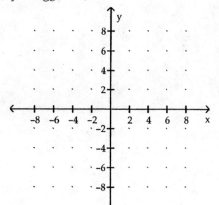

B)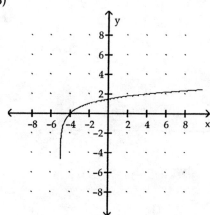

C)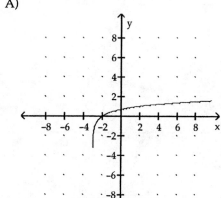

D)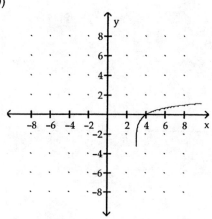

4.6 Logarithmic and Exponential Equations

1 Solve Logarithmic Equations Using the Properties of Logarithms

Solve the equation.

1) $\log_3 x = 4$

   A) $\{64\}$ B) $\{1.26\}$ C) $\{81\}$ D) $\{12\}$

2) $\log_y 10 = 3$

   A) $\{10^{1/3}\}$ B) $\left\{\dfrac{10}{3}\right\}$ C) $\{3^{1/10}\}$ D) $\{10^3\}$

3) $\log_5 (x - 2) = 2$

   A) {27}   B) {34}   C) {23}   D) {30}

4) $\log (x + 4) = \log (4x + 1)$

   A) {1}   B) {-1}   C) $\left\{\dfrac{3}{5}\right\}$ D) $\left\{\dfrac{1}{3}\right\}$

5) $\log (2 + x) - \log (x - 5) = \log 2$

   A) {12}   B) ∅   C) $\left\{\dfrac{5}{2}\right\}$ D) {-12}

6) $\log (4x) = \log 5 + \log (x - 1)$

   A) $\left\{-\dfrac{5}{9}\right\}$ B) {-5}   C) {5}   D) $\left\{\dfrac{4}{3}\right\}$

7) $\log_6 (2x + 5) = \log_6 (2x + 2)$

   A) {0}   B) $\left\{\dfrac{7}{3}\right\}$ C) {3}   D) ∅

8) $\log_3 x + \log_3(x - 24) = 4$

   A) {-3, 27}   B) {27}   C) {53}   D) No real solutions

9) $\dfrac{1}{3} \log_2 (x + 6) = \log_8 (3x)$

   A) ∅   B) {9}   C) {3, 0}   D) {3}

10) $\log_2(3x - 2) - \log_2(x - 5) = 4$

    A) $\left\{\dfrac{38}{5}\right\}$ B) {6}   C) {18}   D) $\left\{\dfrac{3}{13}\right\}$

11) $\log_2 (x - 4) + \log_2 (x - 10) = 4$

    A) {2}   B) {12, 2}   C) {12}   D) {13}

12) $\log_2 (x + 2) = 3 + \log_2 (x - 5)$

    A) {-6}   B) {-1}   C) {6}   D) {1}

13) $\log_{10} (x - 30) = 3 - \log_{10} x$

    A) {-20}   B) {20}   C) {-50}   D) {50}

14) $2 + \log_3(2x + 5) - \log_3 x = 4$

    A) $\left\{\dfrac{5}{4}\right\}$ B) $\left\{\dfrac{5}{7}\right\}$ C) $\left\{\dfrac{1 \pm \sqrt{46}}{9}\right\}$ D) $\left\{\dfrac{1 + \sqrt{46}}{9}\right\}$

15) $\log_a(x - 7) - \log_a(x - 2) = \log_a(x - 4) - \log_a(x + 12)$

Solve the equation. Express irrational answers in exact form and as a decimal rounded to 3 decimal places.

16) $\ln x + \ln(x+7) = 1$

A) $\dfrac{-7 - \sqrt{49 + 4e^1}}{2} \approx -7.369$

B) $-7 + \sqrt{49 + 4e^1} \approx 0.738$

C) $\dfrac{-7 + 2\sqrt{49 + e^1}}{2} \approx 3.692$

D) $\dfrac{-7 + \sqrt{49 + 4e^1}}{2} \approx 0.369$

Solve the problem.

17) The function $f(x) = 1 + 1.4 \ln(x+1)$ models the average number of free-throws a basketball player can make consecutively during practice as a function of time, where x is the number of consecutive days the basketball player has practiced for two hours. After how many days of practice can the basketball player make an average of 10 consecutive free throws?

A) 618 days B) 2585 days C) 2583 days D) 620 days

2 Solve Exponential Equations

Solve the equation.

1) $5^x = 25$

A) {5} B) {3} C) {2} D) {1}

2) $2^{(1 + 2x)} = 32$

A) {16} B) {4} C) {2} D) {-2}

3) $4^{(5 - 3x)} = \dfrac{1}{256}$

A) {-3} B) {3} C) {128} D) $\left\{\dfrac{1}{64}\right\}$

4) $4^{(5 + 3x)} = \dfrac{1}{256}$

A) {3} B) $\left\{\dfrac{1}{64}\right\}$ C) {128} D) {-3}

5) $3 \cdot 5^{2t - 1} = 75$

A) {3} B) $\left\{\dfrac{1}{2}\right\}$ C) $\left\{\dfrac{13}{10}\right\}$ D) $\left\{\dfrac{3}{2}\right\}$

6) $3^{2x} + 3^x - 6 = 0$

7) $9^{7x+3} = 27$

Solve the equation. Express irrational answers in exact form and as a decimal rounded to 3 decimal places.

8) $\left(\dfrac{8}{9}\right)^x = 5^{1-x}$

A) $\dfrac{\ln 40}{\ln 45} \approx 0.969$

B) $\dfrac{\ln 5}{\ln\left(\dfrac{8}{9}\right) + \ln 5} \approx 1.079$

C) $\ln\left(\dfrac{8}{9}\right) - \ln 5 \approx -1.727$

D) $\dfrac{\ln\left(\dfrac{8}{9}\right) + \ln 5}{\ln 5} \approx 0.927$

Solve the exponential equation. Express the solution set in terms of natural logarithms.

9) $e^{4x} = 2$

A) $\left\{\dfrac{\ln 4}{2}\right\}$
B) $\{4 \ln 2\}$
C) $\left\{\dfrac{\ln 2}{4}\right\}$
D) $\left\{\dfrac{1}{2} e\right\}$

10) $e^{x+8} = 6$

A) $\{\ln 14\}$
B) $\{e^6 + 8\}$
C) $\{e^{48}\}$
D) $\{\ln 6 - 8\}$

11) $2^{9x} = 4.1$

A) $\left\{\dfrac{9 \ln 4.1}{\ln 2}\right\}$
B) $\left\{\dfrac{\ln 4.1}{9 \ln 2}\right\}$
C) $\left\{\dfrac{4.1 \ln 9}{\ln 2}\right\}$
D) $\left\{\dfrac{\ln 4.1}{2 \ln 9}\right\}$

12) $3^{x+7} = 4$

A) $\left\{\dfrac{\ln 3}{\ln 4} + \ln 7\right\}$
B) $\left\{\dfrac{\ln 4}{\ln 3} - 7\right\}$
C) $\left\{\dfrac{\ln 3}{\ln 4} + 7\right\}$
D) $\{\ln 4 - \ln 3 - \ln 7\}$

13) $4^{x+4} = 5^{2x+5}$

A) $\{7 \ln 5 - 5 \ln 4\}$
B) $\left\{\ln\left[\dfrac{5^5}{4^4} - \dfrac{4}{5^2}\right]\right\}$
C) $\left\{\dfrac{5 \ln 5 - 4 \ln 4}{\ln 4 - 2 \ln 5}\right\}$
D) $\{\ln 5 - \ln 4\}$

14) $\pi^{x+1} = e^{2x}$

15) $\dfrac{e^x - e^{-x}}{2} = -1$

Solve the exponential equation. Use a calculator to obtain a decimal approximation, correct to two decimal places, for the solution.

16) $6^{3x} = 3.2$

A) $\{0.18\}$
B) $\{1.95\}$
C) $\{0.22\}$
D) $\{1.96\}$

17) $4^{x+7} = 3$

A) $\{-0.97\}$
B) $\{8.26\}$
C) $\{2.11\}$
D) $\{-6.21\}$

18) $\left(\dfrac{1}{2}\right)^x = 17$

A) {4.09} B) {-0.24} C) {-4.09} D) {-8.50}

19) $4^{(x-1)} = 15$

A) {2.95} B) {4.75} C) {0.95} D) {2.12}

20) $3^{(4x-1)} = 16$

A) {0.27} B) {0.38} C) {0.88} D) {1.58}

21) $e^{4x} = 8$

A) {0.17} B) {5.44} C) {8.32} D) {0.52}

22) $e^{x+3} = 7$

A) {-0.99} B) {2.30} C) {-0.70} D) {-1.05}

Solve the problem.

23) Find out how long it takes a $3300 investment to double if it is invested at 7% compounded monthly. Round to the nearest tenth of a year. Use the formula $A = P\left(1 + \dfrac{r}{n}\right)^{nt}$.

A) 9.7 years B) 9.9 years C) 10.1 years D) 10.3 years

24) The formula $A = 236e^{0.028t}$ models the population of a particular city, in thousands, t years after 1998. When will the population of the city reach 271 thousand?

A) 2004 B) 2003 C) 2005 D) 2006

3 Solve Logarithmic and Exponential Equations Using a Graphing Utility

Use a graphing calculator to solve the equation. Round your answer to two decimal places.

1) $\log_3 x + \log_5 x = 2$

A) {1.92} B) {3.69} C) {5.48} D) {0.57}

2) $\log_4(x + 2) - \log_5(x - 1) = 1$

A) {2.05} B) {1.75} C) {1.98} D) {-0.69}

3) $e^x = -x$

A) {0.57} B) {-0.57} C) {1.05} D) {-1.05}

4) $e^x - \ln x = 4$

A) {1.27} B) {2.17} C) {1.48} D) {0.57}

5) $e^x = x^3$

A) {-0.71} B) {2.54} C) {1.86} D) {-0.79}

6) $e^x = x^2 - 1$

A) {-1.15} B) {2.54} C) {-0.71} D) {0}

Solve the given exponential equation. Round answer to three decimal places.

7) $2^x - 4x + 1 = 0$

   A) {14.387}   B) {0.639, 3.847}   C) {3.847}   D) {0.639}

8) $2^x - 3x + 2 = 0$

   A) {4.674}   B) {2.225}   C) {2.000, 2.225}   D) {2.000}

9) $5^{2x} + 5^{(x+1)} - 24 = 0$

   A) {1.292}   B) {1.147}   C) {0.683}   D) {1.099}

4.7 Compound Interest

1 Determine the Future Value of a Lump Sum of Money

Find the amount that results from the investment.

1) $1,000 invested at 8% compounded annually after a period of 8 years

   A) $1999.00   B) $1850.93   C) $1713.82   D) $850.93

2) $1,000 invested at 12% compounded semiannually after a period of 11 years

   A) $3478.55   B) $2603.54   C) $3399.56   D) $3603.54

3) $14,000 invested at 14% compounded semiannually after a period of 13 years

   A) $67,302.94   B) $75,984.06   C) $81,302.94   D) $76,893.76

4) $480 invested at 9% compounded quarterly after a period of 7 years

   A) $875.29   B) $894.98   C) $414.98   D) $877.46

5) $12,000 invested at 4% compounded quarterly after a period of 4 years

   A) $13,931.63   B) $14,038.30   C) $2070.94   D) $14,070.94

Solve the problem.

6) Find the amount owed at the end of 8 years if $5000 is loaned at a rate of 5% compounded monthly.

   A) $8060.16   B) $12,911.25   C) $9093.60   D) $7452.93

7) If $5,000 is invested for 6 years at 5%, compounded continuously, find the future value.

8) John Forgetsalot deposited $100 at a 3% annual interest rate in a savings account fifty years ago, and then he promptly forgot he had done it. Recently, he was cleaning out his home office and discovered the forgotten bank book. How much money is in the account?

9) Meike earned $1565 in tips while working a summer job at a coffee shop. She wants to use this money to take a trip to Europe next summer. If she places the money in an account which pays 6.5% compounded continuously, how much money will she have in nine months?

10) Carla has just inherited a building that is worth $250,000. The building is in a high demand area, and the value of the building is projected to increase at a rate of 25% per year for the next 4 years. How much more money will she make if she waits four years to sell the building instead of selling now?

2 Calculate Effective Rates of Return

Find the effective rate of interest.

1) 10% compounded continuously
   - A) 10.451%
   - B) 10.374%
   - C) 10.089%
   - D) 10.517%

2) 15.3% compounded continuously
   - A) 15.751%
   - B) 16.532%
   - C) 15.674%
   - D) 15.389%

3) 25.09% compounded daily
   - A) 28.507%
   - B) 25.995%
   - C) 25.295%
   - D) 25.213%

4) 8.25% compounded monthly
   - A) 8.455%
   - B) 8.569%
   - C) 8.373%
   - D) 9.155%

5) $6\frac{1}{4}\%$ compounded monthly
   - A) 6.25%
   - B) 6.39%
   - C) 6.43%
   - D) 6.29%

6) 6.5% compounded quarterly
   - A) 7.405%
   - B) 6.623%
   - C) 6.705%
   - D) 6.66%

7) $4\frac{3}{4}\%$ compounded quarterly

Solve the problem.

8) A local bank advertises that it pays interest on savings accounts at the rate of 3% compounded monthly. Find the effective rate. Round answer to two decimal places.
   - A) 36%
   - B) 3.04%
   - C) 3.44%
   - D) 3.40%

9) Which of the two rates would yield the larger amount in 1 year: 4.9% compounded semiannually or 4.8% compounded quarterly?
   - A) 4.8% compounded quarterly
   - B) 4.9% compounded semiannually
   - C) They will yield the same amount.

10) Which of the two rates would yield the larger amount in 1 year: 9% compounded monthly or $9\frac{1}{4}\%$ compounded annually?
    - A) 9% compounded monthly
    - B) $9\frac{1}{4}\%$ compounded annually
    - C) They will yield the same amount.

3 Determine the Present Value of a Lump Sum of Money

Find the present value. Round to the nearest cent.

1) To get $5,600 after 4 years at 10% compounded annually
   - A) $3824.88
   - B) $4207.36
   - C) $6830.13
   - D) $1775.12

2) To get $10,500 after 12 years at 7% compounded annually

   A) $5837.87   B) $4440.12   C) $4662.13   D) $4988.47

3) To get $2,000 after 12 years at 8% compounded semiannually

   A) $780.24   B) $811.45   C) $1219.76   D) $794.23

4) To get $25,000 after 2 years at 13% compounded semiannually

   A) $19,433.08   B) $20,696.23   C) $5566.92   D) $19,578.67

5) To get $6,500 after 11 years at 8% compounded quarterly

   A) $3780.4   B) $2719.60   C) $2774.00   D) $2787.74

6) To get $10,000 after 2 years at 6% compounded monthly

   A) $10,616.78   B) $5000.00   C) $9419.05   D) $8871.86

Solve the problem.

7) What principal invested at 8% compounded continuously for 4 years will yield $1190? Round the answer to two decimal places.

   A) $627.48   B) $864.12   C) $1638.78   D) $1188.62

8) What principal invested at 6%, compounded continuously for 3 years, will yield $1500? Round the answer to two decimal places.

9) How much money needs to be invested now to get $2000 after 4 years at 8% compounded quarterly? Express your answer to the nearest dollar.

   A) $1848   B) $2746   C) $584   D) $1457

4 Determine the Time Required to Double or Triple a Lump Sum of Money

Solve the problem. Round your answer to three decimals.

1) What annual rate of interest is required to double an investment in 6 years?

   A) 6.123%   B) 11.552%   C) 20.094%   D) 12.246%

2) What annual rate of interest is required to triple an investment in 5 years?

   A) 21.972%   B) 14.87%   C) 12.287%   D) 24.573%

3) How long will it take for an investment to double in value if it earns 5.5% compounded continuously?

   A) 19.975 years B) 13.431 years C) 6.301 years   D) 12.603 years

4) How long will it take for an investment to triple in value if it earns 4.5% compounded continuously?

   A) 24.414 years B) 12.207 years C) 27.652 years D) 15.403 years

Solve the problem.

5) How long does it take $1125 to triple if it is invested at 7% interest, compounded quarterly? Round your answer to the nearest tenth.

   A) 15.8 years   B) 18.1 years   C) 15.8 months   D) 18.1 months

6) How long does it take $1700 to double if it is invested at 5% interest, compounded monthly? Round your answer to the nearest tenth.

7) Gillian has $10,000 to invest in a mutual fund. The average annual rate of return for the past five years was 12.25%. Assuming this rate, determine how long it will take for her investment to double.

5 Demonstrate Additional Understanding and Skills

Solve the problem.

1) If Emery has $1800 to invest at 8% per year compounded monthly, how long will it be before he has $3500? If the compounding is continuous, how long will it be? (Round your answers to three decimal places.)

   A) 80.535 yrs, 8.562 yrs
   B) 0.72 yrs, 0.693 yrs
   C) 8.34 yrs, 8.312 yrs
   D) 2.662 yrs, 0.831 yrs

2) Cindy will require $14,000 in 2 years to return to college to get an MBA degree. How much money should she ask her parents for now so that, if she invests it at 9% compounded continuously, she will have enough for school? (Round your answer to the nearest dollar.)

   A) $11,784
   B) $9767
   C) $11,694
   D) $16,761

3) Tracey bought a diamond ring appraised at $1400 at an antique store. If diamonds have appreciated in value at an annual rate of 12%, what was the value of the ring 6 years ago? (Round your answer to the nearest dollar.)

   A) $2763
   B) $332
   C) $709
   D) $681

4) The Feldmans bought their first house for $17,000. Over the years they moved three times into bigger and bigger houses. Now, 42 years later, they are ready to retire and want a smaller house like the first one they bought. If inflation in property values has averaged 3.2% per year during that time, how much will such a house cost them now? (Round your answer to the nearest dollar.)

   A) $63,826
   B) $65,184
   C) $4434
   D) $4528

5) Larry has $1500 to invest and needs $1900 in 11 years. What annual rate of return will he need to get in order to accomplish his goal? (Round your answer to two decimals.)

   A) 1.55%
   B) 3.55%
   C) 2.15%
   D) 2.55%

6) A venture capital firm invested $2,000,000 in a new company in 1995. In 1999, they sold their stake in the company for $10,500,000. What was the average annual rate of return on their investment?

7) Julio figures that he can save $5000 per year. If, at the end of each year, he invests the money in a certificate of deposit (CD) which pays 7% interest annually, how much money will he have saved in six years?

4.8 Exponential Growth and Decay; Newton's Law; Logistic Growth and Decay

1 Find Equations of Populations That Obey the Law of Uninhibited Growth

Solve the problem.

1) The size P of a small herbivore population at time t (in years) obeys the function $P(t) = 600e^{0.24t}$ if they have enough food and the predator population stays constant. After how many years will the population reach 1800?

   A) 8.74 yrs
   B) 4.58 yrs
   C) 29.54 yrs
   D) 10.35 yrs

2) Conservationists tagged 140 black-nosed rabbits in a national forest in 1990. In 1993, they tagged 280 black-nosed rabbits in the same range. If the rabbit population follows the exponential law, how many rabbits will be in the range 10 years from 1990?

   A) 172
   B) 2822
   C) 345
   D) 1411

3) The revenue for a dot.com company is projected to double each year for the first 5 years. If the revenue for the first year is $2 million, write a function showing the revenue R after x years. What is the revenue for the fourth year?

4) In a networking marketing plan for a company, each distributor is expected to recruit 3 new distributors. Jack was the first distributor hired by the company, so he is considered a level 1 distributor. The 3 people he recruits are considered level 2 distributors. The people recruited by the level 2 distributors are considered level 3 distributors, and so on. Write a function that models the number of distributors D at each level L. How many distributors would there be at the fifth level?

5) The bacteria in a container quadruples every day. If there are initially 100 bacteria, write an equation that models the number of bacteria A after d days. How many bacteria will there be after 1 week?

6) The concentration of alcohol in a person's blood is measurable. Suppose that the risk R (given as a percent) of having an accident while driving a car can be modeled by the equation
$$R = 5e^{kx}$$
where x is the variable concentration of alcohol in the blood and k is a constant.

Suppose that a concentration of alcohol in the blood of 0.07 results in a 10% risk (R = 10) of an accident. Find the constant k in the equation.

Using this value of k, what is the risk if the concentration is 0.11?

7) During 1991, 200,000 people visited Rave Amusement Park. During 1997, the number had grown to 834,000. If the number of visitors to the park obeys the law of uninhibited growth, find the exponential growth function that models this data.
   A) $f(t) = 200{,}000e^{0.248t}$ B) $f(t) = 200{,}000e^{0.238t}$ C) $f(t) = 634{,}000e^{0.238t}$ D) $f(t) = 634{,}000e^{0.248t}$

8) A culture of bacteria obeys the law of uninhibited growth. If 140,000 bacteria are present initially and there are 609,000 after 6 hours, how long will it take for the population to reach one million?

9) The size P of a certain insect population at time t (in days) obeys the function $P = 700e^{0.03t}$. After how many days will the population reach 1500?

10) The value of a particular investment follows a pattern of exponential growth. In the year 2000, you invested money in a money market account. The value of your investment t years after 2000 is given by the exponential growth model $A = 5800e^{0.052t}$. How much did you initially invest in the account?
    A) $301.60   B) $5800.00   C) $2900.00   D) $6109.58

11) The value of a particular investment follows a pattern of exponential growth. In the year 2000, you invested money in a money market account. The value of your investment t years after 2000 is given by the exponential growth model $A = 3000e^{0.069t}$. When will the account be worth $4236?
    A) 2006   B) 2007   C) 2004   D) 2005

12) The value of a particular investment follows a pattern of exponential growth. In the year 2000, you invested money in a money market account. The value of your investment t years after 2000 is given by the exponential growth model $A = 1100e^{0.06t}$. By what percentage is the account increasing each year?
    A) 6.2%   B) 6.7%   C) 6.6%   D) 6.4%

13) The population of a particular country was 29 million in 1983; in 1995, it was 39 million. The exponential growth function $A = 29e^{kt}$ describes the population of this country t years after 1983. Use the fact that 12 years after 1983 the population increased by 10 million to find k to three decimal places.

A) 0.025 B) 0.192 C) 0.035 D) 0.586

2 Find Equations of Populations That Obey the Law of Decay

Solve the problem.

1) The half-life of silicon-32 is 710 years. If 80 grams is present now, how much will be present in 100 years? (Round your answer to three decimal places.)

A) 0.056 B) 30.137 C) 72.559 D) 79.223

2) The half-life of plutonium-234 is 9 hours. If 80 milligrams is present now, how much will be present in 6 days? (Round your answer to three decimal places.)

A) 26.39 B) 0.001 C) 0.787 D) 50.396

3) A fossilized leaf contains 12% of its normal amount of carbon 14. How old is the fossil (to the nearest year)? Use 5600 years as the half-life of carbon 14.

A) 20,040 B) 17,099 C) 36,108 D) 1031

4) The half-life of a radioactive element is 130 days, but your sample will not be useful to you after 80% of the radioactive nuclei originally present have disintegrated. About how many days can you use the sample?

A) 287 B) 297 C) 302 D) 312

5) The half-life of carbon-14 is 5700 years. Find the age of a sample in which 8% of the radioactive nuclei originally present have decayed.

6) The half-life of radium is 1690 years. If 150 grams is present now, how long (to the nearest year) till only 100 grams are present?

7) Assume that the half-life of Carbon-14 is 5700 years. Find the age (to the nearest year) of a wooden axe in which the amount of Carbon-14 is 30% of what it originally had.

8) Strontium 90 decays at a constant rate of 2.44% per year. Therefore, the equation for the amount P of strontium 90 after t years is $P = P_0 e^{-0.0244t}$. How long will it take for 15 grams of strontium to decay to 5 grams? Round answer to 2 decimal places.

A) 4.50 years B) 40.50 years C) 45.03 years. D) 450.25 years

9) If a single pane of glass obliterates 15% of the light passing through it, then the percent P of light that passes through n successive panes can be approximated by the equation

$$P = 100e^{-0.15n}$$

How many panes are necessary to block at least 50% of the light?

10) Bob, the incredible shrinking man, loses half of his height each day after he was exposed to a mysterious form of cosmic radiation. How many days before he is literally "knee-high to a grasshopper"? Assume that a grasshopper's knee is 4 millimeters high and that Bob is 2 meters tall. Round your answer to the nearest whole day. (1000 millimeters = 1 meter)

11) The formula

$$D = 8e^{-0.6h}$$

can be used to find the number of milligrams D of a certain drug that is in a patient's bloodstream h hours after the drug has been administered. The drug is to be administered again when the amount in the bloodstream reaches 4 milligrams. What is the time between injections?

12) Between 8:30 a.m. and 9:30 a.m., cars drive through the Cappuccino Express at a rate of 12 cars per hour (0.2 per minute). The following formula from probability can be used to determine the probability that a car will arrive within t minutes of 8:30a.m.

$$F(t) = 1 - e^{-0.2t}$$

Determine how many minutes are needed for the probability to reach 0.6.

13) A rumor is spread at an elementary school with 1200 students according to the model $N = 1200(1 - e^{-0.16d})$ where N is the number of students who have heard the rumor and d is the number of days that have elapsed since the rumor began. How many days must elapse for 500 to have heard the rumor?

14) The function $A = A_0 e^{-0.01386x}$ models the amount in pounds of a particular radioactive material stored in a concrete vault, where x is the number of years since the material was put into the vault. If 800 pounds of the material are initially put into the vault, how many pounds will be left after 70 years?

   A) 303 pounds   B) 286 pounds   C) 488 pounds   D) 560 pounds

15) The function $A = A_0 e^{-0.00693x}$ models the amount in pounds of a particular radioactive material stored in a concrete vault, where x is the number of years since the material was put into the vault. If 600 pounds of the material are placed in the vault, how much time will need to pass for only 227 pounds to remain?

   A) 280 years   B) 145 years   C) 140 years   D) 150 years

16) The amount of a certain drug in the bloodstream is modeled by the function $y = y_0 e^{-0.40t}$, where y_0 is the amount of the drug injected (in milligrams) and t is the elapsed time (in hours). Suppose that 10 milligrams are injected at 10:00 A.M. If a second injection is to be administered when there is 1 milligram of the drug present in the bloodstream, approximately when should the next dose be given? Express your answer to the nearest quarter hour.

   A) 5: 30 P.M   B) 3:45 P.M   C) 5:45 P.M   D) 12:30 P.M

3 Use Newton's Law of Cooling

Solve the problem.

1) Sandy manages a ceramics shop and uses a 800°F kiln to fire ceramic greenware. After turning off her kiln, she must wait until its temperature gauge reaches 175°F before opening it and removing the ceramic pieces. If room temperature is 70°F and the gauge reads 650°F in 9 minutes, how long must she wait before opening the kiln? Assume the kiln cools according to Newton's Law of Cooling:

   $$U = T + (U_0 - T)e^{kt}.$$

   (Round your answer to the nearest whole minute.)

   A) 150 minutes   B) 76 minutes   C) 111 minutes   D) 57 minutes

2) A thermometer reading 93°F is placed inside a cold storage room with a constant temperature of 38°F. If the thermometer reads 88°F in 9 minutes, how long before it reaches 51°F? Assume the cooling follows Newton's Law of Cooling:
   $U = T + (U_0 - T)e^{kt}$.
   (Round your answer to the nearest whole minute.)
   A) -1 minutes B) 136 minutes C) 29 minutes D) -28 minutes

3) A thermometer reading 11°C is brought into a room with a constant temperature of 30°C. If the thermometer reads 16°C after 4 minutes, what will it read after being in the room for 10 minutes? Assume the cooling follows Newton's Law of Cooling:
   $U = T + (U_0 - T)e^{kt}$.
   (Round your answer to two decimal places.)
   A) 38.86°C B) 21.14°C C) 4.67°C D) 29.1°C

4) A thermometer reading 36°F is brought into a room with a constant temperature of 80°F. If the thermometer reads 44°F after 3 minutes, what will it read after being in the room for 9 minutes? Assume the cooling follows Newton's Law of Cooling:
   $U = T + (U_0 - T)e^{kt}$.
   (Round your answer to two decimal places.)
   A) 104.1°F B) 55.9°F C) 72.77°F D) 26.25°F

5) A cup of coffee is heated to 194° and is then allowed to cool in a room whose air temperature is 72°. After 11 minutes, the temperature of the cup of coffee is 140°. Find the time needed for the coffee to cool to a temperature of 102°. Assume the cooling follows Newton's Law of Cooling:
   $U = T + (U_0 - T)e^{kt}$.
   (Round your answer to one decimal place.)
   A) 15.1 minutes B) 41.1 minutes C) 26.4 minutes D) 29.7 minutes

6) A thermometer is taken from a room at 71°F to the outdoors where the temperature is 14°F. Determine what the reading on the thermometer will be after 5 minutes, if the reading drops to 45°F after 1 minute. Assume the cooling follows Newton's Law of Cooling:
   $U = T + (U_0 - T)e^{kt}$.
   (Round your answer to two decimal places.)

7) The temperature (in degrees Fahrenheit) of a dead body that has been cooling in a room set at 70° is measured as 88°. One hour later, the body temperature is 87.5°. How long (to the nearest hour) before the first measurement was the time of death, assuming that the body temperature of the deceased at the time of death was 98.6°. Assume the cooling follows Newton's Law of Cooling:
   $U = T + (U_0 - T)e^{kt}$.

8) A fully cooked turkey is taken out of an oven set at 200°C (Celsius) and placed in a sink of chilled water of temperature 4°C. After 3 minutes, the temperature of the turkey is measured to be 50°C. How long (to the nearest minute) will it take for the temperature of the turkey to reach 15°C? Assume the cooling follows Newton's Law of Cooling:
   $U = T + (U_0 - T)e^{kt}$.
   (Round your answer to the nearest minute.)

4 Use Logistic Models

Solve the problem.

1) The logistic growth model $P(t) = \dfrac{1880}{1 + 36.6e^{-0.335t}}$ represents the population of a bacterium in a culture tube after t hours. What was the initial amount of bacteria in the population?

    A) 51 B) 55 C) 49 D) 50

2) The logistic growth model $P(t) = \dfrac{920}{1 + 17.4e^{-0.332t}}$ represents the population of a bacterium in a culture tube after t hours. When will the amount of bacteria be 630?

    A) 4.92 hours B) 1.44 hours C) 7.46 hours D) 10.94 hours

3) The logistic growth model $P(t) = \dfrac{480}{1 + 79e^{-0.19t}}$ represents the population of a species introduced into a new territory after t years. When will the population be 80?

    A) 14.53 years B) 3.04 years C) 13.57 years D) 4 years

4) The logistic growth model $P(t) = \dfrac{230}{1 + 22e^{-0.157t}}$ represents the population of a species introduced into a new territory after t years. What will the population be in 20 years?

    A) 242 B) 118 C) 127 D) 230

5) The logistic growth model $P(t) = \dfrac{1}{1 + 4.56e^{-0.882t}}$ represents the proportion of the total market of a new product as it penetrates the market t years after introduction. When will the product have 80% of the market?

    A) 1.47 years B) 3.29 years C) 2.47 years D) 4.29 years

6) In 1990, the population of a country was estimated at 4 million. For any subsequent year the population, P(t) (in millions), can be modeled by the equation $P(t) = \dfrac{240}{5 + 54.99e^{-0.0208t}}$, where t is the number of years since 1990. Estimate the year when the population will be 21 million.

    A) approximately the year 2088 B) approximately the year 2016
    C) approximately the year 2093 D) approximately the year 2041

7) In 1992, the population of a country was estimated at 5 million. For any subsequent year, the population, P(t) (in millions), can be modeled using the equation $P(t) = \dfrac{250}{5 + 44.99e^{-0.0208t}}$, where t is the number of years since 1992. Determine the year when the population will be 39 million.

8) In a town whose population is 3000, a disease creates an epidemic. The number of people, N, infected t days after the disease has begun is given by the function $N(t) = \dfrac{3000}{1 + 21.2e^{-0.54t}}$. Find the number of infected people after 10 days.

    A) 1000 people B) 2737 people C) 142 people D) 2000 people.

9) The logistic growth function $f(t) = \dfrac{360}{1 + 3.0e^{-0.2t}}$ describes the population of a species of butterflies t months after they are introduced to a non-threatening habitat. How many butterflies were initially introduced to the habitat?

    A) 360 butterflies   B) 2 butterflies   C) 90 butterflies   D) 3 butterflies

10) The logistic growth function $f(t) = \dfrac{600}{1 + 11.0e^{-0.12t}}$ describes the population of a species of butterflies t months after they are introduced to a non-threatening habitat. What is the limiting size of the butterfly population that the habitat will sustain?

    A) 1200 butterflies   B) 600 butterflies   C) 11 butterflies   D) 50 butterflies

11) The logistic growth function $f(t) = \dfrac{360}{1 + 4.1e^{-0.27t}}$ describes the population of a species of butterflies t months after they are introduced to a non-threatening habitat. How many butterflies are expected in the habitat after 15 months?

    A) 336 butterflies   B) 360 butterflies   C) 5400 butterflies   D) 1050 butterflies

12) The logistic growth function $f(t) = \dfrac{95{,}000}{1 + 1582.3e^{-1.8t}}$ models the number of people who have become ill with a particular infection t weeks after its initial outbreak in a particular community. How many people became ill with this infection when the epidemic began?

    A) 1582 people   B) 95,000 people   C) 1583 people   D) 60 people

13) The logistic growth function $f(t) = \dfrac{80{,}000}{1 + 1999e^{-1.5t}}$ models the number of people who have become ill with a particular infection t weeks after its initial outbreak in a particular community. How many people were ill after 4 weeks?

    A) 160 people   B) 13,434 people   C) 80,005 people   D) 82,000 people

14) The logistic growth function $f(t) = \dfrac{90{,}000}{1 + 2999.0e^{-1.9t}}$ models the number of people who have become ill with a particular infection t weeks after its initial outbreak in a particular community. What is the limiting size of the population that becomes ill?

    A) 2999 people   B) 90,000 people   C) 180,000 people   D) 3000 people

4.9 Building Exponential, Logarithmic, and Logistic Models from Data

1 Use a Graphing Utility to Fit an Exponential Function to Data

Solve the problem.

1) The population (in hundred thousands) for the Colonial United States in ten-year increments for the years 1700-1780 is given in the table. (Source: 1998 Information Please Almanac)

Decade	Population	Decade	Population
0	251	5	1171
1	332	6	1594
2	466	7	2148
3	629	8	2780
4	906		

   State whether the data can be more accurately modeled using an exponential function or a logarithmic function. Using a graphing utility, find a model for population (in hundred thousands) as a function of decades since 1700.

2) A biologist has a bacteria sample. She records the amount of bacteria every week for 8 weeks and finds that the exponential function of best fit to the data is $A = 150 \cdot 1.79^t$. Express the function of best fit in the form $A = A_0 e^{kt}$.

3) A music store manager collected data regarding price and quantity demanded of cassette tapes every week for 10 weeks, and found that the exponential function of best fit to the data was $p = 25 \cdot 0.89^q$. Express the function of best fit in the form $p = p_0 e^{kq}$, and use this expression to predict the quantity demanded if the price is $8.50.

4) A life insurance company uses the following rate table for annual premiums for women for term life insurance. Use a graphing utility to fit an exponential function to the data. Predict the annual premium for a woman aged 70 years.

Age	35	40	45	50	55	60	65
Premium	$103	$133	$190	$255	$360	$503	$818

   A) $y = 6.367e^{0.068x}$, $743
   B) $y = -9306.4 + 2516.3 \ln(x)$, $1723
   C) $y = 8.94e^{0.068x}$, $1044
   D) $y = 0.0000398x^{4.06}$, $1233

5) Money magazine reports that the percentage of trading days in which the Dasdaq loses or gains 2% or more has been increasing since 1995 indicating more volatility in the Dasdaq. Use a graphing utility to fit an exponential function to the data. Predict the percentage of trading days in 2001 having such swings in value.

Year	% of trading days
1995, 0	2
1996, 1	5
1997, 2	8
1998, 3	18
1999, 4	23
2000, 5	49

   A) $y = 1.278e^{0.611x}$, 92 days
   B) $y = 1.849e^{0.531x}$, 76 days
   C) $y = 1.669x^{1.706}$, 46 days
   D) $y = 2.353e^{0.610x}$, 91 days

Precalculus Enhanced with Graphing Utilities

6) A nuclear scientist has a sample of 100 mg of a radioactive material which has a half-life in hours. She monitors the amount of radioactive material over a period of a day and obtains the following data. Use a graphing utility to fit an exponential function to the data. Predict the amount of material remaining at 40 hours.

Hours	0	5	10	15	20	25	30
mg	100	68.3	45.2	31.3	21.5	14.6	9.8

A) $y = 100e^{-0.077x}$, 6.7 mg
B) $y = 92e^{-0.0686x}$, 5.9 mg
C) $y = 100e^{-0.077x}$, 4.6 mg
D) $y = 86e^{-0.071x}$, 5.0 mg

2 Use a Graphing Utility to Fit a Logarithmic Function to Data

Solve the problem.

1) Data representing the price and quantity demanded for hand-held electronic organizers were analyzed every day for 15 days. The logarithmic function of best fit to the data was found to be $p = 398 - 73 \ln(q)$. Use this to predict the number of hand-held electronic organizers that would be demanded if the price were $275.

2) The rates of death (in number of deaths per 100,000 population) for 1-4 year olds in the United States between 1980-1995 are given below. (Source: NCHS Data Warehouse)

Year	Rate of Death
1980	91.4
1985	74.5
1990	69.3
1995	61.3

A logarithmic equation that models this data is $y = 822.99 - 167.55 \ln x$ where x represents the number of years since 1900. Use this equation to predict the rate of death for 1-4 year olds in 2005.

3) The rates of death (in number of deaths per 100,000 population) for 20-24 year olds in the United States between 1985-1993 are given below. (Source: NCHS Data Warehouse)

Year	Rate of Death
1985	134.9
1987	154.7
1989	162.9
1991	174.5
1993	182.2

A logarithmic equation that models this data is $y = 57.76 + 48.56 \ln x$ where x represents the number of years since 1980 and y represents the rate of death in that year. Use this equation to predict the year in which the rate of death for 20-24 year olds first exceeds 200.

4) After introducing an inhibitor into a culture of luminescent bacteria, a scientist monitors the luminosity produced by the culture. Use a graphing utility to fit a logarithmic function to the data. Predict the luminosity after 20 hours.

Time, hrs	2	3	4	5	8	10	15
Luminosity	77.4	60.8	54.5	45.8	30.0	24.3	10.5

A) $y = 100.5 - 32.7 \ln(x)$, 2.54
B) $y = 107.55 - 41 \ln(x)$, -15.27
C) $y = 98.75 - 32.66 \ln(x)$, 0.91
D) $y = 112.97 - 45.97 \ln(x)$, -24.74

5) In a Psychology class, the students were tested at the end of the course on a final exam. Then they were retested with an equivalent test at subsequent time intervals. Their average scores after t months are given in the table.

Time, t (in months)	1	2	3	4	5
Score, y (in percentage)	86.2	85.7	85.4	85.2	85.0

Using a graphing utility, fit a logarithmic function $y = a + b \ln x$ to the data. Using the function you found, estimate how long will it take for the test scores to fall below 84%. Express your answer to the nearest month.

A) 12 months   B) 20 months   C) 8 months   D) 10 months

3 Use a Graphing Utility to Fit a Logistic Function to Data

Solve the problem.

1) A mechanic is testing the cooling system of a boat engine. He measures the engine's temperature over time. Use a graphing utility to fit a logistic function to the data. What is the carrying capacity of the cooling system?

time, min	5	10	15	20	25
temperature, °F	100	180	270	300	305

A) $y = \dfrac{314.79}{1 + 7.86e^{-0.246x}}$, 315°F

B) $y = \dfrac{314.79}{1 + 7.86e^{-1.22x}}$, 315°F

C) $y = \dfrac{306.53}{1 + 7.92e^{-0.254x}}$, 307°F

D) $y = \dfrac{311.63}{1 + 8.1e^{-0.253x}}$, 312°F

Ch. 4 Exponential and Logarithmic Functions
Answer Key

4.1 Composite Functions
1 Form a Composite Function
1) A
2) A
3) D
4) B
5) B
6) D
7) C
8) C
9) A
10) D
11) A
12) B
13) D
14) A
15) C
16) B
17) D
18) A
19) D
20) A
21) C
22) B
23) A
24) A
25) C
26) B
27) A
28) A
29) $S(t) = \frac{1}{18}\pi t^4 + 4\pi t^2$
30) $V(t) = \frac{1}{3}\pi t^4$
31) $C = \frac{\sqrt{1200 - 6p}}{40} + 800$
32) 8 or −8

2 Find the Domain of a Composite Function
1) D
2) B
3) B
4) D
5) A
6) A
7) D
8) C
9) C
10) A
11) A
12) $(f \circ g)(x) = (-1 + 5x)^2$; Domain: all real numbers

4.2 One-to-One Functions; Inverse Functions

1 Determine Whether a Function Is One-to-One
1) B
2) A
3) A
4) B
5) A
6) A
7) A
8) B
9) B
10) A
11) A

2 Determine the Inverse of a Function Defined by a Map or an Ordered Pair
1) A
2) B
3) D
4) A
5) B

3 Obtain the Graph of the Inverse Function from the Graph of the Function
1) B
2) B
3) B
4) A
5) C

4 Find the Inverse of a Function Defined by an Equation
1) B
2) B
3) B
4) A
5) B
6) C
7) D
8) $x(P) = \frac{1}{2}P + 250$
9) (a) $x \geq 0$ is one correct answer; another equally correct answer is $x \leq 0$.
   (b) For the case $x \geq 0$, the inverse is $f^{-1}(x) = x + 5$.
   For the case $x \leq 0$, the inverse is $f^{-1}(x) = -x - 5$.
10) (a) $V(W) = (\frac{W - 1.2}{3.51})^3$
    (b) 0.13

5 Demonstrate Additional Understanding and Skills
1) A
2) A
3) B
4) B
5) B
6) A
7) Inverses. Check to see that $f(g(x)) = x$ and that $g(f(x)) = x$.

4.3 Exponential Functions

1 Evaluate Exponential Functions
1) C
2) A
3) A

4) B
5) C
6) A
7) D
8) B
9) C
10) B
11) 661 students
12) A
13) 14.18 watts
14) 4 mg; 2 mg
15) D
16) C
17) D

2 Graph Exponential Functions

1) D
2) B
3) A
4) B
5) C
6) A
7) C
8) B
9) B
10) B
11) B
12) A
13) A
14) C
15) B
16) B
17) C
18) A
19) A
20) B
21) C
22) C
23) C
24) B
25) C

3 Define the Number e

1) D
2) B
3) A
4) D
5) D
6) B
7) D

4 Solve Exponential Equations

1) A
2) C
3) A
4) D
5) A
6) B

7) D
8) C
9) A
10) C
11) C
12) C

4.4 Logarithmic Functions

1 Change Exponential Expressions to Logarithmic Expressions and Logarithmic Expressions to Exponential Expressions

1) C
2) A
3) B
4) D
5) D
6) D
7) D
8) D
9) $\log_{16} 64 = \dfrac{3}{2}$
10) D
11) D
12) A
13) B
14) B
15) D
16) C
17) B
18) C
19) C
20) A
21) $61^y = x$
22) D
23) C

2 Evaluate Logarithmic Expressions

1) C
2) D
3) C
4) D
5) B
6) A
7) C
8) B
9) C
10) C
11) C
12) B
13) B
14) D
15) A
16) C
17) D
18) B
19) B
20) B
21) C

22) A
3 Determine the Domain of a Logarithmic Function
1) A
2) B
3) D
4) B
5) C
6) D
7) B
8) D
9) D
10) C
11) A
12) B
13) (−2, ∞)

4 Graph Logarithmic Functions
1) C
2) A
3) B
4) A
5) D
6) B
7) A
8) C
9) C
10) B
11) D
12) C
13) D
14) D
15) D
16) D
17) D
18) D
19) C
20) D
21) B
22) B

23)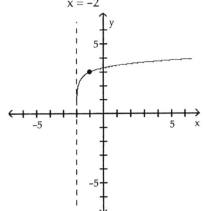

domain $(-2, \infty)$
range $(-\infty, \infty)$
vertical asymptote $x = -2$

5 Solve Logarithmic Equations

1) D
2) B
3) D
4) D
5) A
6) C
7) D
8) C
9) C
10) B
11) A
12) B
13) B
14) D
15) 1.427 decibels
16) 0.20 watt per square meter
17) 117 decibels
18) The intensity of a sound that crosses the human threshold of pain is 10 times as intense as this rock concert.
19) Jim's voice is 10^{10} times as intense as Amy's.
20) B
21) D
22) 2.9731
23) The magnitude of the earthquake measures 6.3 on the Richter scale.
24) The earthquake of greater magnitude has a seismographic reading that is $10^{0.1} \approx 1.26$ times that of the lesser earthquake.
25) A
26) C
27) B
28) (a) $k = 9.90$
    (b) 7%
    (c) 0.28
    (d) 0.07
29) A
30) C

Precalculus Enhanced with Graphing Utilities 409

4.5 Properties of Logarithms

1 Work with the Properties of Logarithms
1) C
2) A
3) C
4) A
5) A
6) A
7) B
8) D
9) B
10) C
11) D
12) A
13) A
14) C
15) B
16) A
17) D

2 Write a Logarithmic Expression as a Sum or Difference of Logarithms
1) C
2) B
3) C
4) C
5) D
6) A
7) C
8) B
9) A
10) B
11) B
12) $\frac{5}{3}\log_b x + \frac{8}{3}\log_b y - \frac{2}{3}\log_b z$
13) A
14) A
15) D

3 Write a Logarithmic Expression as a Single Logarithm
1) C
2) B
3) C
4) B
5) C
6) C
7) B
8) $\log_a \frac{6x^4}{y^3}$
9) B
10) D

4 Evaluate Logarithms Whose Base Is Neither 10 nor e
1) D
2) B
3) A
4) C
5) D

6) B
7) C
8) A
9) A
10) -7.26
11) 2

5 Graph Logarithmic Functions Whose Base Is Neither 10 nor e

1) A
2) A
3) D

4.6 Logarithmic and Exponential Equations

1 Solve Logarithmic Equations Using the Properties of Logarithms

1) C
2) A
3) A
4) A
5) A
6) C
7) D
8) B
9) D
10) B
11) C
12) C
13) D
14) B
15) {8.36}
16) D
17) A

2 Solve Exponential Equations

1) C
2) C
3) B
4) D
5) D
6) $\left\{\dfrac{\ln 2}{\ln 3}\right\}$
7) $\left\{-\dfrac{3}{14}\right\}$
8) B
9) C
10) D
11) B
12) B
13) C
14) $\left\{\dfrac{\ln \pi}{2 - \ln \pi}\right\}$
15) $\{\ln(-1 + \sqrt{2})\}$
16) C
17) D
18) C
19) A
20) C
21) D

22) D
23) B
24) B

3 Solve Logarithmic and Exponential Equations Using a Graphing Utility

1) B
2) C
3) B
4) C
5) C
6) A
7) B
8) C
9) C

4.7 Compound Interest

1 Determine the Future Value of a Lump Sum of Money

1) B
2) D
3) C
4) B
5) D
6) D
7) $6749
8) $438.39
9) $1643.18
10) $360,351.56

2 Calculate Effective Rates of Return

1) D
2) B
3) A
4) B
5) C
6) D
7) 4.84%
8) B
9) B
10) A

3 Determine the Present Value of a Lump Sum of Money

1) A
2) C
3) A
4) A
5) B
6) D
7) B
8) $1252.91
9) D

4 Determine the Time Required to Double or Triple a Lump Sum of Money

1) D
2) A
3) D
4) A
5) A
6) 13.9 years
7) 6 years

5 Demonstrate Additional Understanding and Skills

1) C
2) C
3) C
4) A
5) C
6) 51.37%
7) $35,766.45

4.8 Exponential Growth and Decay; Newton's Law; Logistic Growth and Decay

1 Find Equations of Populations That Obey the Law of Uninhibited Growth

1) B
2) D
3) $R(x) = 2000000 \cdot 2^{x-1}$; $16,000,000
4) $D = 3^{L-1}$; 81 distributors
5) $A = 100 \cdot 4^d$; 1,638,400 bacteria
6) $k = \dfrac{\ln 2}{0.07} \approx 9.90$; about 14.9%
7) B
8) 8.024 hours
9) 26 days
10) B
11) D
12) A
13) A

2 Find Equations of Populations That Obey the Law of Decay

1) C
2) B
3) B
4) C
5) 686 years
6) 989 years
7) 9901 years
8) C
9) 5 panes
10) 9 days
11) about 1.2 hours
12) about 4.6 minutes
13) about 3.4 days
14) A
15) C
16) B

3 Use Newton's Law of Cooling

1) B
2) B
3) B
4) B
5) C
6) 16.71 °F
7) The deceased died approximately 16 hours before the first measurement.
8) 6 minutes

4 Use Logistic Models

1) D
2) D
3) A

  4) B
  5) B
  6) C
  7) in about 166.47 years or approximately the year 2158
  8) B
  9) C
  10) B
  11) A
  12) D
  13) B
  14) B

4.9 Building Exponential, Logarithmic, and Logistic Models from Data

1 Use a Graphing Utility to Fit an Exponential Function to Data

  1) exponential; $y = 252.68 \cdot 1.36^x$

  2) $A = 150e^{0.58t}$

  3) $p = 25e^{-0.12q}$; 9 cassettes

  4) C

  5) D

  6) C

2 Use a Graphing Utility to Fit a Logarithmic Function to Data

  1) 5 electronic organizers

  2) 43.2 deaths per 100,000

  3) 1999

  4) C

  5) B

3 Use a Graphing Utility to Fit a Logistic Function to Data

  1) A

Ch. 5 Trigonometric Functions

5.1 Angles and Their Measure

1 Convert Between Degrees, Minutes, Seconds, and Decimal Forms for Angles

Convert the angle to a decimal in degrees. Round the answer to two decimal places.

1) 19°18'4"
   A) 19.36°   B) 19.26°   C) 19.30°   D) 19.31°

2) 172°30'5"
   A) 172.51°   B) 172.46°   C) 172.56°   D) 172.50°

3) 236°51'36"
   A) 236.86°   B) 236.92°   C) 236.82°   D) 236.87°

4) 23°47'37"
   A) 23.94°   B) 23.79°   C) 23.52°   D) 23.84°

5) 21°17'34"
   A) 21.29°   B) 21.37°   C) 21.22°   D) 21.34°

Convert the angle to D° M' S" form. Round the answer to the nearest second.

6) 15.21°
   A) 15°12'36"   B) 15°12'21"   C) 15°12'42"   D) 15°12'24"

7) 186.61°
   A) 186°36'36"   B) 186°34'61"   C) 186°36'61"   D) 186°37'36"

8) 332.53°
   A) 332°32'47"   B) 332°31'53"   C) 332°31'48"   D) 332°47'53"

2 Find the Arc Length of a Circle

If s denotes the length of the arc of a circle of radius r subtended by a central angle θ, find the missing quantity.

1) r = 8.89 centimeters, θ = 1.2 radians, s = ?
   A) 12.7 cm   B) 11.7 cm   C) 10.7 cm   D) 9.7 cm

2) r = 11.8 inches, θ = 240°, s = ?
   A) 49.6 in.   B) 49.4 in.   C) 49.5 in.   D) 49.7 in.

3) r = $\frac{2}{3}$ feet, s = 8 feet, θ = ?
   A) $\frac{16}{3}$°   B) 12°   C) $\frac{16}{3}$ radians   D) 12 radians

4) s = 4.86 meters, θ = 1.8 radians, r = ?
   A) 0.37 m   B) 3.5 m   C) 1.35 m   D) 2.7 m

Find the length s. Round the answer to three decimal places.

5)

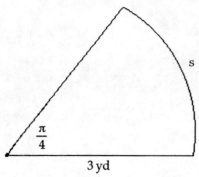

A) 4.189 yd   B) 4.712 yd   C) 2.356 yd   D) 3.82 yd

6)

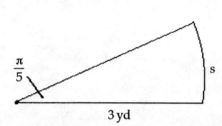

A) 4.775 yd   B) 3.77 yd   C) 5.236 yd   D) 1.885 yd

7)

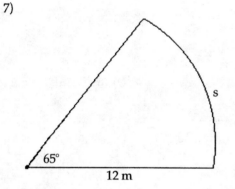

A) 13.614 m   B) 14.975 m   C) 10.891 m   D) 12.253 m

8)

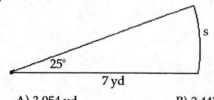

A) 3.054 yd   B) 2.443 yd   C) 3.359 yd   D) 2.749 yd

Solve the problem.

9) For a circle of radius 4 feet, find the arc length s subtended by a central angle of 30°. Round to the nearest hundredth.

A) 4.19 ft   B) 376.99 ft   C) 2.09 ft   D) 6.28 ft

10) For a circle of radius 4 feet, find the arc length s subtended by a central angle of 60°. Round to the nearest hundredth.
   A) 4.25 ft
   B) 4.40 ft
   C) 4.35 ft
   D) 4.19 ft

11) A ship in the Atlantic Ocean measures its position to be 29°40' north latitude. Another ship is reported to be due north of the first ship at 43°44' north latitude. Approximately how far apart are the two ships? Round to the nearest mile. Assume that the radius of the Earth is 3960 miles.
   A) 957 mi
   B) 55,689 mi
   C) 972 mi
   D) 55,704 mi

12) Salt Lake City, Utah, is due north of Flagstaff, Arizona. Find the distance between Salt Lake City (40°45' north latitude) and Flagstaff (35°16' north latitude). Assume that the radius of the Earth is 3960 miles. Round to nearest whole mile.

13) The minute hand of a clock is 6 inches long. How far does the tip of the minute hand move in 20 minutes? If necessary, round the answer to two decimal places.
   A) 13.8 in.
   B) 12.57 in.
   C) 10.83 in.
   D) 15.08 in.

14) A pendulum swings though an angle of 25° each second. If the pendulum is 55 inches long, how far does its tip move each second? If necessary, round the answer to two decimal places.
   A) 22.15 in.
   B) 25.29 in.
   C) 24 in.
   D) 26.43 in.

3 Convert from Degrees to Radians and from Radians to Degrees

Convert the angle in degrees to radians. Express the answer as multiple of π.

1) 90°
   A) $\frac{\pi}{3}$
   B) $\frac{\pi}{8}$
   C) $\frac{\pi}{4}$
   D) $\frac{\pi}{2}$

2) −90°
   A) $-\frac{\pi}{4}$
   B) $-\frac{\pi}{3}$
   C) $-\frac{\pi}{8}$
   D) $-\frac{\pi}{2}$

3) 480°
   A) $\frac{7\pi}{2}$
   B) $\frac{3\pi}{8}$
   C) $\frac{8\pi}{3}$
   D) $\frac{9\pi}{4}$

4) −480°
   A) $-\frac{8\pi}{3}$
   B) $-\frac{3\pi}{8}$
   C) $-\frac{7\pi}{2}$
   D) $-\frac{9\pi}{4}$

5) 87°
   A) $\frac{29\pi}{60}$
   B) $\frac{29\pi}{30}$
   C) $\frac{29\pi}{120}$
   D) $\frac{29\pi}{90}$

6) 6°
   A) $\frac{\pi}{60}$
   B) $\frac{\pi}{18}$
   C) $\frac{\pi}{30}$
   D) $\frac{\pi}{15}$

Convert the angle in degrees to radians. Express the answer in decimal form. If necessary, round to two decimal places.

7) 290°
   A) 5.03
   B) 5.04
   C) 5.06
   D) 5.05

8) −256°
   A) −4.44
   B) −4.45
   C) −4.47
   D) −4.46

Convert the angle in radians to degrees.

9) $\dfrac{4\pi}{5}$
   A) 143°
   B) 146°
   C) 145°
   D) 144°

10) $-\dfrac{11\pi}{10}$
   A) −199°
   B) −198°
   C) −197°
   D) −200°

11) $\dfrac{\pi}{3}$
   A) 60π°
   B) 3°
   C) 1°
   D) 60°

12) $-\dfrac{\pi}{5}$
   A) 1°
   B) −36°
   C) −1°
   D) −36π°

13) $\dfrac{11\pi}{6}$
   A) 330°
   B) 164°
   C) 98π°
   D) 660°

14) $-\dfrac{11}{3}\pi$
   A) −660°
   B) −330°
   C) −1320π°
   D) −12°

15) $\dfrac{\pi}{6}$
   A) 1080°
   B) 15°
   C) 30°
   D) 60°

16) $\dfrac{11\pi}{12}$
   A) 165°
   B) 150°
   C) 210°
   D) 160°

Convert the angle in radians to degrees. Express the answer in decimal form. If necessary, round to two decimal places.

17) 5
   A) 0.09°
   B) 0.22°
   C) 286.48°
   D) 287.41°

18) 2.03
   A) 0.04°
   B) 115.02°
   C) 116.31°
   D) 0.19°

19) $\sqrt{7}$

    A) 0.18°   B) 152.23°   C) 151.59°   D) 0.05°

4 Find the Area of a Sector of a Circle

If A denotes the area of the sector of a circle of radius r formed by the central angle θ, find the missing quantity. If necessary, round the answer to two decimal places.

1) r = 10 inches, θ = $\frac{\pi}{3}$ radians, A = ?

    A) 10.47 in²   B) 5.23 in²   C) 104.67 in²   D) 52.33 in²

2) r = 7 feet, A = 87 square feet, θ = ?

    A) 4263 radians   B) 1.78 radians   C) 3.55 radians   D) 2131.5 radians

3) θ = 4 radians, A = 57 square meters, r = ?

    A) 10.68 m   B) 5.34 m   C) 456 m   D) 114 m

4) r = 5 inches, θ = 45°, A = ?

    A) 19.63 in²   B) 3.93 in²   C) 1.96 in²   D) 9.81 in²

5) r = 5 feet, A = 79 square feet, θ = ?

    A) 56,608.28°   B) 362.29°   C) 181.15°   D) 113,216.56°

6) θ = 120°, A = 74 square meters, r = ?

    A) 309.81 m   B) 8.8 m   C) 8.41 m   D) 77.45 m

7) r = 78.3 centimeters, θ = $\frac{\pi}{11}$ radians, A = ?

    A) 11.2 cm²   B) 278.7 cm²   C) 1751 cm²   D) 875.5 cm²

8) r = 31.3 feet, θ = 38.971°, A = ?

    A) 333.18 ft²   B) 669.36 ft²   C) 666.36 ft²   D) 336.18 ft²

Find the area A. Round the answer to three decimal places.

9)

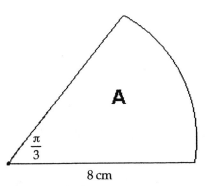

    A) 21.333 cm²   B) 67.021 cm²   C) 4.189 cm²   D) 33.51 cm²

10)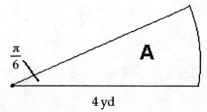

A) 2.667 yd² B) 4.189 yd² C) 1.047 yd² D) 8.378 yd²

11)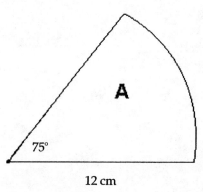

A) 7.854 cm² B) 94.248 cm² C) 30 cm² D) 188.496 cm²

12)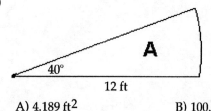

A) 4.189 ft² B) 100.531 ft² C) 50.265 ft² D) 16 ft²

Solve the problem.

13) A circle has a radius of 3 centimeters. Find the area of the sector of the circle formed by an angle of 40°. If necessary, round the answer to two decimal places.

A) 3.14 cm² B) 1 cm² C) 1.05 cm² D) 6.28 cm²

14) An irrigation sprinkler in a field of lettuce sprays water over a distance of 35 feet as it rotates through an angle of 120°. What area of the field receives water? If necessary, round the answer to two decimal places.

A) 2565.63 ft² B) 408.33 ft² C) 36.65 ft² D) 1282.82 ft²

15) As part of an experiment to test different liquid fertilizers, a sprinkler has to be set to cover an area of 120 square yards in the shape of a sector of a circle of radius 40 yards. Through what angle should the sprinkler be set to rotate? If necessary, round the answer to two decimal places.

A) 6.44° B) 27° C) 8.59° D) 4.3°

16) The blade of a windshield wiper sweeps out an angle of 135° in one cycle. The base of the blade is 12 inches from the pivot point and the tip is 32 inches from the pivot point. What area does the wiper cover in one cycle? (Round to the nearest 0.1 square inch.)

A) 1105.3 in² B) 1041.8 in² C) 948.3 in² D) 1036.7 in²

5 Find the Linear Speed of an Object Traveling in Circular Motion

Solve the problem.

1) An object is traveling around a circle with a radius of 10 centimeters. If in 20 seconds a central angle of $\frac{1}{3}$ radian is swept out, what is the linear speed of the object?
   A) 6 cm/sec
   B) 6 radians/sec
   C) $\frac{1}{6}$ radians/sec
   D) $\frac{1}{6}$ cm/sec

2) An object is traveling around a circle with a radius of 20 meters. If in 10 seconds a central angle of $\frac{1}{5}$ radian is swept out, what is the linear speed of the object?
   A) $\frac{2}{5}$ m/sec
   B) $\frac{1}{4}$ m/sec
   C) $\frac{1}{5}$ m/sec
   D) $\frac{1}{8}$ m/sec

3) An object is traveling around a circle with a radius of 10 meters. If in 15 seconds a central angle of 3 radians is swept out, what is the linear speed of the object?
   A) $\frac{1}{3}$ m/sec
   B) 2 m/sec
   C) $\frac{2}{3}$ m/sec
   D) 3 m/sec

4) A weight hangs from a rope 20 feet long. It swings through an angle of 27° each second. How far does the weight travel each second? Round to the nearest 0.1 foot.
   A) 9.4 feet
   B) 8.1 feet
   C) 8.7 feet
   D) 9.0 feet

5) A gear with a radius of 2 centimeters is turning at $\frac{\pi}{11}$ radians/sec. What is the linear speed at a point on the outer edge of the gear?
   A) 22π cm/sec
   B) $\frac{2\pi}{11}$ cm/sec
   C) $\frac{\pi}{22}$ cm/sec
   D) $\frac{11\pi}{2}$ cm/sec

6) A wheel of radius 1.8 feet is moving forward at 12 feet per second. How fast is the wheel rotating?
   A) 0.15 radians/sec
   B) 3.8 radians/sec
   C) 1.75 radians/sec
   D) 6.7 radians/sec

7) A car is traveling at 41 mph. If its tires have a diameter of 25 inches, how fast are the car's tires turning? Express the answer in revolutions per minute. If necessary, round to two decimal places.
   A) 551.26 rpm
   B) 532.26 rpm
   C) 3463.68 rpm
   D) 1102.52 rpm

8) A pick-up truck is fitted with new tires which have a diameter of 43 inches. How fast will the pick-up truck be moving when the wheels are rotating at 395 revolutions per minute? Express the answer in miles per hour rounded to the nearest whole number.
   A) 57 mph
   B) 51 mph
   C) 8 mph
   D) 25 mph

9) The Earth rotates about its pole once every 24 hours. The distance from the pole to a location on Earth 46° north latitude is about 2750.8 miles. Therefore, a location on Earth at 46° north latitude is spinning on a circle of radius 2750.8 miles. Compute the linear speed on the surface of the Earth at 46° north latitude.
   A) 789 mph
   B) 17,284 mph
   C) 115 mph
   D) 720 mph

10) To approximate the speed of a river, a circular paddle wheel with radius 0.68 feet is lowered into the water. If the current causes the wheel to rotate at a speed of 9 revolutions per minute, what is the speed of the current? If necessary, round to two decimal places.

   A) 0.44 mph B) 0.22 mph C) 38.45 mph D) 0.07 mph

11) The four Galilean moons of Jupiter have orbital periods and mean distances from Jupiter given by the following table.

	Distance (km)	Period (Earth hours)
Io	4.214×10^5	42.460
Europa	6.709×10^5	85.243
Ganymeade	1.070×10^6	171.709
Callisto	1.883×10^6	400.536

Find the linear speed of each moon. Which is the fastest (in terms of linear speed)?

12) In a computer simulation, a satellite orbits around Earth at a distance from the Earth's surface of 2.4×10^4 miles. The orbit is circular, and one revolution around Earth takes 10.1 days. Assuming the radius of the Earth is 3960 miles, find the linear speed of the satellite. Express the answer in miles per hour to the nearest whole mile.

   A) 622 mph B) 115 mph C) 15,322 mph D) 725 mph

13) A carousel has a radius of 17 feet and takes 32 seconds to make one complete revolution. What is the linear speed of the carousel at its outside edge? If necessary, round the answer to two decimal places.

   A) 0.53 ft/sec B) 11.83 ft/sec C) 3.34 ft/sec D) 106.81 ft/sec

6 Demonstrate Additional Understanding and Skills

Draw the angle.

1) 60°

A)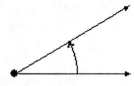

B)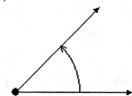

C)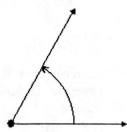

D)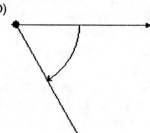

2) 135°

3) $\dfrac{2\pi}{3}$

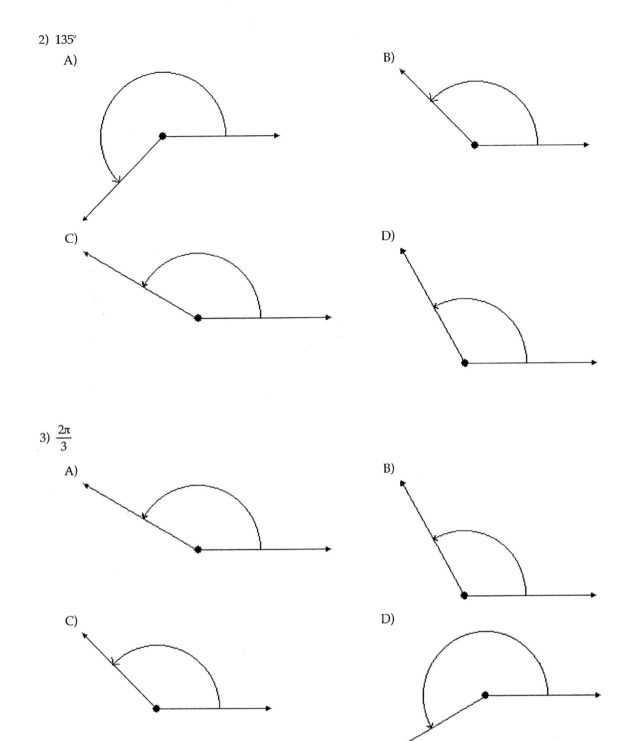

4) $-\dfrac{3\pi}{4}$

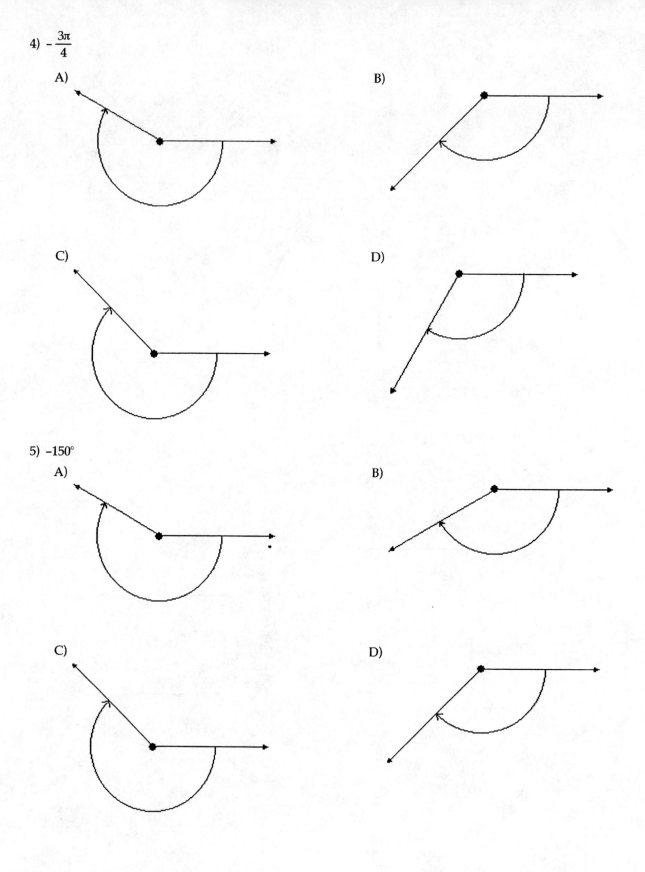

5) −150°

6) 330°

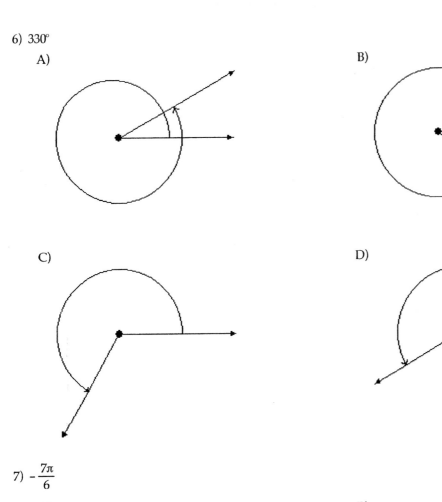

7) $-\dfrac{7\pi}{6}$

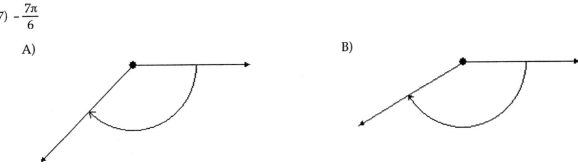

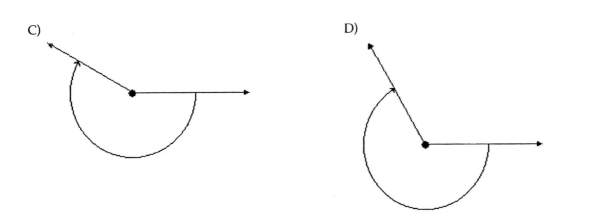

8) $\dfrac{5\pi}{3}$

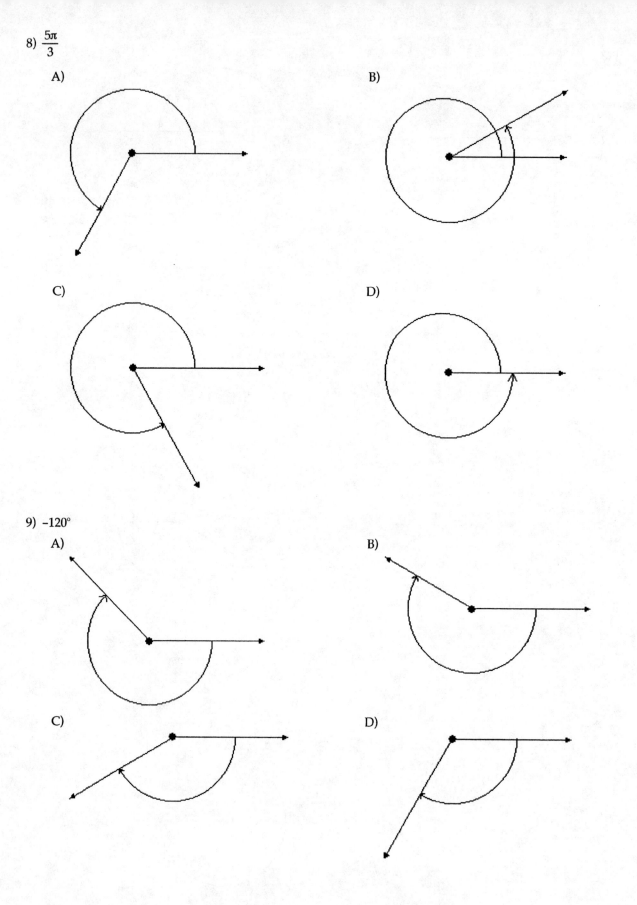

9) $-120°$

10) $\dfrac{7\pi}{4}$

A)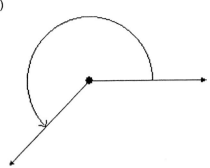

B)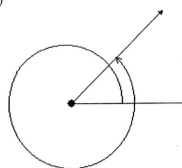

C)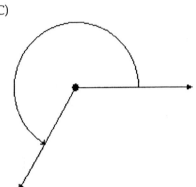

D)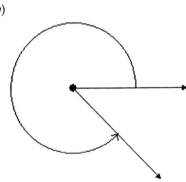

5.2 Trigonometric Functions: Unit Circle Approach

1 Find the Exact Values of the Trigonometric Functions Using a Point on the Unit Circle

In the problem, t is a real number and P = (x, y) is the point on the unit circle that corresponds to t. Find the exact value of the indicated trigonometric function of t.

1) $(\dfrac{4}{7}, \dfrac{\sqrt{33}}{7})$ Find sin t.

   A) $\dfrac{4\sqrt{33}}{33}$ B) $\dfrac{4}{7}$ C) $\dfrac{\sqrt{33}}{7}$ D) $\dfrac{\sqrt{33}}{4}$

2) $(\dfrac{3}{8}, \dfrac{\sqrt{55}}{8})$ Find tan t.

   A) $\dfrac{8}{3}$ B) $\dfrac{3\sqrt{55}}{55}$ C) $\dfrac{\sqrt{55}}{8}$ D) $\dfrac{\sqrt{55}}{3}$

3) $(\dfrac{\sqrt{77}}{9}, \dfrac{2}{9})$ Find sec t.

   A) $\dfrac{9\sqrt{77}}{77}$ B) $\dfrac{\sqrt{77}}{2}$ C) $\dfrac{9}{2}$ D) $\dfrac{2\sqrt{77}}{77}$

4) $(-\dfrac{\sqrt{55}}{8}, \dfrac{3}{8})$ Find cos t.

   A) $-\dfrac{8\sqrt{55}}{55}$ B) $\dfrac{3}{8}$ C) $-\dfrac{\sqrt{55}}{8}$ D) $-\dfrac{\sqrt{55}}{3}$

5) $(-\frac{\sqrt{11}}{6}, \frac{5}{6})$ Find cot t.

   A) $-\frac{\sqrt{11}}{5}$ B) $-\frac{6}{5}$ C) $\frac{\sqrt{11}}{6}$ D) $\frac{5}{6}$

6) $(-\frac{\sqrt{55}}{8}, -\frac{3}{8})$ Find sin t.

   A) $-\frac{8\sqrt{55}}{55}$ B) $-\frac{3}{8}$ C) $-\frac{\sqrt{55}}{8}$ D) $\frac{8}{3}$

7) $(-\frac{\sqrt{55}}{8}, -\frac{3}{8})$ Find cot t.

   A) $-\frac{3\sqrt{55}}{55}$ B) $\frac{\sqrt{55}}{3}$ C) $-\frac{\sqrt{55}}{3}$ D) $\frac{\sqrt{55}}{8}$

8) $(\frac{4}{7}, -\frac{\sqrt{33}}{7})$ Find csc t.

   A) $-\frac{7\sqrt{33}}{33}$ B) $\frac{\sqrt{33}}{7}$ C) $\frac{\sqrt{33}}{4}$ D) $-\frac{\sqrt{33}}{7}$

9) $(\frac{3}{4}, -\frac{\sqrt{7}}{4})$ Find cos t.

   A) $-\frac{\sqrt{7}}{4}$ B) $\frac{\sqrt{7}}{4}$ C) $\frac{3}{4}$ D) $-\frac{3}{4}$

10) $(\frac{1}{5}, -\frac{2\sqrt{6}}{5})$ Find csc t.

   A) $-\frac{5\sqrt{6}}{12}$ B) $\frac{5}{1}$ C) $\frac{1}{5}$ D) $-\frac{\sqrt{6}}{2}$

2 Find the Exact Values of the Trigonometric Functions of Quadrantal Angles

Find the exact value. Do not use a calculator.

1) $\sin 2\pi$

   A) $\frac{\sqrt{2}}{2}$ B) 0   C) 1   D) undefined

2) $\cos 0$

   A) 0   B) $\frac{\sqrt{2}}{2}$ C) 1   D) undefined

3) $\tan 0$

   A) 0   B) $\frac{\sqrt{2}}{2}$ C) 1   D) undefined

4) $\cot 0$

   A) $\frac{\sqrt{2}}{2}$ B) 0   C) 1   D) undefined

5) $\cot \dfrac{\pi}{2}$

A) 1 B) -1 C) 0 D) undefined

6) $\tan \pi$

A) -1 B) 0 C) 1 D) undefined

7) $\cos \pi$

A) -1 B) 1 C) 0 D) undefined

8) $\cot \dfrac{3\pi}{2}$

A) -1 B) 1 C) 0 D) undefined

9) $\tan (17\pi)$

A) -1 B) 0 C) 1 D) undefined

10) $\cos \left(-\dfrac{\pi}{2}\right)$

A) -1 B) 0 C) 1 D) undefined

11) $\sin (-\pi)$

A) -1 B) 0 C) 1 D) undefined

3 Find the Exact Values of the Trigonometric Functions of pi/4 = 45 Degrees

Find the exact value. Do not use a calculator.

1) $\sec \dfrac{\pi}{4}$

A) $\dfrac{\sqrt{2}}{2}$ B) $\sqrt{2}$ C) $-\sqrt{2}$ D) $\dfrac{2\sqrt{3}}{3}$

2) $\cos 45°$

A) $\sqrt{2}$ B) $\dfrac{1}{2}$ C) $\dfrac{\sqrt{2}}{2}$ D) $\dfrac{\sqrt{3}}{2}$

Solve the problem.

3) If friction is ignored, the time t (in seconds) required for a block to slide down an inclined plane is given by the formula

$$t = \sqrt{\dfrac{2a}{g \sin\theta \cos\theta}}$$

where a is the length (in feet) of the base and g ≈ 32 feet per second per second is the acceleration of gravity. How long does it take a block to slide down an inclined plane with base a = 8 when θ = 45°? If necessary, round the answer to the nearest tenth of a second.

A) 1.2 sec B) 1.1 sec C) 1 sec D) 0.3 sec

4) The force acting on a pendulum to bring it to its perpendicular resting point is called the restoring force. The restoring force F, in Newtons, acting on a string pendulum is given by the formula

$F = mg \sin\theta$

where m is the mass in kilograms of the pendulum's bob, $g \approx 9.8$ meters per second per second is the acceleration due to gravity, and θ is angle at which the pendulum is displaced from the perpendicular. What is the value of the restoring force when m = 0.9 kilogram and $\theta = 45°$? If necessary, round the answer to the nearest tenth of a Newton.

A) 6 N   B) 5.9 N   C) 6.2 N   D) 7.5 N

4 Find the Exact Values of the Trigonometric Functions of pi/6 = 30 Degrees and pi/3 = 60 Degrees

Find the exact value. Do not use a calculator.

1) tan 30°

A) 1   B) $\frac{\sqrt{3}}{2}$ C) $\sqrt{3}$ D) $\frac{\sqrt{3}}{3}$

2) tan 60°

A) 2   B) $\frac{\sqrt{3}}{2}$ C) $\sqrt{3}$ D) $\frac{\sqrt{3}}{3}$

3) $\sec \frac{\pi}{6}$

A) $\frac{\sqrt{3}}{2}$ B) 2   C) $\sqrt{2}$ D) $\frac{2\sqrt{3}}{3}$

4) $\sec \frac{\pi}{3}$

A) $\frac{2\sqrt{3}}{3}$ B) 2   C) $\sqrt{2}$ D) $\frac{\sqrt{3}}{2}$

Find the exact value of the expression. Do not use a calculator.

5) cot 45° − cos 30°

A) $\frac{2-\sqrt{2}}{2}$ B) $-\frac{\sqrt{3}}{6}$ C) $\frac{2\sqrt{3}-3\sqrt{2}}{6}$ D) $\frac{2-\sqrt{3}}{2}$

6) sec 30° − cos 45°

A) $\frac{4\sqrt{3}-3\sqrt{2}}{6}$ B) $\frac{4-\sqrt{2}}{2}$ C) $\frac{4\sqrt{2}-3\sqrt{3}}{6}$ D) $\frac{4-\sqrt{3}}{2}$

7) cos 60° + tan 60°

A) $\frac{1+2\sqrt{3}}{2}$ B) $\frac{3\sqrt{3}}{2}$ C) $\frac{1+\sqrt{3}}{2}$ D) $2\sqrt{3}$

8) $\sin \frac{\pi}{3} - \cos \frac{\pi}{6}$

A) $\frac{\sqrt{3}-1}{2}$ B) 0   C) 1   D) $\sqrt{3}$

9) $\tan\frac{\pi}{6} - \cos\frac{\pi}{6}$

A) $\sqrt{3}$ B) $\frac{2\sqrt{3} - 3\sqrt{2}}{6}$ C) $-\frac{\sqrt{6}}{2}$ D) $-\frac{\sqrt{3}}{6}$

Solve the problem.

10) If friction is ignored, the time t (in seconds) required for a block to slide down an inclined plane is given by the formula

$$t = \sqrt{\frac{2a}{g \sin\theta \cos\theta}}$$

where a is the length (in feet) of the base and $g \approx 32$ feet per second per second is the acceleration of gravity. How long does it take a block to slide down an inclined plane with base a = 9 when $\theta = 30°$? If necessary, round the answer to the nearest tenth of a second.

A) 1.1 sec   B) 0.3 sec   C) 1.4 sec   D) 1.9 sec

11) The force acting on a pendulum to bring it to its perpendicular resting point is called the restoring force. The restoring force F, in Newtons, acting on a string pendulum is given by the formula

$$F = mg \sin\theta$$

where m is the mass in kilograms of the pendulum's bob, $g \approx 9.8$ meters per second per second is the acceleration due to gravity, and θ is angle at which the pendulum is displaced from the perpendicular. What is the value of the restoring force when m = 0.7 kilogram and $\theta = 30°$? If necessary, round the answer to the nearest tenth of a Newton.

A) 3.5 N   B) 3.4 N   C) 6.8 N   D) 5.9 N

5 Find the Exact Values for Integer Multiples of pi/6 = 30 Degrees, pi/4 = 45 Degrees, and pi/3 = 60 Degrees

Find the exact value. Do not use a calculator.

1) $\cos\frac{10\pi}{3}$

A) $\frac{1}{2}$ B) $\frac{\sqrt{3}}{2}$ C) $-\frac{\sqrt{3}}{2}$ D) $-\frac{1}{2}$

2) $\sec\frac{21\pi}{4}$

A) $-\frac{2\sqrt{3}}{3}$ B) $\frac{\sqrt{2}}{2}$ C) $-\sqrt{2}$ D) -2

3) $\sin 765°$

A) $-\frac{1}{2}$ B) $-\frac{\sqrt{2}}{2}$ C) $\frac{\sqrt{2}}{2}$ D) $\frac{1}{2}$

4) $\cot 930°$

A) $\sqrt{3}$ B) $\frac{\sqrt{3}}{3}$ C) $-\frac{\sqrt{3}}{3}$ D) $-\sqrt{3}$

Find the exact value of the expression. Do not use a calculator.

5) $\tan\frac{7\pi}{4} + \tan\frac{5\pi}{4}$

A) $\frac{1}{2}$ B) $\frac{\sqrt{2}+1}{2}$ C) $\frac{2\sqrt{2}+1}{6}$ D) 0

6) sin 135° − sin 270°

   A) $\dfrac{\sqrt{2}}{2}$ B) 2   C) $\dfrac{\sqrt{2}+2}{2}$ D) $\dfrac{\sqrt{2}-2}{2}$

7) $\cos\dfrac{\pi}{3} + \tan\dfrac{5\pi}{3}$

   A) $\dfrac{2\sqrt{3}+3}{6}$ B) $\dfrac{\sqrt{3}+1}{2}$ C) $\dfrac{\sqrt{3}+3}{3}$ D) $\dfrac{1-2\sqrt{3}}{2}$

8) cos 120° tan 60°

   A) $-\dfrac{\sqrt{3}}{2}$ B) $\dfrac{3}{2}$ C) $-\dfrac{1}{4}$ D) $\dfrac{\sqrt{3}}{2}$

9) tan 150° cos 210°

   A) $-\dfrac{5\sqrt{3}}{6}$ B) $\dfrac{\sqrt{3}+1}{2}$ C) $\dfrac{2\sqrt{3}+3}{6}$ D) $\dfrac{3\sqrt{3}+2\sqrt{3}}{6}$

10) sin 330° sin 270°

   A) $-\dfrac{1}{2}$ B) $\dfrac{\sqrt{3}}{2}$ C) $-\dfrac{\sqrt{3}}{2}$ D) $\dfrac{1}{2}$

6 Use a Calculator to Approximate the Value of a Trigonometric Function

Use a calculator to find the approximate value of the expression rounded to two decimal places.

1) sin 7°

   A) 0.66   B) 0.12   C) 0.55   D) 0.01

2) cos 71°

   A) −0.19   B) −0.31   C) 0.33   D) 0.45

3) tan 79°

   A) 5.14   B) 0.50   C) 0.42   D) 5.22

4) $\cos\dfrac{2\pi}{5}$

   A) 1.00   B) 1.12   C) 0.43   D) 0.31

5) $\sec\dfrac{\pi}{10}$

   A) 0.91   B) 1.00   C) 0.96   D) 1.05

6) csc 57°

   A) 2.29   B) 1.08   C) 2.18   D) 1.19

7) $\cot\dfrac{\pi}{8}$

   A) 2.41   B) 145.90   C) 145.76   D) 2.55

Precalculus Enhanced with Graphing Utilities 432

8) cot 0.2137

A) 1.02   B) 0.98   C) 0.22   D) 4.61

9) cos 6

A) -0.99   B) -0.96   C) 0.99   D) 0.96

10) cos 8°

A) -0.99   B) 0.15   C) -0.15   D) 0.99

11) tan 37°

Solve the problem.

12) If friction is ignored, the time t (in seconds) required for a block to slide down an inclined plane is given by the formula

$$t = \sqrt{\frac{2a}{g \sin\theta \cos\theta}}$$

where a is the length (in feet) of the base and $g \approx 32$ feet per second per second is the acceleration of gravity. How long does it take a block to slide down an inclined plane with base a = 10 when $\theta = 36°$? If necessary, round the answer to the nearest tenth of a second.

A) 1.1 sec   B) 0.3 sec   C) 1.4 sec   D) 2.2 sec

13) The force acting on a pendulum to bring it to its perpendicular resting point is called the restoring force. The restoring force F, in Newtons, acting on a string pendulum is given by the formula

$F = mg \sin\theta$

where m is the mass in kilograms of the pendulum's bob, $g \approx 9.8$ meters per second per second is the acceleration due to gravity, and θ is angle at which the pendulum is displaced from the perpendicular. What is the value of the restoring force when m = 0.7 kilogram and $\theta = 31°$? If necessary, round the answer to the nearest tenth of a Newton.

A) 5.9 N   B) 2.8 N   C) 3.5 N   D) 3.4 N

14) The strength S of a wooden beam with rectangular cross section is given by the formula

$S = kd^3 \sin^2\theta \cos\theta$

where d is the diagonal length, θ the angle illustrated, and k is a constant that varies with the type of wood used.

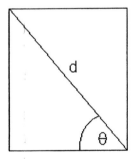

Let d = 1 and express the strength S in terms of the constant k for $\theta = 45°, 50°, 55°, 60°$, and $65°$. Does the strength always increase as θ gets larger?

7 Use a Circle of Radius r to Evaluate the Trigonometric Functions

A point on the terminal side of an angle θ is given. Find the exact value of the indicated trigonometric function of θ.

1) (−3, −4)   Find sin θ.
   A) $-\dfrac{3}{5}$ B) $\dfrac{3}{5}$ C) $\dfrac{4}{5}$ D) $-\dfrac{4}{5}$

2) (−4, −3)   Find cos θ.
   A) $\dfrac{3}{5}$ B) $-\dfrac{3}{5}$ C) $\dfrac{4}{5}$ D) $-\dfrac{4}{5}$

3) $\left(-\dfrac{1}{5}, \dfrac{1}{2}\right)$ Find cos θ.
   A) $-\dfrac{29}{5}$ B) $\dfrac{5\sqrt{29}}{29}$ C) $\dfrac{29}{2}$ D) $-\dfrac{2\sqrt{29}}{29}$

4) (3, −2)   Find tan θ.
   A) $-\dfrac{3}{2}$ B) $-\dfrac{\sqrt{13}}{3}$ C) $-\dfrac{2}{3}$ D) $\dfrac{\sqrt{13}}{3}$

5) (5, 4)   Find cot θ.
   A) $-\dfrac{\sqrt{41}}{5}$ B) $\dfrac{5}{4}$ C) $\dfrac{\sqrt{41}}{5}$ D) $\dfrac{4}{5}$

6) (−2, −1)   Find csc θ.
   A) −2   B) $\sqrt{5}$ C) $-\sqrt{5}$ D) −5

7) (−3, −1)   Find sec θ.
   A) $\dfrac{\sqrt{10}}{3}$ B) $-\sqrt{10}$ C) $-\dfrac{\sqrt{10}}{3}$ D) $-\dfrac{3\sqrt{10}}{10}$

8 Demonstrate Additional Understanding and Skills

Solve the problem.

1) If sin θ = 0.4, find sin (θ + π).
   A) 0.6   B) −0.4   C) 0.4   D) −0.6

2) If sin θ = $\dfrac{1}{3}$, find csc θ.
   A) $\dfrac{2}{3}$ B) 3   C) $-\dfrac{1}{3}$ D) undefined

Find the exact value of the expression if θ = 45°. Do not use a calculator.

3) f(θ) = cot θ   Find f(θ).
   A) −1   B) 0   C) $\sqrt{3}$ D) 1

4) g(θ) = cos θ   Find [g(θ)]².
   A) $-\dfrac{\sqrt{2}}{2}$ B) 2   C) $\sqrt{2}$ D) $\dfrac{1}{2}$

5) $f(\theta) = \cos\theta$ Find $3f(\theta)$.

A) $-\dfrac{\sqrt{2}}{2}$ B) $\dfrac{3\sqrt{2}}{2}$ C) $-\dfrac{3\sqrt{2}}{2}$ D) $\dfrac{\sqrt{2}}{2}$

6) $g(\theta) = \sin\theta$ Find $6g(\theta)$.

A) $6\sqrt{2}$ B) $-3\sqrt{2}$ C) $3\sqrt{2}$ D) $-6\sqrt{2}$

Find the exact value of the expression if $\theta = 30°$. Do not use a calculator.

7) $f(\theta) = \csc\theta$ Find $f(\theta)$.

A) $\sqrt{2}$ B) $\dfrac{1}{2}$ C) 2 D) $\dfrac{2\sqrt{3}}{3}$

8) $g(\theta) = \cos\theta$ Find $g(2\theta)$.

A) 1 B) $\dfrac{\sqrt{3}}{2}$ C) $\sqrt{3}$ D) $\dfrac{1}{2}$

9) $f(\theta) = \sin\theta$ Find $[f(\theta)]^2$.

A) 1 B) $\dfrac{1}{4}$ C) $\dfrac{3}{4}$ D) $\dfrac{1}{2}$

10) $g(\theta) = \sin\theta$ Find $6g(\theta)$.

A) $-\dfrac{1}{2}$ B) $3\sqrt{3}$ C) $-\dfrac{\sqrt{3}}{2}$ D) 3

11) $f(\theta) = \cos\theta$ Find $11f(\theta)$.

A) $\dfrac{11}{2}$ B) $-\dfrac{\sqrt{3}}{2}$ C) $\dfrac{11\sqrt{3}}{2}$ D) $-\dfrac{1}{2}$

Find the exact value of the expression if $\theta = 60°$. Do not use a calculator.

12) $f(\theta) = \tan\theta$ Find $f(\theta)$.

A) $\dfrac{\sqrt{3}}{2}$ B) $\dfrac{\sqrt{3}}{3}$ C) $\sqrt{3}$ D) 2

13) $g(\theta) = \cos\theta$ Find $[g(\theta)]^2$.

A) $\sqrt{3}$ B) $\dfrac{3}{4}$ C) $\dfrac{\sqrt{3}}{2}$ D) $\dfrac{1}{4}$

14) $f(\theta) = \sin\theta$ Find $6f(\theta)$.

A) 3 B) $-\dfrac{\sqrt{3}}{2}$ C) $3\sqrt{3}$ D) $-\dfrac{1}{2}$

15) $g(\theta) = \cos\theta$ Find $7g(\theta)$.

A) $\dfrac{7\sqrt{3}}{2}$ B) $\dfrac{7}{2}$ C) $-\dfrac{1}{2}$ D) $-\dfrac{\sqrt{3}}{2}$

Solve the problem.

16) The displacement d, in inches, from equilibrium of a weight suspended from a spring is given by
$$d = 1 + 2 \sin(15t)°$$
where t in time in seconds. Find the displacement when t = 0, 2, 4, 6, 8, 10, and 12 seconds. Do not use a calculator.

17) The area A of a regular polygon with n sides each of length s is given by the formula
$$A = \frac{ns^2}{4} \cdot \cot \frac{180°}{n}$$
Let s = 1 unit and calculate the area of regular polygons with 3, 4, and 6 sides. Do not use a calculator.

5.3 Properties of the Trigonometric Functions

1 Determine the Domain and the Range of the Trigonometric Functions

Solve the problem.

1) What is the domain of the sine function?

   A) all real numbers, except odd multiples of $\frac{\pi}{2}$ (90°)

   B) all real numbers
   C) all real numbers from −1 to 1, inclusive
   D) all real numbers, except integral multiples of π (180°)

2) For what numbers θ is f(θ) = sec θ not defined?

   A) integral multiples of π (180°)   B) odd multiples of π (180°)

   C) odd multiples of $\frac{\pi}{2}$ (90°)   D) all real numbers

3) For what numbers θ is f(θ) = csc θ not defined?

   A) odd multiples of $\frac{\pi}{2}$ (90°)   B) all real numbers

   C) odd multiples of π (180°)   D) integral multiples of π (180°)

4) What is the range of the cosine function?

   A) all real numbers
   B) all real numbers from −1 to 1, inclusive
   C) all real numbers greater than or equal to 1 or less than or equal to −1
   D) all real numbers greater than or equal to 0

5) What is the range of the tangent function?

   A) all real numbers, except odd multiples of $\frac{\pi}{2}$(90°)

   B) all real numbers from −1 to 1, inclusive
   C) all real numbers
   D) all real numbers greater than or equal to 1 or less than or equal to −1

6) What is the range of the cosecant function?
   A) all real numbers from −1 to 1, inclusive
   B) all real numbers greater than or equal to 1 or less than or equal to −1
   C) all real numbers
   D) all real numbers, except integral multiples of π(180)°

2 Determine the Period of the Trigonometric Functions

Use the fact that the trigonometric functions are periodic to find the exact value of the expression. Do not use a calculator.

1) $\sin 765°$
   A) $\dfrac{\sqrt{2}}{2}$
   B) $-\dfrac{\sqrt{2}}{2}$
   C) $-\dfrac{1}{2}$
   D) $\dfrac{1}{2}$

2) $\tan 390°$
   A) $\sqrt{3}$
   B) $-\sqrt{3}$
   C) $\dfrac{\sqrt{3}}{2}$
   D) $\dfrac{\sqrt{3}}{3}$

3) $\csc 960°$
   A) $-\dfrac{1}{2}$
   B) $-\sqrt{3}$
   C) $-\sqrt{2}$
   D) $-\dfrac{2\sqrt{3}}{3}$

4) $\cot 390°$
   A) $-\dfrac{\sqrt{3}}{3}$
   B) $-\sqrt{3}$
   C) $\dfrac{\sqrt{3}}{3}$
   D) $\sqrt{3}$

5) $\cot 1080°$
   A) $\sqrt{3}$
   B) 1
   C) 0
   D) undefined

6) $\tan 720°$
   A) 0
   B) 1
   C) $\dfrac{\sqrt{3}}{3}$
   D) undefined

7) $\cos \dfrac{14\pi}{3}$
   A) $\dfrac{1}{2}$
   B) $\dfrac{\sqrt{3}}{2}$
   C) $-\dfrac{1}{2}$
   D) $-\dfrac{\sqrt{3}}{2}$

8) $\sin \dfrac{16\pi}{3}$
   A) $\dfrac{\sqrt{3}}{2}$
   B) $-\dfrac{1}{2}$
   C) $-\dfrac{\sqrt{3}}{2}$
   D) −1

9) $\tan \dfrac{17\pi}{4}$
   A) −1
   B) 1
   C) $\dfrac{\sqrt{3}}{3}$
   D) $\sqrt{3}$

10) $\sec \dfrac{21\pi}{4}$

A) -2 B) $-\sqrt{2}$ C) $-\dfrac{2\sqrt{3}}{3}$ D) $\dfrac{\sqrt{2}}{2}$

Solve the problem.

11) If $\sin \theta = -0.6$, find the value of $\sin \theta + \sin(\theta + 2\pi) + \sin(\theta + 4\pi)$.

A) -0.6 B) -1.8 C) 0.2 D) $-1.8 + 6\pi$

12) If $\cot \theta = -9.5$, find the value of $\cot \theta + \cot(\theta + \pi) + \cot(\theta + 2\pi)$.

A) -26.5 B) -28.5 C) $-28.5 + 3\pi$ D) undefined

13) If $f(\theta) = \cos \theta$ and $f(a) = \dfrac{1}{6}$, find the exact value of $f(a) + f(a + 2\pi) + f(a + 4\pi)$.

A) $\dfrac{1}{6}$ B) $\dfrac{1}{2} + 6\pi$ C) $\dfrac{5}{2}$ D) $\dfrac{1}{2}$

14) If $f(\theta) = \cot \theta$ and $f(a) = 6$, find the exact value of $f(a) + f(a + \pi) + f(a + 3\pi)$.

A) 6 B) 18 C) $18 + 4\pi$ D) undefined

15) If $f(\theta) = \cos \theta$ and $f(a) = -\dfrac{1}{12}$, find the exact value of $f(a) + f(a - 2\pi) + f(a + 4\pi)$.

A) $-\dfrac{1}{4}$ B) $-\dfrac{1}{12}$ C) -36 D) -12

16) If $f(\theta) = \sin \theta$ and $f(a) = -\dfrac{1}{9}$, find the exact value of $f(a) + f(a - 4\pi) + f(a - 2\pi)$.

17) If $\sin \theta = 0.1$, find the value of $\sin \theta + \sin(\theta + 2\pi) + \sin(\theta + 4\pi)$.

A) $0.3 + 6\pi$ B) 2.3 C) 0.3 D) 0.1

18) If $\tan \theta = -2.1$, find the value of $\tan \theta + \tan(\theta + \pi) + \tan(\theta + 2\pi)$.

A) -4.3 B) $-6.3 + 3\pi$ C) -6.3 D) undefined

3 Determine the Signs of the Trigonometric Functions in a Given Quadrant

Name the quadrant in which the angle θ lies.

1) $\tan \theta > 0$, $\sin \theta < 0$

A) I B) II C) III D) IV

2) $\cos \theta < 0$, $\csc \theta < 0$

A) I B) II C) III D) IV

3) $\sin \theta > 0$, $\cos \theta < 0$

A) I B) II C) III D) IV

4) $\cot \theta < 0$, $\cos \theta > 0$

A) I B) II C) III D) IV

Precalculus Enhanced with Graphing Utilities 438

5) $\csc\theta > 0$, $\sec\theta > 0$
   A) I   B) II   C) III   D) IV

6) $\sec\theta < 0$, $\tan\theta < 0$
   A) I   B) II   C) III   D) IV

7) $\tan\theta < 0$, $\sin\theta < 0$
   A) I   B) II   C) III   D) IV

8) $\cos\theta > 0$, $\csc\theta < 0$
   A) I   B) II   C) III   D) IV

9) $\cot\theta > 0$, $\sin\theta < 0$
   A) I   B) II   C) III   D) IV

10) $\sin\theta > 0$, $\cos\theta > 0$
    A) I   B) II   C) III   D) IV

Solve the problem.

11) Which of the following trigonometric values are negative?
    I. $\sin(-292°)$
    II. $\tan(-193°)$
    III. $\cos(-207°)$
    IV. $\cot 222°$
    A) I and III   B) III only   C) II, III, and IV   D) II and III

12) Determine the sign of the trigonometric values listed below.
    (i) $\sin 250°$
    (ii) $\tan 330°$
    (iii) $\cos(-40°)$

4 Find the Values of the Trigonometric Functions Utilizing Fundamental Identities

In the problem, sin θ and cos θ are given. Find the exact value of the indicated trigonometric function.

1) $\sin\theta = \frac{2\sqrt{2}}{3}$, $\cos\theta = \frac{1}{3}$ Find $\tan\theta$.
   A) $\frac{\sqrt{2}}{4}$ B) 3   C) $\frac{3\sqrt{2}}{4}$ D) $2\sqrt{2}$

2) $\sin\theta = \frac{2\sqrt{2}}{3}$, $\cos\theta = \frac{1}{3}$ Find $\cot\theta$.
   A) 3   B) $\frac{\sqrt{2}}{4}$ C) $\frac{3\sqrt{2}}{4}$ D) $2\sqrt{2}$

3) $\sin\theta = \frac{\sqrt{5}}{3}$, $\cos\theta = \frac{2}{3}$ Find $\sec\theta$.
   A) $\frac{3\sqrt{5}}{5}$ B) $\frac{2\sqrt{5}}{5}$ C) $\frac{\sqrt{5}}{2}$ D) $\frac{3}{2}$

4) $\sin\theta = \dfrac{2\sqrt{2}}{3}$, $\cos\theta = \dfrac{1}{3}$ Find $\csc\theta$.

  A) $\dfrac{\sqrt{2}}{4}$ B) $\dfrac{3\sqrt{2}}{4}$ C) $2\sqrt{2}$ D) 3

Use the properties of the trigonometric functions to find the exact value of the expression. Do not use a calculator.

5) $\sin^2 70° + \cos^2 70°$

  A) 2   B) 0   C) 1   D) −1

6) $\sec^2 25° - \tan^2 25°$

  A) 1   B) 0   C) 2   D) −1

7) $\cos 20° \sec 20°$

  A) 20   B) −1   C) 1   D) 0

8) $\tan 40° - \dfrac{\sin 40°}{\cos 40°}$

  A) 1   B) 40   C) 0   D) undefined

5 Find the Exact Values of the Trigonometric Functions of an Angle Given One of the Functions and the Quadrant of the Angle

Find the exact value of the indicated trigonometric function of θ.

1) $\tan\theta = -\dfrac{8}{9}$, θ in quadrant II   Find $\cos\theta$.

  A) $\dfrac{9\sqrt{145}}{145}$ B) $-\dfrac{9\sqrt{145}}{145}$ C) $\dfrac{\sqrt{145}}{8}$ D) $-\dfrac{\sqrt{145}}{9}$

2) $\csc\theta = -\dfrac{3}{2}$, θ in quadrant III   Find $\cot\theta$.

  A) $-\dfrac{\sqrt{5}}{3}$ B) $-\dfrac{3\sqrt{5}}{5}$ C) $\dfrac{\sqrt{5}}{2}$ D) $-\dfrac{2\sqrt{5}}{5}$

3) $\sec\theta = \dfrac{7}{4}$, θ in quadrant IV   Find $\tan\theta$.

  A) $-\dfrac{\sqrt{33}}{7}$ B) $-\sqrt{33}$ C) $-\dfrac{\sqrt{33}}{4}$ D) $-\dfrac{7}{4}$

4) $\tan\theta = \dfrac{15}{8}$, $180° < \theta < 270°$ Find $\cos\theta$.

  A) $-\dfrac{8}{17}$ B) $\dfrac{15\sqrt{23}}{23}$ C) $\dfrac{-8\sqrt{23}}{23}$ D) −8

5) $\cos\theta = \dfrac{7}{25}$, $\dfrac{3\pi}{2} < \theta < 2\pi$ Find $\cot\theta$.

  A) $-\dfrac{7}{24}$ B) $\dfrac{25}{7}$ C) $-\dfrac{24}{7}$ D) $\dfrac{-7\sqrt{2}}{6}$

6) $\cos\theta = \dfrac{8}{9}$, $\tan\theta < 0$ Find $\sin\theta$.

   A) $-\dfrac{\sqrt{17}}{8}$ B) $-\dfrac{\sqrt{17}}{9}$ C) $-\sqrt{17}$ D) $-\dfrac{9}{8}$

7) $\sin\theta = -\dfrac{2}{3}$, $\tan\theta > 0$ Find $\sec\theta$.

   A) $-\dfrac{\sqrt{5}}{3}$ B) $-\dfrac{3\sqrt{5}}{5}$ C) $-\dfrac{2\sqrt{5}}{5}$ D) $\dfrac{\sqrt{3}}{2}$

8) $\cot\theta = -\dfrac{5}{2}$, $\cos\theta < 0$ Find $\csc\theta$.

   A) $\dfrac{\sqrt{29}}{2}$ B) $\dfrac{5\sqrt{29}}{29}$ C) $-\dfrac{\sqrt{29}}{5}$ D) $-\dfrac{5\sqrt{29}}{29}$

9) $\sin\theta = \dfrac{1}{2}$, $\sec\theta < 0$ Find $\cos\theta$ and $\tan\theta$.

   A) $\cos\theta = -\dfrac{\sqrt{3}}{2}$, $\tan\theta = -\dfrac{\sqrt{3}}{3}$ B) $\cos\theta = \sqrt{\dfrac{3}{2}}$, $\tan\theta = \dfrac{\sqrt{3}}{3}$

   C) $\cos\theta = -\dfrac{\sqrt{3}}{2}$, $\tan\theta = \dfrac{\sqrt{3}}{3}$ D) $\cos\theta = -\sqrt{3}$, $\tan\theta = -\dfrac{10\sqrt{3}}{3}$

10) $\sin\theta = \dfrac{1}{6}$, $\sec\theta < 0$ Find $\cos\theta$ and $\tan\theta$.

6 Use Even-Odd Properties to Find the Exact Values of the Trigonometric Functions

Use the even-odd properties to find the exact value of the expression. Do not use a calculator.

1) $\sin(-30°)$

   A) $\dfrac{\sqrt{3}}{2}$ B) $-\dfrac{\sqrt{3}}{2}$ C) $-\dfrac{1}{2}$ D) $\dfrac{1}{2}$

2) $\sin(-60°)$

   A) $-\dfrac{\sqrt{3}}{2}$ B) $\dfrac{\sqrt{3}}{2}$ C) $-\dfrac{1}{2}$ D) $\dfrac{1}{2}$

3) $\sec(-60°)$

   A) $-\dfrac{2\sqrt{3}}{3}$ B) $\dfrac{2\sqrt{3}}{3}$ C) 2 D) -2

4) $\cot(-60°)$

   A) $\dfrac{\sqrt{3}}{3}$ B) $\sqrt{3}$ C) $-\sqrt{3}$ D) $-\dfrac{\sqrt{3}}{3}$

5) $\cos(-150°)$

   A) $-\dfrac{\sqrt{3}}{2}$ B) $\dfrac{1}{2}$ C) $\dfrac{-1}{2}$ D) $\dfrac{\sqrt{3}}{2}$

6) $\cos\left(-\dfrac{\pi}{4}\right)$

A) $\dfrac{\sqrt{2}}{2}$ B) $\dfrac{\sqrt{3}}{2}$ C) $-\dfrac{\sqrt{3}}{2}$ D) $-\dfrac{\sqrt{2}}{2}$

7) $\sec\left(-\dfrac{\pi}{6}\right)$

A) $-\dfrac{2\sqrt{3}}{3}$ B) 2 C) -2 D) $\dfrac{2\sqrt{3}}{3}$

8) $\cot\left(-\dfrac{\pi}{6}\right)$

A) $-\dfrac{\sqrt{3}}{3}$ B) $\dfrac{\sqrt{3}}{3}$ C) $\sqrt{3}$ D) $-\sqrt{3}$

9) $\cos\left(-\dfrac{\pi}{2}\right)$

A) -1 B) 1 C) 0 D) undefined

10) $\sin(-\pi)$

A) 1 B) 0 C) -1 D) undefined

11) $\cot\left(-\dfrac{\pi}{4}\right)$

A) 1 B) $-\sqrt{3}$ C) -1 D) $-\dfrac{\sqrt{3}}{3}$

Solve the problem.

12) If $f(\theta) = \sin\theta$ and $f(a) = \dfrac{1}{6}$, find the exact value of $f(-a)$.

A) $-\dfrac{1}{6}$ B) $-\dfrac{5}{6}$ C) $\dfrac{1}{6}$ D) $\dfrac{5}{6}$

13) If $f(\theta) = \cot\theta$ and $f(a) = -6$, find the exact value of $f(-a)$.

A) $\dfrac{1}{6}$ B) 6 C) $-\dfrac{1}{6}$ D) -6

14) Is the function $f(\theta) = \sin\theta + \cos\theta$ even, odd, or neither?

15) Is the function $f(\theta) = \sin\theta + \tan\theta$ even, odd, or neither?

5.4 Graphs of the Sine and Cosine Functions

1 Graph Transformations of the Sine Function

Use transformations to graph the function.

1) $y = 3 \sin x$

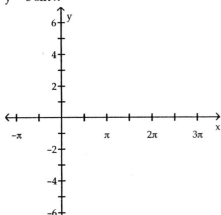

A)
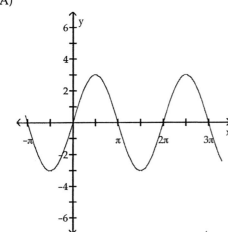

B)
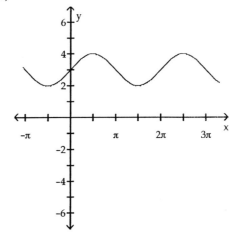

C)
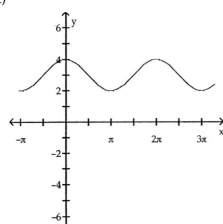

D)
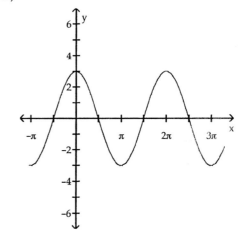

Precalculus Enhanced with Graphing Utilities 443

© 2006 Pearson Education, Inc., Upper Saddle River, NJ. All rights reserved. This material is protected under all copyright laws as they currently exist.
No portion of this material may be reproduced, in any form or by any means, without permission in writing from the publisher.

2) $y = \sin(x + \pi)$

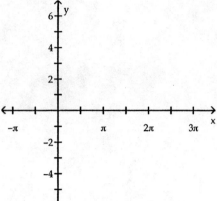

A)
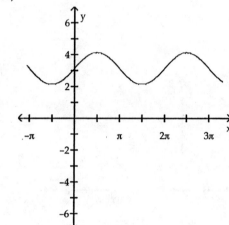

B)
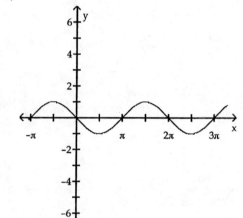

C)
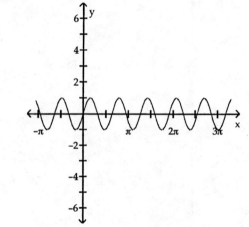

D)
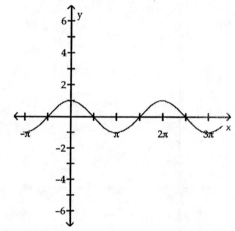

3) $y = \sin x - 5$

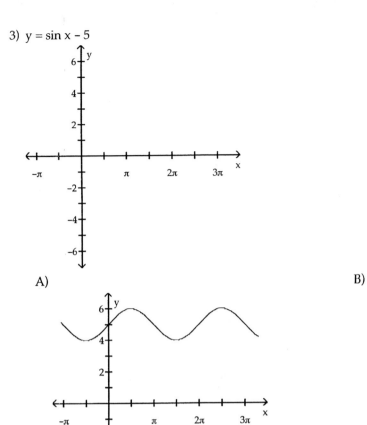

A)

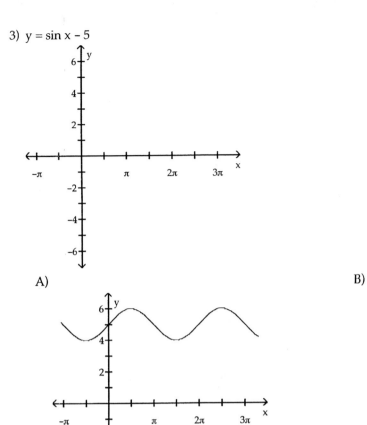

B)

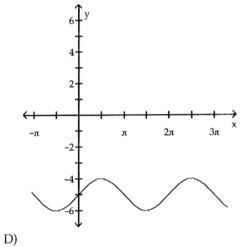

C)

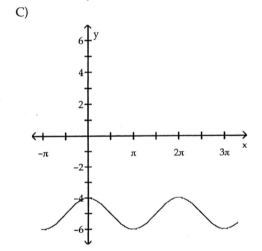

D)

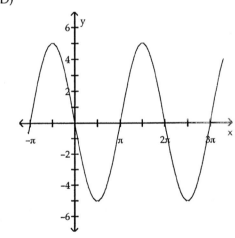

4) $y = -5\sin x$

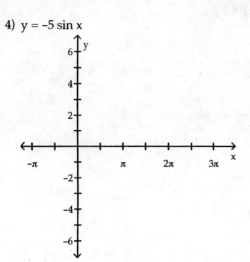

A)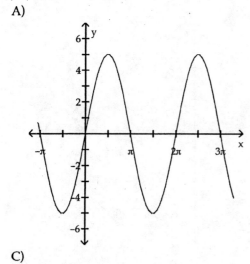

B)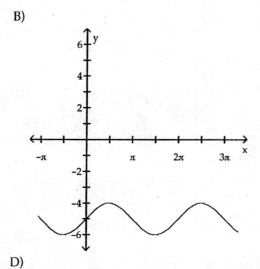

C)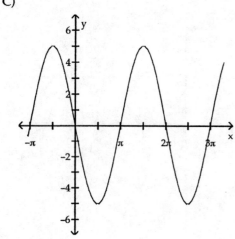

D)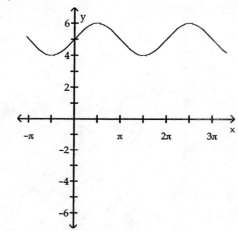

5) $y = \sin(\pi x)$

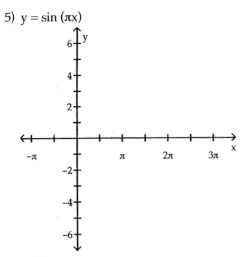

A)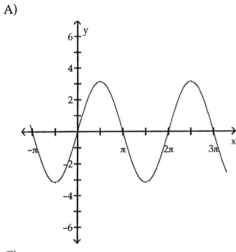

B)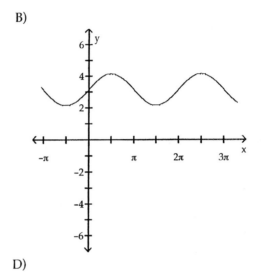

C)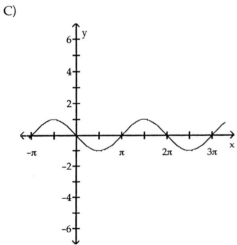

D)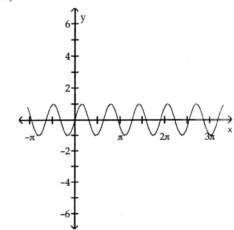

6) $y = 3 \sin x - 2$

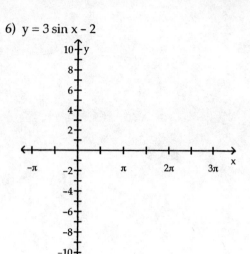

A)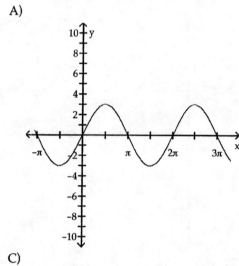

B)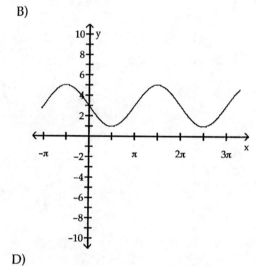

C)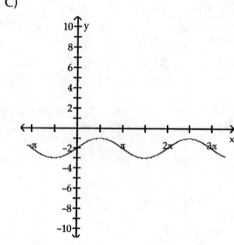

D)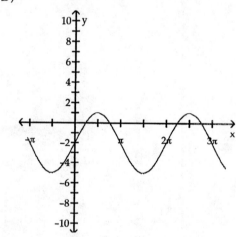

7) $y = -4 \sin\left(x + \dfrac{\pi}{4}\right)$

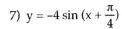

A)
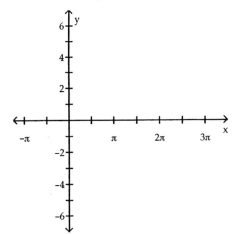

B)
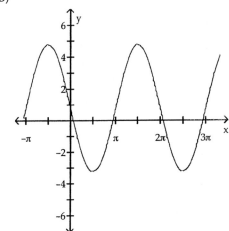

C)
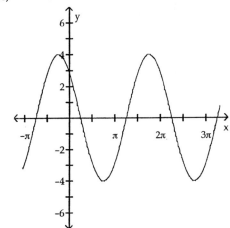

D)
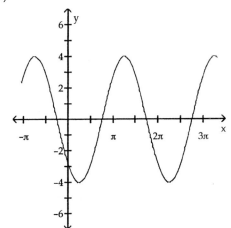

8) $y = 4\sin(\pi - x)$

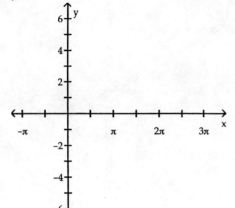

A)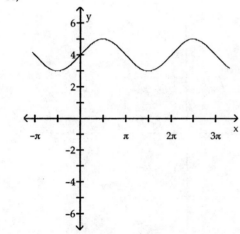

B)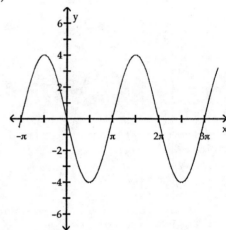

C)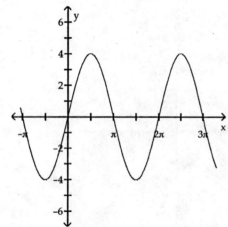

D)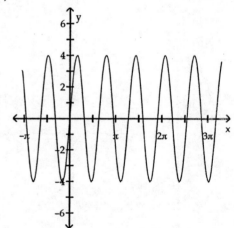

2 Graph Transformations of the Cosine Function

Use transformations to graph the function.

1) $y = 4 \cos x$

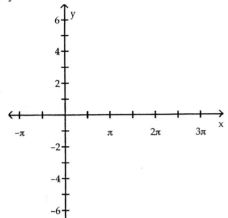

A)
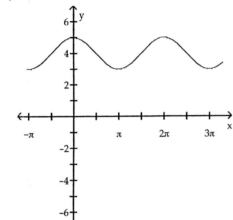

B)
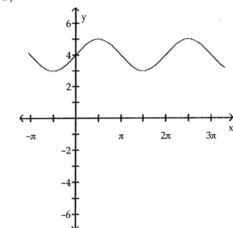

C)
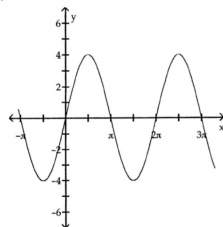

D)
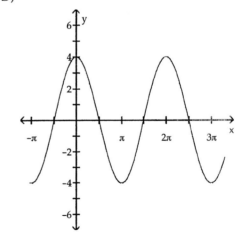

2) $y = \cos\left(x - \dfrac{\pi}{3}\right)$

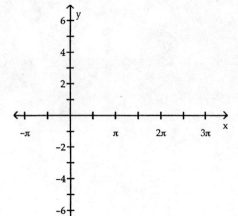

A)
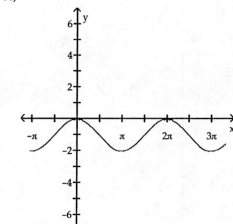

B)
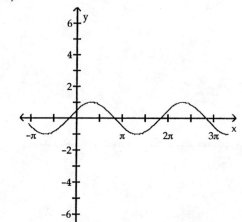

C)
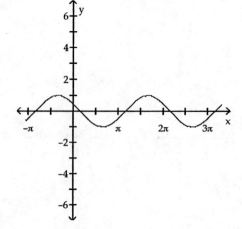

D)
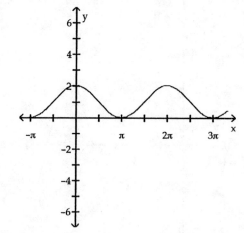

3) y = cos x + 4

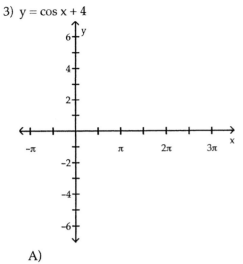

A)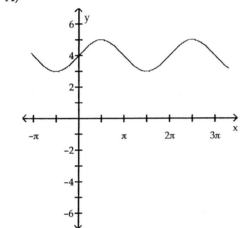

B)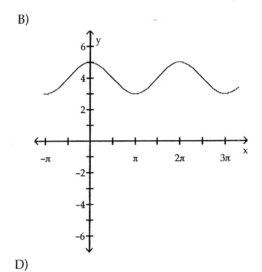

C)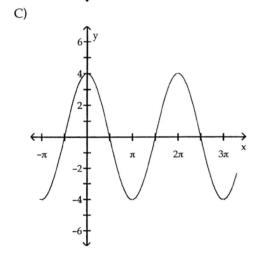

D)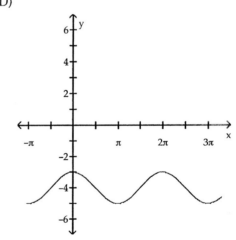

4) $y = -2\cos x$

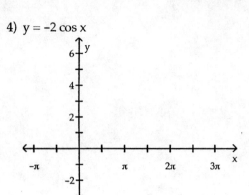

A)
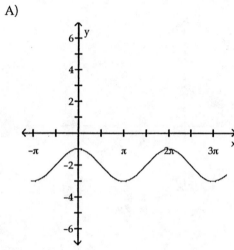

B)
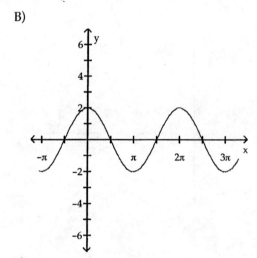

C)
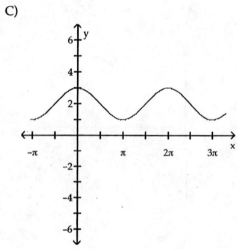

D)
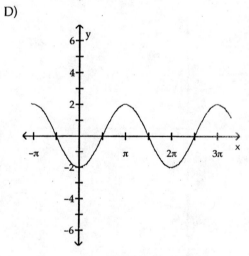

5) $y = \cos\left(\dfrac{\pi}{4}x\right)$

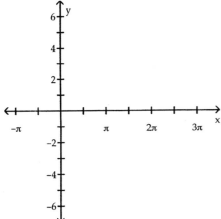

A)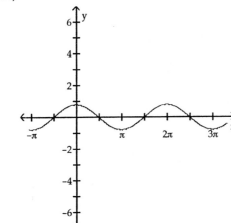

B)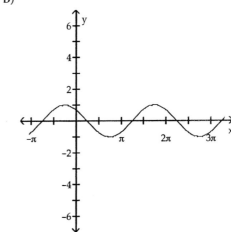

C)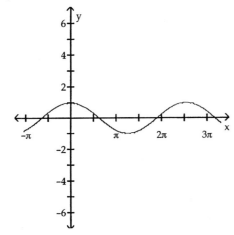

D)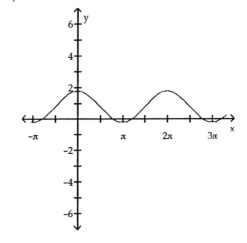

6) $y = 5\cos x + 4$

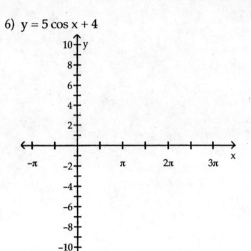

A)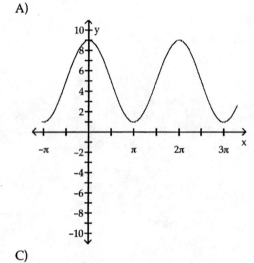

B)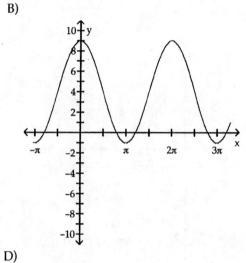

C)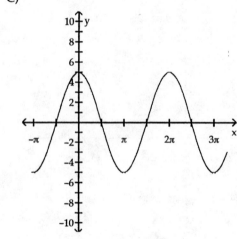

D)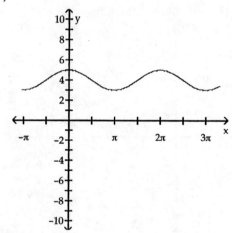

7) $y = -2 \cos\left(x - \dfrac{\pi}{4}\right)$

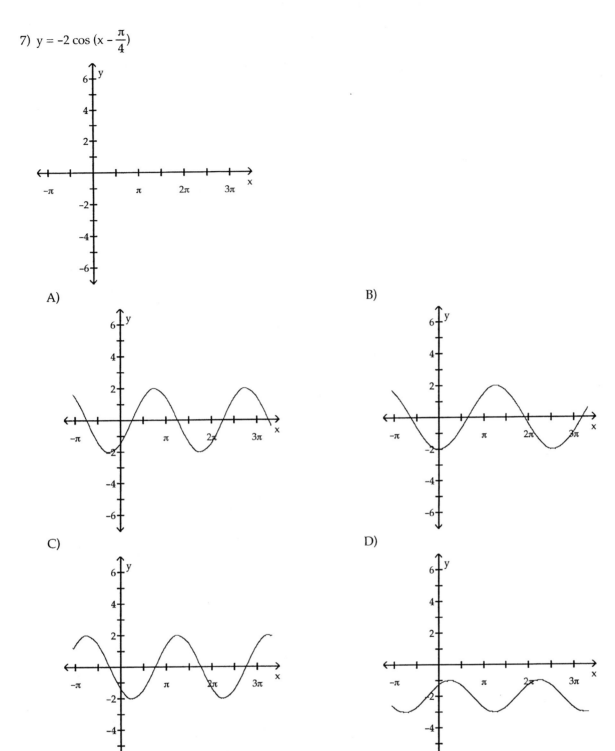

8) $y = 3\cos(\pi - x)$

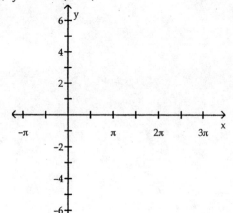

A)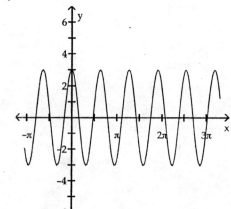

B)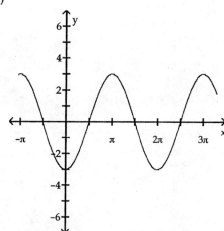

C)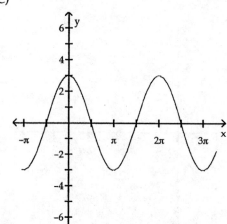

D)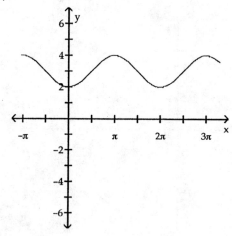

3 Determine the Amplitude and Period of Sinusoidal Functions

Without graphing the function, determine its amplitude or period as requested.

1) $y = -3\sin x$ Find the amplitude.

A) 3   B) -3π C) 2π D) $\dfrac{\pi}{3}$

2) $y = 5 \sin \frac{1}{4}x$ Find the amplitude.

   A) $\frac{\pi}{5}$ B) $\frac{5\pi}{4}$ C) 5   D) 8π

3) $y = -5 \sin 4x$ Find the amplitude.

   A) $\frac{\pi}{5}$ B) $\frac{5}{4}$ C) $\frac{\pi}{4}$ D) 5

4) $y = \sin 5x$ Find the period.

   A) 5   B) 2π C) 1   D) $\frac{2\pi}{5}$

5) $y = -2 \cos \frac{1}{3}x$ Find the amplitude.

   A) $\frac{2\pi}{3}$ B) $\frac{\pi}{2}$ C) 6π D) 2

6) $y = \cos 5x$ Find the period.

   A) 2π B) $\frac{2\pi}{5}$ C) 5   D) 1

7) $y = -5 \cos \frac{1}{4}x$ Find the period.

   A) $\frac{5\pi}{4}$ B) -5 C) $\frac{\pi}{4}$ D) 8π

8) $y = 3 \cos x$ Find the period.

   A) 2π B) π C) 3   D) $\frac{\pi}{3}$

9) $y = \frac{7}{6} \cos (-\frac{8\pi}{3}x)$ Find the period.

   A) $\frac{16\pi}{3}$ B) $\frac{7\pi}{3}$ C) $\frac{3}{4}$ D) $\frac{3}{7}$

10) $y = \frac{7}{4} \sin (-\frac{4\pi}{3}x)$ Find the amplitude.

    A) $\frac{3}{2}$ B) $\frac{4\pi}{3}$ C) $\frac{4\pi}{7}$ D) $\frac{7}{4}$

4 Graph Sinusoidal Functions Using Key Points

Match the given function to its graph.

1) 1) $y = \sin x$ 2) $y = \cos x$
   3) $y = -\sin x$ 4) $y = -\cos x$

A
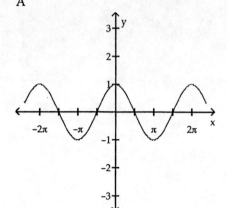

B
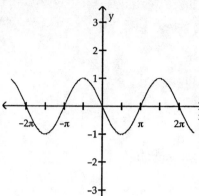

C
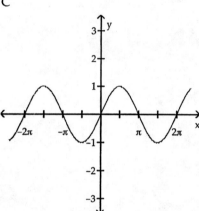

D
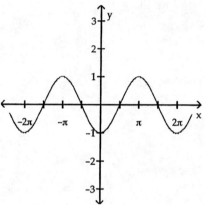

A) 1B, 2D, 3C, 4A   B) 1C, 2A, 3B, 4D   C) 1A, 2D, 3C, 4B   D) 1A, 2B, 3C, 4D

2) 1) $y = \sin 2x$ 2) $y = 2\cos x$
   3) $y = 2\sin x$ 4) $y = \cos 2x$

A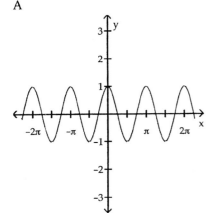

B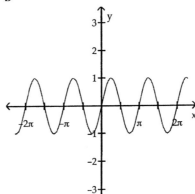

C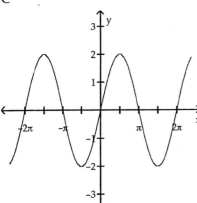

D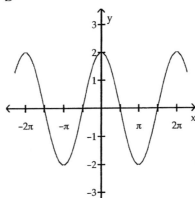

A) 1A, 2C, 3D, 4B   B) 1A, 2D, 3C, 4B   C) 1B, 2D, 3C, 4A   D) 1A, 2B, 3C, 4D

3) 1) $y = \sin(x - \frac{\pi}{3})$ 2) $y = \cos(x + \frac{\pi}{3})$

3) $y = \sin(x + \frac{\pi}{3})$ 4) $y = \cos(x - \frac{\pi}{3})$

A
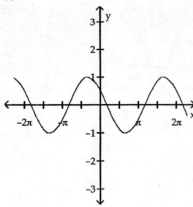

B
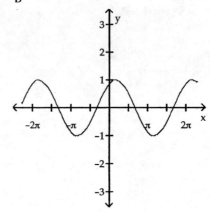

C
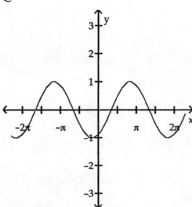

D
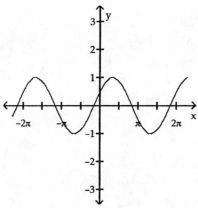

A) 1A, 2D, 3C, 4B   B) 1A, 2B, 3C, 4D   C) 1C, 2A, 3B, 4D   D) 1B, 2D, 3C, 4A

4) 1) y = 1 + sin x   2) y = 1 + cos x
   3) y = −1 + sin x   4) y = −1 + cos x

A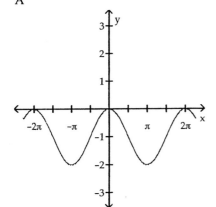

B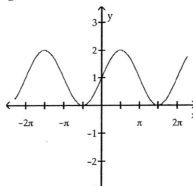

C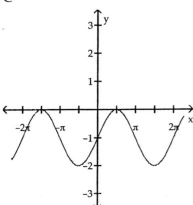

D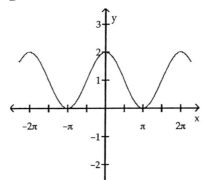

A) 1B, 2D, 3C, 4A   B) 1A, 2C, 3D, 4B   C) 1A, 2B, 3C, 4D   D) 1A, 2D, 3C, 4B

5) 1) $y = \sin(\frac{1}{3}x)$ 2) $y = \frac{1}{3}\cos x$

3) $y = \frac{1}{3}\sin x$ 4) $y = \cos(\frac{1}{3}x)$

A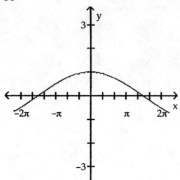

B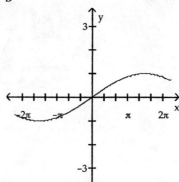

C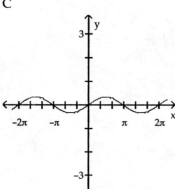

D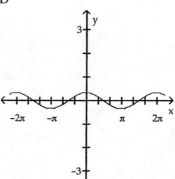

A) 1A, 2C, 3D, 4B   B) 1A, 2D, 3C, 4B   C) 1A, 2B, 3C, 4D   D) 1B, 2D, 3C, 4A

6) 1) $y = 3 \sin(2x)$ 2) $y = 3 \sin(\frac{1}{2}x)$

3) $y = -3 \cos(2x)$ 4) $y = -3 \cos(\frac{1}{2}x)$

A)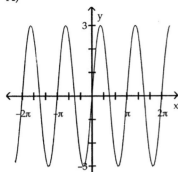

B)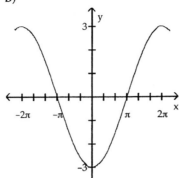

C)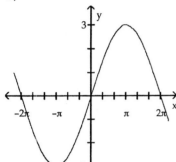

D)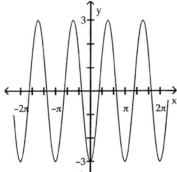

A) 1C, 2A, 3B, 4D   B) 1D, 2B, 3A, 4C   C) 1A, 2C, 3D, 4B   D) 1C, 2A, 3D, 4B

7) 1) $y = 2\sin\left(\dfrac{\pi}{3}x\right)$ 2) $y = 2\sin\left(\dfrac{1}{3}x\right)$

3) $y = 2\cos\left(\dfrac{\pi}{3}x\right)$ 4) $y = 2\cos\left(\dfrac{1}{3}x\right)$

A)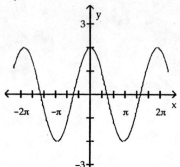

B)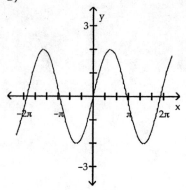

C)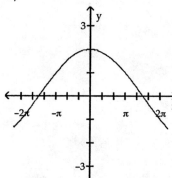

D)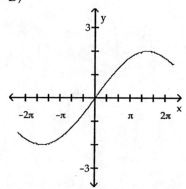

A) 1B, 2D, 3A, 4C   B) 1C, 2A, 3D, 4B   C) 1A, 2C, 3B, 4D   D) 1A, 2C, 3D, 4B

Graph the sinusoidal function.

8) $y = 2\sin(\pi x)$

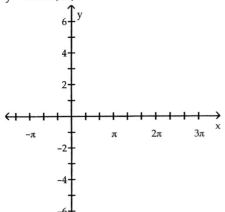

A)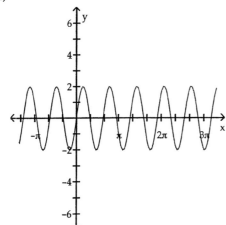

B)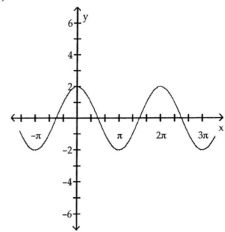

C)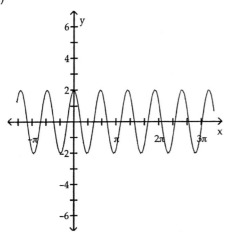

D)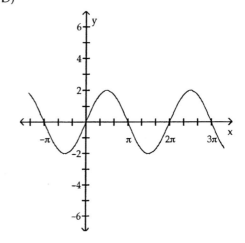

9) $y = -2\cos(\pi x)$

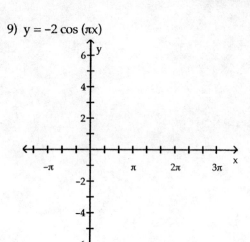

A)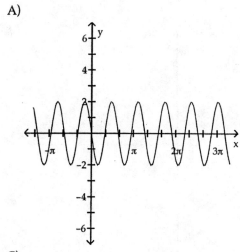

B)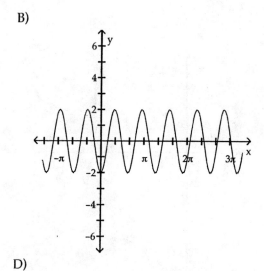

C)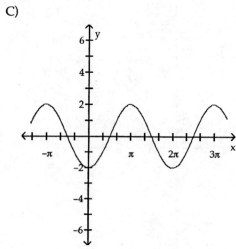

D)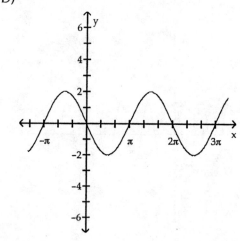

10) $y = 3\sin(2x)$

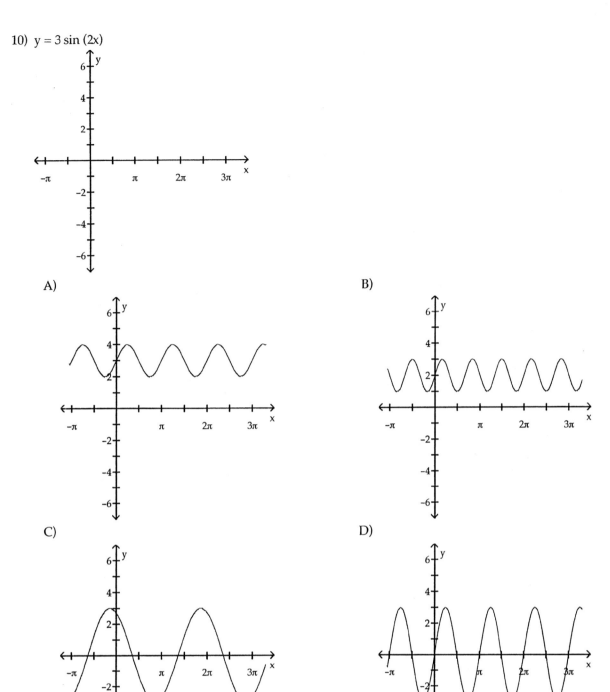

11) $y = 2\cos(\pi x)$

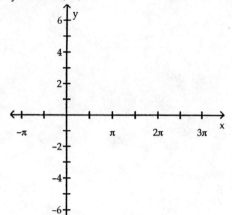

A)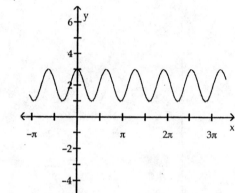

B)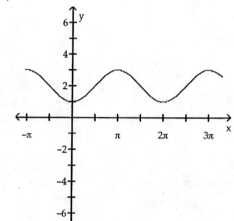

C)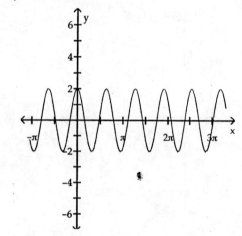

D)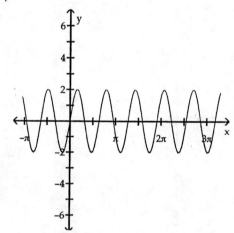

12) $y = -5\sin\left(\frac{1}{4}x\right)$

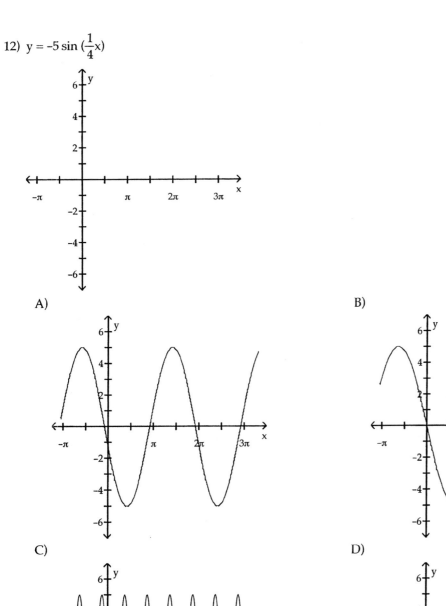

13) $y = \frac{7}{3} \cos\left(-\frac{1}{2}x\right)$

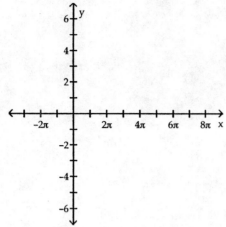

A)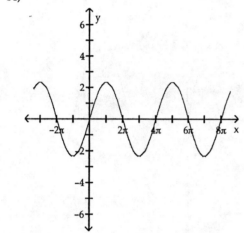

B)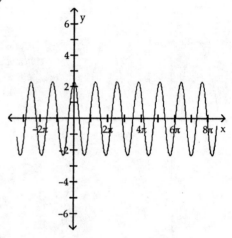

C)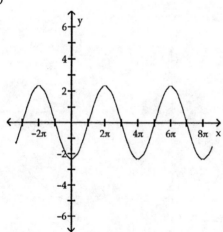

D)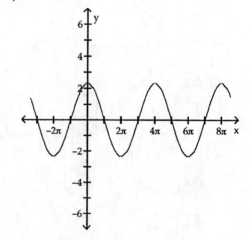

Answer the question.

14) Which one of the equations below matches the graph?

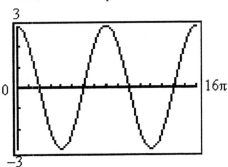

A) $y = 3\sin\left(\frac{1}{4}x\right)$ B) $y = 3\cos\left(\frac{1}{4}x\right)$ C) $y = 3\cos(4x)$ D) $y = -3\sin(4x)$

15) Which one of the equations below matches the graph?

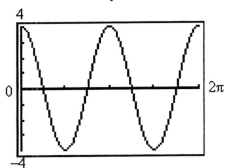

A) $y = 4\sin\left(\frac{1}{2}x\right)$ B) $y = 2\cos\left(\frac{1}{4}x\right)$ C) $y = 4\cos(2x)$ D) $y = 4\cos\left(\frac{1}{2}x\right)$

16) Which one of the equations below matches the graph?

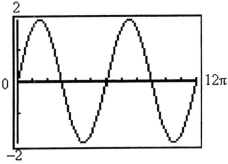

A) $y = 2\cos(3x)$ B) $y = 2\sin\left(\frac{1}{3}x\right)$ C) $y = 2\cos\left(\frac{1}{3}x\right)$ D) $y = -2\sin\left(\frac{1}{3}x\right)$

17) Which one of the equations below matches the graph?

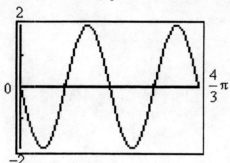

A) $y = -2\cos(3x)$ B) $y = -2\sin\left(\frac{1}{3}x\right)$ C) $y = 2\sin\left(\frac{1}{3}x\right)$ D) $y = -2\sin(3x)$

5 Find an Equation for a Sinusoidal Graph

Write the equation of a sine function that has the given characteristics.

1) Amplitude: 3
   Period: 4π

   A) $y = 3\sin(4x)$ B) $y = \sin(4x) + 3$ C) $y = 3\sin\left(\frac{1}{2}x\right)$ D) $y = 4\sin\left(\frac{2}{3}x\right)$

2) Amplitude: 5
   Period: 6

   A) $y = 6\sin\left(\frac{2}{5}\pi x\right)$ B) $y = \sin(6\pi x) + 5$ C) $y = 5\sin(6x)$ D) $y = 5\sin\left(\frac{1}{3}\pi x\right)$

Find an equation for the graph.

3)

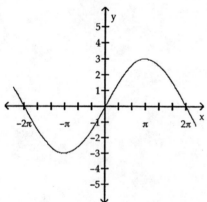

A) $y = 2\sin(3x)$ B) $y = 2\sin\left(\frac{1}{3}x\right)$ C) $y = 3\sin(2x)$ D) $y = 3\sin\left(\frac{1}{2}x\right)$

4)

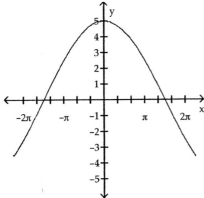

A) $y = 3\cos(5x)$ B) $y = 3\cos\left(\frac{1}{5}x\right)$ C) $y = 5\cos(3x)$ D) $y = 5\cos\left(\frac{1}{3}x\right)$

5)

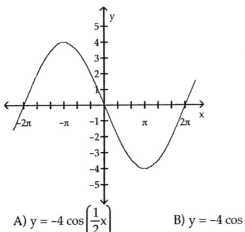

A) $y = -4\cos\left(\frac{1}{2}x\right)$ B) $y = -4\cos(2x)$ C) $y = -4\sin\left(\frac{1}{2}x\right)$ D) $y = -4\sin(2x)$

6)

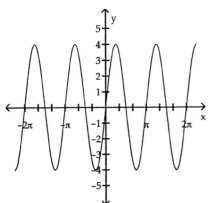

A) $y = 4\sin(2x)$ B) $y = 2\sin(4x)$ C) $y = 4\sin\left(\frac{1}{2}x\right)$ D) $y = 2\sin\left(\frac{1}{4}x\right)$

7)

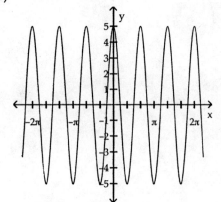

A) $y = 3\cos(5x)$ B) $y = 5\cos\left(\frac{1}{3}x\right)$ C) $y = 3\cos\left(\frac{1}{5}x\right)$ D) $y = 5\cos(3x)$

8)

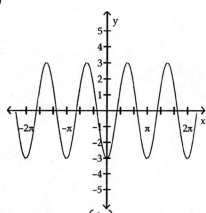

A) $y = -3\cos\left(\frac{1}{2}x\right)$ B) $y = -3\sin\left(\frac{1}{2}x\right)$ C) $y = -3\sin(2x)$ D) $y = -3\cos(2x)$

9)

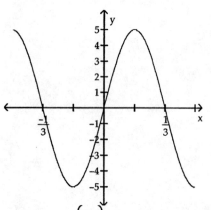

A) $y = 3\sin\left(\frac{\pi}{5}x\right)$ B) $y = 5\sin\left(\frac{\pi}{3}x\right)$ C) $y = 3\sin(5\pi x)$ D) $y = 5\sin(3\pi x)$

10)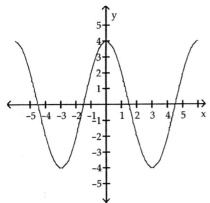

A) $y = 3\cos(4\pi x)$ B) $y = 4\cos(3\pi x)$ C) $y = 3\cos\left(\dfrac{\pi}{4}x\right)$ D) $y = 4\cos\left(\dfrac{\pi}{3}x\right)$

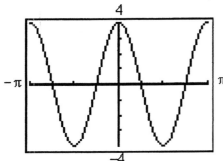

11)
A) $y = -4\cos(2x)$ B) $y = 4\cos(2x)$ C) $y = 4\sin(2x)$ D) $y = 4\cos\left(\dfrac{1}{2}x\right)$

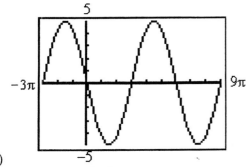

12)
A) $y = -5\sin\left(\dfrac{1}{3}x\right)$ B) $y = -5\sin\left(\dfrac{2}{3}x\right)$ C) $y = -5\sin(3x)$ D) $y = 5\cos\left(\dfrac{1}{3}x\right)$

Precalculus Enhanced with Graphing Utilities 477

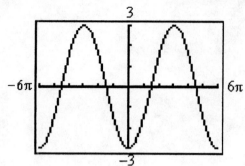

13)
  A) $y = -3\cos(3x)$
  B) $y = 3\cos\left(\dfrac{1}{3}x\right)$
  C) $y = -3\cos\left(\dfrac{1}{3}x\right)$
  D) $y = -3\sin(3x)$

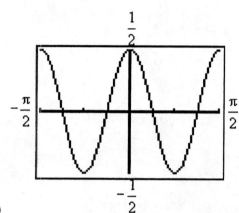

14)
  A) $y = \dfrac{1}{2}\cos\left(\dfrac{1}{2}x\right)$
  B) $y = \dfrac{1}{2}\cos\left(\dfrac{1}{4}x\right)$
  C) $y = \dfrac{1}{2}\cos(4x)$
  D) $y = \cos(4x)$

6 Demonstrate Additional Understanding and Skills

Solve the problem.

1) What is the y-intercept of $y = \sin x$?
  A) 1
  B) $\dfrac{\pi}{2}$
  C) π
  D) 0

2) For what numbers x, $0 \le x \le 2\pi$, does $\sin x = 0$?
  A) 0, 1, 2
  B) $\dfrac{\pi}{2}, \dfrac{3\pi}{2}$
  C) $0, \pi, 2\pi$
  D) 0, 1

3) For what numbers x, $0 \le x \le 2\pi$, does $\cos x = 0$?
  A) 0, 1
  B) 0, 1, 2
  C) $\dfrac{\pi}{2}, \dfrac{3\pi}{2}$
  D) $0, \pi, 2\pi$

4) For what numbers x, $0 \le x \le 2\pi$, does $\cos x = 1$?
  A) $\dfrac{\pi}{2}$
  B) $\dfrac{\pi}{2}, \dfrac{3\pi}{2}$
  C) $0, 2\pi$
  D) none

5) For what numbers x, $0 \le x \le 2\pi$, does $\sin x = 1$?
  A) $0, 2\pi$
  B) $\dfrac{\pi}{2}$
  C) $\dfrac{\pi}{2}, \dfrac{3\pi}{2}$
  D) none

6) For what numbers x, $0 \le x \le 2\pi$, does $\sin x = -1$?

   A) $\dfrac{3\pi}{2}$ B) π C) $\dfrac{\pi}{2}, \dfrac{3\pi}{2}$ D) none

7) For what numbers x, $0 \le x \le 2\pi$, does $\cos x = -1$?

   A) $\dfrac{\pi}{2}, \dfrac{3\pi}{2}$ B) $\dfrac{\pi}{2}$ C) π D) none

8) Wildlife management personnel use predator-prey equations to model the populations of certain predators and their prey in the wild. Suppose the population M of a predator after t months is given by

   $$M = 750 + 125 \sin \dfrac{\pi}{6} t$$

   while the population N of its primary prey is given by

   $$N = 12{,}250 + 3050 \cos \dfrac{\pi}{6} t$$

   Find the period for each of these functions.

9) The average daily temperature T of a city in the United States is approximated by

   $$T = 55 - 23 \cos \dfrac{2\pi}{365}(t - 30)$$

   where t is in days, $1 \le t \le 365$, and $t = 1$ corresponds to January 1. Find the period of T.

10) The current I, in amperes, flowing through a particular ac (alternating current) circuit at time t seconds is
    $$I = 220 \sin(50\pi t)$$
    What is the period and amplitude of the current?

    A) period $= \dfrac{1}{200}$ second, amplitude $= 200$
    B) period $= \dfrac{\pi}{220}$ second, amplitude $= 50$
    C) period $= 50\pi$ seconds, amplitude $= \dfrac{1}{25}$
    D) period $= \dfrac{1}{25}$ second, amplitude $= 220$

11) The current I, in amperes, flowing through a particular ac (alternating current) circuit at time t seconds is
    $$I = 110 \sin(40\pi t)$$
    Using Ohm's Law, $V = IR$, if a resistance of $R = 60$ ohms is present, what is the voltage V?

    A) $V = 110 \sin(40\pi t)$ B) $V = 6600 \sin(40\pi t)$ C) $V = -50 \sin(\dfrac{2}{3}\pi t)$ D) $V = 110 \sin(2400\pi t)$

12) The current I, in amperes, flowing through an ac (alternating current) circuit at time t, in seconds, is
$$I = 30 \sin(50\pi t)$$
What is the amplitude? What is the period?
Graph this function over two periods beginning at t = 0.

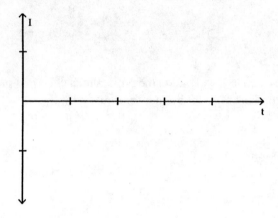

13) A mass hangs from a spring which oscillates up and down. The position P of the mass at time t is given by
$$P = 4 \cos(4t)$$
What is the amplitude? What is the period?
Graph this function over two periods beginning at t = 0.

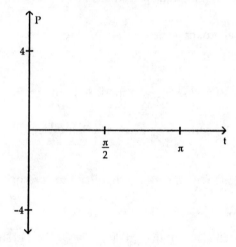

14) Before exercising, an athlete measures her air flow and obtains
$$a = 0.65 \sin\left(\frac{2\pi}{5}t\right)$$
where a is measured in liters per second and t is the time in seconds. If a > 0, the athlete is inhaling; if a < 0, the athlete is exhaling. The time to complete one complete inhalation/exhalation sequence is a respiratory cycle.
What is the amplitude? What is the period? What is the respiratory cycle?
Graph a over two periods beginning at t = 0.

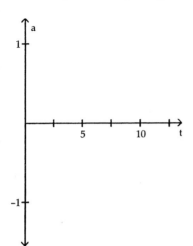

15) A boy is flying a model airplane while standing on a straight line. The plane, at the end of a twenty-five foot wire, flies in circles around the boy. The directed distance of the plane from the straight line is found to be
$$d = 25 \cos\left(\frac{3\pi}{4}t\right)$$
where d is measured in feet and t is the time in seconds. If d > 0, the plane is in front of the boy; if d < 0, the plane is behind him.
What is the amplitude? What is the period?
Graph d over two periods beginning at t = 0.

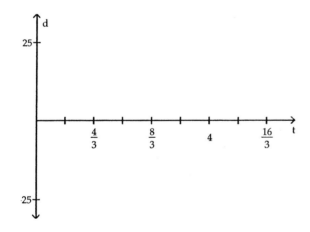

16) At a certain port, the height of a tide at any time during the day is given by

$$h = 6.4 \sin\left(\frac{\pi}{6}t\right)$$

Here, h is the height in feet above or below a central line and t is the hour of the day with t = 0 corresponding to midnight.

What is the amplitude? What is the period?

Graph h over two periods beginning at t = 0.

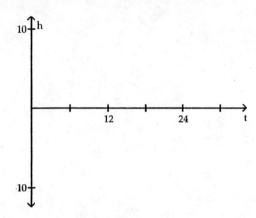

5.5 Graphs of the Tangent, Cotangent, Cosecant, and Secant Functions

1 Graph Transformations of the Tangent Function and Cotangent Function

Match the function to its graph.

1) $y = -\tan x$

A)
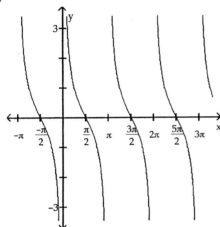

B)
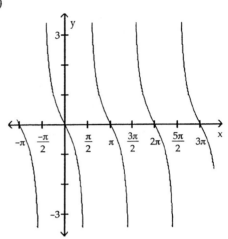

C)
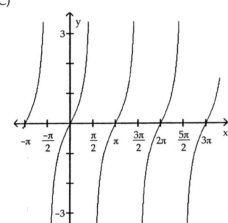

D)
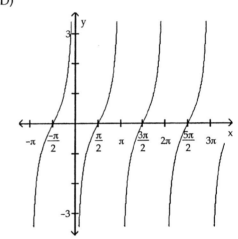

2) $y = \tan\left(x + \dfrac{\pi}{2}\right)$

A)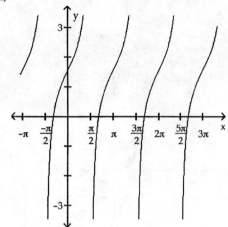

B)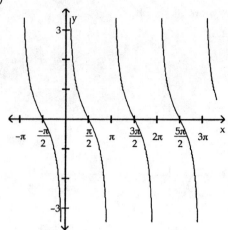

C)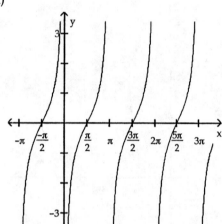

D)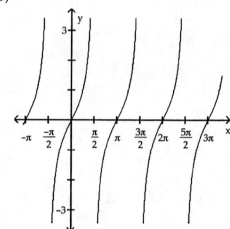

3) $y = \tan(x + \pi)$

A)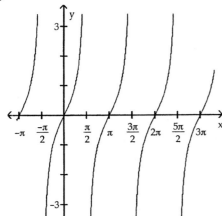

B)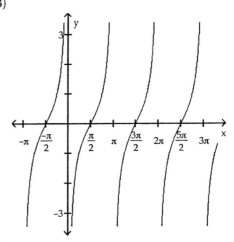

C)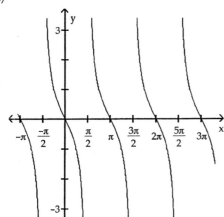

D)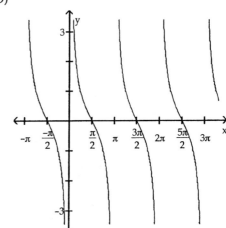

4) $y = \tan\left(x - \dfrac{\pi}{2}\right)$

A)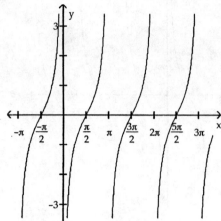

B)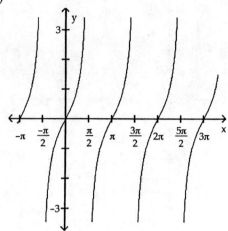

C)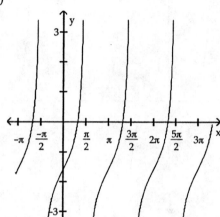

D)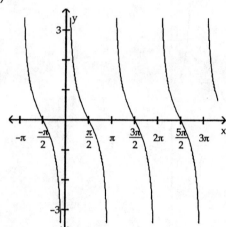

Graph the function.

5) $y = \cot x$

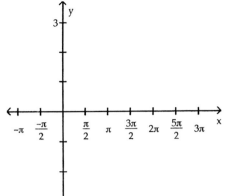

A)
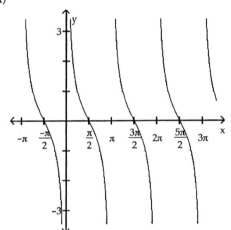

B)
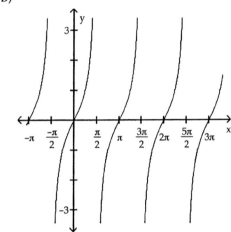

C)
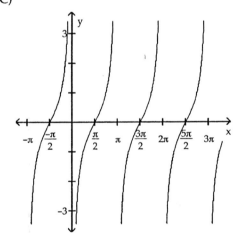

D)
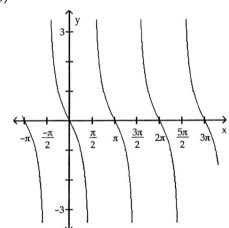

6) $y = -\tan(x - \pi)$

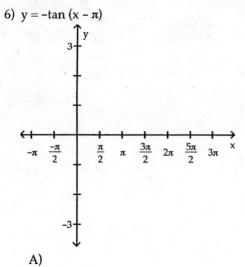

A)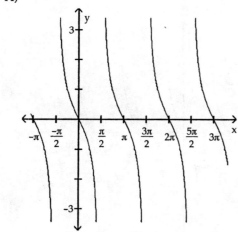

B)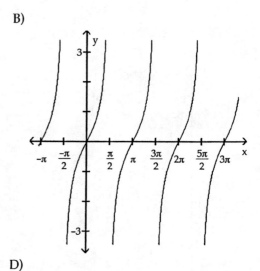

C)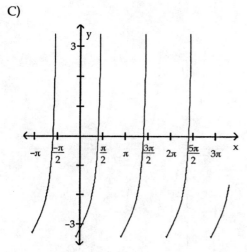

D)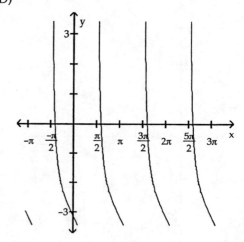

7) $y = 4\tan(2x)$

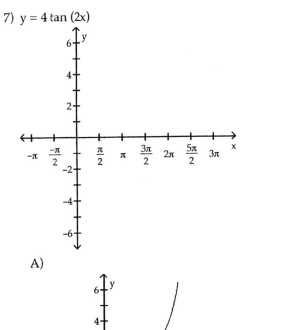

A)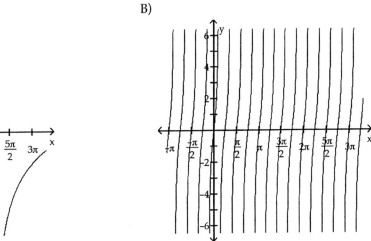

B)

C)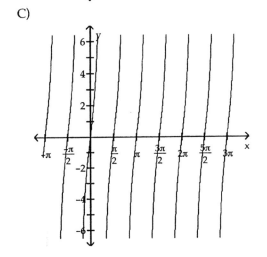

D)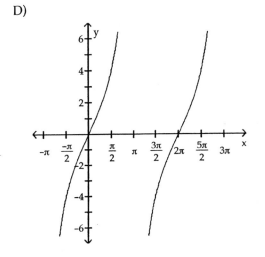

8) $y = -3\tan\left(\dfrac{1}{4}x\right)$

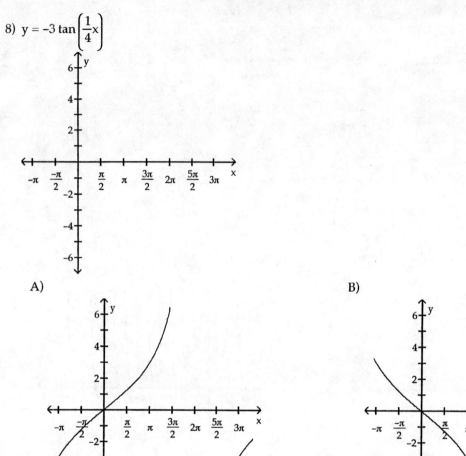

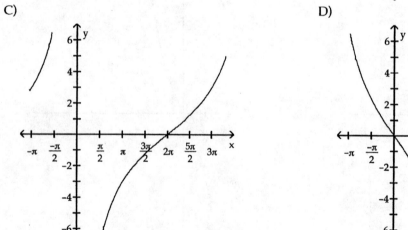

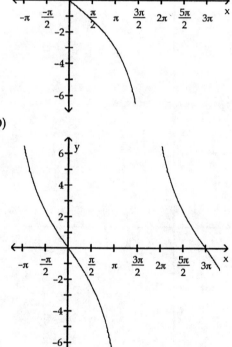

9) $y = -\cot(\pi x)$

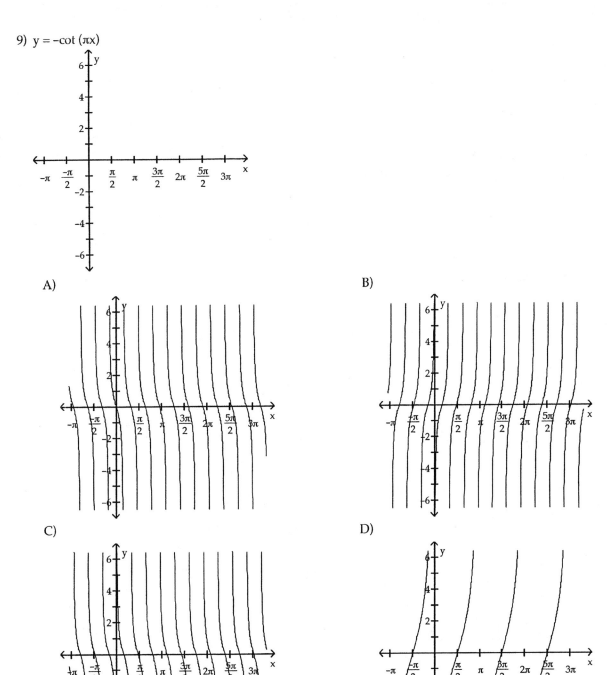

10) $y = -\cot(2x)$

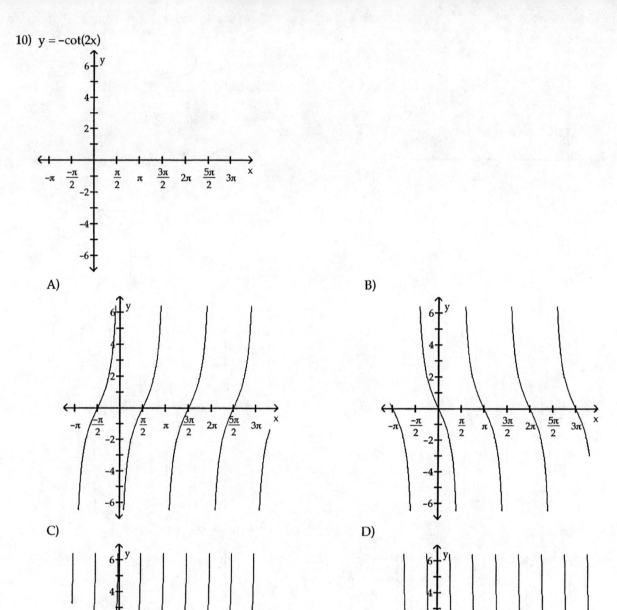

11) y = -2 cot (4x)

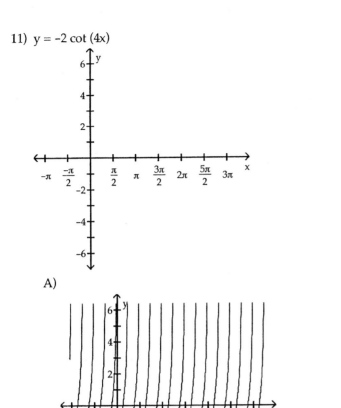

A)
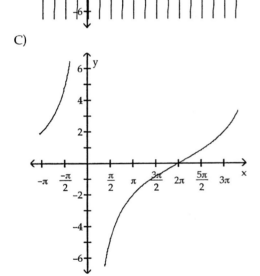

B)
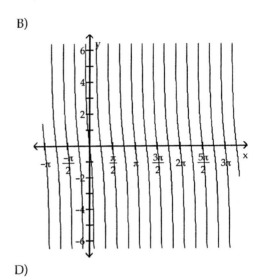

C)
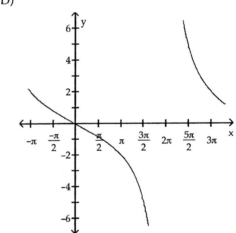

D)

12) $y = -4\tan\left(x + \dfrac{\pi}{4}\right)$

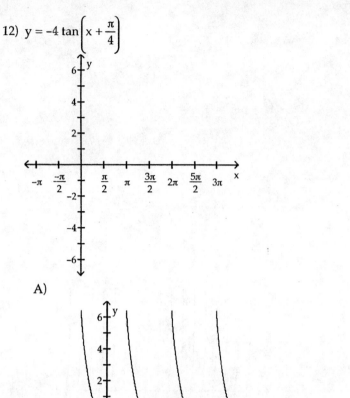

A)

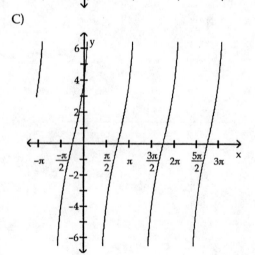

B)

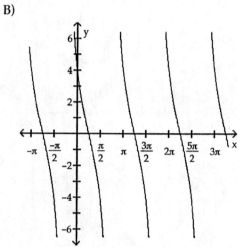

C)

D)

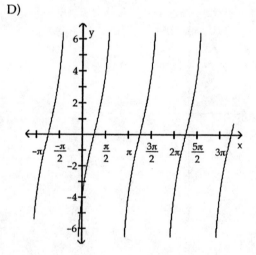

13) $y = \frac{1}{3} \cot\left(x + \frac{\pi}{4}\right)$

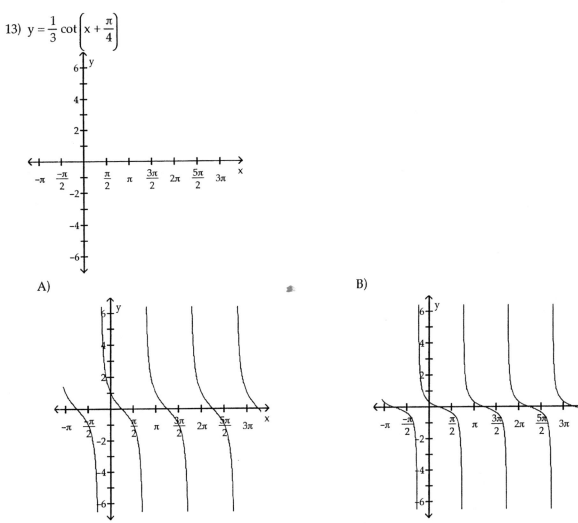

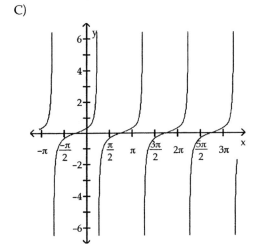

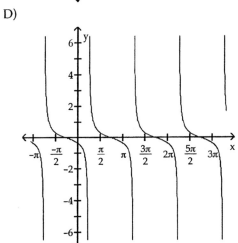

14) $y = -\tan\left(x - \dfrac{\pi}{4}\right)$

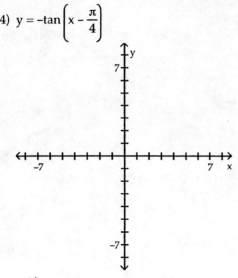

A)
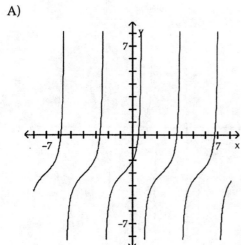

B)
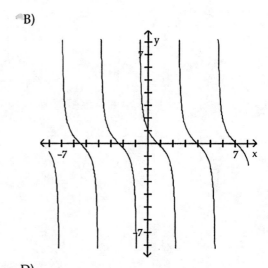

C)
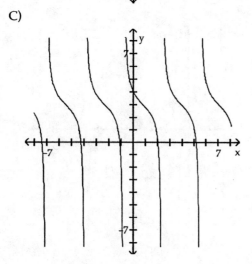

D)
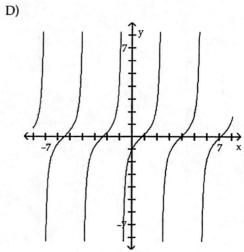

15) $y = \tan\left(x - \dfrac{\pi}{4}\right)$

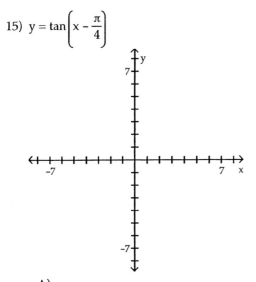

A)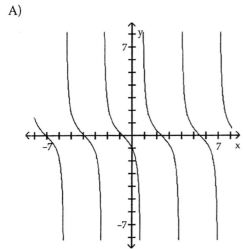

B)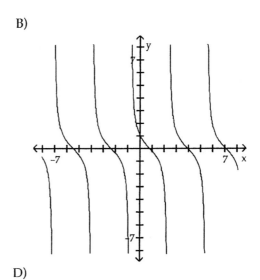

C)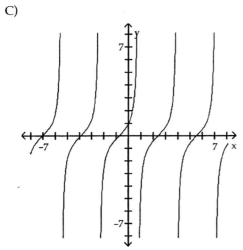

D)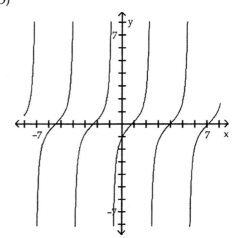

16) $y = -\cot\left(x - \dfrac{\pi}{4}\right)$

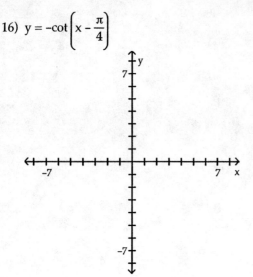

A)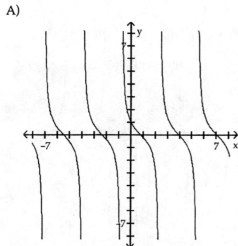

B)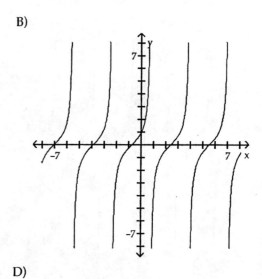

C)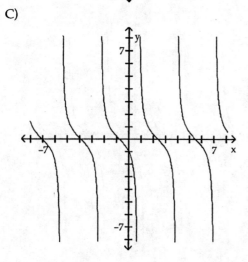

D)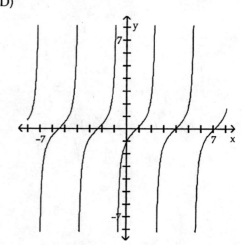

2 Graph Transformations of the Cosecant Function and Secant Function

Graph the function.

1) $y = \csc\left(x - \dfrac{\pi}{3}\right)$

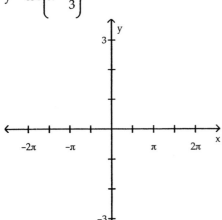

A)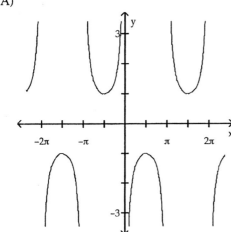

B)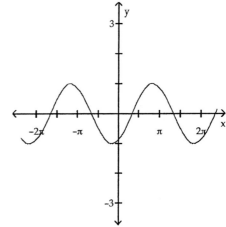

C)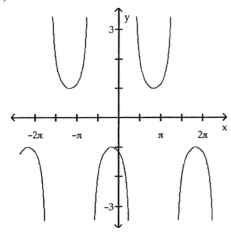

D)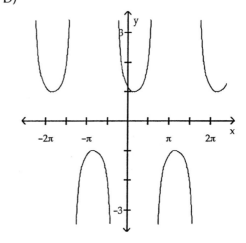

Precalculus Enhanced with Graphing Utilities 499

2) $y = \sec\left(x - \dfrac{\pi}{4}\right)$

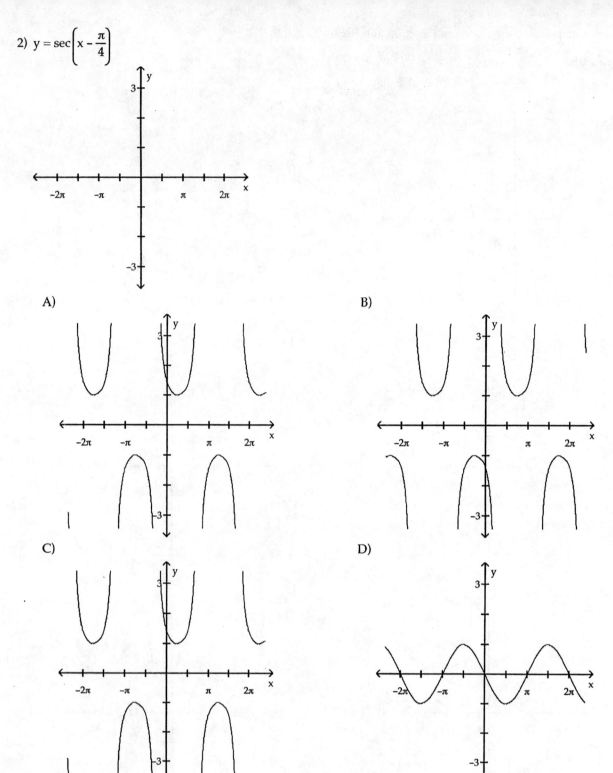

3) y = -sec x

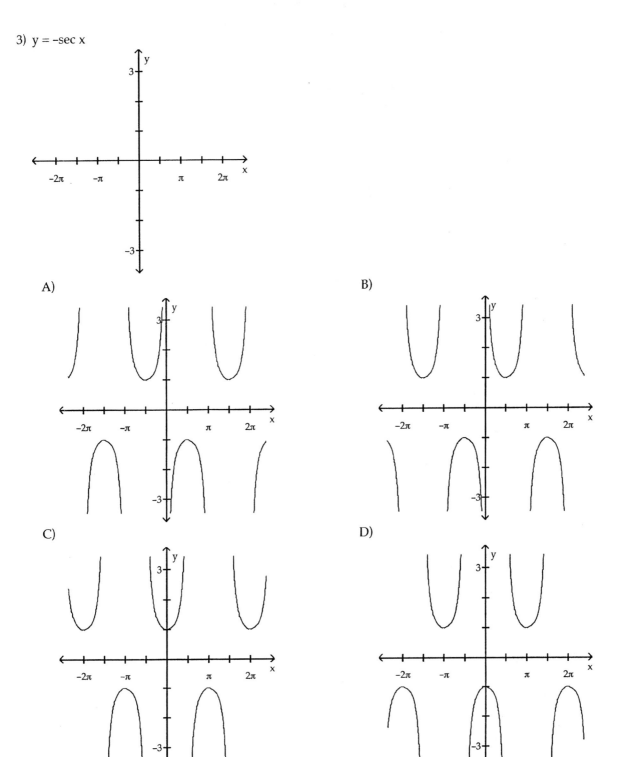

4) $y = \csc(2x)$

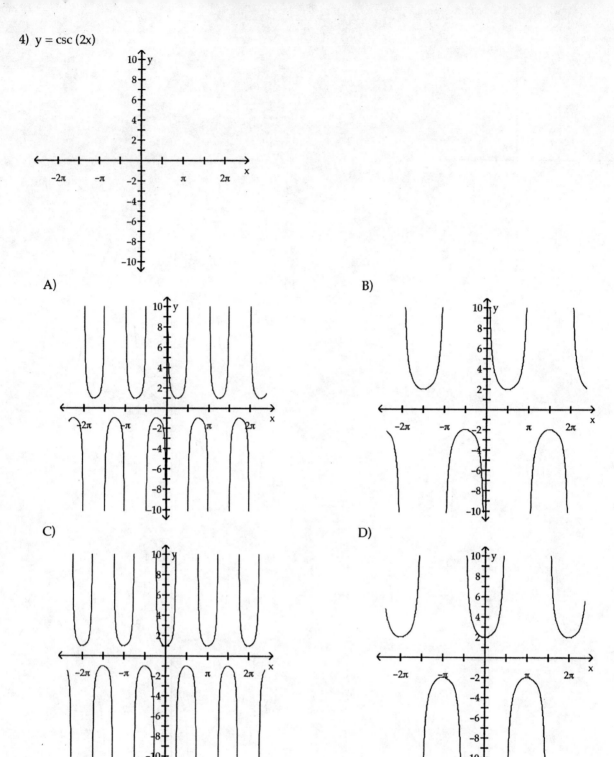

5) $y = \sec\left(\dfrac{1}{2}x\right)$

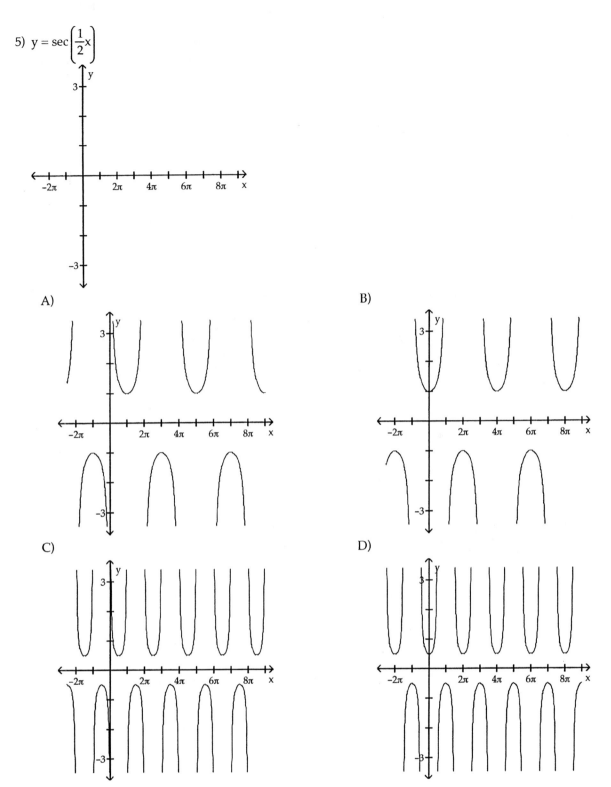

6) $y = 2\csc\left(\dfrac{1}{4}x\right)$

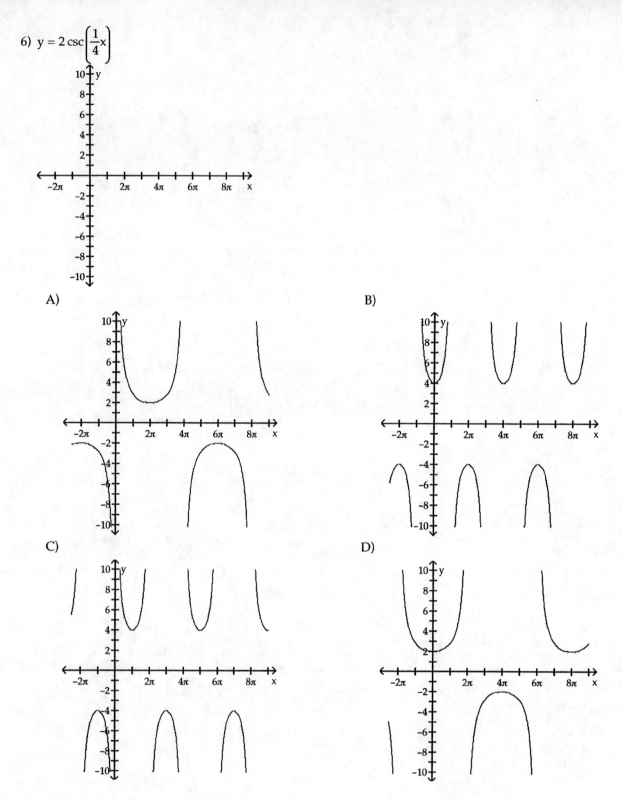

7) $y = 6\csc(3x)$

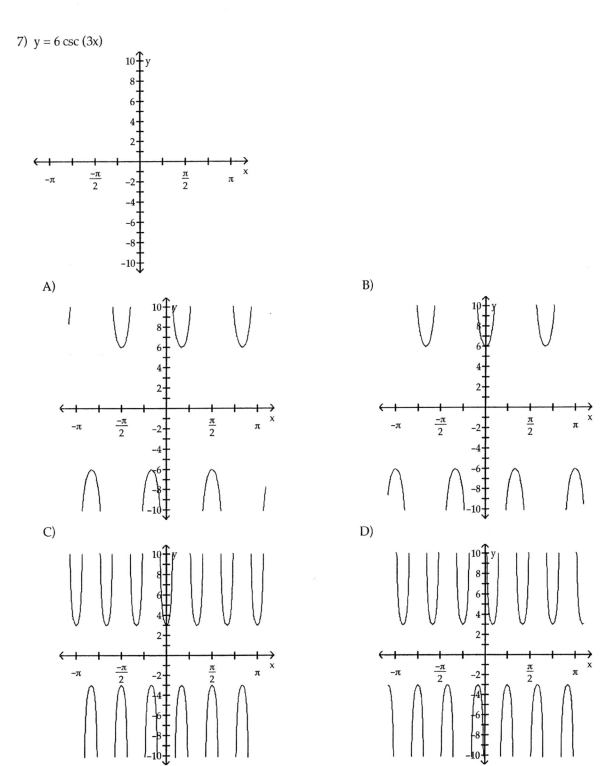

8) $y = -5\sec\left(x + \dfrac{\pi}{4}\right)$

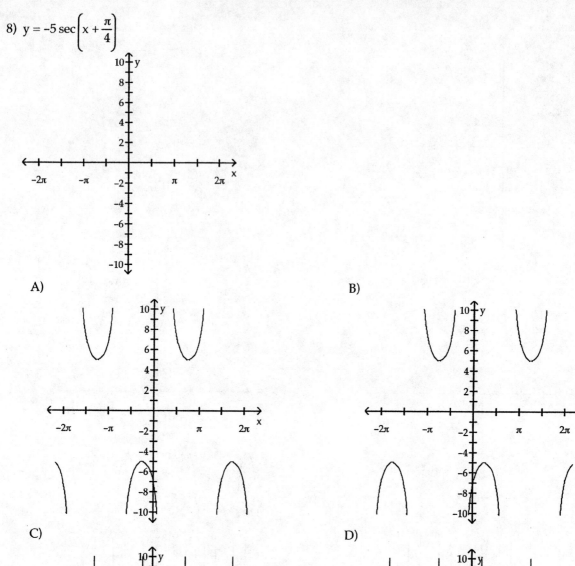

9) $y = 3\csc\left(x - \dfrac{\pi}{4}\right)$

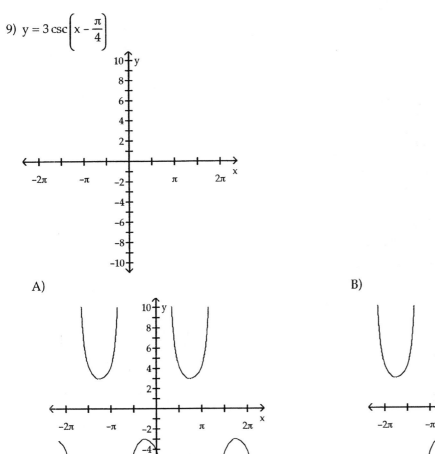

A)

B)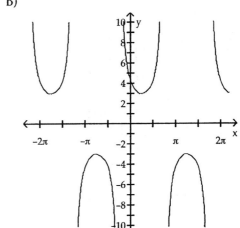

C)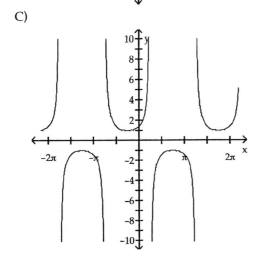

D)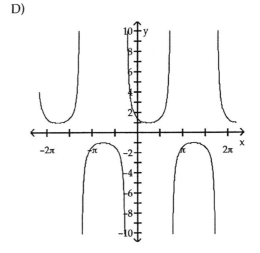

10) $y = \csc\left(\dfrac{\pi x}{5} + \dfrac{3\pi}{5}\right)$

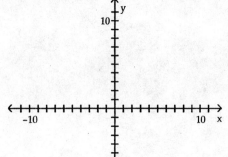

A)

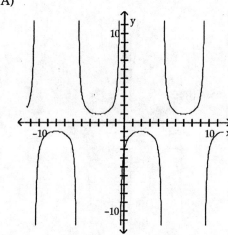

B)

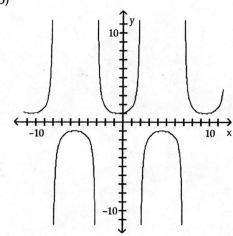

C)

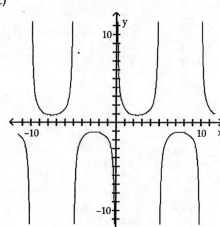

D)

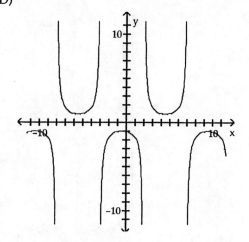

3 Demonstrate Additional Understanding and Skills

Solve the problem.

1) What is the y-intercept of $y = \tan x$?

   A) 0   B) $\dfrac{\pi}{2}$ C) 1   D) none

Precalculus Enhanced with Graphing Utilities

2) What is the y-intercept of $y = \csc x$?

A) 0   B) 1   C) $\dfrac{\pi}{2}$ D) none

3) For what numbers x, $-2\pi \le x \le 2\pi$, does the graph of $y = \sec x$ have vertical asymptotes.?

A) -2, -1, 0, 1, 2   B) $-\dfrac{3\pi}{2}, -\dfrac{\pi}{2}, \dfrac{\pi}{2}, \dfrac{3\pi}{2}$ C) $-2\pi, -\pi, 0, \pi, 2\pi$ D) none

4) For what numbers x, $-2\pi \le x \le 2\pi$, does the graph of $y = \cot x$ have vertical asymptotes?

A) -2, -1, 0, 1, 2   B) $-\dfrac{3\pi}{2}, -\dfrac{\pi}{2}, \dfrac{\pi}{2}, \dfrac{3\pi}{2}$ C) $-2\pi, -\pi, 0, \pi, 2\pi$ D) none

5) For what numbers x, $-2\pi \le x \le 2\pi$, does $\sec x = 1$?

A) $-\dfrac{3\pi}{2}, \dfrac{\pi}{2}$ B) $-\pi, \pi$ C) $-2\pi, 0, 2\pi$ D) none

6) For what numbers x, $-2\pi \le x \le 2\pi$, does $\csc x = 1$?

A) $-\pi, \pi$ B) $-\dfrac{3\pi}{2}, \dfrac{\pi}{2}$ C) $-2\pi, 0, 2\pi$ D) none

7) For what numbers x, $-2\pi \le x \le 2\pi$, does $\sec x = -1$?

A) $-\dfrac{\pi}{2}, \dfrac{3\pi}{2}$ B) $-2\pi, 0, 2\pi$ C) $-\pi, \pi$ D) none

8) For what numbers x, $-2\pi \le x \le 2\pi$, does $\csc x = -1$?

A) $-\pi, \pi$ B) $-\dfrac{\pi}{2}, \dfrac{3\pi}{2}$ C) $-2\pi, 0, 2\pi$ D) none

5.6 Phase Shift; Sinusoidal Curve Fitting

1 Graph Sinusoidal Functions of the Form y = A sin (omega x - phi) Using the Amplitude, Period, and Phase Shift

Find the phase shift of the function.

1) $y = 2 \sin\left(x - \dfrac{\pi}{4}\right)$

A) 2 units down   B) $\dfrac{\pi}{4}$ units to the right   C) 2 units up   D) $\dfrac{\pi}{4}$ units to the left

2) $y = -4 \cos\left(x + \dfrac{\pi}{2}\right)$

A) -4 units down   B) -4 units up   C) $\dfrac{\pi}{2}$ units to the left   D) $\dfrac{\pi}{2}$ units to the right

3) $y = -2 \sin\left(2x - \dfrac{\pi}{2}\right)$

A) $\dfrac{\pi}{2}$ units to the left   B) 2π units down   C) $\dfrac{\pi}{4}$ units to the right   D) 2π units up

4) $y = -4\cos(6x + \pi)$

A) $\dfrac{\pi}{6}$ units to the left B) 4π units to the right C) $\dfrac{\pi}{4}$ units to the left D) 6π units to the right

5) $y = -2\sin\left(\dfrac{1}{2}x - \dfrac{\pi}{2}\right)$

A) $\dfrac{\pi}{2}$ units to the right B) $\dfrac{\pi}{4}$ units to the left C) $\dfrac{\pi}{2}$ units to the left D) π units to the right

6) $y = -3\cos\left(\dfrac{1}{2}x + \dfrac{\pi}{2}\right)$

A) π units to the left B) 3π units to the right C) $\dfrac{\pi}{2}$ units to the left D) $\dfrac{\pi}{4}$ units to the right

7) $y = 4\sin(3\pi x - 2)$

A) $\dfrac{2}{3}$ units to the left

B) 2 units to the right

C) 2 units to the left

D) $\dfrac{2}{3\pi}$ units to the right

8) $y = 2\sin\left(-3x - \dfrac{\pi}{2}\right)$

A) $\dfrac{\pi}{2}$ units to the right B) $\dfrac{\pi}{6}$ units to the left C) $\dfrac{\pi}{2}$ units to the left D) $\dfrac{\pi}{6}$ units to the right

Graph the function. Show at least one period.

9) $y = 4\sin(3\pi x + 2)$

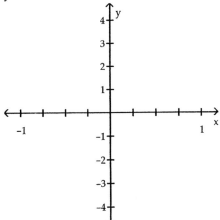

A)
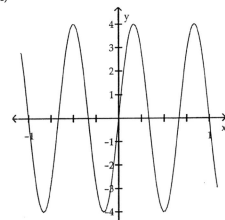

B)
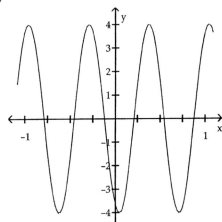

C)
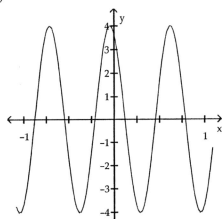

D)
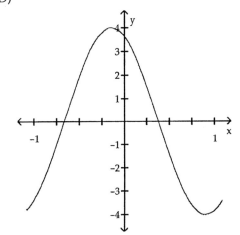

10) $y = 5\sin(2x - \pi)$

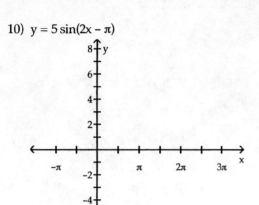

A)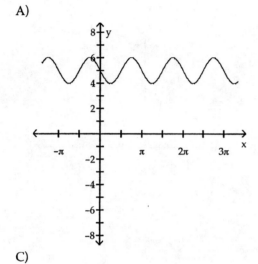

B)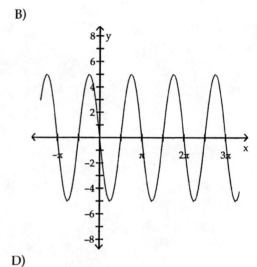

C)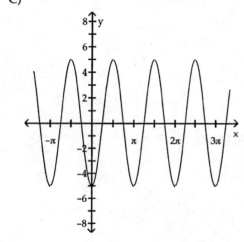

D)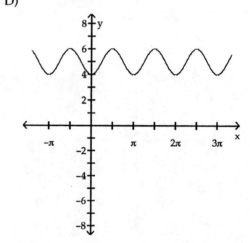

11) $y = 4\cos\left(2x + \dfrac{\pi}{2}\right)$

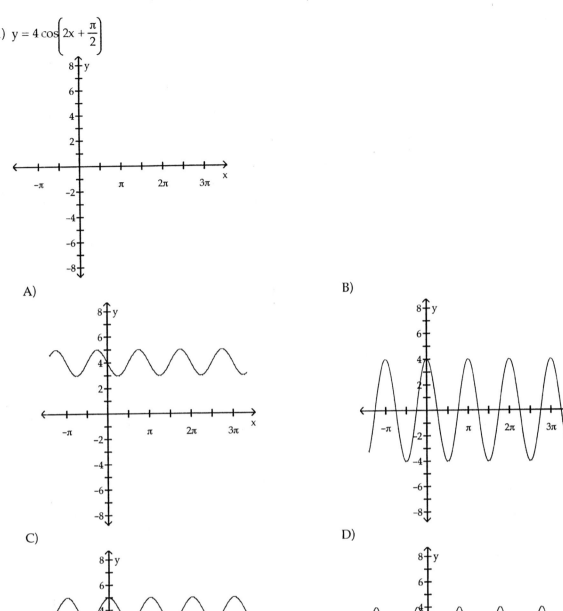

12) $y = -4\sin\left(3x + \dfrac{\pi}{2}\right)$

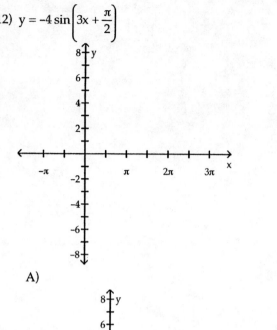

A)
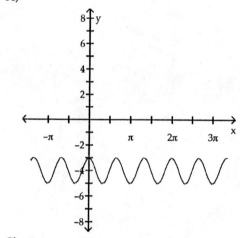

B)
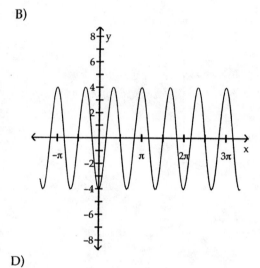

C)
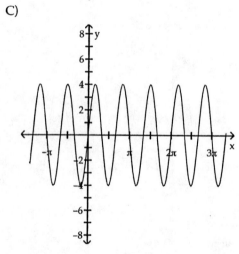

D)
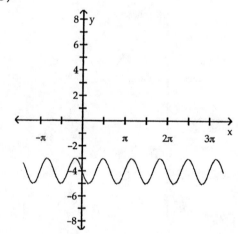

13) $y = 3\sin(\pi x + 4)$

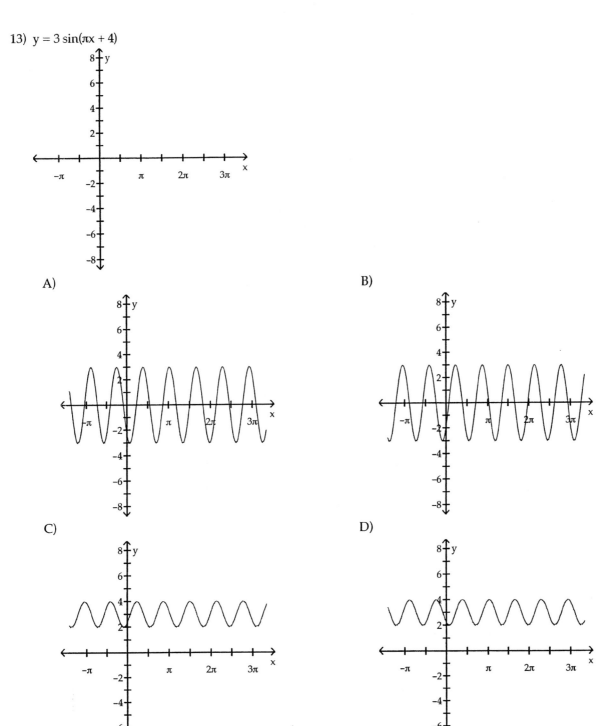

14) $y = 4\cos\left(-3x + \dfrac{\pi}{2}\right)$

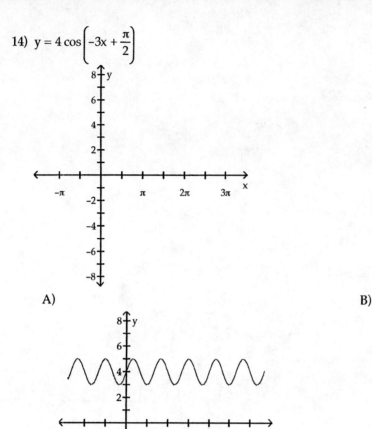

A)

B)
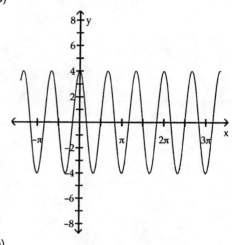

C)
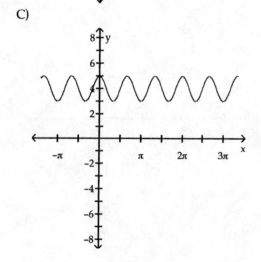

D)
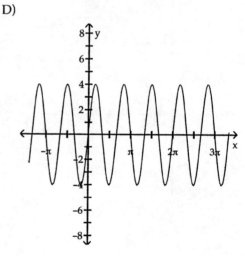

15) $y = 4\sin(-2x - \pi)$

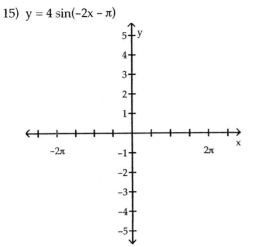

A)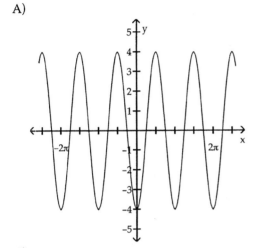

B)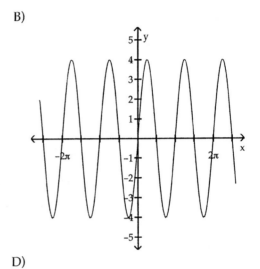

C)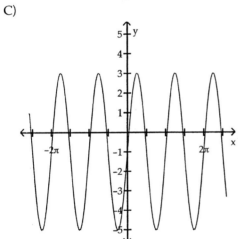

D)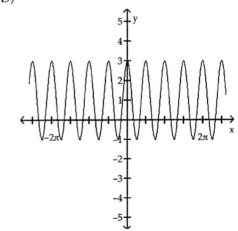

16) $y = -3\cos(x - \pi)$

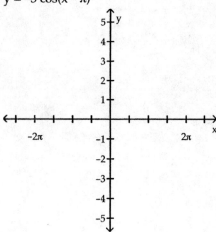

Solve the problem.

17) For the equation $y = -\dfrac{1}{2}\sin(4x + 3\pi)$, identify (i) the amplitude, (ii) the phase shift, and (iii) the period.

A) (i) 2   (ii) 3π (iii) $\dfrac{\pi}{2}$

B) (i) $-\dfrac{1}{2}$ (ii) $-\dfrac{4\pi}{3}$ (iii) 4

C) (i) $\dfrac{1}{2}$ (ii) $-\dfrac{3\pi}{4}$ (iii) $\dfrac{\pi}{2}$

D) (i) $\dfrac{1}{2}$ (ii) $-\dfrac{3\pi}{4}$ (iii) 4

18) For the equation $y = -\dfrac{1}{2}\cos(2x - 2\pi)$, identify (i) the amplitude, (ii) the phase shift, and (iii) the period.

A) (i) $\dfrac{1}{2}$ (ii) π (iii) π

B) (i) 2   (ii) π (iii) π

C) (i) 2   (ii) 2π (iii) 2π

D) (i) $\dfrac{1}{2}$ (ii) $\dfrac{\pi}{2}$ (iii) π

2 Find a Sinusoidal Function from Data

Solve the problem.

1) An experiment in a wind tunnel generates cyclic waves. The following data is collected for 52 seconds:

Time (in seconds)	Wind speed (in feet per second)
0	21
13	44
26	67
39	44
52	21

Let V represent the wind speed (velocity) in feet per second and let t represent the time in seconds. Write a sine equation that describes the wave.

A) $V = 67\sin(52t - 26) + 21$

B) $V = 67\sin\left(\dfrac{\pi}{26}t - \dfrac{\pi}{2}\right) + 21$

C) $V = 23\sin\left(\dfrac{\pi}{26}t - \dfrac{\pi}{2}\right) + 44$

D) $V = 46\sin(52t - 26) + 23$

2) A town's average monthly temperature data is represented in the table below:

Month, x	Average Monthly Temperature, °F
January, 1	27.6
February, 2	30.6
March, 3	41.6
April, 4	56.1
May, 5	70.1
June, 6	78.6
July, 7	83.3
August, 8	78.6
September, 9	80.6
October, 10	56.5
November, 11	41.4
December, 12	30.7

Find a sinusoidal function of the form $y = A \sin(\omega x - \phi) + B$ that fits the data.

A) $y = 27.6 \sin\left(\dfrac{\pi}{6}x - \dfrac{\pi}{4}\right) + 83.3$

B) $y = 27.85 \sin\left(\dfrac{\pi}{6}x - \dfrac{2\pi}{3}\right) + 55.45$

C) $y = 83.3 \sin\left(\dfrac{\pi}{6}x - \dfrac{2\pi}{3}\right) + 27.6$

D) $y = 55.45 \sin\left(\dfrac{\pi}{6}x - \dfrac{\pi}{4}\right) + 27.85$

3) The number of hours of sunlight in a day can be modeled by a sinusoidal function. In the northern hemisphere, the longest day of the year occurs at the summer solstice and the shortest day occurs at the winter solstice. In 2000, these dates were June 22 (the 172nd day of the year) and December 21 (the 356th day of the year), respectively.

A town experiences 11.68 hours of sunlight at the summer solstice and 7.31 hours of sunlight at the winter solstice. Find a sinusoidal function $y = A \sin(\omega x - \phi) + B$ that fits the data, where x is the day of the year. (Note: There are 366 days in the year 2000.)

A) $y = 11.68 \sin\left(\pi x - \dfrac{2\pi}{3}\right) + 7.31$

B) $y = 2.185 \sin\left(\dfrac{\pi}{183}x - \dfrac{161\pi}{366}\right) + 9.495$

C) $y = 11.68 \sin\left(\dfrac{172\pi}{356}x - \dfrac{2\pi}{3}\right) + 9.495$

D) $y = 2.185 \sin\left(\dfrac{\pi}{183}x - \dfrac{2\pi}{3}\right) + 9.495$

4) The data below represent the average monthly cost of natural gas in an Oregon home.

month	Aug	Sep	Oct	Nov	Dec	Jan
cost	21.20	28.24	44.73	67.25	89.77	106.26

month	Feb	Mar	Apr	May	Jun	Jul
cost	111.30	106.26	89.77	67.25	43.73	28.24

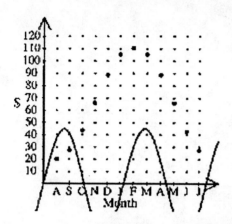

Above is the graph of 45.05 sin x superimposed over a scatter diagram of the data. Find the sinusoidal function of the form y = A sin (ωx − φ) + B which best fits the data.

A) $y = 45.05 \sin\left(\dfrac{\pi}{4}x - \dfrac{2\pi}{3}\right) + 21.20$

B) $y = 45.05 \sin\left(\dfrac{\pi}{6}x - \dfrac{\pi}{12}\right) + 66.25$

C) $y = 45.05 \sin\left(\dfrac{\pi}{6}x - \dfrac{2\pi}{3}\right) + 66.25$

D) $y = 45.05 \sin\left(\dfrac{\pi}{8}t + 12\right) + 21.20$

5) The data below represent the average monthly cost of natural gas in an Oregon home.

month	Aug	Sep	Oct	Nov	Dec	Jan
cost	18.90	24.24	44.58	68.25	91.92	109.26

month	Feb	Mar	Apr	May	Jun	Jul
cost	113.60	106.26	91.92	68.25	42.58	24.24

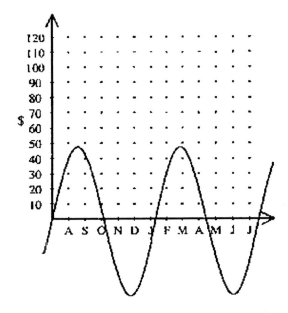

Above is the graph of 47.35 sin x. Make a scatter diagram of the data. Find the sinusoidal function of the form $y = A \sin(\omega x - \phi) + B$ which fits the data.

6) The following data represents the normal monthly precipitation for a certain city in California.

Month, x	Normal Monthly Precipitation, inches
January, 1	6.06
February, 2	4.45
March, 3	4.38
April, 4	2.08
May, 5	1.27
June, 6	0.56
July, 7	0.17
August, 8	0.46
September, 9	0.91
October, 10	2.24
November, 11	5.21
December, 12	5.51

Draw a scatter diagram of the data for one period. Find a sinusoidal function of the form $y = A \sin(\omega x - \phi) + B$ that fits the data. Draw the sinusoidal function on the scatter diagram. Use a graphing utility to find the sinusoidal function of best fit. Draw the sinusoidal function of best fit on the scatter diagram.

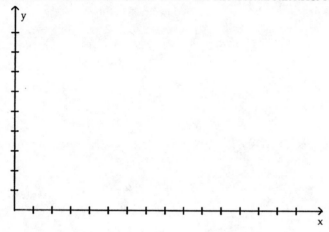

7) The following data represents the normal monthly precipitation for a certain city in Arkansas.

Month, x	Normal Monthly Precipitation, inches
January, 1	3.91
February, 2	4.36
March, 3	5.31
April, 4	6.21
May, 5	7.02
June, 6	7.84
July, 7	8.19
August, 8	8.06
September, 9	7.41
October, 10	6.30
November, 11	5.21
December, 12	4.28

Draw a scatter diagram of the data for one period. Find the sinusoidal function of the form
$y = A \sin(\omega x - \phi) + B$ that fits the data. Draw the sinusoidal function on the scatter diagram. Use a graphing utility to find the sinusoidal function of best fit. Draw the sinusoidal function of best fit on the scatter diagram.

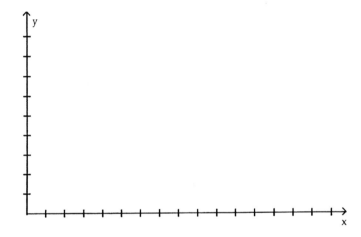

Precalculus Enhanced with Graphing Utilities

8) The following data represents the average monthly minimum temperature for a certain city in California.

Month, x	Average Monthly Minimum Temperature, °F
January, 1	49.6
February, 2	50.8
March, 3	55.6
April, 4	57.5
May, 5	60.7
June, 6	63.6
July, 7	65.9
August, 8	65.6
September, 9	64.4
October, 10	62.1
November, 11	54.2
December, 12	50.1

Draw a scatter diagram of the data for one period. Find a sinusoidal function of the form $y = A \sin(\omega x - \phi) + B$ that fits the data. Draw the sinusoidal function on the scatter diagram. Use a graphing utility to find the sinusoidal function of best fit. Draw the sinusoidal function of best fit on the scatter diagram.

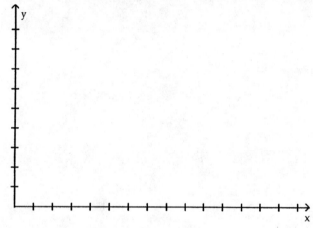

9) The following data represents the average percent of possible sunshine for a certain city in Indiana.

Month, x	Average Percent of Possible Sunshine
January, 1	46
February, 2	51
March, 3	55
April, 4	60
May, 5	68
June, 6	73
July, 7	75
August, 8	74
September, 9	68
October, 10	62
November, 11	41
December, 12	38

Draw a scatter diagram of the data for one period. Find the sinusoidal function of the form $y = A \sin(\omega x - \phi) + B$ that fits the data. Draw the sinusoidal function on the scatter diagram. Use a graphing utility to find the sinusoidal function of best fit. Draw the sinusoidal function of best fit on the scatter diagram.

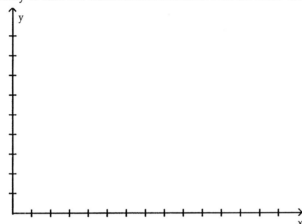

3 Demonstrate Additional Understanding and Skills

Write the equation of a sine function that has the given characteristics.

1) Amplitude: 2
   Period: 3π
   Phase Shift: $\dfrac{\pi}{3}$

   A) $y = 2 \sin\left(\dfrac{2}{3}x - \dfrac{2}{9}\pi\right)$
   B) $y = 2 \sin\left(\dfrac{2}{3}x + \dfrac{2}{9}\pi\right)$
   C) $y = 2 \sin\left(3x + \dfrac{\pi}{3}\right)$
   D) $y = 2 \sin\left(\dfrac{3}{2}x + \dfrac{2}{9}\pi\right)$

2) Amplitude: 5
   Period: 4π
   Phase Shift: $-\dfrac{\pi}{4}$

   A) $y = 5 \sin\left(\dfrac{1}{2}x - \dfrac{1}{8}\pi\right)$
   B) $y = 5 \sin\left(4x - \dfrac{\pi}{4}\right)$
   C) $y = 5 \sin\left(2x - \dfrac{1}{8}\pi\right)$
   D) $y = 5 \sin\left(\dfrac{1}{2}x + \dfrac{1}{8}\pi\right)$

3) Amplitude: 2
   Period: π
   Phase Shift: -6

   A) $y = 2\sin(2x+12)$ B) $y = 2\sin\left(\dfrac{1}{2}x - 12\right)$ C) $y = \sin(2x+6)$ D) $y = 2\sin(x-6)$

4) Amplitude: 4
   Period: π
   Phase Shift: $\dfrac{3}{2}$

   A) $y = 4\sin\left(2x + \dfrac{3}{2}\right)$ B) $y = 4\sin\left(\dfrac{1}{2}x - 6\right)$ C) $y = 4\sin(2x-3)$ D) $y = \sin(4x+3)$

Ch. 5 Trigonometric Functions
Answer Key

5.1 Angles and Their Measure
1 Convert Between Degrees, Minutes, Seconds, and Decimal Forms for Angles
1) C
2) D
3) A
4) B
5) A
6) A
7) A
8) C

2 Find the Arc Length of a Circle
1) C
2) B
3) D
4) D
5) C
6) D
7) A
8) A
9) C
10) D
11) C
12) 379 mi
13) B
14) C

3 Convert from Degrees to Radians and from Radians to Degrees
1) D
2) D
3) C
4) A
5) A
6) C
7) C
8) C
9) D
10) B
11) D
12) B
13) A
14) A
15) C
16) A
17) C
18) C
19) C

4 Find the Area of a Sector of a Circle
1) D
2) C
3) B
4) D
5) B
6) C

7) D
8) A
9) D
10) B
11) B
12) C
13) A
14) D
15) C
16) D

5 Find the Linear Speed of an Object Traveling in Circular Motion
1) D
2) A
3) B
4) A
5) B
6) D
7) A
8) B
9) D
10) A
11) 6.24×10^4 kmp; 4.95×10^4 kmp; 3.92×10^4 kmp; 2.95×10^4 kmp; Io
12) D
13) C

6 Demonstrate Additional Understanding and Skills
1) C
2) B
3) B
4) B
5) B
6) B
7) C
8) C
9) D
10) D

5.2 Trigonometric Functions: Unit Circle Approach
1 Find the Exact Values of the Trigonometric Functions Using a Point on the Unit Circle
1) C
2) D
3) A
4) C
5) A
6) B
7) B
8) A
9) C
10) A

2 Find the Exact Values of the Trigonometric Functions of Quadrantal Angles
1) B
2) C
3) A
4) D
5) C
6) B

7) A
8) C
9) B
10) B
11) B

3 Find the Exact Values of the Trigonometric Functions of pi/4 = 45 Degrees

1) B
2) C
3) C
4) C

4 Find the Exact Values of the Trigonometric Functions of pi/6 = 30 Degrees and pi/3 = 60 Degrees

1) D
2) C
3) D
4) B
5) D
6) A
7) A
8) B
9) D
10) A
11) B

5 Find the Exact Values for Integer Multiples of pi/6 = 30 Degrees, pi/4 = 45 Degrees, and pi/3 = 60 Degrees

1) D
2) C
3) C
4) A
5) D
6) C
7) D
8) A
9) A
10) D

6 Use a Calculator to Approximate the Value of a Trigonometric Function

1) B
2) C
3) A
4) D
5) D
6) D
7) A
8) D
9) D
10) D
11) 0.75
12) A
13) C
14) 0.354k; 0.377k; 0.385k; 0.375k and 0.347k; No, it reaches a maximum near 55°.

7 Use a Circle of Radius r to Evaluate the Trigonometric Functions

1) D
2) D
3) D
4) C
5) B
6) C

7) C
8 Demonstrate Additional Understanding and Skills
1) B
2) B
3) D
4) D
5) B
6) C
7) C
8) D
9) B
10) D
11) C
12) C
13) D
14) C
15) B
16) 1 in.; 2 in.; $(1+\sqrt{3})$ in.; 3 in.; $(1+\sqrt{3})$ in.; 2 in.; 1 in.
17) $\frac{\sqrt{3}}{4}$ sq units; 1 sq unit; $\frac{3\sqrt{3}}{2}$ sq units

5.3 Properties of the Trigonometric Functions
1 Determine the Domain and the Range of the Trigonometric Functions
1) B
2) C
3) D
4) B
5) C
6) B

2 Determine the Period of the Trigonometric Functions
1) A
2) D
3) D
4) D
5) D
6) A
7) C
8) C
9) B
10) B
11) B
12) B
13) D
14) B
15) A
16) $-\frac{1}{3}$
17) C
18) C

3 Determine the Signs of the Trigonometric Functions in a Given Quadrant
1) C
2) C
3) B
4) D
5) A

6) B
7) D
8) D
9) C
10) A
11) D
12) (i) negative
   (ii) negative
   (iii) positive

4 Find the Values of the Trigonometric Functions Utilizing Fundamental Identities
1) D
2) B
3) D
4) B
5) C
6) A
7) C
8) C

5 Find the Exact Values of the Trigonometric Functions of an Angle Given One of the Functions and the Quadrant of the Angle
1) B
2) C
3) C
4) A
5) A
6) B
7) B
8) A
9) A
10) $\cos\theta = -\dfrac{\sqrt{35}}{6}$, $\tan\theta = -\dfrac{\sqrt{35}}{35}$

6 Use Even–Odd Properties to Find the Exact Values of the Trigonometric Functions
1) C
2) A
3) C
4) D
5) A
6) A
7) D
8) D
9) C
10) B
11) C
12) A
13) B
14) neither
15) odd

5.4 Graphs of the Sine and Cosine Functions
1 Graph Transformations of the Sine Function
1) A
2) B
3) B
4) C
5) D

- 6) D
- 7) D
- 8) C

2 Graph Transformations of the Cosine Function
- 1) D
- 2) B
- 3) B
- 4) D
- 5) C
- 6) B
- 7) C
- 8) B

3 Determine the Amplitude and Period of Sinusoidal Functions
- 1) A
- 2) C
- 3) D
- 4) D
- 5) D
- 6) B
- 7) D
- 8) A
- 9) C
- 10) D

4 Graph Sinusoidal Functions Using Key Points
- 1) B
- 2) C
- 3) C
- 4) A
- 5) D
- 6) C
- 7) A
- 8) A
- 9) B
- 10) D
- 11) C
- 12) D
- 13) D
- 14) B
- 15) C
- 16) B
- 17) D

5 Find an Equation for a Sinusoidal Graph
- 1) C
- 2) D
- 3) D
- 4) D
- 5) C
- 6) A
- 7) D
- 8) D
- 9) D
- 10) D
- 11) B
- 12) A
- 13) C

14) C
6 Demonstrate Additional Understanding and Skills
1) D
2) C
3) C
4) C
5) B
6) A
7) C
8) 12, 12
9) 365 days
10) D
11) B
12) amplitude = 30, period = $\frac{1}{25}$

$I = 30\sin(50\pi t)$

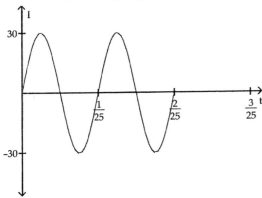

13) amplitude = 4, period = $\frac{\pi}{2}$

$P = 4\cos(4t)$

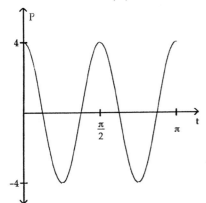

14) amplitude = 0.65, period = 5, respiratory cycle = 5 seconds

$$a = 0.65\sin\left(\frac{2\pi}{5}t\right)$$

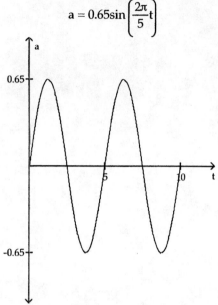

15) amplitude = 25, period = 8/3

$$d = 25\cos\left(\frac{3\pi}{4}t\right)$$

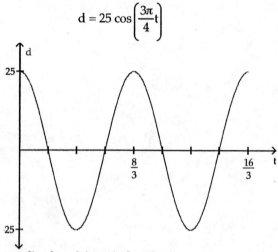

16) amplitude = 6.4, period = 12

$$h = 6.4\sin\left(\frac{\pi}{6}t\right)$$

Precalculus Enhanced with Graphing Utilities 534

5.5 Graphs of the Tangent, Cotangent, Cosecant, and Secant Functions

1 Graph Transformations of the Tangent Function and Cotangent Function

1) B
2) C
3) A
4) A
5) A
6) A
7) C
8) B
9) B
10) C
11) A
12) A
13) B
14) B
15) D
16) B

2 Graph Transformations of the Cosecant Function and Secant Function

1) C
2) C
3) D
4) A
5) B
6) A
7) A
8) A
9) A
10) B

3 Demonstrate Additional Understanding and Skills

1) A
2) D
3) B
4) C
5) C
6) B
7) C
8) B

5.6 Phase Shift; Sinusoidal Curve Fitting

1 Graph Sinusoidal Functions of the Form y = A sin (omega x – phi) Using the Amplitude, Period, and Phase Shift

1) B
2) C
3) C
4) A
5) D
6) A
7) D
8) B
9) C
10) B
11) D
12) B
13) A
14) D

15) B
16)

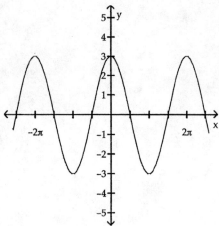

17) C
18) A

2 Find a Sinusoidal Function from Data

1) C
2) B
3) B
4) C
5)

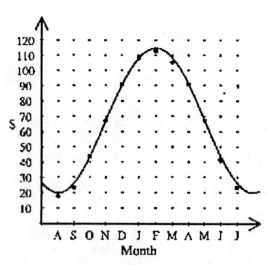

$y = 47.35 \sin(\frac{\pi}{6}x - \frac{2\pi}{3}) + 66.25$

6) $y = 3.14 \sin(0.46x + 1.52) + 3.16$

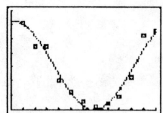

7) $y = 2.17 \sin(0.49x - 1.88) + 6.02$

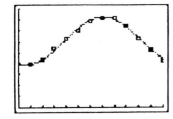

8) $y = 8.33 \sin(0.50x - 2.06) + 57.97$

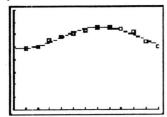

9) $y = 15.99 \sin(0.57x - 2.29) + 60.62$

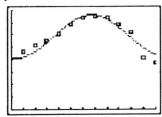

3 Demonstrate Additional Understanding and Skills
1) A
2) D
3) A
4) C

Ch. 6 Analytic Trigonometry

6.1 The Inverse Sine, Cosine, and Tangent Functions

1 Find the Exact Value of the Inverse Sine, Cosine, and Tangent Functions

Find the exact value of the expression.

1) $\sin^{-1}\dfrac{\sqrt{3}}{2}$

   A) $\dfrac{3\pi}{4}$ B) $\dfrac{\pi}{4}$ C) $\dfrac{2\pi}{3}$ D) $\dfrac{\pi}{3}$

2) $\cos^{-1}\dfrac{\sqrt{2}}{2}$

   A) $\dfrac{\pi}{4}$ B) $\dfrac{7\pi}{4}$ C) $\dfrac{11\pi}{6}$ D) $\dfrac{\pi}{6}$

3) $\cos^{-1}\left(-\dfrac{\sqrt{3}}{2}\right)$

   A) $\dfrac{\pi}{3}$ B) $\dfrac{5\pi}{6}$ C) $\dfrac{2\pi}{3}$ D) $\dfrac{\pi}{6}$

4) $\tan^{-1}(1)$

   A) $\dfrac{7\pi}{4}$ B) $\dfrac{5\pi}{4}$ C) $\dfrac{3\pi}{4}$ D) $\dfrac{\pi}{4}$

5) $\sin^{-1}(-0.5)$

   A) $\dfrac{\pi}{3}$ B) $\dfrac{\pi}{6}$ C) $\dfrac{7\pi}{3}$ D) $-\dfrac{\pi}{6}$

6) $\tan^{-1}\dfrac{\sqrt{3}}{3}$

   A) $\dfrac{7\pi}{6}$ B) $\dfrac{\pi}{3}$ C) $\dfrac{\pi}{6}$ D) $\dfrac{\pi}{4}$

Find the exact value of the expression. Do not use a calculator.

7) $\cos[\cos^{-1}(-0.9372)]$

   A) -0.4686 B) 0.9372 C) -0.9372 D) 0.4686

8) $\sin[\sin^{-1}(0.7)]$

   A) 0.7 B) 1.4286 C) 0.3 D) 1.5523

9) $\tan[\tan^{-1}(-0.8)]$

   A) -0.2 B) -0.9712 C) -1.25 D) -0.8

10) $\cos^{-1}\left[\cos\left(-\dfrac{3\pi}{5}\right)\right]$

    A) $-\dfrac{3\pi}{5}$ B) $\dfrac{3\pi}{5}$ C) $\dfrac{2\pi}{5}$ D) $-\dfrac{2\pi}{5}$

11) $\tan^{-1}\left[\tan\left(\frac{3\pi}{5}\right)\right]$

A) $\frac{3\pi}{5}$ B) $-\frac{3\pi}{5}$ C) $-\frac{2\pi}{5}$ D) $\frac{2\pi}{5}$

12) $\sin^{-1}\left[\sin\left(\frac{5\pi}{7}\right)\right]$

A) $\frac{7}{5\pi}$ B) $\frac{5\pi}{7}$ C) $\frac{2\pi}{7}$ D) $\frac{7}{2\pi}$

13) $\sin[\sin^{-1}(-0.3)]$

A) -0.3 B) -2.7 C) 0.3 D) 2.7

14) $\cos^{-1}\left[\cos\left(\frac{\pi}{10}\right)\right]$

A) $-\frac{\pi}{10}$ B) $\frac{\pi}{10}$ C) 10π D) -10π

15) $\sin^{-1}\left[\sin\left(\frac{\pi}{5}\right)\right]$

A) -5π B) $\frac{\pi}{5}$ C) 5π D) $-\frac{\pi}{5}$

16) $\tan^{-1}\left[\tan\left(-\frac{\pi}{8}\right)\right]$

A) -8π B) $-\frac{\pi}{8}$ C) $\frac{\pi}{8}$ D) 8π

2 Find an Approximate Value of the Inverse Sine, Cosine, and Tangent Functions

Use a calculator to find the value of the expression rounded to two decimal places.

1) $\sin^{-1}(0.4)$

A) 0.41 B) 23.58 C) 66.42 D) 1.16

2) $\cos^{-1}(-0.8)$

A) -53.13 B) 2.50 C) 143.13 D) -0.93

3) $\tan^{-1}(-0.3)$

A) -16.70 B) -1.28 C) -0.29 D) -73.30

4) $\sin^{-1}\left(-\frac{1}{5}\right)$

A) 1.77 B) -11.54 C) 101.54 D) -0.20

5) $\cos^{-1}\left(-\frac{4}{5}\right)$

A) -0.93 B) 143.13 C) 2.50 D) -53.13

6) $\sin^{-1}\left(\dfrac{\sqrt{5}}{5}\right)$

A) 63.44 B) 1.11 C) 0.46 D) 26.57

7) $\cos^{-1}\left(-\dfrac{\sqrt{2}}{5}\right)$

A) −16.43 B) 106.43 C) −0.29 D) 1.86

Solve the problem.

8) The formula
$$D = 24\left[1 - \dfrac{\cos^{-1}(\tan i \tan \theta)}{\pi}\right]$$
can be used to approximate the number of hours of daylight when the declination of the sun is $i°$ at a location $\theta°$ north latitude for any date between the vernal equinox and autumnal equinox. To use this formula, $\cos^{-1}(\tan i \tan \theta)$ must be expressed in radians. Approximate the number of hours of daylight in Fargo, North Dakota, (46°52' north latitude) for vernal equinox ($i = 0°$).

9) The formula
$$D = 24\left[1 - \dfrac{\cos^{-1}(\tan i \tan \theta)}{\pi}\right]$$
can be used to approximate the number of hours of daylight when the declination of the sun is $i°$ at a location $\theta°$ north latitude for any date between the vernal equinox and autumnal equinox. To use this formula, $\cos^{-1}(\tan i \tan \theta)$ must be expressed in radians. Approximate the number of hours of daylight in Flagstaff, Arizona, (35°13' north latitude) for summer solstice ($i = 23.5°$).

10) When light travels from one medium to another—from air to water, for instance—it changes direction. (This is why a pencil, partially submerged in water, looks as though it is bent.) The angle of incidence θ_i is the angle in the first medium; the angle of refraction θ_r is the second medium. (See illustration.) Each medium has an index of refraction—n_i and n_r, respectively—which can be found in tables. Snell's law relates these quantities in the formula
$$n_i \sin\theta_i = n_r \sin\theta_r$$
Solving for θ_r, we obtain
$$\theta_r = \sin^{-1}\left(\dfrac{n_i}{n_r}\sin\theta_i\right)$$
Find θ_r for crown glass ($n_i = 1.52$), water ($n_r = 1.33$), and $\theta_i = 38°$.

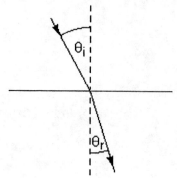

11) When light travels from one medium to another—from air to water, for instance—it changes direction. (This is why a pencil, partially submerged in water, looks as though it is bent.) The angle of incidence θ_i is the angle in the first medium; the angle of refraction θ_r is the second medium. (See illustration.) Each medium has an index of refraction—n_i and n_r, respectively—which can be found in tables. Snell's law relates these quantities in the formula
$$n_i \sin\theta_i = n_r \sin\theta_r$$
Solving for θ_r, we obtain
$$\theta_r = \sin^{-1}\left(\frac{n_i}{n_r}\sin\theta_i\right)$$
Find θ_r for air ($n_i = 1.0003$), methylene iodide ($n_r = 1.74$), and $\theta_i = 14.7°$.

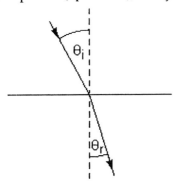

12) When light travels from one medium to another—from air to water, for instance—it changes direction. (This is why a pencil, partially submerged in water, looks as though it is bent.) The angle of incidence θ_i is the angle in the first medium; the angle of refraction θ_r is the second medium. (See illustration.) Each medium has an index of refraction—n_i and n_r, respectively—which can be found in tables. Snell's law relates these quantities in the formula
$$n_i \sin\theta_i = n_r \sin\theta_r$$
Solving for θ_r, we obtain
$$\theta_r = \sin^{-1}\left(\frac{n_i}{n_r}\sin\theta_i\right)$$
Find θ_r for fused quartz ($n_i = 1.46$), ethyl alcohol ($n_r = 1.36$), and $\theta_i = 8.5°$.

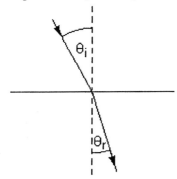

6.2 The Inverse Trigonometric Functions (Continued)

1 Find the Exact Value of Expressions Involving the Inverse Sine, Cosine, and Tangent Functions

Find the exact value of the expression.

1) $\sin(\tan^{-1} 2)$

   A) $\dfrac{2\sqrt{5}}{5}$ B) $\dfrac{5\sqrt{2}}{2}$ C) $5\sqrt{2}$ D) $2\sqrt{5}$

2) $\cos\left(\sin^{-1}\dfrac{1}{4}\right)$

   A) $\dfrac{\sqrt{15}}{2}$ B) $\dfrac{\sqrt{15}}{4}$ C) $\dfrac{4\sqrt{15}}{15}$ D) $\dfrac{2\sqrt{15}}{15}$

3) $\cos\left(2\sin^{-1}\dfrac{1}{4}\right)$

   A) $\dfrac{1}{8}$ B) $\dfrac{7}{8}$ C) $\dfrac{5}{8}$ D) $\dfrac{3}{8}$

4) $\cos\left(\sin^{-1}\dfrac{4}{5}\right)$

   A) $\dfrac{1}{5}$ B) $-\dfrac{4}{5}$ C) $\dfrac{3}{5}$ D) $-\dfrac{3}{5}$

5) $\cos^{-1}\left(\cos\dfrac{7\pi}{6}\right)$

   A) $\dfrac{4\pi}{5}$ B) $\dfrac{\pi}{6}$ C) $\dfrac{\pi}{3}$ D) $\dfrac{5\pi}{6}$

6) $\cos^{-1}\left(\sin\dfrac{7\pi}{6}\right)$

   A) $\dfrac{2\pi}{3}$ B) $\dfrac{\pi}{6}$ C) $\dfrac{4\pi}{5}$ D) $\dfrac{\pi}{3}$

7) $\tan\left(\cos^{-1}\dfrac{1}{3}\right)$

   A) $-2\sqrt{3}$ B) $2\sqrt{3}$ C) $2\sqrt{2}$ D) $\dfrac{2\sqrt{2}}{3}$

8) $\cos\left[\tan^{-1}\left(\dfrac{\sqrt{3}}{3}\right)\right]$

   A) $\dfrac{\sqrt{3}}{3}$ B) $\dfrac{\pi}{3}$ C) $\dfrac{1}{2}$ D) $\dfrac{\sqrt{3}}{2}$

9) $\cos^{-1}[\cos(-0.9372)]$

   A) -0.4686 B) 0.9372 C) -0.9372 D) 0.4686

10) $\sin^{-1}\left[\sin\left(\frac{3\pi}{5}\right)\right]$

   A) $\frac{3\pi}{5}$ B) $\frac{5}{3\pi}$ C) $\frac{5}{2\pi}$ D) $\frac{2\pi}{5}$

11) $\tan^{-1}\left[\tan\left(\frac{3\pi}{5}\right)\right]$

   A) $-\frac{3\pi}{5}$ B) $\frac{2\pi}{5}$ C) $\frac{3\pi}{5}$ D) $-\frac{2\pi}{5}$

12) $\sec\left[\sin^{-1}\left(-\frac{\sqrt{3}}{2}\right)\right]$

   A) $\frac{\sqrt{2}}{2}$ B) 0   C) 2   D) 1

13) $\csc\left(\tan^{-1}\frac{\sqrt{3}}{3}\right)$

   A) $\frac{1}{2}$ B) $\sqrt{3}$ C) $\frac{2\sqrt{3}}{3}$ D) 2

14) $\tan\left(\sin^{-1}\frac{\sqrt{2}}{2}\right)$

   A) 2   B) $\sqrt{2}$ C) 1   D) $\frac{\sqrt{2}}{2}$

2   Know the Definition of the Inverse Secant, Cosecant, and Cotangent Functions

Find the exact value of the expression.

1) $\cot^{-1}(-1)$

   A) $\frac{\pi}{4}$ B) $\frac{3\pi}{4}$ C) $\frac{\pi}{3}$ D) $\frac{2\pi}{3}$

2) $\csc^{-1}\left(\frac{2\sqrt{3}}{3}\right)$

   A) $\frac{2\pi}{3}$ B) $-\frac{\pi}{3}$ C) $\frac{\pi}{3}$ D) $\frac{\pi}{6}$

3) $\sec^{-1}(-\sqrt{2})$

   A) $\frac{\pi}{3}$ B) $-\frac{3\pi}{4}$ C) $-\frac{\pi}{4}$ D) $\frac{3\pi}{4}$

4) $\sec^{-1}(-2)$

   A) $-\frac{2\pi}{3}$ B) $\frac{4\pi}{3}$ C) $\frac{2\pi}{3}$ D) $-\frac{\pi}{3}$

3 Use a Calculator to Evaluate sec^-1 x, csc^-1 x, and cot^-1 x

Use a calculator to find the value of the expression in radian measure rounded to 2 decimal places.

1) $\csc^{-1}\left(-\dfrac{7}{2}\right)$

    A) 106.60   B) -0.29   C) 1.86   D) -16.60

2) $\sec^{-1}(-7)$

    A) -8.21   B) 1.71   C) 98.21   D) -0.14

3) $\cot^{-1}\left(\dfrac{10}{19}\right)$

    A) 0.48   B) 1.09   C) 27.76   D) 62.24

Using a calculator, approximate the value of the expression. Round answer to three decimal places.

4) $\sec^{-1}\left(-\dfrac{7}{3}\right)$

    A) -2.014   B) 1.128   C) 2.014   D) 0.497

6.3 Trigonometric Identities

1 Use Algebra to Simplify Trigonometric Expressions

Simplify the trigonometric expression by following the indicated direction.

1) Rewrite in terms of sine and cosine: $\cot\theta \cdot \tan\theta$

2) Multiply $\dfrac{\sin\theta}{1-\cos\theta}$ by $\dfrac{1+\cos\theta}{1+\cos\theta}$

3) Rewrite over a common denominator: $\dfrac{1}{1-\sin\theta} + \dfrac{1}{1+\sin\theta}$

4) Multiply and simplify: $\dfrac{(\tan\theta+1)(\tan\theta+1) - \sec^2\theta}{\tan\theta}$

5) Factor and simplify: $\dfrac{7\sin^2\theta + 8\sin\theta + 1}{\sin^2\theta - 1}$

Simplify the expression as far as possible.

6) $\dfrac{\cos\theta}{1+\sin\theta} + \tan\theta$

    A) 1   B) $\sin^2\theta$ C) $\cos\theta + \sin\theta$ D) $\sec\theta$

7) $(1+\cot\theta)(1-\cot\theta) - \csc^2\theta$

    A) $2\cot^2\theta$ B) $-2\cot^2\theta$ C) 0   D) 2

2 Establish Identities

Complete the identity.

1) $\sec\theta - \dfrac{1}{\sec\theta} = ?$

   A) $\sec\theta \csc\theta$ B) $-2\tan^2\theta$ C) $\sin\theta \tan\theta$ D) $1 + \cot\theta$

2) $\csc\theta(\sin\theta + \cos\theta) = ?$

   A) $1 + \cot\theta$ B) $\sin\theta \tan\theta$ C) $\sec\theta \csc\theta$ D) $-2\tan^2\theta$

3) $\dfrac{\sin\theta}{1 + \sin\theta} - \dfrac{\sin\theta}{1 - \sin\theta} = ?$

   A) $\sin\theta \tan\theta$ B) $1 + \cot\theta$ C) $\sec\theta \csc\theta$ D) $-2\tan^2\theta$

4) $\dfrac{\sin\theta}{\cos\theta} + \dfrac{\cos\theta}{\sin\theta} = ?$

   A) $-2\tan^2\theta$ B) $1 + \cot\theta$ C) $\sin\theta \tan\theta$ D) $\sec\theta \csc\theta$

5) $\tan^2\theta - 3\sin\theta \tan\theta \sec\theta = ?$

   A) $\sin\theta \tan\theta$ B) $1 + \cot\theta$ C) $-2\tan^2\theta$ D) $\sec\theta \csc\theta$

6) $\dfrac{(\sin\theta + \cos\theta)^2}{1 + 2\sin\theta \cos\theta} = ?$

   A) 1 B) 0 C) $1 - \sin\theta$ D) $-\sec^2\theta$

7) $\dfrac{1}{\cos^2\theta} - \dfrac{1}{\cot^2\theta} = ?$

   A) $1 - \sin\theta$ B) $-\sec^2\theta$ C) 1 D) 0

8) $2\tan\theta - (1 + \tan\theta)^2 = ?$

   A) $1 - \sin\theta$ B) $-\sec^2\theta$ C) 1 D) 0

9) $\tan\theta(\cot\theta - \cos\theta) = ?$

   A) $-\sec^2\theta$ B) 1 C) $1 - \sin\theta$ D) 0

10) $\dfrac{\sec\theta \sin\theta}{\tan\theta} - 1 = ?$

    A) $-\sec^2\theta$ B) $1 - \sin\theta$ C) 1 D) 0

11) $\dfrac{\sin^2\theta - 1}{\sin\theta + 1} = ?$

    A) $\sin\theta$ B) $\sin\theta + 1$ C) $\cos^2\theta$ D) $\sin\theta - 1$

12) $\sec^4\theta - 2\sec^2\theta \tan^2\theta + \tan^4\theta = ?$

    A) $\sec^2\theta(1 + \tan^2\theta)$ B) 1 C) $\sec^2\theta + \tan^2\theta$ D) 2

13) $\dfrac{\csc\theta\,(\sin^2\theta + \cos^2\theta\,\tan\theta)}{\sin\theta + \cos\theta} = ?$

    A) $\csc^2\theta + \tan^2\theta$ B) 1 C) $\csc\theta\,\tan\theta$ D) $\sin\theta + \cos\theta$

14) $\sin^2\theta + \sin^2\theta\,\cot^2\theta = ?$

    A) 1 B) $\sin^2\theta + 1$ C) $\cot^2\theta - 1$ D) $\cot^2\theta + 1$

15) $\sec^4\theta + \sec^2\theta\,\tan^2\theta - 2\tan^4\theta = ?$

    A) $\sec^2\theta + 2$ B) $3\sec^2\theta - 2$ C) $4\sec^2\theta$ D) $\tan^2\theta - 1$

16) $\tan^4\theta - \sec^4\theta = ?$

    A) $-2\tan^2\theta - 1$ B) $\sec^2\theta + \tan^2\theta$ C) $\tan^2\theta - \sec^2\theta$ D) $\sec^2\theta$

17) $\dfrac{\csc\theta\,\cot\theta}{\sec\theta} = ?$

    A) $\csc^2\theta$ B) $\sec^2\theta$ C) 1 D) $\cot^2\theta$

18) $\dfrac{1}{\cot^2\theta} + \sec\theta\,\cos\theta = ?$

    A) $\csc^2\theta$ B) $\sec^2\theta$ C) 1 D) $\tan^2\theta$

19) $\cos\theta - \cos\theta\,\sin^2\theta = ?$

    A) $\sec^2\theta$ B) $\tan^2\theta$ C) $\cos^3\theta$ D) $\sin\theta$

20) $\sin^2\theta + \tan^2\theta + \cos^2\theta = ?$

    A) $\tan^2\theta$ B) $\sec^2\theta$ C) $\sin\theta$ D) $\cos^3\theta$

21) $\dfrac{7 + 7\sin\theta}{-5\cos\theta} =$

    A) $\dfrac{7\sin\theta}{-5(1 + \cos\theta)}$ B) $\dfrac{7\cos\theta}{-5(1 - \sin\theta)}$ C) $\dfrac{7\cos\theta}{-5(1 + \sin\theta)}$ D) $\dfrac{7\sin\theta}{-5(1 - \cos\theta)}$

22) $\ln|\cot\theta| = ?$

    A) $\ln|\sin\theta| - \ln|\cos\theta|$ B) $\dfrac{\ln|\cos\theta|}{\ln|\sin\theta|}$ C) $\dfrac{\ln|\sin\theta|}{\ln|\cos\theta|}$ D) $\ln|\cos\theta| - \ln|\sin\theta|$

23) $\ln|\csc\theta| = ?$

    A) $-\ln|\cos\theta|$ B) $\ln|\sin\theta|$ C) $\ln|\cos\theta|$ D) $-\ln|\sin\theta|$

3 Demonstrate Additional Understanding and Skills

Use a right triangle to write the expression as an algebraic expression. Assume that v is positive and in the domain of the given inverse trigonometric function.

1) $\sin(\tan^{-1} v)$

    A) $v\sqrt{v^2 + 1}$ B) $\dfrac{v\sqrt{v^2 + 1}}{v^2 + 1}$ C) $\dfrac{v\sqrt{v^2 - 1}}{v^2 - 1}$ D) $\dfrac{\sqrt{v^2 + 1}}{v^2 + 1}$

2) $\cos(\tan^{-1} v)$

   A) $\dfrac{v\sqrt{v^2+1}}{v^2+1}$
   B) $\dfrac{\sqrt{v^2+1}}{v^2+1}$
   C) $v\sqrt{v^2+1}$
   D) $\dfrac{\sqrt{v^2-1}}{v^2-1}$

3) $\cos(\sin^{-1} v)$

   A) $\dfrac{\sqrt{v^2+1}}{v}$
   B) $\sqrt{v^2+1}$
   C) $\sqrt{v^2-1}$
   D) $\sqrt{1-v^2}$

4) $\sin\left(\tan^{-1}\dfrac{v}{\sqrt{2}}\right)$

   A) $\dfrac{\sqrt{v^2+2}}{v^2+2}$
   B) $\dfrac{v\sqrt{v^2+2}}{v^2+2}$
   C) $v\sqrt{v^2+2}$
   D) $\dfrac{v\sqrt{v^2-2}}{v^2-2}$

5) $\sin\left(\sin^{-1}\dfrac{v}{\sqrt{3}}\right)$

   A) $\dfrac{v\sqrt{v^2-3}}{v^2-3}$
   B) $\dfrac{\sqrt{v^2+3}}{v^2+3}$
   C) $\dfrac{v\sqrt{3}}{3}$
   D) $v\sqrt{3}$

6) $\tan\left(\sec^{-1}\dfrac{\sqrt{v^2+25}}{v}\right)$

   A) $5v$
   B) $\dfrac{\sqrt{v^2+5}}{v^2+5}$
   C) $\dfrac{v\sqrt{v^2+25}}{v^2+25}$
   D) $\dfrac{5}{v}$

7) $\sin\left(\sec^{-1}\dfrac{\sqrt{v^2+9}}{v}\right)$

   A) $\dfrac{v\sqrt{v^2+3}}{v^2+3}$
   B) $\dfrac{\sqrt{v^2+3}}{v^2+3}$
   C) $v\sqrt{3}$
   D) $\dfrac{3\sqrt{v^2+9}}{v^2+9}$

6.4 Sum and Difference Formulas

1 Use Sum and Difference Formulas to Find Exact Values

Find the exact value of the trigonometric function.

1) $\sin\left(-\dfrac{11\pi}{12}\right)$

   A) $-\dfrac{\sqrt{6}+\sqrt{2}}{4}$
   B) $\dfrac{\sqrt{2}+\sqrt{6}}{4}$
   C) $\dfrac{\sqrt{2}-\sqrt{6}}{4}$
   D) $\dfrac{\sqrt{6}-\sqrt{2}}{4}$

2) $\sin\dfrac{11\pi}{12}$

   A) $-\sqrt{2}(\sqrt{3}-1)$
   B) $-\dfrac{\sqrt{2}(\sqrt{3}-1)}{4}$
   C) $\sqrt{2}(\sqrt{3}-1)$
   D) $\dfrac{\sqrt{2}(\sqrt{3}-1)}{4}$

3) $\cos\dfrac{5\pi}{12}$

   A) $\dfrac{\sqrt{2}(\sqrt{3}-1)}{4}$
   B) $\sqrt{2}(\sqrt{3}-1)$
   C) $-\sqrt{2}(\sqrt{3}-1)$
   D) $-\dfrac{\sqrt{2}(\sqrt{3}-1)}{4}$

4) $\tan \dfrac{13\pi}{12}$

A) $-2-\sqrt{3}$ B) $2+\sqrt{3}$ C) $2-\sqrt{3}$ D) $\sqrt{3}-2$

5) $\sin 15°$

A) $\dfrac{\sqrt{2}(\sqrt{3}+1)}{4}$ B) $\dfrac{\sqrt{2}(\sqrt{3}-1)}{4}$ C) $-\dfrac{\sqrt{2}(\sqrt{3}-1)}{4}$ D) $-\dfrac{\sqrt{2}(\sqrt{3}+1)}{4}$

6) $\cos 285°$

A) $-\dfrac{\sqrt{2}(\sqrt{3}-1)}{4}$ B) $-\sqrt{2}(\sqrt{3}-1)$ C) $-\sqrt{2}(\sqrt{3}+1)$ D) $\dfrac{\sqrt{2}(\sqrt{3}-1)}{4}$

7) $\cot 15°$

A) $-\sqrt{3}+2$ B) $\sqrt{3}+2$ C) $-\sqrt{3}-2$ D) $\sqrt{3}-2$

8) $\sin 255°$

A) $\dfrac{\sqrt{2}(\sqrt{3}-1)}{4}$ B) $\dfrac{\sqrt{2}(\sqrt{3}+1)}{4}$ C) $\dfrac{-\sqrt{2}(\sqrt{3}+1)}{4}$ D) $\dfrac{-\sqrt{2}(\sqrt{3}-1)}{4}$

9) $\tan 105°$

A) $\dfrac{2+\sqrt{3}}{4}$ B) $-2-\sqrt{3}$ C) $\dfrac{2-\sqrt{3}}{4}$ D) $2+\sqrt{3}$

Find the exact value of the expression.

10) $\sin 20° \cos 40° + \cos 20° \sin 40°$

A) $\dfrac{1}{2}$ B) $\dfrac{\sqrt{3}}{3}$ C) $\dfrac{\sqrt{3}}{2}$ D) $\dfrac{1}{3}$

11) $\sin 250° \cos 10° - \cos 250° \sin 10°$

A) $-\dfrac{25}{6}$ B) $-\dfrac{1}{2}$ C) $\dfrac{\sqrt{3}}{2}$ D) $-\dfrac{\sqrt{3}}{2}$

12) $\sin 220° \cos 100° - \cos 220° \sin 100°$

A) $-\dfrac{\sqrt{3}}{2}$ B) $-\dfrac{1}{2}$ C) $\dfrac{\sqrt{3}}{2}$ D) $-\dfrac{11}{3}$

13) $\sin 15° \cos 105° + \cos 15° \sin 105°$

A) $\dfrac{\sqrt{3}}{2}$ B) $\dfrac{1}{4}$ C) $-\dfrac{\sqrt{3}}{2}$ D) $-\dfrac{1}{2}$

14) $\cos 10° \cos 50° - \sin 10° \sin 50°$

A) $\dfrac{1}{2}$ B) $\dfrac{1}{4}$ C) $\sqrt{3}$ D) $\dfrac{\sqrt{3}}{2}$

15) $\cos\left(\frac{7\pi}{12}\right)\cos\left(\frac{5\pi}{12}\right) + \sin\left(\frac{7\pi}{12}\right)\sin\left(\frac{5\pi}{12}\right)$

   A) $\frac{1}{2}$ B) 1   C) $\frac{\sqrt{3}}{2}$ D) $\frac{1}{4}$

16) $\frac{\tan 25° + \tan 5°}{1 - \tan 25° \tan 5°}$

   A) 2   B) $\frac{\sqrt{3}}{3}$ C) $\sqrt{3}$ D) $\frac{1}{2}$

17) $\frac{\tan 80° + \tan 70°}{1 - \tan 80° \tan 70°}$

   A) $-\sqrt{3}$ B) $-\frac{\sqrt{3}}{3}$ C) -2 D) $-\frac{1}{2}$

18) $\frac{\tan 175° - \tan 55°}{1 + \tan 175° \tan 55°}$

   A) $-\frac{\sqrt{3}}{3}$ B) -2 C) $-\sqrt{3}$ D) $-\frac{1}{2}$

19) $\frac{\tan 65° - \tan(-55°)}{1 + \tan 65° \tan(-55°)}$

   A) -2 B) $-\sqrt{3}$ C) $-\frac{\sqrt{3}}{3}$ D) $-\frac{1}{2}$

20) $\frac{1 - \tan 80° \tan 70°}{\tan 80° + \tan 70°}$

   A) $-\frac{\sqrt{3}}{3}$ B) $\frac{\sqrt{3}}{3}$ C) $\sqrt{3}$ D) $-\sqrt{3}$

Find the exact value under the given conditions.

21) $\sin\alpha = \frac{12}{13}$, $0 < \alpha < \frac{\pi}{2}$; $\cos\beta = \frac{21}{29}$, $0 < \beta < \frac{\pi}{2}$ Find $\cos(\alpha + \beta)$.

   A) $\frac{352}{377}$ B) $-\frac{135}{377}$ C) $\frac{345}{377}$ D) $\frac{152}{377}$

22) $\sin\alpha = \frac{8}{17}$, $\frac{\pi}{2} < \alpha < \pi$; $\cos\beta = \frac{21}{29}$, $0 < \beta < \frac{\pi}{2}$ Find $\sin(\alpha - \beta)$.

   A) $\frac{155}{493}$ B) $\frac{475}{493}$ C) $-\frac{132}{493}$ D) $\frac{468}{493}$

23) $\tan\alpha = \frac{20}{21}$, $\pi < \alpha < \frac{3\pi}{2}$; $\cos\beta = -\frac{24}{25}$, $\frac{\pi}{2} < \beta < \pi$ Find $\sin(\alpha + \beta)$.

   A) $\frac{644}{725}$ B) $\frac{333}{725}$ C) $\frac{627}{725}$ D) $\frac{364}{725}$

24) $\sin \alpha = -\dfrac{20}{29}$, $\dfrac{3\pi}{2} < \alpha < 2\pi$; $\tan \beta = -\dfrac{8}{15}$, $\dfrac{\pi}{2} < \beta < \pi$ Find $\cos(\alpha + \beta)$.

A) $\dfrac{132}{493}$ B) $-\dfrac{155}{493}$ C) $\dfrac{468}{493}$ D) $-\dfrac{475}{493}$

25) $\sin \alpha = \dfrac{3}{5}$, $\dfrac{\pi}{2} < \alpha < \pi$; $\cos \beta = \dfrac{2}{5}$, $0 < \beta < \dfrac{\pi}{2}$ Find $\cos(\alpha - \beta)$.

A) $\dfrac{8 - 3\sqrt{21}}{25}$ B) $\dfrac{-8 + 3\sqrt{21}}{25}$ C) $\dfrac{6 + 4\sqrt{21}}{25}$ D) $\dfrac{6 - 4\sqrt{21}}{25}$

26) $\sin \alpha = -\dfrac{24}{25}$, $\dfrac{3\pi}{2} < \alpha < 2\pi$; $\cos \beta = -\dfrac{\sqrt{21}}{5}$, $\pi < \beta < \dfrac{3\pi}{2}$ Find $\sin(\alpha - \beta)$.

A) $\dfrac{-48 - 7\sqrt{21}}{125}$ B) $\dfrac{14 + 24\sqrt{21}}{125}$ C) $\dfrac{-14 + 24\sqrt{21}}{125}$ D) $\dfrac{-48 + 7\sqrt{21}}{125}$

27) $\sin \alpha = -\dfrac{7}{25}$, $\pi < \alpha < \dfrac{3\pi}{2}$; $\tan \beta = -\dfrac{2\sqrt{21}}{21}$, $\dfrac{\pi}{2} < \beta < \pi$ Find $\cos(\alpha + \beta)$.

A) $\dfrac{14 - 24\sqrt{21}}{125}$ B) $\dfrac{14 + 24\sqrt{21}}{125}$ C) $\dfrac{-48 - 7\sqrt{21}}{125}$ D) $\dfrac{48 - 7\sqrt{21}}{125}$

28) $\cos \alpha = -\dfrac{24}{25}$, $\dfrac{\pi}{2} < \alpha < \pi$; $\sin \beta = -\dfrac{\sqrt{21}}{5}$, $\pi < \beta < \dfrac{3\pi}{2}$ Find $\cos(\alpha + \beta)$.

A) $\dfrac{-14 + 24\sqrt{21}}{125}$ B) $\dfrac{-48 - 7\sqrt{21}}{125}$ C) $\dfrac{48 + 7\sqrt{21}}{125}$ D) $\dfrac{14 - 24\sqrt{21}}{125}$

2 Use Sum and Difference Formulas to Establish Identities

Complete the identity.

1) $\cos(\alpha + \beta) \cos(\alpha - \beta) = ?$
   A) $\cos^2 \beta - \sin^2 \alpha$
   B) $\cos^2 \beta - 2\sin^2 \alpha \sin^2 \beta$
   C) $2 - \sin^2 \alpha - \sin^2 \beta$
   D) $\cos(\alpha^2)\cos(\beta^2) + \sin(\alpha^2)\sin(\beta^2)$

2) $\csc(\alpha - \beta) = ?$
   A) $\csc \alpha \sec \beta - \sec \alpha \csc \beta$
   B) $\dfrac{\sec \alpha \sec \beta}{\tan \alpha \tan \beta - 1}$
   C) $\csc \alpha - \csc \beta$
   D) $\dfrac{\sec \alpha \sec \beta}{\tan \alpha - \tan \beta}$

3) $\sin(\alpha + \beta) \cos \beta - \cos(\alpha + \beta) \sin \beta$
   A) $\sin \alpha \cos \beta - \cos \alpha \sin \beta$
   B) $2\sin \beta \cos \beta (\sin \alpha - \cos \alpha)$
   C) $\sin \alpha$
   D) $\sin \alpha \cos^2 \beta - \sin \alpha \sin^2 \beta$

4) $\cos\left(\dfrac{\pi}{2} + \theta\right) = ?$
   A) $\cos \theta$ B) $-\cos \theta$ C) $-\sin \theta$ D) $\sin \theta$

5) $\dfrac{\cos(\alpha - \beta)}{\sin \alpha \cos \beta} = ?$

  A) $\cos \alpha - \sin \beta$ B) $\cos \alpha + \sin \beta$ C) $\cot \alpha - 1$ D) $\cot \alpha + \tan \beta$

6) $\dfrac{\sin(\alpha - \beta)}{\sin(\alpha + \beta)} = ?$

  A) $\dfrac{\sin \alpha - \sin \beta}{\sin \alpha + \sin \beta}$ B) $\dfrac{\tan \alpha - \tan \beta}{\tan \alpha + \tan \beta}$ C) $\dfrac{\tan \alpha \tan \beta - 1}{\tan \alpha \tan \beta + 1}$ D) -1

7) $\sin(\alpha + \beta) \cos(\alpha - \beta) = ?$

  A) $\sin \alpha \cos \alpha + \sin \beta \cos \beta$
  B) $\sin^2 \alpha \cos^2 \beta - \sin^2 \beta \cos^2 \alpha$
  C) $\sin^2 \alpha - \cos^2 \beta$
  D) $\sin \alpha \cos \alpha - \sin \alpha \cos \beta + \sin \beta \cos \alpha + \sin \beta \cos \beta$

8) $\sec(\alpha + \beta) \sec(\alpha - \beta) = ?$

  A) $\dfrac{\sec^2 \alpha \sec^2 \beta}{1 - \tan^2 \alpha \tan^2 \beta}$ B) $\dfrac{1}{\cos^2 \alpha - \cos^2 \beta}$ C) $\sec^2 \alpha - \sec^2 \beta$ D) $\dfrac{\sec \alpha \sec \beta}{\cos^2 \alpha - \sin^2 \beta}$

9) $\sin(\pi - \theta) = ?$

  A) $\sin \theta$ B) $-\sin \theta$ C) $\cos \theta$ D) $-\cos \theta$

10) $\tan(\pi - \theta) = ?$

  A) $\tan \theta$ B) $\cot \theta$ C) $-\cot \theta$ D) $-\tan \theta$

3 Use Sum and Difference Formulas Involving Inverse Trigonometric Functions

Find the exact value of the expression.

1) $\sin\left(\cos^{-1} \dfrac{1}{2} - \sin^{-1} \dfrac{\sqrt{3}}{2}\right)$

  A) $\dfrac{\sqrt{3}}{3}$ B) $\dfrac{2\sqrt{3}}{2}$ C) 1 D) 0

2) $\cos\left(\tan^{-1} \dfrac{4}{3} - \sin^{-1} \dfrac{3}{5}\right)$

  A) $\dfrac{24}{25}$ B) 1 C) $\dfrac{2\sqrt{3}}{5}$ D) $\dfrac{2\sqrt{6}}{5}$

3) $\sin\left(\sin^{-1} \dfrac{2}{3} + \cos^{-1} \dfrac{1}{3}\right)$

  A) $\dfrac{2\sqrt{6}}{5}$ B) $\dfrac{2 + 2\sqrt{10}}{9}$ C) $\dfrac{2\sqrt{3} + 2\sqrt{10}}{9}$ D) $\dfrac{2\sqrt{3}}{5}$

4) $\tan\left(\tan^{-1} \dfrac{3}{4} + \sin^{-1} \dfrac{1}{2}\right)$

  A) $\dfrac{2\sqrt{3} + 2\sqrt{10}}{9}$ B) $\dfrac{9 + 4\sqrt{3}}{12 - 3\sqrt{3}}$ C) $\dfrac{2\sqrt{6}}{5}$ D) $\dfrac{2\sqrt{3}}{5}$

5) $\cos\left(\sin^{-1}\dfrac{1}{3} - \tan^{-1}\dfrac{1}{2}\right)$

A) $\dfrac{2\sqrt{3}+4}{3\sqrt{5}}$ B) $\dfrac{2\sqrt{3}+1}{5}$ C) $\dfrac{2\sqrt{6}}{5}$ D) $\dfrac{4\sqrt{10}+\sqrt{5}}{15}$

6) $\cos\left(\tan^{-1}\dfrac{5}{12} - \cos^{-1}\dfrac{4}{5}\right)$

A) $\dfrac{63}{65}$ B) $\dfrac{13}{24}$ C) $\dfrac{52}{65}$ D) $\dfrac{7}{13}$

Write the trigonometric expression as an algebraic expression containing u and v.

7) $\cos(\sin^{-1} u - \cos^{-1} v)$

A) $v\sqrt{1-u^2} - u\sqrt{1-v^2}$ B) $uv - (\sqrt{1-u^2})(\sqrt{1-v^2})$
C) $v\sqrt{1-u^2} + u\sqrt{1-v^2}$ D) $uv + (\sqrt{1-u^2})(\sqrt{1-v^2})$

8) $\cos(\tan^{-1} u + \tan^{-1} v)$

A) $\dfrac{\sqrt{u^2+1} \cdot \sqrt{v^2+1}}{1-uv}$ B) $\dfrac{1-uv}{\sqrt{u^2+1}\cdot\sqrt{v^2+1}}$ C) $\dfrac{1+uv}{\sqrt{u^2+1}\cdot\sqrt{v^2+1}}$ D) $\dfrac{u+v}{\sqrt{u^2+1}\cdot\sqrt{v^2+1}}$

9) $\sin(\tan^{-1} u + \tan^{-1} v)$

A) $\dfrac{\sqrt{u^2+1} \cdot \sqrt{v^2+1}}{1-uv}$ B) $\dfrac{1+uv}{\sqrt{u^2+1}\cdot\sqrt{v^2+1}}$ C) $\dfrac{u+v}{\sqrt{u^2+1}\cdot\sqrt{v^2+1}}$ D) $\dfrac{1-uv}{\sqrt{u^2+1}\cdot\sqrt{v^2+1}}$

10) $\sin(\tan^{-1} u - \tan^{-1} v)$

A) $\dfrac{1-uv}{\sqrt{u^2+1}\cdot\sqrt{v^2+1}}$ B) $\dfrac{u-v}{\sqrt{u^2+1}\cdot\sqrt{v^2+1}}$ C) $\dfrac{\sqrt{u^2+1}\cdot\sqrt{v^2+1}}{1-uv}$ D) $\dfrac{1+uv}{\sqrt{u^2+1}\cdot\sqrt{v^2+1}}$

11) $\cos(\sin^{-1} u + \cos^{-1} v)$

A) $v\sqrt{1-u^2} - u\sqrt{1-v^2}$ B) $uv - (\sqrt{1-u^2})(\sqrt{1-v^2})$
C) $v\sqrt{1-u^2} + u\sqrt{1-v^2}$ D) $uv + (\sqrt{1-u^2})(\sqrt{1-v^2})$

6.5 Double-angle and Half-angle Formulas

1 Use Double-Angle Formulas to Find Exact Values

Use the information given about the angle θ, $0 \le \theta \le 2\pi$, to find the exact value of the indicated trigonometric function.

1) $\sin\theta = \dfrac{24}{25}$, $0 < \theta < \dfrac{\pi}{2}$ Find $\cos(2\theta)$.

A) $\dfrac{336}{625}$ B) $\dfrac{527}{625}$ C) $-\dfrac{527}{625}$ D) $-\dfrac{21}{25}$

2) $\cos\theta = \dfrac{12}{13}$, $\dfrac{3\pi}{2} < \theta < 2\pi$ Find $\sin(2\theta)$.

A) $-\dfrac{120}{169}$ B) $-\dfrac{119}{169}$ C) $\dfrac{120}{169}$ D) $\dfrac{119}{169}$

3) $\tan \theta = \dfrac{24}{7}$, $\pi < \theta < \dfrac{3\pi}{2}$ Find sin (2θ).

   A) $-\dfrac{336}{625}$ B) $-\dfrac{527}{625}$ C) $\dfrac{527}{625}$ D) $\dfrac{336}{625}$

4) $\csc \theta = \dfrac{25}{7}$, $\dfrac{\pi}{2} < \theta < \pi$ Find cos (2θ).

   A) $-\dfrac{527}{625}$ B) $\dfrac{527}{625}$ C) $-\dfrac{336}{625}$ D) $\dfrac{336}{625}$

5) $\csc \theta = -\dfrac{5}{2}$, $\tan \theta > 0$ Find cos (2θ).

   A) $\dfrac{-4\sqrt{21}}{25}$ B) $\dfrac{4\sqrt{21}}{25}$ C) $\dfrac{17}{25}$ D) $-\dfrac{17}{25}$

6) $\sec \theta = -\dfrac{3\sqrt{5}}{5}$, $\csc \theta > 0$ Find sin (2θ).

   A) $\dfrac{-4\sqrt{5}}{9}$ B) $-\dfrac{1}{9}$ C) $\dfrac{4\sqrt{5}}{9}$ D) $\dfrac{1}{9}$

7) $\sin \theta = \dfrac{2\sqrt{6}}{7}$, $\tan \theta < 0$ Find sin (2θ).

   A) $\dfrac{20\sqrt{6}}{49}$ B) $-\dfrac{1}{49}$ C) $\dfrac{-20\sqrt{6}}{49}$ D) $\dfrac{1}{49}$

8) $\cos \theta = -\dfrac{1}{5}$, $\csc \theta < 0$ Find cos (2θ).

   A) $-\dfrac{23}{25}$ B) $\dfrac{23}{25}$ C) $\dfrac{4\sqrt{6}}{25}$ D) $\dfrac{-4\sqrt{6}}{25}$

9) $\sin \theta = -\dfrac{4}{5}$, $\dfrac{3\pi}{2} < \theta < 2\pi$ Find cos (2θ).

   A) $-\dfrac{24}{25}$ B) $\dfrac{7}{25}$ C) $\dfrac{24}{25}$ D) $-\dfrac{7}{25}$

10) $\cos \theta = \dfrac{\sqrt{5}}{5}$, $0 < \theta < \dfrac{\pi}{2}$ Find sin (2θ).

    A) $\dfrac{3}{5}$ B) $\dfrac{2}{5}$ C) $\dfrac{1}{5}$ D) $\dfrac{4}{5}$

2 Use Double-Angle Formulas to Establish Identities

Complete the identity.

1) cos (4θ) = ?

   A) $4 \sin \theta \cos^3 \theta - 4 \sin^3 \theta \cos \theta$
   B) $\cos^3 \theta - 3 \sin^2 \theta \cos \theta$
   C) $\cos^4 \theta - 6 \sin^2 \theta \cos^2 \theta + \sin^4 \theta$
   D) $3 \sin \theta - 4 \sin^3 \theta$

2) $\sin(4\theta) = ?$

A) $3\sin\theta - 4\sin^3\theta$

B) $\cos^3\theta - 3\sin^2\theta\cos\theta$

C) $1 - 8\sin^2\theta\cos^2\theta$

D) $4\sin\theta\cos^3\theta - 4\sin^3\theta\cos\theta$

3) $\csc(2\theta) - \sec(2\theta)(\tan\theta - 1) = ?$

A) $\dfrac{\sin\theta + \cos\theta}{\sin(2\theta)(\sin\theta - \cos\theta)}$

B) $\dfrac{\csc\theta + \sec\theta}{\sin(2\theta)(\sin\theta - \cos\theta)}$

C) $\dfrac{\cos\theta - \sin\theta}{\sin(2\theta)(\sin\theta + \cos\theta)}$

D) $\dfrac{\csc\theta - \sec\theta}{\sin(2\theta)(\sin\theta + \cos\theta)}$

4) $\dfrac{\sin(2\theta) + 2\sin^2\theta}{\cos(2\theta)} = ?$

A) $\dfrac{2\tan\theta}{1 - \tan\theta}$

B) $2\tan\theta + 2$

C) $\tan(2\theta) + 2\tan^2\theta$

D) $\dfrac{\tan(2\theta) + \tan^2\theta}{\sec(2\theta)}$

5) $\sin\theta\cos^3\theta + \sin^3\theta\cos\theta = ?$

A) $\dfrac{1}{4}\sin^2(2\theta)$

B) $\dfrac{1}{2}\sin(2\theta)\cos(2\theta)$

C) $\dfrac{1}{2}\sin(2\theta)$

D) $\dfrac{1}{6}\sin(3\theta)\sin(2\theta)$

6) $\dfrac{1}{2}\sec\theta\csc\theta = ?$

A) $\dfrac{\cot(2\theta)}{\cos\theta}$

B) $\csc(2\theta)$

C) $\sec(2\theta)$

D) $\dfrac{\tan(2\theta)}{\sin\theta}$

7) $1 - \dfrac{1}{2}\sin(2\theta) = ?$

A) $\dfrac{\sin^3\theta - \cos^3\theta}{\sin\theta + \cos\theta}$

B) $\dfrac{\sin^3\theta + \cos^3\theta}{\sin\theta + \cos\theta}$

C) $\dfrac{\sin\theta\cos\theta\cos(2\theta)}{\sin\theta + \cos\theta}$

D) $\dfrac{\sin\theta\cos(2\theta)}{\sin\theta + \cos\theta}$

8) $\dfrac{\sin(3\theta)}{\sin\theta} = ?$

A) $2\sin\theta\cos\theta$

B) $2\cos(2\theta)$

C) $\cos(2\theta) + \sin\theta\cos\theta$

D) $4\cos^2\theta - 1$

9) $(4\sin\theta\cos\theta)(1 - 2\sin^2\theta) = ?$

A) $\sin(4\theta)$

B) $\sin(2\theta) + \cos(2\theta)$

C) $\sin(2\theta)\cos(2\theta)$

D) $\dfrac{\cos(4\theta)}{\sin(2\theta)}$

10) $\sin(2\theta)\tan\theta + \cos(2\theta) = ?$

A) $\cos(3\theta)$

B) 1

C) $2\cos(2\theta)$

D) $\sec(2\theta)$

11) $\pm\sqrt{\dfrac{1 + \cos(36\theta)}{2}} = ?$

A) $\cos(72\theta)$

B) $\sin(72\theta)$

C) $\sin(18\theta)$

D) $\cos(18\theta)$

12) $\pm\sqrt{\dfrac{1-\cos(22\theta)}{2}} = ?$

   A) $\sin(44\theta)$ B) $\cos(44\theta)$ C) $\cos(11\theta)$ D) $\sin(11\theta)$

13) $\csc^2 \dfrac{\theta}{2} = ?$

   A) $\dfrac{2}{1+\sin\theta}$ B) $\dfrac{2}{1-\cos\theta}$ C) $\dfrac{1+\sec\theta}{2}$ D) $\dfrac{1-\sec\theta}{2}$

14) $\csc^2 \dfrac{\theta}{2} - \sec^2 \dfrac{\theta}{2} = ?$

   A) 0 B) $4\cot\theta\csc\theta$ C) $2\sin^2\theta\sec\theta$ D) $\dfrac{2}{1-\sec\theta}$

15) $\sin\dfrac{\theta}{2}\cos\dfrac{\theta}{2} = ?$

   A) $\dfrac{1}{2}\cos\theta$ B) $\dfrac{1+\sin\theta}{2}$ C) $\dfrac{1}{2}\sin\theta$ D) $\dfrac{1-\cos\theta}{2}$

16) $\dfrac{\sin^2 \dfrac{\theta}{2}}{\sin^2\theta} = ?$

   A) $\dfrac{1}{2+2\cos\theta}$ B) $\dfrac{1}{2-2\cos\theta}$ C) $\sin^2\dfrac{\theta}{2}$ D) $\cos^2\dfrac{\theta}{2}$

17) $\csc(2\theta) + \cot(2\theta) = ?$

   A) $\cot\theta$ B) $\cot\dfrac{\theta}{2}$ C) $\tan\dfrac{\theta}{2}$ D) $\tan\theta$

Rewrite each expression as an equivalent expression that does not contain powers of trigonometric functions greater than 1.

18) $8\cos^2 x$

   A) $4 + 4\cos 2x$ B) $16\cos x$ C) $4 - 4\cos 2x$ D) $1 + \cos 2x$

19) $\sin^4 x$

   A) $\dfrac{3}{2} - \dfrac{3}{2}\cos 2x$ B) $\dfrac{3}{2} - 2\cos 2x + \dfrac{1}{2}\cos 4x$

   C) $\dfrac{3}{8} - \dfrac{1}{2}\cos 2x + \dfrac{1}{8}\cos 4x$ D) $\dfrac{3}{8} + \dfrac{5}{8}\cos 2x$

Simplify the expression as far as possible.

20) $4\sin\theta\cos^3\theta - 4\sin^3\theta\cos\theta$

   A) $\sin 2\theta\cos 2\theta$ B) $\sin 4\theta$ C) $4\sin\theta\cos\theta\cos 2\theta$ D) $4\sin\theta\cos\theta$

21) $\sin 8x\cos 8x$

   A) $\cos 8x$ B) $\cos 4x$ C) $\dfrac{1}{2}\sin 16x$ D) $2\sin 4x$

22) $\cos^2 4x - \sin^2 4x$

   A) $\cos 8x$ B) $\cos 4x$ C) $\frac{1}{2}\sin 16x$ D) $2\sin 4x$

23) $4\sin 2x \cos 2x$

   A) $\frac{1}{2}\sin 16x$ B) $\cos 8x$ C) $2\sin 4x$ D) $\cos 4x$

24) $1 - 2\sin^2 2x$

   A) $2\sin 4x$ B) $\cos 8x$ C) $\frac{1}{2}\sin 16x$ D) $\cos 4x$

25) $2\cos^2 4x - 1$

   A) $2\sin 4x$ B) $\cos 8x$ C) $\cos 4x$ D) $\frac{1}{2}\sin 16x$

3 Use Half-Angle Formulas to Find Exact Values

Use the information given about the angle θ, $0 \le \theta \le 2\pi$, to find the exact value of the indicated trigonometric function.

1) $\sin\theta = \frac{1}{4}$, $0 < \theta < \frac{\pi}{2}$ Find $\sin\frac{\theta}{2}$.

   A) $\frac{\sqrt{10}}{4}$ B) $\frac{\sqrt{6}}{4}$ C) $\frac{\sqrt{8 - 2\sqrt{15}}}{4}$ D) $\frac{\sqrt{8 + 2\sqrt{15}}}{4}$

2) $\sin\theta = \frac{1}{4}$, $\tan\theta > 0$ Find $\cos\frac{\theta}{2}$.

   A) $\frac{\sqrt{8 - 2\sqrt{15}}}{4}$ B) $\frac{\sqrt{6}}{4}$ C) $\frac{\sqrt{10}}{4}$ D) $\frac{\sqrt{8 + 2\sqrt{15}}}{4}$

3) $\cos\theta = \frac{1}{4}$, $\csc\theta > 0$ Find $\sin\frac{\theta}{2}$.

   A) $\frac{\sqrt{10}}{4}$ B) $\frac{\sqrt{8 - 2\sqrt{15}}}{4}$ C) $\frac{\sqrt{6}}{4}$ D) $\frac{\sqrt{8 + 2\sqrt{15}}}{4}$

4) $\sec\theta = 4$, $0 < \theta < \frac{\pi}{2}$ Find $\cos\frac{\theta}{2}$.

   A) $\frac{\sqrt{6}}{4}$ B) $\frac{\sqrt{8 + 2\sqrt{15}}}{4}$ C) $\frac{\sqrt{10}}{4}$ D) $\frac{\sqrt{8 - 2\sqrt{15}}}{4}$

5) $\cos\theta = -\frac{3}{5}$, $\pi < \theta < \frac{3\pi}{2}$ Find $\cos\frac{\theta}{2}$.

   A) $\frac{\sqrt{30}}{10}$ B) $-\frac{\sqrt{5}}{5}$ C) $-\frac{\sqrt{30}}{10}$ D) $\frac{\sqrt{5}}{5}$

6) $\cos\theta = -\frac{3}{5}$, $\sin\theta > 0$ Find $\cos\frac{\theta}{2}$.

   A) $\frac{\sqrt{30}}{10}$ B) $-\frac{\sqrt{5}}{5}$ C) $\frac{\sqrt{5}}{5}$ D) $-\frac{\sqrt{30}}{10}$

7) $\sin\theta = -\dfrac{3}{5}$, $\dfrac{3\pi}{2} < \theta < 2\pi$ Find $\sin\dfrac{\theta}{2}$.

   A) $-\dfrac{\sqrt{5}}{5}$ B) $-\dfrac{\sqrt{30}}{10}$ C) $\dfrac{\sqrt{5}}{5}$ D) $-\dfrac{\sqrt{10}}{10}$

8) $\sec\theta = -\dfrac{17}{15}$, $\dfrac{\pi}{2} < \theta < \pi$ Find $\sin\dfrac{\theta}{2}$.

   A) $-\dfrac{\sqrt{17}}{17}$ B) $\dfrac{4\sqrt{17}}{17}$ C) $-\dfrac{4}{17}$ D) $\dfrac{\sqrt{17}}{17}$

9) $\sin\theta = -\dfrac{20}{29}$, $\dfrac{3\pi}{2} < \theta < 2\pi$ Find $\cos\dfrac{\theta}{2}$.

   A) $-\dfrac{5\sqrt{29}}{29}$ B) $\dfrac{5\sqrt{29}}{29}$ C) $-\dfrac{2\sqrt{29}}{29}$ D) $\dfrac{21}{58}$

10) $\csc\theta = -\dfrac{6}{5}$, $\tan\theta > 0$ Find $\cos\dfrac{\theta}{2}$.

    A) $\dfrac{\sqrt{11}}{12}$ B) $\dfrac{\sqrt{18+3\sqrt{11}}}{6}$ C) $-\dfrac{\sqrt{18-3\sqrt{11}}}{6}$ D) $-\dfrac{\sqrt{6+\sqrt{11}}}{12}$

11) $\cos\theta = \dfrac{4}{5}$, $\dfrac{3\pi}{2} \le \theta \le 2\pi$ Find $\cos\dfrac{\theta}{2}$.

    A) $-\dfrac{3\sqrt{10}}{10}$ B) $\dfrac{2}{5}$ C) $\dfrac{3\sqrt{10}}{10}$ D) $-\dfrac{2}{5}$

Use the Half-angle Formulas to find the exact value of the trigonometric function.

12) $\sin 22.5°$

    A) $\dfrac{1}{2}\sqrt{2+\sqrt{2}}$ B) $-\dfrac{1}{2}\sqrt{2-\sqrt{2}}$ C) $\dfrac{1}{2}\sqrt{2-\sqrt{2}}$ D) $-\dfrac{1}{2}\sqrt{2+\sqrt{2}}$

13) $\cos 22.5°$

    A) $\dfrac{1}{2}\sqrt{2-\sqrt{2}}$ B) $-\dfrac{1}{2}\sqrt{2-\sqrt{2}}$ C) $-\dfrac{1}{2}\sqrt{2+\sqrt{2}}$ D) $\dfrac{1}{2}\sqrt{2+\sqrt{2}}$

14) $\sin 75°$

    A) $\dfrac{1}{2}\sqrt{2-\sqrt{3}}$ B) $-\dfrac{1}{2}\sqrt{2+\sqrt{3}}$ C) $-\dfrac{1}{2}\sqrt{2-\sqrt{3}}$ D) $\dfrac{1}{2}\sqrt{2+\sqrt{3}}$

15) $\cos 75°$

    A) $\dfrac{1}{2}\sqrt{2+\sqrt{3}}$ B) $\dfrac{1}{2}\sqrt{2-\sqrt{3}}$ C) $-\dfrac{1}{2}\sqrt{2-\sqrt{3}}$ D) $-\dfrac{1}{2}\sqrt{2+\sqrt{3}}$

16) $\tan 75°$

    A) $2-\sqrt{3}$ B) $-2-\sqrt{3}$ C) $-2+\sqrt{3}$ D) $2+\sqrt{3}$

17) $\sin \dfrac{5\pi}{12}$

   A) $-\dfrac{1}{2}\sqrt{2-\sqrt{3}}$ B) $\dfrac{1}{2}\sqrt{2+\sqrt{3}}$ C) $\dfrac{1}{2}\sqrt{2-\sqrt{3}}$ D) $-\dfrac{1}{2}\sqrt{2+\sqrt{3}}$

18) $\cos \dfrac{5\pi}{12}$

   A) $-\dfrac{1}{2}\sqrt{2-\sqrt{3}}$ B) $-\dfrac{1}{2}\sqrt{2+\sqrt{3}}$ C) $\dfrac{1}{2}\sqrt{2-\sqrt{3}}$ D) $\dfrac{1}{2}\sqrt{2+\sqrt{3}}$

19) $\sin 165°$

   A) $-\dfrac{1}{2}\sqrt{2+\sqrt{3}}$ B) $\dfrac{1}{2}\sqrt{2-\sqrt{3}}$ C) $\dfrac{1}{2}\sqrt{2+\sqrt{3}}$ D) $-\dfrac{1}{2}\sqrt{2-\sqrt{3}}$

20) $\cos 165°$

   A) $\dfrac{1}{2}\sqrt{2-\sqrt{3}}$ B) $\dfrac{1}{2}\sqrt{2+\sqrt{3}}$ C) $-\dfrac{1}{2}\sqrt{2-\sqrt{3}}$ D) $-\dfrac{1}{2}\sqrt{2+\sqrt{3}}$

21) $\tan 165°$

   A) $-2+\sqrt{3}$ B) $2-\sqrt{3}$ C) $-2-\sqrt{3}$ D) $2+\sqrt{3}$

22) $\sin 15°$

4 Demonstrate Additional Understanding and Skills

Find the exact value of the expression.

1) $\sin\left[2\cos^{-1}\left(-\dfrac{3}{5}\right)\right]$

   A) $-\dfrac{12}{25}$ B) $\dfrac{24}{25}$ C) $-\dfrac{24}{25}$ D) $\dfrac{12}{25}$

2) $\cos\left[2\sin^{-1}\left(-\dfrac{5}{13}\right)\right]$

   A) $\dfrac{119}{169}$ B) $-\dfrac{12}{13}$ C) $\dfrac{2\sqrt{5}+10}{13}$ D) $\dfrac{10}{13}$

3) $\cos\left[\sin^{-1}\dfrac{2}{3} + 2\sin^{-1}\left(-\dfrac{1}{3}\right)\right]$

   A) $\dfrac{2\sqrt{3}}{5}$ B) $\dfrac{2\sqrt{6}}{5}$ C) $\dfrac{7\sqrt{5}+8\sqrt{2}}{27}$ D) $\dfrac{2\sqrt{3}+2\sqrt{10}}{9}$

Solve the problem.

4) Draw a triangle so that $\tan \frac{\theta}{2} = u$. The hypotenuse of the triangle with have length $\sqrt{1 + u^2}$. Use the illustration and the double angle formulas to write $\sin \theta$ and $\cos \theta$ in terms of u.

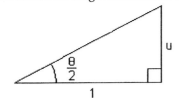

5) The path of a projectile fired at an inclination θ (in degrees) to the horizontal with an initial speed v_0 is a parabola. The range R of the projectile, that is, the horizontal distance that the projectile travels, is found by using the formula

$$R = \frac{v_0^2}{g} \sin(2\theta)$$

where g is the acceleration due to gravity. The maximum height H of the projectile is

$$H = \frac{v_0^2}{4g}(1 - \cos(2\theta))$$

Find the range R and the maximum height H in terms of g if the projectile is fired with an initial speed of 200 meters per second at an angle of 15° and then at an angle of 22.5°. Do not use a calculator, but simplify the answers.

6) The two equal sides of an isosceles triangle measure three feet. Let the angle between the sides measure θ. Find the area A of the triangle as a function of $\frac{\theta}{2}$. The answer may include more than one trigonometric function.

7) An important result in analytic geometry concerns simplifying general second-degree equations of the form $Ax^2 + Bxy + Cy^2 + Dx + Ey + F = 0$ by means of a procedure known as rotation of axes. This is done by calculating the angle θ where

$$\cot(2\theta) = \frac{A - C}{B}, \quad -\frac{\pi}{2} < \theta < \frac{\pi}{2}$$

Determine the exact value of θ for the equation $x^2 - 4xy - 3y^2 + 2x + 5 = 0$.

8) An important result in analytic geometry concerns simplifying general second-degree equations of the form $Ax^2 + Bxy + Cy^2 + Dx + Ey + F = 0$ by means of a procedure known as rotation of axes. This is done by calculating the angle θ where

$$\cot 2\theta = \frac{A - C}{B}, \quad -\frac{\pi}{2} < \theta < \frac{\pi}{2}$$

Determine the exact value of θ for the equation $5x^2 + 3\sqrt{3} xy + 2y^2 - 8 = 0$.

6.6 Product-to-Sum and Sum-to-Product Formulas

1 Express Products as Sums

Express the product as a sum containing only sines or cosines.

1) $\sin(6\theta)\cos(4\theta)$

   A) $\frac{1}{2}[\cos(10\theta) - \cos(2\theta)]$
   B) $\frac{1}{2}[\sin(10\theta) + \sin(2\theta)]$
   C) $\frac{1}{2}[\sin(10\theta) + \cos(2\theta)]$
   D) $\sin\cos(24\theta^2)$

2) $\sin(5\theta)\sin(3\theta)$

   A) $\frac{1}{2}[\sin(8\theta) + \cos(2\theta)]$
   B) $\frac{1}{2}[\cos(2\theta) - \cos(8\theta)]$
   C) $\sin^2(15\theta^2)$
   D) $\frac{1}{2}[\cos(8\theta) - \cos(2\theta)]$

3) $\cos(3\theta)\cos(2\theta)$

   A) $\cos^2(6\theta^2)$
   B) $\frac{1}{2}[\cos(5\theta) - \sin\theta]$
   C) $\frac{1}{2}[\cos(5\theta) - \cos\theta]$
   D) $\frac{1}{2}[\cos\theta + \cos(5\theta)]$

4) $\sin(3\theta)\sin(7\theta)$

   A) $\frac{1}{2}[\cos(10\theta) - \sin(4\theta)]$
   B) $\frac{1}{2}[-\cos(4\theta) - \cos(10\theta)]$
   C) $\sin^2(21\theta^2)$
   D) $\frac{1}{2}[\cos(4\theta) - \cos(10\theta)]$

5) $\sin(5\theta)\cos(6\theta)$

   A) $\frac{1}{2}[\cos(11\theta) - \cos\theta]$
   B) $\frac{1}{2}[\sin(11\theta) - \sin\theta]$
   C) $\frac{1}{2}[\cos(11\theta) + \sin\theta]$
   D) $\sin\cos(30\theta^2)$

6) $\cos(4\theta)\cos(7\theta)$

   A) $\frac{1}{2}[\cos(11\theta) - \sin(3\theta)]$
   B) $\frac{1}{2}[\cos(3\theta) + \cos(11\theta)]$
   C) $\cos^2(19\theta^2)$
   D) $\frac{1}{2}[\cos(11\theta) - \cos(3\theta)]$

7) $\cos\dfrac{5\theta}{2}\cos\dfrac{\theta}{2}$

   A) $\frac{1}{4}[\cos(6\theta) - \sin(4\theta)]$
   B) $\frac{1}{2}[\cos(2\theta) + \cos(3\theta)]$
   C) $\frac{1}{4}\cos^2(5\theta)$
   D) $\frac{1}{2}[\cos(3\theta) - \sin(2\theta)]$

8) $\sin \frac{\theta}{2} \cos \frac{9\theta}{2}$

   A) $\frac{1}{4}[\cos(10\theta) - \sin(8\theta)]$ B) $\frac{1}{2}[\cos(5\theta) + \sin(4\theta)]$

   C) $\frac{1}{4} \sin \cos(9\theta)$ D) $\frac{1}{2}[\sin(5\theta) - \sin(4\theta)]$

9) $-2 \sin(5\theta) \sin \theta$

   A) $\cos(6\theta) + \cos(4\theta)$ B) $\cos(6\theta) - \cos(4\theta)$ C) $\cos(7\theta) - \cos(3\theta)$ D) $\cos(7\theta) + \cos(3\theta)$

10) $2 \cos(7\theta) \cos \theta$

    A) $\cos(8\theta) + \sin(6\theta)$ B) $\cos(14\theta) + \cos(2\theta)$ C) $\cos(10\theta) + \sin(4\theta)$ D) $\cos(8\theta) + \cos(6\theta)$

Solve the problem.

11) A product of two oscillations with different frequencies such as
    $f(t) = \sin(10t) \sin(t)$
    is important in acoustics. The result is an oscillation with "oscillating amplitude."
    (i) Write the product f(t) of the two oscillations as a sum of two cosines and call it g(t).
    (ii) Using a graphing utility, graph the function g(t) on the interval $0 \le t \le 2\pi$.
    (iii) On the same system as your graph, graph $y = \sin t$ and $y = -\sin t$.
    (iv) The last two functions constitute an "envelope" for the function g(t). For certain values of t, the two cosine functions in g(t) cancel each other out and near-silence occurs; between these values, the two functions combine in varying degrees. The phenomenon is known (and heard) as "beats." For what values of t do the functions cancel each other?

2 Express Sums as Products

Express the sum or difference as a product of sines and/or cosines.

1) $\sin(8\theta) + \sin(2\theta)$

   A) $2 \cos(5\theta) \sin(3\theta)$ B) $2 \sin(5\theta) \sin(3\theta)$ C) $2 \sin(5\theta) \cos(3\theta)$ D) $2 \sin(10\theta)$

2) $\cos(8\theta) - \cos(4\theta)$

   A) $-2 \cos(6\theta) \sin(2\theta)$ B) $2 \cos(2\theta)$ C) $-2 \sin(6\theta) \sin(2\theta)$ D) $2 \cos(6\theta) \cos(2\theta)$

3) $\cos(6\theta) + \cos(4\theta)$

   A) $2 \cos(5\theta) \sin \theta$ B) $2 \sin(5\theta) \sin \theta$ C) $2 \cos(5\theta) \cos \theta$ D) $2 \cos(5\theta)$

4) $\cos(4\theta) - \cos(6\theta)$

   A) $-2 \sin(5\theta) \sin \theta$ B) $2 \sin(5\theta) \sin \theta$ C) $\cos(-2\theta)$ D) $-2 \cos(5\theta) \sin \theta$

5) $\sin(7\theta) - \sin(3\theta)$

   A) $2 \sin(2\theta)$ B) $2 \sin(5\theta) \cos(2\theta)$ C) $2 \cos(3\theta) \cos(5\theta)$ D) $2 \sin(2\theta) \cos(5\theta)$

6) $\sin(2\theta) - \sin(4\theta)$

   A) $2 \sin(3\theta) \cos \theta$ B) $2 \cos(2\theta) \cos(3\theta)$ C) $-2 \sin \theta$ D) $-2 \sin \theta \cos(3\theta)$

7) $\cos \dfrac{7\theta}{2} + \cos \dfrac{5\theta}{2}$

A) $2 \sin (3\theta) \sin \dfrac{\theta}{2}$
B) $2 \cos (3\theta)$
C) $2 \cos (3\theta) \cos \dfrac{\theta}{2}$
D) $2 \sin (3\theta) \sin \theta$

8) $\sin \dfrac{11\theta}{2} + \sin \dfrac{7\theta}{2}$

A) $2 \sin \dfrac{9\theta}{2} \sin \theta$
B) $2 \sin \dfrac{9\theta}{2} \cos \theta$
C) $2 \sin (9\theta)$
D) $2 \cos (9\theta) \sin \theta$

9) $\sin (6\theta) - \sin (4\theta)$

A) $-2 \sin \theta \cos (5\theta)$
B) $2 \sin \theta \cos (5\theta)$
C) $-2 \sin (5\theta) \cos \theta$
D) $2 \sin (5\theta) \cos \theta$

10) $\sin (4\theta) - \sin (2\theta)$

A) $2 \sin (3\theta) \cos \theta$
B) $2 \sin \theta \cos (3\theta)$
C) $\sin (3\theta) \cos \theta$
D) $\sin \theta \cos (3\theta)$

Solve the problem.

11) On a Touch-Tone phone, each button produces a unique sound. The sound produced is the sum of two tones, given by
$$y = \sin (2\pi l t) \text{ and } y = \sin (2\pi h t)$$
where l and h are the low and high frequencies (cycles per second) shown on the illustration.

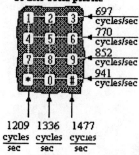

The sound produced is thus given by
$$y = \sin (2\pi l t) + \sin (2\pi h t)$$
Write the sound emitted by touching the 4 key as a product of sines and cosines.

A) $y = 2 \sin (2106\pi t) \cos (566\pi t)$
B) $y = 2 \sin (566\pi t) \cos (2106\pi t)$
C) $y = 2 \sin (439\pi t) \cos (1979\pi t)$
D) $y = 2 \sin (1979\pi t) \cos (439\pi t)$

12) If two sound sources at the same volume are equidistant from a microphone, the pressure on the microphone is given by
$$p = a \cos \omega_1 t + a \cos \omega_2 t$$
where a, ω_1, ω_2 are constants and t is time. Write p as a product of cosine functions.

3 Demonstrate Additional Understanding and Skills

Complete the identity.

1) $\dfrac{\sin (2\theta) + \sin (8\theta)}{\cos (2\theta) + \cos (8\theta)} = ?$

A) $\tan (5\theta) \cot (3\theta)$
B) $2 \tan (5\theta) \tan (3\theta)$
C) $\tan (5\theta)$
D) $\tan (2\theta) + \tan (8\theta)$

Precalculus Enhanced with Graphing Utilities 562

2) $\dfrac{\cos(4\theta) - \cos(8\theta)}{\cos(4\theta) + \cos(8\theta)} = ?$

   A) $-\tan(6\theta)$ B) $\cot(6\theta)$ C) $\tan(2\theta)\tan(6\theta)$ D) 0

3) $\sin(4\theta)\sin(7\theta)\cos(4\theta)\cos(7\theta) = ?$

   A) $\dfrac{\cos^2(11\theta) + \cos^2(3\theta)}{4}$ B) $\dfrac{\sin^2(56\theta)}{4}$

   C) $\dfrac{\cos^2(3\theta) - \cos^2(11\theta)}{4}$ D) $\cos^2(56\theta)$

4) $\sin\theta\,[\sin(4\theta) + \sin(6\theta)] = ?$

   A) $\cos\theta\,[\cos(4\theta) + \cos(6\theta)]$ B) $\dfrac{1}{2}\cos\theta\,[\cos(4\theta) - \cos(6\theta)]$

   C) $\cos\theta\,[\cos(4\theta) - \cos(6\theta)]$ D) $\dfrac{1}{2}\cos\theta\,[\cos(4\theta) + \cos(6\theta)]$

5) $\dfrac{\sin(4\theta) + \sin(6\theta)}{\cos(3\theta) + \cos(5\theta)} = ?$

   A) $\tan(5\theta) + \tan(4\theta)$ B) $\dfrac{\cos(10\theta)}{\sin(8\theta)}$ C) $\dfrac{\sin(5\theta)}{\cos(4\theta)}$ D) $\dfrac{\sin(10\theta)}{\cos(8\theta)}$

6) $\dfrac{\sin\alpha + \sin\beta}{\cos\alpha + \cos\beta} = ?$

   A) $\cot\dfrac{\alpha - \beta}{2}$ B) $\tan\dfrac{\alpha + \beta}{2}$ C) $\tan\alpha + \tan\beta$ D) $\tan\dfrac{\alpha - \beta}{2}$

7) $\tan\alpha + \cot\beta = ?$

   A) $\dfrac{\cos(\alpha - \beta)}{\cos\alpha\sin\beta}$ B) $\dfrac{\sin(\alpha + \beta)}{\cos\alpha\sin\beta}$ C) $\dfrac{\cos(\alpha + \beta)}{\cos\alpha\sin\beta}$ D) $\dfrac{\sin(\alpha - \beta)}{\cos\alpha\sin\beta}$

8) $1 - \cos(2\theta) + \cos(4\theta) - \cos(6\theta) = ?$

   A) $4\cos\theta\cos(2\theta)\cos(3\theta)$ B) $4\sin\theta\sin(2\theta)\sin(3\theta)$
   C) $4\cos\theta\cos(2\theta)\sin(3\theta)$ D) $4\sin\theta\cos(2\theta)\sin(3\theta)$

6.7 Trigonometric Equations (I)

1 Solve Equations Involving a Single Trigonometric Function

Solve the equation on the interval $0 \leq \theta < 2\pi$.

1) $\sin(4\theta) = \dfrac{\sqrt{3}}{2}$

   A) $0, \dfrac{\pi}{4}, \pi$

   B) $\dfrac{\pi}{12}, \dfrac{\pi}{6}, \dfrac{2\pi}{3}, \dfrac{7\pi}{12}, \dfrac{7\pi}{6}, \dfrac{13\pi}{12}, \dfrac{5\pi}{3}, \dfrac{19\pi}{12}$

   C) 0

   D) $\dfrac{\pi}{4}, \dfrac{5\pi}{4}$

2) $2\sqrt{3}\sin(4\theta) = 3$

   A) $0, \dfrac{\pi}{4}, \pi$

   B) $\dfrac{\pi}{12}, \dfrac{\pi}{6}, \dfrac{2\pi}{3}, \dfrac{7\pi}{12}, \dfrac{7\pi}{6}, \dfrac{13\pi}{12}, \dfrac{5\pi}{3}, \dfrac{19\pi}{12}$

   C) $\dfrac{\pi}{4}, \dfrac{5\pi}{4}$

   D) 0

3) $\csc(3\theta) = 0$

   A) $0, \dfrac{2\pi}{3}, \pi, \dfrac{4\pi}{3}$

   B) $\dfrac{\pi}{8}, \dfrac{9\pi}{8}$

   C) $\dfrac{\pi}{4}, \dfrac{3\pi}{4}, \dfrac{5\pi}{4}, \dfrac{7\pi}{4}$

   D) No solution

4) $\sqrt{2}\cos(2\theta) = 1$

   A) $\dfrac{\pi}{4}, \dfrac{3\pi}{4}, \dfrac{5\pi}{4}, \dfrac{7\pi}{4}$

   B) $\dfrac{\pi}{8}, \dfrac{7\pi}{8}, \dfrac{9\pi}{8}, \dfrac{15\pi}{8}$

   C) $0, \dfrac{2\pi}{3}, \pi, \dfrac{4\pi}{3}$

   D) No solution

5) $\cos\theta - 1 = 0$

   A) $\dfrac{\pi}{2}$

   B) 0

   C) π

   D) $\dfrac{3\pi}{2}$

6) $5\csc\theta - 3 = 2$

   A) $\dfrac{3\pi}{2}$

   B) π

   C) 2π

   D) $\dfrac{\pi}{2}$

7) $2\cos(2\theta) = \sqrt{3}$

   A) $\dfrac{3\pi}{2}$

   B) $\dfrac{\pi}{2}$

   C) $\dfrac{\pi}{6}, \dfrac{11\pi}{6}$

   D) $\dfrac{\pi}{12}, \dfrac{11\pi}{12}, \dfrac{13\pi}{12}, \dfrac{23\pi}{12}$

8) $2\cos\theta + 1 = 0$

   A) $\dfrac{3\pi}{2}$

   B) $\dfrac{\pi}{2}, \dfrac{3\pi}{2}$

   C) $\dfrac{2\pi}{3}, \dfrac{4\pi}{3}$

   D) $\dfrac{\pi}{3}, \dfrac{5\pi}{3}$

9) $\cot\left(2\theta - \dfrac{\pi}{2}\right) = 1$

   A) $\dfrac{3\pi}{8}, \dfrac{7\pi}{8}, \dfrac{11\pi}{8},$ and $\dfrac{15\pi}{8}$

   B) $\dfrac{\pi}{4}, \dfrac{5\pi}{4}, \dfrac{9\pi}{4},$ and $\dfrac{13\pi}{4}$

   C) $\dfrac{3\pi}{8}, \dfrac{7\pi}{8}$

   D) $\dfrac{3\pi}{8}$

Solve the equation. Give a general formula for all the solutions.

10) $\cos\theta = 1$

    A) $\theta = \pi + 2k\pi$

    B) $\theta = 0 + 2k\pi$

    C) $\theta = \dfrac{3\pi}{2} + 2k\pi$

    D) $\theta = \dfrac{\pi}{2} + 2k\pi$

11) $\sin\theta = 1$

    A) $\theta = \dfrac{\pi}{2} + 2k\pi$

    B) $\theta = \dfrac{3\pi}{2} + 2k\pi$

    C) $\theta = 0 + 2k\pi$

    D) $\theta = \pi + 2k\pi$

12) $\sin\theta = 0$

    A) $\theta = 0 + k\pi$

    B) $\theta = \dfrac{\pi}{2} + 2k\pi$

    C) $\theta = 0 + 2k\pi$

    D) $\theta = \dfrac{\pi}{2} + k\pi$

13) $\cos\theta = 0$

   A) $\theta = 0 + 2k\pi$
   B) $\theta = \dfrac{\pi}{2} + 2k\pi$
   C) $\theta = \dfrac{\pi}{2} + k\pi$
   D) $\theta = 0 + k\pi$

14) $\sin\theta = \dfrac{\sqrt{3}}{2}$

   A) $\theta = \dfrac{\pi}{3} + 2k\pi,\ \theta = \dfrac{2\pi}{3} + 2k\pi$
   B) $\theta = \dfrac{\pi}{6} + 2k\pi,\ \theta = \dfrac{5\pi}{6} + 2k\pi$
   C) $\theta = \dfrac{\pi}{3} + k\pi,\ \theta = \dfrac{2\pi}{3} + k\pi$
   D) $\theta = \dfrac{\pi}{6} + k\pi,\ \theta = \dfrac{5\pi}{6} + k\pi$

15) $\tan\theta = -1$

   A) $\theta = \dfrac{\pi}{4} + k\pi$
   B) $\theta = \dfrac{3\pi}{4} + k\pi$
   C) $\theta = \dfrac{\pi}{4} + 2k\pi$
   D) $\theta = \dfrac{3\pi}{4} + 2k\pi$

16) $\cos\theta - 1 = 0$

   A) $\theta = \pi + 2k\pi$
   B) $\theta = \dfrac{\pi}{2} + 2k\pi$
   C) $\theta = \dfrac{3\pi}{2} + 2k\pi$
   D) $\theta = 2k\pi$

17) $2\cos\theta + 1 = 0$

   A) $\theta = \dfrac{3\pi}{2} + k\pi$
   B) $\theta = \dfrac{2\pi}{3} + k\pi,\ \theta = \dfrac{4\pi}{3} + k\pi$
   C) $\theta = \dfrac{2\pi}{3} + 2k\pi,\ \theta = \dfrac{4\pi}{3} + 2k\pi$
   D) $\theta = \dfrac{\pi}{2} + 2k\pi,\ \theta = \dfrac{3\pi}{2} + 2k\pi$

18) $\cos(2\theta) = \dfrac{\sqrt{2}}{2}$

   A) $\theta = \dfrac{2\pi}{3} + k\pi,\ \theta = \dfrac{4\pi}{3} + k\pi$
   B) $\theta = \dfrac{\pi}{8} + k\pi,\ \theta = \dfrac{7\pi}{8} + k\pi$
   C) $\theta = \dfrac{\pi}{4} + k\pi,\ \theta = \dfrac{3\pi}{4} + k\pi$
   D) $\theta = \dfrac{\pi}{8} + 2k\pi,\ \theta = \dfrac{7\pi}{8} + 2k\pi$

19) $\csc\dfrac{\theta}{3} = \dfrac{2\sqrt{3}}{3}$

   A) $\theta = \dfrac{\pi}{9} + 2k\pi$
   B) $\theta = \dfrac{\pi}{18} + 2k\pi$
   C) $\theta = \pi + 6k\pi$
   D) $\theta = \dfrac{\pi}{2} + 6k\pi$

Use a calculator to solve the equation on the interval $0 \le \theta < 2\pi$. Round the answer to two decimal places.

20) $\sin\theta = 0.26$

   A) 0.26, 2.88
   B) 0.26, 6.02
   C) 0.26, 1.83
   D) 0.26, 3.40

21) $\cos\theta = 0.62$

   A) 0.90, 2.24
   B) 0.90, 2.47
   C) 0.90, 5.38
   D) 0.90, 4.04

22) $\tan\theta = 4.2$

   A) 1.34, 2.91
   B) 1.34, 1.80
   C) 1.34, 4.95
   D) 1.34, 4.48

23) $\sin\theta = -0.44$

   A) 0.46, 3.60　　B) 0.46, 5.83　　C) 0.46, 2.03　　D) 3.60, 5.83

24) $\cos\theta = -0.84$

   A) 2.57, 5.71　　B) 2.57, 3.72　　C) 0.57, 2.57　　D) 0.57, 3.72

Using a graphing utility, solve the equation on the interval $0 \le \theta < 2\pi$. Round answer to three decimal places.

25) $2\csc\theta = 5$

   A) 0.412　　B) 0.201, 2.941　　C) 0.201　　D) 0.412, 2.730

Solve the equation on the interval $[0, 2\pi)$.

26) Suppose $f(x) = \cos\theta - 1$. Solve $f(x) = 0$.

   A) 0　　B) $\dfrac{\pi}{2}$　　C) $\dfrac{3\pi}{2}$　　D) π

27) Suppose $f(x) = 6\csc\theta - 2$. Solve $f(x) = 4$.

   A) $\dfrac{3\pi}{2}$　　B) π　　C) $\dfrac{\pi}{2}$　　D) 2π

28) Suppose $f(x) = 2\cos\theta + 1$. Solve $f(x) = 0$.

   A) $\dfrac{3\pi}{2}$　　B) $\dfrac{2\pi}{3}, \dfrac{4\pi}{3}$　　C) $\dfrac{\pi}{2}, \dfrac{3\pi}{2}$　　D) $\dfrac{\pi}{3}, \dfrac{5\pi}{3}$

Solve the problem using Snell's Law: $\dfrac{\sin\theta_1}{\sin\theta_2} = \dfrac{v_1}{v_2}$.

29) A light beam in air travels at 2.99×10^8 meters per second. If its angle of incidence to a second medium is 49° and its angle of refraction in the second medium is 39°, what is its speed in the second medium (to two decimal places)?

   A) 2.49×10^8 mps　　B) 2.26×10^8 mps　　C) 3.59×10^8 mps　　D) 1.88×10^8 mps

30) A ray of light near the horizon with an angle of incidence of 84° enters a pool of water and strikes a fish's eye. If the index of refraction is 1.33, what is the angle of refraction (to two decimal places)?

   A) 43.72°　　B) 48.40°　　C) 41.60°　　D) 46.28°

31) The index of refraction of light passing from air into a second medium is 1.42. If the angle of incidence is 82°, what is the angle of refraction (to two decimal places)?

   A) 42.05°　　B) 47.95°　　C) 44.22°　　D) 45.78°

32) A light beam traveling through air makes an angle of incidence of 37° upon a second medium. The refracted beam makes an angle of refraction of 27°. What is the index of refraction of the material of the second medium? Give the answer to two decimal places.

   A) 1.33　　B) 0.45　　C) 0.60　　D) 0.75

33) A light beam in air travels at 2.99×10^8 meters per second. If its angle of incidence to a second medium is 85° and its angle of refraction in the second medium is 76°, what is its speed in the second medium (to two decimal places)?

   A) 2.90×10^8 mps　　B) 2.91×10^8 mps　　C) 2.98×10^8 mps　　D) 3.07×10^8 mps

Precalculus Enhanced with Graphing Utilities

Solve the problem.

34) The function
$$I(t) = 40 \sin\left(60\pi t - \frac{\pi}{2}\right)$$
represents the amperes of current produced by an electric generator as a function of time t, where t is measured in seconds. Find the smallest value of t for which the current is 20 amperes. Round your answer to three decimal places, if necessary.

A) 0.011   B) 0.017   C) 0.033   D) 0.008

35) A weight suspended from a spring is vibrating vertically with up being the positive direction. The function
$$f(t) = 10 \sin\left(\frac{3\pi t}{4} - \frac{\pi}{4}\right)$$
represents the distance in centimeters of the weight from its rest position as a function of time t, where t is measured in seconds. Find the smallest positive value of t for which the displacement of the weight above its rest position is 5 cm. Round answer to thre decimal places, if necessary.

A) 2.293   B) 0.222   C) 0.556   D) 1.586

36) You are flying a kite and want to know its angle of elevation. The string on the kite is 43 meters long and the kite is level with the top of a building that you know is 28 meters high. Use an inverse trigonometric function to find the angle of elevation of the kite. Round to two decimal places.

37) Before exercising, an athlete measures her air flow and obtains
$$a = 0.65 \sin\left(\frac{2\pi}{5}t\right)$$
where a is measured in liters per second and t is the time in seconds. If a > 0, the athlete is inhaling; if a < 0, the athlete is exhaling. The time to complete one complete inhalation/exhalation sequence is a respiratory cycle. Find the values of t for which the athlete's air flow is zero. Find all values of t for t < 20 seconds.

38) A mass hangs from a spring which oscillates up and down. The position P (in feet) of the mass at time t (in seconds) is given by
$$P = 4 \cos(4t)$$
For what values of t, $0 \le t < \pi$, will the position be $2\sqrt{2}$ feet? Find the exact values. Do not use a calculator.

39) The path of a projectile fired at an inclination θ (in degrees) to the horizontal with an initial velocity v_0 is a parabola. The range R of the projectile, that is, the horizontal distance that the projectile travels, is found by using the formula
$$R = \frac{v_0^2}{g} \sin(2\theta)$$
where g is the acceleration due to gravity. Suppose the projectile is fired with an initial velocity of 400 feet per seconds and g = 32 feet per second². What angle θ, $0° \le \theta < 90°$, would you select for the range to be 2500 feet? (There should be two values of θ.)

40) Wildlife management personnel use predator-prey equations to model the populations of certain predators and their prey in the wild. Suppose the population M of a predator after t months is given by

$$M = 750 + 125 \sin \frac{\pi}{6} t$$

while the population N of its primary prey is given by

$$N = 12{,}250 + 3050 \cos \frac{\pi}{6} t$$

Find the values of t, $0 \le t < 12$, for which the predator population is 875. Find the values of t, $0 \le t < 12$, for which the prey population is 10,725.

41) A consumer notes the sinusoidal nature of her monthly power bills. In winter when she uses electricity to heat her home and in summer when she cools her home, the bills are high. In spring and fall, significantly less electricity is used and the bills are much smaller. The following function models this behavior.

$$C = 60 + 40 \cos\left(\frac{\pi}{3} t - \frac{\pi}{3}\right)$$

Here C is the cost of power in dollars for the month t, $1 \le t \le 12$, with $t = 1$ corresponding to January. For what values of t, $1 \le t \le 12$, is the cost exactly $80?

42) The average daily temperature T of a city in the United States is approximated by

$$T = 55 - 23 \cos \frac{2\pi}{365}(t - 30)$$

where t is in days, $1 \le t \le 365$, and $t = 1$ corresponds to January 1. For what range of values of t is the average daily temperature above 70°F? Use a calculator and round answers to the nearest whole number.

6.8 Trigonometric Equations (II)

1 Solve Trigonometric Equations Quadratic in Form

Solve the equation on the interval $0 \le \theta < 2\pi$.

1) $\cos^2 \theta + 2 \cos \theta + 1 = 0$

   A) $\frac{\pi}{4}, \frac{7\pi}{4}$ B) 2π C) π D) $\frac{\pi}{2}, \frac{3\pi}{2}$

2) $2 \sin^2 \theta = \sin \theta$

   A) $\frac{\pi}{6}, \frac{5\pi}{6}$ B) $\frac{\pi}{2}, \frac{3\pi}{2}, \frac{\pi}{3}, \frac{2\pi}{3}$ C) $\frac{\pi}{3}, \frac{2\pi}{3}$ D) $0, \pi, \frac{\pi}{6}, \frac{5\pi}{6}$

3) $\cos \theta = \sin \theta$

   A) $\frac{\pi}{4}, \frac{5\pi}{4}$ B) $\frac{3\pi}{4}, \frac{5\pi}{4}$ C) $\frac{\pi}{4}, \frac{7\pi}{4}$ D) $\frac{3\pi}{4}, \frac{7\pi}{2}$

4) $\sec^2 \theta - 2 = \tan^2 \theta$

   A) $\frac{\pi}{6}$ B) $\frac{\pi}{4}$ C) $\frac{\pi}{3}$ D) No solution

5) $\tan \theta + \sec \theta = 1$

   A) $\frac{\pi}{4}$ B) 0 C) $\frac{5\pi}{4}$ D) No solution

6) $\csc^5 \theta - 4 \csc \theta = 0$

   A) $\frac{\pi}{4}, \frac{5\pi}{4}, \frac{\pi}{3}, \frac{5\pi}{3}$
   B) $\frac{\pi}{4}, \frac{3\pi}{4}, \frac{\pi}{6}, \frac{5\pi}{6}$
   C) $\frac{\pi}{4}, \frac{3\pi}{4}, \frac{\pi}{3}, \frac{5\pi}{6}$
   D) $\frac{\pi}{4}, \frac{3\pi}{4}, \frac{5\pi}{4}, \frac{7\pi}{4}$

7) $\sin^2 \theta - \cos^2 \theta = 0$

   A) $\frac{\pi}{4}, \frac{\pi}{6}$
   B) $\frac{\pi}{4}, \frac{3\pi}{4}, \frac{5\pi}{4}, \frac{7\pi}{4}$
   C) $\frac{\pi}{4}$
   D) $\frac{\pi}{4}, \frac{\pi}{3}$

8) $\sin^2 \theta + \sin \theta = 0$

   A) $0, \pi, \frac{3\pi}{2}$
   B) $0, \pi, \frac{\pi}{3}, \frac{2\pi}{3}$
   C) $0, \pi, \frac{4\pi}{3}, \frac{5\pi}{3}$
   D) $0, \pi, \frac{\pi}{3}, \frac{5\pi}{3}$

Solve the problem.

9) A water wheel rotates through the angle θ, the water level L behind the wheel changes according to the equation
   $$L = 1 - \sin \theta - 2 \cos^2 \theta$$
   where L is measured in inches. Determine the values of θ for which the water level is zero. Find the exact values. Do not use a calculator.

10) A weight is suspended on a system of springs and oscillates up and down according to
    $$P = \frac{1}{10}[\sin (2t) + \sin t]$$
    where P is the position in meters above or below the point of equilibrium (P = 0) and t is time in seconds. Find the time when the weight is at equilibrium. Find the exact values. Do not use a calculator.

11) A mass hanging from two wires oscillates according to the equation
    $$P = \sin\left(\frac{\pi}{3}t\right) - \cos\left(\frac{\pi}{6}t\right)$$
    where t is measured in seconds and P > 0 is interpreted as the weight being right of center, and if P < 0 it is left of center. Find the values of t (0 < t < 12) when the mass is at the center position. Find the exact values. Do not use a calculator. $\left[\text{Hint: } \frac{\pi}{3}t = 2\left(\frac{\pi}{6}t\right)\right]$

2 Solve Trigonometric Equations Using Identities

Solve the equation on the interval $0 \leq \theta < 2\pi$.

1) $\tan (2\theta) - \tan \theta = 0$

   A) $0, \pi$
   B) $\frac{\pi}{12}, \frac{\pi}{6}, \frac{2\pi}{3}, \frac{7\pi}{12}, \frac{7\pi}{6}, \frac{13\pi}{12}, \frac{5\pi}{3}$
   C) $\frac{\pi}{4}, \frac{5\pi}{4}$
   D) 0

2) $\sec \frac{\theta}{2} = \cos \frac{\theta}{2}$

   A) $\frac{\pi}{12}, \frac{\pi}{6}, \frac{2\pi}{3}, \frac{7\pi}{12}, \frac{7\pi}{6}, \frac{13\pi}{12}, \frac{5\pi}{3}$
   B) $0, \frac{\pi}{4}, \pi, \frac{5\pi}{3}$
   C) 0
   D) $\frac{\pi}{4}, \frac{5\pi}{4}$

3) csc (3θ) = 0

   A) $\frac{\pi}{8}, \frac{9\pi}{8}$
   B) $\frac{\pi}{4}, \frac{3\pi}{4}, \frac{5\pi}{4}, \frac{7\pi}{4}$
   C) $0, \frac{2\pi}{3}, \pi, \frac{4\pi}{3}$
   D) No solution

4) $\sin^2(2\theta) = 1$

   A) $0, \frac{2\pi}{3}, \pi, \frac{4\pi}{3}$
   B) $\frac{\pi}{4}, \frac{3\pi}{4}, \frac{5\pi}{4}, \frac{7\pi}{4}$
   C) $\frac{\pi}{8}, \frac{9\pi}{8}$
   D) No solution

5) $\cos(2\theta) = \sqrt{2} - \cos(2\theta)$

   A) $0, \frac{2\pi}{3}, \pi, \frac{4\pi}{3}$
   B) $\frac{\pi}{8}, \frac{7\pi}{8}, \frac{9\pi}{8}, \frac{15\pi}{8}$
   C) $\frac{\pi}{4}, \frac{3\pi}{4}, \frac{5\pi}{4}, \frac{7\pi}{4}$
   D) No solution

6) $\sin(2\theta) + \sin\theta = 0$

   A) $0, \frac{2\pi}{3}, \pi, \frac{4\pi}{3}$
   B) $\frac{\pi}{4}, \frac{3\pi}{4}, \frac{5\pi}{4}, \frac{7\pi}{4}$
   C) $\frac{\pi}{8}, \frac{9\pi}{8}$
   D) No solution

7) $3\cot^2\theta - 4\csc\theta = 1$

   A) $\frac{\pi}{6}$
   B) $\frac{7\pi}{6}$
   C) $\frac{\pi}{6}, \frac{5\pi}{6}$
   D) $\frac{7\pi}{6}, \frac{11\pi}{6}$

Solve the problem.

8) The altitude of a projectile in feet (neglecting air resistance) is given by

$$y = (\tan\theta)x - \frac{16}{v^2\cos^2\theta}x^2,$$

where x is the horizontal distance covered in feet and v is the initial velocity of the projectile at an angle θ from the horizontal. Find the firing angle (in degrees) of a projectile fired at an initial velocity of 100 feet per second so that it strikes the ground 312.5 feet from the firing point.

   A) 45°
   B) 22.5°
   C) 50°
   D) 30°

3 Solve Trigonometric Equations Linear in Sine and Cosine

Solve the equation on the interval $0 \le \theta < 2\pi$.

1) $\cos\theta - \sin\theta = 0$

   A) $\frac{\pi}{4}, \frac{5\pi}{4}$
   B) $\frac{\pi}{4}$
   C) $\frac{\pi}{6}, \frac{\pi}{3}$
   D) $\frac{\pi}{2}$

2) $\sqrt{3}\sin\theta - \cos\theta = -1$

   A) $0, \frac{2\pi}{3}$
   B) $\frac{\pi}{2}, \frac{7\pi}{6}$
   C) $\frac{3\pi}{2}, \frac{\pi}{6}$
   D) $0, \frac{4\pi}{3}$

3) $\sin\theta = \sqrt{2} - \cos\theta$

   A) $\frac{\pi}{4}$
   B) $\frac{\pi}{2}$
   C) $\frac{5\pi}{4}$
   D) $\frac{3\pi}{2}$

4) $\sin\theta + \sqrt{3}\cos\theta = -1$

   A) $\frac{3\pi}{2}, \frac{5\pi}{6}$
   B) $0, \frac{2\pi}{3}$
   C) $\frac{\pi}{2}, \frac{7\pi}{6}$
   D) $\frac{3\pi}{2}, \frac{\pi}{6}$

Solve the problem.

5) The path of a projectile fired at an inclination θ (in degrees) to the horizontal with an initial velocity v_0 is a parabola. The range R of the projectile, that is, the horizontal distance that the projectile travels, is found by using the formula

$$R = \frac{v_0^2}{g} \sin 2\theta$$

where g is the acceleration due to gravity. Suppose the projectile is fired with an initial velocity of 400 feet per seconds and $g = 32$ feet per second^2. What angle θ, $0° \leq \theta < 90°$, would you select for the range to be 2500 feet? (There should be two values of θ.)

6) An electric generator produces a 30-cycle alternating current described by the equation

$$I(t) = 40 \sin 60\pi \left(t - \frac{1}{120}\right),$$

where $I(t)$ amperes is the current at t seconds. Find the smallest value of t for which the current is 20 amperes.

A) $\frac{1}{90}$ B) $\frac{1}{120}$ C) $\frac{1}{60}$ D) $\frac{1}{30}$

7) A weight suspended from a spring is vibrating vertically according to the equation

$$y = 10 \sin \frac{3\pi}{4}\left(t - \frac{1}{3}\right),$$

where y centimeters is the distance of the weight from its rest position with up being the positive direction. Find the smallest positive value of t for which the displacement of the weight above its rest position is 5 cm.

A) $3 - \sqrt{2}$ B) $\frac{5}{9}$ C) $\frac{6 - \sqrt{2}}{2}$ D) $\frac{2}{9}$

4 Solve Trigonometric Equations Using a Graphing Utility

Use a graphing utility to solve the equation on the interval $0° \leq x < 360°$. Express the solution(s) rounded to one decimal place.

1) $2 + 13 \sin x = 14 \cos^2 x$

A) 34.9°, 145.2°   B) 34.9°, 214.9°   C) 55.2°, 124.9°   D) 214.9°, 325.2°

2) $-11 + 24 \sin x = 16 \cos^2 x$

3) $\sin^2 x - 8 \sin x + 16 = 0$

A) 28.2°, 151.8°
B) 208.2°, 331.8°
C) 28.2°, 151.8°, 208.2°, 331.8°
D) No solution

4) $\sin^2 x + 8 \sin x + 16 = 0$

A) 208.2°, 331.8°
B) 28.2°, 151.8°
C) 28.2°, 151.8°, 208.2°, 331.8°
D) No solution

5) $\sin^2 x - 8 \sin x - 4 = 0$

A) 28.2°, 151.8°, 208.2°, 331.8°
B) 208.2°, 331.8°
C) 28.2°, 151.8°
D) No solution

6) $\sin^2 x + 8 \sin x - 4 = 0$
   A) 28.2°, 151.8°, 208.2°, 331.8°
   B) 28.2°, 151.8°
   C) 208.2°, 331.8°
   D) No solution

7) $\tan^2 x + 5 \tan x + 3 = 0$
   A) 51.8°, 128.2°
   B) 49.8°, 130.2°, 229.8°, 310.2°
   C) 103.1°, 145.1°, 283.1°, 325.1°
   D) 70.5°, 109.5°, 180.0°

8) $3 \cos^2 x + 2 \cos x = 1$
   A) 103.2°, 145.2°, 283.2°, 325.2°
   B) 70.5°, 180.0°, 289.5°
   C) 51.8°, 128.2°
   D) 49.8°, 130.2°, 229.8°, 310.2°

9) $7 \cot^2 x - 5 = 0$
   A) 70.5°, 109.5°, 180.0°
   B) 49.8°, 130.2°, 229.8°, 310.2°
   C) 51.8°, 128.2°
   D) 103.2°, 145.2°, 283.2°, 325.2°

10) $\cos^2 x + \cos x - 1 = 0$
    A) 103.2°, 145.2°, 283.2°, 325.2°
    B) 51.8°, 308.2°
    C) 49.8°, 130.2°, 229.8°, 310.2°
    D) 70.5°, 109.5°, 180.0°

Use a calculator to solve the equation on the interval $0 \le x < 2\pi$. Round the answer to one decimal place if necessary.

11) $x + 3 \sin x = 1$

12) $2x - 3 \cos x = 0$

13) $e^x = \cos x$

14) $2x^2 - 3x \sin x = 2$

Solve the problem.

15) A weight is suspended on a system of spring and oscillates up and down according to
    $$P = 0.1[3 \cos(8t) - \sin(8t)]$$
    where P is the position in meters above or below the point of equilibrium (P = 0) and t is time in seconds. Find the time when the weight is at equilibrium. Find all values of t, $0 \le t \le 1$, rounded to the nearest 0.01 second.

16) The ground movement of an earthquake near a fault line is modeled by the equation
    $$d = D \tan\left[\frac{\pi}{2}\left(1 - \frac{2M}{S}\right)\right]$$
    where M is the horizontal movement (in meters) at a distance d (in kilometers) from the earthquake, D is the depth (also in kilometers) below the surface of the center of the earthquake, and S is the total horizontal displacement (also in meters) at the fault line. What is the horizontal movement 5 kilometers from an earthquake centered 3 kilometers below the surface with a total horizontal displacement of 4 meters? Round the answer to the nearest 0.01 meter.

17) The seasonal variation in the length of daylight can be represented by a sine function. For example, the daily number of hours of daylight in a certain city in the U.S. can be given by $h = \frac{41}{4} + \frac{5}{3} \sin \frac{2\pi x}{365}$, where x is the number of days after March 21 (disregarding leap year). On what day(s) will there be about 10 hours of daylight?

Precalculus Enhanced with Graphing Utilities 572

Ch. 6 Analytic Trigonometry
Answer Key

6.1 The Inverse Sine, Cosine, and Tangent Functions
1 Find the Exact Value of the Inverse Sine, Cosine, and Tangent Functions
1) D
2) A
3) B
4) D
5) D
6) C
7) C
8) A
9) D
10) B
11) C
12) C
13) A
14) B
15) B
16) B

2 Find an Approximate Value of the Inverse Sine, Cosine, and Tangent Functions
1) A
2) B
3) C
4) D
5) C
6) C
7) D
8) 12 hr
9) 14.38 hr
10) $\theta_r = 44.72°$
11) $\theta_r = 8.39°$
12) $\theta_r = 9.13°$

6.2 The Inverse Trigonometric Functions (Continued)
1 Find the Exact Value of Expressions Involving the Inverse Sine, Cosine, and Tangent Functions
1) A
2) B
3) B
4) C
5) D
6) A
7) C
8) D
9) B
10) D
11) D
12) C
13) D
14) C

2 Know the Definition of the Inverse Secant, Cosecant, and Cotangent Functions
1) B
2) C
3) D

       4) C
3 Use a Calculator to Evaluate sec^-1 x, csc^-1 x, and cot^-1 x
       1) B
       2) B
       3) B
       4) C

6.3 Trigonometric Identities

1 Use Algebra to Simplify Trigonometric Expressions
       1) 1

       2) $\dfrac{1 + \cos\theta}{\sin\theta}$

       3) $\dfrac{2}{\cos^2\theta}$

       4) 2

       5) $\dfrac{7\sin\theta + 1}{\sin\theta - 1}$

       6) D
       7) B

2 Establish Identities
       1) C
       2) A
       3) D
       4) D
       5) C
       6) A
       7) C
       8) B
       9) C
       10) D
       11) D
       12) B
       13) B
       14) A
       15) B
       16) A
       17) D
       18) B
       19) C
       20) B
       21) B
       22) D
       23) D

3 Demonstrate Additional Understanding and Skills
       1) B
       2) B
       3) D
       4) B
       5) C
       6) D
       7) D

6.4 Sum and Difference Formulas

1 Use Sum and Difference Formulas to Find Exact Values
       1) C
       2) D

3) A
4) C
5) B
6) D
7) B
8) C
9) B
10) C
11) D
12) C
13) A
14) A
15) C
16) B
17) B
18) C
19) B
20) D
21) B
22) D
23) B
24) B
25) B
26) B
27) B
28) C

2 Use Sum and Difference Formulas to Establish Identities

1) A
2) D
3) C
4) C
5) D
6) B
7) A
8) A
9) A
10) D

3 Use Sum and Difference Formulas Involving Inverse Trigonometric Functions

1) D
2) A
3) B
4) B
5) D
6) A
7) C
8) B
9) C
10) B
11) A

6.5 Double-angle and Half-angle Formulas

1 Use Double-Angle Formulas to Find Exact Values

1) C
2) A
3) D
4) B

5) C
6) A
7) C
8) A
9) D
10) D

2 Use Double-Angle Formulas to Establish Identities
1) C
2) D
3) C
4) A
5) C
6) B
7) B
8) D
9) A
10) B
11) D
12) D
13) B
14) B
15) C
16) A
17) A
18) A
19) C
20) B
21) C
22) A
23) C
24) D
25) B

3 Use Half-Angle Formulas to Find Exact Values
1) C
2) D
3) C
4) C
5) B
6) C
7) D
8) B
9) B
10) C
11) A
12) C
13) D
14) D
15) B
16) D
17) B
18) C
19) B
20) D
21) A

22) $\dfrac{\sqrt{2-\sqrt{3}}}{2}$ or $\dfrac{1}{4}(\sqrt{6}-\sqrt{2})$

4 Demonstrate Additional Understanding and Skills
1) C
2) A
3) C
4) $\sin\theta = \dfrac{2u}{1+u^2}$; $\cos\theta = \dfrac{1-u^2}{1+u^2}$
5) $\theta = 15°$: $R = \dfrac{20{,}000}{g}$, $H = \dfrac{5000(2-\sqrt{3})}{g}$;

   $\theta = 22.5°$: $R = \dfrac{20{,}000(\sqrt{2})}{g}$, $H = \dfrac{5000(2-\sqrt{2})}{g}$
6) $A = 9\sin\dfrac{\theta}{2}\cos\dfrac{\theta}{2}$
7) $-22.5°$
8) $30°$

6.6 Product-to-Sum and Sum-to-Product Formulas
1 Express Products as Sums
1) B
2) B
3) D
4) D
5) B
6) B
7) B
8) D
9) B
10) D
11) (i) $g(t) = \dfrac{1}{2}\cos(9t) - \dfrac{1}{2}\cos(11t)$

   (ii), (iii)

   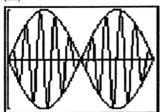

   (iv) $t = \dfrac{n\pi}{10}$, n any integer

2 Express Sums as Products
1) C
2) C
3) C
4) B
5) D
6) D
7) C
8) B
9) B
10) B
11) D

12) $p = 2a \cos(\frac{\omega_1 + \omega_2}{2}t) \cos(\frac{\omega_1 - \omega_2}{2}t)$

3 Demonstrate Additional Understanding and Skills
1) C
2) C
3) C
4) C
5) C
6) B
7) A
8) D

6.7 Trigonometric Equations (I)
1 Solve Equations Involving a Single Trigonometric Function
1) B
2) B
3) D
4) B
5) B
6) D
7) D
8) C
9) A
10) B
11) A
12) A
13) C
14) A
15) B
16) D
17) C
18) B
19) C
20) A
21) C
22) D
23) D
24) B
25) D
26) A
27) C
28) B
29) A
30) B
31) C
32) A
33) B
34) A
35) C
36) 40.63°
37) t = 0, 2.5, 5, 7.5, 10, 12.5, 15, 17.5
38) $t = \frac{\pi}{16}, \frac{7\pi}{16}, \frac{9\pi}{16}, \frac{15\pi}{16}$
39) 15°, 75°
40) M = 875, t = 3; N = 10,725, t = 4, 8

41) t = 0, 2, 6, 8, 12
42) 163 ≤ t ≤ 262

6.8 Trigonometric Equations (II)

1 Solve Trigonometric Equations Quadratic in Form

1) C
2) D
3) A
4) D
5) B
6) D
7) B
8) A
9) $\theta = \dfrac{\pi}{2}, \dfrac{7\pi}{6}, \dfrac{11\pi}{6}$
10) $0 \text{ sec}, \dfrac{2\pi}{3} \text{ sec}, \dfrac{4\pi}{3} \text{ sec}, \pi \text{ sec}$
11) 1, 3, 5, 9

2 Solve Trigonometric Equations Using Identities

1) A
2) C
3) D
4) B
5) B
6) A
7) C
8) A

3 Solve Trigonometric Equations Linear in Sine and Cosine

1) A
2) D
3) A
4) A
5) 15°, 75°
6) A
7) B

4 Solve Trigonometric Equations Using a Graphing Utility

1) A
2) 48.6°, 131.4°
3) D
4) D
5) B
6) B
7) C
8) B
9) B
10) B
11) 0.3
12) 0.9
13) 0
14) 1.9
15) 0.16, 0.55, 0.94
16) 1.38 m
17) 191 days after March 21 and 356 days after March 21 (i.e., September 28 and March 12)

Ch. 7 Applications of Trigonometric Functions

7.1 Right Triangle Trigonometry; Applications

1 Find the Values of Trigonometric Functions of Acute Angles

Two sides of a right triangle ABC (C is the right angle) are given. Find the indicated trigonometric function of the given angle. Give exact answers with rational denominators.

1) Find sin A when a = 4 and b = 9.

   A) $\dfrac{\sqrt{97}}{9}$ B) $\dfrac{4\sqrt{97}}{97}$ C) $\dfrac{9\sqrt{97}}{97}$ D) $\dfrac{\sqrt{97}}{4}$

2) Find sin B when b = 3 and c = 8.

   A) $\dfrac{8\sqrt{55}}{55}$ B) $\dfrac{3\sqrt{55}}{55}$ C) $\dfrac{3}{8}$ D) $\dfrac{\sqrt{55}}{8}$

3) Find cos A when a = 4 and b = 5.

   A) $\dfrac{4\sqrt{41}}{41}$ B) $\dfrac{\sqrt{41}}{5}$ C) $\dfrac{5\sqrt{41}}{41}$ D) $\dfrac{\sqrt{41}}{4}$

4) Find cos A when a = $\sqrt{11}$ and c = 12

   A) $\dfrac{\sqrt{133}}{12}$ B) $\dfrac{133}{11}$ C) $\dfrac{\sqrt{11}}{12}$ D) $\dfrac{11}{12}$

5) Find csc B when a = 9 and b = 4.

   A) $\dfrac{\sqrt{97}}{4}$ B) $\dfrac{9\sqrt{97}}{97}$ C) $\dfrac{\sqrt{97}}{9}$ D) $\dfrac{4\sqrt{97}}{97}$

6) Find sec B when a = 5 and b = 4.

   A) $\dfrac{\sqrt{41}}{5}$ B) $\dfrac{4\sqrt{41}}{41}$ C) $\dfrac{5\sqrt{41}}{41}$ D) $\dfrac{5\sqrt{41}}{4}$

7) Find tan A when a = 5 and b = 6.

   A) $\dfrac{\sqrt{61}}{6}$ B) $\dfrac{6}{5}$ C) $\dfrac{\sqrt{61}}{5}$ D) $\dfrac{5}{6}$

8) Find tan B when a = 2 and b = 9.

   A) $\dfrac{9\sqrt{85}}{85}$ B) $\dfrac{9}{2}$ C) $\dfrac{2}{9}$ D) $\dfrac{2\sqrt{85}}{85}$

9) Find cot A when a = 3 and c = 8.

   A) $\dfrac{\sqrt{55}}{3}$ B) $\dfrac{3\sqrt{55}}{55}$ C) $\dfrac{\sqrt{55}}{8}$ D) $\dfrac{8\sqrt{55}}{55}$

10) Find cot A when b = 9 and c = 10.

   A) $\dfrac{9\sqrt{19}}{19}$ B) $\dfrac{\sqrt{19}}{9}$ C) $\dfrac{\sqrt{19}}{10}$ D) $\dfrac{10\sqrt{19}}{19}$

Solve the problem.

11) Find tan P.

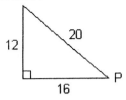

A) $\tan P = \dfrac{4}{3}$ B) $\tan P = \dfrac{3}{4}$ C) $\tan P = \dfrac{4}{5}$ D) $\tan P = \dfrac{3}{5}$

12) Find the exact value of each of the six trigonometric functions of the angle P.

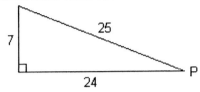

2 Use the Complementary Angle Theorem

Find the exact value of the expression. Do not use a calculator.

1) $-\dfrac{\sec 50°}{\csc 40°}$

A) 1 B) −1 C) 0 D) undefined

2) $\csc^2 5° - \tan^2 85°$

A) 0 B) 1 C) 2 D) −1

3) $\tan 45° - \dfrac{\cos 45°}{\cos 45°}$

A) 0 B) 2 C) −1 D) 1

4) $\cos 80° \sin 10° + \sin 80° \cos 10°$

A) 1 B) −1 C) 0 D) 2

3 Solve Right Triangles

Solve the right triangle using the information given. Round answers to two decimal places, if necessary.

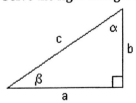

1) $b = 8$, $\alpha = 20°$; find a, c, and β

A) a = 2.91
  c = 8.51
  β = 70°

B) a = 2.91
  c = 9.51
  β = 70°

C) a = 3.91
  c = 8.51
  β = 70°

D) a = 3.91
  c = 9.51
  β = 70°

Precalculus Enhanced with Graphing Utilities 581

2) $a = 3$, $\alpha = 25°$; find b, c, and β

   A) $b = 6.43$
      $c = 8.1$
      $\beta = 75°$

   B) $b = 6.43$
      $c = 8.1$
      $\beta = 65°$

   C) $b = 6.43$
      $c = 7.1$
      $\beta = 75°$

   D) $b = 6.43$
      $c = 7.1$
      $\beta = 65°$

3) $a = 4$, $b = 5$; find c, α, and β

   A) $c = 6.4$
      $\alpha = 39.66°$
      $\beta = 50.34°$

   B) $c = 4.58$
      $\alpha = 38.66°$
      $\beta = 51.34°$

   C) $c = 4.58$
      $\alpha = 39.66°$
      $\beta = 50.34°$

   D) $c = 6.4$
      $\alpha = 38.66°$
      $\beta = 51.34°$

4) $b = 2$, $c = 8$; find a, β, and α

   A) $a = 7.75$
      $\beta = 14.48°$
      $\alpha = 75.52°$

   B) $a = 7.75$
      $\beta = 75.52°$
      $\alpha = 14.48°$

   C) $a = 8.25$
      $\beta = 15.48°$
      $\alpha = 74.52°$

   D) $a = 8.25$
      $\beta = 14.48°$
      $\alpha = 75.52°$

4 Solve Applied Problems

Solve the problem.

1) A surveyor is measuring the distance across a small lake. He has set up his transit on one side of the lake 80 feet from a piling that is directly across from a pier on the other side of the lake. From his transit, the angle between the piling and the pier is 60°. What is the distance between the piling and the pier to the nearest foot?

   A) 139 ft   B) 46 ft   C) 40 ft   D) 69 ft

2) A radio transmission tower is 210 feet tall. How long should a guy wire be if it is to be attached 12 feet from the top and is to make an angle of 24° with the ground? Give your answer to the nearest tenth of a foot.

   A) 486.8 ft   B) 216.7 ft   C) 229.9 ft   D) 516.3 ft

3) A straight trail with a uniform inclination of 11° leads from a lodge at an elevation of 600 feet to a mountain lake at an elevation of 8900 feet. What is the length of the trail (to the nearest foot)?

   A) 9067 ft   B) 43,499 ft   C) 8455 ft   D) 46,644 ft

4) A building 160 feet tall casts a 50 foot long shadow. If a person looks down from the top of the building, what is the measure of the angle between the end of the shadow and the vertical side of the building (to the nearest degree)? (Assume the person's eyes are level with the top of the building.)

   A) 72°   B) 17°   C) 73°   D) 18°

5) John (whose line of sight is 6 ft above horizontal) is trying to estimate the height of a tall oak tree. He first measures the angle of elevation from where he is standing as 35°. He walks 30 feet closer to the tree and finds that the angle of elevation has increased by 12°. Estimate the height of the tree rounded to the nearest whole number.

   A) 90 ft   B) 61 ft   C) 67 ft   D) 86 ft

6) A photographer points a camera at a window in a nearby building forming an angle of 42° with the camera platform. If the camera is 52 m from the building, how high above the platform is the window, to the nearest hundredth of a meter?

   A) 0.9 m   B) 1.11 m   C) 46.82 m   D) 57.75 m

7) A tree casts a shadow of 26 meters when the angle of elevation of the sun is 24°. Find the height of the tree to the nearest meter.

   A) 10 m   B) 11 m   C) 13 m   D) 12 m

8) A twenty-five foot ladder just reaches the top of a house and forms an angle of 41.5° with the wall of the house. How tall is the house? Round your answer to the nearest 0.1 foot.

   A) 19 ft   B) 18.6 ft   C) 18.8 ft   D) 18.7 ft

9) In 1838, the German mathematician and astronomer Friedrich Wilhelm Bessel was the first person to calculate the distance to a star other than the Sun. He accomplished this by first determining the parallax of the star, 61 Cygni, at 0.314 arc seconds (Parallax is the change in position of the star measured against background stars as Earth orbits the Sun. See illustration.) If the distance from Earth to the Sun is about 150,000,000 km and

$$\theta = 0.314 \text{ seconds} = \frac{0.314}{60} \text{ minutes} = \frac{0.314}{60 \cdot 60} \text{ degrees}$$

determine the distance d from Earth to 61 Cygni using Bessel's figures.

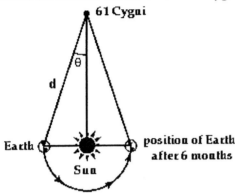

10) Two hikers on opposite sides of a canyon each stand precisely 525 meters above the canyon floor. They each sight a landmark on the canyon floor on a line directly between them. The angles of depression from each hiker to the landmark meter are 37° and 21°. How far apart are the hikers? Round your answer to the nearest whole meter.

    A) 1064 m   B) 2063 m   C) 2064 m   D) 2065 m

11) Yosemite Falls in California consists of three sections: Upper Yosemite Fall (by itself one of the ten highest waterfalls in the world), the Middle Cascade, and Lower Yosemite Fall. From a footbridge across the creek 2500 feet from the falls, the angles of elevation to the top and bottom of Upper Yosemite Fall are 45.74° and 24.42°, respectively. How high is the total series of three falls? How high is Upper Yosemite Fall? Round your answers to the whole foot.

12) A forest ranger at Lookout A sights a fire directly north of her position. Another ranger at Lookout B, exactly 2 kilometers directly west of A, sights the same fire at a bearing of N41.2°E. How far is the fire from Lookout A? Round your answer to the nearest 0.01 km.

    A) 2.18 km   B) 2.28 km   C) 2.32 km   D) 2.25 km

13) A sailboat leaves port on a bearing of S72°W. After sailing for two hours at 12 knots, the boat turns 90° toward the south. After sailing for three hours at 9 knots on this course, what is the bearing to the ship from port? Round your answer to the nearest 0.1°.

    A) S24.6°W   B) S23.6°W   C) N24.6°E   D) N23.6°E

14) From the edge of a 1000-foot cliff, the angles of depression to two cars in the valley below are 21° and 28°. How far apart are the cars? Round your answers to the nearest 0.1 ft.

    A) 714.4 ft   B) 724.4 ft   C) 713.4 ft   D) 724.5 ft

7.2 The Law of Sines

1 Solve SAA or ASA Triangles

Solve the triangle.

1)
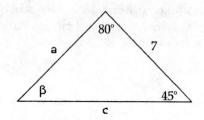

A) $a = 6.04$, $c = 8.42$, $\beta = 60°$
B) $a = 8.42$, $c = 6.04$, $\beta = 50°$
C) $a = 6.04$, $c = 8.42$, $\beta = 55°$
D) $a = 8.42$, $c = 6.04$, $\beta = 55°$

2)
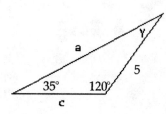

A) $a = 7.55$, $c = 3.68$, $\gamma = 25°$
B) $a = 7.55$, $c = 3.68$, $\gamma = 30°$
C) $a = 3.68$, $c = 7.55$, $\gamma = 25°$
D) $a = 3.68$, $c = 7.55$, $\gamma = 20°$

3) $\alpha = 70°$, $\beta = 10°$, $a = 4$
A) $\gamma = 100°$, $b = 4.19$, $c = 0.74$
B) $\gamma = 100°$, $b = 4.19$, $c = -0.26$
C) $\gamma = 100°$, $b = 1.74$, $c = 4.19$
D) $\gamma = 100°$, $b = 0.74$, $c = 4.19$

4) $\alpha = 40°$, $\beta = 90°$, $c = 2$
A) $\gamma = 50°$, $a = 3.61$, $b = 1.68$
B) $\gamma = 50°$, $a = 2.68$, $b = 2.61$
C) $\gamma = 50°$, $a = 1.68$, $b = 2.61$
D) $\gamma = 50°$, $a = 2.61$, $b = 1.68$

2 Solve SSA Triangles

Two sides and an angle are given. Determine whether the given information results in one triangle, two triangles, or no triangle at all. Solve any triangle(s) that results.

1) $a = 7$, $b = 9$, $\beta = 49°$

A) two triangles
$\alpha_1 = 76.01°$, $\gamma_1 = 54.99°$, $c_1 = 7.60$ or
$\alpha_2 = 103.99°$, $\gamma_2 = 27.01$, $c_2 = 12.14$

B) one triangle
$\alpha = 35.94°$, $\gamma = 95.06°$, $c = 11.88$

C) one triangle
$\alpha = 76.01°$, $\gamma = 54.99°$, $c = 7.60$

D) no triangle

Precalculus Enhanced with Graphing Utilities 584

2) $a = 4$, $b = 5$, $\alpha = 80°$

A) one triangle
$\beta = 39°$, $\gamma = -50°$, $c = 11$

B) one triangle
$\beta = 41°$, $\gamma = 59°$, $c = 13$

C) one triangle
$\alpha = 40°$, $\gamma = 60°$, $c = 9$

D) no triangle

3) $a = 17$, $b = 14$, $\beta = 25°$

A) one triangle
$\alpha = 149.12°$, $\gamma = 5.88°$, $c = 3.39$

B) one triangle
$\alpha = 30.88°$, $\gamma = 124.12°$, $c = 27.42$

C) two triangles
$\alpha_1 = 30.88°$, $\gamma_1 = 124.12°$, $c_1 = 27.42$ or
$\alpha_2 = 149.12°$, $\gamma_2 = 5.88°$, $c_2 = 3.39$

D) no triangle

Solve the problem.

4) Given a triangle with $a = 9$, $b = 11$, $\alpha = 31°$, what is (are) the possible length(s) of c? Round your answer to two decimal places.

A) 16.42 or 3.41 B) 6.61 C) 16.42 or 2.44 D) 14.21

3 Solve Applied Problems

Solve the problem.

1) An airplane is sighted at the same time by two ground observers who are 5 miles apart and both directly west of the airplane. They report the angles of elevation as 12° and 22°. How high is the airplane?

A) 2.24 mi B) 1.87 mi C) 4.04 mi D) 1.04 mi

2) A surveyor standing 54 meters from the base of a building measures the angle to the top of the building and finds it to be 35°. The surveyor then measures the angle to the top of the radio tower on the building and finds that it is 47°. How tall is the radio tower?

A) 20.1 m B) 11.48 m C) 8.52 m D) 7.41 m

3) A ship sailing parallel to shore sights a lighthouse at an angle of 14° from its direction of travel. After traveling 3 miles farther, the angle is 22°. At that time, how far is the ship from the lighthouse?

A) 3 mi B) 1.94 mi C) 8.07 mi D) 5.21 mi

4) A rocket tracking station has two telescopes A and B placed 2.0 miles apart. The telescopes lock onto a rocket and transmit their angles of elevation to a computer after a rocket launch. What is the distance to the rocket from telescope B at the moment when both tracking stations are directly east of the rocket telescope A reports an angle of elevation of 27° and telescope B reports an angle of elevation of 49°?

A) 1.2 mi B) 3.32 mi C) 2.42 mi D) 4.03 mi

5) A guy wire to the top of a tower makes an angle of 54° with the level ground. At a point 29 feet farther from the base of the tower and in line with the base of the wire, the angle of elevation to the top of the tower is 23°. What is the length of the guy wire?

A) 14.01 ft B) 22 ft C) 45.55 ft D) 60.05 ft

6) A ship at sea, the Admiral, spots two other ships, the Barstow and the Cauldrew and measures the angle between them at be 45°. They radio the Barstow and by comparing known landmarks, the distance between the the Admiral and the Barstow is found to be 323 meters. The Barstow reports an angle of 59° between the Admiral and the Cauldrew. To the nearest meter, what is the distance between the Barstow and the Cauldrew?

   A) 49 m   B) 235 m   C) 266 m   D) 81 m

7) Two surveyors 180 meters apart on the same side of a river measure their respective angles to a point between them on the other side of the river and obtain 54° and 68°. How far from the point (line-of-sight distance) is each surveyor? Round your answer to the nearest 0.1 meter.

8) A flagpole is perpendicular to the horizontal but is on a slope that rises 10° from the horizontal. The pole casts a 43-foot shadow down the slope and angle of elevation of the sun measured from the slope is 36°. How tall is the pole? Round your answer to the nearest 0.1 foot.

   A) 36.2 ft   B) 35.4 ft   C) 36.4 ft   D) 33.5 ft

9) It is 4.7 km from Lighthouse A to Port B. The bearing of the port from the lighthouse is N73°E. A ship has sailed due west from the port and its bearing from the lighthouse is N31°E. How far has the ship sailed from the port? Round your answer to the nearest 0.1 km.

   A) 3.5 km   B) 3.1 km   C) 3.7 km   D) 2.7 km

10) A pier 1250 meters long extends at an angle from the shoreline. A surveyor walks to a point 1500 meters down the shoreline from the pier and measures the angle formed by the ends of the pier. If is found to be 53°. What acute angle (correct to the nearest 0.1°) does the pier form with the shoreline? Is there more than one possibility? If so, how can we know which is the correct one?

7.3 The Law of Cosines

1 Solve SAS Triangles

Solve the triangle.

1)
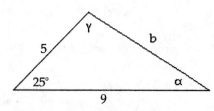

A) $b = 3.94, \alpha = 129.7°, \gamma = 25.3°$
B) $b = 4.94, \alpha = 25.3°, \gamma = 129.7°$
C) $b = 5.94, \alpha = 25.3°, \gamma = 129.7°$
D) $b = 4.94, \alpha = 129.7°, \gamma = 25.3°$

2)
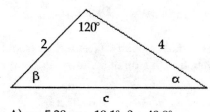

A) $c = 5.29, \alpha = 19.1°, \beta = 40.9°$
B) $c = 6.29, \alpha = 19.1°, \beta = 40.9°$
C) $c = 5.29, \alpha = 40.9°, \beta = 19.1°$
D) $c = 4.29, \alpha = 40.9°, \beta = 19.1°$

3) $b = 4, c = 5, \alpha = 70°$
   A) $a = 4.23, \beta = 64°, \gamma = 46°$
   B) $a = 6.23, \beta = 46°, \gamma = 64°$
   C) $a = 5.23, \beta = 46°, \gamma = 64°$
   D) $a = 5.23, \beta = 64°, \gamma = 46°$

4) $a = 6, b = 8, \gamma = 70°$
   A) $c = 9, \alpha = 52.8°, \beta = 57.2°$
   B) $c = 6.3, \alpha = 28.6°, \beta = 81.4°$
   C) $c = 10, \alpha = 56.9°, \beta = 53.1°$
   D) $c = 8.2, \alpha = 43.5°, \beta = 66.5°$

5) $a = 60, b = 12, \gamma = 120°$
   A) $c = 69.71, \alpha = 53.1°, \beta = 6.9°$
   B) $c = 66.81, \alpha = 51.1°, \beta = 8.9°$
   C) $c = 72.61, \alpha = 49.1°, \beta = 10.9°$
   D) no triangle

6) $a = 6, c = 5, \beta = 90°$
   A) $b = 7.81, \alpha = 39.8°, \gamma = 50.2°$
   B) $b = 8.81, \alpha = 50.2°, \gamma = 39.8°$
   C) $b = 7.81, \alpha = 50.2°, \gamma = 39.8°$
   D) $b = 6.81, \alpha = 39.8°, \gamma = 50.2°$

2   Solve SSS Triangles

Solve the triangle. Find the angles α and β first.

1)

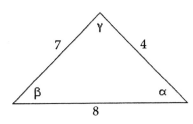

   A) $\alpha = 30°, \beta = 61°, \gamma = 89°$
   B) $\alpha = 61°, \beta = 30°, \gamma = 89°$
   C) $\alpha = 30°, \beta = 89°, \gamma = 61°$
   D) $\alpha = 61°, \beta = 89°, \gamma = 30°$

2)

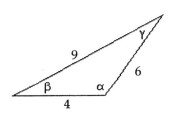

   A) $\alpha = 32.1°, \beta = 20.7°, \gamma = 127.2°$
   B) $\alpha = 127.2°, \beta = 20.7°, \gamma = 32.1°$
   C) $\alpha = 32.1°, \beta = 127.2°, \gamma = 20.7°$
   D) $\alpha = 127.2°, \beta = 32.1°, \gamma = 20.7°$

3) $a = 10, b = 10, c = 8$
   A) $\alpha = 66.4°, \beta = 47.2°, \gamma = 66.4°$
   B) $\alpha = 67.4°, \beta = 67.4°, \gamma = 45.2°$
   C) $\alpha = 66.4°, \beta = 66.4°, \gamma = 47.2°$
   D) $\alpha = 47.2°, \beta = 66.4°, \gamma = 66.4°$

4) $a = 8, b = 6, c = 4$
   A) $\alpha = 46.6°, \beta = 28.9°, \gamma = 104.5°$
   B) $\alpha = 104.5°, \beta = 46.6°, \gamma = 28.9°$
   C) $\alpha = 46.6°, \beta = 104.5°, \gamma = 28.9°$
   D) $\alpha = 104.5°, \beta = 28.9°, \gamma = 46.6°$

5) a = 7, b = 13, c = 17

   A) α = 24.3°, β = 42.8°, γ = 112.9°
   B) α = 20.3°, β = 44.8°, γ = 114.9°
   C) α = 22.3°, β = 44.8°, γ = 112.9°
   D) no triangle

6) a = 19, b = 16, c = 11

   A) α = 57.3°, β = 87.4°, γ = 35.3°
   B) α = 35.3°, β = 57.3°, γ = 87.4°
   C) α = 87.4°, β = 57.3°, γ = 35.3°
   D) α = 87.4°, β = 35.3°, γ = 57.3°

3 Solve Applied Problems

Solve the problem.

1) In flying the 78 miles from Champaign to Peoria, a student pilot sets a heading that is 11° off course and maintains an average speed of 116 miles per hour. After 15 minutes, the instructor notices the course error and tells the student to correct his heading. Through what angle will the plane move to correct the heading and how many miles away is Peoria when the plane turns?

   A) 17.4°; 49.84 mi   B) 162.6°; 68.59 mi   C) 17.4°; 68.59 mi   D) 162.6°; 49.84 mi

2) Two points A and B are on opposite sides of a building. A surveyor selects a third point C to place a transit. Point C is 52 feet from point A and 64 feet from point B. The angle ACB is 52°. How far apart are points A and B?

   A) 104.4 ft   B) 94.1 ft   C) 68.9 ft   D) 52 ft

3) The distance from home plate to dead center field in a certain baseball stadium is 408 feet. A baseball diamond is a square with a distance from home plate to first base of 90 feet. How far is it from first base to dead center field?

   A) 385.5 ft   B) 333.1 ft   C) 475.9 ft   D) 350.2 ft

4) A famous golfer tees off on a long, straight 453 yard par 4 and slices his drive 13° to the right of the line from tee to the hole. If the drive went 284 yards, how many yards will the golfer's second shot have to be to reach the hole?

   A) 400.6 yd   B) 187.5 yd   C) 732.5 yd   D) 641.3 yd

5) A famous golfer tees off on a straight 390 yard par 4 and slices his drive to the right. The drive goes 250 yards from the tee. Using a 7-iron on his second shot, he hits the ball 190 yards and it lands inches from the hole. How many degrees (to the nearest degree) to the right of the line from the tee to the hole did he slice his drive?

   A) 52°   B) 32°   C) 124°   D) 24°

6) Island A is 150 miles from island B. A ship captain travels 250 miles from island A and then finds that he is off course and 160 miles from island B. What angle, in degrees, must he turn through to head straight for island B? Round the answer to two decimal places. (Hint: Be careful to properly identify which angle is the turning angle.)

   A) 55.08°   B) 145.08°   C) 34.92°   D) 110.17°

7) A ladder leans against a building that has a wall slanting away from the ladder at an angle of 96° with the ground. If the bottom of the ladder is 23 feet from the base of the wall and it reaches a point 52 feet up the wall, how tall is the ladder to the nearest foot?

   A) 60 ft   B) 59 ft   C) 58 ft   D) 61 ft

8) A plane takes off from an airport on the bearing S29°W. It continues for 20 minutes then changes to bearing S52°W and flies for 2 hours 20 minutes on this course then lands at a second airport. If the plane's speed is 420 mph, how far from the first airport is the second airport? Round your answer correct to the nearest mile.

   A) 1110 mi B) 1010 mi C) 1011 mi D) 1111 mi

9) A box has dimensions 2" × 3" × 4". (See illustration.)

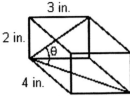

Determine the angle θ formed by the diagonal of the 2" × 3" side and the diagonal of the 3" × 4" side. Round your answer to the nearest degree.

   A) 62° B) 50° C) 60° D) 65°

7.4 Area of a Triangle

1 Find the Area of SAS Triangles

Find the area of the triangle. If necessary, round the answer to two decimal places.

1)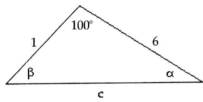

   A) 98.48 B) 2.95 C) 0.52 D) 11.82

2)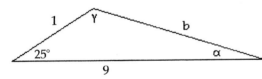

   A) 3.8 B) 4.08 C) 7.61 D) 1.9

3) α = 30°, b = 13, c = 8

   A) 24 B) 47.03 C) 26 D) 45.03

4) α = 83°, b = 9, c = 6

   A) 27.01 B) 53.60 C) 26.80 D) 3.29

5) α = 23°, b = 9, c = 2

   A) 3.52 B) 3.50 C) 3.51 D) 3.53

6) a = 12, b = 15, γ = 52°

   A) 141.84 B) 70.92 C) 88.80 D) 35.46

2 Find the Area of SSS Triangles

Find the area of the triangle. If necessary, round the answer to two decimal places.

1)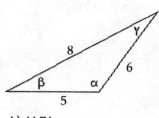

   A) 44.74   B) 14.98   C) 4.86   D) 195.03

2)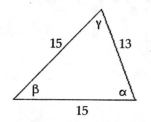

   A) 153.36   B) 1005.66   C) 87.87   D) 18.95

3) a = 11, b = 14, c = 16

   A) 84.47   B) 78.47   C) 81.47   D) 75.47

4) a = 4, b = 5, c = 7

   A) 16.01   B) 10.01   C) 9.80   D) 3.46

5) a = 6, b = 6, c = 7

   A) 21.14   B) 18.25   C) 15.54   D) 17.06

6) a = 14, b = 32, c = 26

   A) 181.99   B) 5280.01   C) 177.99   D) 3219.69

3 Solve Applied Problems

Solve the problem.

1) Find the area of the shaded portion (see illustration) of a circle of radius 25 cm, formed by a central angle of 115°. Round your answer to the nearest square cm.

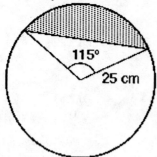

[Hint: Subtract the area of the triangle from the area of the sector of the circle to obtain the area of the shaded portion.]

2) A room in the shape of a triangle has sides of length 7 yd, 9 yd, and 14 yd. If carpeting costs $16.50 a square yard and padding costs $2.25 a square yard, how much to the nearest dollar will it cost to carpet the room, assuming that there is no waste?
   A) $443
   B) $503
   C) $490
   D) $483

3) A new homeowner has a triangular-shaped back yard. Two of the three sides measure 65 ft and 80 ft and form an included angle of 125°. The owner wants to approximate the area of the yard, so that he can determine the amount of fertilizer and grass seed to be purchased. Find the area of the yard rounded to the nearest square foot.
   A) 5200 sq. ft
   B) 2130 sq. ft
   C) 2129 sq. ft
   D) 4260 sq. ft

4) Find the area of the Bermuda Triangle if the sides of the triangle have the approximate lengths 847 miles, 923 miles, and 1312 miles.
   A) 389,039 mi
   B) 514,174 mi
   C) 1,556,156 mi
   D) 493,670 mi

5) Penrose tiles are formed from a rhombus WXYZ with sides of length 1 and interior angles 72° and 108°. (Refer to the illustration.) A point O is chosen on the diagonal 1 unit from Y. Line segments OX and OZ are drawn to the other vertices. The two resulting tiles are called a kite (figure OXYZ) and a dart (figure OXWZ). Find the area of the kite tile and the dart tile, correct to the nearest 0.01.

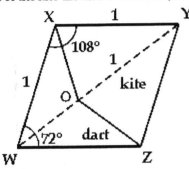

7.5 Simple Harmonic Motion; Damped Motion; Combining Waves

1 Find an Equation for an Object in Simple Harmonic Motion

An object attached to a coiled spring is pulled down a distance a from its rest position and then released. Assuming that the motion is simple harmonic with period T, write an equation that relates the displacement d of the object from its rest position after t seconds. Also assume that the positive direction of the motion is up.

1) $a = 6$; $T = 4$ seconds
   A) $d = -4\cos\left(\frac{1}{3}\pi t\right)$
   B) $d = -6\cos\left(\frac{1}{2}\pi t\right)$
   C) $d = -6\cos\left(\frac{1}{4}\pi t\right)$
   D) $d = -6\sin\left(\frac{1}{2}\pi t\right)$

2) $a = 16$; $T = 6$ seconds
   A) $d = -16\cos\left(\frac{1}{3}\pi t\right)$
   B) $d = -6\cos\left(\frac{1}{8}\pi t\right)$
   C) $d = -16\cos\left(\frac{1}{6}\pi t\right)$
   D) $d = -16\sin\left(\frac{1}{3}\pi t\right)$

3) $a = 14$; $T = 4\pi$ seconds
   A) $d = -14\sin\left(\frac{1}{2}\pi t\right)$
   B) $d = -4\cos\left(\frac{1}{7}t\right)$
   C) $d = -14\cos\left(\frac{1}{2}t\right)$
   D) $d = -14\cos\left(\frac{1}{2}\pi t\right)$

4) a = 5;   T = 10 seconds

   A) d = 5 cos (10t)   B) d = -5 cos (10t)   C) d = 5 cos$\left(\frac{\pi}{5}t\right)$ D) d = -5 cos$\left(\frac{\pi}{5}t\right)$

2 Analyze Simple Harmonic Motion

The displacement d (in meters) of an object at time t (in seconds) is given. Describe the motion of the object. What is the maximum displacement from its resting position, the time required for one oscillation, and the frequency?

1) d = 6 sin (3t)

   A) simple harmonic; -6 m; $\frac{2}{3}\pi$ sec; $\frac{3}{2\pi}$ oscillations/sec

   B) simple harmonic; 6 m; $\frac{3}{2\pi}$ sec; $\frac{2}{3}\pi$ oscillations/sec

   C) simple harmonic; 6 m; $\frac{2}{3}\pi$ sec; $\frac{3}{2\pi}$ oscillations/sec

   D) simple harmonic; 6 m; 3π sec; $\frac{3}{\pi}$ oscillations/sec

2) d = -4 sin (5t)

   A) simple harmonic; -4 m; 5π sec; $\frac{5}{\pi}$ oscillations/sec

   B) simple harmonic; 4 m; $\frac{2}{5}\pi$ sec; $\frac{5}{2\pi}$ oscillations/sec

   C) simple harmonic; -4 m; $\frac{2}{5}\pi$ sec; $\frac{5}{2\pi}$ oscillations/sec

   D) simple harmonic; 4 m; $\frac{5}{2\pi}$ sec; $\frac{2}{5}\pi$ oscillations/sec

3) d = 4 cos (5t)

   A) simple harmonic; 4 m; 5π sec; $\frac{5}{\pi}$ oscillations/sec

   B) simple harmonic; 4 m; $\frac{5}{2\pi}$ sec; $\frac{2}{5}\pi$ oscillations/sec

   C) simple harmonic; 4 m; $\frac{2}{5}\pi$ sec; $\frac{5}{2\pi}$ oscillations/sec

   D) simple harmonic; -4 m; $\frac{2}{5}\pi$ sec; $\frac{5}{2\pi}$ oscillations/sec

4) d = 6 cos$\left(\frac{\pi}{2}t\right)$

   A) simple harmonic; 6 m; 2 sec; $\frac{1}{2}$ oscillation/sec   B) simple harmonic; 6 m; 4 sec; $\frac{1}{4}$ oscillation/sec

   C) simple harmonic; -6 m; 4 sec; $\frac{1}{4}$ oscillation/sec   D) simple harmonic; 6 m; $\frac{1}{4}$ sec; 4 oscillations/sec

5) $d = 7 + 2\cos(2\pi t)$

A) simple harmonic; −2 m; 1 sec; 1 oscillation/sec
B) simple harmonic; 7 m; 1 sec; 1 oscillation/sec
C) simple harmonic; 2 m; $\frac{1}{2}$ sec; 2 oscillations/sec
D) simple harmonic; 2 m; 1 sec; 1 oscillation/sec

3 Analyze an Object in Damped Motion

Graph the damped vibration curve for $0 \leq t \leq 2\pi$.

1) $d(t) = e^{-t/4\pi} \cos t$

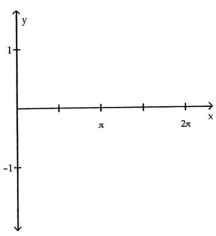

A)

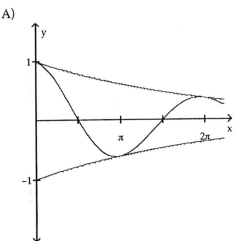

B)

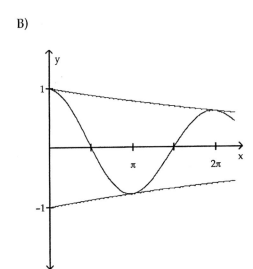

Solve the problem.

2) An object is suspended from a coiled spring. It is pulled downward and released. The position P above or below its rest position after t seconds is given by

$$P = -e^{-x/2\pi} \cos x$$

Find all values of t ($0 \le t < 4\pi$) for which the object is at the rest position and graph the equation for $0 \le t < 4\pi$.

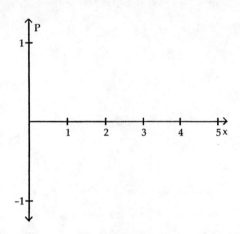

3) An object of mass m attached to a coiled spring with damping factor b is pulled down a distance a from its rest position and then released. Assume the positive direction of the motion is up and the period of the first oscillation is T. Write an equation that relates the distance d of the object from its rest position after t seconds.
m = 15 g; a = 11 cm; b = 0.7 g/sec; T = 3 sec

4) The distance d (in meters) of the bob of a pendulum from its rest position at time t (in seconds) is given by:

$$d = -9e^{-0.6t/60} \cos\left(\sqrt{\left(\frac{2\pi}{7}\right)^2 - \frac{0.36}{3600}}\, t\right)$$

What is the maximum displacement of the bob after the first oscillation?

A) about 5.18 m B) about 8.39 m C) about 8.91 m D) about 0.13 m

4 **Graph the Sum of Two Functions**

Use the method of adding y-coordinates to graph the function.

1) $f(x) = x + \cos(2x)$

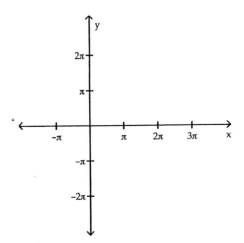

A)

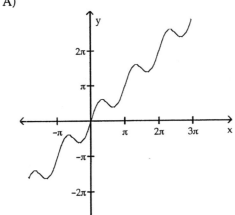

B)

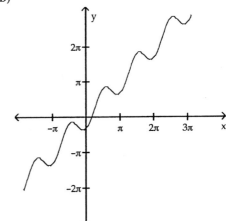

C)

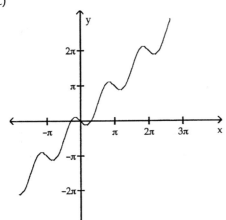

D)

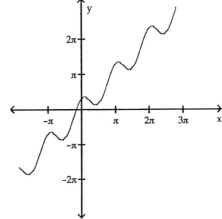

Precalculus Enhanced with Graphing Utilities 595

2) $g(x) = \sin x + \cos(2x)$

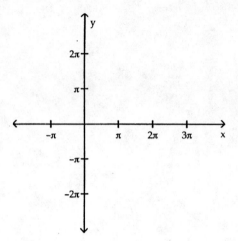

A)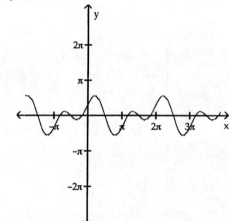

B)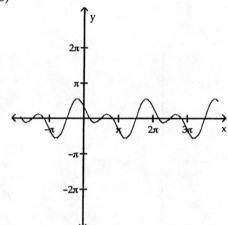

C)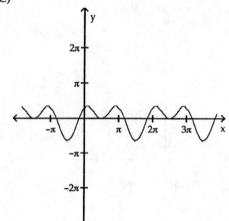

D)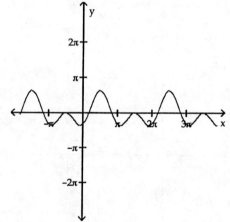

Solve the problem.

3) A square wave is built up from sinusoidal curves of varying periods and amplitudes. Graph the following function, which can be used to approximate the square wave.

$$f(x) = \frac{4}{\pi}\left[\sin(\pi x) + \frac{1}{3}\sin(3\pi x)\right] \quad 0 \le x \le 4$$

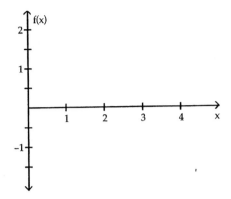

A better approximation to the square wave is given by

$$f(x) = \frac{4}{\pi}\left[\sin(\pi x) + \frac{1}{3}\sin(3\pi x) + \frac{1}{5}\sin(5\pi x)\right] \quad 0 \le x \le 4$$

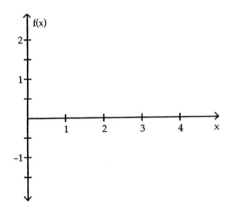

Graph this function and compare the result to the previous graph. Adding another term will improve the approximation even more. Write this new function with four terms.

Ch. 7 Applications of Trigonometric Functions
Answer Key

7.1 Right Triangle Trigonometry; Applications
1 Find the Values of Trigonometric Functions of Acute Angles
1) B
2) C
3) C
4) A
5) A
6) A
7) D
8) B
9) A
10) A
11) B
12) $\sin P = \frac{7}{25}$, $\cos P = \frac{24}{25}$, $\tan P = \frac{7}{24}$, $\csc P = \frac{25}{7}$, $\sec P = \frac{25}{24}$, and $\cot P = \frac{24}{7}$

2 Use the Complementary Angle Theorem
1) B
2) B
3) A
4) A

3 Solve Right Triangles
1) A
2) D
3) D
4) A

4 Solve Applied Problems
1) A
2) A
3) B
4) B
5) C
6) C
7) D
8) D
9) 9.85×10^{13} km. (More recent observations have refined this value to about 1.08×10^{14} km.)
10) C
11) 2565 ft; 1430 ft
12) B
13) B
14) B

7.2 The Law of Sines
1 Solve SAA or ASA Triangles
1) C
2) A
3) D
4) C

2 Solve SSA Triangles
1) B
2) D
3) C
4) C

3 Solve Applied Problems
1) A
2) A
3) D
4) C
5) B
6) B
7) 196.8 m, 171.7 m
8) C
9) C
10) 53.6° or 20.4°; yes; direct observation

7.3 The Law of Cosines
1 Solve SAS Triangles
1) B
2) A
3) C
4) D
5) B
6) C

2 Solve SSS Triangles
1) B
2) D
3) C
4) B
5) C
6) C

3 Solve Applied Problems
1) A
2) D
3) D
4) B
5) D
6) B
7) B
8) A
9) C

7.4 Area of a Triangle
1 Find the Area of SAS Triangles
1) B
2) D
3) C
4) C
5) A
6) B

2 Find the Area of SSS Triangles
1) B
2) C
3) D
4) C
5) D
6) C

3 Solve Applied Problems
1) 344 sq. cm
2) B
3) B

4) A
5) 0.59 sq. units; 0.36 sq. units

7.5 Simple Harmonic Motion; Damped Motion; Combining Waves

1 Find an Equation for an Object in Simple Harmonic Motion

1) B
2) A
3) C
4) D

2 Analyze Simple Harmonic Motion

1) C
2) B
3) C
4) B
5) D

3 Analyze an Object in Damped Motion

1) B
2) $t = \dfrac{\pi}{2}; \dfrac{3\pi}{2}; \dfrac{5\pi}{2}; \dfrac{7\pi}{2}$

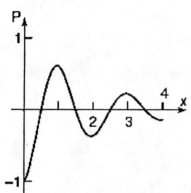

3) $d = -11e^{-0.7t/30} \cos\left(\sqrt{\left(\dfrac{2\pi}{3}\right)^2 - \dfrac{0.49}{900}}\, t\right)$

4) B

4 Graph the Sum of Two Functions

1) D
2) C
3)

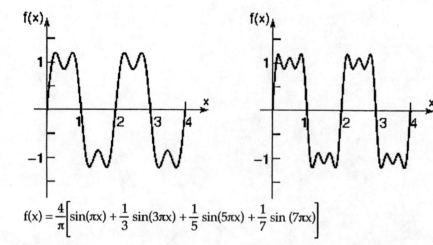

$f(x) = \dfrac{4}{\pi}\left[\sin(\pi x) + \dfrac{1}{3}\sin(3\pi x) + \dfrac{1}{5}\sin(5\pi x) + \dfrac{1}{7}\sin(7\pi x)\right]$

Ch. 8 Polar Coordinates; Vectors

8.1 Polar Coordinates

1 Plot Points Using Polar Coordinates

Match the point in polar coordinates with either A, B, C, or D on the graph.

1) $\left(-3, \dfrac{\pi}{3}\right)$

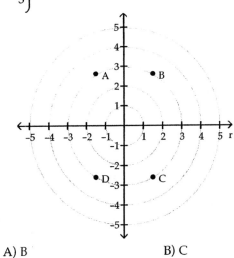

A) B B) C C) A D) D

2) $\left(3, -\dfrac{5\pi}{3}\right)$

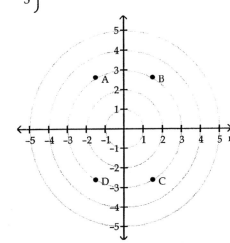

A) A B) C C) D D) B

3) $\left(-3, -\dfrac{\pi}{3}\right)$

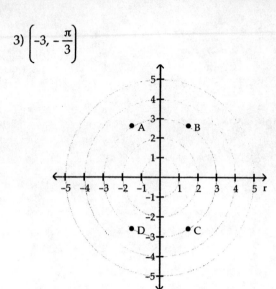

A) C B) D C) B D) A

Plot the point given in polar coordinates.

4) $\left(2, -\dfrac{\pi}{4}\right)$

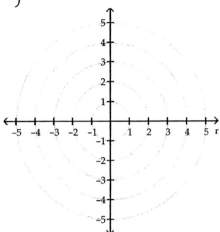

A)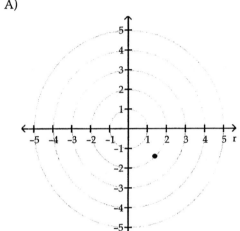

B)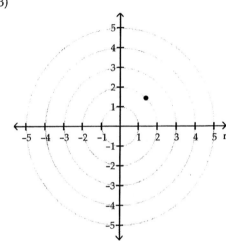

C)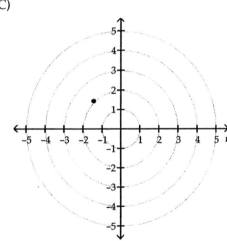

D)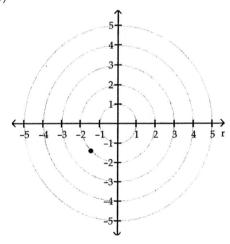

5) (−2, 45°)

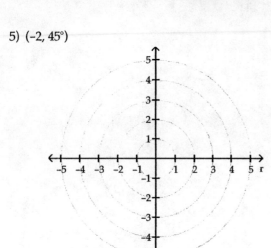

A)
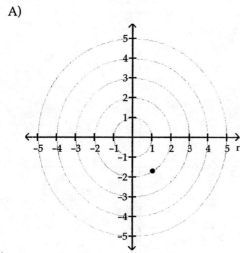

B)
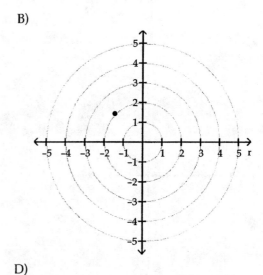

C)
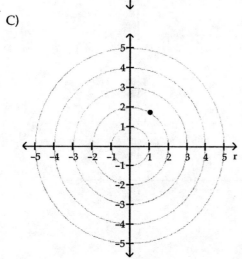

D)
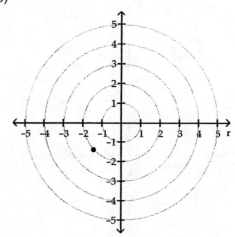

Precalculus Enhanced with Graphing Utilities 604

6) (2, 45°)

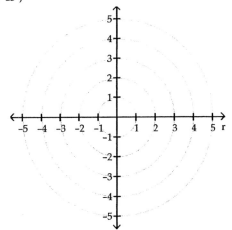

A)
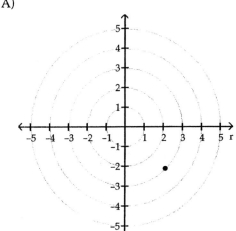

B)
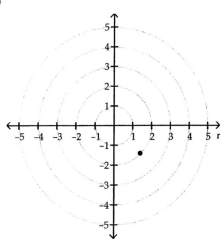

C)
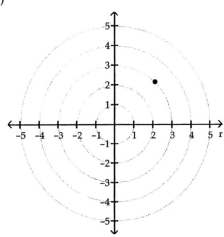

D)
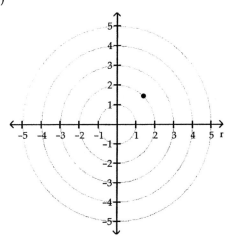

7) $\left(3, \dfrac{7\pi}{6}\right)$

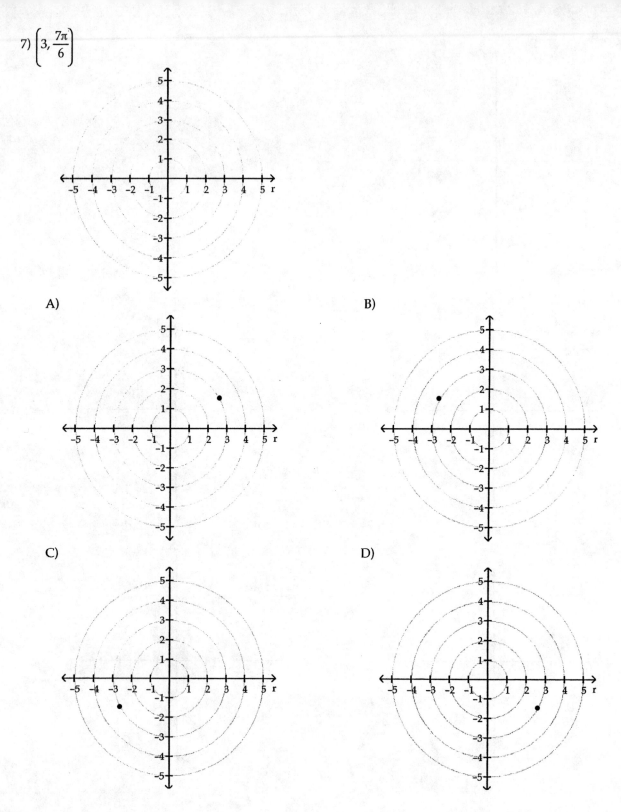

8) $(2, 0°)$

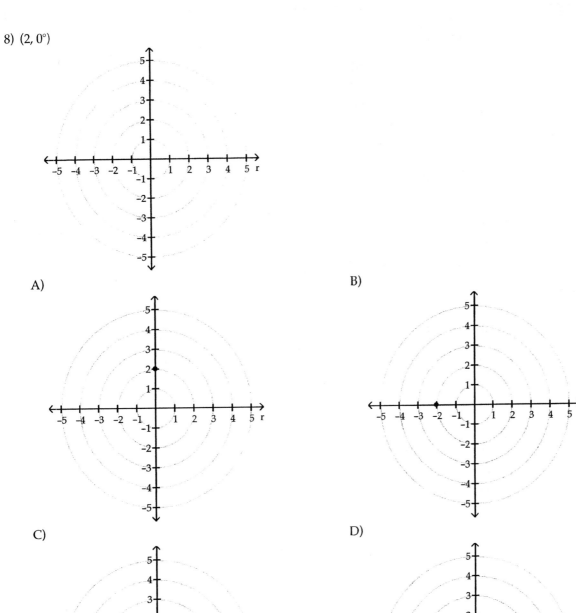

A)

B)

C)

D)

9) (2, 360°)

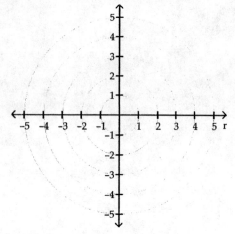

A)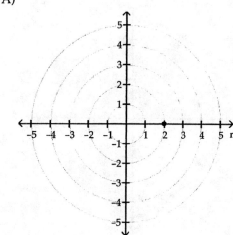

B)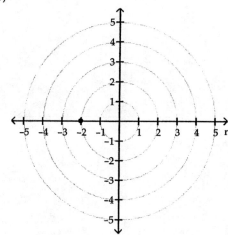

C)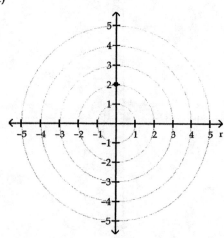

D)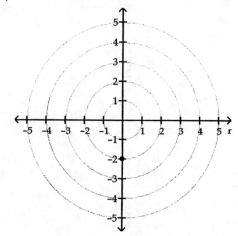

10) $\left(-3, -\dfrac{\pi}{4}\right)$

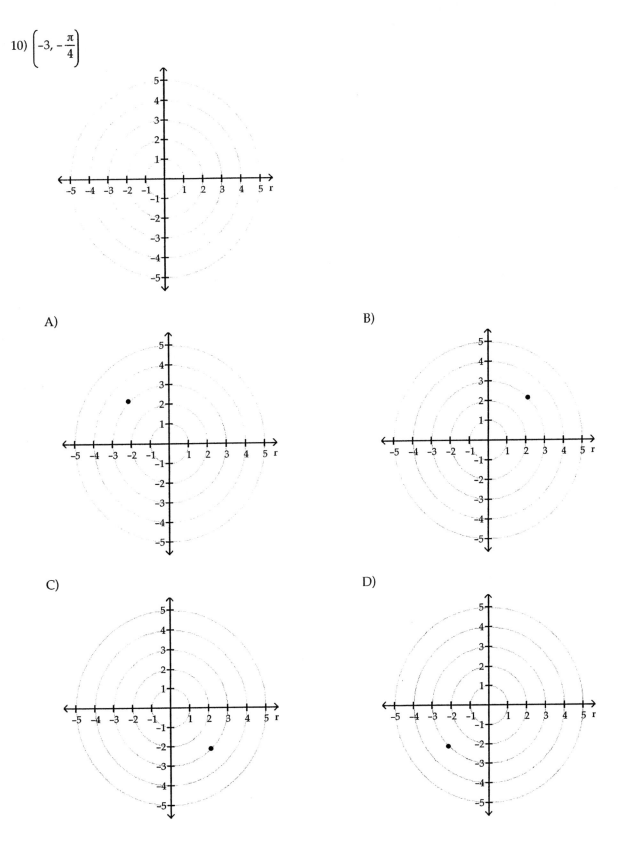

A)

B)

C)

D)

11) $\left(5, \dfrac{5\pi}{3}\right)$

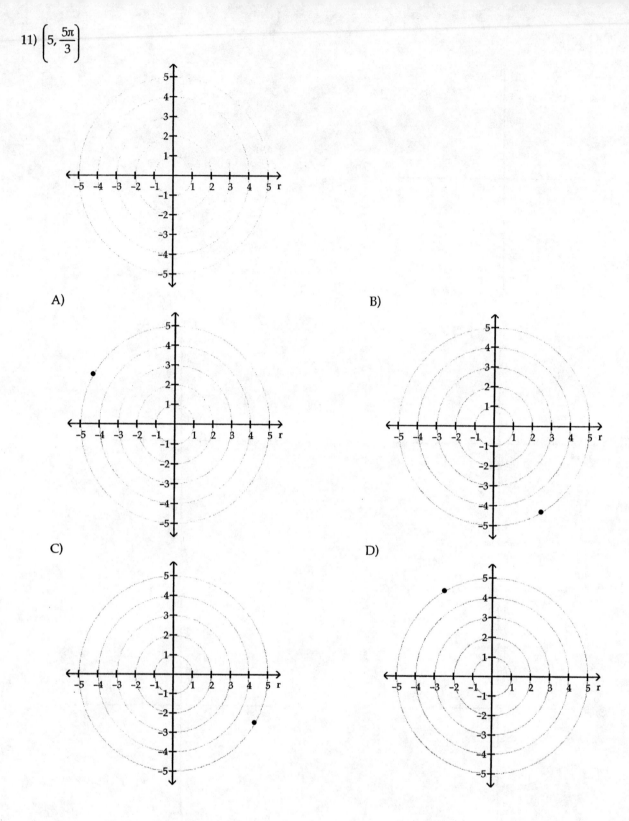

A)

B)

C)

D)

12) $\left(-3, -\dfrac{\pi}{3}\right)$

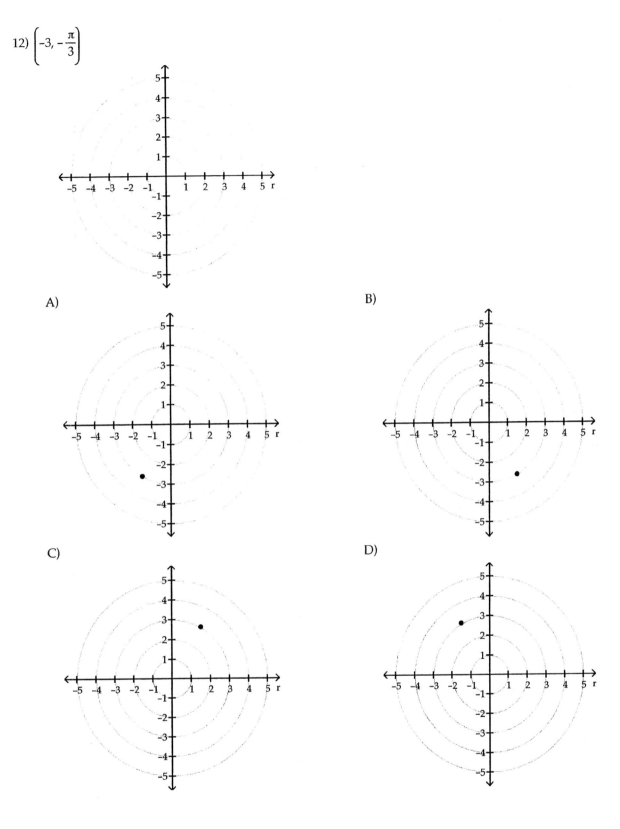

Solve the problem.

13) Plot the point $\left(4, \dfrac{\pi}{6}\right)$ and find other polar coordinates (r, θ) of the point for which:

   (a) $r > 0$, $\quad -2\pi \le \theta < 0$
   (b) $r < 0$, $\quad 0 \le \theta < 2\pi$
   (c) $r > 0$ $\quad 2\pi \le \theta < 4\pi$

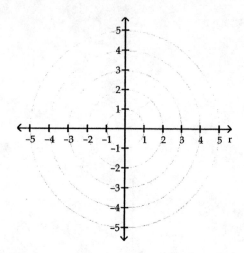

2 Convert from Polar Coordinates to Rectangular Coordinates

The polar coordinates of a point are given. Find the rectangular coordinates of the point.

1) $\left(7, \dfrac{2\pi}{3}\right)$

   A) $\left(-\dfrac{7}{2}, \dfrac{-7\sqrt{3}}{2}\right)$
   B) $\left(\dfrac{7}{2}, \dfrac{-7\sqrt{3}}{2}\right)$
   C) $\left(-\dfrac{7}{2}, \dfrac{7\sqrt{3}}{2}\right)$
   D) $\left(\dfrac{7}{2}, \dfrac{7\sqrt{3}}{2}\right)$

2) $\left(-3, \dfrac{2\pi}{3}\right)$

   A) $\left(\dfrac{3}{2}, \dfrac{3\sqrt{3}}{2}\right)$
   B) $\left(\dfrac{3}{2}, \dfrac{-3\sqrt{3}}{2}\right)$
   C) $\left(-\dfrac{3}{2}, \dfrac{-3\sqrt{3}}{2}\right)$
   D) $\left(-\dfrac{3}{2}, \dfrac{3\sqrt{3}}{2}\right)$

3) $\left(-5, \dfrac{3\pi}{4}\right)$

   A) $\left(\dfrac{-5\sqrt{2}}{2}, \dfrac{-5\sqrt{2}}{2}\right)$
   B) $\left(\dfrac{5\sqrt{2}}{2}, \dfrac{-5\sqrt{2}}{2}\right)$
   C) $\left(\dfrac{-5\sqrt{2}}{2}, \dfrac{5\sqrt{2}}{2}\right)$
   D) $\left(\dfrac{5\sqrt{2}}{2}, \dfrac{5\sqrt{2}}{2}\right)$

4) $\left(5, \dfrac{3\pi}{4}\right)$

   A) $\left(\dfrac{5\sqrt{2}}{2}, \dfrac{-5\sqrt{2}}{2}\right)$
   B) $\left(\dfrac{-5\sqrt{2}}{2}, \dfrac{-5\sqrt{2}}{2}\right)$
   C) $\left(\dfrac{5\sqrt{2}}{2}, \dfrac{5\sqrt{2}}{2}\right)$
   D) $\left(\dfrac{-5\sqrt{2}}{2}, \dfrac{5\sqrt{2}}{2}\right)$

5) $\left(5, -\dfrac{4\pi}{3}\right)$

   A) $\left(-\dfrac{5\sqrt{3}}{2}, -\dfrac{5}{2}\right)$
   B) $\left(\dfrac{5}{2}, -\dfrac{5\sqrt{3}}{2}\right)$
   C) $\left(-\dfrac{5}{2}, \dfrac{5\sqrt{3}}{2}\right)$
   D) $\left(\dfrac{5\sqrt{3}}{2}, \dfrac{5}{2}\right)$

Precalculus Enhanced with Graphing Utilities 612

6) (-9, 120°)

   A) $\left(-\dfrac{9}{2}, \dfrac{9\sqrt{3}}{2}\right)$
   B) $\left(\dfrac{9}{2}, \dfrac{9\sqrt{3}}{2}\right)$
   C) $\left(-\dfrac{9}{2}, \dfrac{-9\sqrt{3}}{2}\right)$
   D) $\left(\dfrac{9}{2}, \dfrac{-9\sqrt{3}}{2}\right)$

7) (-3, -135°)

   A) $\left(\dfrac{-3\sqrt{2}}{2}, \dfrac{-3\sqrt{2}}{2}\right)$
   B) $\left(\dfrac{-3\sqrt{2}}{2}, \dfrac{3\sqrt{2}}{2}\right)$
   C) $\left(\dfrac{3\sqrt{2}}{2}, \dfrac{-3\sqrt{2}}{2}\right)$
   D) $\left(\dfrac{3\sqrt{2}}{2}, \dfrac{3\sqrt{2}}{2}\right)$

8) (4, 90°)

   A) (4, 0)   B) (0, -4)   C) (0, 4)   D) (-4, 0)

9) (400, 130°) Round the rectangular coordinates to two decimal places.

   A) (-257.12, 306.42)   B) (306.42, 257.12)   C) (306.42, -257.12)   D) (-257.12, -306.42)

10) (4, 70°) Round the rectangular coordinates to two decimal places.

   A) (1.59, 4.01)   B) (3.76, 1.37)   C) (4.01, 1.59)   D) (1.37, 3.76)

The letters r and θ represent polar coordinates. Write the equation using rectangular coordinates (x, y).

11) $r = \cos \theta$

   A) $x^2 + y^2 = x$ B) $(x + y)^2 = x$ C) $x^2 + y^2 = y$ D) $(x + y)^2 = y$

12) $r = 1 + 2 \sin \theta$

   A) $\sqrt{x^2 + y^2} = x^2 + y^2 + 2x$
   B) $x^2 + y^2 = \sqrt{x^2 + y^2} + 2y$
   C) $\sqrt{x^2 + y^2} = x^2 + y^2 + 2y$
   D) $x^2 + y^2 = \sqrt{x^2 + y^2} + 2x$

13) $r = 10 \sin \theta$

   A) $\sqrt{x^2 + y^2} = 10y$ B) $x^2 + y^2 = 10y$ C) $\sqrt{x^2 + y^2} = 10x$ D) $x^2 + y^2 = 10x$

14) $r = 2(\sin \theta - \cos \theta)$

   A) $x^2 + y^2 = 2x - 2y$ B) $x^2 + y^2 = 2y - 2x$ C) $2x^2 + 2y^2 = x - y$ D) $2x^2 + 2y^2 = y - x$

15) $r = 5$

   A) $x^2 + y^2 = 25$ B) $x^2 - y^2 = 25$ C) $x + y = 5$ D) $x + y = 25$

16) $r = \dfrac{5}{1 + \cos \theta}$

   A) $x^2 = 10y - 25$ B) $x^2 = 25 - 10y$ C) $y^2 = 10x - 25$ D) $y^2 = 25 - 10x$

17) $r \sin \theta = 10$

   A) $x = 10y$ B) $x = 10$ C) $y = 10$ D) $y = 10x$

18) $r(1 - 2 \cos \theta) = 1$

   A) $\sqrt{x^2 + y^2} = 2 + x$ B) $\sqrt{x^2 + y^2} = 1 + 2x$ C) $x^2 + y^2 = 1 + 2x$ D) $x^2 + y^2 = 2 + x$

3 Convert from Rectangular Coordinates to Polar Coordinates

The rectangular coordinates of a point are given. Find polar coordinates for the point.

1) $(-7, 0)$
   A) $(-7, \pi)$
   B) $(7, \pi)$
   C) $\left(7, \dfrac{3\pi}{2}\right)$
   D) $\left(7, \dfrac{\pi}{2}\right)$

2) $(0, 8)$
   A) $\left(8, -\dfrac{\pi}{2}\right)$
   B) $\left(8, \dfrac{\pi}{2}\right)$
   C) $(8, \pi)$
   D) $(8, 0)$

3) $(3, -3)$
   A) $\left(-3\sqrt{2}, -\dfrac{3\pi}{4}\right)$
   B) $\left(-3\sqrt{2}, \dfrac{\pi}{4}\right)$
   C) $\left(3\sqrt{2}, -\dfrac{\pi}{4}\right)$
   D) $\left(3\sqrt{2}, \dfrac{\pi}{4}\right)$

4) $(-\sqrt{3}, -1)$
   A) $\left(2, \dfrac{\pi}{6}\right)$
   B) $\left(2, \dfrac{5\pi}{6}\right)$
   C) $\left(2, -\dfrac{\pi}{6}\right)$
   D) $\left(2, -\dfrac{5\pi}{6}\right)$

5) $(-3, 0.6)$ Round the polar coordinates to two decimal places, if necessary, with θ in radians.
   A) $(-3.06, 1.37)$
   B) $(3.06, -1.37)$
   C) $(3.06, 2.94)$
   D) $(3.06, 1.37)$

6) $(100, -30)$ Round the polar coordinates to two decimal places, with θ in degrees.
   A) $(104.40, -106.70°)$
   B) $(104.40, 106.70°)$
   C) $(104.40, 16.70°)$
   D) $(104.40, -16.70°)$

7) $(0.6, -1.1)$ Round the polar coordinates to two decimal places, with θ in degrees.
   A) $(1.25, 57.93°)$
   B) $(1.25, 61.39°)$
   C) $(1.25, -61.39°)$
   D) $(1.25, -57.93°)$

The letters x and y represent rectangular coordinates. Write the equation using polar coordinates (r, θ).

8) $x^2 + 4y^2 = 4$
   A) $\cos^2\theta + 4\sin^2\theta = 4r$
   B) $4\cos^2\theta + \sin^2\theta = 4r$
   C) $r^2(4\cos^2\theta + \sin^2\theta) = 4$
   D) $r^2(\cos^2\theta + 4\sin^2\theta) = 4$

9) $x^2 + y^2 - 4x = 0$
   A) $r\sin^2\theta = 4\cos\theta$
   B) $r = 4\cos\theta$
   C) $r\cos^2\theta = 4\sin\theta$
   D) $r = 4\sin\theta$

10) $x^2 = 4y$
    A) $r\cos^2\theta = 4\sin\theta$
    B) $4\cos^2\theta = r\sin\theta$
    C) $r\sin^2\theta = 4\cos\theta$
    D) $4\sin^2\theta = r\cos\theta$

11) $y^2 = 16x$
    A) $\sin^2\theta = 16r\cos\theta$
    B) $r^2\sin^2\theta = 16\cos\theta$
    C) $r\sin^2\theta = 16\cos\theta$
    D) $\sin^2\theta = 16r^2\cos\theta$

12) $xy = 1$
    A) $r\sin 2\theta = 2$
    B) $r^2\sin 2\theta = 2$
    C) $2r\sin\theta\cos\theta = 1$
    D) $2r^2\sin\theta\cos\theta = 1$

13) $2x + 3y = 6$
    A) $2\sin\theta + 3\cos\theta = 6r$
    B) $2\cos\theta + 3\sin\theta = 6r$
    C) $r(2\sin\theta + 3\cos\theta) = 6$
    D) $r(2\cos\theta + 3\sin\theta) = 6$

Precalculus Enhanced with Graphing Utilities

14) x = -3
- A) r cos θ = 3
- B) r sin θ = 3
- C) r cos θ = -3
- D) r sin θ = -3

15) y = 5
- A) sin θ cos θ = 5
- B) r cos θ = 5
- C) r = 5
- D) r sin θ = 5

16) y = x
- A) r = sin θ
- B) sin θ = - cos θ
- C) r = cos θ
- D) sin θ = cos θ

8.2 Polar Equations and Graphs

1 Graph and Identify Polar Equations by Converting to Rectangular Equations

Transform the polar equation to an equation in rectangular coordinates. Then identify and graph the equation.

1) $r = 2\sin\theta$

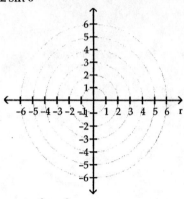

A) $(x-1)^2 + y^2 = 1$; circle, radius 1, center at $(1, 0)$ in rectangular coordinates

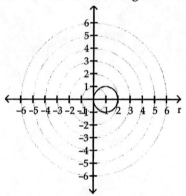

B) $x^2 + (y+1)^2 = 1$; circle, radius 1, center at $(0, -1)$ in rectangular coordinates

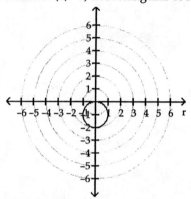

C) $(x+1)^2 + y^2 = 1$; circle, radius 1, center at $(-1, 0)$ in rectangular coordinates

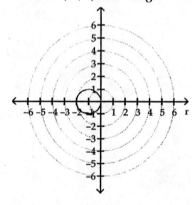

D) $x^2 + (y-1)^2 = 1$; circle, radius 1, center at $(0, 1)$ in rectangular coordinates

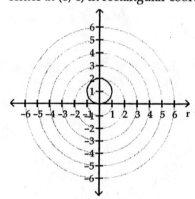

2) $r = 6 \cos \theta$

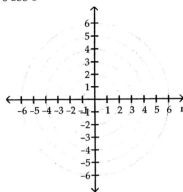

A) $(x + 3)^2 + y^2 = 9$; circle, radius 3, center at (−3, 0) in rectangular coordinates

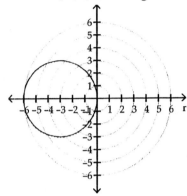

B) $x^2 + (y + 3)^2 = 9$; circle, radius 3, center at (0, −3) in rectangular coordinates

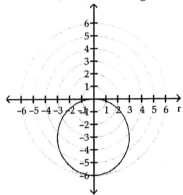

C) $(x − 3)^2 + y^2 = 9$; circle, radius 3, center at (3, 0) in rectangular coordinates

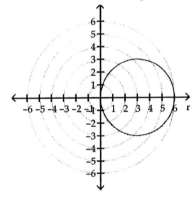

D) $x^2 + (y − 3)^2 = 9$; circle, radius 3, center at (0, 3) in rectangular coordinates

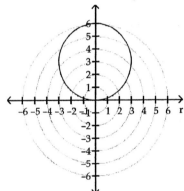

3) $r \sin \theta = 5$

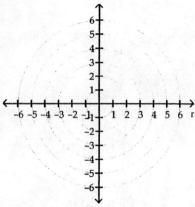

A) $y = 5$; horizontal line 5 units above the pole

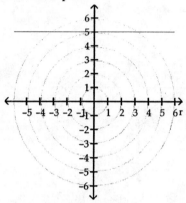

B) $x = -5$; vertical line 5 units to the left of the pole

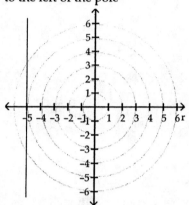

C) $y = -5$; horizontal line 5 units below the pole

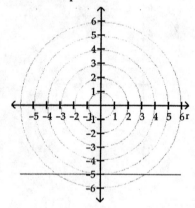

D) $x = 5$; vertical line 5 units to the right of the pole

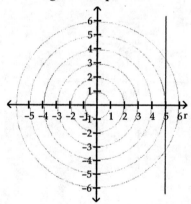

4) $\theta = \dfrac{\pi}{3}$

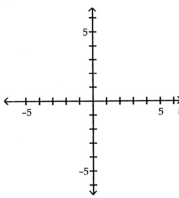

A) $y = -\dfrac{\sqrt{3}}{3}x$; line through the pole making an angle of $\dfrac{\pi}{3}$ with the polar axis

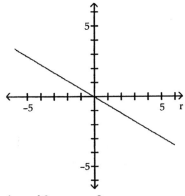

B) $y = \sqrt{3}x$; line through the pole making an angle of $\dfrac{\pi}{3}$ with the polar axis

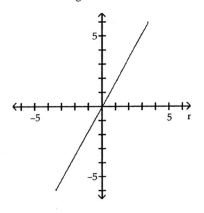

C) $\left(x - \dfrac{\pi}{3}\right)^2 + y^2 = \dfrac{\pi^2}{9}$; circle, radius $\dfrac{\pi}{3}$, center at $\left(\dfrac{\pi}{3}, 0\right)$ in rectangular coordinates

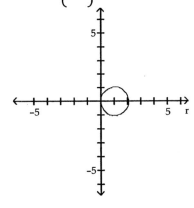

D) $y = -\dfrac{\pi}{3}$; horizontal line $\dfrac{\pi}{3}$ units below the pole

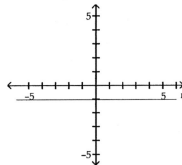

5) $r \sec \theta = -6$

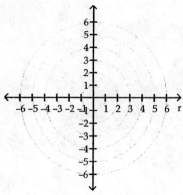

A) $y = -6$; horizontal line 6 units below the pole

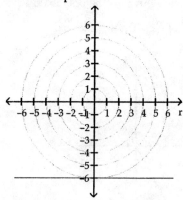

C) $x^2 + (y + 3)^2 = 9$; circle, radius 3, center at $(0, -3)$ in rectangular coordinates

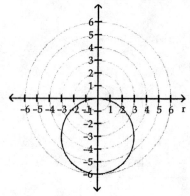

B) $x = -6$; vertical line 6 units to the left of the pole

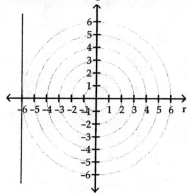

D) $(x + 3)^2 + y^2 = 9$; circle, radius 3, center at $(-3, 0)$ in rectangular coordinates

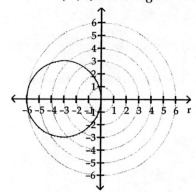

Match the graph to one of the polar equations.

6)

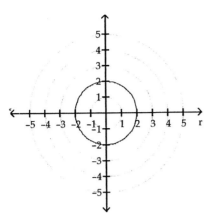

A) $r = 4 \cos \theta$ B) $r = 2$ C) $r = 4 \sin \theta$ D) $r \sin \theta = 2$

7)

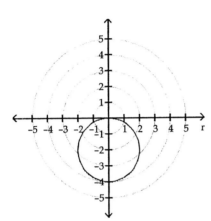

A) $r = -2$ B) $r = -4 \sin \theta$ C) $r = -4 \cos \theta$ D) $r \sin \theta = -2$

8)

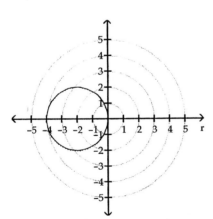

A) $r = -4 \sin \theta$ B) $r = -4 \cos \theta$ C) $r = -2$ D) $r \sin \theta = -2$

9)

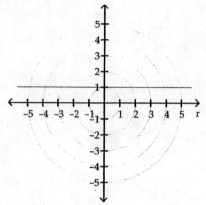

A) $r \sin \theta = 1$ B) $r = 2 \sin \theta$ C) $r = 1$ D) $r = 2 \cos \theta$

10)

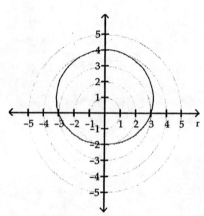

A) $r = 6 \cos \theta$ B) $r = 3 + \cos \theta$ C) $r = 3 + \sin \theta$ D) $r = 6 \sin \theta$

11)

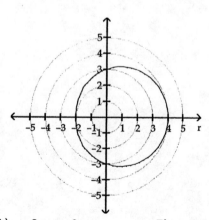

A) $r = 3 + \cos \theta$ B) $r = 6 \sin \theta$ C) $r = 6 \cos \theta$ D) $r = 3 + \sin \theta$

2 Graph Polar Equations Using a Graphing Utility

Use a graphing utility to graph the polar equation.

1) $r = 3$

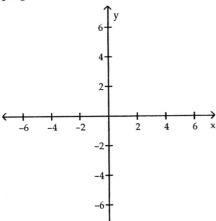

A)
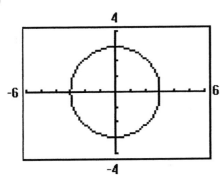

B)
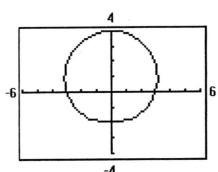

C)
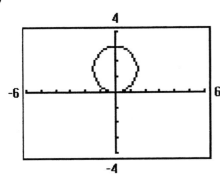

D)
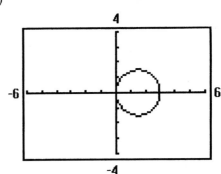

2) $r = 3 + \sin\theta$

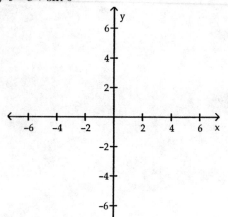

A)
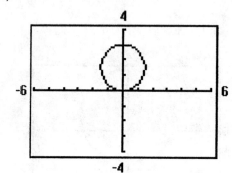

B)
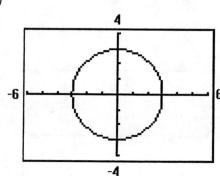

C)
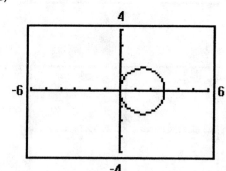

D)
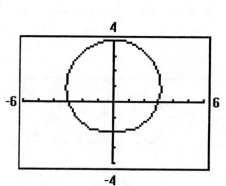

3 Test Polar Equations for Symmetry

Test the equation for symmetry with respect to the given axis, line, or pole.

1) $r = -4\cos\theta$; the polar axis

   A) May or may not be symmetric with respect to the polar axis
   B) Symmetric with respect to the polar axis

2) $r = -4\cos\theta$; the line $\theta = \dfrac{\pi}{2}$

   A) Symmetric with respect to the line $\theta = \dfrac{\pi}{2}$
   B) May or may not be symmetric with respect to the line $\theta = \dfrac{\pi}{2}$

Precalculus Enhanced with Graphing Utilities 624

3) r = -2 sin θ; the pole
   A) May or may not be symmetric with respect to the pole
   B) Symmetric with respect to the pole

4) r = 2 - 2 sin θ; polar axis
   A) Symmetric with respect to the polar axis
   B) May or may not be symmetric with respect to the polar axis

5) r = 2 - 2 cos θ; the line $\theta = \frac{\pi}{2}$
   A) Symmetric with respect to the line $\theta = \frac{\pi}{2}$
   B) May or may not be symmetric with respect to the line $\theta = \frac{\pi}{2}$

6) r = 6 + 2 sin θ; the line $\theta = \frac{\pi}{2}$
   A) May or may not be symmetric with respect to the line $\theta = \frac{\pi}{2}$
   B) Symmetric with respect to the line $\theta = \frac{\pi}{2}$

7) r = 6 + 2 cos θ; the pole
   A) May or may not be symmetric with respect to the pole
   B) Symmetric with respect to the pole

8) r = 3 + 6 sin θ; the polar axis
   A) May or may not be symmetric with respect to the polar axis
   B) Symmetric with respect to the polar axis

9) r^2 = sin(2θ); the pole
   A) May or may not be symmetric with respect to the pole
   B) Symmetric with respect to the pole

10) r = 3 sin(3θ); the line $\theta = \frac{\pi}{2}$
    A) Symmetric with respect to the line $\theta = \frac{\pi}{2}$
    B) May or may not be symmetric with respect to the line $\theta = \frac{\pi}{2}$

4 Graph Polar Equations by Plotting Points

Identify and graph the polar equation.

1) $r = 2 - 2\sin\theta$

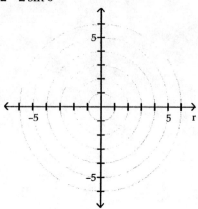

A)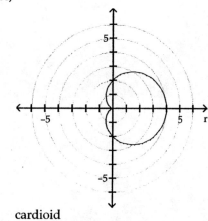
cardioid

B)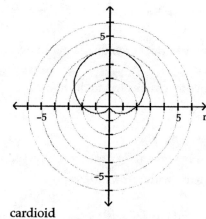
cardioid

C)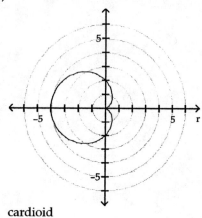
cardioid

D)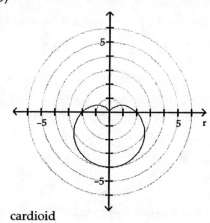
cardioid

2) $r = 6 - 5\sin\theta$

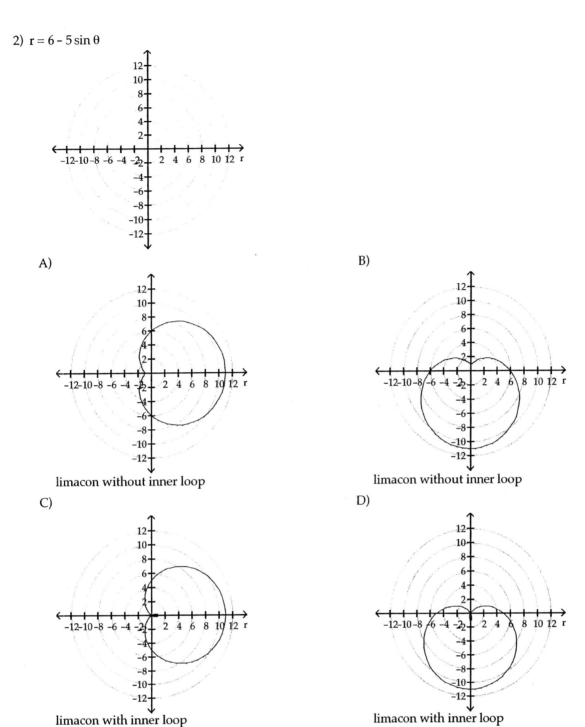

A) limacon without inner loop

B) limacon without inner loop

C) limacon with inner loop

D) limacon with inner loop

3) $r = 2 - 3\sin\theta$

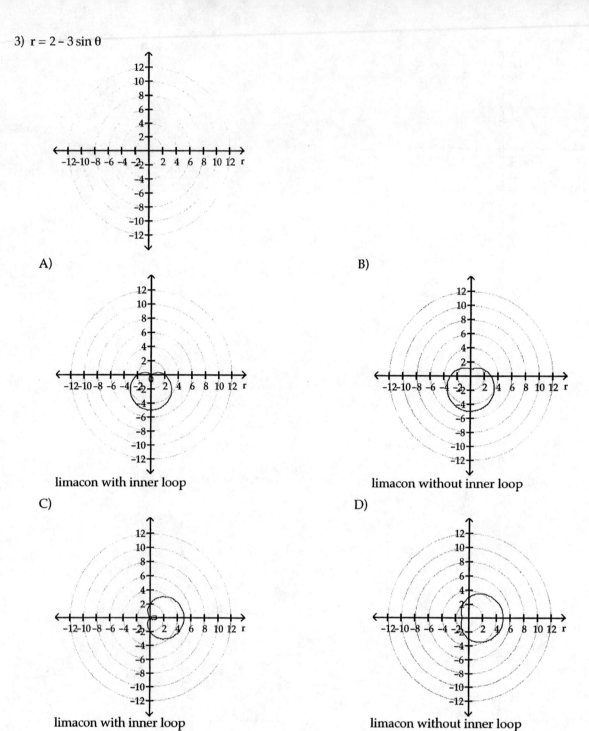

A) limacon with inner loop

B) limacon without inner loop

C) limacon with inner loop

D) limacon without inner loop

4) $r = 4\sin(2\theta)$

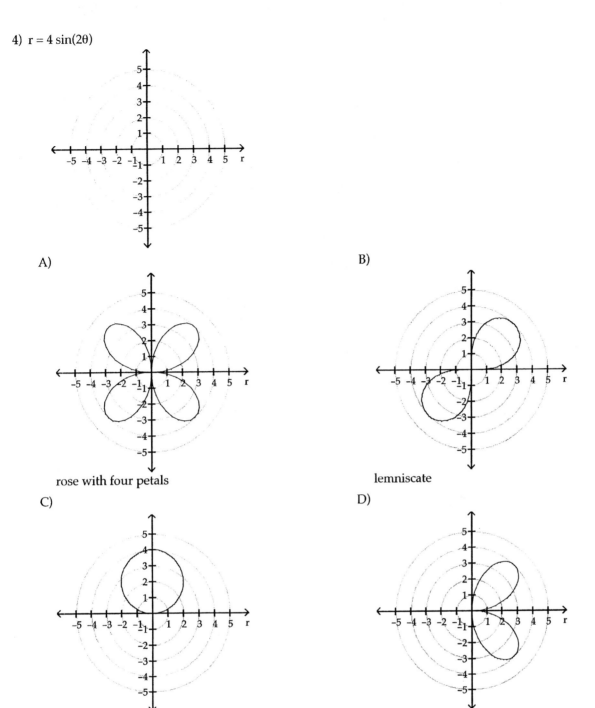

A) rose with four petals

B) lemniscate

C) circle

D) rose with two petals

5) $r^2 = 3\cos(2\theta)$

A)

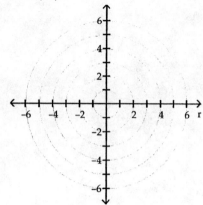

rose with four petals

B)

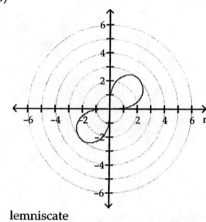

lemniscate

C)

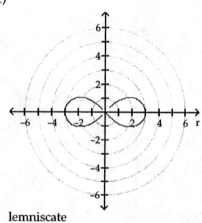

lemniscate

D)

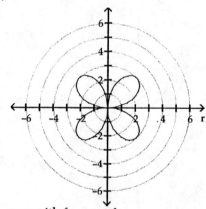

rose with four petals

6) $r = 4^\theta$

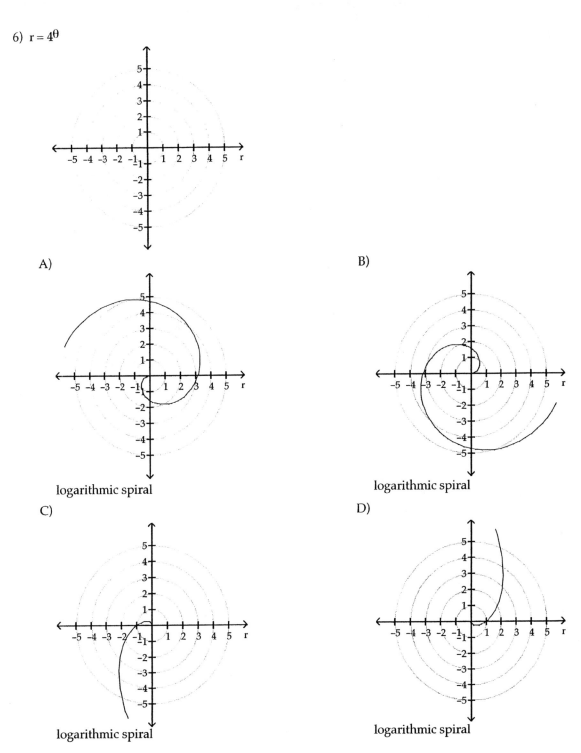

A) logarithmic spiral

B) logarithmic spiral

C) logarithmic spiral

D) logarithmic spiral

Graph the polar equation.

7) $r = 1 + \cos\theta$

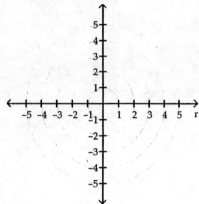

A)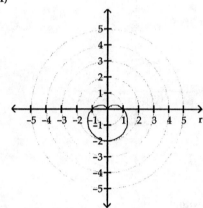

B)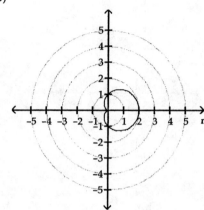

C)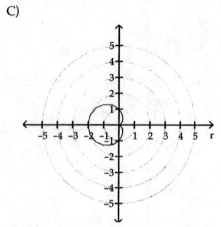

D)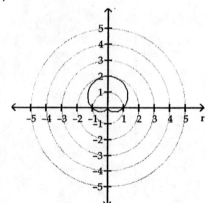

8) $r = \dfrac{4}{1 - \cos\theta}$

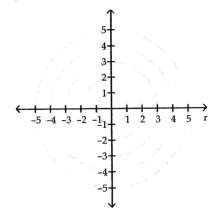

A)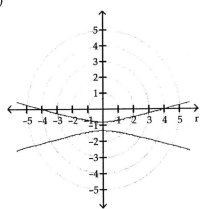

B)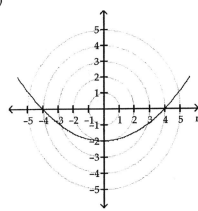

C)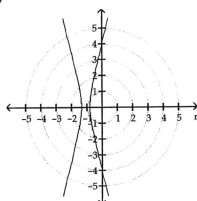

D)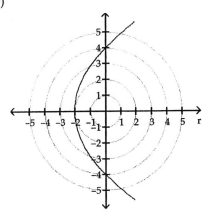

9) $r = \dfrac{2}{1 - \sin\theta}$

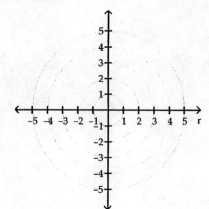

A)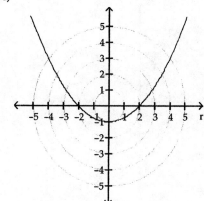

B)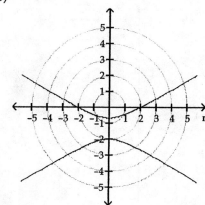

C)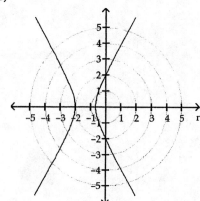

D)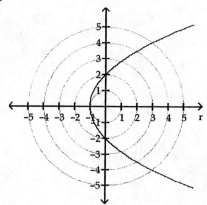

10) $r = \dfrac{5}{1 - 5\sin\theta}$

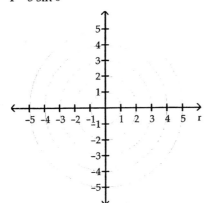

A)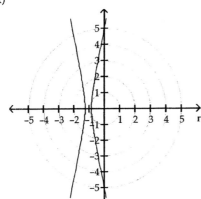

B)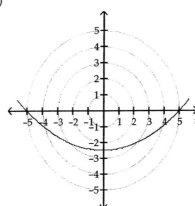

C)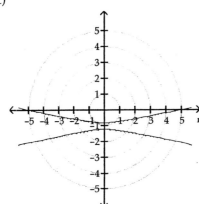

D)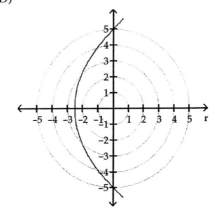

11) $r = \dfrac{5}{5 - 4\cos\theta}$

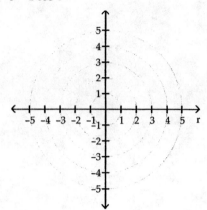

A)
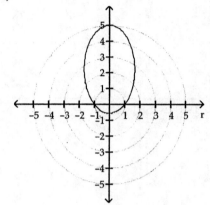

B)
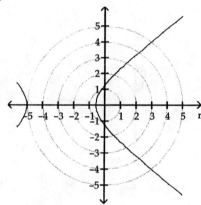

C)
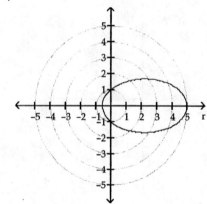

D)
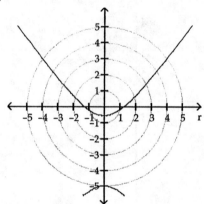

12) $r = \dfrac{5}{5 - 4\sin\theta}$

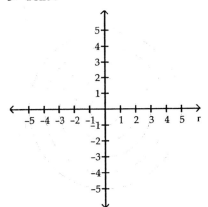

A)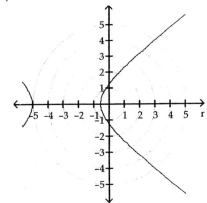

B)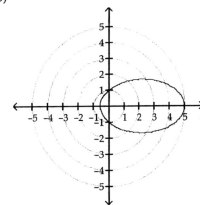

C)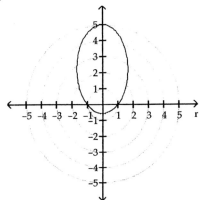

D)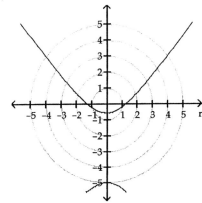

13) $r = \theta$, $\theta \geq 0$

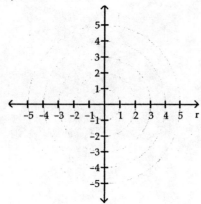

A)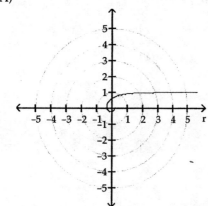

B)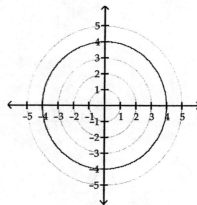

C)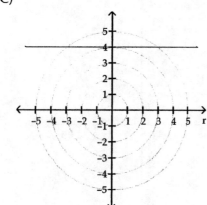

D)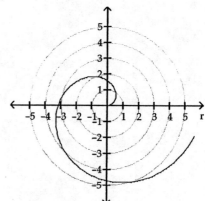

14) $r = \dfrac{3}{\theta}$

A)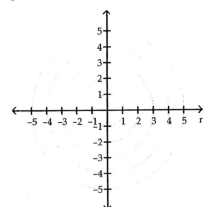

B)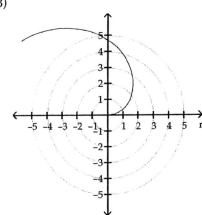

C)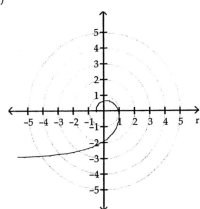

D)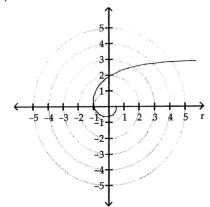

15) $r = \csc\theta - 4$, $0 < \theta < \pi$

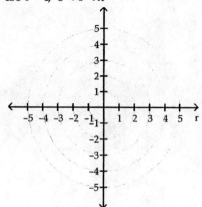

A)

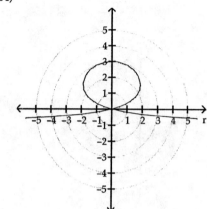

B)

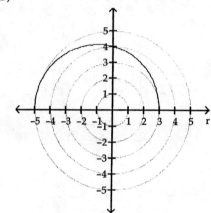

C)

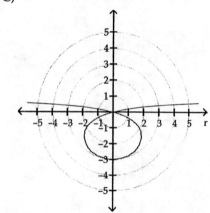

D)

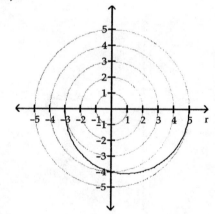

16) r = sin θ tan θ

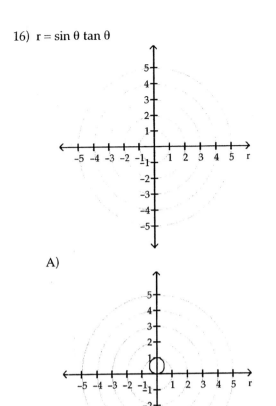

A)
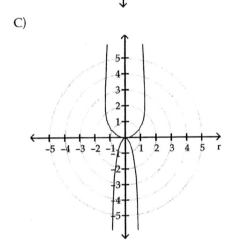

B)
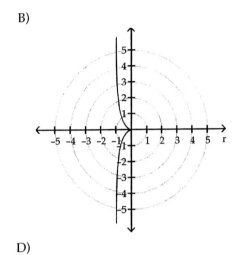

C)

D)
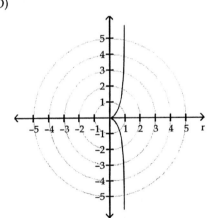

17) $r = \tan\theta,\ -\dfrac{\pi}{2} < \theta < \dfrac{\pi}{2}$

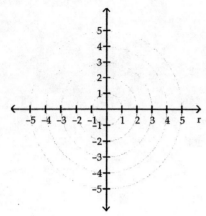

A)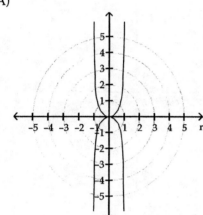

B)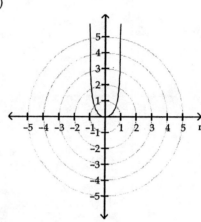

C)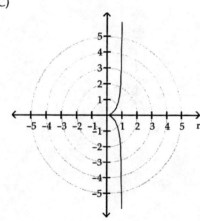

D)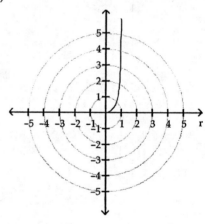

8.3 The Complex Plane; De Moivre's Theorem

1 Convert a Complex Number from Rectangular Form to Polar Form

Write the complex number in polar form. Express the argument in degrees, rounded to the nearest tenth, if necessary.

1) $1 - \sqrt{3}i$

   A) $2(\cos 300° + i \sin 300°)$　　　　B) $4(\cos 300° + i \sin 300°)$
   C) $2(\cos 330° + i \sin 330°)$　　　　D) $4(\cos 330° + i \sin 330°)$

2) 3 − 3i
   A) 9(cos 300° + i sin 300°)
   B) 9(cos 315° + i sin 315°)
   C) 3√2(cos 300° + i sin 300°)
   D) 3√2(cos 315° + i sin 315°)

3) −5
   A) 5(cos 270° + i sin 270°)
   B) 5(cos 90° + i sin 90°)
   C) 5(cos 180° + i sin 180°)
   D) 5(cos 0° + i sin 0°)

4) −3i
   A) 3(cos 180° + i sin 180°)
   B) 3(cos 0° + i sin 0°)
   C) 3(cos 270° + i sin 270°)
   D) 3(cos 90° + i sin 90°)

5) −12 − 16i
   A) 20(cos 306.9° + i sin 306.9°)
   B) 20(cos 126.9° + i sin 126.9°)
   C) 20(cos 233.1° + i sin 233.1°)
   D) 20(cos 53.1° + i sin 53.1°)

2 Plot Points in the Complex Plane

Plot the complex number in the complex plane.

1) $6 + 4i$

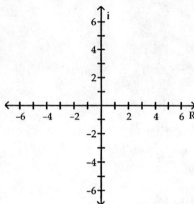

A)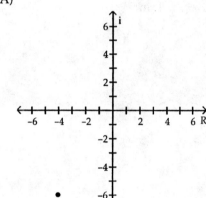

B)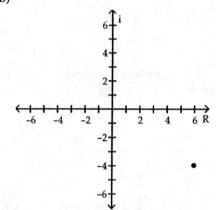

C)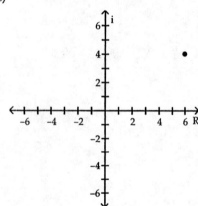

D)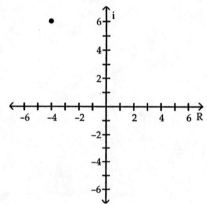

2) 5i

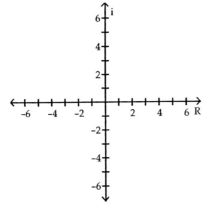

A)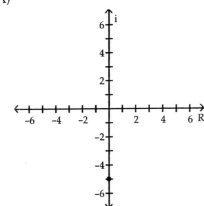

B)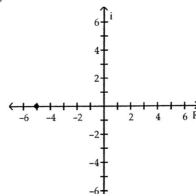

C)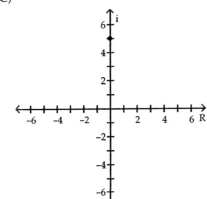

D)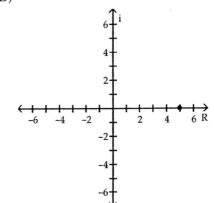

3) $-2 + i$

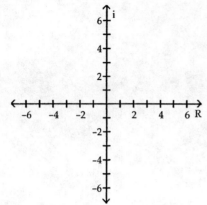

A)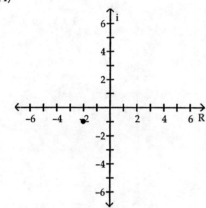

B)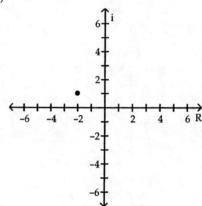

C)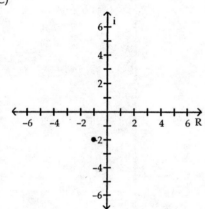

D)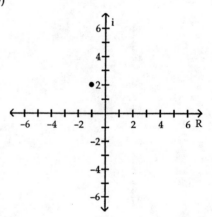

4) −6 − i

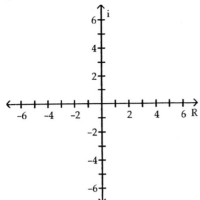

A)
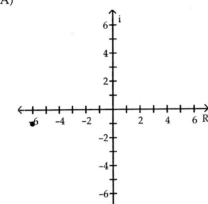

B)
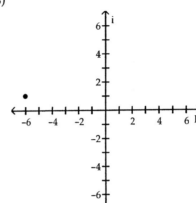

C)
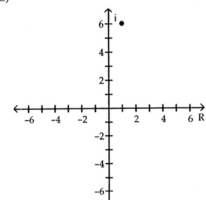

D)
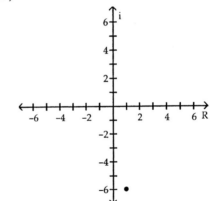

5) $2 + \sqrt{6}i$

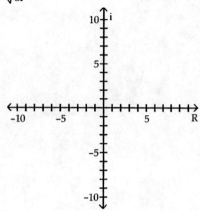

A)
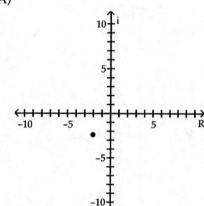

B)
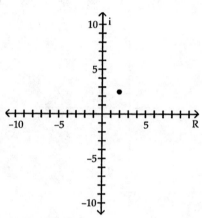

C)
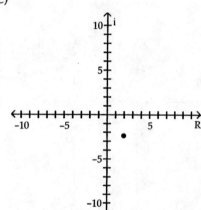

D)
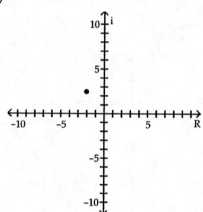

3 Find Products and Quotients of Complex Numbers in Polar Form

Solve the problem. Leave your answer in polar form.

1) $z = 10(\cos 30° + i \sin 30°)$
   $w = 5(\cos 10° + i \sin 10°)$
   Find zw.

   A) $15(\cos 40° + i \sin 40°)$
   B) $50(\cos 40° + i \sin 40°)$
   C) $50(\cos 300° + i \sin 300°)$
   D) $15(\cos 300° + i \sin 300°)$

2) $z = 10(\cos 45° + i \sin 45°)$
   $w = 5(\cos 15° + i \sin 15°)$
   Find zw.
   A) $5(\cos 60° + i \sin 60°)$
   B) $50(\cos 30° + i \sin 30°)$
   C) $5(\cos 30° + i \sin 30°)$
   D) $50(\cos 60° + i \sin 60°)$

3) $z = 5(\cos 35° + i \sin 35°)$
   $w = 2(\cos 40° + i \sin 40°)$
   Find zw.
   A) $10(\cos 50.9° + i \sin 50.9°)$
   B) $7(\cos 75° + i \sin 75°)$
   C) $7(\cos 50.9° + i \sin 50.9°)$
   D) $10(\cos 75° + i \sin 75°)$

4) $z = 8\left(\cos \frac{\pi}{6} + i \sin \frac{\pi}{6}\right)$
   $w = 3\left(\cos \frac{\pi}{2} + i \sin \frac{\pi}{2}\right)$
   Find zw.
   A) $24\left(\cos \frac{\pi}{3} + i \sin \frac{\pi}{3}\right)$
   B) $12\left(\cos \frac{\pi}{3} + i \sin \frac{\pi}{3}\right)$
   C) $24\left(\cos \frac{2\pi}{3} + i \sin \frac{2\pi}{3}\right)$
   D) $12\left(\cos \frac{2\pi}{3} + i \sin \frac{2\pi}{3}\right)$

5) $z = 6\left(\cos \frac{3\pi}{2} + i \sin \frac{3\pi}{2}\right)$
   $w = 12\left(\cos \frac{5\pi}{6} + i \sin \frac{5\pi}{6}\right)$
   Find zw.
   A) $72\left(\cos \frac{\pi}{6} + i \sin \frac{\pi}{6}\right)$
   B) $36\left(\cos \frac{\pi}{6} + i \sin \frac{\pi}{6}\right)$
   C) $72\left(\cos \frac{\pi}{3} + i \sin \frac{\pi}{3}\right)$
   D) $36\left(\cos \frac{\pi}{3} + i \sin \frac{\pi}{3}\right)$

6) $z = 2 + 2i$
   $w = \sqrt{3} - i$
   Find zw.
   A) $4\sqrt{2}\left(\cos \frac{23\pi}{12} + i \sin \frac{23\pi}{12}\right)$
   B) $4\sqrt{2}\left(\cos \frac{\pi}{12} + i \sin \frac{\pi}{12}\right)$
   C) $4\left(\cos \frac{23\pi}{12} + i \sin \frac{23\pi}{12}\right)$
   D) $4\left(\cos \frac{\pi}{12} + i \sin \frac{\pi}{12}\right)$

7) $z = 10(\cos 30° + i \sin 30°)$
   $w = 5(\cos 10° + i \sin 10°)$
   Find $\frac{z}{w}$.
   A) $5(\cos 20° + i \sin 20°)$
   B) $5(\cos 3° + i \sin 3°)$
   C) $2(\cos 20° + i \sin 20°)$
   D) $2(\cos 3° + i \sin 3°)$

8) $z = 10(\cos 45° + i \sin 45°)$
   $w = 5(\cos 15° + i \sin 15°)$
   Find $\dfrac{z}{w}$.

   A) $\dfrac{1}{2}(\cos 30° + i \sin 30°)$ B) $2(\cos 45° + i \sin 45°)$
   C) $\dfrac{1}{2}(\cos 45° + i \sin 45°)$ D) $2(\cos 30° + i \sin 30°)$

9) $z = 5(\cos 200° + i \sin 200°)$
   $w = 4(\cos 50° + i \sin 50°)$
   Find $\dfrac{z}{w}$.

   A) $\dfrac{4}{5}(\cos 150° + i \sin 150°)$ B) $\dfrac{5}{4}(\cos 150° + i \sin 150°)$
   C) $\dfrac{5}{4}(\cos 40° + i \sin 40°)$ D) $\dfrac{4}{5}(\cos 40° + i \sin 40°)$

10) $z = 8\left(\cos \dfrac{\pi}{2} + i \sin \dfrac{\pi}{2}\right)$
    $w = 3\left(\cos \dfrac{\pi}{6} + i \sin \dfrac{\pi}{6}\right)$
    Find $\dfrac{z}{w}$.

    A) $5\left(\cos \dfrac{\pi}{3} + i \sin \dfrac{\pi}{3}\right)$ B) $\dfrac{8}{3}\left(\cos \dfrac{\pi}{12} + i \sin \dfrac{\pi}{12}\right)$
    C) $\dfrac{8}{3}\left(\cos \dfrac{\pi}{3} + i \sin \dfrac{\pi}{3}\right)$ D) $5\left(\cos \dfrac{\pi}{12} + i \sin \dfrac{\pi}{12}\right)$

11) $z = \sqrt{3}\left(\cos \dfrac{7\pi}{4} + i \sin \dfrac{7\pi}{4}\right)$
    $w = \sqrt{6}\left(\cos \dfrac{9\pi}{4} + i \sin \dfrac{9\pi}{4}\right)$
    Find $\dfrac{z}{w}$.

    A) $\dfrac{\sqrt{2}}{2}\left(\cos \dfrac{\pi}{2} + i \sin \dfrac{\pi}{2}\right)$ B) $\dfrac{\sqrt{2}}{2}\left(\cos \dfrac{3\pi}{2} + i \sin \dfrac{3\pi}{2}\right)$
    C) $3\sqrt{2}\left(\cos \dfrac{3\pi}{2} + i \sin \dfrac{3\pi}{2}\right)$ D) $3\sqrt{2}\left(\cos \dfrac{\pi}{2} + i \sin \dfrac{\pi}{2}\right)$

Precalculus Enhanced with Graphing Utilities 650

12) $z = 6\left(\cos\dfrac{3\pi}{2} + i\sin\dfrac{3\pi}{2}\right)$

$w = 12\left(\cos\dfrac{5\pi}{6} + i\sin\dfrac{5\pi}{6}\right)$

Find $\dfrac{z}{w}$.

A) $\dfrac{1}{6}\left(\cos\dfrac{2\pi}{3} + i\sin\dfrac{2\pi}{3}\right)$

B) $\dfrac{1}{2}\left(\cos\dfrac{2\pi}{3} + i\sin\dfrac{2\pi}{3}\right)$

C) $\dfrac{1}{2}\left(\cos\dfrac{\pi}{3} + i\sin\dfrac{\pi}{3}\right)$

D) $\dfrac{1}{6}\left(\cos\dfrac{\pi}{3} + i\sin\dfrac{\pi}{3}\right)$

13) $z = 1 - i$

$w = 1 - \sqrt{3}i$

Find $\dfrac{z}{w}$.

A) $\dfrac{\sqrt{2}}{2}(\cos 15° + i\sin 15°)$

B) $\dfrac{\sqrt{2}}{2}(\cos 75° + i\sin 75°)$

C) $\dfrac{1}{2}(\cos 75° + i\sin 75°)$

D) $\dfrac{1}{2}\cos 15° + i\sin 15°$

4 Use De Moivre's Theorem

Write the expression in the standard form a + bi.

1) $[2(\cos 15° + i\sin 15°)]^3$

A) $3\sqrt{2} + 3\sqrt{2}i$ B) $4 + 4i$ C) $3 + 3i$ D) $4\sqrt{2} + 4\sqrt{2}i$

2) $\left[\sqrt{2}\left(\cos\dfrac{3\pi}{4} + i\sin\dfrac{3\pi}{4}\right)\right]^4$

A) -4 B) $4i$ C) $-4i$ D) 4

3) $\left[\sqrt{3}\left(\cos\dfrac{5\pi}{6} + i\sin\dfrac{5\pi}{6}\right)\right]^4$

A) $-\dfrac{9}{2} - \dfrac{9\sqrt{3}}{2}i$ B) $-\dfrac{9\sqrt{3}}{2} + \dfrac{9}{2}i$ C) $-\dfrac{9}{2} + \dfrac{9\sqrt{3}}{2}i$ D) $-\dfrac{9\sqrt{3}}{2} - \dfrac{9}{2}i$

4) $(1 + i)^{20}$

A) -1024 B) 1024 C) $-1024i$ D) $1024i$

5) $(1 - i)^{10}$

A) $-32i$ B) $32 - 32i$ C) 32 D) $-32 + 32i$

6) $(-\sqrt{3} + i)^6$

A) $64 - 64\sqrt{3}i$ B) $-64\sqrt{3} + 64i$ C) -64 D) $64i$

7) $(\sqrt{3} + i)^5$

A) $16\sqrt{3} - 16i$ B) $9\sqrt{3} + 5i$ C) $16 - 16\sqrt{3}i$ D) $-16\sqrt{3} + 16i$

8) $\left(-\dfrac{1}{2}-\dfrac{\sqrt{3}}{2}i\right)^{10}$

A) $\dfrac{1}{2}-\dfrac{\sqrt{3}}{2}i$ B) $\dfrac{1}{2}+\dfrac{\sqrt{3}}{2}i$ C) $-\dfrac{1}{2}-\dfrac{\sqrt{3}}{2}i$ D) $-\dfrac{1}{2}+\dfrac{\sqrt{3}}{2}i$

5 Find Complex Roots

Find all the complex roots. Leave your answers in polar form with the argument in degrees.

1) The complex cube roots of $-8i$

A) $8(\cos 90° + i \sin 90°)$, $8(\cos 210° + i \sin 210°)$, $8(\cos 330° + i \sin 330°)$

B) $512(\cos 90° + i \sin 90°)$, $512(\cos 210° + i \sin 210°)$, $512(\cos 330° + i \sin 330°)$

C) $2(\cos 90° + i \sin 90°)$, $2(\cos 210° + i \sin 210°)$, $2(\cos 330° + i \sin 330°)$

D) $2(\cos 180° + i \sin 180°)$, $2(\cos 300° + i \sin 300°)$, $2(\cos 60° + i \sin 60°)$

2) The complex fourth roots of -16

A) $\sqrt[4]{2}(\cos 45° + i \sin 45°)$, $\sqrt[4]{2}(\cos 135° + i \sin 135°)$, $\sqrt[4]{2}(\cos 225° + i \sin 225°)$, $\sqrt[4]{2}(\cos 315° + i \sin 315°)$

B) $2(\cos 45° + i \sin 45°)$, $2(\cos 135° + i \sin 135°)$, $2(\cos 225° + i \sin 225°)$, $16(\cos 315° + i \sin 315°)$

C) $16(\cos 45° + i \sin 45°)$, $16(\cos 135° + i \sin 135°)$, $16(\cos 225° + i \sin 225°)$, $16(\cos 315° + i \sin 315°)$

D) $2(\cos 90° + i \sin 90°)$, $2(\cos 180° + i \sin 180°)$, $2(\cos 270° + i \sin 270°)$, $2(\cos 360° + i \sin 360°)$

3) The complex fifth roots of $\sqrt{3}+i$

A) $32(\cos 30° + i \sin 30°)$, $32(\cos 102° + i \sin 102°)$, $32(\cos 174° + i \sin 174°)$, $32(\cos 246° + i \sin 246°)$, $32(\cos 318° + i \sin 318°)$

B) $\sqrt[5]{2}(\cos 6° + i \sin 6°)$, $\sqrt[5]{2}(\cos 78° + i \sin 78°)$, $\sqrt[5]{2}(\cos 150° + i \sin 150°)$, $\sqrt[5]{2}(\cos 222° + i \sin 222°)$, $\sqrt[5]{2}(\cos 294° + i \sin 294°)$

C) $32(\cos 6° + i \sin 6°)$, $32(\cos 78° + i \sin 78°)$, $32(\cos 150° + i \sin 150°)$, $32(\cos 222° + i \sin 222°)$, $32(\cos 294° + i \sin 294°)$

D) $\sqrt[5]{2}(\cos 30° + i \sin 30°)$, $\sqrt[5]{2}(\cos 102° + i \sin 102°)$, $\sqrt[5]{2}(\cos 174° + i \sin 174°)$, $\sqrt[5]{2}(\cos 246° + i \sin 246°)$, $\sqrt[5]{2}(\cos 318° + i \sin 318°)$

4) The complex fifth roots of $-2i$

A) $\sqrt[4]{2}(\cos 54° + i \sin 54°)$, $\sqrt[4]{2}(\cos 126° + i \sin 126°)$, $\sqrt[4]{2}(\cos 198° + i \sin 198°)$, $\sqrt[4]{2}(\cos 270° + i \sin 270°)$, $\sqrt[4]{2}(\cos 342° + i \sin 342°)$

B) $\sqrt[5]{2}(\cos 45° + i \sin 45°)$, $\sqrt[5]{2}(\cos 117° + i \sin 117°)$, $\sqrt[5]{2}(\cos 189° + i \sin 189°)$, $\sqrt[5]{2}(\cos 261° + i \sin 261°)$, $\sqrt[5]{2}(\cos 333° + i \sin 333°)$

C) $32(\cos 54° + i \sin 54°)$, $32(\cos 126° + i \sin 126°)$, $32(\cos 198° + i \sin 198°)$, $32(\cos 270° + i \sin 270°)$, $32(\cos 342° + i \sin 342°)$

D) $\sqrt[5]{2}(\cos 54° + i \sin 54°)$, $\sqrt[5]{2}(\cos 126° + i \sin 126°)$, $\sqrt[5]{2}(\cos 198° + i \sin 198°)$, $\sqrt[5]{2}(\cos 270° + i \sin 270°)$, $\sqrt[5]{2}(\cos 342° + i \sin 342°)$

Precalculus Enhanced with Graphing Utilities

6 Demonstrate Additional Understanding and Skills

Write the complex number in rectangular form.

1) $8\left(\cos \frac{\pi}{6} + i \sin \frac{\pi}{6}\right)$

   A) $\frac{\sqrt{3}}{4} + \frac{1}{4}i$ B) $\frac{1}{4} + \frac{\sqrt{3}}{4}i$ C) $4 + 4\sqrt{3}i$ D) $4\sqrt{3} + 4i$

2) $3\left(\cos \frac{\pi}{3} + i \sin \frac{\pi}{3}\right)$

   A) $\frac{3}{2} + \frac{3\sqrt{3}}{2}i$ B) $\frac{\sqrt{3}}{6} + \frac{\sqrt{3}}{6}i$ C) $\sqrt{3} + i$ D) $\frac{\sqrt{3}}{2} + \frac{3\sqrt{3}}{2}i$

3) $6(\cos 330° + i \sin 330°)$

   A) $3\sqrt{3} + 3i$ B) $-3\sqrt{3} + 3i$ C) $-3\sqrt{3} - 3i$ D) $3\sqrt{3} - 3i$

4) $9(\cos 180° + i \sin 180°)$

   A) $9i$ B) -9 C) $-9i$ D) 9

8.4 Vectors

1 Graph Vectors

Use the vectors in the figure below to graph the following vector.

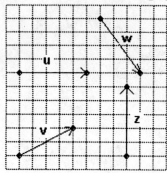

1) **u + z**

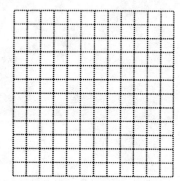

A)

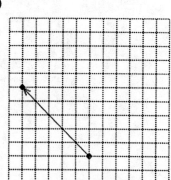

B)

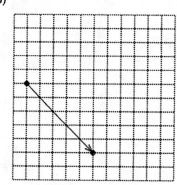

C)

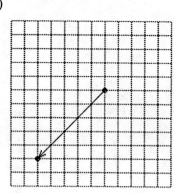

D)

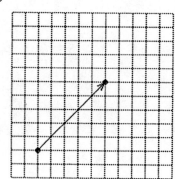

2) 3**w**

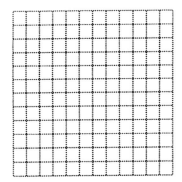

A)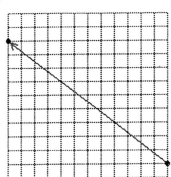
B)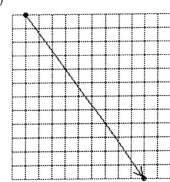

C)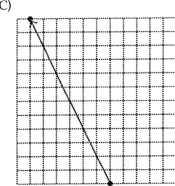
D)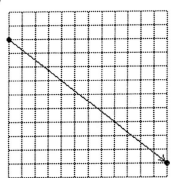

3) **v − w**

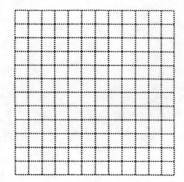

A)

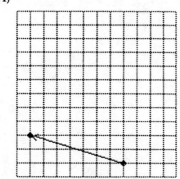

B)

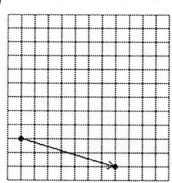

C)

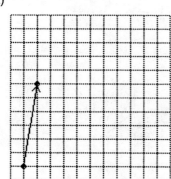

D)

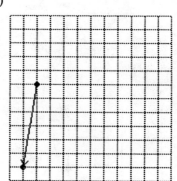

Precalculus Enhanced with Graphing Utilities 656

4) z − v

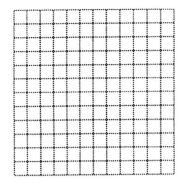

A)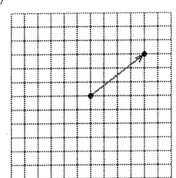

B)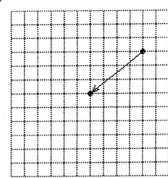

C)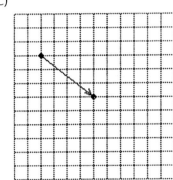

D)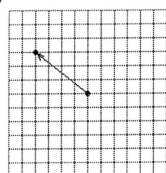

5) 2u − z − w

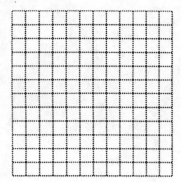

A)

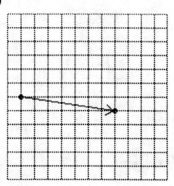

B)

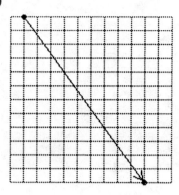

C)

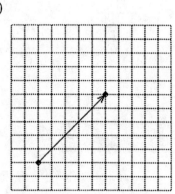

D)

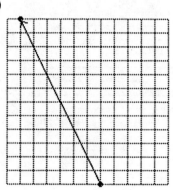

Use the figure below. Determine whether the given statement is true or false.

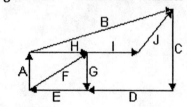

6) A + H = F

   A) True B) False

7) G + H = F

   A) True B) False

8) H + I + J = B

   A) True B) False

9) A + B + C + D + E = 0

A) True   B) False

10) C + D + G + I + J = 0

A) True   B) False

2   Find a Position Vector

The vector v has initial position P and terminal point Q. Write v in the form ai + bj; that is, find its position vector.

1) P = (0, 0); Q = (-6, 3)

A) v = 3i + 3j   B) v = -3i + 6j   C) v = -6i + 3j   D) v = 6i - 3j

2) P = (5, 3); Q = (-6, -6)

A) v = -11i - 9j   B) v = 9i + 11j   C) v = -9i - 11j   D) v = 11i + 9j

3) P = (6, -3); Q = (-6, -1)

A) v = -7i - 3j   B) v = -12i + 2j   C) v = 2i - 12j   D) v = -3i - 7j

3   Add and Subtract Vectors

Solve the problem.

1) If **u** = 12i - 3j and **v** = -4i + 8j, find **u** + **v**.

A) 16i + 7j   B) 8i + 5j   C) -16i + 5j   D) 7i + 5j

2) If **u** = 3i - 2j and **v** = -6i + 7j, find **u** - **v**.

A) -3i + 5j   B) 9i - 9j   C) 8i + 5j   D) 7i + 5j

4   Find a Scalar Product and the Magnitude of a Vector

Solve the problem.

1) If **w** = 9i + 3j, find 3**w**.

A) 12i + 3j   B) 12i + 6j   C) 27i + 9j   D) 27i + 3j

2) If **v** = 3i - 5j and **w** = -7i + 4j, find 3**v** - 4**w**.

A) -19i + j   B) 37i - 31j   C) 17i - 10j   D) -4i - j

3) If **v** = 6i - 8j, find $\|\mathbf{v}\|$.

A) 10   B) $\sqrt{10}$ C) 100   D) 14

4) If **v** = -3i + 5j, find $\|\mathbf{v}\|$.

A) 8   B) 34   C) $\sqrt{34}$ D) $2\sqrt{2}$

5) If **v** = -i - j, find $\|\mathbf{v}\|$.

A) 0   B) $\sqrt{2}$ C) 2   D) 1

6) If **v** = -5i + 7j, what is $\|-9\mathbf{v}\|$?

A) $18\sqrt{6}$ B) $-9\sqrt{74}$ C) $18i\sqrt{6}$ D) $9\sqrt{74}$

7) If **v** = 9**i** + **j** and **w** = -2**i** + **j**, find $\|\mathbf{v} + \mathbf{w}\|$.
  A) 11
  B) $5\sqrt{2}$
  C) $\sqrt{87}$
  D) $\sqrt{53}$

Find the quantity if v = 5i - 7j and w = 3i + 2j.

8) $\|\mathbf{v}\| + \|\mathbf{w}\|$
  A) $\sqrt{39}$
  B) $\sqrt{87}$
  C) $2\sqrt{6} + \sqrt{13}$
  D) $\sqrt{74} + \sqrt{13}$

9) $\|\mathbf{v}\| - \|\mathbf{w}\|$
  A) $\sqrt{11}$
  B) $\sqrt{74} - \sqrt{13}$
  C) $2\sqrt{6} - \sqrt{13}$
  D) $\sqrt{85}$

10) $\|\mathbf{v} + \mathbf{w}\|$
  A) $\sqrt{39}$
  B) $\sqrt{74} + \sqrt{13}$
  C) $\sqrt{89}$
  D) $2\sqrt{6} + \sqrt{13}$

11) $\|\mathbf{v} - \mathbf{w}\|$
  A) $2\sqrt{6} - \sqrt{13}$
  B) $\sqrt{85}$
  C) $\sqrt{77}$
  D) $\sqrt{74} - \sqrt{13}$

5 Find a Unit Vector

Find the unit vector having the same direction as v.

1) **v** = 6**i**
  A) $\mathbf{u} = \frac{1}{6}\mathbf{i}$
  B) $\mathbf{u} = \mathbf{i}$
  C) $\mathbf{u} = 36\mathbf{i}$
  D) $\mathbf{u} = 6\mathbf{i}$

2) **v** = -8**j**
  A) $\mathbf{u} = -\mathbf{j}$
  B) $\mathbf{u} = 64\mathbf{j}$
  C) $\mathbf{u} = -8\mathbf{j}$
  D) $\mathbf{u} = -\frac{1}{8}\mathbf{j}$

3) **v** = -4**i** - 3**j**
  A) $\mathbf{u} = \frac{3}{5}\mathbf{i} + \frac{4}{5}\mathbf{j}$
  B) $\mathbf{u} = -20\mathbf{i} - 15\mathbf{j}$
  C) $\mathbf{u} = -\frac{5}{4}\mathbf{i} - \frac{5}{3}\mathbf{j}$
  D) $\mathbf{u} = -\frac{4}{5}\mathbf{i} - \frac{3}{5}\mathbf{j}$

4) **v** = 12**i** - 5**j**
  A) $\mathbf{u} = \frac{5}{13}\mathbf{i} - \frac{12}{13}\mathbf{j}$
  B) $\mathbf{u} = 156\mathbf{i} - 65\mathbf{j}$
  C) $\mathbf{u} = \frac{12}{13}\mathbf{i} - \frac{5}{13}\mathbf{j}$
  D) $\mathbf{u} = \frac{13}{12}\mathbf{i} - \frac{13}{5}\mathbf{j}$

5) **v** = -3**i** + **j**
  A) $\mathbf{u} = -3\sqrt{10}\mathbf{i} + \sqrt{10}\mathbf{j}$
  B) $\mathbf{u} = -\frac{3\sqrt{10}}{10}\mathbf{i} + \frac{\sqrt{10}}{10}\mathbf{j}$
  C) $\mathbf{u} = -\frac{3\sqrt{10}}{10}\mathbf{i} - \frac{\sqrt{10}}{10}\mathbf{j}$
  D) $\mathbf{u} = -\frac{\sqrt{10}}{3}\mathbf{i} + \sqrt{10}\mathbf{j}$

Precalculus Enhanced with Graphing Utilities

6 Find a Vector from its Direction and Magnitude

Solve the problem.

1) Find a vector **v** whose magnitude is 4 and whose component in the **i** direction is twice the component in the **j** direction.

  A) $v = -\frac{8}{5}\sqrt{5}\,i + \frac{4}{5}\sqrt{5}\,j$ or $v = \frac{8}{5}\sqrt{5}\,i - \frac{4}{5}\sqrt{5}\,j$ B) $v = \frac{4}{5}\sqrt{5}\,i - \frac{8}{5}\sqrt{5}\,j$ or $v = -\frac{4}{5}\sqrt{5}\,i + \frac{8}{5}\sqrt{5}\,j$

  C) $v = \frac{4}{5}\sqrt{5}\,i + \frac{8}{5}\sqrt{5}\,j$ or $v = -\frac{4}{5}\sqrt{5}\,i - \frac{8}{5}\sqrt{5}\,j$ D) $v = \frac{8}{5}\sqrt{5}\,i + \frac{4}{5}\sqrt{5}\,j$ or $v = -\frac{8}{5}\sqrt{5}\,i - \frac{4}{5}\sqrt{5}\,j$

2) If $P = (-4, -9)$ and $Q = (x, -45)$, find all numbers x such that the vector represented by $\overrightarrow{PQ}$ has length –60.

  A) {44, –56}   B) {52, 44}   C) {52, –52}   D) {44, –52}

Write the vector v in the form ai + bj, given its magnitude ‖v‖ and the angle α it makes with the positive x-axis.

3) ‖v‖ = 13, α = 60°

  A) $v = 13\left(\frac{\sqrt{2}}{2}i + \frac{\sqrt{2}}{2}j\right)$ B) $v = 13\left(\frac{13\sqrt{3}}{2}i + \frac{13}{2}j\right)$

  C) $v = 13\left(\frac{13}{2}i + \frac{13\sqrt{3}}{2}j\right)$ D) $v = 13\left(-\frac{13}{2}i - \frac{13\sqrt{3}}{2}j\right)$

4) ‖v‖ = 5, α = 135°

  A) $v = 5\left(-\frac{5}{2}i + \frac{5\sqrt{3}}{2}j\right)$ B) $v = 5\left(\frac{\sqrt{2}}{2}i - \frac{\sqrt{2}}{2}j\right)$ C) $v = 5\left(\frac{5\sqrt{3}}{2}i - \frac{5}{2}j\right)$ D) $v = 5\left(-\frac{\sqrt{2}}{2}i + \frac{\sqrt{2}}{2}j\right)$

5) ‖v‖ = 9, α = 30°

  A) $v = 9\left(-\frac{9\sqrt{3}}{2}i + \frac{9}{2}j\right)$ B) $v = 9\left(\frac{9}{2}i + \frac{9\sqrt{3}}{2}j\right)$ C) $v = 9\left(\frac{9\sqrt{3}}{2}i + \frac{1}{2}j\right)$ D) $v = 9\left(\frac{\sqrt{2}}{2}i + \frac{\sqrt{2}}{2}j\right)$

6) ‖v‖ = 4, α = 90°

  A) v = 4i   B) v = 4j   C) v = 4i – 4j)   D) v = 4i + 4j

7) ‖v‖ = 6, α = 180°

  A) v = –6i   B) v = –6j   C) v = –6i – 6j   D) v = 6j

Solve the problem.

8) A truck pushes a load of 45 tons up a hill with an inclination of 35°. Express the force vector **F** in terms of **i** and **j**. Round the components of **F** to two decimal places.

9) Two forces, F_1 of magnitude 35 newtons (N) and F_2 of magnitude 55 newtons, act on an object at angles of 45° and –60° (respectively) with the positive x-axis. Find the direction and magnitude of the resultant force; that is, find $F_1 + F_2$. Round the direction and magnitude to two decimal places.

10) Two forces, F_1 of magnitude 60 newtons (N) and F_2 of magnitude 70 newtons, act on an object at angles of 40° and 130° (respectively) with the positive x-axis. Find the direction and magnitude of the resultant force; that is, find $F_1 + F_2$. Round the direction and magnitude to two decimal places.

11) Two forces of magnitude 25 pounds and 40 pounds act on an object. The force of 40 lb acts along the positive x-axis, and the force of 25 lb acts at an angle of 80° with the positive x-axis. Find the direction and magnitude of the resultant force. Round the direction and magnitude to the nearest whole number.

   A) Direction: 29°; magnitude: 51 lb
   B) Direction: 4°; magnitude: 65 lb
   C) Direction: 51°; magnitude: 51 lb
   D) Direction: 40°; magnitude: 47 lb

7 Work with Objects in Static Equilibrium

Solve the problem.

1) An audio speaker that weighs 50 pounds hangs from the ceiling of a restaurant from two cables as shown in the figure. To two decimal places, what is the tension in the two cables?

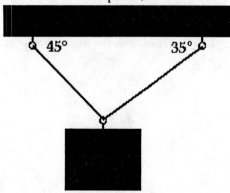

2) A box of supplies that weighs 1500 kilograms is suspended by two cables as shown in the figure. To two decimal places, what is the tension in the two cables?

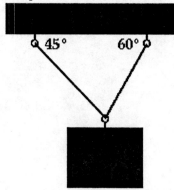

8.5 The Dot Product

1 Find the Dot Product of Two Vectors

Find the dot product v · w.

1) $v = -i - j$, $w = -i - j$

   A) 2   B) 1   C) −1   D) 0

2) $v = i$, $w = -4j$

   A) −4   B) $\sqrt{17}$ C) −3   D) 0

3) $v = -6i + 4j$, $w = 12i - 4j$

   A) −72   B) −88   C) −16   D) −56

4) $v = 11i - 11j$, $w = -13i - 12j$

A) -275   B) 132   C) -11   D) -143

2 Find the Angle between Two Vectors

Find the angle between v and w. Round your answer to one decimal place, if necessary.

1) $v = -3i - 5j$, $w = 3i - 2j$

A) 97.3°   B) 43.7°   C) 33.7°   D) 87.3°

2) $v = -5i + 7j$, $w = -6i - 4j$

A) 88.2°   B) 110.8°   C) 20.7°   D) 90.9°

3) $v = 3i$, $w = j$

A) 90°   B) 180°   C) 270°   D) 0°

3 Determine Whether Two Vectors Are Parallel

Solve the problem.

1) Which of the following vectors is parallel to $v = -10i - 8j$?

A) $w = 3i - 5j$ B) $w = 20i + 16j$ C) $w = 4i + 4j$ D) $w = -20i + 25j$

2) Which of the following vectors is parallel to $v = 9i + 5j$?

A) $w = 45i - 25j$ B) $w = -\frac{15}{2}i + \frac{25}{6}j$ C) $w = \frac{3}{2}i + \frac{5}{6}j$ D) $w = \frac{15}{2}i - \frac{25}{6}j$

4 Determine Whether Two Vectors Are Orthogonal

Solve the problem.

1) Which of the following vectors is orthogonal to $20i - 8j$?

A) $w = 20i + 4j$ B) $w = 15i - 6j$ C) $w = 4i + 3j$ D) $w = -10i - 25j$

State whether the vectors are parallel, orthogonal, or neither.

2) $v = 3i + j$, $w = i - 3j$

A) Orthogonal   B) Parallel   C) Neither

3) $v = 2i + 4j$, $w = 4i - 2j$

A) Orthogonal   B) Parallel   C) Neither

4) $v = 4i - j$, $w = 8i - 2j$

A) Parallel   B) Orthogonal   C) Neither

5) $v = 3i + 4j$, $w = 6i + 8j$

A) Orthogonal   B) Parallel   C) Neither

6) $v = 4i - 2j$, $w = 4i + 2j$

A) Parallel   B) Orthogonal   C) Neither

7) $v = i + \sqrt{2}j$, $w = i - 4j$

A) Orthogonal   B) Parallel   C) Neither

Solve the problem.

8) Find a so that the vectors $\mathbf{v} = \mathbf{i} + a\mathbf{j}$ and $\mathbf{w} = -3\mathbf{i} + 4\mathbf{j}$ are orthogonal.

A) $-\frac{4}{3}$ B) $\frac{4}{3}$ C) $\frac{3}{4}$ D) $-\frac{3}{4}$

5 Decompose a Vector into Two Orthogonal Vectors

Decompose $\mathbf{v}$ into two vectors $\mathbf{v}_1$ and $\mathbf{v}_2$, where $\mathbf{v}_1$ is parallel to $\mathbf{w}$ and $\mathbf{v}_2$ is orthogonal to $\mathbf{w}$.

1) $\mathbf{v} = \mathbf{i} + 9\mathbf{j}$, $\mathbf{w} = \mathbf{i} + \mathbf{j}$

A) $\mathbf{v}_1 = \frac{11}{2}\mathbf{i} + \frac{11}{2}\mathbf{j}$, $\mathbf{v}_2 = -\frac{9}{2}\mathbf{i} + \frac{7}{2}\mathbf{j}$ B) $\mathbf{v}_1 = 5\mathbf{i} + 5\mathbf{j}$, $\mathbf{v}_2 = 4\mathbf{i} - 4\mathbf{j}$

C) $\mathbf{v}_1 = 5\mathbf{i} + 5\mathbf{j})$, $\mathbf{v}_2 = -4\mathbf{i} + 4\mathbf{j}$ D) $\mathbf{v}_1 = 10\mathbf{i} + 10\mathbf{j}$, $\mathbf{v}_2 = -8\mathbf{i} + 8\mathbf{j}$

2) $\mathbf{v} = \mathbf{i} - 2\mathbf{j}$, $\mathbf{w} = -3\mathbf{i} + \mathbf{j}$

A) $\mathbf{v}_1 = \frac{3}{2}\mathbf{i} - \frac{1}{2}\mathbf{j}$, $\mathbf{v}_2 = \frac{7}{10}\mathbf{i} - \frac{19}{10}\mathbf{j}$ B) $\mathbf{v}_1 = \frac{3}{2}\mathbf{i} - \frac{1}{2}\mathbf{j}$, $\mathbf{v}_2 = \frac{3}{2}\mathbf{i} - \frac{3}{2}\mathbf{j}$

C) $\mathbf{v}_1 = \frac{5}{3}\mathbf{i} - \frac{5}{9}\mathbf{j}$, $\mathbf{v}_2 = -\frac{2}{3}\mathbf{i} - \frac{13}{9}\mathbf{j}$ D) $\mathbf{v}_1 = \frac{3}{2}\mathbf{i} - \frac{1}{2}\mathbf{j}$, $\mathbf{v}_2 = -\frac{1}{2}\mathbf{i} - \frac{3}{2}\mathbf{j}$

3) $\mathbf{v} = -2\mathbf{i} - 5\mathbf{j}$, $\mathbf{w} = -2\mathbf{i} + \mathbf{j}$

A) $\mathbf{v}_1 = +\frac{1}{2}\mathbf{i} - \frac{1}{4}\mathbf{j}$, $\mathbf{v}_2 = -\frac{5}{2}\mathbf{i} - \frac{19}{4}\mathbf{j}$ B) $\mathbf{v}_1 = +\frac{2}{5}\mathbf{i} - \frac{1}{5}\mathbf{j}$, $\mathbf{v}_2 = -\frac{12}{5}\mathbf{i} - \frac{24}{5}\mathbf{j}$

C) $\mathbf{v}_1 = +\frac{2}{5}\mathbf{i} - \frac{1}{5}\mathbf{j}$, $\mathbf{v}_2 = \frac{8}{5}\mathbf{i} - \frac{34}{5}\mathbf{j}$ D) $\mathbf{v}_1 = +\frac{2}{5}\mathbf{i} - \frac{1}{5}\mathbf{j}$, $\mathbf{v}_2 = -\frac{9}{5}\mathbf{i} - \frac{18}{5}\mathbf{j}$

4) $\mathbf{v} = -2\mathbf{i} + 2\mathbf{j}$, $\mathbf{w} = -3\mathbf{i} - \mathbf{j}$

A) $\mathbf{v}_1 = -\frac{4}{3}\mathbf{i} + \frac{4}{9}\mathbf{j}$, $\mathbf{v}_2 = -\frac{2}{3}\mathbf{i} + \frac{14}{9}\mathbf{j}$ B) $\mathbf{v}_1 = -\frac{6}{5}\mathbf{i} + \frac{2}{5}\mathbf{j}$, $\mathbf{v}_2 = \frac{2}{5}\mathbf{i} + \frac{6}{5}\mathbf{j}$

C) $\mathbf{v}_1 = -\frac{6}{5}\mathbf{i} + \frac{2}{5}\mathbf{j}$, $\mathbf{v}_2 = -\frac{12}{5}\mathbf{i} + \frac{8}{5}\mathbf{j}$ D) $\mathbf{v}_1 = -\frac{6}{5}\mathbf{i} + \frac{2}{5}\mathbf{j}$, $\mathbf{v}_2 = -\frac{4}{5}\mathbf{i} + \frac{8}{5}\mathbf{j}$

6 Compute Work

Solve the problem. Round your answer to the nearest tenth.

1) Find the work done by a force of 7 pounds acting in the direction of 42° to the horizontal in moving an object 5 feet from (0, 0) to (5, 0).

A) 28.0 ft-lb   B) 23.4 ft-lb   C) 52.0 ft-lb   D) 26.0 ft-lb

2) A wagon is pulled horizontally by exerting a force of 60 pounds on the handle at an angle of 25° to the horizontal. How much work is done in moving the wagon 50 feet?" .

A) 1267.9 ft-lb   B) 2718.9 ft-lb   C) 2110.8 ft-lb   D) 1617.4 ft-lb

3) A person is pulling a freight cart with a force of 51 pounds. How much work is done in moving the cart 20 feet if the cart's handle makes an angle of 18° with the ground?

A) 970.1 ft-lb   B) 993.9 ft-lb   C) 31.5 ft-lb   D) 315.2 ft-lb

4) Find the work done by a force of 200 pounds acting in the direction $-\mathbf{i} + 2\mathbf{j}$ in moving an object 75 feet from (0, 0) to (-75, 0).

A) 13,416.1 ft-lb   B) 15,000.0 ft-lb   C) 6708.2 ft-lb   D) 8944.9 ft-lb

7 Demonstrate Additional Understanding and Skills

Solve the problem.

1) An airplane has an air speed of 550 miles per hour bearing N30°W. The wind velocity is 50 kilometers per hour in the direction N30°E. Find the resultant vector (with exact components) representing the path of the plane relative to the ground. To the nearest tenth, what is the ground speed of the plane? What is its direction?

2) A DC-10 jumbo jet maintains an airspeed of 600 miles per hour in a southeasterly direction. The velocity of the jet stream is a constant 50 miles per hour from the west. Find the actual speed and direction of the aircraft. (Round the speed and direction to the nearest tenth.)

3) A plane is headed due south with an airspeed of 210 miles per hour. A wind from a direction of S30°W is blowing at 20 miles per hour. Find the groundspeed and resulting direction of the plane, rounded to the nearest whole number.
   A) 193 mph; S3°W   B) 193 mph; S3°E   C) 211 mph; S15°W   D) 201 mph; S5°E

4) An SUV weighing 5000 pounds is parked on a street which has an incline of 16°. Find the force required to keep the SUV from rolling down the hill and the force of the SUV perpendicular to the hill. Round the forces to the nearest hundredth.
   A) 689.09 lb and 2403.15 lb
   B) 1378.19 lb and 4806.31 lb
   C) 1124.76 lb and 4871.85 lb
   D) 1627.84 lb and 4727.59 lb

8.6 Vectors in Space

1 Find the Distance between Two Points in Space

Find the distance between the two points.

1) $P_1 = (0, 0, 0)$ and $P_2 = (3, -1, -3)$
   A) $\sqrt{7}$ B) $\sqrt{17}$ C) $\sqrt{19}$ D) $\sqrt{13}$

2) $P_1 = (2, 4, 3)$ and $P_2 = (-1, 0, -4)$
   A) $\sqrt{42}$ B) $\sqrt{74}$ C) $\sqrt{61}$ D) $\sqrt{58}$

3) $P_1 = (3, -1, 2)$ and $P_2 = (1, 2, -3)$
   A) $\sqrt{35}$ B) $\sqrt{46}$ C) $2\sqrt{11}$ D) $\sqrt{38}$

4) $P_1 = (1, 0, 3)$ and $P_2 = (-2, -1, 4)$
   A) $\sqrt{10}$ B) 3   C) $\sqrt{11}$ D) $\sqrt{35}$

5) $P_1 = (0, 2, -4)$ and $P_2 = (-3, 1, -3)$
   A) $\sqrt{59}$ B) $2\sqrt{2}$ C) $\sqrt{2}$ D) $\sqrt{11}$

2 Find Position Vectors in Space

Find the position vector for the vector having initial point P and terminal point Q.

1) $P = (0, 0, 0)$ and $Q = (-2, -3, 3)$
   A) $\mathbf{v} = -2\mathbf{i} - 3\mathbf{j} + 3\mathbf{k}$ B) $\mathbf{v} = -2\mathbf{i} - 3\mathbf{j} - 3\mathbf{k}$ C) $\mathbf{v} = 2\mathbf{i} + 3\mathbf{j} - 3\mathbf{k}$ D) $\mathbf{v} = 3\mathbf{i} - 3\mathbf{j} - 2\mathbf{k}$

2) $P = (1, -3, 0)$ and $Q = (-3, -1, -4)$
   A) $\mathbf{v} = 4\mathbf{i} - 2\mathbf{j} + 4\mathbf{k}$ B) $\mathbf{v} = -5\mathbf{i} + 2\mathbf{j} - 3\mathbf{k}$ C) $\mathbf{v} = -4\mathbf{i} + 2\mathbf{j} - 4\mathbf{k}$ D) $\mathbf{v} = -3\mathbf{i} + 2\mathbf{j} + 3\mathbf{k}$

3) P = (-4, 3, -3) and Q = (4, 0, -1)
   A) v = -8i + 3j - 2k   B) v = 7i - 3j + 5k   C) v = 8i - 3j + 2k   D) v = 3i - 3j + 7k

4) P = (4, 1, -1) and Q = (-4, -2, 1)
   A) v = -8i - 3j + 2k   B) v = -3i - 3j - 3k   C) v = -3i - 3j - 5k   D) v = 8i + 3j - 2k

3 Perform Operations on Vectors

Perform the operation.

1) $v = 4i - 6j - 2k$ Find $\|v\|$.
   A) $2\sqrt{13}$ B) $2\sqrt{14}$ C) $2\sqrt{5}$ D) $2\sqrt{3}$

2) $v = -5i - 5j + 5k$ Find $\|v\|$.
   A) 5   B) $\sqrt{55}$ C) $5\sqrt{3}$ D) $5\sqrt{2}$

3) $v = 3i + 2j + 6k$ and $w = 3i + 2j + 5k$
   Find $3v + 2w$
   A) $12i + 8j + 23k$ B) $6i + 4j + 11k$ C) $15i + 10j + 28k$ D) $9i + 6j + 16k$

4) $v = -4i + 3j - 5k$ and $w = -4i + 5j - 2k$
   Find $3v - 4w$
   A) $12i - 17j + 3k$ B) $-8i + 8j - 7k$ C) $-16i + 14j - 17k$ D) $4i - 11j - 7k$

5) $v = 5i + 2j + 4k$ and $w = 2i + 6j - 3k$ Find $\|v - w\|$.
   A) $3\sqrt{10}$ B) $3\sqrt{5} - 7$ C) $\sqrt{74}$ D) $2\sqrt{21}$

6) $v = 6i + 3j + 2k$ and $w = 4i - 2j + 5k$ Find $\|v\| - \|w\|$.
   A) $\sqrt{70}$ B) $\sqrt{43} - \sqrt{39}$ C) $7 - 3\sqrt{5}$ D) $\sqrt{38}$

4 Find the Dot Product

Find the dot product v · w.

1) $v = i + j$ and $w = i + j - k$
   A) 2   B) 1   C) -2   D) 3

2) $v = 2i + j - 3k$ and $w = i + 3j - 3k$
   A) 8   B) 14   C) -14   D) 41

3) $v = 3i + 2j + k$ and $w = 2i + 5j + 2k$
   A) 18   B) 53   C) -18   D) -2

5 Find the Angle between Two Vectors

Find the angle between v and w. Round to one decimal place, if necessary.

1) $v = i + j$ and $w = i + j - k$
   A) 0°   B) 35.3°   C) 66°   D) 90°

2) $v = 3i + j - 3k$ and $w = i + 2j + 2k$
   A) 92.9°   B) 94.4°   C) 92°   D) -4.4°

3) $v = 3i + 4j + k$ and $w = 6i + 4j + 3k$

   A) 63.5° B) 58.5° C) 68.3° D) 21.7°

Solve the problem.

4) A sail on a sailboat is supported by a mast and a boom. The length and direction of the mast and boom can be described by the vectors $m = i - 4j + 5k$ and $b = -2i + 3j + k$, respectively, where the components are in meters. Find the angle (to the nearest tenth of a degree) between the mast and boom.

6 Find the Direction Angles of a Vector

Find the direction angles of each vector. Round to the nearest degree, if necessary.

1) $v = 3i + 6j - 2k$

   A) $\alpha = 65°, \beta = 31°, \gamma = 107°$
   B) $\alpha = 62°, \beta = 18°, \gamma = 34°$
   C) $\alpha = 68°, \beta = 41°, \gamma = 104°$
   D) $\alpha = 86°, \beta = 83°, \gamma = 92°$

2) $v = -12i + 6j - 4k$

   A) $\alpha = 161°, \beta = 62°, \gamma = 108°$
   B) $\alpha = 94°, \beta = 88°, \gamma = 91°$
   C) $\alpha = 157°, \beta = 63°, \gamma = 108°$
   D) $\alpha = 149°, \beta = 65°, \gamma = 107°$

3) $v = i - j + 2k$

   A) $\alpha = 80°, \beta = 100°, \gamma = 71°$
   B) $\alpha = 69°, \beta = 111°, \gamma = 45°$
   C) $\alpha = 66°, \beta = 114°, \gamma = 35°$
   D) $\alpha = 63°, \beta = 117°, \gamma = 26°$

4) $v = 2i - 4j - 3k$

   A) $\alpha = 61°, \beta = 166°, \gamma = 137°$
   B) $\alpha = 68°, \beta = 138°, \gamma = 124°$
   C) $\alpha = 65°, \beta = 147°, \gamma = 129°$
   D) $\alpha = 67°, \beta = 140°, \gamma = 125°$

5) $v = 2i - 3j - 4k$

   A) $\alpha = 61°, \beta = 137°, \gamma = 166°$
   B) $\alpha = 68°, \beta = 124°, \gamma = 138°$
   C) $\alpha = 67°, \beta = 125°, \gamma = 140°$
   D) $\alpha = 65°, \beta = 129°, \gamma = 147°$

7 Demonstrate Additional Understanding and Skills

Find the equation of a sphere with radius r and center P_0.

1) $r = 5; P_0 = (1, 0, 3)$

   A) $(x - 1)^2 + y^2 + (z - 3)^2 = 125$
   B) $(x + 1)^2 + y^2 + (z + 3)^2 = 25$
   C) $(x - 1)^2 + y^2 + (z - 3)^2 = 10$
   D) $(x - 1)^2 + y^2 + (z - 3)^2 = 25$

2) $r = 4; P_0 = (3, -4, -8)$

   A) $(x - 3)^2 + (y + 4)^2 + (z + 8)^2 = 8$
   B) $(x + 3)^2 + (y - 4)^2 + (z - 8)^2 = 16$
   C) $(x - 3)^2 + (y + 4)^2 + (z + 8)^2 = 16$
   D) $(x - 3)^2 + (y + 4)^2 + (z + 8)^2 = 64$

Find the center P_0 and radius r of the sphere with the given equation.

3) $(x - 7)^2 + (y + 5)^2 + (z - 6)^2 = 112$

   A) $P_0 = (7, -5, 6); r = 28$
   B) $P_0 = (-7, 5, -6); r = 4\sqrt{7}$
   C) $P_0 = (7, -5, 6); r = 4\sqrt{7}$
   D) $P_0 = (-7, 5, -6); r = 28$

4) $x^2 + 6x + y^2 - 10y + z^2 + 7z = \dfrac{3}{4}$

A) $P_0 = (-3, 5, -\dfrac{7}{2})$; $r = \sqrt{47}$
B) $P_0 = (-3, 5, -\dfrac{7}{2})$; $r = \sqrt{86}$
C) $P_0 = (3, -5, \dfrac{7}{2})$; $r = \sqrt{86}$
D) $P_0 = (3, -5, \dfrac{7}{2})$; $r = \sqrt{47}$

8.7 The Cross Product

1 Find the Cross Product of Two Vectors

Find the indicated cross product.

1) $\mathbf{v} = 6\mathbf{i} + 6\mathbf{j} + \mathbf{k}$, $\mathbf{w} = 3\mathbf{i} - 2\mathbf{j} - \mathbf{k}$
Find $\mathbf{v} \times \mathbf{w}$.

A) $4\mathbf{i} - 9\mathbf{j} + 42\mathbf{k}$
B) $-8\mathbf{i} + 3\mathbf{j} + 6\mathbf{k}$
C) $-4\mathbf{i} + 9\mathbf{j} - 30\mathbf{k}$
D) $-30\mathbf{i} + 4\mathbf{j} - 9\mathbf{k}$

2) $\mathbf{v} = \mathbf{i} + 3\mathbf{j} + 2\mathbf{k}$, $\mathbf{w} = 5\mathbf{i} - 4\mathbf{j} - \mathbf{k}$
Find $\mathbf{v} \times \mathbf{w}$.

A) $-19\mathbf{i} - 5\mathbf{j} - 11\mathbf{k}$
B) $-11\mathbf{i} - 9\mathbf{j} + 11\mathbf{k}$
C) $2\mathbf{i} - 7\mathbf{j} + 23\mathbf{k}$
D) $5\mathbf{i} + 11\mathbf{j} - 19\mathbf{k}$

3) $\mathbf{v} = 6\mathbf{i} + 5\mathbf{j}$, $\mathbf{w} = 4\mathbf{i} - 4\mathbf{j} - 3\mathbf{k}$
Find $\mathbf{v} \times \mathbf{w}$.

A) $-12\mathbf{i} - 12\mathbf{j} + 46\mathbf{k}$
B) $-44\mathbf{i} + 15\mathbf{j} - 18\mathbf{k}$
C) $-15\mathbf{i} + 18\mathbf{j} - 44\mathbf{k}$
D) $-15\mathbf{i} + 18\mathbf{j} - 4\mathbf{k}$

4) $\mathbf{v} = -5\mathbf{i} + 6\mathbf{j} - 5\mathbf{k}$, $\mathbf{w} = 3\mathbf{i} + 2\mathbf{k}$
Find $\mathbf{v} \times \mathbf{w}$.

A) $-30\mathbf{i} - 19\mathbf{j} - 30\mathbf{k}$
B) $-18\mathbf{i} - 12\mathbf{j} + 5\mathbf{k}$
C) $12\mathbf{i} + 25\mathbf{j} + 18\mathbf{k}$
D) $12\mathbf{i} - 5\mathbf{j} - 18\mathbf{k}$

5) $\mathbf{v} = -4\mathbf{i} - 3\mathbf{j}$, $\mathbf{w} = -3\mathbf{i} - 5\mathbf{k}$
Find $\mathbf{v} \times \mathbf{w}$.

A) $-15\mathbf{j} - 12\mathbf{k}$
B) $-9\mathbf{i} - 15\mathbf{j} + 20\mathbf{k}$
C) $15\mathbf{i} - 20\mathbf{j} + 9\mathbf{k}$
D) $15\mathbf{i} - 20\mathbf{j} - 9\mathbf{k}$

6) $\mathbf{v} = -5\mathbf{i} + 6\mathbf{j} - 4\mathbf{k}$, $\mathbf{w} = -3\mathbf{i} + 4\mathbf{j} - 4\mathbf{k}$
Find $\mathbf{v} \times \mathbf{w}$.

A) $-8\mathbf{i} - 8\mathbf{j} - 18\mathbf{k}$
B) $-40\mathbf{i} - 32\mathbf{j} - 38\mathbf{k}$
C) $-8\mathbf{i} - 8\mathbf{j} - 2\mathbf{k}$
D) $-2\mathbf{i} + 8\mathbf{j} + 8\mathbf{k}$

7) $\mathbf{v} = 2\mathbf{i} - 4\mathbf{j} + \mathbf{k}$, $\mathbf{w} = -5\mathbf{i} - 2\mathbf{j} - \mathbf{k}$
Find $\mathbf{w} \times \mathbf{v}$.

A) $24\mathbf{i} + 6\mathbf{j} - 6\mathbf{k}$
B) $6\mathbf{i} - 3\mathbf{j} + 18\mathbf{k}$
C) $-6\mathbf{i} + 3\mathbf{j} + 24\mathbf{k}$
D) $-3\mathbf{i} + 6\mathbf{j} + 24\mathbf{k}$

8) $\mathbf{v} = \mathbf{i} + 6\mathbf{j} - 5\mathbf{k}$, $\mathbf{w} = -4\mathbf{i} + 6\mathbf{j} - \mathbf{k}$
Find $\mathbf{w} \times \mathbf{v}$.

A) $24\mathbf{i} - 9\mathbf{j} - 30\mathbf{k}$
B) $-24\mathbf{i} - 21\mathbf{j} - 30\mathbf{k}$
C) $-30\mathbf{i} + 24\mathbf{j} - 24\mathbf{k}$
D) $21\mathbf{i} + 24\mathbf{j} - 30\mathbf{k}$

9) $\mathbf{v} = 2\mathbf{i} + 3\mathbf{j}$, $\mathbf{w} = 5\mathbf{i} + 4\mathbf{j} + 5\mathbf{k}$
Find $\mathbf{w} \times \mathbf{v}$.

A) $7\mathbf{i} + 15\mathbf{j} - 15\mathbf{k}$
B) $20\mathbf{i} - 25\mathbf{j} + 14\mathbf{k}$
C) $-10\mathbf{i} + 15\mathbf{j} + 7\mathbf{k}$
D) $-15\mathbf{i} + 10\mathbf{j} + 7\mathbf{k}$

10) $v = 4i - 4j + 6k$, $w = -4i + 2k$
   Find $w \times v$.
   A) $24i + 32j + 16k$ B) $8i + 32j + 16k$ C) $-32i - 8j + 16k$ D) $16i - 8j + 8k$

11) $v = 4i + 6j$, $w = -6i + 3k$
   Find $w \times v$.
   A) $-36i + 18j - 18k$ B) $-12i + 18j - 36k$ C) $18j - 24k$ D) $-18i + 12j - 36k$

12) $v = 5i + 5j - 4k$, $w = 5i + 6j + 6k$
   Find $w \times v$.
   A) $56i - 50j + 5k$ B) $-50i + 54j - 5k$ C) $-5i + 54j - 54k$ D) $-54i + 50j - 5k$

2 Know Algebraic Properties of the Cross Product

Use the given vectors to find the requested expression.

1) $v = -4i - 5j + k$, $w = -2i - 4j - k$
   Find $(3v) \times w$.
   A) $17i - 10j - 14k$ B) $-27i + 18j - 18k$ C) $19i - 14j + 38k$ D) $27i - 18j + 18k$

2) $v = 2i + 2j - 5k$, $w = -3i + 3j - 4k$
   Find $v \times (4w)$.
   A) $-17i + 47j + 30k$ B) $-28i - 20j - 48k$ C) $28i + 92j + 48k$ D) $52i + 68j + 30k$

3) $v = 3i + 2j - 3k$, $u = -4i - 4j - 2k$
   Find $v \times (-2u)$.
   A) $4i - 48j - 28k$ B) $-32i + 36j - 8k$ C) $-16i + 24j - 4k$ D) $32i - 36j + 8k$

4) $v = -2i - 3j - 3k$, $w = 5i - 4j + 4k$, $u = -2i - 4j - 2k$
   Find $w \cdot (v \times u)$.
   A) -94 B) -30 C) 0 D) -48

5) $v = 5i - 4j + 5k$, $w = 4i + 3j - 3k$, $u = -5i + 4j + 3k$
   Find $u \cdot (v \times w)$.
   A) -228 B) 225 C) 273 D) 248

6) $v = 2i + 3j - 3k$, $w = 3i - 2j - 2k$, $u = 3i + 3j - 4k$
   Find $v \cdot (v \times w)$.
   A) -64 B) 0 C) 46 D) 64

3 Know Geometric Properties of the Cross Product

Solve the problem.

1) Find a unit vector normal to the plane containing $u = -i + j + 4k$ and $v = 2i - 3j + k$.

2) Find a unit vector normal to the plane containing $u = -i + 3j - 5k$ and $v = 2i - j + 6k$.

4 Find a Vector Orthogonal to Two Given Vectors

Find the requested vector.

1) $v = -2i - 2j + k$, $w = -5i - 4j - k$
   Find a vector orthogonal to both **v** and **w**.
   A) $6i - 5j + 18k$
   B) $-6i + 7j - 7k$
   C) $6i - 7j - 2k$
   D) $16i - 30j + 40k$

2) $v = 2i + 4j + k$, $u = -4i + 4j + 2k$
   Find a vector orthogonal to both **v** and **u**.
   A) $-6i + 5j - 8k$
   B) $4i - 8j + 24k$
   C) $6i - 5j + 5k$
   D) $-8i + 2j - 20k$

3) $v = 5i - 5j + k$
   Find a vector orthogonal to both **v** and $i + j$.
   A) $-i + j + 10k$
   B) $i + j + 10k$
   C) $i - j - 7k$
   D) $-5i + 5j - 6k$

4) $v = 3i - 2j + k$
   Find a vector orthogonal to both **v** and $i + k$.
   A) $-5i + 2j + 5k$
   B) $i - j + 2k$
   C) $-2i - 2j + 2k$
   D) $3i - 6j - 3k$

5) $w = 4i - 3j - k$
   Find a vector orthogonal to both **w** and $j + k$.
   A) $-4i - 2j + 4k$
   B) $4i - 4j - 2k$
   C) $-2i - 4j + 4k$
   D) $5i + 4j - 4k$

5 Find the Area of a Parallelogram

Find the area of the parallelogram.

1) One corner at P_1 and adjacent sides $\overrightarrow{P_1P_2}$ and $\overrightarrow{P_1P_3}$
   $P_1(0, 0, 0)$, $P_2(2, -3, 1)$, $P_3(-1, 4, -1)$
   A) $\sqrt{14}$
   B) $3\sqrt{3}$
   C) $3\sqrt{2}$
   D) $\sqrt{6}$

2) One corner at P_1 and adjacent sides $\overrightarrow{P_1P_2}$ and $\overrightarrow{P_1P_3}$
   $P_1(0, 0, 0)$, $P_2(4, 3, 1)$, $P_3(-2, 2, 1)$
   A) 3
   B) $\sqrt{293}$
   C) $\sqrt{26}$
   D) $\sqrt{233}$

3) One corner at P_1 and adjacent sides $\overrightarrow{P_1P_2}$ and $\overrightarrow{P_1P_3}$
   $P_1(1, 2, 0)$, $P_2(-2, 4, 2)$, $P_3(0, -3, 3)$
   A) $\sqrt{35}$
   B) $\sqrt{17}$
   C) $3\sqrt{57}$
   D) $3\sqrt{66}$

4) One corner at P_1 and adjacent sides $\overrightarrow{P_1P_2}$ and $\overrightarrow{P_1P_3}$
   $P_1(-1, 0, 1)$, $P_2(2, 1, -2)$, $P_3(2, -1, 2)$
   A) $4\sqrt{2}$
   B) $\sqrt{11}$
   C) $2\sqrt{46}$
   D) $\sqrt{19}$

Ch. 8 Polar Coordinates; Vectors
Answer Key

8.1 Polar Coordinates
1 Plot Points Using Polar Coordinates
1) D
2) D
3) D
4) A
5) D
6) D
7) C
8) D
9) A
10) A
11) B
12) D
13)

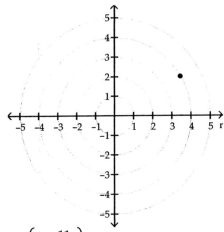

(a) $\left(4, -\dfrac{11\pi}{6}\right)$

(b) $\left(-4, \dfrac{7\pi}{6}\right)$

(c) $\left(4, \dfrac{13\pi}{6}\right)$

2 Convert from Polar Coordinates to Rectangular Coordinates
1) C
2) B
3) B
4) D
5) C
6) D
7) D
8) C
9) A
10) D
11) A
12) B
13) B
14) B

Precalculus Enhanced with Graphing Utilities 671

15) A
16) D
17) C
18) B

3 Convert from Rectangular Coordinates to Polar Coordinates
1) B
2) B
3) C
4) D
5) C
6) D
7) C
8) D
9) B
10) A
11) C
12) B
13) D
14) C
15) D
16) D

8.2 Polar Equations and Graphs
1 Graph and Identify Polar Equations by Converting to Rectangular Equations
1) D
2) C
3) A
4) B
5) D
6) B
7) B
8) B
9) A
10) C
11) A

2 Graph Polar Equations Using a Graphing Utility
1) A
2) D

3 Test Polar Equations for Symmetry
1) B
2) B
3) A
4) B
5) B
6) B
7) A
8) A
9) B
10) A

4 Graph Polar Equations by Plotting Points
1) D
2) B
3) A
4) A
5) C
6) D

7) B
8) D
9) A
10) C
11) C
12) C
13) D
14) D
15) C
16) D
17) B

8.3 The Complex Plane; De Moivre's Theorem
1 Convert a Complex Number from Rectangular Form to Polar Form
1) A
2) D
3) C
4) C
5) C

2 Plot Points in the Complex Plane
1) C
2) C
3) B
4) A
5) B

3 Find Products and Quotients of Complex Numbers in Polar Form
1) B
2) D
3) D
4) C
5) C
6) B
7) C
8) D
9) B
10) C
11) B
12) B
13) A

4 Use De Moivre's Theorem
1) D
2) A
3) A
4) A
5) A
6) C
7) D
8) C

5 Find Complex Roots
1) C
2) B
3) B
4) D

6 Demonstrate Additional Understanding and Skills
1) D
2) A

3) D
4) B

8.4 Vectors

1 Graph Vectors

1) D
2) B
3) C
4) D
5) A
6) A
7) B
8) A
9) A
10) B

2 Find a Position Vector

1) C
2) A
3) B

3 Add and Subtract Vectors

1) B
2) B

4 Find a Scalar Product and the Magnitude of a Vector

1) C
2) B
3) A
4) C
5) B
6) D
7) D
8) D
9) B
10) C
11) B

5 Find a Unit Vector

1) B
2) A
3) D
4) C
5) B

6 Find a Vector from its Direction and Magnitude

1) D
2) D
3) C
4) D
5) C
6) B
7) A
8) 36.86i + 25.81j
9) Direction: −23.65°; magnitude: 57.04 N
10) Direction: 89.40°; magnitude: 92.20 N
11) A

7 Work with Objects in Static Equilibrium

1) Tension in right cable: 35.90 lb; tension in left cable: 41.59 lb
2) Tension in right cable: 1098.08 kg; tension in left cable: 776.46 kg

8.5 The Dot Product
1 Find the Dot Product of Two Vectors
- 1) A
- 2) D
- 3) B
- 4) C

2 Find the Angle between Two Vectors
- 1) D
- 2) A
- 3) A

3 Determine Whether Two Vectors Are Parallel
- 1) B
- 2) C

4 Determine Whether Two Vectors Are Orthogonal
- 1) D
- 2) A
- 3) A
- 4) A
- 5) B
- 6) C
- 7) C
- 8) C

5 Decompose a Vector into Two Orthogonal Vectors
- 1) C
- 2) D
- 3) B
- 4) D

6 Compute Work
- 1) D
- 2) B
- 3) A
- 4) C

7 Demonstrate Additional Understanding and Skills
- 1) $-250i + 300\sqrt{3}j$; 576.6 mph; N25.7°W
- 2) 636.3 mph; S48.2°E
- 3) B
- 4) B

8.6 Vectors in Space
1 Find the Distance between Two Points in Space
- 1) C
- 2) B
- 3) D
- 4) C
- 5) D

2 Find Position Vectors in Space
- 1) A
- 2) C
- 3) C
- 4) A

3 Perform Operations on Vectors
- 1) B
- 2) C
- 3) C
- 4) D
- 5) C

6) C
4 Find the Dot Product
1) A
2) B
3) A

5 Find the Angle between Two Vectors
1) B
2) B
3) D
4) θ ≈ 111.8°

6 Find the Direction Angles of a Vector
1) A
2) D
3) C
4) B
5) B

7 Demonstrate Additional Understanding and Skills
1) D
2) C
3) C
4) A

8.7 The Cross Product
1 Find the Cross Product of Two Vectors
1) C
2) D
3) C
4) D
5) D
6) C
7) C
8) B
9) D
10) B
11) D
12) D

2 Know Algebraic Properties of the Cross Product
1) D
2) C
3) D
4) B
5) D
6) B

3 Know Geometric Properties of the Cross Product
1) $\dfrac{13\sqrt{251}}{251}\mathbf{i} + \dfrac{9\sqrt{251}}{251}\mathbf{j} + \dfrac{\sqrt{251}}{251}\mathbf{k}$
2) $\dfrac{13\sqrt{210}}{210}\mathbf{i} - \dfrac{2\sqrt{210}}{105}\mathbf{j} - \dfrac{\sqrt{210}}{42}\mathbf{k}$

4 Find a Vector Orthogonal to Two Given Vectors
1) C
2) B
3) A
4) C
5) C

5 Find the Area of a Parallelogram
1) B
2) D
3) D
4) C

Ch. 9   Analytic Geometry

9.1   Conics

1   Know the Names of Conics

Name the conic.

1)

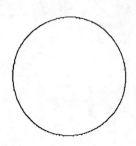

A) ellipse   B) circle   C) hyperbola   D) parabola

2)

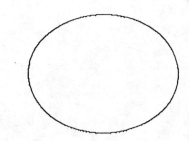

A) circle   B) parabola   C) hyperbola   D) ellipse

3)

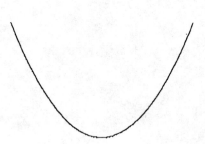

A) hyperbola   B) ellipse   C) circle   D) parabola

4)

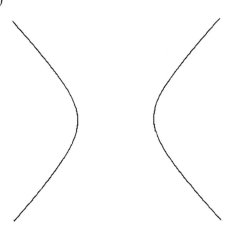

A) parabola   B) circle   C) ellipse   D) hyperbola

9.2 The Parabola

1 Work with Parabolas with Vertex at the Origin

Match the equations to the graph.

1) $y^2 = 4x$

A)
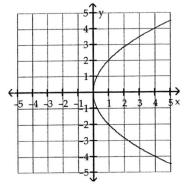

B)
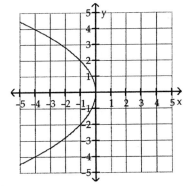

C)
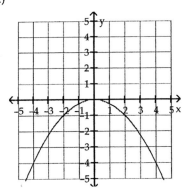

D)
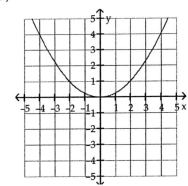

2) $y^2 = -2x$

A)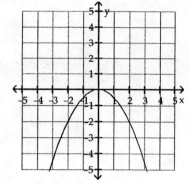

B)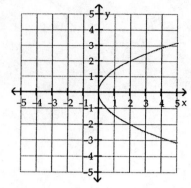

C)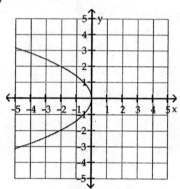

D)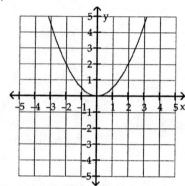

3) $x^2 = 13y$

A)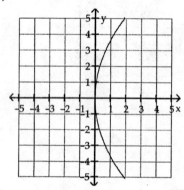

B)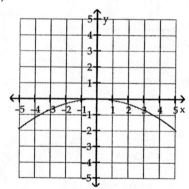

C)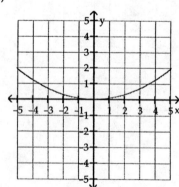

D)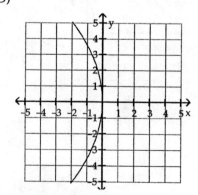

4) $x^2 = -8y$

A)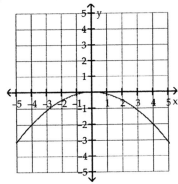

B)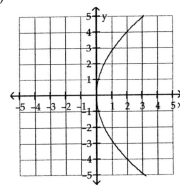

C)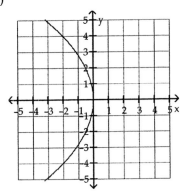

D)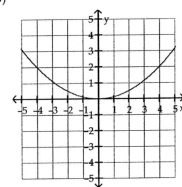

5) $y^2 = 20x$

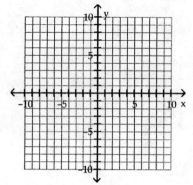

A)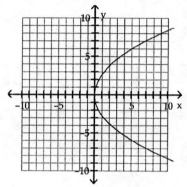

B)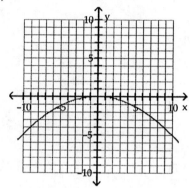

C)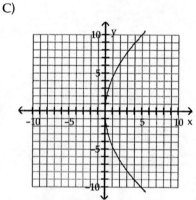

D)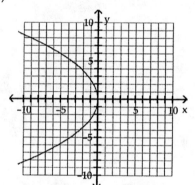

6) $y^2 = -16x$

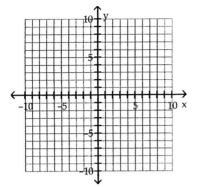

A)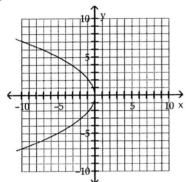

B)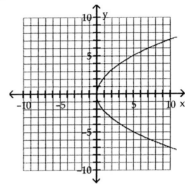

C)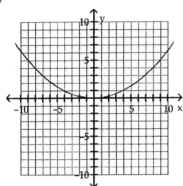

D)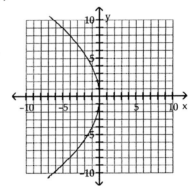

7) $x^2 = 20y$

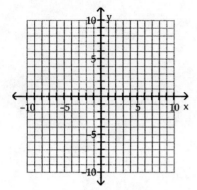

A)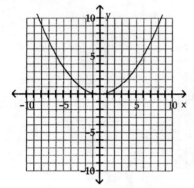

B)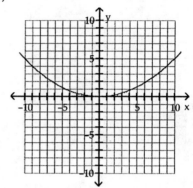

C)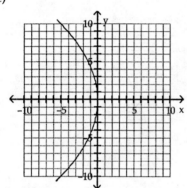

D)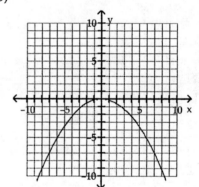

8) $x^2 = -9y$

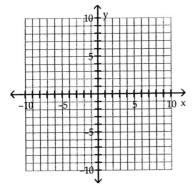

A)
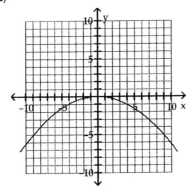

B)
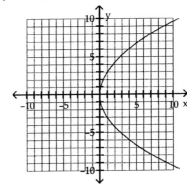

C)
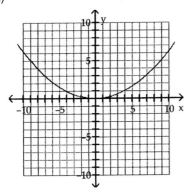

D)
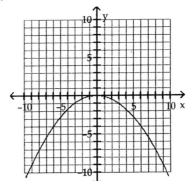

Write an equation for the parabola.

9)

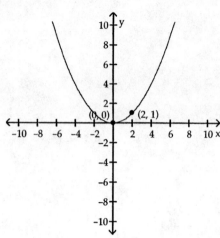

A) $y^2 = 4x$ B) $x^2 = 4y$ C) $x^2 = -4y$ D) $y^2 = -4x$

10)

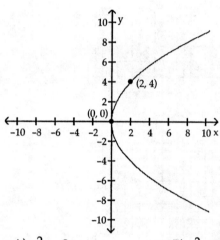

A) $x^2 = -8y$ B) $x^2 = 8y$ C) $y^2 = 8x$ D) $y^2 = -8x$

Find an equation of the parabola satisfying the given conditions.

11) Focus: (-7, 0); Directrix: x = 7
   A) $x^2 = -28y$ B) $y^2 = -28x$ C) $y^2 = 28x$ D) $y^2 = -7x$

12) Focus: (0, 9); Directrix: y = -9
   A) $x^2 = 36y$ B) $y^2 = 9x$ C) $x^2 = -36y$ D) $y^2 = 36x$

13) Focus at (5, 0); vertex at (0, 0)
   A) $y^2 = 5x$ B) $x^2 = 20y$ C) $x^2 = 5y$ D) $y^2 = 20x$

14) Directrix the line y = 3; vertex at (0, 0)
   A) $x = 3y^2$ B) $x = -\frac{1}{12}y^2$ C) $y = -12x^2$ D) $y = -\frac{1}{12}x^2$

15) Focus at (5, 0); vertex at (0, 0)
   A) $x^2 = 20y$ B) $y^2 = 20x$ C) $y = 20x^2$ D) $x = 20y^2$

16) Vertex at (0, 0); axis of symmetry the x-axis; containing the point (6, 5)

   A) $y^2 = \dfrac{25}{24}x$
   B) $y^2 = \dfrac{25}{6}x$
   C) $x^2 = \dfrac{25}{24}y$
   D) $x^2 = \dfrac{25}{6}y$

Find an equation of the parabola satisfying the stated conditions and state the two points that define the latus rectum.

17) Focus: (0, 1); Directrix the line y = −1

   A) $x^2 = 4y$; latus rectum: (1, 2) and (−1, 2)
   B) $x^2 = 4y$; latus rectum: (2, 1) and (−2, 1)
   C) $y^2 = 8x$; latus rectum: (3, 4) and (−3, 4)
   D) $x^2 = 8y$; latus rectum: (4, 1) and (−4, 1)

Find the vertex, focus, and directrix of the parabola.

18) $x^2 = 8y$

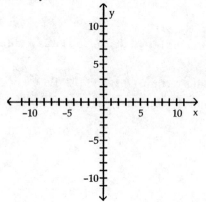

A)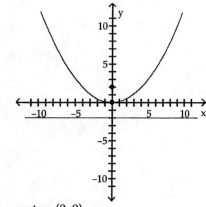

vertex: $(0, 0)$
focus: $(0, 2)$
directrix: $y = -2$

B)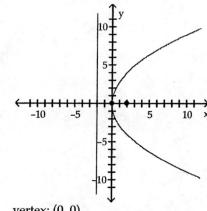

vertex: $(0, 0)$
focus: $(2, 0)$
directrix: $x = -2$

C)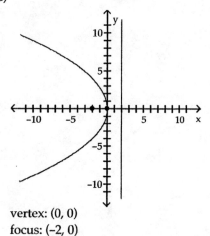

vertex: $(0, 0)$
focus: $(-2, 0)$
directrix: $x = 2$

D)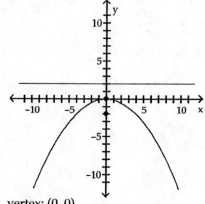

vertex: $(0, 0)$
focus: $(0, -2)$
directrix: $y = 2$

19) $y^2 = 8x$

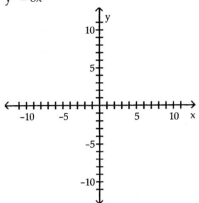

A)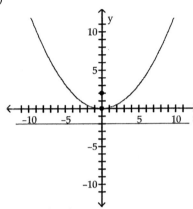
vertex: (0, 0)
focus: (0, 2)
directrix: $y = -2$

B)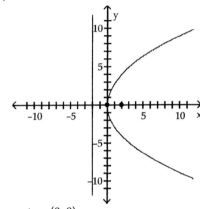
vertex: (0, 0)
focus: (2, 0)
directrix: $x = -2$

C)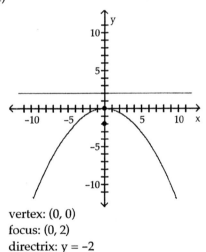
vertex: (0, 0)
focus: (0, 2)
directrix: $y = -2$

D)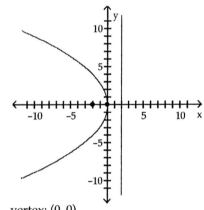
vertex: (0, 0)
focus: (-2, 0)
directrix: $x = 2$

2 Work with Parabolas with Vertex at (h, k)

Find the vertex, focus, and directrix of the parabola with the given equation.

1) $(y + 2)^2 = 4(x - 3)$

A) vertex: (-3, 2)
focus: (-2, 2)
directrix: $x = -4$

B) vertex: (3, -2)
focus: (2, -2)
directrix: $x = 4$

C) vertex: (3, -2)
focus: (4, -2)
directrix: $x = 2$

D) vertex: (-2, 3)
focus: (-1, 3)
directrix: $x = -3$

2) $(y + 1)^2 = -12(x - 4)$

    A) vertex: (4, -1)
       focus: (7, -1)
       directrix: x = 1

    B) vertex: (-1, 4)
       focus: (-4, 4)
       directrix: x = 2

    C) vertex: (4, -1)
       focus: (1, -1)
       directrix: x = 7

    D) vertex: (-4, 1)
       focus: (-7, 1)
       directrix: x = -1

3) $(x + 1)^2 = 8(y - 3)$

    A) vertex: (3, -1)
       focus: (3, 1)
       directrix: y = -3

    B) vertex: (1, -3)
       focus: (1, -1)
       directrix: y = -5

    C) vertex: (-1, 3)
       focus: (-1, 5)
       directrix: y = 1

    D) vertex: (-1, 3)
       focus: (-1, 1)
       directrix: x = 5

4) $(x + 4)^2 = -12(y + 2)$

    A) vertex: (-4, -2)
       focus: (-4, -5)
       directrix: y = 1

    B) vertex: (-2, -4)
       focus: (-2, -7)
       directrix: y = -1

    C) vertex: (4, 2)
       focus: (4, -1)
       directrix: y = 5

    D) vertex: (-4, -2)
       focus: (-4, 1)
       directrix: x = -5

Graph the equation.

5) $(y-1)^2 = 5(x-2)$

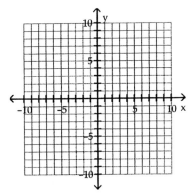

A)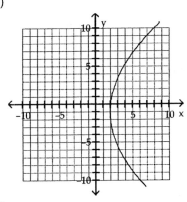

B)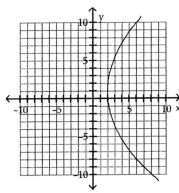

C)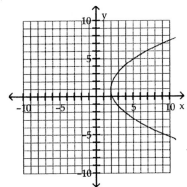

D)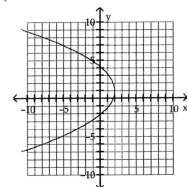

6) $(y+2)^2 = -5(x-1)$

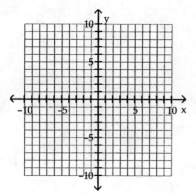

A)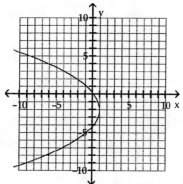

B)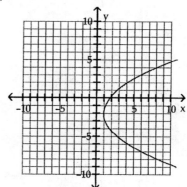

C)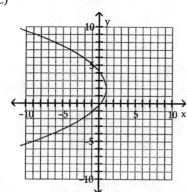

D)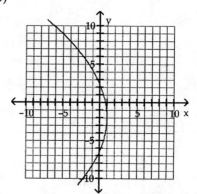

7) $(x+1)^2 = 6(y-1)$

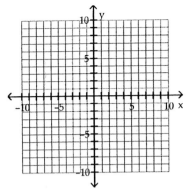

A)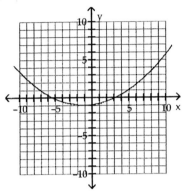

B)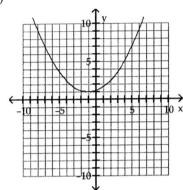

C)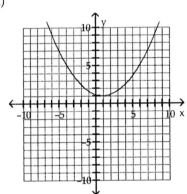

D)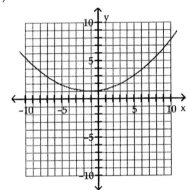

8) $(x-2)^2 = -7(y+2)$

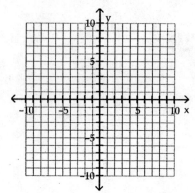

A)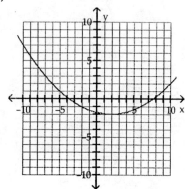

B)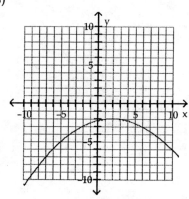

C)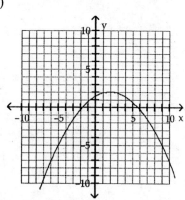

D)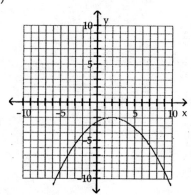

Find an equation for the parabola satisfying the stated conditions.

9) Vertex at (5, 3) and focus at (4, 3)

   A) $(x-3)^2 = -4(y-3)$ B) $(y-3)^2 = 4(x-5)$ C) $(y-3)^2 = -4(x-5)$ D) $(x-3)^2 = 4(y-3)$

10) Vertex at (1, 7) and focus at (1, 5)

   A) $(x-1)^2 = 8(y-7)$ B) $(x-1)^2 = -8(y-7)$ C) $(y-7)^2 = 16(x-1)$ D) $(y-7)^2 = -16(x-1)$

11) Vertex at (4, -4); focus at (4, 1)

   A) $(x-4)^2 = 20(y+4)$ B) $(x+4)^2 = 5(y-4)$ C) $(x-4)^2 = 5(y+4)$ D) $(x+4)^2 = 20(y-4)$

12) Vertex at (3, 4); focus at (3, 2)

   A) $(y-4)^2 = 8(x-4)$ B) $(x-3)^2 = 8(y-4)$ C) $(y-4)^2 = -8(x-4)$ D) $(x-3)^2 = -8(y-4)$

13) Vertex at (9, -3); focus at (9, -5)

   A) $(y - 3)^2 = 16(x + 9)$ B) $(y - 3)^2 = -16(x + 9)$ C) $(x - 9)^2 = 8(y + 3)$ D) $(x - 9)^2 = -8(y + 3)$

14) Vertex at (7, -1); focus at (2, -1)

   A) $(y + 1)^2 = -20(x - 7)$ B) $(x + 7)^2 = -4(y - 1)$ C) $(x + 7)^2 = 4(y - 1)$ D) $(y + 1)^2 = 20(x - 7)$

Find the vertex, focus, and directrix of the parabola. Graph the equation.

15) $(y - 1)^2 = -(x + 3)$

A)

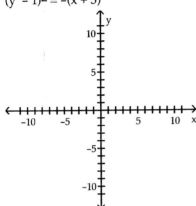

vertex: (-1, 3)
focus: (-1.25, 3)
directrix: x = -0.75

B)

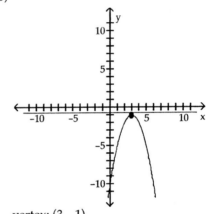

vertex: (3, -1)
focus: (3, -1.25)
directrix: y = -0.75

C)

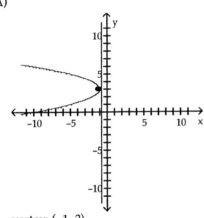

vertex: (-3, 1)
focus: (-3, 0.75)
directrix: y = 1.25

D)

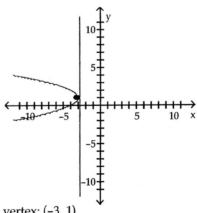

vertex: (-3, 1)
focus: (-3.25, 1)
directrix: x = -2.75

16) $(x-1)^2 = -4(y-2)$

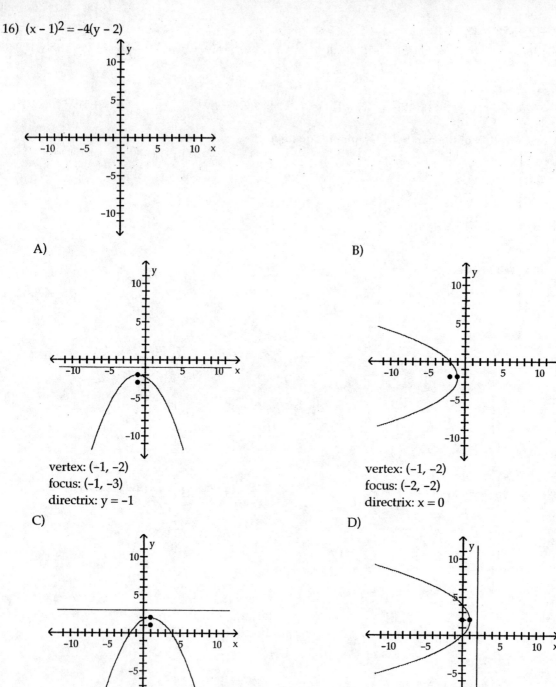

A) vertex: (−1, −2)
focus: (−1, −3)
directrix: y = −1

B) vertex: (−1, −2)
focus: (−2, −2)
directrix: x = 0

C) vertex: (1, 2)
focus: (1, 1)
directrix: y = 3

D) vertex: (1, 2)
focus: (0, 2)
directrix: x = 2

17) $x^2 - 6x = 8y - 49$

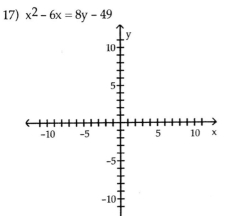

A)
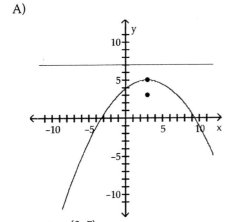
vertex: (3, 5)
focus: (3, 3)
directrix: $y = 7$

B)
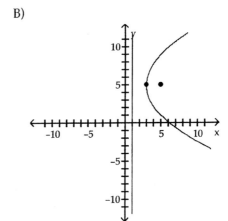
vertex: (3, 5)
focus: (5, 5)
directrix: $x = 1$

C)
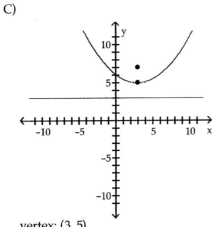
vertex: (3, 5)
focus: (3, 7)
directrix: $y = 3$

D)
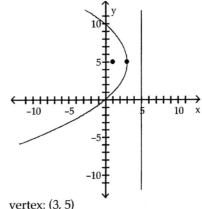
vertex: (3, 5)
focus: (1, 5)
directrix: $x = 5$

18) $y^2 + 14y = 4x - 25$

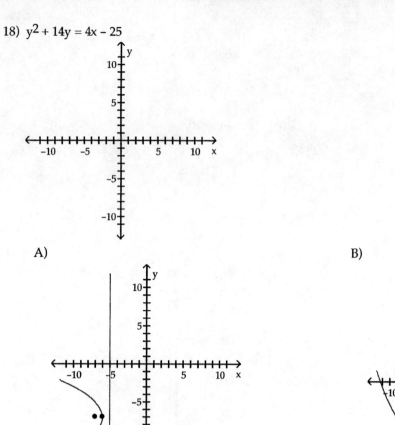

A)
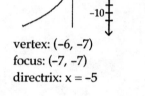
vertex: (-6, -7)
focus: (-7, -7)
directrix: x = -5

B)
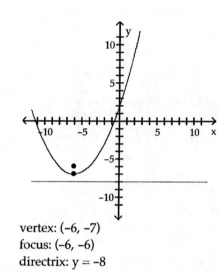
vertex: (-6, -7)
focus: (-6, -6)
directrix: y = -8

C)
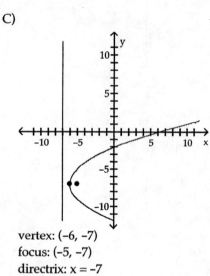
vertex: (-6, -7)
focus: (-5, -7)
directrix: x = -7

D)
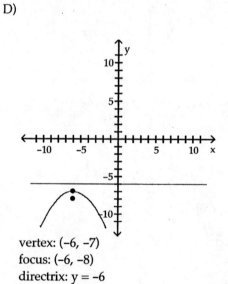
vertex: (-6, -7)
focus: (-6, -8)
directrix: y = -6

3 Solve Applied Problems Involving Parabolas

Solve the problem.

1) A reflecting telescope contains a mirror shaped like a paraboloid of revolution. If the mirror is 18 inches across at its opening and is 5 feet deep, where will the light be concentrated?
   A) 0.7 inches from the vertex
   B) 4.1 inches from the vertex
   C) 1.4 inches from the vertex
   D) 0.3 inches from the vertex

2) A searchlight is shaped like a paraboloid of revolution. If the light source is located 5 feet from the base along the axis of symmetry and the opening is 14 feet across, how deep should the searchlight be?
   A) 12.3 feet   B) 9.8 feet   C) 0.9 feet   D) 2.5 feet

3) A bridge is built in the shape of a parabolic arch. The bridge arch has a span of 166 feet and a maximum height of 40 feet. Find the height of the arch at 25 feet from its center.
   A) 14.5 feet   B) 32.4 feet   C) 0.9 feet   D) 36.4 feet

4) A reflecting telescope has a mirror shaped like a paraboloid of revolution. If the distance of the vertex to the focus is 30 feet and the distance across the top of the mirror is 50 inches, how deep is the mirror in the center?
   A) $\frac{125}{3456}$ in.   B) $\frac{125}{24}$ in.   C) $\frac{125}{288}$ in.   D) 9 in.

5) An experimental model for a suspension bridge is built in the shape of a parabolic arch. In one section, cable runs from the top of one tower down to the roadway, just touching it there, and up again to the top of a second tower. The towers are both 12.25 inches tall and stand 70 inches apart. Find the vertical distance from the roadway to the cable at a point on the road 14 inches from the lowest point of the cable.
   A) 1.96 in.   B) 1.76 in.   C) 7.84 in.   D) 2.16 in.

6) An experimental model for a suspension bridge is built in the shape of a parabolic arch. In one section, cable runs from the top of one tower down to the roadway, just touching it there, and up again to the top of a second tower. The towers are both 9 inches tall and stand 60 inches apart. At some point along the road from the lowest point of the cable, the cable is 1.44 inches above the roadway. Find the distance between that point and the base of the nearest tower.
   A) 11.8 in.   B) 18 in.   C) 12.2 in.   D) 18.2 in.

7) An experimental model for a suspension bridge is built in the shape of a parabolic arch. In one section, cable runs from the top of one tower down to the roadway, just touching it there, and up again to the top of a second tower. The towers stand 50 inches apart. At a point between the towers and 17.5 inches along the road from the base of one tower, the cable is 0.56 inches above the roadway. Find the height of the towers.
   A) 8.25 in.   B) 5.75 in.   C) 6.25 in.   D) 6.75 in.

8) A satellite dish is shaped like a paraboloid of revolution. The signals that emanate from a satellite strike the surface of the dish and are reflected to a single point, where the receiver is located. If the dish is 8 feet across at its opening and is 2 feet deep at its center, at what position should the receiver be placed?

9) A sealed-beam headlight is in the shape of a paraboloid of revolution. The bulb, which is placed at the focus, is 3 centimeters from the vertex. If the depth is to be 6 centimeters, what is the diameter of the headlight at its opening?

10) A spotlight has a parabolic cross section that is 6 ft wide at the opening and 2.5 ft deep at the vertex. How far from the vertex is the focus? Round answer to two decimal places.

A) 0.21 ft B) 0.52 ft C) 0.26 ft D) 0.90 ft

9.3 The Ellipse

1 Work with Ellipses with Center at the Origin

Match the graph to the equation.

1)

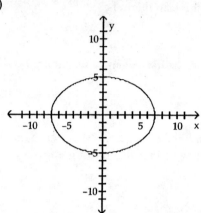

A) $\dfrac{y^2}{25} + \dfrac{x^2}{49} = 1$ B) $\dfrac{x^2}{25} + \dfrac{y^2}{49} = 1$ C) $\dfrac{x^2}{49} - \dfrac{y^2}{25} = 1$ D) $\dfrac{y^2}{25} - \dfrac{x^2}{49} = 1$

2)

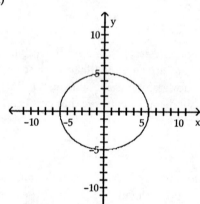

A) $\dfrac{y^2}{25} - \dfrac{x^2}{36} = 1$ B) $\dfrac{y^2}{25} + \dfrac{x^2}{36} = 1$ C) $\dfrac{x^2}{25} + \dfrac{y^2}{36} = 1$ D) $-\dfrac{y^2}{25} + \dfrac{x^2}{36} = 1$

Find the center, foci, and vertices of the ellipse.

3) $\dfrac{x^2}{49} + \dfrac{y^2}{25} = 1$

A) center at (0, 0)
foci at $(-2\sqrt{6}, 0)$ and $(2\sqrt{6}, 0)$
vertices at (-7, 0), (7, 0)

B) center at (0, 0)
foci at (0, -5) and (0, 5)
vertices at (0, -25), (0, 25)

C) center at (0, 0)
foci at $(0, -2\sqrt{6})$ and $(0, 2\sqrt{6})$
vertices at (0, -7), (0, 7)

D) center at (0, 0)
foci at (-7, 0) and (7, 0)
vertices at (-49, 0), (49, 0)

Precalculus Enhanced with Graphing Utilities 700

4) $\dfrac{x^2}{4} + \dfrac{y^2}{36} = 1$

   A) center at (0, 0)
      foci at $(-4\sqrt{2}, 0)$ and $(4\sqrt{2}, 0)$
      vertices at (-6, 0), (6, 0)

   B) center at (0, 0)
      foci at $(0, -4\sqrt{2})$ and $(0, 4\sqrt{2})$
      vertices at (0, -6), (0, 6)

   C) center at (0, 0)
      foci at (0, 6) and (2, 0)
      vertices at (0, 36), (4, 0)

   D) center at (0, 0)
      foci at (0, -6) and (0, 6)
      vertices at (0, -36), (0, 36)

5) $4x^2 + 81y^2 = 324$

   A) center at (0, 0)
      foci at $(-\sqrt{77}, 0)$ and $(\sqrt{77}, 0)$
      vertices at (-9, 0), (9, 0)

   B) center at (0, 0)
      foci at $(0, -\sqrt{77})$ and $(0, \sqrt{77})$
      vertices at (0, -9), (0, 9)

   C) center at (0, 0)
      foci at (0, -2) and (0, 2)
      vertices at (0, -4), (0, 4)

   D) center at (0, 0)
      foci at (-9, 0) and (9, 0)
      vertices at (-81, 0), (81, 0)

6) $25x^2 + 4y^2 = 100$

   A) center at (0, 0)
      foci at (0, 5) and (2, 0)
      vertices at (0, 25) and (4, 0)

   B) center at (0, 0)
      foci at (0, -5) and (0, 5)
      vertices at (0, -25), (0, 25)

   C) center at (0, 0)
      foci at $(0, -\sqrt{21})$ and $(0, \sqrt{21})$
      vertices at (0, -5), (0, 5)

   D) center at (0, 0)
      foci at $(-\sqrt{21}, 0)$ and $(\sqrt{21}, 0)$
      vertices at (-5, 0), (5, 0)

Find an equation for the ellipse satisfying the given conditions.

7) Foci: (-4, 0), (4, 0)   vertices: (-5, 0), (5, 0)

   A) $\dfrac{x^2}{16} + \dfrac{y^2}{25} = 1$
   B) $\dfrac{x^2}{9} + \dfrac{y^2}{25} = 1$
   C) $\dfrac{x^2}{16} + \dfrac{y^2}{9} = 1$
   D) $\dfrac{x^2}{25} + \dfrac{y^2}{9} = 1$

8) Foci: (-2, 0), (2, 0)   x-intercepts: (-6, 0), (6, 0)

   A) $\dfrac{x^2}{4} + \dfrac{y^2}{36} = 1$
   B) $\dfrac{x^2}{36} + \dfrac{y^2}{32} = 1$
   C) $\dfrac{x^2}{32} + \dfrac{y^2}{36} = 1$
   D) $\dfrac{x^2}{4} + \dfrac{y^2}{32} = 1$

9) Foci: (0, -7), (0, 7)   vertices: (0, -8), (0, 8)

   A) $\dfrac{x^2}{15} + \dfrac{y^2}{64} = 1$
   B) $\dfrac{x^2}{49} + \dfrac{y^2}{15} = 1$
   C) $\dfrac{x^2}{64} + \dfrac{y^2}{15} = 1$
   D) $\dfrac{x^2}{49} + \dfrac{y^2}{64} = 1$

10) Foci: (0, -5), (0, 5)   y-intercepts: (0, -7), (0, 7)

    A) $\dfrac{x^2}{49} + \dfrac{y^2}{24} = 1$
    B) $\dfrac{x^2}{25} + \dfrac{y^2}{24} = 1$
    C) $\dfrac{x^2}{25} + \dfrac{y^2}{49} = 1$
    D) $\dfrac{x^2}{24} + \dfrac{y^2}{49} = 1$

11) Major axis horizontal with length 18; length of minor axis = 16; center (0, 0)

    A) $\dfrac{x^2}{324} + \dfrac{y^2}{256} = 1$
    B) $\dfrac{x^2}{18} + \dfrac{y^2}{64} = 1$
    C) $\dfrac{x^2}{64} + \dfrac{y^2}{81} = 1$
    D) $\dfrac{x^2}{81} + \dfrac{y^2}{64} = 1$

12) Major axis vertical with length 12; length of minor axis = 6; center (0, 0)

A) $\dfrac{x^2}{36} + \dfrac{y^2}{144} = 1$ B) $\dfrac{x^2}{9} + \dfrac{y^2}{36} = 1$ C) $\dfrac{x^2}{6} + \dfrac{y^2}{36} = 1$ D) $\dfrac{x^2}{36} + \dfrac{y^2}{9} = 1$

13) Center at (0, 0); focus at (5, 0); vertex at (7, 0)

A) $\dfrac{x^2}{25} + \dfrac{y^2}{49} = 1$ B) $\dfrac{x^2}{49} + \dfrac{y^2}{24} = 1$ C) $\dfrac{x^2}{49} - \dfrac{y^2}{24} = 1$ D) $\dfrac{x^2}{24} + \dfrac{y^2}{7} = 1$

14) Center at (0, 0); focus at (0, 3); vertex at (0, 5)

A) $\dfrac{x^2}{16} + \dfrac{y^2}{25} = 1$ B) $\dfrac{x^2}{16} - \dfrac{y^2}{25} = 1$ C) $\dfrac{x^2}{3} + \dfrac{y^2}{5} = 1$ D) $\dfrac{x^2}{25} + \dfrac{y^2}{16} = 1$

Graph the ellipse and locate the foci.

15) $\dfrac{x^2}{9} + \dfrac{y^2}{4} = 1$

A) foci at $(0, \sqrt{5})$ and $(0, -\sqrt{5})$

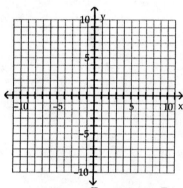

B) foci at $(2\sqrt{3}, 0)$ and $(-2\sqrt{3}, 0)$

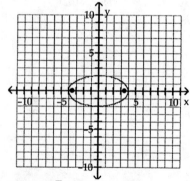

C) foci at $(\sqrt{13}, 0)$ and $(-\sqrt{13}, 0)$

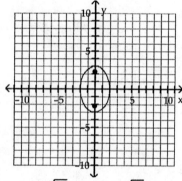

D) foci at $(\sqrt{5}, 0)$ and $(-\sqrt{5}, 0)$

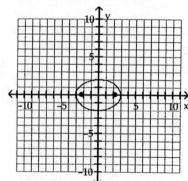

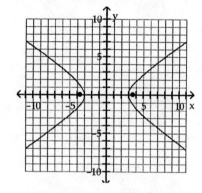

16) $\dfrac{x^2}{4} + \dfrac{y^2}{16} = 1$

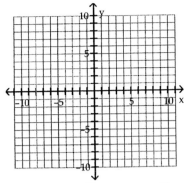

A) foci at $(2\sqrt{3}, 0)$ and $(-2\sqrt{3}, 0)$

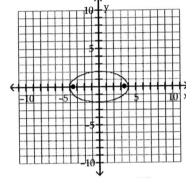

B) foci at $(0, 2\sqrt{3})$ and $(0, -2\sqrt{3})$

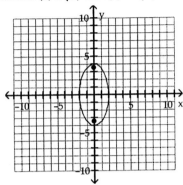

C) foci at $(\sqrt{21}, 0)$ and $(-\sqrt{21}, 0)$

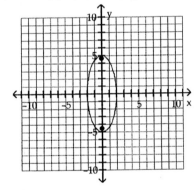

D) foci at $(2\sqrt{5}, 0)$ and $(-2\sqrt{5}, 0)$

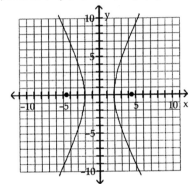

17) $9x^2 + 16y^2 = 144$

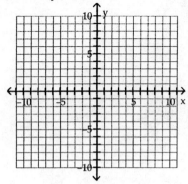

A) foci at (5, 0) and (−5, 0)

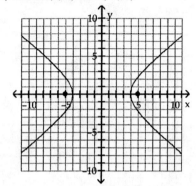

B) foci at $(0, \sqrt{7})$ and $(0, -\sqrt{7})$

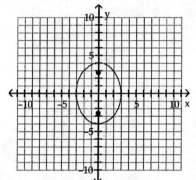

C) foci at (4, 0) and (−4, 0)

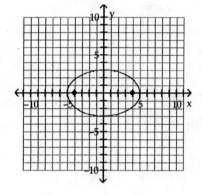

D) foci at $(\sqrt{7}, 0)$ and $(-\sqrt{7}, 0)$

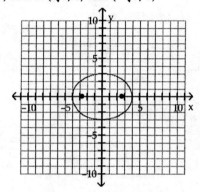

18) $16x^2 + 4y^2 = 64$

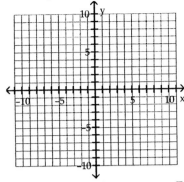

A) foci at $(0, 2\sqrt{3})$ and $(0, -2\sqrt{3})$

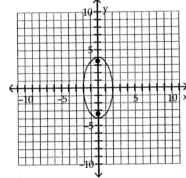

B) foci at $(2\sqrt{5}, 0)$ and $(-2\sqrt{5}, 0)$

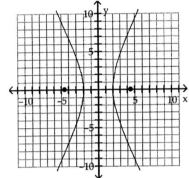

C) foci at $(2\sqrt{3}, 0)$ and $(-2\sqrt{3}, 0)$

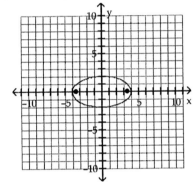

D) foci at $(\sqrt{21}, 0)$ and $(-\sqrt{21}, 0)$

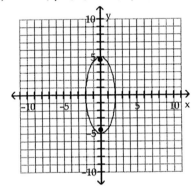

2 Work with Ellipses with Center at (h, k)

Write an equation for the graph.

1)
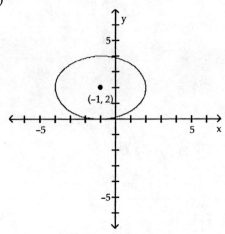

A) $\dfrac{(x+1)^2}{4} + \dfrac{(y-2)^2}{9} = 1$

B) $\dfrac{(x+1)^2}{9} + \dfrac{(y-2)^2}{4} = 1$

C) $\dfrac{(x-1)^2}{9} + \dfrac{(y+2)^2}{4} = 1$

D) $\dfrac{(x-2)^2}{9} + \dfrac{(y+1)^2}{4} = 1$

Graph the equation.

2) $\dfrac{(x-1)^2}{9} + \dfrac{(y+1)^2}{4} = 1$

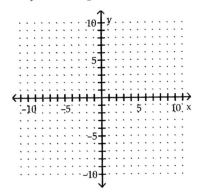

A)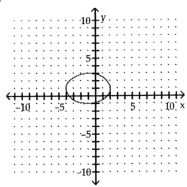

B)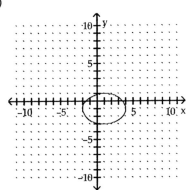

C)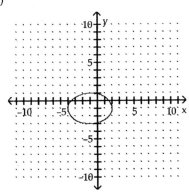

D)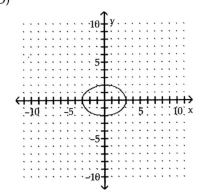

3) $\dfrac{(x+1)^2}{4} + \dfrac{(y+2)^2}{9} = 1$

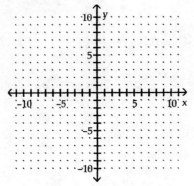

A)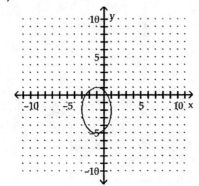

B)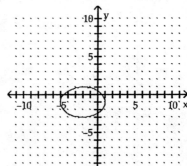

C)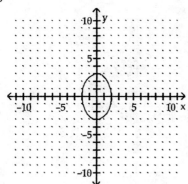

D)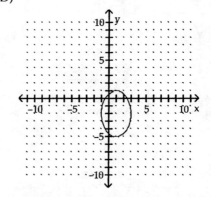

4) $4(x+1)^2 + 16(y-1)^2 = 64$

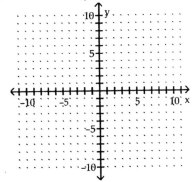

A)
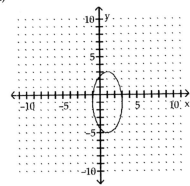

B)
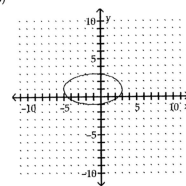

C)
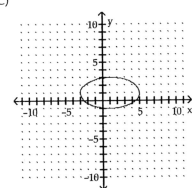

D)
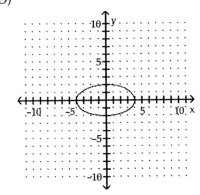

5) $9(x-2)^2 + 4(y+1)^2 = 36$

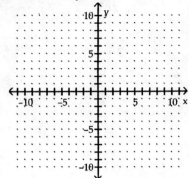

A)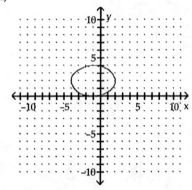

B)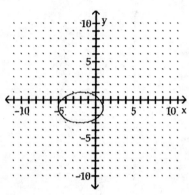

C)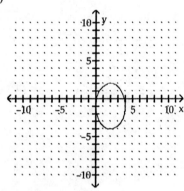

D)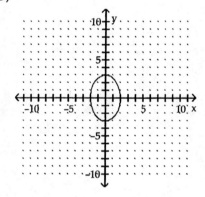

Find the center, foci, and vertices of the ellipse.

6) $\dfrac{(x+2)^2}{16} + \dfrac{(y-1)^2}{9} = 1$

    A) center at $(1, -2)$
        foci at $(1+\sqrt{7}, -2)$, $(1-\sqrt{7}, -2)$
        vertices at $(-6, 1)$, $(2, 1)$

    B) center at $(-2, 1)$
        foci at $(-2+\sqrt{7}, 1)$, $(-2-\sqrt{7}, 1)$
        vertices at $(-6, 1)$, $(2, 1)$

    C) center at $(-2, 1)$
        foci at $(-\sqrt{7}, 1)$, $(\sqrt{7}, 1)$
        vertices at $(4, 1)$, $(-4, 1)$

    D) center at $(-2, 1)$
        foci at $(-2+\sqrt{7}, -2)$, $(-2-\sqrt{7}, -2)$
        vertices at $(4, 1)$, $(-4, 1)$

7) $36(x - 2)^2 + 25(y + 2)^2 = 900$

   A) center at (3, -2)
      foci at $(3, -2 - \sqrt{11})$, $(3, -2 + \sqrt{11})$
      vertices at (3, 4), (3, -8)

   B) center at (2, -2)
      foci at $(2, -2 - \sqrt{11})$, $(2, -2 + \sqrt{11})$
      vertices at (2, 4), (2, -8)

   C) center at (-2, 2)
      foci at $(-2, 2 - \sqrt{11})$, $(-2, 2 + \sqrt{11})$
      vertices at (-2, 4), (-2, -8)

   D) center at (-2, -2)
      foci at $(-2, -2 - \sqrt{11})$, $(-2, -2 + \sqrt{11})$
      vertices at (-2, 4), (-2, -8)

8) $2x^2 + 3y^2 - 24x + 36y + 174 = 0$

   A) $\dfrac{(x - 6)^2}{2} + \dfrac{(y + 6)^2}{3} = 1$
      center: (6, -6); foci: (7, -6), (5, -6); vertices: (7.7, -6), (4.3, -6)

   B) $\dfrac{(x - 6)^2}{3} + \dfrac{(y + 6)^2}{2} = 1$
      center: (6, -6); foci: (7, -6), (5, -6); vertices: (7.7, -6), (4.3, -6)

   C) $\dfrac{(x - 6)^2}{3} + \dfrac{(y + 6)^2}{2} = 1$
      center: (-6, 6); foci: (-5, 6), (-7, 6); vertices: (-7.7, 6), (-4.3, 6)

   D) $\dfrac{(x - 6)^2}{2} + \dfrac{(y + 6)^2}{3} = 1$
      center: (-6, 6); foci: (-5, 6), (-7, 6); vertices: (-7.7, 6), (-4.3, 6)

9) $64x^2 + y^2 - 512x + 960 = 0$

   A) $\dfrac{x^2}{64} + (y - 4)^2 = 1$
      center: (4, 0); foci: $(4, 3\sqrt{7})$, $(4, -3\sqrt{7})$; vertices: (4, 8), (4, -8)

   B) $(x - 8)^2 + \dfrac{y^2}{16} = 1$
      center: (8, 0); foci: $(8, \sqrt{15})$, $(8, -\sqrt{15})$; vertices: (8, 4), (8, -4)

   C) $\dfrac{x^2}{16} + (y - 8)^2 = 1$
      center: (8, 0); foci: $(8, \sqrt{15})$, $(8, -3\sqrt{7})$; vertices: (8, 4), (8, -4)

   D) $(x - 4)^2 + \dfrac{y^2}{64} = 1$
      center: (4, 0); foci: $(4, 3\sqrt{7})$, $(4, -3\sqrt{7})$; vertices: (4, 8), (4, -8)

Find an equation for the ellipse satisfying the stated conditions.

10) Center at (6, 6); focus at (9, 6); vertex at (11, 6)

   A) $\dfrac{(x + 6)^2}{9} - \dfrac{(y - 6)^2}{13} = 1$

   B) $\dfrac{(x - 6)^2}{64} + \dfrac{(y + 6)^2}{7} = 2$

   C) $\dfrac{(x + 6)^2}{25} + \dfrac{(y + 6)^2}{16} = 1$

   D) $\dfrac{(x - 6)^2}{25} + \dfrac{(y - 6)^2}{16} = 1$

11) Vertices at (−2, 4) and (12, 4) ; focus at (10, 4)

A) $\dfrac{(x+5)^2}{25} + \dfrac{(y+4)^2}{24} = 1$

B) $\dfrac{(x-4)^2}{36} + \dfrac{(y-5)^2}{23} = 1$

C) $\dfrac{(x-5)^2}{49} + \dfrac{(y-4)^2}{24} = 1$

D) $\dfrac{(x-5)^2}{81} - \dfrac{(y+4)^2}{29} = 1$

Find an equation for the ellipse satisfying the given conditions. Graph the equation.

12) Foci at (2, −1) and (2, −7); length of major axis is 10

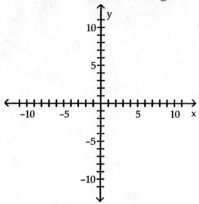

A)
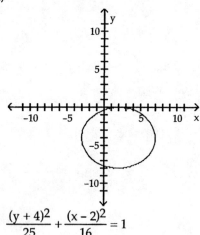

$\dfrac{(y+4)^2}{25} + \dfrac{(x-2)^2}{16} = 1$

B)
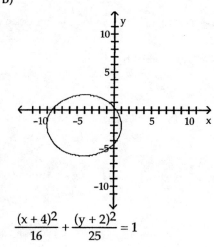

$\dfrac{(x+4)^2}{16} + \dfrac{(y+2)^2}{25} = 1$

C)
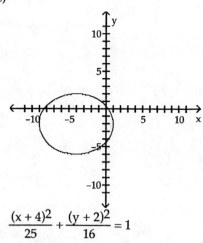

$\dfrac{(x+4)^2}{25} + \dfrac{(y+2)^2}{16} = 1$

D)
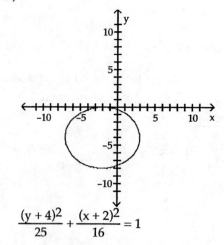

$\dfrac{(y+4)^2}{25} + \dfrac{(x+2)^2}{16} = 1$

13) Foci at (−7, −4) and (−1, −4); length of major axis is 10

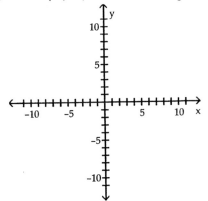

A)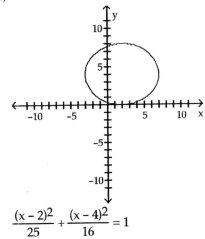

$$\frac{(x-2)^2}{25} + \frac{(x-4)^2}{16} = 1$$

B)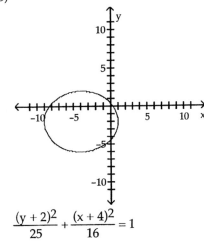

$$\frac{(y+2)^2}{25} + \frac{(x+4)^2}{16} = 1$$

C)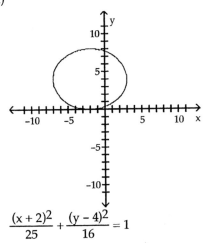

$$\frac{(x+2)^2}{25} + \frac{(y-4)^2}{16} = 1$$

D)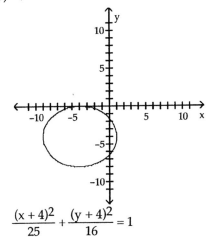

$$\frac{(x+4)^2}{25} + \frac{(y+4)^2}{16} = 1$$

14) Vertices at (5, −4) and (5, 8); length of minor axis is 6

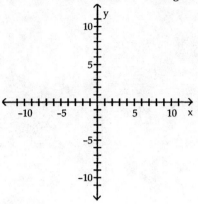

A)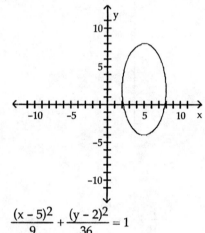

$$\frac{(x-5)^2}{9} + \frac{(y-2)^2}{36} = 1$$

B)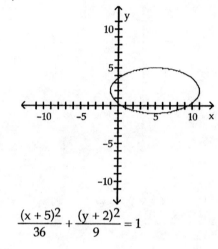

$$\frac{(x+5)^2}{36} + \frac{(y+2)^2}{9} = 1$$

C)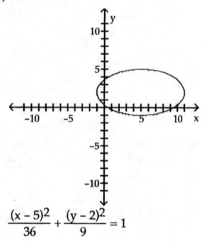

$$\frac{(x-5)^2}{36} + \frac{(y-2)^2}{9} = 1$$

D)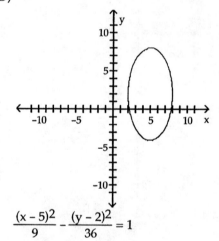

$$\frac{(x-5)^2}{9} - \frac{(y-2)^2}{36} = 1$$

15) Foci at (–1, 4), and (–5, 4); vertex at (–6, 4)

A) $\dfrac{(x-4)^2}{9} + \dfrac{(y+3)^2}{5} = 1$

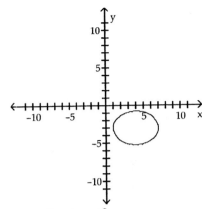

B) $\dfrac{(x+3)^2}{9} + \dfrac{(y-4)^2}{5} = 1$

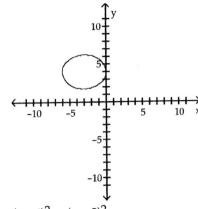

C) $\dfrac{(x+3)^2}{5} + \dfrac{(y-4)^2}{9} = 1$

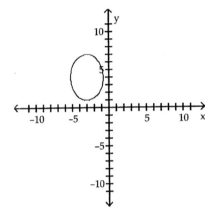

D) $\dfrac{(x-4)^2}{5} + \dfrac{(y+3)^2}{9} = 1$

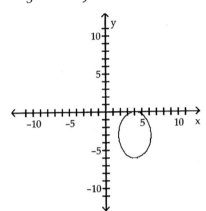

16) center at (-4, 5); focus at (-9, 5); contains the point (-10, 5)

A) $\dfrac{(x+4)^2}{11} + \dfrac{(y-5)^2}{36} = 1$

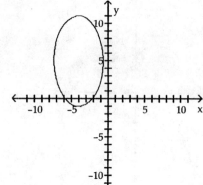

B) $\dfrac{(x+4)^2}{36} + \dfrac{(y-5)^2}{11} = 1$

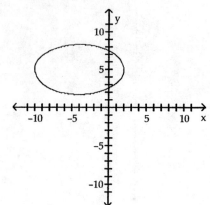

C) $\dfrac{(x+5)^2}{11} + \dfrac{(y-4)^2}{36} = 1$

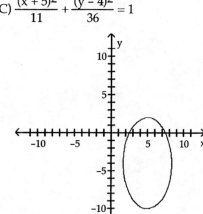

D) $\dfrac{(x+5)^2}{36} + \dfrac{(y-4)^2}{11} = 1$

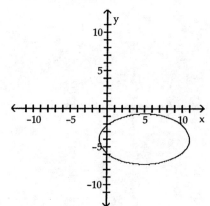

3 Solve Applied Problems Involving Ellipses

Solve the problem.

1) A bridge is built in the shape of a semielliptical arch. It has a span of 110 feet. The height of the arch 30 feet from the center is to be 12 feet. Find the height of the arch at its center.

   A) 30.74 feet   B) 22 feet   C) 12.47 feet   D) 14.32 feet

2) An arch for a bridge over a highway is in the form of a semiellipse. The top of the arch is 30 feet above ground (the major axis). What should the span of the bridge be (the length of its minor axis) if the height 28 feet from the center is to be 10 feet above ground?

   A) 29.7 feet   B) 55.71 feet   C) 59.4 feet   D) 168 feet

3) The orbit of a planet around a sun is an ellipse with the sun at one focus. The aphelion of a planet is its greatest distance from the sun, its perihelion is its shortest distance, and its mean distance is the length of the semimajor axis of the elliptical orbit. If a planet has a perihelion of 461.4 million miles and a mean distance of 463 million miles, write an equation for the orbit of the planet around the sun.

   A) $\dfrac{x^2}{463^2} + \dfrac{y^2}{1.6^2} = 1$

   B) $\dfrac{x^2}{463^2} + \dfrac{y^2}{462.997^2} = 1$

   C) $\dfrac{x^2}{463^2} + \dfrac{y^2}{461.4^2} = 1$

   D) $\dfrac{x^2}{463.003^2} + \dfrac{y^2}{463^2} = 1$

4) An arch in the form of a semiellipse is 52 ft wide at the base and has a height of 20 ft. How wide is the arch at a height of 12 ft above the base?

   A) 35.5 feet   B) 41.6 feet   C) 20.8 feet   D) 17.7 feet

5) A hall 130 feet in length was designed as a whispering gallery. If the ceiling is 25 feet high at the center, how far from the center are the foci located?

6) A race track is in the shape of an ellipse 80 feet long and 60 feet wide. What is the width 32 feet from the center?

9.4 The Hyperbola

1 Work with Hyperbolas with Center at the Origin

Write an equation for the hyperbola.

1)

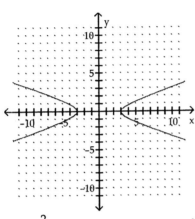

A) $x^2 - \dfrac{y^2}{9} = 1$ B) $y^2 - \dfrac{x^2}{9} = 1$ C) $\dfrac{x^2}{9} - y^2 = 1$ D) $\dfrac{y^2}{9} - x^2 = 1$

2)

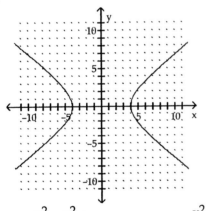

A) $\dfrac{y^2}{9} - \dfrac{x^2}{16} = 1$ B) $\dfrac{y^2}{16} - \dfrac{x^2}{9} = 1$ C) $\dfrac{x^2}{9} - \dfrac{y^2}{16} = 1$ D) $\dfrac{x^2}{16} - \dfrac{y^2}{9} = 1$

3)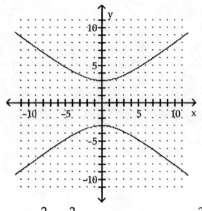

A) $\dfrac{y^2}{9} - \dfrac{x^2}{16} = 1$ B) $\dfrac{x^2}{9} - \dfrac{y^2}{16} = 1$ C) $\dfrac{y^2}{16} - \dfrac{x^2}{9} = 1$ D) $\dfrac{x^2}{16} - \dfrac{y^2}{9} = 1$

Match the equation to the graph.

4) $\dfrac{x^2}{4} - \dfrac{y^2}{9} = 1$

A)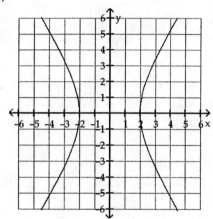

B)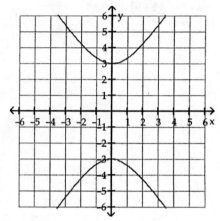

C)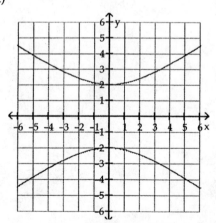

D)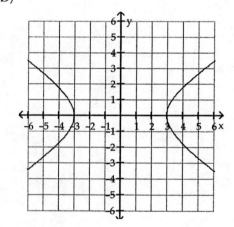

5) $\dfrac{y^2}{4} - \dfrac{x^2}{9} = 1$

A)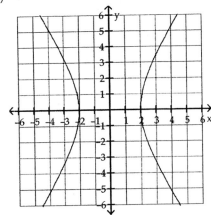

B)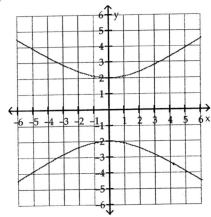

C)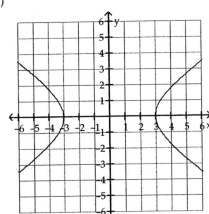

D)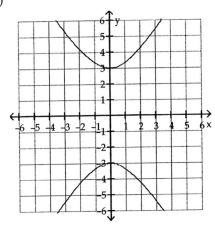

Find an equation for the hyperbola satisfying the stated conditions.

6) Vertices at $(0, \pm 10)$; asymptotes at $y = \pm \dfrac{5}{9}x$

   A) $\dfrac{y^2}{100} - \dfrac{x^2}{81} = 1$
   B) $\dfrac{y^2}{81} - \dfrac{x^2}{25} = 1$
   C) $\dfrac{y^2}{324} - \dfrac{x^2}{100} = 1$
   D) $\dfrac{y^2}{100} - \dfrac{x^2}{324} = 1$

7) Vertices at $(\pm 6, 0)$; foci at $(\pm 7, 0)$

   A) $\dfrac{x^2}{36} - \dfrac{y^2}{13} = 1$
   B) $\dfrac{x^2}{36} - \dfrac{y^2}{49} = 1$
   C) $\dfrac{x^2}{13} - \dfrac{y^2}{36} = 1$
   D) $\dfrac{x^2}{49} - \dfrac{y^2}{36} = 1$

Find an equation for the hyperbola satisfying the stated conditions. Graph the equation.

8) Center at (0, 0); focus at ($\sqrt{34}$, 0); vertex at (5, 0)

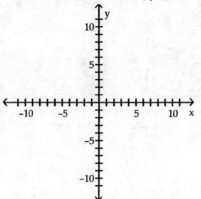

A)
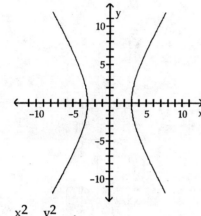
$$\frac{x^2}{9} - \frac{y^2}{25} = 1$$

B)
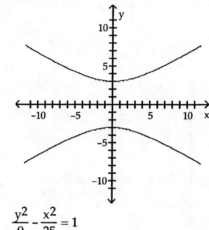
$$\frac{y^2}{9} - \frac{x^2}{25} = 1$$

C)
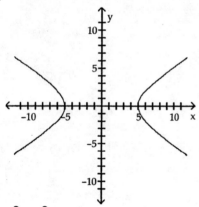
$$\frac{x^2}{25} - \frac{y^2}{9} = 1$$

D)
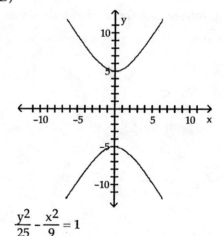
$$\frac{y^2}{25} - \frac{x^2}{9} = 1$$

Precalculus Enhanced with Graphing Utilities 720

9) Center at (0, 0); vertex at (0, 8); focus at (0, 10)

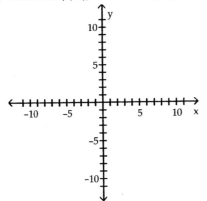

A)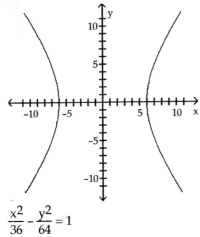

$$\frac{x^2}{36} - \frac{y^2}{64} = 1$$

B)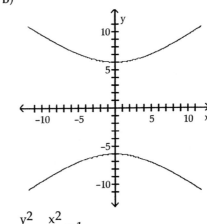

$$\frac{y^2}{36} - \frac{x^2}{64} = 1$$

C)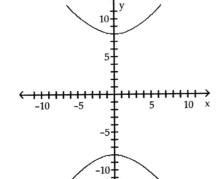

$$\frac{y^2}{64} - \frac{x^2}{36} = 1$$

D)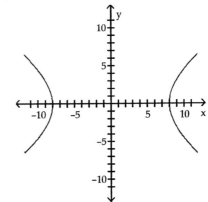

$$\frac{x^2}{64} - \frac{y^2}{36} = 1$$

Find the center, transverse axis, vertices, foci, and asymptotes of the hyperbola.

10) $\dfrac{x^2}{64} - \dfrac{y^2}{121} = 1$

A) center at (0, 0)
   transverse axis is x-axis
   vertices at (-11, 0) and (11, 0)
   foci at $(-\sqrt{185}, 0)$ and $(\sqrt{185}, 0)$
   asymptotes of $y = -\dfrac{11}{8}$ and $y = \dfrac{11}{8}$

B) center at (0, 0)
   transverse axis is x-axis
   vertices at (-8, 0) and (8, 0)
   foci at (-11, 0) and (11, 0)
   asymptotes of $y = -\dfrac{11}{8}$ and $y = \dfrac{11}{8}$

C) center at (0, 0)
   transverse axis is y-axis
   vertices at (0, -8) and (0, 8)
   foci at $(-\sqrt{185}, 0)$ and $(\sqrt{185}, 0)$
   asymptotes of $y = -\dfrac{11}{8}$ and $y = \dfrac{11}{8}$

D) center at (0, 0)
   transverse axis is x-axis
   vertices at (-8, 0) and (8, 0)
   foci at $(-\sqrt{185}, 0)$ and $(\sqrt{185}, 0)$
   asymptotes of $y = -\dfrac{11}{8}$ and $y = \dfrac{11}{8}$

11) $9y^2 - 100x^2 = 900$

A) center at (0, 0)
   transverse axis is y-axis
   vertices at (0, -10) and (0, 10)
   foci at $(0, -\sqrt{109})$ and $(0, \sqrt{109})$
   asymptotes of $y = -\dfrac{10}{3}$ and $y = \dfrac{10}{3}$

B) center at (0, 0)
   transverse axis is x-axis
   vertices: (-3, 0), (3, 0)
   foci: $(-\sqrt{109}, 0)$, $(\sqrt{109}, 0)$
   asymptotes of $y = -\dfrac{10}{3}$ and $y = \dfrac{10}{3}$

C) center at (0, 0)
   transverse axis is x-axis
   vertices: (-10, 0), (10, 0)
   foci: (-3, 0), (3, 0)
   asymptotes of $y = -\dfrac{10}{3}$ and $y = \dfrac{10}{3}$

D) center at (0, 0)
   transverse axis is y-axis
   vertices: (0, -10), (0, 10)
   foci: $(-\sqrt{109}, 0)$, $(\sqrt{109}, 0)$
   asymptotes of $y = -\dfrac{10}{3}$ and $y = \dfrac{10}{3}$

Write an equation for the hyperbola.

12)

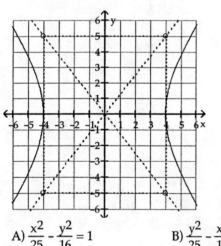

A) $\dfrac{x^2}{25} - \dfrac{y^2}{16} = 1$ B) $\dfrac{y^2}{25} - \dfrac{x^2}{16} = 1$ C) $\dfrac{y^2}{16} - \dfrac{x^2}{25} = 1$ D) $\dfrac{x^2}{16} - \dfrac{y^2}{25} = 1$

13)

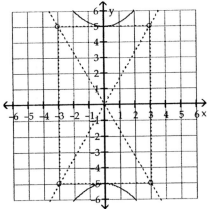

A) $\dfrac{x^2}{9} - \dfrac{y^2}{25} = 1$ B) $\dfrac{x^2}{25} - \dfrac{y^2}{9} = 1$ C) $\dfrac{y^2}{9} - \dfrac{x^2}{25} = 1$ D) $\dfrac{y^2}{25} - \dfrac{x^2}{9} = 1$

Graph the hyperbola.

14) $\dfrac{x^2}{9} - \dfrac{y^2}{16} = 1$

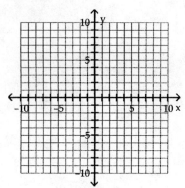

A)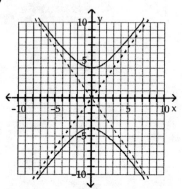

B)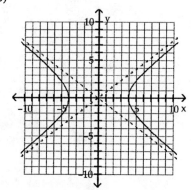

C)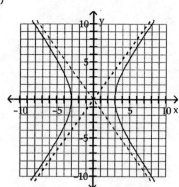

D)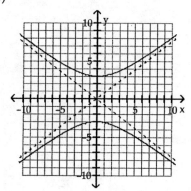

15) $\dfrac{y^2}{9} - \dfrac{x^2}{36} = 1$

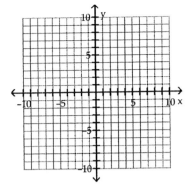

A)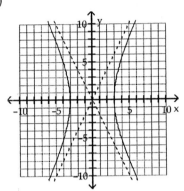

B)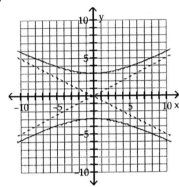

C)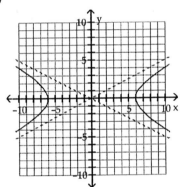

D)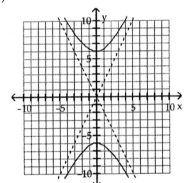

16) $36x^2 - 9y^2 = 324$

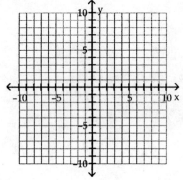

A)
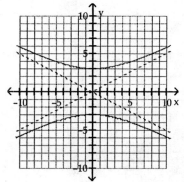

B)
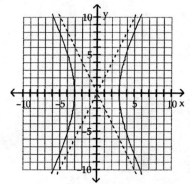

C)
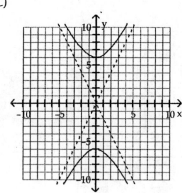

D)
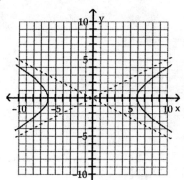

17) $25y^2 - 4x^2 = 100$

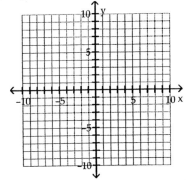

A)
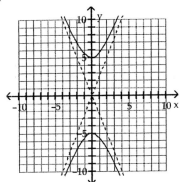

B)
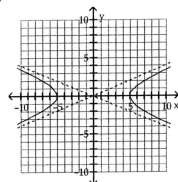

C)
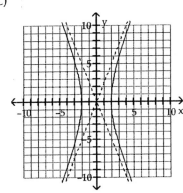

D)
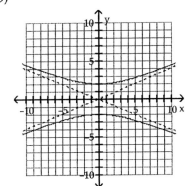

18) $36x^2 = 9y^2 + 324$

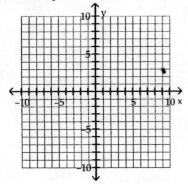

A)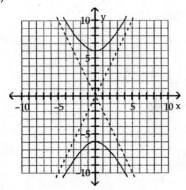

B)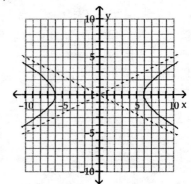

C)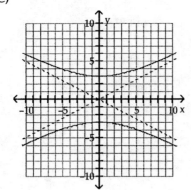

D)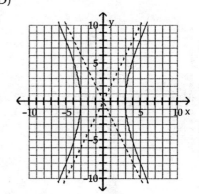

19) $9y^2 = 4x^2 + 36$

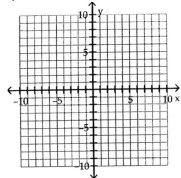

A)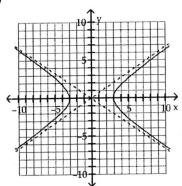

B)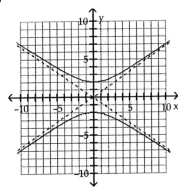

C)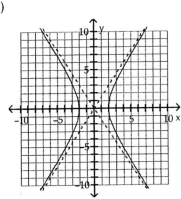

D)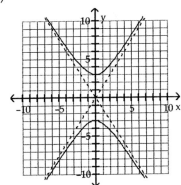

2 Find the Asymptotes of a Hyperbola

Find the asymptotes of the hyperbola.

1) $\dfrac{x^2}{25} - \dfrac{y^2}{16} = 1$

   A) $y = \dfrac{5}{4}x$ and $y = -\dfrac{5}{4}x$ 　　　　B) $y = \dfrac{25}{16}x$ and $y = -\dfrac{25}{16}x$

   C) $y = \dfrac{16}{25}x$ and $y = -\dfrac{16}{25}x$ 　　　D) $y = \dfrac{4}{5}x$ and $y = -\dfrac{4}{5}x$

2) $y^2 - x^2 = 25$

   A) $y = 5x$ and $y = -5x$ 　　　　　B) $y = \dfrac{1}{25}x$ and $y = -\dfrac{1}{25}x$

   C) $y = \dfrac{1}{5}x$ and $y = -\dfrac{1}{5}x$ 　　　　D) $y = x$ and $y = -x$

Precalculus Enhanced with Graphing Utilities 729

3) $\dfrac{(x-3)^2}{25} - \dfrac{(y-1)^2}{9} = 1$

A) $y - 1 = \dfrac{5}{3}(x - 3)$ and $y - 1 = -\dfrac{5}{3}(x - 3)$

B) $y - 3 = \dfrac{3}{5}(x - 1)$ and $y - 3 = -\dfrac{3}{5}(x - 1)$

C) $y - 1 = \dfrac{3}{5}(x - 3)$ and $y - 1 = -\dfrac{3}{5}(x - 3)$

D) $y = \dfrac{3}{5}(x - 3)$ and $y = -\dfrac{3}{5}(x - 3)$

4) $x^2 - y^2 + 6x + 4y - 20 = 0$

A) $y - 2 = (x + 3)$ and $y - 2 = -(x + 3)$

B) $y + 3 = (x - 2)$ and $y + 3 = -(x - 2)$

C) $y - 3 = (x + 2)$ and $y - 3 = -(x + 2)$

D) $y - 2 = \dfrac{1}{5}(x + 3)$ and $y - 2 = -\dfrac{1}{5}(x + 3)$

3 Work with Hyperbolas with Center at (h, k)

Find an equation for the hyperbola satisfying the stated conditions.

1) Vertices $(\dfrac{1}{2}, -3)$ and $(-\dfrac{9}{2}, -3)$;   asymptotes $y + 3 = \pm\dfrac{6}{5}(x + 2)$

A) $\dfrac{4(x - 2)^2}{25} - \dfrac{(y - 3)^2}{9} = 1$

B) $\dfrac{(x + 2)^2}{9} - \dfrac{4(y + 3)^2}{25} = 1$

C) $\dfrac{(y + 3)^2}{9} - \dfrac{4(x + 2)^2}{25} = 1$

D) $\dfrac{4(x + 2)^2}{25} - \dfrac{(y + 3)^2}{9} = 1$

2) center at (7, 6); focus at (2, 6); vertex at (6, 6)

A) $\dfrac{(x - 7)^2}{24} - (y - 6)^2 = 1$

B) $(x - 7)^2 - \dfrac{(y - 6)^2}{24} = 1$

C) $\dfrac{(x - 6)^2}{24} - (y - 7)^2 = 1$

D) $(x - 6)^2 - \dfrac{(y - 7)^2}{24} = 1$

3) Vertices at (0, ±6); asymptotes at $y = \pm\dfrac{3}{2}x$

A) $\dfrac{y^2}{16} - \dfrac{x^2}{36} = 1$

B) $\dfrac{y^2}{4} - \dfrac{x^2}{9} = 1$

C) $\dfrac{y^2}{36} - \dfrac{x^2}{4} = 1$

D) $\dfrac{y^2}{36} - \dfrac{x^2}{16} = 1$

4) Vertices at (±2, 0); foci at (±11, 0)

A) $\dfrac{x^2}{4} - \dfrac{y^2}{117} = 1$

B) $\dfrac{x^2}{117} - \dfrac{y^2}{4} = 1$

C) $\dfrac{x^2}{121} - \dfrac{y^2}{4} = 1$

D) $\dfrac{x^2}{4} - \dfrac{y^2}{121} = 1$

Find the center, transverse axis, vertices, foci, and asymptotes of the hyperbola.

5) $\dfrac{(x-3)^2}{25} - \dfrac{(y-4)^2}{9} = 1$

    A) center at (4, 3)
       transverse axis is parallel to x-axis
       vertices at (-1, 3) and (9, 3)
       foci at $(4 - \sqrt{34}, 3)$ and $(4 + \sqrt{34}, 3)$
       asymptotes of $y - 3 = -\dfrac{3}{5}(x - 4)$ and $y - 3 = \dfrac{3}{5}(x - 4)$

    B) center at (3, 4)
       transverse axis is parallel to x-axis
       vertices at (-2, 4) and (8, 4)
       foci at $(3 - \sqrt{34}, 4)$ and $(3 + \sqrt{34}, 4)$
       asymptotes of $y - 4 = -\dfrac{3}{5}(x - 3)$ and $y - 4 = \dfrac{3}{5}(x - 3)$

    C) center at (3, 4)
       transverse axis is parallel to x-axis
       vertices at (0, 4) and (6, 4)
       foci at $(3 - \sqrt{34}, 4)$ and $(3 + \sqrt{34}, 4)$
       asymptotes of $y - 4 = -\dfrac{5}{3}(x - 3)$ and $y - 4 = \dfrac{5}{3}(x - 3)$

    D) center at (3, 4)
       transverse axis is parallel to y-axis
       vertices at (3, -1) and (3, 9)
       foci at $(3, 4 - \sqrt{34})$ and $(3, 4 + \sqrt{34})$
       asymptotes of $y + 4 = -\dfrac{5}{3}(x + 3)$ and $y + 4 = \dfrac{5}{3}(x + 3)$

6) $(x - 2)^2 - 25(y - 3)^2 = 25$

A) center at (3, 2)
transverse axis is parallel to x-axis
vertices at (-2, 2) and (8, 2)
foci at $(3 - \sqrt{26}, 2)$ and $(3 + \sqrt{26}, 2)$
asymptotes of $y - 2 = -\frac{1}{5}(x - 3)$ and $y - 2 = \frac{1}{5}(x - 3)$

B) center at (2, 3)
transverse axis is parallel to x-axis
vertices at (1, 3) and (3, 3)
foci at $(2 - \sqrt{26}, 3)$ and $(2 + \sqrt{26}, 3)$
asymptotes of $y - 3 = -5(x - 2)$ and $y - 3 = 5(x - 2)$

C) center at (2, 3)
transverse axis is parallel to y-axis
vertices at (2, -2) and (2, 8),
foci at $(2, 3 - \sqrt{26})$ and $(2, 3 + \sqrt{26})$,
asymptotes of $y + 3 = -5(x + 2)$ and $y + 3 = 5(x + 2)$

D) center at (2, 3)
transverse axis is parallel to x-axis
vertices at (-3, 3) and (7, 3)
foci at $(2 - \sqrt{26}, 3)$ and $(2 + \sqrt{26}, 3)$
asymptotes of $y - 3 = -\frac{1}{5}(x - 2)$ and $y - 3 = \frac{1}{5}(x - 2)$

7) $x^2 - 9y^2 + 6x + 54y - 81 = 0$

A) center at (3, -3)
transverse axis is parallel to x-axis
vertices at (0, -3) and (6, -3)
foci at $(3 - \sqrt{10}, -3)$ and $(3 + \sqrt{10}, -3)$
asymptotes of $y + 3 = -\frac{1}{3}(x - 3)$ and $y + 3 = \frac{1}{3}(x - 3)$

B) center at (-3, 3)
transverse axis is parallel to y-axis
vertices at (-3, 0) and (-3, 6)
foci at $(-3, 3 - \sqrt{10})$ and $(-3, 3 + \sqrt{10})$
asymptotes of $y + 3 = -3(x - 3)$ and $y + 3 = 3(x - 3)$

C) center at (-3, 3)
transverse axis is parallel to x-axis
vertices at (-6, 3) and (0, 3)
foci at $(-3 - \sqrt{10}, 3)$ and $(-3 + \sqrt{10}, 3)$
asymptotes of $y - 3 = -\frac{1}{3}(x + 3)$ and $y - 3 = \frac{1}{3}(x + 3)$

D) center at (-3, 3)
transverse axis is parallel to x-axis
vertices at (-4, 3) and (-2, 3)
foci at $(-3 - \sqrt{10}, 3)$ and $(-3 + \sqrt{10}, 3)$
asymptotes of $y - 3 = -3(x + 3)$ and $y - 3 = 3(x + 3)$

Graph the hyperbola.

8) $\dfrac{(x+2)^2}{4} - \dfrac{(y-1)^2}{9} = 1$

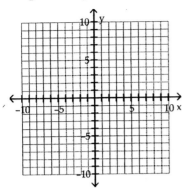

A)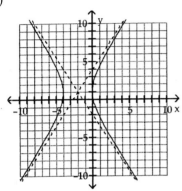

B)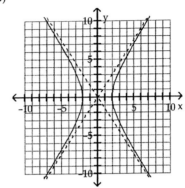

C)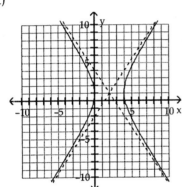

D)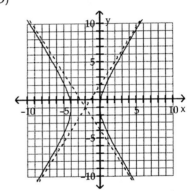

9) $\dfrac{(y-1)^2}{9} - \dfrac{(x+1)^2}{4} = 1$

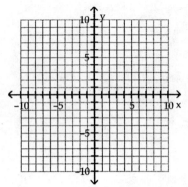

A)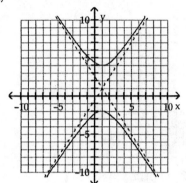

B)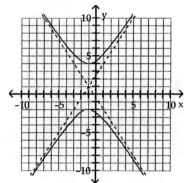

C)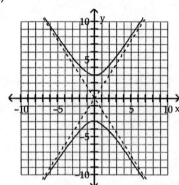

D)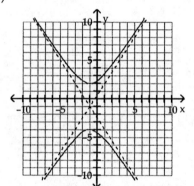

10) $(x-1)^2 - 9(y+1)^2 = 9$

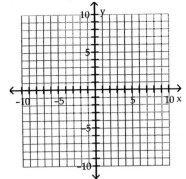

A)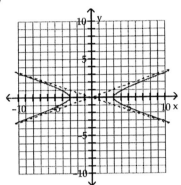

B)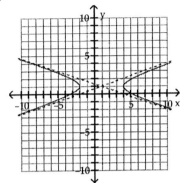

C)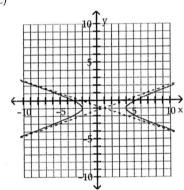

D)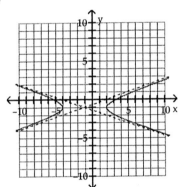

11) $(y+4)^2 - 9(x+3)^2 = 9$

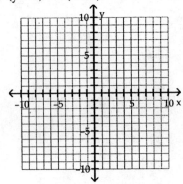

A)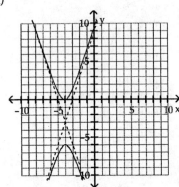

B)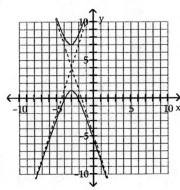

C)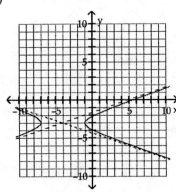

D)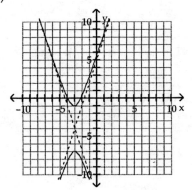

4 Solve Applied Problems Involving Hyperbolas

Solve the problem.

1) Two recording devices are set 3200 feet apart, with the device at point A to the west of the device at point B. At a point on a line between the devices, 400 feet from point B, a small amount of explosive is detonated. The recording devices record the time the sound reaches each one. How far directly north of site B should a second explosion be done so that the measured time difference recorded by the devices is the same as that for the first detonation?

A) 4098.78 feet B) 1763.83 feet C) 933.33 feet D) 1549.19 feet

9.5 Rotation of Axes; General Form of a Conic

1 Identify a Conic

Identify the equation without completing the square.

1) $3x^2 - 3x + y - 1 = 0$

A) ellipse B) parabola C) hyperbola D) not a conic

Precalculus Enhanced with Graphing Utilities 736

2) $3y^2 - 3x + 3y = 0$
   A) hyperbola   B) ellipse   C) parabola   D) not a conic

3) $4x^2 + 3y^2 - 2x - 2y = 0$
   A) hyperbola   B) parabola   C) ellipse   D) not a conic

4) $4x^2 + 2y^2 + 4x - 4 = 0$
   A) hyperbola   B) parabola   C) ellipse   D) not a conic

5) $2x^2 - 4y^2 + 5x + 2y + 1 = 0$
   A) parabola   B) hyperbola   C) ellipse   D) not a conic

6) $y^2 - 2x^2 + 8x + 3y + 3 = 0$
   A) ellipse   B) hyperbola   C) parabola   D) not a conic

2 Use a Rotation of Axes to Transform Equations

Determine the appropriate rotation formulas to use so that the new equation contains no xy-term.

1) $x^2 + 2xy + y^2 - 8x + 8y = 0$

   A) $x = \dfrac{\sqrt{2+\sqrt{2}}}{2}x' - \dfrac{\sqrt{2-\sqrt{2}}}{2}y'$ and $y = \dfrac{\sqrt{2-\sqrt{2}}}{2}x' + \dfrac{\sqrt{2+\sqrt{2}}}{2}y'$

   B) $x = -y'$ and $y = x'$

   C) $x = \dfrac{\sqrt{2}}{2}(x' - y')$ and $y = \dfrac{\sqrt{2}}{2}(x' + y')$

   D) $x = \dfrac{1}{2}x' - \dfrac{\sqrt{3}}{2}y'$ and $y = \dfrac{\sqrt{3}}{2}x' + \dfrac{1}{2}y'$

2) $3x^2 + 3xy + 3y^2 - 8x + 8y = 0$

   A) $x = \dfrac{\sqrt{2+\sqrt{2}}}{2}x' - \dfrac{\sqrt{2-\sqrt{2}}}{2}y'$ and $y = \dfrac{\sqrt{2-\sqrt{2}}}{2}x' + \dfrac{\sqrt{2+\sqrt{2}}}{2}y'$

   B) $x = \dfrac{\sqrt{2}}{2}(x' - y')$ and $y = \dfrac{\sqrt{2}}{2}(x' + y')$

   C) $x = \dfrac{1}{2}x' - \dfrac{\sqrt{3}}{2}y'$ and $y = \dfrac{\sqrt{3}}{2}x' + \dfrac{1}{2}y'$

   D) $x = -y'$ and $y = x'$

3) $10x^2 - 4xy + 6y^2 - 8x + 8y = 0$

   A) $x = \dfrac{1}{2}x' - \dfrac{\sqrt{3}}{2}y'$ and $y = \dfrac{\sqrt{3}}{2}x' + \dfrac{1}{2}y'$

   B) $x = \dfrac{\sqrt{2-\sqrt{2}}}{2}x' - \dfrac{\sqrt{2+\sqrt{2}}}{2}y'$ and $y = \dfrac{\sqrt{2+\sqrt{2}}}{2}x' + \dfrac{\sqrt{2-\sqrt{2}}}{2}y'$

   C) $x = \dfrac{\sqrt{2}}{2}(x' - y')$ and $y = \dfrac{\sqrt{2}}{2}(x' + y')$

   D) $x = -y'$ and $y = x'$

3 Discuss an Equation Using a Rotation of Axes

Rotate the axes so that the new equation contains no xy-term. Discuss the new equation.

1) $24xy - 7y^2 + 36 = 0$

A) $\theta = 36.9°$
$$\frac{y'^2}{9} - \frac{x'^2}{16} = 1$$
hyperbola
center at (0, 0)
transverse axis is the y'-axis
vertices at (0, ±3)

B) $\theta = 53.1°$
$$\frac{y'^2}{4} - \frac{4x'^2}{9} = 1$$
hyperbola
center at (0, 0)
transverse axis is the y'-axis
vertices at (0, ±2)

C) $\theta = 36.9°$
$$\frac{4y'^2}{9} - \frac{x'^2}{4} = 1$$
hyperbola
center at (0, 0)
transverse axis is the y'-axis
vertices at $(0, \pm\frac{3}{2})$

D) $\theta = 36.9°$
$$\frac{y'^2}{4} - \frac{4x'^2}{9} = 1$$
hyperbola
center at (0, 0)
transverse axis is the y'-axis
vertices at (0, ±2)

2) $x^2 + 2xy + y^2 - 8x + 8y = 0$

A) $\theta = 36.9°$
$$\frac{x'^2}{4} + \frac{y'^2}{4} = 1$$
ellipse
center (0, 0)
major axis is x'-axis
vertices at (±2, 0)

B) $\theta = 45°$
$y'^2 = -4\sqrt{2}x'$
parabola
vertex at (0, 0)
focus at $(-\sqrt{2}, 0)$

C) $\theta = 45°$
$x'^2 = -4\sqrt{2}y'$
parabola
vertex at (0, 0)
focus at $(0, -\sqrt{2})$

D) $\theta = 36.9°$
$$\frac{x'^2}{4} + \frac{y'^2}{2} = 1$$
ellipse
center (0, 0)
major axis is x'-axis
vertices at (±2, 0)

3) $31x^2 + 10\sqrt{3}xy + 21y^2 - 144 = 0$

A) $\theta = 45°$
$x'^2 = -4\sqrt{2}y'$
parabola
vertex at (0, 0)
focus at $(0, -\sqrt{2})$

B) $\theta = 30°$
$$\frac{x'^2}{4} + \frac{y'^2}{9} = 1$$
ellipse
center at (0, 0)
major axis is y'-axis
vertices at (0, ±3)

C) $\theta = 36.9°$
$$\frac{x'^2}{9} + \frac{y'^2}{4} = 1$$
ellipse
center at (0, 0)
major axis is x'-axis
vertices at (±3, 0)

D) $\theta = 45°$
$y'^2 = -4\sqrt{2}x'$
parabola
vertex at (0, 0)
focus at $(-\sqrt{2}, 0)$

4) $xy + 16 = 0$

A) $\theta = 36.9°$
$$\frac{x'^2}{4} + \frac{y'^2}{2} = 1$$
ellipse
center at (0, 0)
major axis is the x'-axis
vertices at (±2, 0)

B) $\theta = 45°$
$y'^2 = -32x'$
parabola
vertex at (0, 0)
focus at (-8, 0)

C) $\theta = 45°$
$$\frac{y'^2}{32} + \frac{x'^2}{32} = 1$$
ellipse
center at (0, 0)
major axis is y'-axis
vertices at $(0, \pm 4\sqrt{2})$

D) $\theta = 45°$
$$\frac{y'^2}{32} - \frac{x'^2}{32} = 1$$
hyperbola
center at (0, 0)
transverse axis is y'-axis
vertices at $(0, \pm 4\sqrt{2})$

5) $x^2 + xy + y^2 - 3y - 6 = 0$

A) $\theta = 45°$
$$\frac{x'^2}{3} + \frac{y'^2}{4} = 1$$
ellipse
center at (0, 0)
major axis is y'-axis
vertices at (0, ±2)

B) $\theta = 45°$
$$\frac{\left(x' - \frac{\sqrt{2}}{2}\right)^2}{5} + \frac{\left(y' - \frac{3\sqrt{2}}{2}\right)^2}{15} = 1$$
ellipse
center at $(\frac{\sqrt{2}}{2}, \frac{3\sqrt{2}}{2})$
major axis is y'-axis
vertices at $(\frac{\sqrt{2}}{2}, -\frac{3\sqrt{2}}{2})$ and $(\frac{\sqrt{2}}{2}, \frac{9\sqrt{2}}{2})$

C) $\theta = 45°$
$y'^2 = -18x'$
parabola
vertex at (0, 0)
focus at $(-\frac{9}{2}, 0)$

D) $\theta = 45°$
$$\frac{x'^2}{6} - \frac{y'^2}{8} = 1$$
hyperbola
center at (0, 0)
transverse axis is the x'-axis
vertices at $(\pm\sqrt{6}, 0)$

6) $17x^2 - 12xy + 8y^2 - 68x + 24y - 12 = 0$

A) $\theta = 63.4°$
$x'^2 = -16y'$
parabola
vertex at $(0, 0)$
focus at $(0, -4)$

B) $\theta = 63.4°$
$\dfrac{x'^2}{16} - \dfrac{y'^2}{4} = 1$
hyperbola
center at $(0, 0)$
transverse axis is the x'-axis
vertices at $(\pm 4, 0)$

C) $\theta = 26.6°$
$\dfrac{x'^2}{4} + \dfrac{y'^2}{16} = 1$
ellipse
center at $(0, 0)$
major axis is y'-axis
vertices at $(0, \pm 4)$

D) $\theta = 63.4°$
$\dfrac{\left(x' - \dfrac{2\sqrt{5}}{5}\right)^2}{16} + \dfrac{\left(y' + \dfrac{4\sqrt{5}}{5}\right)^2}{4} = 1$
ellipse
center at $\left(\dfrac{2\sqrt{5}}{5}, -\dfrac{4\sqrt{5}}{5}\right)$
major axis is x'-axis
vertices at $\left(4 + \dfrac{2\sqrt{5}}{5}, -\dfrac{4\sqrt{5}}{5}\right)$ and $\left(-4 + \dfrac{2\sqrt{5}}{5}, -\dfrac{4\sqrt{5}}{5}\right)$

7) $5x^2 - 6xy + 5y^2 - 8 = 0$

A) $\theta = 45°$
$x'^2 = -4y'$
parabola
vertex at $(0, 0)$
focus at $(0, -1)$

B) $\theta = 45°$
$\dfrac{x'^2}{4} + y'^2 = 1$
ellipse
center at $(0, 0)$
major axis is the x'-axis
vertices at $(\pm 2, 0)$

C) $\theta = 45°$
$y'^2 = -4x'$
parabola
vertex at $(0, 0)$
focus at $(-1, 0)$

D) $\theta = 45°$
$\dfrac{x'^2}{4} - y'^2 = 1$
hyperbola
center at $(0, 0)$
transverse axis is the x'-axis
vertices at $(\pm 2, 0)$

Rotate the axes so that the new equation contains no xy–term. Graph the new equation.

8) $24xy - 7y^2 + 36 = 0$

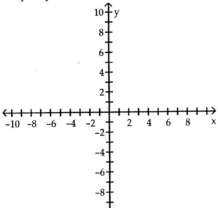

A)

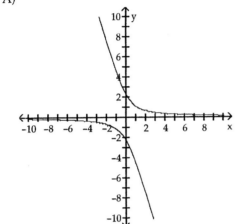

B)

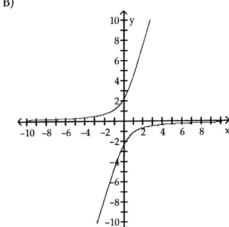

9) $x^2 + 2xy + y^2 - 8x + 8y = 0$

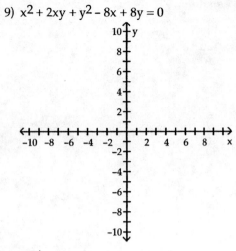

A)

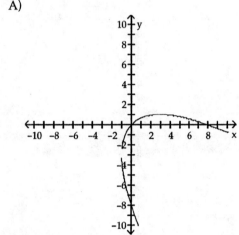

B)

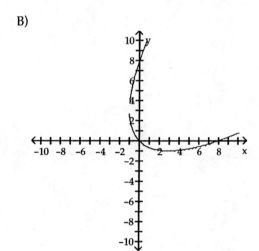

10) $31x^2 + 10\sqrt{3}xy + 21y^2 - 144 = 0$

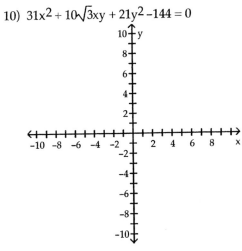

A)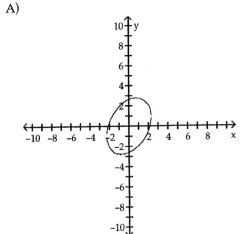

B)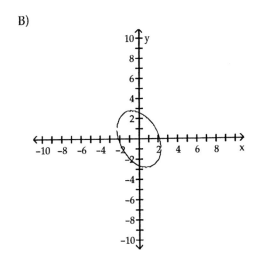

11) $xy + 16 = 0$

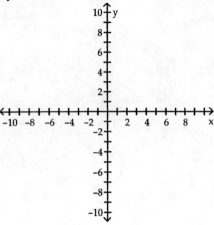

A)
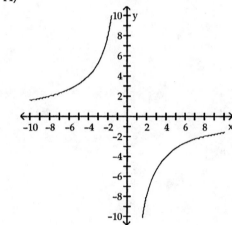

B)
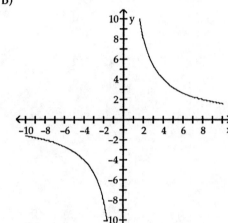

12) $x^2 + xy + y^2 - 3y - 6 = 0$

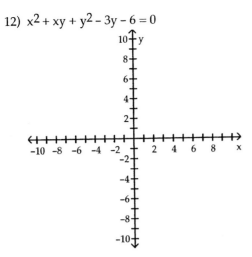

A)
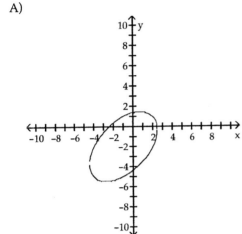

B)
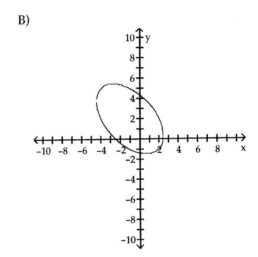

13) $17x^2 - 12xy + 8y^2 - 68x + 24y - 12 = 0$

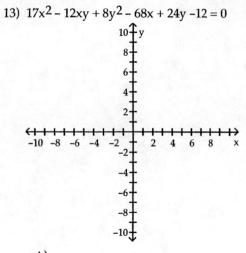

A)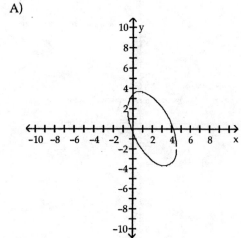

B)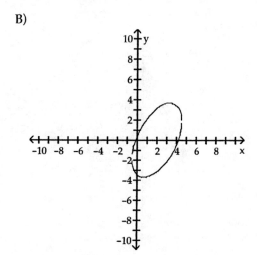

14) $5x^2 - 6xy + 5y^2 - 8 = 0$

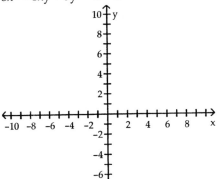

A)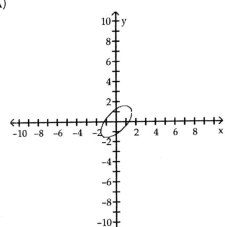

B)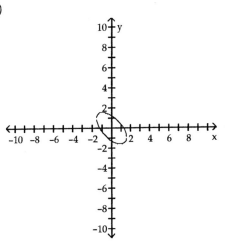

4 Identify a Conic without a Rotation of Axes

Identify the equation without applying a rotation of axes.

1) $x^2 + 12xy + 36y^2 - 2x - 3y - 1 = 0$
   A) ellipse B) hyperbola C) parabola D) not a conic

2) $2x^2 + 6xy + 9y^2 - 3x + 4y - 5 = 0$
   A) hyperbola B) parabola C) ellipse D) not a conic

3) $3x^2 - 7xy + 3y^2 - 4x + 4y + 10 = 0$
   A) parabola B) ellipse C) hyperbola D) not a conic

4) $x^2 + 2xy - 4y^2 + 4x + 2y + 8 = 0$
   A) hyperbola B) parabola C) ellipse D) not a conic

5) $5x^2 - 7xy + 3y^2 + 4x + 4y + 1 = 0$
   A) hyperbola B) ellipse C) parabola D) not a conic

6) $9x^2 + 7xy + 4y^2 - 3x + 3y + 6 = 0$
   A) hyperbola B) parabola C) ellipse D) not a conic

9.6 Polar Equations of Conics

1 Discuss and Graph Polar Equations of Conics

Identify the conic that the polar equation represents. Also, give the position of the directrix.

1) $r = \dfrac{3}{1 - 3\cos\theta}$

   A) ellipse, directrix perpendicular to the polar axis 1 left of the pole
   B) hyperbola, directrix perpendicular to the polar axis 1 left of the pole
   C) hyperbola, directrix perpendicular to the polar axis 1 right of the pole
   D) ellipse, directrix perpendicular to the polar axis 1 right of the pole

2) $r = \dfrac{2}{2 + 2\sin\theta}$

   A) hyperbola, directrix perpendicular to the polar axis 1 right of the pole
   B) parabola, directrix perpendicular to the polar axis 1 right of the pole
   C) parabola, directrix parallel to the polar axis 1 above the pole
   D) hyperbola, directrix parallel to the polar axis 1 above the pole

3) $r = \dfrac{3}{4 - 2\sin\theta}$

   A) ellipse, directrix perpendicular to the polar axis $\dfrac{3}{2}$ right of the pole
   B) ellipse, directrix perpendicular to the polar axis $\dfrac{3}{2}$ left of the pole
   C) ellipse, directrix parallel to the polar axis $\dfrac{3}{2}$ below the pole
   D) ellipse, directrix parallel to the polar axis $\dfrac{3}{2}$ above the pole

4) $r = \dfrac{6}{3 - 4\cos\theta}$

   A) ellipse, directrix is perpendicular to the polar axis at a distance 3 units to the left of the pole
   B) hyperbola, directrix is perpendicular to the polar axis at a distance $\dfrac{3}{2}$ units to the left of the pole
   C) hyperbola, directrix is perpendicular to the polar axis at a distance 3 units to the right of the pole
   D) ellipse, directrix is perpendicular to the polar axis at a distance $\dfrac{3}{2}$ units to the right of the pole

Discuss the equation and graph it.

5) $r = \dfrac{9}{3 - 3\cos\theta}$

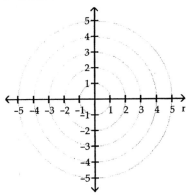

A) directrix parallel to polar axis 3 above pole

focus $(0, 0)$, vertex $\left(\dfrac{3}{2}, \dfrac{\pi}{2}\right)$

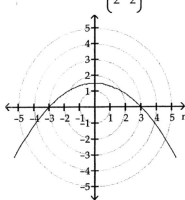

B) directrix parallel to polar axis 3 below pole

focus $(0, 0)$, vertex $\left(\dfrac{3}{2}, \dfrac{3\pi}{2}\right)$

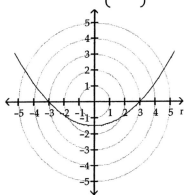

C) directrix perpendicular to polar axis 3 left of pole

focus $(0, 0)$, vertex $\left(\dfrac{3}{2}, \pi\right)$

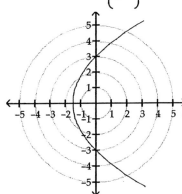

D) directrix perpendicular to polar axis 3 right of pole

focus $(0, 0)$, vertex $\left(\dfrac{3}{2}, 0\right)$

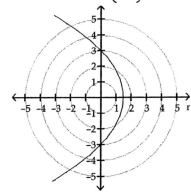

Precalculus Enhanced with Graphing Utilities 749

6) $r = \dfrac{3}{3 - \sin\theta}$

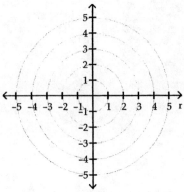

A) directrix perpendicular to polar axis 3 right of pole
center $\left(-\dfrac{3}{8}, 0\right)$
vertices $\left(\dfrac{3}{2}, \pi\right), \left(\dfrac{3}{4}, 0\right)$

B) directrix perpendicular to polar axis 3 left of pole
center $\left(\dfrac{3}{8}, 0\right)$
vertices $\left(\dfrac{3}{4}, \pi\right), \left(\dfrac{3}{2}, 0\right)$

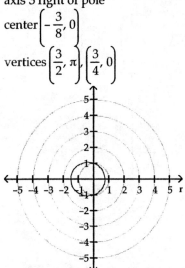
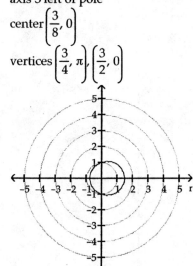

C) directrix parallel to polar axis 3 above pole
center $\left(-\dfrac{3}{8}, \dfrac{\pi}{2}\right)$
vertices $\left(-\dfrac{3}{4}, \dfrac{3\pi}{2}\right), \left(\dfrac{3}{2}, \dfrac{3\pi}{2}\right)$

D) directrix parallel to polar axis 3 below pole
center $\left(\dfrac{3}{8}, \dfrac{\pi}{2}\right)$
vertices $\left(\dfrac{3}{2}, \dfrac{\pi}{2}\right), \left(\dfrac{3}{4}, \dfrac{3\pi}{2}\right)$

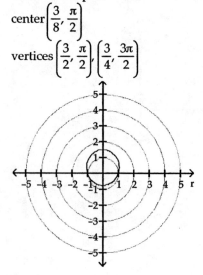

7) $r = \dfrac{3}{2 + 4 \sin \theta}$

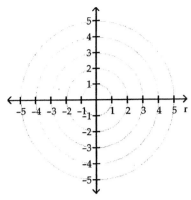

A) hyperbola; directrix parallel to the polar axis $\dfrac{3}{4}$ unit above the pole

vertices $(\dfrac{1}{2}, \dfrac{\pi}{2})$, $(-\dfrac{3}{2}, \dfrac{3\pi}{2})$

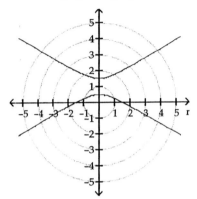

B) ellipse, directrix parallel to the polar axis $\dfrac{3}{2}$ unit below the pole

vertices $(\dfrac{3}{2}, \dfrac{\pi}{2})$, $(\dfrac{1}{2}, \dfrac{3\pi}{2})$

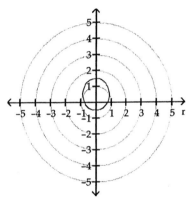

C) hyperbola, directrix perpendicular to the polar axis $\dfrac{3}{4}$ unit right of the pole

vertices $(\dfrac{1}{2}, 0)$, $(-\dfrac{3}{2}, \pi)$

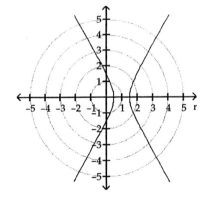

D) ellipse, directrix perpendicular to the polar axis $\dfrac{3}{2}$ unit left of the pole

vertices $(\dfrac{3}{2}, 0)$, $(\dfrac{1}{2}, \pi)$

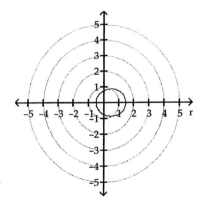

2 Convert a Polar Equation of a Conic to a Rectangular Equation

Convert the polar equation to a rectangular equation.

1) $r = \dfrac{3}{3 - 3\cos\theta}$

   A) $y^2 = 2x + 1$ B) $x^2 = -2y + 1$ C) $y^2 = -2x + 1$ D) $x^2 = 2y + 1$

2) $r = \dfrac{3}{3 + \cos\theta}$

   A) $10x^2 + 9y^2 - 6x - 9 = 0$ B) $9x^2 + 9y^2 + 6x - 9 = 0$
   C) $9x^2 + 8y^2 + 6y - 9 = 0$ D) $8x^2 + 9y^2 + 6x - 9 = 0$

3) $r = \dfrac{4\sec\theta}{4\sec\theta + 1}$

   A) $16x^2 + 15y^2 + 8y - 16 = 0$ B) $16x^2 + 16y^2 + 8x - 16 = 0$
   C) $17x^2 + 16y^2 - 8x - 16 = 0$ D) $15x^2 + 16y^2 + 8x - 16 = 0$

4) $r = \dfrac{4\sec\theta}{\sec\theta + 2}$

   A) $3x^2 - y^2 + 16 = 0$ B) $3y^2 - x^2 + 16 = 0$
   C) $3y^2 - x^2 - 16y + 16 = 0$ D) $3x^2 - y^2 - 16x + 16 = 0$

Find a polar equation for the conic. A focus is at the pole.

5) $e = 1$; directrix is parallel to the polar axis 4 above the pole

   A) $r = \dfrac{4}{1 - \cos\theta}$ B) $r = \dfrac{4}{1 + \sin\theta}$ C) $r = \dfrac{4}{1 + \cos\theta}$ D) $r = \dfrac{4}{1 - \sin\theta}$

6) $e = \dfrac{1}{2}$; directrix is perpendicular to the polar axis 3 to the left of the pole

   A) $r = \dfrac{6}{4 + 2\cos\theta}$ B) $r = \dfrac{12}{4 + 2\cos\theta}$ C) $r = \dfrac{6}{4 - 2\cos\theta}$ D) $r = \dfrac{12}{4 - 2\cos\theta}$

7) $e = 5$; directrix is perpendicular to the polar axis 1 to the right of the pole

   A) $r = \dfrac{5}{1 - 5\sin\theta}$ B) $r = \dfrac{5}{1 + 5\sin\theta}$ C) $r = \dfrac{5}{1 + 5\cos\theta}$ D) $r = \dfrac{5}{1 - 5\cos\theta}$

9.7 Plane Curves and Parametric Equations

1 Graph Parametric Equations by Hand

Graph the curve whose parametric equations are given.

1) $x = 2t, y = t + 2; -2 \leq t \leq 3$

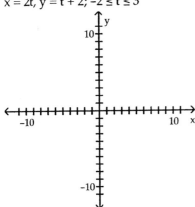

A)
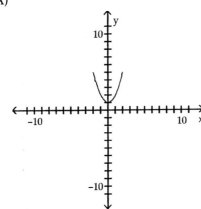

B)
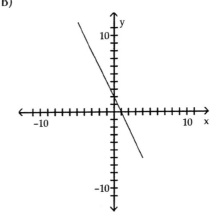

C)
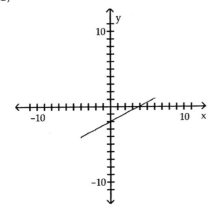

D)
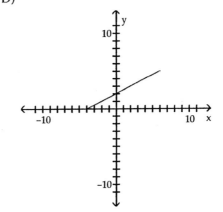

2) $x = 2t - 1, y = t^2 + 5; -4 \leq t \leq 4$

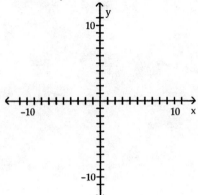

A)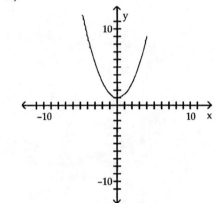

B)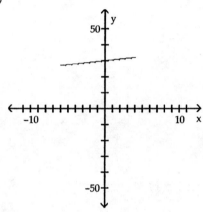

C)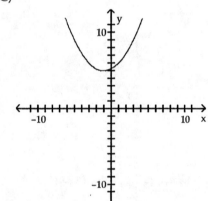

D)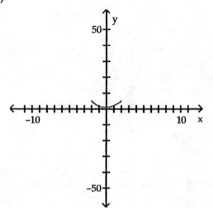

3) $x = t^3 + 1$, $y = t^3 - 20$; $-2 \le t \le 2$

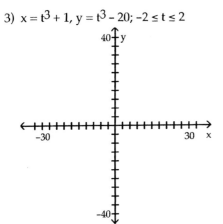

A)
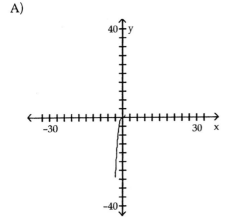

B)
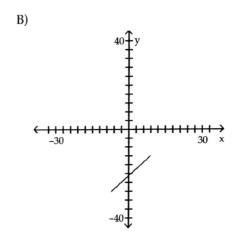

C)
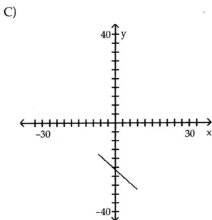

D)
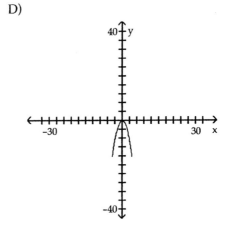

4) $x = 8 \sin t$, $y = 8 \cos t$; $0 \leq t \leq 2\pi$

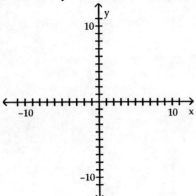

A)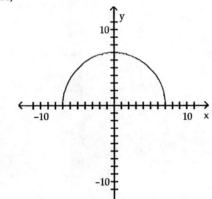

B)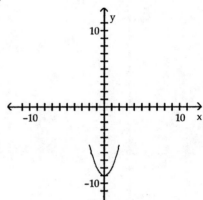

C)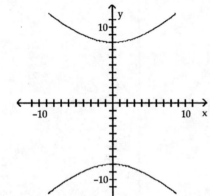

D)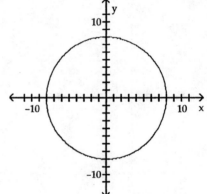

5) $x = 2\tan t,\ y = 3\sec t;\ 0 \le t \le 2\pi$

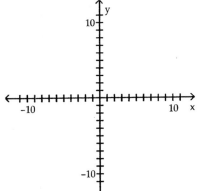

A)
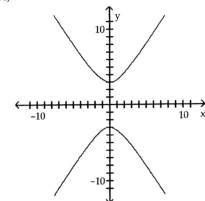

B)
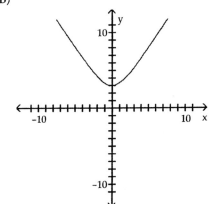

C)
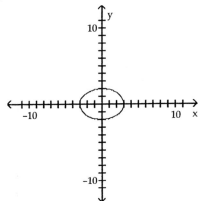

D)
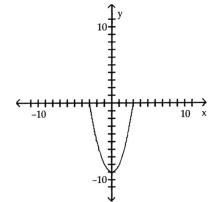

6) $x = -\sec t,\ y = \tan t;\ -\dfrac{\pi}{2} < t < \dfrac{\pi}{2}$

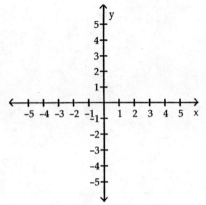

A)
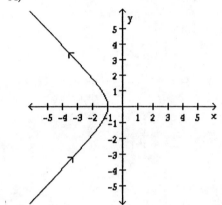

B)
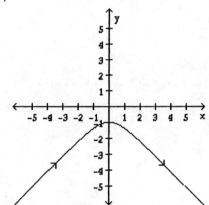

C)
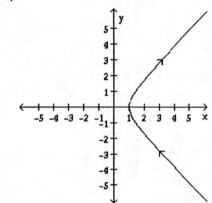

D)
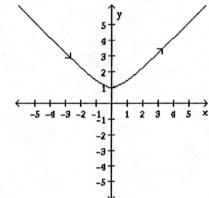

7) $x = 3\cos t, y = -3\sin t; \quad \dfrac{\pi}{2} \le t \le \dfrac{3\pi}{2}$

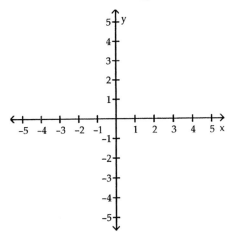

A)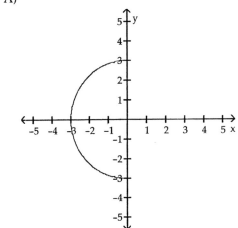

B)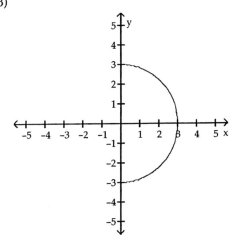

C)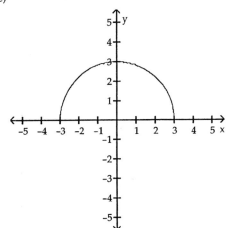

D)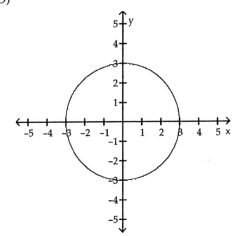

The parametric equations of four curves are given. Graph each of them, indicating the orientation.

8) C_1: $x = 7\sin t$, $y = 7 - 7\cos^2 t$; $\frac{\pi}{2} \le t \le \frac{3\pi}{2}$

C_2: $x = \ln t$, $y = \ln t^2$; $e^{-4} \le t \le e^3$

C_3: $x = t^2 - 8$, $y = t - 3$; $-4 \le t \le 4$

C_4: $x = t - 5$, $y = t + 2$; $-4 \le t \le 7$

2 Graph Parametric Equations Using a Graphing Utility

Use a graphing utility to graph the curve defined by the given parametric equations.

1) $x = t + 2$, $y = 3t - 1$; $0 \le t \le 3$

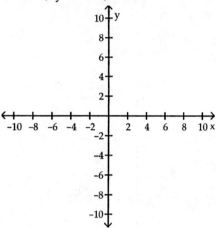

A)

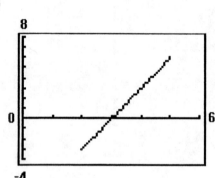

B)

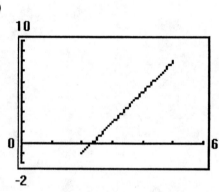

C)

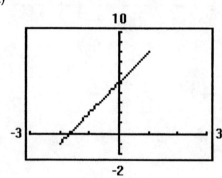

D)

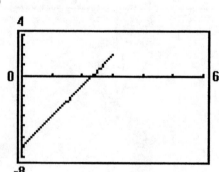

2) $x = 2t^2$, $y = t + 2$; $-\infty < t < \infty$

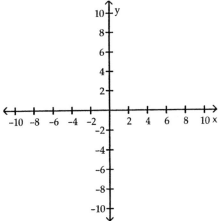

A)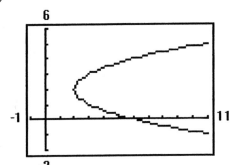

B)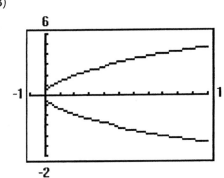

C)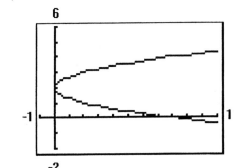

D)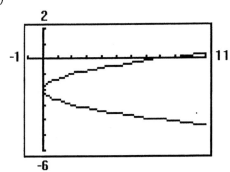

3) $x = 3\cos t,\ y = 2\sin t;\ 0 \le t \le 2\pi$

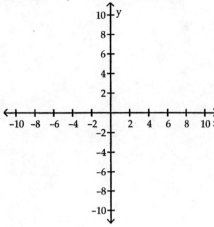

A)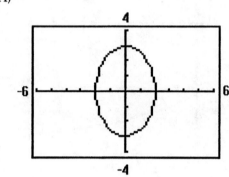

B)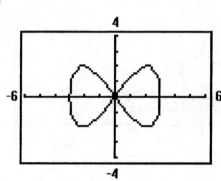

C)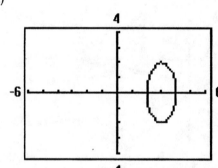

D)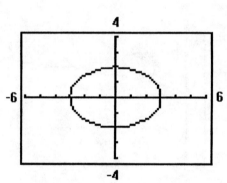

3 Find a Rectangular Equation for a Curve Defined Parametrically

Find a rectangular equation for the plane curve defined by the parametric equations.

1) $x = 3t,\ y = t + 4;\ -2 \le t \le 3$

A) $y = \frac{1}{3}x + 4$; for x in $-6 \le x \le 9$

B) $y = -3x + 4$; for x in $-\infty < x < \infty$

C) $y = x^2 + 1$; for x in $-2 \le x \le 2$

D) $y = \frac{1}{3}x - 4$; for x in $-\infty < x < \infty$

2) $x = 2t - 1, y = t^2 + 6; -4 \le t \le 4$

A) $y = -\frac{1}{2}x + 30$; for x in $-6 \le x \le 4$

B) $y = \frac{1}{4}x^2 + \frac{1}{2}x + \frac{25}{4}$; for x in $-9 \le x \le 7$

C) $y = x^2 + 1$; for x in $-2 \le x \le 2$

D) $y = \frac{1}{2}x^2 + 1$; for x in $-6 \le x \le 4$

3) $x = t^3 + 1, y = t^3 - 5; -2 \le t \le 2$

A) $y = x^3$; for x in $-3 \le x \le 1$

B) $y = x - 6$; for x in $-7 \le x \le 9$

C) $y = -x - 6$; for x in $-7 \le x \le 9$

D) $y = -x^2$; for x in $-4 \le x \le 4$

4) $x = 5 \sin t, y = 5 \cos t; 0 \le t \le 2\pi$

A) $y = \sqrt{a^2 - x^2} = 25$; for x in $-\infty < x < \infty$

B) $y^2 - x^2 = 25$; for x in $-\infty < x < \infty$

C) $x^2 + y^2 = 25$; for x in $-5 \le x \le 5$

D) $y = x^2 - 9$; for x in $-2 \le x \le 2$

5) $x = 2 \tan t, y = 3 \sec t; 0 \le t \le 2\pi$

A) $\frac{y^2}{9} + \frac{x^2}{4} = 1$; for x in $-\infty < x < \infty$

B) $y = x^2 - 9$; for x in $-3 \le x \le 3$

C) $\frac{y^2}{9} - \frac{x^2}{4} = 1$; for x in $-\infty < x < \infty$

D) $y = 3\sqrt{1 + \frac{x^2}{4}}$; for x in $-\infty < x < \infty$

6) $x = 5 \cos t, y = -2 \sin t; 0 \le t \le 2\pi$

A) $4x^2 + 25y^2 = 1; -\frac{1}{5} \le x \le \frac{1}{5}$

B) $4x^2 + 25y^2 = 100; -5 \le x \le 5$

C) $4x^2 - 25y^2 = 1; x \ge \frac{1}{2}$

D) $4x^2 - 25y^2 = 100; x \ge 5$

4 Use Time as a Parameter in Parametric Equations

Solve.

1) Ron throws a ball straight up with an initial speed of 60 feet per second from a height of 7 feet. Find parametric equations that describe the motion of the ball as a function of time. How long is the ball in the air? When is the ball at its maximum height? What is the maximum height of the ball?

A) $x = 0$ and $y = -16t^2 + 60t + 7$
3.863 sec, 1.875 sec,
63.25 feet

B) $x = 0$ and $y = -16t^2 + 60t + 7$
7.726 sec, 1.875 sec,
56.25 feet

C) $x = 0$ and $y = -16t^2 + 60t + 7$
7.259 sec, 1.875 sec,
428.454 feet

D) $x = 0$ and $y = -16t^2 + 60t + 7$
3.629 sec, 1.875 sec,
7.026 feet

2) A baseball pitcher throws a baseball with an initial speed of 125 feet per second at an angle of 20° to the horizontal. The ball leaves the pitcher's hand at a height of 5 feet. Find parametric equations that describe the motion of the ball as a function of time. How long is the ball in the air? When is the ball at its maximum height? What is the maximum height of the ball?

A) $x = 117.46t$ and $y = -16t^2 + 42.75t + 5$
2.784 sec, 1.336 sec,
33.556 feet

B) $x = 117.46t$ and $y = -16t^2 + 42.75t + 5$
5.099 sec, 1.336 sec,
233.445 feet

C) $x = 117.46t$ and $y = -16t^2 + 42.75t + 5$
2.549 sec, 1.336 sec,
5.011 feet

D) $x = 117.46t$ and $y = -16t^2 + 42.75t + 5$
5.568 sec, 1.336 sec,
28.556 feet

3) A baseball player hit a baseball with an initial speed of 190 feet per second at an angle of 40° to the horizontal. The ball was hit at a height of 4 feet off the ground. Find parametric equations that describe the motion of the ball as a function of time. How long is the ball in the air? When is the ball at its maximum height? What is the distance the ball traveled?

A) $x = 145.54t$ and $y = -16t^2 + 122.17t + 4$
7.603 sec, 3.818 sec,
1106.541 feet

B) $x = 145.54t$ and $y = -16t^2 + 122.17t + 4$
7.668 sec, 3.818 sec,
1116.001 feet

C) $x = 145.54t$ and $y = -16t^2 + 122.17t + 4$
7.668 sec, 3.818 sec,
1877.571 feet

D) $x = 145.54t$ and $y = -16t^2 + 122.17t + 4$
15.336 sec, 3.818 sec,
2232.001 feet

5 Find Parametric Equations for Curves Defined by Rectangular Equations

Find parametric equations for the rectangular equation.

1) $y = 2x + 5$

A) $x = t, y = 2t + 5; 0 \le t < \infty$

B) $x = \frac{t}{2}, y = t + \frac{5}{2}; 0 \le t < \infty$

C) $y = 2t, 2x = t - 5; 0 \le t < \infty$

D) $x = t, y = 2t^2 + 5; 0 \le t < \infty$

2) $y = x^4 + 3$

A) $x = t, y = t^4 + 3; 0 \le t < \infty$

B) $x = t^2, y = t^4 + 3; 0 \le t < \infty$

C) $x = t, y = t^2 + 3; 0 \le t < \infty$

D) $x = t^2, y = t^2 + 3; 0 \le t < \infty$

3) $y = 3x^2 + 6$

A) $x = t; y = 3t^2 + 6; 0 \le t < \infty$

B) $x = \sqrt{t}; y = 3t + 6; t \ge 0$

C) $x = t^2; y = 3t + 6; 0 \le t < \infty$

D) $y = t; x = 3t^2 + 6; 0 \le t < \infty$

Solve the problem.

4) Find parametric equations for an object that moves along the ellipse $\frac{x^2}{9} + \frac{y^2}{16} = 1$ with the motion described.

The motion begins at (3, 0), is counterclockwise, and requires 9 seconds for a complete revolution.

A) $x = 3\cos(\frac{2\pi}{9}t), y = 4\sin(\frac{2\pi}{9}t), 0 \le t \le 9$

B) $x = 3\cos(\frac{2\pi}{9}t), y = -4\sin(\frac{2\pi}{9}t), 0 \le t \le 9$

C) $x = -3\cos(\frac{2\pi}{9}t), y = 4\sin(\frac{2\pi}{9}t), 0 \le t \le 9$

D) $x = -3\cos(\frac{2\pi}{9}t), y = -4\sin(\frac{2\pi}{9}t), 0 \le t \le 9$

Ch. 9 Analytic Geometry
Answer Key

9.1 Conics
1 Know the Names of Conics
1) B
2) D
3) D
4) D

9.2 The Parabola
1 Work with Parabolas with Vertex at the Origin
1) A
2) C
3) C
4) A
5) C
6) D
7) B
8) D
9) B
10) C
11) B
12) A
13) D
14) D
15) B
16) B
17) B
18) A
19) B

2 Work with Parabolas with Vertex at (h, k)
1) C
2) C
3) C
4) A
5) C
6) A
7) B
8) D
9) C
10) B
11) A
12) D
13) D
14) A
15) D
16) C
17) C
18) C

3 Solve Applied Problems Involving Parabolas
1) D
2) D
3) D
4) C
5) A

6) B
7) C
8) The receiver should be located 2 feet from the base of the dish, along its axis of symmetry.
9) about 17 centimeters
10) D

9.3 The Ellipse
1 Work with Ellipses with Center at the Origin
1) A
2) B
3) A
4) B
5) A
6) C
7) D
8) B
9) A
10) D
11) D
12) B
13) B
14) A
15) D
16) B
17) D
18) A

2 Work with Ellipses with Center at (h, k)
1) B
2) B
3) A
4) B
5) C
6) B
7) B
8) B
9) D
10) D
11) C
12) A
13) D
14) A
15) B
16) B

3 Solve Applied Problems Involving Ellipses
1) D
2) C
3) B
4) B
5) 60 feet
6) 36 feet

9.4 The Hyperbola
1 Work with Hyperbolas with Center at the Origin
1) C
2) D
3) A

  4) A
  5) B
  6) D
  7) A
  8) C
  9) C
  10) D
  11) A
  12) D
  13) D
  14) C
  15) B
  16) B
  17) D
  18) D
  19) B
2 Find the Asymptotes of a Hyperbola
  1) D
  2) D
  3) C
  4) A
3 Work with Hyperbolas with Center at (h, k)
  1) D
  2) B
  3) D
  4) A
  5) B
  6) D
  7) C
  8) A
  9) B
  10) C
  11) D
4 Solve Applied Problems Involving Hyperbolas
  1) C

9.5 Rotation of Axes; General Form of a Conic
1 Identify a Conic
  1) B
  2) C
  3) C
  4) C
  5) B
  6) B
2 Use a Rotation of Axes to Transform Equations
  1) C
  2) B
  3) B
3 Discuss an Equation Using a Rotation of Axes
  1) D
  2) C
  3) B
  4) D
  5) B
  6) D
  7) B

8) B
9) A
10) B
11) A
12) B
13) B
14) A

4 Identify a Conic without a Rotation of Axes

1) C
2) C
3) C
4) A
5) B
6) C

9.6 Polar Equations of Conics

1 Discuss and Graph Polar Equations of Conics

1) B
2) C
3) C
4) B
5) C
6) D
7) A

2 Convert a Polar Equation of a Conic to a Rectangular Equation

1) A
2) D
3) D
4) D
5) B
6) C
7) C

9.7 Plane Curves and Parametric Equations

1 Graph Parametric Equations by Hand

1) D
2) C
3) B
4) D
5) A
6) A
7) A

8)

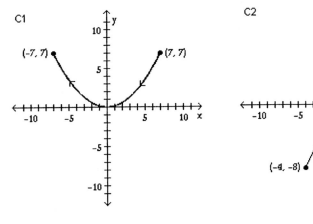

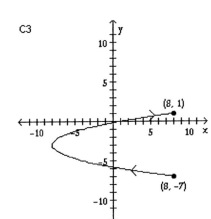 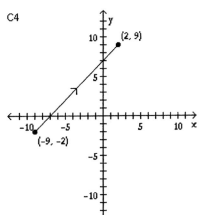

2 **Graph Parametric Equations Using a Graphing Utility**
   1) B
   2) C
   3) D
3 **Find a Rectangular Equation for a Curve Defined Parametrically**
   1) A
   2) B
   3) B
   4) C
   5) C
   6) B
4 **Use Time as a Parameter in Parametric Equations**
   1) A
   2) A
   3) B
5 **Find Parametric Equations for Curves Defined by Rectangular Equations**
   1) A
   2) A
   3) A
   4) B

Ch. 10 Systems of Equations and Inequalities

10.1 Systems of Linear Equations: Substitution and Elimination

1 Solve Systems of Equations by Substitution

Solve the system of equations by using substitution.

1) $\begin{cases} x + 2y = 2 \\ 7x - 9y = -9 \end{cases}$
   A) $x = 1, y = 0$ B) $x = 0, y = 0$ C) $x = 0, y = 1$ D) $x = 1, y = 1$

2) $\begin{cases} 3x + y = 13 \\ 2x + 9y = -8 \end{cases}$
   A) $x = 5, y = 2$ B) $x = -5, y = -2$ C) $x = -5, y = 2$ D) $x = 5, y = -2$

3) $\begin{cases} 5x - 2y = -1 \\ x + 4y = 35 \end{cases}$
   A) $x = 3, y = 9$ B) $x = 2, y = 9$ C) $x = 3, y = 8$ D) $x = 2, y = 8$

4) $\begin{cases} 5x + 3y = 80 \\ 2x + y = 30 \end{cases}$
   A) $x = 10, y = 0$ B) $x = 10, y = 10$ C) $x = 0, y = 10$ D) $x = 0, y = 0$

5) $\begin{cases} x + 7y = -2 \\ 3x + y = 34 \end{cases}$
   A) $x = 3, y = 7$ B) $x = -2, y = 3$ C) $x = 12, y = -2$ D) $x = 7, y = 12$

2 Solve Systems of Equations by Elimination

Use the elimination method to solve the system.

1) $\begin{cases} x + y = -9 \\ x - y = 19 \end{cases}$
   A) $x = 9, y = 5$ B) $x = 5, y = -14$ C) $x = 9, y = -14$ D) $x = 5, y = 14$

2) $\begin{cases} 4x + 10y = 10 \\ 3x - 2y = -2 \end{cases}$
   A) $x = 1, y = 1$ B) $x = 1, y = 0$ C) $x = 0, y = 1$ D) $x = 0, y = 0$

3) $\begin{cases} 2x + 6y = -20 \\ 8x + 2y = 30 \end{cases}$
   A) $x = -2, y = 5$ B) $x = -5, y = 5$ C) $x = 5, y = -5$ D) $x = 8, y = -8$

3 Identify Inconsistent Systems of Equations Containing Two Variables

Solve the system.

1) $\begin{cases} x - 4y = -10 \\ 2x - 8y = -17 \end{cases}$
   A) dependent (many solutions) B) (2, 4)
   C) inconsistent (no solution) D) (2, 3)

2) $\begin{cases} x + y = -8 \\ x + y = 7 \end{cases}$

A) (0, −1)   B) inconsistent (no solution)
C) (−8, 7)   D) dependent (many solutions)

3) $\begin{cases} 5x - 6y = 7 \\ 5x - 6y = 8 \end{cases}$

A) $\left(\dfrac{7}{6}, \dfrac{8}{5}\right)$ B) (7, 8)

C) consistent (many solutions)   D) inconsistent (no solution)

4) $\begin{cases} 7x - 8y = 1 \\ 21x - 24y = 2 \end{cases}$

A) inconsistent (no solution)   B) consistent (many solutions)

C) $\left(\dfrac{3}{28}, -\dfrac{3}{32}\right)$ D) (3, 2)

4 Express the Solution of a System of Dependent Equations Containing Two Variables

Solve the system.

1) $\begin{cases} 5x + y = 5 \\ -15x - 3y = -15 \end{cases}$

A) x = −5y + 5, where y is any real number   B) inconsistent (no solution)
C) y = −5x + 5, where x is any real number   D) y = 5x + 5, where x is any real number

2) $\begin{cases} x - 3y = -7 \\ 3x - 9y = -21 \end{cases}$

A) (0, 0)   B) inconsistent (no solution)

C) $y = -\dfrac{x}{-3} - 7$, where x is any real number   D) (−7, 0)

5 Solve Systems of Three Equations Containing Three Variables

Solve the system of equations.

1) $\begin{cases} x + y + z = 2 \\ x - y + 5z = 12 \\ 4x + y + z = -4 \end{cases}$

A) inconsistent (no solution)   B) x = 3, y = −2, z = 1
C) x = 3, y = 1, z = −2   D) x = −2, y = 1, z = 3

2) $\begin{cases} x - y + 4z = 7 \\ 4x + z = 2 \\ x + 3y + z = 5 \end{cases}$

A) x = 2, y = 0, z = 1   B) x = 0, y = 1, z = 2
C) x = 2, y = 1, z = 0   D) inconsistent (no solution)

Precalculus Enhanced with Graphing Utilities 771

3) $\begin{cases} x - y + 3z = -2 \\ 4x + z = 0 \\ x + 4y + z = 8 \end{cases}$

A) $x = 0, y = 2, z = -2$ B) $x = 0, y = 2, z = 0$ C) $x = 0, y = 0, z = 2$ D) Inconsistent, $\emptyset$

4) $\begin{cases} x + y + z = 7 \\ x - y + 2z = 7 \\ 5x + y + z = 11 \end{cases}$

A) $x = 4, y = 2, z = 1$ B) $x = 4, y = 1, z = 2$ C) $x = 1, y = 4, z = 2$ D) $x = 1, y = 2, z = 4$

5) $\begin{cases} x - y + z = 8 \\ x + y + z = 6 \\ x + y - z = -12 \end{cases}$

A) $x = -2, y = -1, z = 9$ B) $x = 2, y = -1, z = -9$
C) $x = -2, y = -1, z = -9$ D) $x = 2, y = -1, z = 9$

6 Identify Inconsistent Systems of Equations Containing Three Variables

Solve the system of equations.

1) $\begin{cases} x + y + z = 9 \\ x - y + 5z = 19 \\ 4x + 4y + 4z = 15 \end{cases}$

A) $x = 4, y = 3, z = 2$ B) $x = 4, y = 2, z = 3$
C) inconsistent (no solution)   D) $x = 2, y = 3, z = 4$

2) $\begin{cases} x - y + 3z = 2 \\ 5x + z = 0 \\ -x + y - 3z = -6 \end{cases}$

A) $x = 3, y = -2, z = 0$ B) inconsistent (no solution)
C) $x = 0, y = 0, z = -2$ D) $x = 0, y = -2, z = 0$

7 Express the Solution of a System of Dependent Equations Containing Three Variables

Solve the system of equations.

1) $\begin{cases} x + 4y - z = 3 \\ x + 5y - 2z = 5 \\ 3x + 12y - 3z = 9 \end{cases}$

A) $x = z - 2$
   $y = -3z - 5$
   $z =$ any real number

B) inconsistent (no solution)

C) $x = -3z - 5$
   $y = z + 2$
   $z =$ any real number

D) $x = 3z + 5$
   $y = z - 2$
   $z =$ any real number

2) $\begin{cases} -x + y + 2z = 0 \\ x + 2y + z = 6 \\ -2x - y + z = -6 \end{cases}$

A) inconsistent (no solution)

B) $x = z + 2$
   $y = 2 - z$
   $z = $ any real number

C) $x = z + 2$
   $y = z - 2$
   $z = $ any real number

D) $x = 2 - z$
   $y = z + 2$
   $z = $ any real number

3) $\begin{cases} 2x - y + 5z = -7 \\ x + y - 2z = -2 \\ x - y + 4z = 8 \end{cases}$

A) $x = -3 - z$
   $y = 3z + 1$
   $z = $ any real number

B) $x = 3z + 1$
   $y = z - 3$
   $z = $ any real number

C) inconsistent (no solution)

D) $x = z + 3$
   $y = 3z + 1$
   $z = $ any real number

8 Demonstrate Additional Understanding and Skills

Verify that the values of the variables listed are solutions of the system of equations.

1) $\begin{cases} x + y = -3 \\ x - y = -7 \end{cases}$
   $x = 5, y = 2$

   A) not a solution
   B) solution

2) $\begin{cases} 4x + y = -16 \\ 3x + 4y = -25 \end{cases}$
   $x = -3, y = -4$

   A) solution
   B) not a solution

3) $\begin{cases} 4x + y = -29 \\ 3x + 4y = -38 \end{cases}$
   $x = -6, y = 5$

   A) solution
   B) not a solution

4) $\begin{cases} x + y = 1 \\ x - y = 5 \end{cases}$
   $x = 3, y = -2$

   A) not a solution
   B) solution

5) $\begin{cases} x + y + z = -5 \\ x - y + 2z = -12 \\ 3x + y + z = -13 \end{cases}$
   $x = -4, y = 2, z = -3$

   A) not a solution
   B) solution

6) $\begin{cases} x + y + z = 1 \\ x - y + 4z = 11 \\ 2x + y + z = -3 \end{cases}$
$x = 4, y = 1, z = -4$

A) not a solution   B) solution

7) $\begin{cases} x - y + 5z = -16 \\ 5x + z = -4 \\ x + 3y + z = -16 \end{cases}$
$x = 0, y = -4, z = -4$

A) solution   B) not a solution

8) $\begin{cases} x - y + 3z = -20 \\ 4x + z = -5 \\ x + 4y + z = 15 \end{cases}$
$x = -5, y = 5, z = 0$

A) solution   B) not a solution

Solve the problem.

9) A flat rectangular piece of aluminum has a perimeter of 68 inches. The length is 10 inches longer than the width. Find the width.

A) 12 inches   B) 32 inches   C) 34 inches   D) 22 inches

10) The Family Fine Arts Center charges $22 per adult and $14 per senior citizen for its performances. On a recent weekend evening when 460 people paid admission, the total receipts were $7600. How many who paid were senior citizens?

A) 225 senior citizens   B) 145 senior citizens   C) 315 senior citizens   D) 235 senior citizens

11) A retired couple has $150,000 to invest to obtain annual income. They want some of it invested in safe Certificates of Deposit yielding 7%. The rest they want to invest in AA bonds yielding 10% per year. How much should they invest in each to realize exactly $13,500 per year?

A) $90,000 at 7% and $60,000 at 10%   B) $100,000 at 10% and $50,000 at 7%
C) $110,000 at 10% and $40,000 at 7%   D) $100,000 at 7% and $50,000 at 10%

12) A tour group split into two groups when waiting in line for food at a fast food counter. The first group bought 7 slices of pizza and 5 soft drinks for $34.99. The second group bought 5 slices of pizza and 6 soft drinks for $29.68. How much does one slice of pizza cost?

A) $1.93 per slice of pizza   B) $3.62 per slice of pizza
C) $3.12 per slice of pizza   D) $2.43 per slice of pizza

13) A movie theater charges $8.00 for adults and $5.00 for children. If there were 40 people altogether and the theater collected $272.00 at the end of the day, how many of them were adults?

A) 24 adults   B) 29 adults   C) 10 adults   D) 16 adults

14) An 8-cylinder Crown Victoria gives 18 miles per gallon in city driving and 21 miles per gallon in highway driving. A 300-mile trip required 15.5 gallons of gasoline. How many whole miles were driven in the city?

A) 153 miles   B) 168 miles   C) 132 miles   D) 147 miles

15) The Family Arts Center charges $21 for adults, $15 for senior citizens, and $10 for children under 12 for their live performances on Sunday afternoon. This past Sunday, the paid revenue was $13,167 for 874 tickets sold. There were 48 more children than adults. How many children attended?

A) 335 children B) 345 children C) 232 children D) 297 children

16) Lexie wants to have an income of $9000 per year from investments. To that end she is going to invest $90,000 in three different accounts. These accounts pay 7%, 10%, and 14% simple interest. If she wants to have $10,000 more in the account paying 7% simple interest than she has in the account paying 14% simple interest, how much should go into each account?

17) Find real numbers a, b, and c such that the graph of the function $y = ax^2 + bx + c$ contains the points (1, 1), (2, 4), and (-3, 29).

18) A company has sales (measured in millions of dollars) of 50, 60, and 75 during the first three consecutive years. Find a quadratic function that fits these data, and use the result to predict the sales during the fourth year. Assume that the quadratic function is of the form $y = ax^2 + bx + c$

A) $y = \frac{15}{2}x^2 - \frac{25}{2}x + \frac{325}{4}$; sales during the fourth year = $151.25 million

B) $y = \frac{5}{2}x^2 + \frac{5}{2}x + 45$; sales during the fourth year = $95 million

C) $y = -5x^2 + 40x + 15$; sales during the fourth year = $95 million

D) $y = 5x^2 + 5x + 40$; sales during the fourth year = $180 million

10.2 Systems of Linear Equations: Matrices

1 Write Augmented Matrix of a System of Linear Equations

Write the augmented matrix for the system.

1) $\begin{cases} 2x + 3y = 34 \\ 4x + 7y = 74 \end{cases}$

A) $\begin{bmatrix} 2 & 3 & | & 74 \\ 7 & 4 & | & 34 \end{bmatrix}$
B) $\begin{bmatrix} 34 & 3 & | & 2 \\ 74 & 4 & | & 7 \end{bmatrix}$
C) $\begin{bmatrix} 2 & 3 & | & 34 \\ 4 & 7 & | & 74 \end{bmatrix}$
D) $\begin{bmatrix} 2 & 4 & | & 34 \\ 3 & 7 & | & 74 \end{bmatrix}$

2) $\begin{cases} 2x + 4y = 42 \\ 6y = 36 \end{cases}$

A) $\begin{bmatrix} 6 & 0 & | & 36 \\ 2 & 4 & | & 4 \end{bmatrix}$
B) $\begin{bmatrix} 2 & 4 & | & 42 \\ 0 & 6 & | & 36 \end{bmatrix}$
C) $\begin{bmatrix} 2 & 4 & | & 42 \\ 6 & 36 & | & 0 \end{bmatrix}$
D) $\begin{bmatrix} 42 & 4 & | & 2 \\ 36 & 0 & | & 6 \end{bmatrix}$

3) $\begin{cases} -2x + 9y + 7z = 2 \\ 5x + 6y + 7z = 48 \\ -2x + 4y + 8z = -2 \end{cases}$

A) $\begin{bmatrix} -2 & 9 & | & 7 \\ 5 & 6 & | & 7 \\ -2 & 4 & | & 8 \end{bmatrix}$
B) $\begin{bmatrix} 2 & 7 & 9 & | & -2 \\ 48 & 7 & 6 & | & 5 \\ -2 & 8 & 4 & | & -2 \end{bmatrix}$
C) $\begin{bmatrix} -2 & 5 & -2 & | & 2 \\ 9 & 6 & 4 & | & 48 \\ 7 & 7 & 8 & | & -2 \end{bmatrix}$
D) $\begin{bmatrix} -2 & 9 & 7 & | & 2 \\ 5 & 6 & 7 & | & 48 \\ -2 & 4 & 8 & | & -2 \end{bmatrix}$

4) $\begin{cases} 8x + 3z = 17 \\ -2y + 3z = 7 \\ 7x - 2y + 9z = 32 \end{cases}$

A) $\begin{bmatrix} 8 & 0 & | & 3 \\ 0 & -2 & | & 3 \\ 7 & -2 & | & 9 \end{bmatrix}$
B) $\begin{bmatrix} 8 & 3 & 0 & | & 17 \\ -2 & 3 & 0 & | & 7 \\ 7 & -2 & 9 & | & 32 \end{bmatrix}$
C) $\begin{bmatrix} 8 & 0 & 3 & | & 17 \\ 0 & -2 & 3 & | & 7 \\ 7 & -2 & 9 & | & 32 \end{bmatrix}$
D) $\begin{bmatrix} 8 & 0 & 7 & | & 17 \\ 0 & -2 & -2 & | & 7 \\ 3 & 3 & 9 & | & 32 \end{bmatrix}$

2 Write the System from the Augmented Matrix

Write a system of equations associated with the augmented matrix. Do not solve.

1) $\begin{bmatrix} 1 & 0 & 0 & | & 1 \\ 0 & 1 & 0 & | & -5 \\ 0 & 0 & 1 & | & 2 \end{bmatrix}$

A) $\begin{cases} x = 0 \\ y = -4 \\ z = 3 \end{cases}$
B) $\begin{cases} x = 1 \\ y = -5 \\ z = 2 \end{cases}$
C) $\begin{cases} x = -1 \\ y = -7 \\ z = 0 \end{cases}$
D) $\begin{cases} x = -1 \\ y = 5 \\ z = -2 \end{cases}$

2) $\begin{bmatrix} 5 & 5 & 8 & | & -2 \\ 7 & 0 & 8 & | & 4 \\ 6 & 9 & 0 & | & 2 \end{bmatrix}$

A) $\begin{cases} 5x + 5y + 8z = -2 \\ 7x + 8z = 4 \\ 6x + 9z = 2 \end{cases}$
B) $\begin{cases} 5x + 5y + 8z = -2 \\ 7x + 8z = 4 \\ 6x + 9y = 2 \end{cases}$
C) $\begin{cases} 5x - 5y + 8z = -2 \\ 7x + 8z = -4 \\ 6x + 9y = -2 \end{cases}$

3 Perform Row Operations on a Matrix

Perform the indicated operation.

1) Perform the row operation $R_3 = 4r_1 + r_3$ on the following augmented matrix:

$\begin{bmatrix} -7 & -5 & -1 & | & -10 \\ 6 & -2 & 9 & | & 5 \\ 28 & -6 & 6 & | & 18 \end{bmatrix}$

Perform in order (a), (b), and (c) on the augmented matrix.

2) (a) $R_2 = -3r_1 + r_2$
   (b) $R_3 = -2r_1 + r_3$
   (c) $R_3 = 5r_2 + r_3$

$\begin{bmatrix} 1 & -3 & -5 & | & -2 \\ 3 & -5 & -4 & | & 5 \\ 2 & 5 & 4 & | & 6 \end{bmatrix}$

A) $\begin{bmatrix} 1 & -3 & -5 & | & -2 \\ 0 & 14 & 19 & | & 4 \\ 0 & 81 & 109 & | & 30 \end{bmatrix}$
B) $\begin{bmatrix} 1 & -3 & -5 & | & -2 \\ 0 & 4 & 11 & | & 11 \\ 0 & 31 & 69 & | & 65 \end{bmatrix}$
C) $\begin{bmatrix} 1 & -3 & -5 & | & -2 \\ 0 & -8 & -9 & | & 3 \\ 0 & -29 & -31 & | & -5 \end{bmatrix}$
D) $\begin{bmatrix} 1 & -3 & -5 & | & -2 \\ 0 & 4 & 11 & | & 11 \\ 0 & 15 & 25 & | & 21 \end{bmatrix}$

3) (a) $R_2 = 2r_1 + r_2$
   (b) $R_3 = 2r_1 + r_3$
   (c) $R_3 = 4r_2 + r_3$

$\begin{bmatrix} 1 & -3 & -5 & | & 2 \\ -2 & -5 & 2 & | & 5 \\ -2 & -5 & 4 & | & 6 \end{bmatrix}$

A) $\begin{bmatrix} 1 & -3 & -5 & | & 2 \\ 0 & -1 & -8 & | & 9 \\ 0 & -12 & -14 & | & 19 \end{bmatrix}$
B) $\begin{bmatrix} 1 & -3 & -5 & | & 2 \\ 0 & -11 & -8 & | & 9 \\ 0 & -55 & -38 & | & 46 \end{bmatrix}$
C) $\begin{bmatrix} 1 & -3 & -5 & | & 2 \\ 0 & -11 & -8 & | & 9 \\ 0 & -43 & -18 & | & 46 \end{bmatrix}$
D) $\begin{bmatrix} 1 & -3 & -5 & | & 2 \\ 0 & -11 & -8 & | & 9 \\ 0 & -7 & 42 & | & 46 \end{bmatrix}$

4 Solve a System of Linear Equations Using Matrices

Solve each system of equations using matrices (row operations). If the system has no solution, say that it is inconsistent.

1) $\begin{cases} 6x + 3y = 3 \\ 5x + 6y = 20 \end{cases}$

   A) Inconsistent, ∅   B) x = -2, y = 5   C) x = 5, y = -2   D) x = -2, y = -5

2) $\begin{cases} 2x - 9y - z = -78 \\ x - 8y - 8z = -93 \\ 8x + y + z = 36 \end{cases}$

   A) x = 3, y = 3, z = 9   B) x = -3, y = 9, z = 6   C) x = 3, y = 9, z = 3   D) Inconsistent, ∅

3) $\begin{cases} -4x - y + 2z = -9 \\ 6x - 7z = -11 \\ 5y + z = 20 \end{cases}$

   A) x = -4, y = 3, z = 8   B) x = 4, y = 5, z = 3   C) Inconsistent, ∅   D) x = 4, y = 3, z = 5

Solve the problem using matrices.

4) Find real numbers a, b, and c such that the graph of the function $y = ax^2 + bx + c$ contains the points (-2, -4), (1, -1), and (3, -19).

5) Melody has $45,000 to invest and wishes to receive an annual income of $4290 from this money. She has chosen investments that pay 5%, 8%, and 12% simple interest. Melody wants to have the amount invested at 12% to be double the amount invested at 8%. How much should she invest at each rate?

6) A company manufactures three types of wooden chairs: the Kitui, the Goa, and the Santa Fe. To make a Kitui chair requires 1 hour of cutting time, 1.5 hours of assembly time, and 1 hour of finishing time. A Goa chair requires 1.5 hours of cutting time, 2.5 hours of assembly time and 2 hours of finishing time. A Santa Fe chair requires 1.5 hours of cutting time, 3 hours of assembly time, and 3 hours of finishing time. If 41 hours of cutting time, 70 hours of assembly time, and 58 hours of finishing time were used one week, how many of each type of chair were produced?

Solve the problem.

7) Find the function $f(x) = ax^3 + bx^2 + cx + d$ for which $f(0) = -2$, $f(1) = 5$, $f(-1) = 3$, $f(2) = 4$.

   A) $f(x) = \frac{8}{3}x^3 + 4x^2 + \frac{5}{3}x - 2$

   B) $f(x) = -\frac{10}{3}x^3 + 6x^2 + \frac{13}{3}x - 2$

   C) $f(x) = 10x^3 - 18x^2 - 13x + 6$

   D) $f(x) = -8x^3 + 12x^2 + 5x - 6$

5 Demonstrate Additional Understanding and Skills

Use a graphing utility to solve the problem.

1) $\begin{cases} 5x + 5y + z = 1 \\ 5x - 2y - z = 34 \\ 2x + y + 5z = 10 \end{cases}$

   A) x = 1, y = -5, z = 5   B) Inconsistent, ∅   C) x = 5, y = -5, z = 1   D) x = 5, y = 1, z = -5

2) Ron attends a cocktail party (with his graphing calculator in his pocket). He wants to limit his food intake to 150 g protein, 132 g fat, and 189 g carbohydrate. According to the health conscious hostess, the marinated mushroom caps have 3 g protein, 5 g fat, and 9 g carbohydrate; the spicy meatballs have 14 g protein, 7 g fat, and 15 g carbohydrate; and the deviled eggs have 13 g protein, 15 g fat, and 6 g carbohydrate. How many of each snack can he eat to obtain his goal?

   A) 3 mushrooms, 9 meatballs, 6 eggs   B) 10 mushrooms, 7 meatballs, 4 eggs
   C) 6 mushrooms, 3 meatballs, 9 eggs   D) 9 mushrooms, 6 meatballs, 3 eggs

Solve the problem.

3) Rob bought 2 pairs of shorts, 3 shirts and a pair of shoes for $146.64. Jessie bought 3 pairs of shorts, 5 shirts and 2 pairs of shoes for $256.35. Allen bought a pair of shorts and 4 shirts for $104.07. What is the price of a pair of shorts? Express answer rounded to two decimal places.

   A) $50.40   B) $10.30   C) $14.55   D) $22.38

4) Jenny receives $1270 per year from three different investments totaling $20,000. One of the investments pays 6%, the second one pays 8%, and the third one pays 5%. If the money invested at 8% is $1500 less than the amount invested at 5%, how much money has Jenny invested in the investment that pays 6%?

   A) $4500   B) $8500   C) $10,000   D) $1500

10.3 Systems of Linear Equations: Determinants

1 Evaluate 2 by 2 Determinants

Find the value of the determinant.

1) $\begin{vmatrix} 3 & 7 \\ 5 & 9 \end{vmatrix}$

   A) −24   B) 62   C) −8   D) 8

2) $\begin{vmatrix} 9 & -6 \\ -4 & 7 \end{vmatrix}$

   A) 87   B) −26   C) −39   D) 39

3) $\begin{vmatrix} -1 & 2 \\ 2 & 1 \end{vmatrix}$

   A) −4   B) 5   C) −5   D) 3

4) $\begin{vmatrix} 12 & -7 \\ -4 & 3 \end{vmatrix}$

   A) 64   B) 4   C) 8   D) −8

2 Use Cramer's Rule to Solve a System of Two Equations Containing Two Variables

Use Cramer's rule to solve the linear system.

1) $\begin{cases} 3x + 2y = -8 \\ 4x + y = -14 \end{cases}$

   A) $x = 2, y = -4$ B) $x = -2, y = -4$ C) $x = -4, y = 2$ D) $x = 4, y = -2$

2) $\begin{cases} 3x + 4y = 36 \\ -2x + 5y = 22 \end{cases}$

   A) $x = 6, y = 4$ B) $x = -6, y = 4$ C) $x = 4, y = 6$ D) $x = -4, y = -6$

3) $\begin{cases} 3x + 2y = 29 \\ 3x - 3y = 24 \end{cases}$

A) $x = -1, y = 9$ B) $x = 9, y = 1$ C) $x = -9, y = -1$ D) $x = 1, y = 9$

4) $\begin{cases} 4x - 7y = 5 \\ 2x + 5y = -3 \end{cases}$

A) $x = -\frac{2}{17}, y = \frac{11}{17}$ B) $x = \frac{2}{3}, y = \frac{1}{3}$ C) $x = \frac{23}{3}, y = -\frac{11}{3}$ D) $x = \frac{2}{17}, y = -\frac{11}{17}$

3 Evaluate 3 by 3 Determinants

Find the value of the determinant.

1) $\begin{vmatrix} -2 & 3 & -4 \\ 3 & 5 & -4 \\ -3 & -2 & -3 \end{vmatrix}$

A) 107 B) 1 C) 73 D) −73

2) $\begin{vmatrix} 1 & 5 & 3 \\ 2 & 2 & 4 \\ 3 & 3 & 2 \end{vmatrix}$

A) −32 B) 132 C) 32 D) −88

3) $\begin{vmatrix} -2 & 5 & 4 \\ 3 & -2 & 1 \\ 1 & 6 & -3 \end{vmatrix}$

A) −12 B) −90 C) 80 D) 130

4 Use Cramer's Rule to Solve a System of Three Equations Containing Three Variables

Use Cramer's rule to solve the linear system.

1) $\begin{cases} 5x - 9y - z = -21 \\ x - 7y - 6z = -58 \\ 6x + y + z = 64 \end{cases}$

A) $x = 9, y = 7, z = 3$ B) $x = 7, y = 3, z = 7$ C) $x = 10, y = 5, z = 3$ D) $x = 9, y = -7, z = -3$

2) $\begin{cases} 6x - 7y - 4z = -50 \\ -6x + 4y + 8z = 62 \\ 5x - 4y + 7z = 66 \end{cases}$

A) $x = 7, y = 8, z = 9$ B) $x = 8, y = 9, z = 8$ C) $x = 8, y = 6, z = 9$ D) $x = 7, y = -8, z = -9$

3) $\begin{cases} 4x - 5z = -16 \\ -3x + 3y + 8z = 50 \\ -4x - 4y = -32 \end{cases}$

A) $x = 2, y = 5, z = 4$ B) $x = 7, y = 4, z = 7$ C) $x = 1, y = -7, z = -4$ D) $x = 1, y = 7, z = 4$

5 Know Properties of Determinants

Use the properties of determinants to find the value of the second determinant, given the value of the first.

1) $\begin{vmatrix} x & y & z \\ u & v & w \\ 1 & -3 & 2 \end{vmatrix} = 68 \qquad \begin{vmatrix} 1 & -3 & 2 \\ u & v & w \\ x & y & z \end{vmatrix} = ?$

   A) Can't determine   B) −68   C) 0   D) 68

2) $\begin{vmatrix} x & y & z \\ u & v & w \\ 1 & 1 & -3 \end{vmatrix} = 23 \qquad \begin{vmatrix} x & y & z \\ u & v & w \\ -3 & -3 & 9 \end{vmatrix} = ?$

   A) −69   B) 23   C) 69   D) −23

3) $\begin{vmatrix} x & y & z \\ u & v & w \\ 1 & 3 & 2 \end{vmatrix} = 33 \qquad \begin{vmatrix} u & v & w \\ 3 & 9 & 6 \\ x & y & z \end{vmatrix} = ?$

   A) −99   B) −33   C) 33   D) 99

4) $\begin{vmatrix} x & y & z \\ u & v & w \\ 1 & -1 & -1 \end{vmatrix} = -14 \qquad \begin{vmatrix} 1 & -1 & -1 \\ 2u & 2v & 2w \\ x-1 & y+1 & z+1 \end{vmatrix} = ?$

   A) −28   B) 28   C) −14   D) 14

5) $\begin{vmatrix} x & y & z \\ u & v & w \\ 1 & 2 & 3 \end{vmatrix} = 22 \qquad \begin{vmatrix} x & y & z-x \\ u & v & w-u \\ 1 & 2 & 2 \end{vmatrix} = ?$

   A) 0   B) Can't determine   C) 22   D) −22

6) $\begin{vmatrix} x & y & z \\ u & v & w \\ 1 & -1 & -4 \end{vmatrix} = 16 \qquad \begin{vmatrix} x-3 & y+3 & z+12 \\ -3u-1 & -3v+1 & -3w+4 \\ 1 & -1 & -4 \end{vmatrix} = ?$

   A) −16   B) 16   C) 48   D) −48

7) Given $\begin{vmatrix} s & t & u \\ v & w & x \\ 4 & 2 & 8 \end{vmatrix} = 3$, find the value of $\begin{vmatrix} 32-s & 16-t & 64-u \\ v & w & x \\ 4 & 2 & 8 \end{vmatrix}$.

   A) −3   B) 24   C) −24   D) 3

Solve the problem.

8) Given that $\begin{vmatrix} x & y & z \\ a & b & c \\ 2 & 4 & 5 \end{vmatrix} = 3$, find the value of the determinant $\begin{vmatrix} 2 & 4 & 5 \\ 3a & 3b & 3c \\ x-2 & y-4 & z-5 \end{vmatrix}$.

   A) 6   B) 0   C) 9   D) −9

6 Demonstrate Additional Understanding and Skills

Solve for x.

1) $\begin{vmatrix} 8 & x \\ 2 & 5 \end{vmatrix} = 32$

2) $\begin{vmatrix} x & -4 & -1 \\ -2 & 2 & 0 \\ -1 & -2 & 8 \end{vmatrix} = 10$

Solve the problem.

3) The equation of the line passing through the distinct points (x_1, y_1) and (x_2, y_2) is given by $\begin{vmatrix} x & y & 1 \\ x_1 & y_1 & 1 \\ x_2 & y_2 & 1 \end{vmatrix} = 0$.

Find the equation of the line passing through the points (3, 5) and (-1, 4).

A) $x + 7y + 17 = 0$ B) $x - 4y + 17 = 0$ C) $-x + 4y - 17 = 0$ D) $x + 4y + 17 = 0$

10.4 Matrix Algebra

1 Find the Sum and Difference of Two Matrices

Perform the indicated operation, whenever possible.

1) $\begin{bmatrix} 4 & 3 \\ -4 & 1 \\ -1 & -9 \end{bmatrix} + \begin{bmatrix} 2 & 1 \\ -3 & -2 \\ -2 & 3 \end{bmatrix}$

A) $\begin{bmatrix} 6 & 4 \\ -7 & -1 \\ -3 & -6 \end{bmatrix}$ B) $\begin{bmatrix} 6 & 1 \\ -7 & -1 \\ -3 & -6 \end{bmatrix}$ C) $\begin{bmatrix} 2 & 2 \\ -1 & 3 \\ 1 & -12 \end{bmatrix}$ D) $\begin{bmatrix} 6 & 4 \\ 7 & 1 \\ -3 & 6 \end{bmatrix}$

2) $\begin{bmatrix} -1 & 0 \\ 5 & 3 \end{bmatrix} - \begin{bmatrix} -1 & 5 \\ 3 & 1 \end{bmatrix}$

A) $\begin{bmatrix} 0 & 5 \\ -2 & -2 \end{bmatrix}$ B) $[-1]$ C) $\begin{bmatrix} -2 & 5 \\ 8 & 4 \end{bmatrix}$ D) $\begin{bmatrix} 0 & -5 \\ 2 & 2 \end{bmatrix}$

3) If $A = \begin{bmatrix} 7 & -4 & 8 \\ -6 & 5 & -1 \\ 0 & 6 & -3 \end{bmatrix}$ and $B = \begin{bmatrix} -2 & -6 & -1 \\ -7 & -4 & 3 \\ -3 & -9 & -5 \end{bmatrix}$, find $A - B$.

A) $\begin{bmatrix} 9 & 2 & 9 \\ 1 & 9 & -4 \\ 3 & 15 & 2 \end{bmatrix}$ B) $\begin{bmatrix} 5 & -10 & 7 \\ -13 & 1 & 2 \\ -3 & -3 & -8 \end{bmatrix}$ C) $\begin{bmatrix} 9 & 2 & 9 \\ 1 & 9 & 2 \\ 3 & 15 & -4 \end{bmatrix}$ D) $\begin{bmatrix} 5 & -10 & 7 \\ -13 & 1 & -8 \\ -3 & -3 & 2 \end{bmatrix}$

4) Let $A = \begin{bmatrix} -4 & 6 & 7 \\ 3 & -5 & 12 \\ 7 & -11 & 14 \end{bmatrix}$ and $B = \begin{bmatrix} 6 & 10 & -4 \\ -5 & 6 & -8 \\ 3 & 11 & 7 \end{bmatrix}$. Find $A - B$.

A) $\begin{bmatrix} 10 & 4 & -11 \\ -8 & 11 & -20 \\ -4 & 22 & -7 \end{bmatrix}$ B) $\begin{bmatrix} -10 & -4 & 11 \\ 8 & -11 & 20 \\ 4 & -22 & 7 \end{bmatrix}$ C) $\begin{bmatrix} -10 & -4 & 3 \\ -2 & -11 & 4 \\ 4 & -22 & 7 \end{bmatrix}$ D) $\begin{bmatrix} 2 & 16 & 3 \\ -2 & 1 & 4 \\ 10 & 0 & 21 \end{bmatrix}$

2 Find Scalar Multiples of a Matrix

Perform the indicated matrix operations.

1) Let $A = \begin{bmatrix} -3 & 5 \\ 0 & 2 \end{bmatrix}$. Find $2A$.

A) $\begin{bmatrix} -6 & 5 \\ 0 & 2 \end{bmatrix}$ B) $\begin{bmatrix} -6 & 10 \\ 0 & 2 \end{bmatrix}$ C) $\begin{bmatrix} -6 & 10 \\ 0 & 4 \end{bmatrix}$ D) $\begin{bmatrix} -1 & 7 \\ 2 & 4 \end{bmatrix}$

2) Let B = [-1 2 4 -3]. Find -4B.

   A) [4 2 4 -3]   B) [4 -8 -16 12]   C) [-4 8 16 -12]   D) [-3 0 2 -5]

3) Let $A = \begin{bmatrix} 3 & 3 \\ 2 & 4 \end{bmatrix}$ and $B = \begin{bmatrix} 0 & 4 \\ -1 & 6 \end{bmatrix}$. Find 4A + B.

   A) $\begin{bmatrix} 12 & 16 \\ 7 & 22 \end{bmatrix}$ B) $\begin{bmatrix} 12 & 16 \\ 1 & 10 \end{bmatrix}$ C) $\begin{bmatrix} 12 & 7 \\ 7 & 10 \end{bmatrix}$ D) $\begin{bmatrix} 12 & 28 \\ 4 & 40 \end{bmatrix}$

4) Let $C = \begin{bmatrix} 1 \\ -3 \\ 2 \end{bmatrix}$ and $D = \begin{bmatrix} -1 \\ 3 \\ -2 \end{bmatrix}$. Find C - 3D.

   A) $\begin{bmatrix} 4 \\ -6 \\ 4 \end{bmatrix}$ B) $\begin{bmatrix} -2 \\ 6 \\ -4 \end{bmatrix}$ C) $\begin{bmatrix} -4 \\ 12 \\ -8 \end{bmatrix}$ D) $\begin{bmatrix} 4 \\ -12 \\ 8 \end{bmatrix}$

5) Let A = [-2 2] and B = [1 0]. Find 3A + 4B.

   A) [1 2]   B) [-2 6]   C) [-3 4]   D) [-6 4]

6) Let $A = \begin{bmatrix} -9 & 3 & 5 \\ 5 & 5 & 4 \\ -6 & 2 & 5 \end{bmatrix}$ and $B = \begin{bmatrix} 5 & 2 & 2 \\ -9 & 8 & 5 \\ -8 & 8 & 9 \end{bmatrix}$. Find -4A + 3B.

   A) $\begin{bmatrix} 41 & -10 & -18 \\ -29 & -12 & -11 \\ 16 & 0 & -11 \end{bmatrix}$ B) $\begin{bmatrix} 51 & -6 & -14 \\ -47 & 4 & -1 \\ 0 & 16 & 7 \end{bmatrix}$ C) $\begin{bmatrix} -4 & 5 & 7 \\ -4 & 13 & 9 \\ -14 & 10 & 14 \end{bmatrix}$ D) $\begin{bmatrix} -4 & -4 & -14 \\ 5 & 13 & 10 \\ 7 & 9 & 14 \end{bmatrix}$

3 Find the Product of Two Matrices

Perform the matrix multiplication.

1) Given $A = \begin{bmatrix} 0 & -3 & 1 \\ 5 & -1 & 0 \end{bmatrix}$ and $B = \begin{bmatrix} 1 & 2 \\ 0 & 1 \\ 1 & -1 \end{bmatrix}$, find AB.

   A) $\begin{bmatrix} 10 & -5 & 1 \\ 5 & -1 & 0 \\ -5 & -2 & 0 \end{bmatrix}$ B) $\begin{bmatrix} 1 & -4 \\ 5 & 9 \end{bmatrix}$ C) $\begin{bmatrix} 1 & 5 \\ -4 & 9 \end{bmatrix}$ D) $\begin{bmatrix} 0 & 10 \\ 0 & -1 \\ 0 & 0 \end{bmatrix}$

2) Given $A = \begin{bmatrix} -2 \\ 1 \\ -1 \end{bmatrix}$ and B = [1 -1 0], find AB.

3) $\begin{bmatrix} -2 & 3 \\ 4 & 2 \end{bmatrix} \begin{bmatrix} -2 & 0 \\ -1 & 4 \end{bmatrix}$

   A) $\begin{bmatrix} 1 & 12 \\ -10 & 8 \end{bmatrix}$ B) $\begin{bmatrix} 4 & -6 \\ -6 & 5 \end{bmatrix}$ C) $\begin{bmatrix} 12 & 1 \\ 8 & -10 \end{bmatrix}$ D) $\begin{bmatrix} 4 & 0 \\ -4 & 8 \end{bmatrix}$

4) $\begin{bmatrix} -3 & 8 & -2 \\ -7 & 6 & 6 \end{bmatrix} \begin{bmatrix} 7 & 6 & 7 \\ 8 & -7 & 4 \\ -9 & 8 & 5 \end{bmatrix}$

A) $\begin{bmatrix} 61 & -90 & 1 \\ -55 & -36 & 5 \end{bmatrix}$
B) $\begin{bmatrix} 61 & -55 \\ -90 & -36 \\ 1 & 5 \end{bmatrix}$
C) $\begin{bmatrix} -21 & 48 & -14 \\ -56 & -42 & 24 \\ 63 & 48 & 30 \end{bmatrix}$
D) $\begin{bmatrix} -3 & 8 & -2 \\ -7 & 6 & 6 \\ 7 & 6 & 6 \\ 8 & -7 & 4 \\ -9 & -9 & 5 \end{bmatrix}$

Perform the indicated operations and simplify.

5) Let $A = \begin{bmatrix} 3 & -4 \\ -2 & 5 \end{bmatrix}$, $B = \begin{bmatrix} 5 & -2 & 8 \\ 1 & 0 & -3 \end{bmatrix}$, and $C = \begin{bmatrix} 7 & -9 & 0 \\ 3 & -5 & 1 \\ -1 & 6 & 2 \end{bmatrix}$. Find AB + BC.

A) $\begin{bmatrix} 32 & 7 & 50 \\ 5 & -23 & -37 \end{bmatrix}$
B) $\begin{bmatrix} 32 & 19 & 40 \\ -15 & 31 & -37 \end{bmatrix}$
C) $\begin{bmatrix} 68 & 3 & 31 \\ 8 & -2 & -5 \end{bmatrix}$
D) $\begin{bmatrix} -10 & -19 & 12 \\ -15 & 31 & -25 \end{bmatrix}$

4 Find the Inverse of a Matrix

Find the inverse of the matrix.

1) $\begin{bmatrix} 0 & -5 \\ 1 & -5 \end{bmatrix}$

A) $\begin{bmatrix} -1 & -1 \\ \frac{1}{5} & 0 \end{bmatrix}$
B) $\begin{bmatrix} 0 & 1 \\ -\frac{1}{5} & -1 \end{bmatrix}$
C) $\begin{bmatrix} -\frac{1}{5} & 0 \\ -1 & 1 \end{bmatrix}$
D) $\begin{bmatrix} -1 & 1 \\ -\frac{1}{5} & 0 \end{bmatrix}$

2) $\begin{bmatrix} 4 & 0 \\ 1 & -5 \end{bmatrix}$

A) $\begin{bmatrix} \frac{1}{4} & 0 \\ -\frac{1}{20} & -\frac{1}{5} \end{bmatrix}$
B) No inverse
C) $\begin{bmatrix} \frac{1}{4} & 0 \\ \frac{1}{20} & -\frac{1}{5} \end{bmatrix}$
D) $\begin{bmatrix} -\frac{1}{5} & 0 \\ \frac{1}{20} & \frac{1}{4} \end{bmatrix}$

3) $\begin{bmatrix} 6 & 2 \\ 2 & 0 \end{bmatrix}$

A) $\begin{bmatrix} -\frac{3}{2} & \frac{1}{2} \\ \frac{1}{2} & 0 \end{bmatrix}$
B) $\begin{bmatrix} 0 & \frac{1}{2} \\ \frac{1}{2} & -\frac{3}{2} \end{bmatrix}$
C) $\begin{bmatrix} 0 & -\frac{1}{2} \\ -\frac{1}{2} & -\frac{3}{2} \end{bmatrix}$
D) $\begin{bmatrix} \frac{1}{2} & -\frac{3}{2} \\ 0 & \frac{1}{2} \end{bmatrix}$

4) $\begin{bmatrix} 3 & 3 \\ 0 & 6 \end{bmatrix}$

A) $\begin{bmatrix} \frac{1}{6} & -\frac{1}{6} \\ 0 & \frac{1}{3} \end{bmatrix}$
B) $\begin{bmatrix} \frac{1}{3} & -\frac{1}{6} \\ 0 & \frac{1}{6} \end{bmatrix}$
C) $\begin{bmatrix} \frac{1}{3} & \frac{1}{6} \\ 0 & \frac{1}{6} \end{bmatrix}$
D) $\begin{bmatrix} 0 & \frac{1}{6} \\ \frac{1}{3} & -\frac{1}{6} \end{bmatrix}$

5) $\begin{bmatrix} 3 & 3 \\ 5 & 5 \end{bmatrix}$

A) No inverse
B) $\begin{bmatrix} -\frac{5}{16} & -\frac{3}{16} \\ -\frac{5}{16} & -\frac{3}{16} \end{bmatrix}$
C) $\begin{bmatrix} -\frac{5}{16} & \frac{3}{16} \\ \frac{5}{16} & -\frac{3}{16} \end{bmatrix}$
D) $\begin{bmatrix} \frac{5}{16} & -\frac{3}{16} \\ -\frac{5}{16} & \frac{3}{16} \end{bmatrix}$

6) $\begin{bmatrix} -6 & -3 \\ 1 & 2 \end{bmatrix}$

A) $\begin{bmatrix} \frac{1}{9} & \frac{2}{3} \\ -\frac{2}{9} & -\frac{1}{3} \end{bmatrix}$
B) $\begin{bmatrix} \frac{2}{3} & -\frac{1}{3} \\ \frac{1}{9} & -\frac{2}{9} \end{bmatrix}$
C) $\begin{bmatrix} -\frac{2}{9} & -\frac{1}{3} \\ \frac{1}{9} & \frac{2}{3} \end{bmatrix}$
D) $\begin{bmatrix} -\frac{2}{9} & \frac{1}{3} \\ -\frac{1}{9} & \frac{2}{3} \end{bmatrix}$

7) $\begin{bmatrix} 1 & 0 & 0 \\ -9 & 1 & 0 \\ 0 & -7 & 1 \end{bmatrix}$

A) $\begin{bmatrix} 1 & 0 & 0 \\ 9 & 1 & 0 \\ 63 & 7 & 1 \end{bmatrix}$
B) $\begin{bmatrix} 1 & 0 & 0 \\ -7 & -1 & 0 \\ -63 & -9 & 1 \end{bmatrix}$
C) $\begin{bmatrix} 1 & 0 & 0 \\ -9 & 1 & 0 \\ 0 & 0 & -7 \end{bmatrix}$
D) $\begin{bmatrix} 1 & -7 & -63 \\ 0 & 1 & 1 \\ 0 & 0 & 1 \end{bmatrix}$

8) $A = \begin{bmatrix} 1 & 0 & 0 \\ -1 & 1 & 0 \\ 1 & 1 & 1 \end{bmatrix}$

A) $\begin{bmatrix} 1 & 1 & 1 \\ 0 & 1 & 1 \\ 0 & 0 & 1 \end{bmatrix}$
B) $\begin{bmatrix} -1 & 0 & 0 \\ -1 & -1 & 0 \\ -1 & -1 & -1 \end{bmatrix}$
C) $\begin{bmatrix} 1 & -1 & 1 \\ 0 & 1 & -1 \\ 0 & 0 & 1 \end{bmatrix}$
D) $\begin{bmatrix} 1 & 0 & 0 \\ 1 & 1 & 0 \\ -2 & -1 & 1 \end{bmatrix}$

5 Solve a System of Linear Equations Using Inverse Matrices

Solve the system using the inverse method.

1) $\begin{cases} 2x + 6y = 2 \\ 2x - y = -5 \end{cases}$

A) $x = 1, y = -2$
B) $x = 2, y = -1$
C) $x = -1, y = 2$
D) $x = -2, y = 1$

2) $\begin{cases} x + 3y = -8 \\ 21x + 6y = 3 \end{cases}$

   A) $x = 3, y = -1$ B) $x = -1, y = 3$ C) $x = 1, y = -3$ D) $x = -3, y = 1$

3) $\begin{cases} 4x + 2z = -20 \\ x - y - 5z = 5 \\ -3x - 2y - z = 12 \end{cases}$

4) $\begin{cases} 2x + 4y - 5z = -8 \\ x + 5y + 2z = -1 \\ 3x + 3y + 3z = 15 \end{cases}$

   A) $x = -5, y = -2, z = -2$ B) $x = 5, y = -2, z = 2$

   C) $x = 2, y = 5, z = 2$ D) $x = 5, y = 2, z = -2$

5) $\begin{cases} x + 2y + 3z = 11 \\ x + y + z = -9 \\ 2x + 2y + z = 5 \end{cases}$

   A) $x = -20, y = 99, z = -12$ B) $x = -20, y = -24, z = -13$

   C) $x = -52, y = 66, z = -23$ D) $x = 44, y = 18, z = 5$

6) $\begin{cases} x + 2y + 3z = -5 \\ x + y + z = -6 \\ x - 2z = -9 \end{cases}$

   A) $x = -43, y = -63, z = -26$ B) $x = -5, y = -3, z = 2$

   C) $x = 1, y = -8, z = 2$ D) $x = -5, y = 0, z = 0$

7) $\begin{cases} x + 2y + 3z = -11 \\ x + y + z = -12 \\ -x + y + 2z = -3 \end{cases}$

   A) $x = 4, y = -33, z = 17$ B) $x = 19, y = -40, z = -10$

   C) $x = -26, y = -99, z = -61$ D) $x = -11, y = -48, z = 6$

6 Demonstrate Additional Understanding and Skills

Show that the matrix has no inverse.

1) $A = \begin{bmatrix} 10 & -6 \\ -5 & 3 \end{bmatrix}$

Use a graphing utility to find the inverse of the matrix, if it exists. Round answers to two decimal places, if necessary.

2) $\begin{bmatrix} -16 & 3 & 28 \\ 5 & -14 & 15 \\ 34 & 25 & 2 \end{bmatrix}$

A) $\begin{bmatrix} -0.01 & 0.03 & 0.02 \\ 0.02 & -0.03 & 0.01 \\ 0.02 & 0.02 & 0.01 \end{bmatrix}$ B) $\begin{bmatrix} -0.02 & 0.03 & 0.02 \\ 0.02 & -0.04 & 0.02 \\ 0.02 & 0.02 & 0.01 \end{bmatrix}$

C) $\begin{bmatrix} -0.02 & 0.03 & 0.02 \\ 0.02 & -0.04 & 0.01 \\ 0.02 & 0.02 & 0.00 \end{bmatrix}$ D) $\begin{bmatrix} -0.01 & 0.02 & 0.01 \\ 0.02 & -0.03 & 0.01 \\ 0.02 & 0.02 & 0.01 \end{bmatrix}$

3) $\begin{bmatrix} 1 & 7 & 0 & 0 \\ 0 & 1 & -2 & 0 \\ 0 & 0 & 1 & 6 \\ 0 & 0 & 0 & 1 \end{bmatrix}$

A) $\begin{bmatrix} 1 & 0 & 0 & 0 \\ -6 & 1 & 0 & 0 \\ 12 & 2 & 1 & 0 \\ 84 & 14 & -7 & 1 \end{bmatrix}$
B) $\begin{bmatrix} 1 & 0 & 0 & 0 \\ -7 & 1 & 0 & 0 \\ -14 & 2 & 1 & 0 \\ 42 & -12 & -6 & 1 \end{bmatrix}$
C) $\begin{bmatrix} 1 & -7 & -14 & 84 \\ 0 & 1 & 2 & -12 \\ 0 & 0 & 1 & -6 \\ 0 & 0 & 0 & 1 \end{bmatrix}$
D) $\begin{bmatrix} 1 & -6 & 12 & 84 \\ 0 & 1 & 2 & 14 \\ 0 & 0 & 1 & -7 \\ 0 & 0 & 0 & 1 \end{bmatrix}$

10.5 Partial Fraction Decomposition

1 Decompose P/Q, Where Q Has Only Nonrepeated Linear Factors

Write the partial fraction decomposition of the rational expression.

1) $\dfrac{x - 13}{(x - 3)(x - 5)}$

A) $\dfrac{5}{x - 3} + \dfrac{4}{x - 5}$
B) $\dfrac{-4}{x - 3} + \dfrac{5}{x - 5}$
C) $\dfrac{4}{x - 3} + \dfrac{-5}{x - 5}$
D) $\dfrac{5}{x - 3} + \dfrac{-4}{x - 5}$

2) $\dfrac{3x^2 - x - 20}{x(x + 1)(x - 1)}$

A) $\dfrac{20}{x} + \dfrac{8}{x + 1} + \dfrac{-9}{x - 1}$
B) $\dfrac{20}{x} + \dfrac{-8}{x + 1} + \dfrac{-9}{x - 1}$
C) $\dfrac{20}{x} + \dfrac{-8}{x + 1} + \dfrac{9}{x - 1}$
D) $\dfrac{20}{x} + \dfrac{-9}{x + 1} + \dfrac{8}{x - 1}$

3) $\dfrac{12x^2 + 162x + 384}{(x + 8)(x + 2)(x + 11)}$

A) $\dfrac{8}{x + 8} + \dfrac{2}{x + 2} + \dfrac{2}{x + 11}$
B) $\dfrac{8}{x + 8} + \dfrac{2}{x + 2} - \dfrac{2}{x + 11}$
C) $-\dfrac{8}{x + 8} - \dfrac{2}{x + 2} - \dfrac{2}{x + 11}$
D) $-\dfrac{8}{x + 8} + \dfrac{2}{x + 2} + \dfrac{2}{x + 11}$

4) $\dfrac{2x - 5}{x^2 - 5x - 6}$

A) $\dfrac{9}{x + 2} + \dfrac{1}{x - 3}$
B) $\dfrac{1}{x - 3} + \dfrac{1}{x - 2}$
C) $\dfrac{17}{x + 6} - \dfrac{3}{x - 1}$
D) $\dfrac{1}{x - 6} + \dfrac{1}{x + 1}$

2 Decompose P/Q, Where Q Has Repeated Linear Factors

Write the partial fraction decomposition of the rational expression.

1) $\dfrac{2x^2 + 6x + 9}{(x + 2)(x + 1)^2}$

A) $\dfrac{5}{x + 2} + \dfrac{-3}{x + 1} + \dfrac{-5}{(x + 1)^2}$
B) $\dfrac{5}{x + 2} + \dfrac{3}{x + 1} + \dfrac{5}{(x + 1)^2}$
C) $\dfrac{5}{x + 2} + \dfrac{-3}{x + 1} + \dfrac{5}{(x + 1)^2}$
D) $\dfrac{-5}{x + 2} + \dfrac{-3}{x + 1} + \dfrac{-5}{(x + 1)^2}$

2) $\dfrac{7x^3 - 2}{x^2(x + 1)^3}$

Precalculus Enhanced with Graphing Utilities 786

3) $\dfrac{x+5}{x^3-2x^2+x}$

A) $\dfrac{5}{x}+\dfrac{-5}{x-1}+\dfrac{11}{(x-1)^2}$

B) $\dfrac{-5}{x}+\dfrac{5}{x-1}+\dfrac{6}{(x-1)^2}$

C) $\dfrac{5}{x}+\dfrac{-5}{x-1}+\dfrac{6}{(x-1)^2}$

D) $\dfrac{5}{x}+\dfrac{6}{x-1}+\dfrac{-5}{(x-1)^2}$

4) $\dfrac{x+1}{(x-2)^2(x+4)}$

A) $\dfrac{\frac{1}{2}}{(x-2)^2}+\dfrac{-\frac{1}{12}}{x+4}$

B) $\dfrac{12}{x-2}+\dfrac{2}{(x-2)^2}+\dfrac{-12}{x+4}$

C) $\dfrac{-1}{x-2}+\dfrac{\frac{1}{4}x}{(x-2)^2}+\dfrac{-\frac{1}{4}}{x+4}$

D) $\dfrac{\frac{1}{12}}{x-2}+\dfrac{\frac{1}{2}}{(x-2)^2}+\dfrac{-\frac{1}{12}}{x+4}$

3 Decompose P/Q, Where Q Has Nonrepeated Irreducible Quadratic Factors

Write the partial fraction decomposition of the rational expression.

1) $\dfrac{8x+1}{(x-1)(x^2+x+1)}$

A) $\dfrac{3}{x-1}+\dfrac{-3x+2}{x^2+x+1}$

B) $\dfrac{-3}{x-1}+\dfrac{3x+2}{x^2+x+1}$

C) $\dfrac{3}{x-1}+\dfrac{2x-3}{x^2+x+1}$

D) $\dfrac{3}{x-1}+\dfrac{-3}{x+1}+\dfrac{2}{x-1}$

2) $\dfrac{x^2-111}{x^4-x^2-72}$

A) $\dfrac{1}{x+3}-\dfrac{1}{x-3}-\dfrac{7}{x^2+8}$

B) $\dfrac{1}{x+3}+\dfrac{1}{x-3}-\dfrac{7}{x^2+8}$

C) $\dfrac{1}{x+3}-\dfrac{1}{x-3}+\dfrac{7}{x^2+8}$

D) $\dfrac{1}{x+3}+\dfrac{1}{x-3}+\dfrac{7}{x^2+8}$

3) $\dfrac{x^2-56}{x^4+5x^2-36}$

4) $\dfrac{3x-2}{x^3-1}$

A) $\dfrac{\frac{1}{2}}{x-1}+\dfrac{\frac{5}{2}}{x+1}$

B) $\dfrac{3}{(x-1)^2}+\dfrac{1}{(x-1)^3}$

C) $\dfrac{\frac{1}{3}}{x-1}+\dfrac{-\frac{1}{3}x+\frac{7}{3}}{(x^2+x+1)}$

D) $\dfrac{3}{x-1}+\dfrac{-3(x-7)}{x^2+x+1}$

4 Decompose P/Q, Where Q Has Repeated Irreducible Quadratic Factors

Write the partial fraction decomposition of the rational expression.

1) $\dfrac{2x^3 + 3x^2}{(x^2 + 5)^2}$

   A) $\dfrac{2x + 3}{x^2 + 5} + \dfrac{10x - 15}{(x^2 + 5)^2}$
   B) $\dfrac{2x + 3}{x^2 + 5} + \dfrac{-10x - 15}{(x^2 + 5)^2}$
   C) $\dfrac{2x + 3}{x^2 + 5} + \dfrac{10x + 15}{(x^2 + 5)^2}$
   D) $\dfrac{2x - 3}{x^2 + 5} + \dfrac{-10x + 15}{(x^2 + 5)^2}$

2) $\dfrac{2x^3 - 2x^2 + 10x - 3}{(x^2 + 3)^3}$

   A) $\dfrac{2x - 2}{(x^2 + 3)^2} + \dfrac{4x + 3}{(x^2 + 3)^3}$
   B) $\dfrac{2x + 2}{(x^2 + 3)^2} + \dfrac{4x - 3}{(x^2 + 3)^3}$
   C) $\dfrac{x}{x^2 + 3} + \dfrac{2x - 2}{(x^2 + 3)^2} + \dfrac{4x + 3}{(x^2 + 3)^3}$
   D) $\dfrac{x + 1}{x^2 + 3} + \dfrac{2x - 2}{(x^2 + 3)^2} + \dfrac{4x + 3}{(x^2 + 3)^3}$

10.6 Systems of Nonlinear Equations

1 Solve a System of Nonlinear Equations Using Substitution

Solve the system of equations using substitution.

1) $\begin{cases} x^2 + y^2 = 61 \\ x + y = -11 \end{cases}$

   A) $x = 5, y = 6$; $x = 6, y = 5$
   B) $x = 5, y = -6$; $x = 6, y = -5$
   C) $x = -5, y = -6$; $x = -6, y = -5$
   D) $x = -5, y = 6$; $x = -6, y = 5$

2) $\begin{cases} xy = 20 \\ x + y = -9 \end{cases}$

   A) $x = 4, y = -5$; $x = 5, y = -4$
   B) $x = 4, y = 5$; $x = 5, y = 4$
   C) $x = -4, y = -5$; $x = -5, y = -4$
   D) $x = -4, y = 5$; $x = -5, y = 4$

3) $\begin{cases} x^2 + y^2 = 13 \\ x - y = 1 \end{cases}$

   A) $x = -2, y = -3$; $x = 3, y = 2$
   B) $x = -2, y = 3$; $x = -3, y = 2$
   C) $x = 2, y = 3$; $x = 3, y = 2$
   D) $x = 2, y = -3$; $x = 3, y = -2$

4) $\begin{cases} y = x^2 - 14x + 49 \\ x + y = 27 \end{cases}$

   A) $x = 7, y = 20$
   B) $x = -2, y = 29$; $x = -11, y = 38$
   C) $x = 2, y = 29$; $x = 11, y = 16$
   D) $x = 2, y = 25$; $x = 11, y = 16$

5)
$$\begin{cases} y = -x^2 + 13 \\ x^2 + y^2 = 25 \end{cases}$$
   A) $x = 3, y = 4; x = -3, y = 4$
   B) $x = 4, y = -3; x = -4, y = -3$
   C) $x = 4, y = -3; x = 3, y = 4; x = -3, y = 4; x = -4, y = -3$
   D) $x = 9, y = 94; x = 16, y = 269$

6)
$$\begin{cases} x^2 + y^2 = 4 \\ x + y = 2 \end{cases}$$
   A) $x = 0, y = 2; x = 2, y = 0$ B) $x = 0, y = -2; x = -2, y = 0$
   C) $x = 2, y = -2; x = -2, y = -2$ D) $x = 0, y = 0; x = 2, y = -2$

7)
$$\begin{cases} xy = 20 \\ x + y = 9 \end{cases}$$
   A) $x = 6, y = 3; x = 3, y = 6$ B) $x = 10, y = 2; x = 2, y = 10$
   C) $x = 20, y = 1; x = 1, y = 20$ D) $x = 5, y = 4; x = 4, y = 5$

8)
$$\begin{cases} x^2 + y^2 = 169 \\ x + y = 17 \end{cases}$$
   A) $x = -12, y = 5; x = -5, y = 12$ B) $x = -12, y = -5; x = -5, y = -12$
   C) $x = 12, y = 5; x = 5, y = 12$ D) $x = 12, y = -5; x = 5, y = -12$

9)
$$\begin{cases} xy - x^2 = -20 \\ x - 2y = 3 \end{cases}$$
   A) $x = -5, y = -1; x = 8, y = \dfrac{11}{2}$ B) $x = -5, y = -1; x = \dfrac{11}{2}, y = 8$
   C) $x = 5, y = 1; x = -\dfrac{11}{2}, y = -8$ D) $x = 5, y = 1; x = -8, y = -\dfrac{11}{2}$

10)
$$\begin{cases} x^2 - y^2 = 39 \\ x - y = 3 \end{cases}$$
   A) $x = 8, y = 5$ B) $x = -8, y = -5$ C) $x = 8, y = -5$ D) $x = -8, y = 5$

11)
$$\begin{cases} y = 6x^2 - 5x \\ y = 2x + 3 \end{cases}$$
   A) $x = \dfrac{1}{6}, y = \dfrac{10}{3}; x = 1, y = 5$ B) $x = \dfrac{1}{3}, y = \dfrac{11}{3}; x = -\dfrac{3}{2}, y = 0$
   C) $x = -\dfrac{1}{2}, y = 2; x = 1, y = 5$ D) $x = \dfrac{3}{2}, y = 6; x = -\dfrac{1}{3}, y = -\dfrac{7}{3}$

12)
$$\begin{cases} \ln x = 3\ln y \\ 3^x = 27y \end{cases}$$

A) $x = \sqrt{3}, y = 3\sqrt{3}$ B) $x = 9, y = \sqrt{3}$ C) $x = \sqrt{3}, y = 9$ D) $x = 3\sqrt{3}, y = \sqrt{3}$

2 Solve a System of Nonlinear Equations Using Elimination

Solve the system of equations using elimination.

1)
$$\begin{cases} x^2 + y^2 = 125 \\ x^2 - y^2 = -75 \end{cases}$$

A) $x = -5, y = -10$; $x = -10, y = -5$
B) $x = 5, y = -10$; $x = 5, y = 10$
C) $x = 5, y = 10$; $x = 10, y = 5$; $x = -5, y = -10$; $x = -10, y = -5$
D) $x = 5, y = 10$; $x = -5, y = 10$; $x = 5, y = -10$; $x = -5, y = -10$

2)
$$\begin{cases} 4x^2 - 5y^2 = -44 \\ 2x^2 + 4y^2 = 82 \end{cases}$$

A) $x = 3, y = 4$; $x = -3, y = 4$; $x = 3, y = -4$; $x = -3, y = -4$
B) $x = 3, y = -4$; $x = 3, y = 4$
C) $x = -3, y = -4$; $x = -4, y = -3$
D) $x = 3, y = 4$; $x = 4, y = 3$; $x = -3, y = -4$; $x = -4, y = -3$

3)
$$\begin{cases} x^2 + y^2 = 4 \\ x^2 - y^2 = 4 \end{cases}$$

A) $x = -2, y = 0$; $x = -2, y = 2$ B) $x = 2, y = 2$; $x = -2, y = 2$
C) $x = 2, y = 0$; $x = 2, y = 2$ D) $x = 2, y = 0$; $x = -2, y = 0$

4)
$$\begin{cases} x^2 + y^2 = 64 \\ \dfrac{x^2}{64} + \dfrac{y^2}{9} = 1 \end{cases}$$

A) $x = 0, y = -8$; $x = 0, y = 8$ B) $x = -8, y = 0$; $x = 8, y = 0$
C) $x = 0, y = -3$; $x = 0, y = 3$ D) inconsistent

5)
$$\begin{cases} 3x^2 + 2y^2 = 89 \\ x^2 - 2y^2 = -21 \end{cases}$$

6)
$$\begin{cases} 2x^2 + y^2 = 17 \\ 3x^2 - 2y^2 = -6 \end{cases}$$

A) $x = 1, y = 3$; $x = 1, y = -3$; $x = -1, y = 3$; $x = -1, y = -3$
B) $x = 1, y = 3$; $x = -1, y = -3$
C) $x = 2, y = 3$; $x = 2, y = -3$; $x = -2, y = 3$; $x = -2, y = -3$
D) $x = 2, y = -3$; $x = -2, y = 3$

7)
$$\begin{cases} 2x^2 + xy - y^2 = 3 \\ x^2 + 2xy + y^2 = 3 \end{cases}$$

A) $x = \frac{2\sqrt{3}}{3}, y = \frac{\sqrt{3}}{3}$; $x = -\frac{2\sqrt{3}}{3}, y = -\frac{\sqrt{3}}{3}$
B) $x = \frac{-2}{3}, y = \frac{1}{3}$; $x = \frac{2\sqrt{3}}{3}, y = -\frac{1}{3}$
C) $x = \frac{-2}{3}, y = -\frac{1}{3}$; $x = \frac{2\sqrt{3}}{3}, y = \frac{1}{3}$
D) $x = \frac{2\sqrt{3}}{3}, y = -\frac{\sqrt{3}}{3}$; $x = -\frac{2\sqrt{3}}{3}, y = \frac{\sqrt{3}}{3}$

3 Demonstrate Additional Understanding and Skills

Use a graphing utility to solve the system of equations. Express the solution rounded to two decimal places.

1) $$\begin{cases} 3x^3 + y^2 = 6 \\ x^4 y = 2 \end{cases}$$

A) no solution
B) $x = 1.81, y = 0.28$; $x = 0.20, y = 2.45$; $x = -0.20, y = -2.45$
C) $x = 1.18, y = 1.03$; $x = 1.07, y = 1.52$; $x = -0.91, y = 2.88$
D) $x = 1.99, y = 0.13$; $x = 1.04, y = 1.70$; $x = -0.91, y = 2.95$

2) $$\begin{cases} x^3 + y^2 = 2 \\ x^2 y = 4 \end{cases}$$

A) $x = 1.37, y = 2.14$ B) $x = -1.37, y = 2.14$ C) $x = 2.14, y = -1.37$ D) $x = 2.14, y = 1.37$

Solve the problem.

3) The sum of the squares of two numbers is 37. The sum of the two numbers is 5. Find the two numbers.
   A) −1 and 6; or −6 and 1
   B) −6 and 1
   C) −1 and 6
   D) −6 and −1; or 1 and 6

4) The sum of the squares of two numbers is 73. The difference of the two numbers is 5. Find the two numbers.
   A) −3 and 8; or −8 and 3
   B) 3 and 8
   C) 3 and 8; or −8 and −3
   D) −8 and −3

5) The difference of two numbers is 5 and the difference of their squares is 55. Find the numbers.

6) A right triangle has an area of 30 square inches. The square of the hypotenuse is 136. Find the lengths of the legs of the triangle. Round your answer to the nearest inch.
   A) 3 in. and 20 in.   B) 6 in. and 10 in.   C) 36 in. and 100 in.   D) 12 in. and 5 in.

7) The perimeter of a rectangle is 36 inches and its area is 56 square inches. What are its dimensions?

   A) 5 in. by 13 in.   B) 3 in. by 15 in.   C) 3 in. by 13 in.   D) 4 in. by 14 in.

8) A rectangular piece of tin has an area of 684 square inches. A square of 3 inches is cut from each corner, and an open box is made by turning up the ends and sides. If the volume of the box is 1170 cubic inches, what were the original dimensions of the piece of tin?

   A) 19 in. by 36 in.   B) 16 in. by 33 in.   C) 13 in. by 27 in.   D) 22 in. by 39 in.

9) The diagonal of the floor of a rectangular office cubicle is 2 ft longer than the length of the cubicle and 5 ft longer than twice the width. Find the dimensions of the cubicle. Round to the nearest tenth, if necessary.

   A) width = 2 ft, length = 9 ft
   B) width = 9.7 ft, length = 22.4 ft
   C) width = 4 ft, length = 11 ft
   D) width = 3.9 ft, length = 9.7 ft

10) In a 1-mile race, the winner crosses the finish line 14 feet ahead of the second-place runner and 24 feet ahead of the third-place runner. Assuming that each runner maintains a constant speed throughout the race, by how many feet does the second-place runner beat the third-place runner? (5280 feet in 1 mile.)

   A) -14.06 ft   B) -4.01 ft   C) -10.05 ft   D) 10.03 ft

11) A person at the top of a 600 foot tall building drops a yellow ball. The height of the yellow ball is given by the equation $h = -16t^2 + 600$ where h is measured in feet and t is the number of seconds since the yellow ball was dropped. A second person, in the same building but on a lower floor that is 516 feet from the ground, drops a white ball 1.5 seconds after the yellow ball was dropped. The height of the white ball is given by the equation $h = -16(t - 1.5)^2 + 516$ where h is measured in feet and t is the number of seconds since the yellow ball was dropped. Find the time that the balls are the same distance above the ground and find this distance.

   A) 2 sec; 536 ft   B) 3 sec; 456 ft   C) 1.5 sec; 564 ft   D) 2.5 sec; 500 ft

10.7 Systems of Inequalities

1 Graph an Inequality by Hand

Graph the inequality.

1) $x > 6$

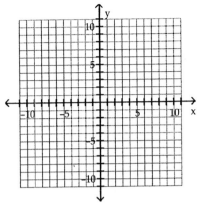

A)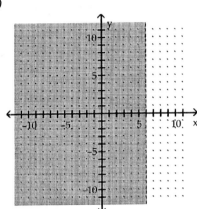

B)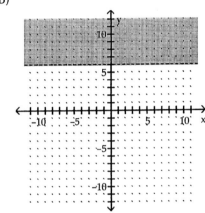

C)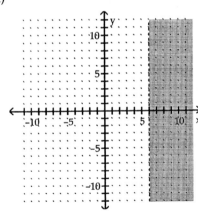

D)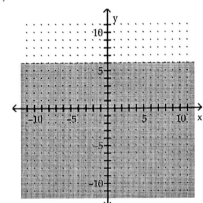

2) $y \leq 2$

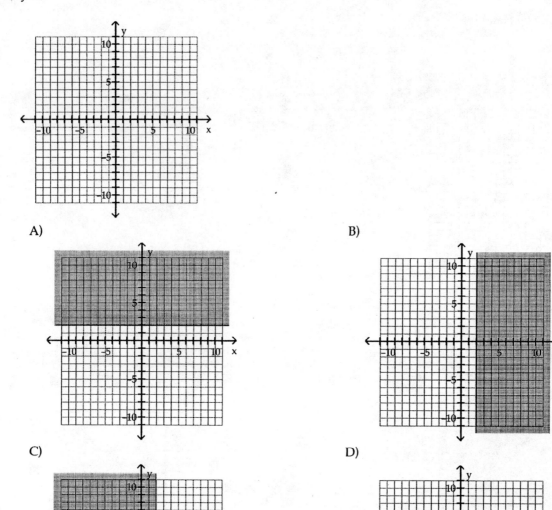

3) $-2x - 4y \leq 8$

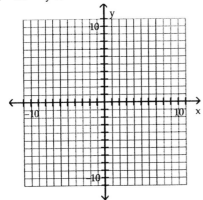

A)
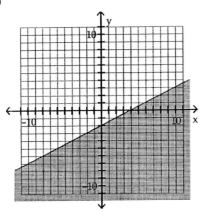

B)
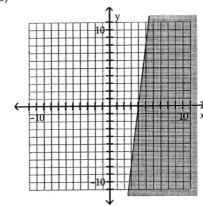

C)
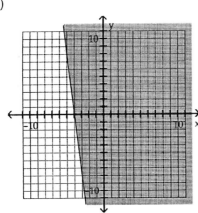

D)
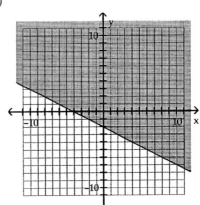

4) $3x + 4y \leq 12$

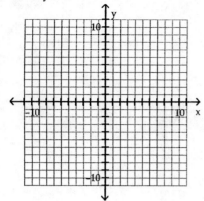

A)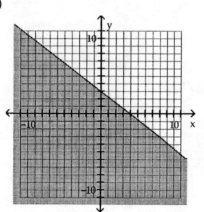

B)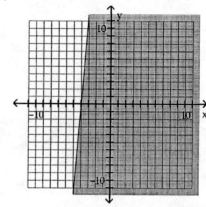

C)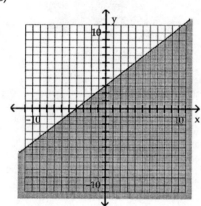

D)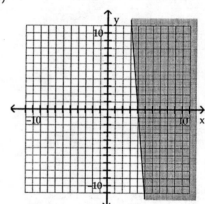

5) $-3x - 5y \leq -15$

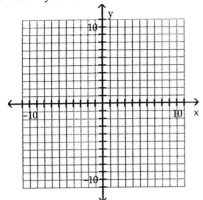

A)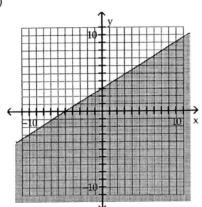

B)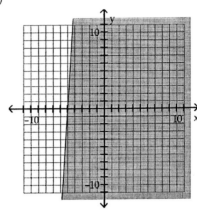

C)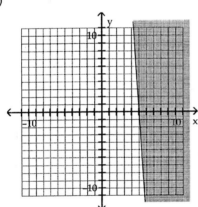

D)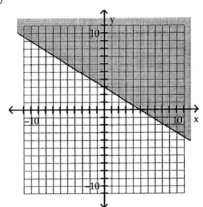

6) $x - y > -4$

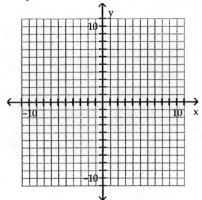

A)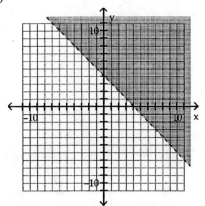

B)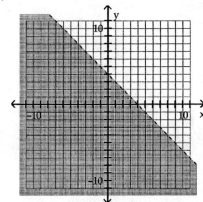

C)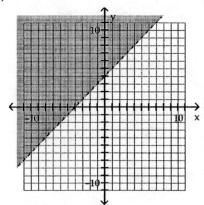

D)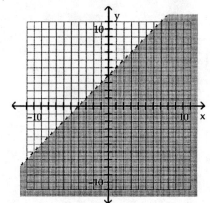

7) $x + y < -5$

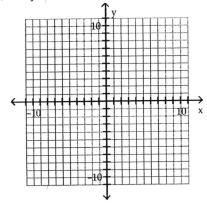

A)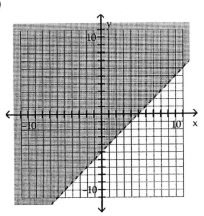

B)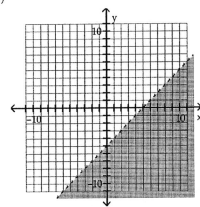

C)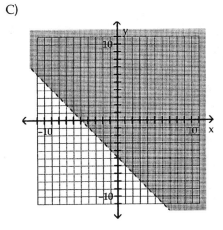

D)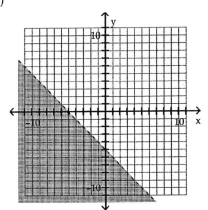

8) $x - y < -5$

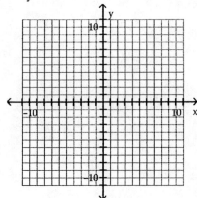

A)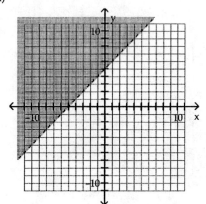

B)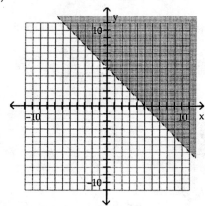

C)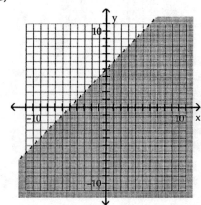

D)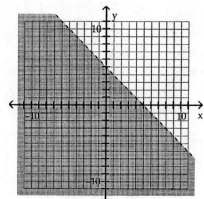

9) $x^2 + y^2 \leq 1$

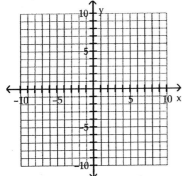

A)
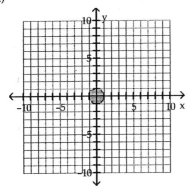

B)
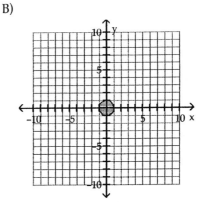

C)
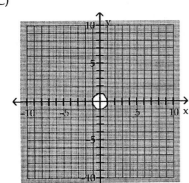

D)
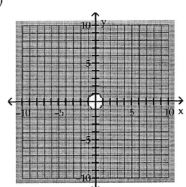

10) $x^2 + y^2 > 16$

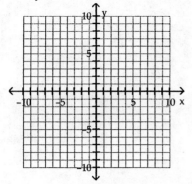

A)
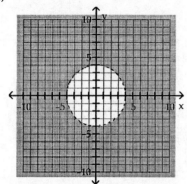

B)
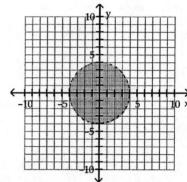

C)
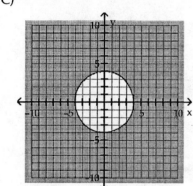

D)
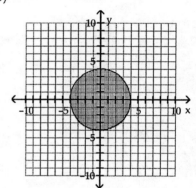

11) $y > x^2 - 2$

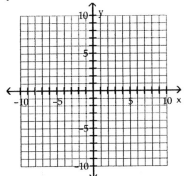

A)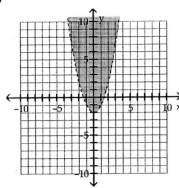

B)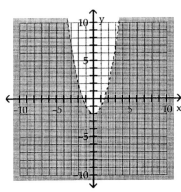

C)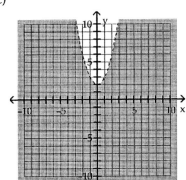

D)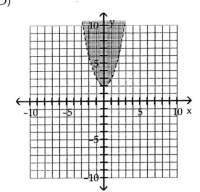

12) $y \leq x^2 + 1$

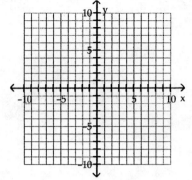

A)
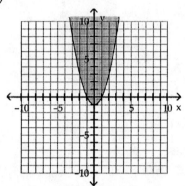

B)
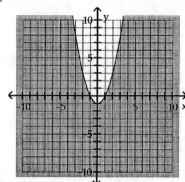

C)
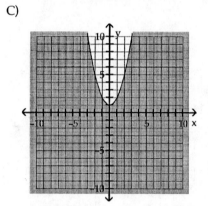

D)
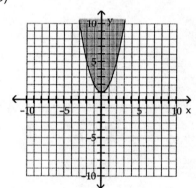

2 Graph an Inequality Using a Graphing Utility

Graph the inequality using a graphing utility.

1) $x + y \leq 5$

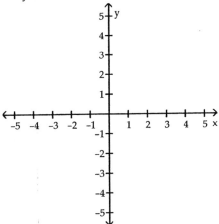

A)
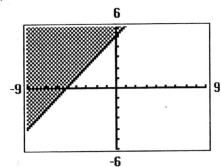

B)
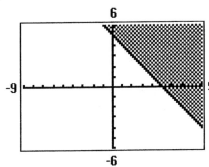

C)
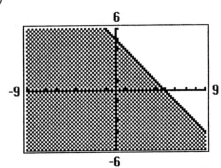

D)
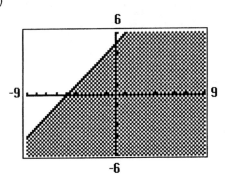

2) $x + 2y \geq 4$

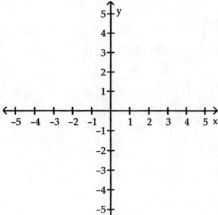

A)
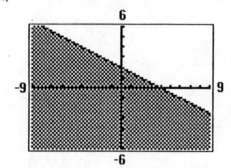

B)
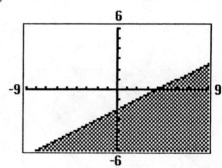

C)
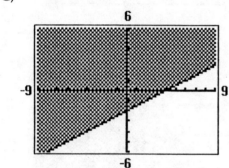

D)
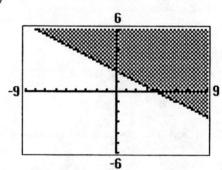

3 Graph a System of Inequalities

Graph the system of inequalities.

1) $\begin{cases} 2x + 3y \le 6 \\ x - y \le 3 \end{cases}$

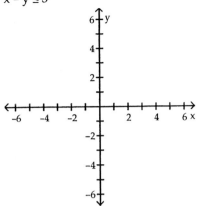

A)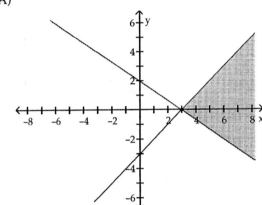

B)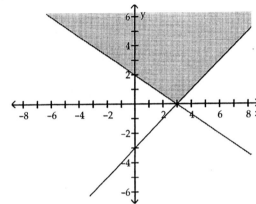

C)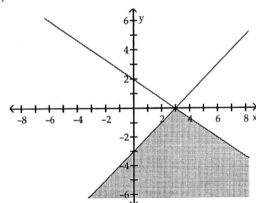

D)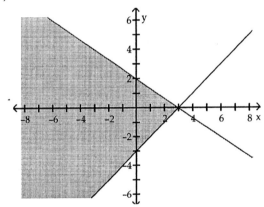

2) $\begin{cases} 4x + 3y \geq 12 \\ x \geq y \end{cases}$

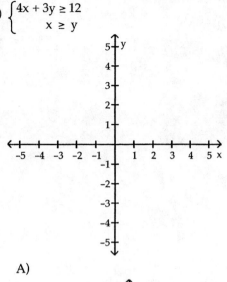

A)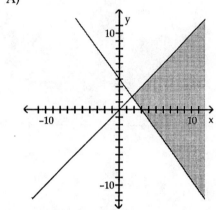

B)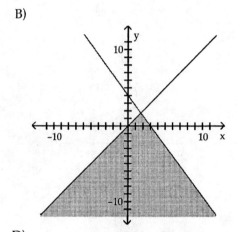

C)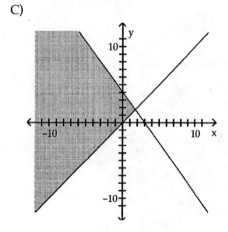

D)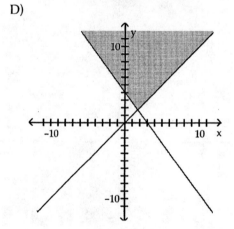

3) $\begin{cases} 2x + 3y \geq 6 \\ x - y \leq 3 \\ y \leq 2 \end{cases}$

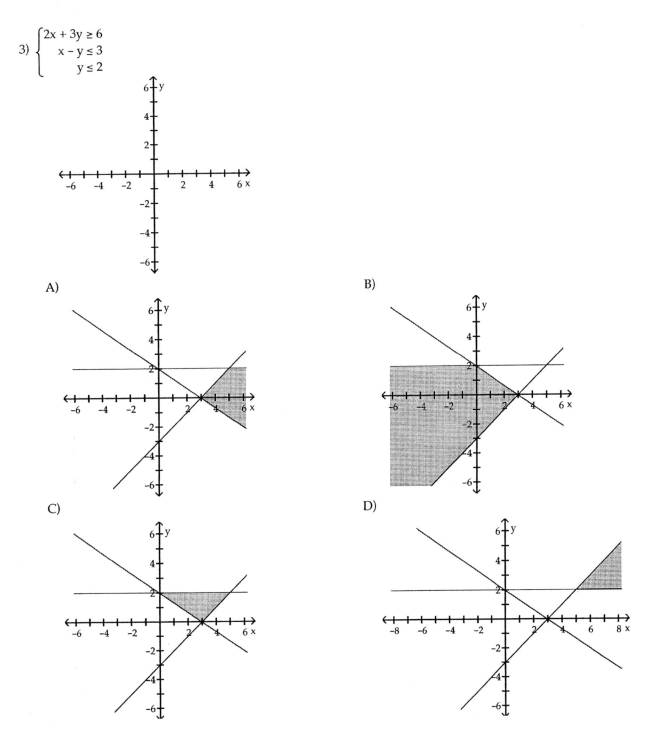

4) $\begin{cases} 2x + 3y \geq 6 \\ x - y \geq 3 \\ y \leq 2 \end{cases}$

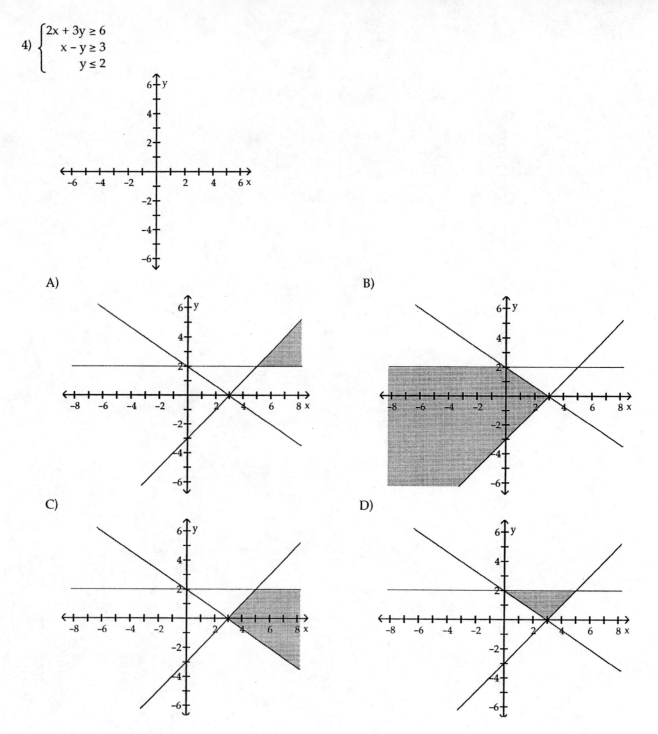

5) $\begin{cases} -x + 2y \leq -6 \\ 3x + 2y > -18 \end{cases}$

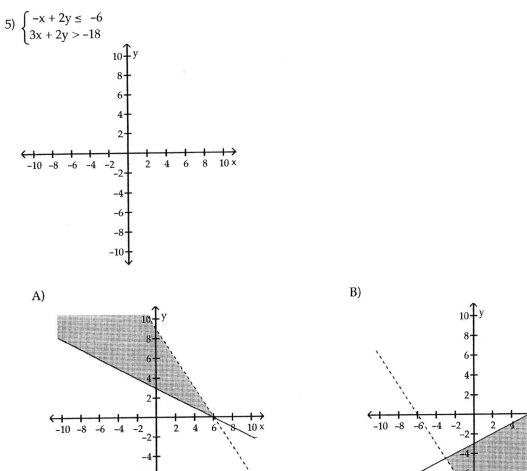

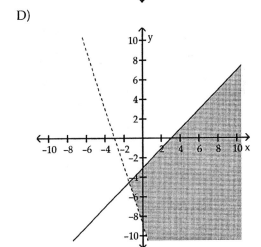

6) $\begin{cases} x^2 + y^2 \leq 49 \\ -4x + 4y \leq -16 \end{cases}$

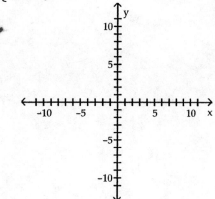

A)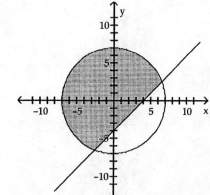

B)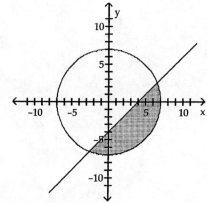

C)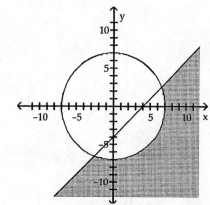

D)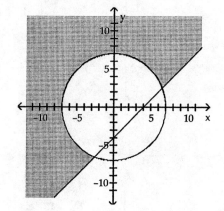

7)
$$\begin{cases} y > x^2 \\ 8x + 2y \le 16 \end{cases}$$

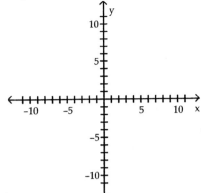

A)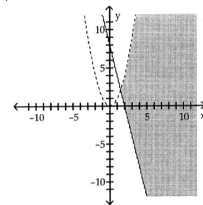

B)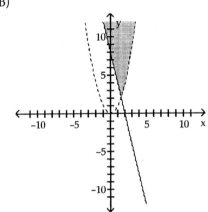

C)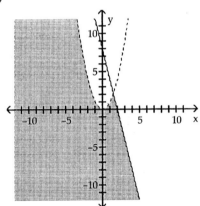

D)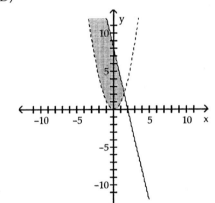

8) $y > x^2$
$8x + 7y \leq 56$

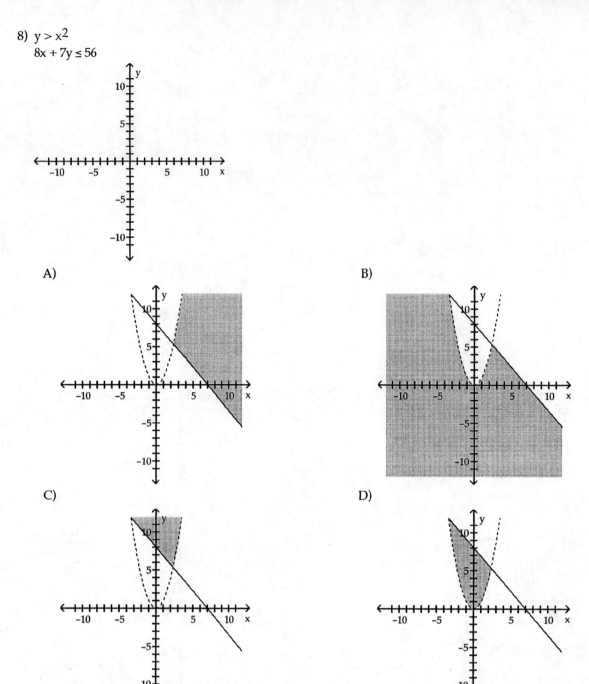

9) $x^2 + y^2 \le 81$
$9x + 7y \le 63$

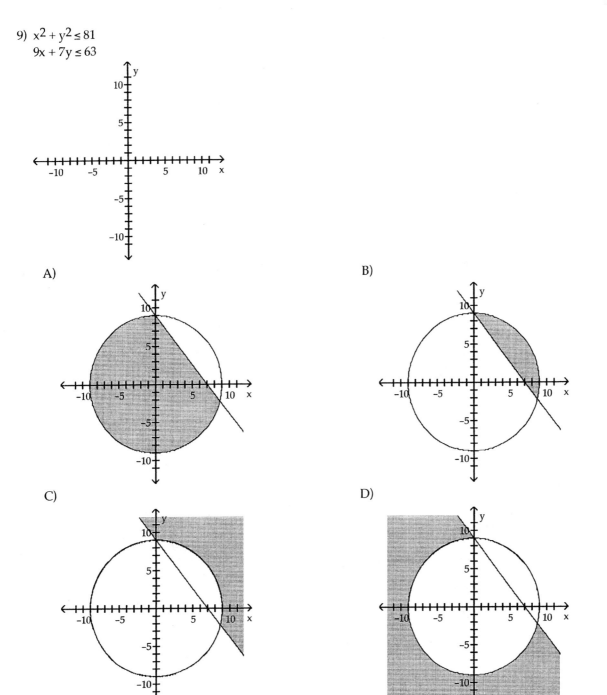

10) $x^2 + y^2 \leq 49$
$x^2 + y^2 \geq 1$

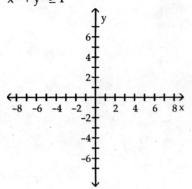

A)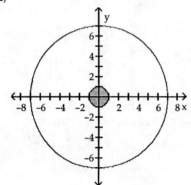

B) no solution

C)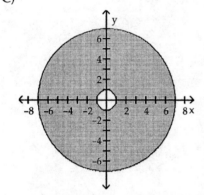

D)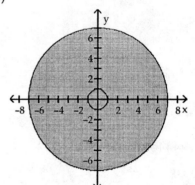

10.8 Linear Programming

1 Set Up a Linear Programming Problem

Write a system of linear inequalities that has the given graph.

1)

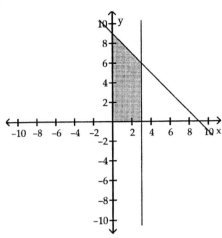

A) $y \geq 0$, $x \geq 0$, $x \leq 3$, and $y + x \leq 9$

B) $x \leq 3$, and $y + x \leq 9$

C) $y \geq 0$, $x \geq 0$, $x \leq 9$, and $y + x \leq 3$

D) $y \geq 0$, $x \geq 0$, $x \leq 3$, and $y + x \geq 9$

2)

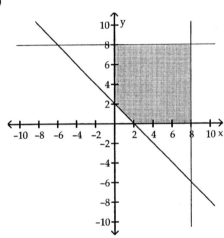

A) $y \geq 0$, $x \geq 0$, $x \leq 8$, and $y + x \geq 2$

B) $y \geq 0$, $x \geq 0$, $y \leq 8$, and $y + x \geq 2$

C) $x \leq 8$, $y \leq 8$, and $y + x \geq 2$

D) $y \geq 0$, $x \geq 0$, $x \leq 8$, $y \leq 8$, and $y + x \geq 2$

Set up the linear programming problem.

3) The Jillson's have up to $75,000 to invest. They decide that they want to have at least $40,000 invested in stable bonds yielding 6% and that no more than $20,000 should be invested in more volatile bonds yielding 12%.
   (a) Using x to denote the amount of money invested in the stable bonds and y the amount invested in the more volatile bonds, write a system of linear inequalities that describe the possible amounts of each investment.
   (b) Graph the system and label the corner points.

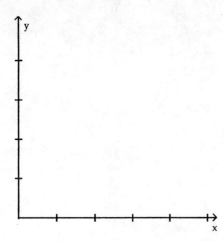

Solve the problem.

4) The liquid portion of a diet is to provide at least 300 calories, 36 units of vitamin A, and 90 units of vitamin C daily. A cup of dietary drink X provides 60 calories, 12 units of vitamin A, and 10 units of vitamin C. A cup of dietary drink Y provides 60 calories, 6 units of vitamin A, and 30 units of vitamin C. Set up a system of linear inequalities that describes the minimum daily requirements for calories and vitamins. Let x = number of cups of dietary drink X, and y = number of cups of dietary drink Y. Write all the constraints as a system of linear inequalities.

A) $\begin{cases} 60x + 60y \geq 300 \\ 12x + 6y > 36 \\ 10x + 30y \geq 90 \end{cases}$
B) $\begin{cases} 60x + 60y \leq 300 \\ 12x + 6y \leq 36 \\ 10x + 30y \leq 90 \end{cases}$
C) $\begin{cases} 60x + 60y \geq 300 \\ 12x + 6y \geq 36 \\ 10x + 30y \geq 90 \\ x \geq 0 \\ y \geq 0 \end{cases}$
D) $\begin{cases} 60x + 60y > 300 \\ 12x + 6y > 36 \\ 10x + 30y > 90 \\ x > 0 \\ y > 0 \end{cases}$

2 Solve a Linear Programming Problem

Find the maximum or minimum value of the objective function, subject to the constraints graphed in this feasible region.

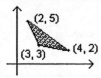

1) $z = x + 8y$. Find maximum.
   A) Maximum 34   B) Maximum 42   C) Maximum 20   D) Maximum 27

2) $z = x + 7y + 8$ Find minimum.
   A) Minimum 33   B) Minimum 45   C) No Minimum   D) Minimum 26

Find the value(s) of the function, subject to the system of inequalities.

3) Find the maximum and minimum of
   $z = 6x + 4y$ subject to $x \geq 0$, $y \geq 0$, $2x + 3y \geq 6$, $x \leq 10$, $y \leq 5$.
   A) 80; 8   B) 60; 12   C) 80; 60   D) 20; 12

4) Find the maximum and minimum of
z = 5x + 25y subject to x ≥ 0, y ≥ 0, 4x + 5y ≤ 30, 4x + 3y ≤ 20, x ≤ 5, y ≤ 8.

A) 25; 0 B) −118.75; −150 C) 25; −150 D) −150; 0

Solve the problem.

5) The Jillson's have up to $75,000 to invest. They decide that they want to have at least $25,000 invested in stable bonds yielding 6% and that no more than $45,000 should be invested in more volatile bonds yielding 12%. How much should they invest in each type of bond to maximize income if the amount in the more volatile bond should not exceed the amount in the more stable bond? What is the maximum income?

6) A doctor has told a patient to take vitamin pills. The patient needs at least 45 units of vitamin A, at least 9 units of vitamin C, and at least 63 units of vitamin D. The red vitamin pills cost 20¢ each and contain 8 units of A, 1 unit of C, and 5 units of D. The blue vitamin pills cost 40¢ each and contain 3 units of A, 1 unit of C, and 11 units of D. How many pills should the patient take each day to minimize costs?

A) 3 red and 6 blue B) 7 red and 2 blue C) 6 red and 3 blue D) 9 red and 0 blue

Ch. 10 Systems of Equations and Inequalities
Answer Key

10.1 Systems of Linear Equations: Substitution and Elimination

1 Solve Systems of Equations by Substitution
1) C
2) D
3) C
4) B
5) C

2 Solve Systems of Equations by Elimination
1) B
2) C
3) C

3 Identify Inconsistent Systems of Equations Containing Two Variables
1) C
2) B
3) D
4) A

4 Express the Solution of a System of Dependent Equations Containing Two Variables
1) C
2) C

5 Solve Systems of Three Equations Containing Three Variables
1) D
2) B
3) B
4) D
5) A

6 Identify Inconsistent Systems of Equations Containing Three Variables
1) C
2) B

7 Express the Solution of a System of Dependent Equations Containing Three Variables
1) C
2) B
3) A

8 Demonstrate Additional Understanding and Skills
1) A
2) A
3) B
4) B
5) B
6) A
7) A
8) B
9) A
10) C
11) B
12) B
13) A
14) A
15) B
16) $40,000 at 7%, $20,000 at 10%, $30,000 at 14%
17) $a = 2, b = -3, c = 2$
18) B

10.2 Systems of Linear Equations: Matrices

1 Write Augmented Matrix of a System of Linear Equations
1) C
2) B
3) D
4) C

2 Write the System from the Augmented Matrix
1) B
2) B

3 Perform Row Operations on a Matrix
1) $\begin{bmatrix} -7 & -5 & -1 & | & -10 \\ 6 & -2 & 9 & | & 5 \\ 0 & -26 & 2 & | & -22 \end{bmatrix}$
2) B
3) B

4 Solve a System of Linear Equations Using Matrices
1) B
2) C
3) D
4) $a = -2$, $b = -1$, $c = 2$
5) $9000 at 5%, $12,000 at 8%, and $24,000 at 12%
6) 14 Kitui chairs, 10 Goa chairs, 8 Santa Fe chairs
7) B

5 Demonstrate Additional Understanding and Skills
1) C
2) D
3) C
4) D

10.3 Systems of Linear Equations: Determinants

1 Evaluate 2 by 2 Determinants
1) C
2) D
3) C
4) C

2 Use Cramer's Rule to Solve a System of Two Equations Containing Two Variables
1) C
2) C
3) B
4) D

3 Evaluate 3 by 3 Determinants
1) C
2) C
3) D

4 Use Cramer's Rule to Solve a System of Three Equations Containing Three Variables
1) A
2) A
3) D

5 Know Properties of Determinants
1) B
2) A
3) D
4) B
5) C
6) D
7) D

8) D
6 Demonstrate Additional Understanding and Skills
- 1) 4
- 2) 5
- 3) B

10.4 Matrix Algebra
1 Find the Sum and Difference of Two Matrices
- 1) A
- 2) D
- 3) A
- 4) B

2 Find Scalar Multiples of a Matrix
- 1) C
- 2) B
- 3) A
- 4) D
- 5) B
- 6) B

3 Find the Product of Two Matrices
- 1) B
- 2) $\begin{bmatrix} -2 & 2 & 0 \\ 1 & -1 & 0 \\ -1 & 1 & 0 \end{bmatrix}$
- 3) A
- 4) A
- 5) A

4 Find the Inverse of a Matrix
- 1) D
- 2) C
- 3) B
- 4) B
- 5) A
- 6) C
- 7) A
- 8) D

5 Solve a System of Linear Equations Using Inverse Matrices
- 1) D
- 2) C
- 3) $x = -4, y = 1, z = -2$
- 4) B
- 5) C
- 6) B
- 7) A

6 Demonstrate Additional Understanding and Skills

1) $[A \mid I_2] = \begin{bmatrix} 10 & -6 & | & 1 & 0 \\ -5 & 3 & | & 0 & 1 \end{bmatrix} \xrightarrow{R_1 = \frac{1}{10}r_1} \begin{bmatrix} 1 & -\frac{3}{5} & | & \frac{1}{10} & 0 \\ -5 & 3 & | & 0 & 1 \end{bmatrix} \xrightarrow{R_2 = 5r_1 + r_2} \begin{bmatrix} 1 & -\frac{3}{5} & | & \frac{1}{10} & 0 \\ 0 & 0 & | & \frac{1}{2} & 1 \end{bmatrix}$

The matrix $[A \mid I_2]$ is sufficiently reduced for us to see that the identity matrix cannot appear to the left of the vertical bar. We conclude that A is singular and so has no inverse.

- 2) B
- 3) C

10.5 Partial Fraction Decomposition

1 Decompose P/Q, Where Q Has Only Nonrepeated Linear Factors

1) D
2) B
3) A
4) D

2 Decompose P/Q, Where Q Has Repeated Linear Factors

1) C
2) $\dfrac{6}{x} - \dfrac{2}{x^2} - \dfrac{6}{x+1} + \dfrac{3}{(x+1)^2} - \dfrac{9}{(x+1)^3}$
3) C
4) D

3 Decompose P/Q, Where Q Has Nonrepeated Irreducible Quadratic Factors

1) A
2) C
3) $\dfrac{1}{x+2} - \dfrac{1}{x-2} + \dfrac{5}{x^2+9}$
4) C

4 Decompose P/Q, Where Q Has Repeated Irreducible Quadratic Factors

1) B
2) A

10.6 Systems of Nonlinear Equations

1 Solve a System of Nonlinear Equations Using Substitution

1) C
2) C
3) A
4) D
5) C
6) A
7) D
8) C
9) D
10) A
11) D
12) D

2 Solve a System of Nonlinear Equations Using Elimination

1) D
2) A
3) D
4) B
5) $x = \sqrt{17}, y = \sqrt{19};\ x = -\sqrt{17}, y = \sqrt{19};\ x = \sqrt{17}, y = -\sqrt{19};\ x = -\sqrt{17}, -y = \sqrt{19}$
6) C
7) A

3 Demonstrate Additional Understanding and Skills

1) C
2) B
3) C
4) C
5) 8 and 3
6) B
7) D
8) A
9) B
10) D

Precalculus Enhanced with Graphing Utilities

11) D
10.7 Systems of Inequalities
1 Graph an Inequality by Hand
1) C
2) D
3) D
4) A
5) D
6) D
7) D
8) A
9) B
10) A
11) A
12) C

2 Graph an Inequality Using a Graphing Utility
1) C
2) D

3 Graph a System of Inequalities
1) D
2) D
3) C
4) C
5) B
6) B
7) D
8) D
9) A
10) C

10.8 Linear Programming
1 Set Up a Linear Programming Problem
1) A
2) D
3) (a)
$$x + y \leq 75{,}000$$
$$x \geq 40{,}000$$
$$y \leq 20{,}000$$
$$x \geq 0$$
$$y \geq 0$$

(b)

4) C
2 Solve a Linear Programming Problem
1) B
2) D
3) A
4) C
5) $37,500 in the stable bonds and $37,500 in the volatile bonds; maximum income $6750
6) C

Ch. 11 Sequences; Induction; the Binomial Theorem

11.1 Sequences

1 Write the First Several Terms of a Sequence

Write out the first five terms of the sequence.

1) $\{n - 1\}$
   A) –4, –3, –2, –1, 0
   B) 0, 1, 2, 3, 4
   C) –1, 0, 1, 2, 3
   D) 1, 2, 3, 4, 5

2) $\{3n - 2\}$
   A) 1, 2, 3, 4, 5
   B) 5, 8, 11, 14, 17
   C) –1, –4, –7, –10, –13
   D) 1, 4, 7, 10, 13

3) $\{2(2n - 3)\}$
   A) –2, –4, –6, –8, –10
   B) –6, –2, 2, 6, 10
   C) –1, 1, 3, 5, 7
   D) –2, 2, 6, 10, 14

4) $\{n^2 - n\}$
   A) 2, 6, 12, 20, 30
   B) 0, 3, 8, 15, 24
   C) 0, 2, 6, 12, 20
   D) 1, 4, 9, 16, 25

5) $\{2^n\}$
   A) 2, 4, 8, 16, 32
   B) 1, 4, 9, 16, 25
   C) 4, 8, 16, 32, 64
   D) 1, 2, 4, 8, 16

6) $\left\{(-1)^{n-1}\left(\dfrac{n+3}{2n-1}\right)\right\}$
   A) $4, \dfrac{5}{3}, \dfrac{6}{5}, 1, \dfrac{8}{9}$
   B) $4, -\dfrac{5}{3}, \dfrac{6}{5}, -1, \dfrac{8}{9}$
   C) $-4, \dfrac{5}{3}, -\dfrac{6}{5}, 1, -\dfrac{8}{9}$
   D) $-4, \dfrac{5}{3}, \dfrac{6}{5}, -1, \dfrac{8}{9}$

7) $\left\{\dfrac{n}{n^2 + 2}\right\}$
   A) $\dfrac{1}{4}, \dfrac{1}{3}, \dfrac{3}{8}, \dfrac{2}{5}, \dfrac{5}{12}$
   B) $\dfrac{1}{3}, \dfrac{1}{3}, \dfrac{3}{11}, \dfrac{2}{9}, \dfrac{5}{27}$
   C) $\dfrac{1}{3}, \dfrac{1}{3}, \dfrac{3}{8}, \dfrac{2}{5}, \dfrac{5}{12}$
   D) $\dfrac{1}{2}, \dfrac{1}{3}, \dfrac{3}{8}, \dfrac{2}{5}, \dfrac{5}{12}$

8) $\left\{\dfrac{(2n-1)!}{n!}\right\}$
   A) 2, 3, 20, 210, 3024
   B) 2, 6, 40, 1260, 72,576
   C) 1, 3, 20, 210, 3024
   D) 1, 6, 40, 1260, 72,576

9) $\left\{\dfrac{4}{n^2}\right\}$
   A) $1, \dfrac{1}{2}, \dfrac{1}{3}, \dfrac{1}{4}, \dfrac{1}{5}$
   B) $1, \dfrac{1}{4}, \dfrac{1}{9}, \dfrac{1}{16}, \dfrac{1}{25}$
   C) $\dfrac{1}{4}, \dfrac{2}{9}, \dfrac{3}{16}, \dfrac{4}{25}, \dfrac{5}{36}$
   D) $4, 1, \dfrac{4}{9}, \dfrac{1}{4}, \dfrac{4}{25}$

Write the first four terms of the sequence whose general term is given.

10) $\left\{\dfrac{3n}{(n+3)!}\right\}$
    A) $\dfrac{3}{7}, \dfrac{9}{8}, 3, \dfrac{81}{10}$
    B) $\dfrac{3}{4}, \dfrac{9}{5}, \dfrac{9}{2}, \dfrac{81}{7}$
    C) $\dfrac{1}{8}, \dfrac{3}{40}, \dfrac{3}{80}, \dfrac{9}{560}$
    D) $\dfrac{1}{8}, \dfrac{3}{40}, \dfrac{3}{40}, \dfrac{9}{280}$

11) $\left\{\dfrac{2(n+1)!}{n!}\right\}$

A) $4, 3, \dfrac{8}{3}, \dfrac{5}{2}$ B) 4, 6, 8, 10   C) 3, 4, 5, 6   D) $4, 3, \dfrac{4}{3}, \dfrac{5}{12}$

The given pattern continues. Write down the nth term of the sequence suggested by the pattern.

12) 4, 10, 16, 22, 28, ...

A) $a_n = 4(6)^{n-1}$ B) $a_n = 2n - 6$ C) $a_n = 6n - 1$ D) $a_n = 2(3n - 1)$

13) 2, 6, 10, 14, 18, ...

A) $a_n = n + 4$ B) $a_n = 2(4)^{n-1}$ C) $a_n = 2n - 4$ D) $a_n = 4n - 2$

14) 0, 2, 6, 12, 20, ...

A) $a_n = 4n - 6$ B) $a_n = 2n - 2$ C) $a_n = n^2 - n$ D) $a_n = 2^{n-1} - 1$

15) 4, 16, 64, 256, 1024, ...

A) $a_n = 4^n$ B) $a_n = 12n$ C) $a_n = 4^{n-1} + 3$ D) $a_n = 4 + 12(n - 1)$

16) $\dfrac{1}{1}, \dfrac{1}{4}, \dfrac{1}{9}, \dfrac{1}{16}, \dfrac{1}{25}, ...$

A) $a_n = \left(\dfrac{1}{2}\right)^{n-1}$ B) $a_n = \dfrac{1}{n^{n-1}}$ C) $a_n = \dfrac{1}{3n - 2}$ D) $a_n = \dfrac{1}{n^2}$

2 Write the Terms of a Sequence Defined by a Recursive Formula

The sequence is defined recursively. Write the first four terms.

1) $a_1 = 3$ and $a_n = a_{n-1} - 3$ for $n \geq 2$

A) 3, 6, 9, 12   B) 3, 0, -3, -6   C) -3, -6, -9, -12   D) 3, 2, -1, -4

2) $a_1 = 2$ and $a_n = 3a_{n-1}$ for $n \geq 2$

A) 2, 6, 18, 54   B) 3, 9, 27, 54   C) 2, 5, 4, 3   D) 2, 8, 20, 56

3) $a_1 = 3$ and $a_n = 3a_{n-1} + 4$ for $n \geq 2$

A) 3, 13, 31, 85   B) 3, 5, 11, 29   C) 3, 9, 27, 81   D) 3, 13, 43, 133

4) $a_1 = 2$, $a_2 = 5$ and $a_n = a_{n-2} - 3a_{n-1}$ for $n \geq 3$

A) 2, 5, 1, 2   B) 2, 5, -1, -16   C) 2, 5, -13, 44   D) 2, 5, 17, -46

5) $a_1 = 135$ and $a_{n+1} = \dfrac{1}{3}(a_n)$ for $n \geq 2$

6) $a_1 = w$; $a_n = a_{n-1} + T$

A) T, T + w, T + 2w, T + 3w   B) w, w + T, w + 2T, w + 3T
C) w, T, 2T, 3T   D) w, w - T, w - 2T, w - 3T

7) $a_1 = \sqrt{5}; a_n = \sqrt{5a_{n-1}}$

A) $\sqrt{5}, \sqrt{5\sqrt{5}}, \sqrt{5\sqrt{5\sqrt{5}}}, \sqrt{5\sqrt{5\sqrt{5\sqrt{5}}}}$

B) $\sqrt{5}, \sqrt{\sqrt{5}}, \sqrt{\sqrt{\sqrt{5}}}, \sqrt{\sqrt{\sqrt{\sqrt{5}}}}$

C) $\sqrt{5}, 5\sqrt{5}, 25\sqrt{5}, 125\sqrt{5}$

D) $\sqrt{5}, 5, 5\sqrt{5}, 25$

3 Use Summation Notation

Write out the sum. Do not evaluate.

1) $\sum_{k=1}^{4} (k+6)$

A) $7 + 8 + 9 + 10$ B) $7 + 8 + 9 + 10 + ...$ C) $6 + 7 + 8 + 9$ D) $10 + 11 + 12 + 13$

2) $\sum_{k=1}^{5} \frac{k+1}{k+2}$

A) $\frac{2}{3} + \frac{3}{4} + \frac{4}{5} + \frac{5}{6} + \frac{6}{7}$ B) $\frac{1}{3} + \frac{2}{4} + \frac{3}{5} + \frac{4}{6}$ C) $\frac{1}{2} + \frac{2}{3} + \frac{3}{4} + \frac{4}{5} + \frac{5}{6}$ D) $\frac{6}{7}$

3) $\sum_{k=1}^{6} (5k-2)$

A) $3 + 8 + 13 + 18 + 23 + 28 + ...$

B) $3 + 8 + 13 + 18 + 23 + 28$

C) 65

D) $1 + 3 + 8 + 13 + 18 + 23$

4) $\sum_{k=1}^{4} (k^2 - 5k - 3)$

A) $-7 - 9 - 9 - 7 + ...$ B) $-7 - 9 - 9 - 7$ C) $-7 + 9 - 9 + 7$ D) $1 - 7 - 9 - 9$

5) $\sum_{k=1}^{n} 8^{k+1}$

A) $8 + 8^2 + 8^3 + ... + 8^n$

B) $8^2 + 8^3 + 8^4 + ... + 8^n$

C) $8^2 + 8^3 + 8^4 + ... + 8^{n+1}$

D) $8 + 8^2 + 8^3 + ... + 8^{n+1}$

Express the sum using summation notation.

6) $2 + 4 + 6 + \cdots + 10$

A) $\sum_{k=0}^{5} 2k$ B) $\sum_{k=1}^{5} k^2$ C) $\sum_{k=1}^{5} 2k^2$ D) $\sum_{k=1}^{5} 2k$

Precalculus Enhanced with Graphing Utilities

7) $2^2 + 3^2 + 4^2 + \cdots + 9^2$

A) $\sum_{k=2}^{n} k^2$　　B) $\sum_{k=2}^{9} k^2$　　C) $\sum_{k=1}^{9} k^2$　　D) $\sum_{k=3}^{9} (k-1)^2$

8) $2 + 8 + 18 + \ldots + 50$

A) $\sum_{k=1}^{5} k^2$　　B) $\sum_{k=1}^{5} 2^{2k}$　　C) $\sum_{k=1}^{5} 2k^2$　　D) $\sum_{k=0}^{5} 2k^2$

9) $\frac{1}{4} + \frac{2}{5} + \frac{1}{2} + \cdots + \frac{13}{16}$

A) $\sum_{k=1}^{n} \frac{k}{k+3}$　　B) $\sum_{k=3}^{13} \frac{k}{k+1}$　　C) $\sum_{k=0}^{13} \frac{k}{k+3}$　　D) $\sum_{k=1}^{13} \frac{k}{k+3}$

10) $\frac{4}{5} - \frac{16}{25} + \frac{64}{125} - \cdots + (-1)^{(11+1)} \left(\frac{4}{5}\right)^{11}$

A) $\sum_{k=1}^{11} (-1)^{(k+1)} \left(\frac{4}{5}\right)^k$

B) $\sum_{k=1}^{n} (-1)^{(k+1)} \left(\frac{4}{5}\right)^k$

C) $\sum_{k=1}^{12} (-1)^{(k+1)} \left(\frac{4}{5}\right)^k$

D) $\sum_{k=1}^{11} \left(\frac{4}{5}\right)^k$

11) $\frac{4^9}{9^4} + \frac{4^8}{9^5} + \frac{4^7}{9^6} + \ldots + \frac{4^2}{9^{11}}$

A) $\sum_{k=1}^{4} \frac{4^{10+k}}{9^{3+k}}$　　B) $\sum_{k=1}^{8} \frac{4^{10-k}}{9^{3+k}}$　　C) $\sum_{k=1}^{8} \frac{4^{-k}}{9^{3+k}}$　　D) $\sum_{k=1}^{8} \frac{4^{10-k}}{9^k}$

12) $4^2 + 8^3 + 12^4 + \ldots + 32^9$

A) $\sum_{k=1}^{8} 2(k-1)^{k+1}$　　B) $\sum_{k=1}^{8} (4k)^{k+1}$　　C) $\sum_{k=1}^{8} 4k^{2k-1}$　　D) $\sum_{k=1}^{8} (4k)^k$

13) $\frac{1}{y} + \frac{r}{2y} + \frac{r^2}{3y} + \ldots + \frac{r^{n-1}}{ny}$

A) $\sum_{k=0}^{n} \frac{r^{k-1}}{ky}$　　B) $\sum_{k=0}^{n} \frac{<b>k}{ky}$　　C) $\sum_{k=1}^{n} \frac{r^{k-1}}{ky}$　　D) $\sum_{k=1}^{n} \frac{r^k}{ky}$

Express the sum using summation notation with a lower limit of summation not necessarily 1 and with k for the index of summation.

14) $5 + 7 + 9 + 11 + \ldots + 25$

A) $\sum_{k=0}^{10} 5+2k$ B) $\sum_{k=1}^{10} 5k+2$ C) $\sum_{k=0}^{16} 5+2k$ D) $\sum_{k=1}^{10} 5+2k$

15) $11 + 15 + 19 + 23 + \ldots + 43$

A) $\sum_{k=2}^{10} 3+4k$ B) $\sum_{k=0}^{32} 3+4k$ C) $\sum_{k=2}^{32} 3+4k$ D) $\sum_{k=1}^{10} 3+4k$

16) $\frac{5}{6} + \frac{6}{7} + \frac{7}{8} + \frac{8}{9} + \ldots + \frac{18}{19}$

A) $\sum_{k=5}^{18} \frac{k+1}{k}$ B) $\sum_{k=5}^{18} \frac{k}{k \cdot 1.5}$ C) $\sum_{k=6}^{18} \frac{k}{k+1}$ D) $\sum_{k=5}^{18} \frac{k}{k+1}$

17) $4 + \frac{9}{2} + 5 + \frac{11}{2} + \ldots + 9$

A) $\sum_{k=8}^{18} \frac{k}{2}$ B) $\sum_{k=8}^{12} \frac{k}{2}$ C) $\sum_{k=1}^{18} \frac{k}{2}$ D) $\sum_{k=2}^{18} \frac{k}{2}$

4 Find the Sum of a Sequence Algebraically and Using a Graphing Utility

Find the sum of the sequence.

1) $\sum_{k=1}^{9} 7$

A) 2 B) 63 C) 7 D) 9

2) $\sum_{k=1}^{7} (-k)$

A) −42 B) −7 C) −6 D) −28

3) $\sum_{k=3}^{6} 4k$

A) 48 B) 36 C) 24 D) 72

4) $\sum_{k=1}^{5} (k+10)$

A) 26 B) 50 C) 65 D) 15

5) $\sum_{k=2}^{5}(4k-5)$

A) 23   B) 30   C) 36   D) 33

6) $\sum_{k=1}^{16}(2k+7)$

A) 272   B) 384   C) 53   D) 400

7) $\sum_{k=1}^{4} 2^k$

A) 14   B) 30   C) 20   D) 18

8) $\sum_{k=3}^{5}(k^2-5)$

A) 9   B) 35   C) 30   D) -3

9) $\sum_{k=2}^{4} k(k-6)$

A) -25   B) -30   C) 0   D) -16

10) $\sum_{k=1}^{4}\left(-\frac{1}{2}\right)^k$

A) $-\frac{5}{16}$ B) $-\frac{1}{16}$ C) $\frac{15}{16}$ D) $\frac{5}{16}$

11) $\sum_{k=1}^{4}(-1)^k(k+13)$

A) -62   B) 2   C) 62   D) 54

12) $\sum_{k=2}^{5}(-1)^{k+1}(k+9)^2$

A) 50   B) 510,870   C) -630   D) 630

13) $\sum_{k=1}^{4} (-1)^k \cdot 4k$

A) -40   B) 8   C) 256   D) 40

14) $\sum_{k=5}^{8} \frac{4}{k}$

A) $-\frac{197}{210}$ B) $\frac{692}{15}$ C) $\frac{624}{625}$ D) $\frac{533}{210}$

15) $\sum_{k=1}^{4} \frac{1}{9k}$

A) $\frac{11}{54}$ B) $\frac{25}{108}$ C) $\frac{1}{36}$ D) $\frac{5}{36}$

16) $\sum_{k=7}^{10} \frac{1}{k-6}$

A) $\frac{25}{12}$ B) $-\frac{2400}{2401}$ C) $\frac{487}{420}$ D) 10

5 Solve Annuity and Amortization Problems

Solve the problem.

1) Karen has a balance of $1000 on a department store credit card that charges 1.5% interest per month on any unpaid balance. She can afford to pay $150 toward the balance each month. Her balance each month, after making a $150 payment, is given by the recursively defined sequence

$B_0 = \$1000 \qquad B_n = 1.015 B_{n-1} - 150$

Determine Karen's balance after making the first payment. That is, determine B_1.

2) Jake bought a truck by taking out a loan for $26,500 at 0.25% interest per month. Jake's regular monthly payment is $567, but he decides to pay an extra $75 toward the balance each month. His balance each month, after making his payment, is given by the recursively defined sequence

$B_0 = \$26,500 \qquad B_n = 1.0025 B_{n-1} - 642$

Determine Jake's balance after making the first payment. That is, determine B_1.

3) A wildlife refuge currently has 100 deer in it. A local wildlife society decides to add an additional 2 deer each month. It is already known that the deer population is growing 12% per year. The size of the population is given by the recursively defined sequence

$p_0 = 100 \qquad p_n = 1.01 p_{n-1} + 2$

How many deer are in the wildlife refuge at the end of the second month? That is, what is p_2?

4) Lexington Reservoir has 300 million gallons of water. About 0.07% of the water is lost to evaporation every week. About 150,000 gallons of water enter the reservoir every week. The amount of water in the reservoir at the end of each week is given by the recursively defined sequence

$$w_0 = 300 \qquad w_n = (0.9993)w_{n-1} + 0.15$$

Determine the amount of water in the reservoir at the beginning of the second week. That is, determine w_2.

5) Maria deposited $1500 in an independent retirement account at her bank. She earns 0.5% interest per month on the balance. Each month, she deposits $100 in the account. Her balance each month after making a $100 deposit, is given by the recursively defined sequence

$$B_0 = \$1500 \qquad B_n = 1.005 B_{n-1} + 100$$

If she made the initial deposit on September 30, and makes each monthly deposit on the last day of the month, how much money will be in the account at the end of the year for Maria to count as a deduction for that year's federal income taxes? That is, determine B_3.

6) Susie and Doug borrowed $180,000 at 7.5% per annum compounded monthly for 30 years to purchase a home. Their monthly payment is determined to be $1350. Find a recursive formula for their balance after each monthly payment has been made and then use a graphing utility to determine when their balance will be below $175,000.

A) $B_0 = 180{,}000 \quad B_n = \left(1 + \dfrac{0.0075}{12}\right)A(n-1) - 1350$; after 6 payments

B) $B_0 = 180{,}000 \quad B_n = \left(1 + \dfrac{0.075}{12}\right)A(n-1) - 1350$; after 21 payments

C) $B_0 = 180{,}000 \quad B_n = \left(1 + \dfrac{0.075}{30}\right)A(n-1) - 1350$; after 7 payments

D) $B_0 = 180{,}000 \quad B_n = \left(1 + \dfrac{7.5}{12}\right)A(n-1) - 1350$; after 24 payments

7) Jack decided to put $600 into an IRA account every 3 months at a rate of 6% compounded quarterly. Find a recursive formula that represents his balance at the end of each quarter. How long will it be before the value of the account is $100,000? What will be the balance in 30 years when Jack retires?

A) $B_0 = 600 \quad B_n = \left(1 + \dfrac{0.06}{4}\right)A(n-1) + 600$; 21 years; $202,355

B) $B_0 = 600 \quad B_n = \left(1 + \dfrac{0.06}{12}\right)A(n-1) + 600$; 10 years; $6.8 million

C) $B_0 = 600 \quad B_n = \left(1 + \dfrac{0.06}{3}\right)A(n-1) + 600$; 18 years; $151,860

D) $B_0 = 600 \quad B_n = \left(1 + \dfrac{0.6}{4}\right)A(n-1) + 600$; 7 years; $4 million

6 Demonstrate Additional Understanding and Skills

Evaluate the factorial expression.

1) $\dfrac{10!}{8!}$

A) 10 B) 90 C) $\dfrac{10}{8}$ D) 2!

2) $\dfrac{7!}{9!}$

  A) 72   B) 2!   C) $\dfrac{1}{2!}$ D) $\dfrac{1}{72}$

3) $\dfrac{6!}{4!\,2!}$

  A) 1   B) 0!   C) 15   D) 6

4) $\dfrac{5!}{4!}$

  A) 5!   B) 5   C) $\dfrac{5}{4}$ D) 1

5) $\dfrac{12!}{6!6!}$

  A) 462   B) 1848   C) 924   D) 665,280

6) $\dfrac{(n+9)!}{n+9}$

  A) 9!   B) n + 9!   C) (n + 8)!   D) 1

7) $\dfrac{n(n+9)!}{(n+10)!}$

  A) $\dfrac{n}{(n+10)!}$ B) $\dfrac{n}{n+10}$ C) $\dfrac{n}{10}$ D) $\dfrac{1}{n+10}$

11.2 Arithmetic Sequences

1 Determine If a Sequence Is Arithmetic

Determine whether the sequence is arithmetic.

1) 4, 12, 36, 108, 972, ...

  A) Arithmetic   B) Not arithmetic

2) 2, −3, −8, −13, −18, ...

  A) Arithmetic   B) Not arithmetic

3) 2, 4, 6, 10, 12, ...

  A) Arithmetic   B) Not arithmetic

4) 2, −1, −4, −7, −10, ...

  A) Arithmetic   B) Not arithmetic

5) 5, −15, 45, −135, 405, ...

  A) Arithmetic   B) Not arithmetic

6) 2, 6, 10, 14, 18, ...

  A) Arithmetic   B) Not arithmetic

An arithmetic sequence is given. Find the common difference and write out the first four terms.

7) {6 − 3n}

   A) d = 3; 3, 6, 9, 12
   B) d = −3; 3, 0, −3, −6
   C) d = −3; 3, 2, −1, −4
   D) d = −3; −3, −6, −9, −12

8) {8n + 7}

   A) d = 8; 15, 23, 31, 39
   B) d = 7; 8, 15, 23, 31
   C) d = 8; 8, 15, 23, 31
   D) d = 7; 15, 23, 31, 39

9) $\left\{\dfrac{1}{5} + \dfrac{n}{9}\right\}$

   A) $d = \dfrac{1}{5}; \dfrac{14}{45}, \dfrac{19}{45}, \dfrac{8}{15}, \dfrac{29}{45}$
   B) $d = \dfrac{1}{9}; \dfrac{1}{5}, \dfrac{14}{45}, \dfrac{19}{45}, \dfrac{8}{15}$
   C) $d = \dfrac{1}{9}; \dfrac{14}{45}, \dfrac{19}{45}, \dfrac{8}{15}, \dfrac{29}{45}$
   D) $d = \dfrac{1}{5}; \dfrac{1}{5}, \dfrac{14}{45}, \dfrac{19}{45}, \dfrac{8}{15}$

2 Find a Formula for an Arithmetic Sequence

Find the nth term and the indicated term of the arithmetic sequence whose initial term, a, and common difference, d, are given.

1) a = 4; d = 8
   a_n = ?; a_{13} = ?

   A) $a_n = -4 + 8n$; $a_{13} = 100$
   B) $a_n = -4 - 8n$; $a_{13} = 100$
   C) $a_n = 4 + 8n$; $a_{13} = 100$
   D) $a_n = -4 + 8n$; $a_{13} = 60$

2) a = 88; d = −8
   a_n = ?; a_6 = ?

   A) $a_n = 88 - 8n$; $a_6 = 48$
   B) $a_n = 88 - 8n$; $a_6 = -32$
   C) $a_n = 96 - 8n$; $a_6 = 48$
   D) $a_n = 96 - 8n$; $a_6 = -32$

3) a = 5; d = −3
   a_n = ?; a_{15} = ?

   A) $a_n = 5 - 3n$; $a_{15} = -37$
   B) $a_n = 8 + 3n$; $a_{15} = -37$
   C) $a_n = 8 - 3n$; $a_{15} = -22$
   D) $a_n = 8 - 3n$; $a_{15} = -37$

4) a = −6; d = 10
   a_n = ?; a_7 = ?

   A) $a_n = -16 + 10n$; $a_7 = 124$
   B) $a_n = -6 + 10n$; $a_7 = 54$
   C) $a_n = -16 + 10n$; $a_7 = 54$
   D) $a_n = -16 - 10n$; $a_7 = 54$

Find the indicated term of the sequence.

5) The twenty-fifth term of the arithmetic sequence −15, −12, −9, ...

   A) 57
   B) −90
   C) −87
   D) 60

6) The tenth term of the arithmetic sequence 0, 5, 10, ...

   A) 55
   B) 36
   C) 45
   D) 50

7) The tenth term of the arithmetic sequence 20, 14, 8, ...

A) 74 B) -54 C) -40 D) -34

8) The twenty-first term of the arithmetic sequence $3\sqrt{3}, 9\sqrt{3}, 15\sqrt{3}, ...$

A) $-123\sqrt{3}$ B) $123\sqrt{3}$ C) $129\sqrt{3}$ D) $-117\sqrt{3}$

Find the first term, the common difference, and give a recursive formula for the arithmetic sequence.

9) 8th term is 44; 13th term is 24

A) $a_1 = 76$, $d = 4$, $a_n = a_{n-1} + 4$
B) $a_1 = 72$, $d = -4$, $a_n = a_{n-1} - 4$
C) $a_1 = 76$, $d = -4$, $a_n = a_{n-1} - 4$
D) $a_1 = 72$, $d = 4$, $a_n = a_{n-1} + 4$

10) 6th term is -20; 14th term is -60

A) $a_1 = 10$, $d = -5$, $a_n = a_{n-1} - 5$
B) $a_1 = 5$, $d = -5$, $a_n = a_{n-1} - 5$
C) $a_1 = 5$, $d = 5$, $a_n = a_{n-1} + 5$
D) $a_1 = 10$, $d = 5$, $a_n = a_{n-1} + 5$

11) 10th term is 68; 13th term is 92

A) $a_1 = -12$, $d = 8$, $a_n = a_{n-1} + 8$
B) $a_1 = -12$, $d = 8$, $a_n = a_{n-1} - 8$
C) $a_1 = -4$, $d = 8$, $a_n = a_{n-1} - 8$
D) $a_1 = -4$, $d = 8$, $a_n = a_{n-1} + 8$

12) 6th term is -10; 15th term is -46

A) $a_1 = -30$, $d = 4$, $a_n = a_{n-1} + 4$
B) $a_1 = -10$, $d = -4$, $a_n = a_{n-1} - 4$
C) $a_1 = 10$, $d = -4$, $a_n = a_{n-1} - 4$
D) $a_1 = 10$, $d = 4$, $a_n = a_{n-1} + 4$

Find the indicated term using the given information.

13) $a = -4$, $d = -5$; a_8

A) 36 B) 31 C) -44 D) -39

14) $a = 9$, $d = 1$; a_{35}

A) 44 B) -25 C) 43 D) -26

15) $a = 6$, $d = 4$; a_{13}

A) -42 B) -46 C) 58 D) 54

16) -12, -7, -2, ...; a_{33}

A) -177 B) 148 C) 153 D) -172

17) $a_{17} = 87$, $a_{15} = 75$; a_1

A) -3 B) 3 C) 6 D) -9

18) $a_{18} = 88$, $a_{12} = 52$; a_3

A) 6 B) 4 C) -2 D) -14

19) $a_{19} = -\frac{33}{4}$, $a_{38} = -13$; a_3

A) $-\frac{9}{2}$ B) 15   C) $-\frac{1}{4}$ D) $-\frac{17}{4}$

Solve the problem.

20) The population of a town is increasing by 200 inhabitants each year. If its population at the beginning of 1990 was 20,167, what was its population at the beginning of 1999?

A) 362,862 inhabitants   B) 21,767 inhabitants   C) 181,431 inhabitants   D) 21,967 inhabitants

3 Find the Sum of an Arithmetic Sequence

Find the sum of the arithmetic sequence.

1) $-3 + 1 + 5 + 9 + 13 + \ldots + (4n - 7)$

A) $n(4n + 7)$ B) $n(4n - 7)$ C) $n(2n - 5)$ D) $n(2n + 5)$

2) $(-6) + (-1) + 4 + 9 + \ldots + 39$

A) 160   B) 170   C) 165   D) 330

3) $(-9) + (-1) + 7 + 15 + \ldots + 71$

4) $\{5n + 5\}$, $n = 31$

A) 2852   B) 2743.5   C) 2635   D) 2557.5

5) $\{2n - 1\}$, $n = 42$

A) 1764   B) 1953   C) 1722   D) 2100

6) $\{-3n + 5\}$, $n = 40$

A) -2080   B) -2260   C) -1940   D) -2200

7) $\{-4n - 7\}$, $n = 26$

A) -1404   B) -1495   C) -1534   D) -1586

Solve the problem.

8) A theater has 24 rows with 23 seats in the first row, 28 in the second row, 33 in the third row, and so forth. How many seats are in the theater?

A) 3984 seats   B) 1932 seats   C) 1992 seats   D) 3864 seats

9) A brick staircase has a total of 15 steps The bottom step requires 126 bricks. Each successive step requires 5 less bricks than the prior one. How many bricks are required to build the staircase?

A) 1365 bricks   B) 1327.5 bricks   C) 2415 bricks   D) 2730 bricks

10) Suppose you just received a job offer with a starting salary of $37,000 per year and a guaranteed raise of $1500 per year. How many years will it be before you've made a total (or aggregate) salary of $1,025,000?

11) A local civic theater has 22 seats in the first row and 21 rows in all. Each successive row contains 3 additional seats. How many seats are in the civic theater?

A) 1010 seats   B) 1070 seats   C) 1092 seats   D) 790 seats

11.3 Geometric Sequences; Geometric Series

1 Determine If a Sequence Is Geometric

Determine whether the sequence is geometric.

1) 4, 12, 36, 108, 324, ...
   A) Geometric   B) Not geometric

2) 3, 1, -1, -3, -5, ...
   A) Geometric   B) Not geometric

3) 3, 5, 7, 11, 13, ...
   A) Geometric   B) Not geometric

4) 4, -12, 36, -108, 324, ...
   A) Geometric   B) Not geometric

5) 7, 4, 1, -2, -5, ...
   A) Geometric   B) Not geometric

6) 4, 8, 16, 32, 64, ...
   A) Geometric   B) Not geometric

7) $-\frac{1}{4}, -\frac{1}{9}, -\frac{1}{14}, -\frac{1}{19}, ...$
   A) Geometric   B) Not geometric

If the sequence is geometric, find the common ratio. If the sequence is not geometric, say so.

8) 4, 8, 16, 32, 64
   A) 8   B) 1/2   C) 2   D) Not geometric

9) 3, -9, 27, -81, 243
   A) 3   B) -3   C) -12   D) Not geometric

10) $\frac{4}{3}, \frac{8}{3}, \frac{16}{3}, \frac{32}{3}, \frac{64}{3}$
   A) 2   B) 6   C) 4   D) Not geometric

11) 31, 15.5, 7.75, 3.875, 1.94
   A) -0.5   B) 2   C) 0.5   D) 1.94

12) $\frac{3}{4}, \frac{3}{16}, \frac{3}{64}, \frac{3}{256}, \frac{3}{1024}$
   A) $\frac{1}{4}$ B) 4   C) 40   D) $\frac{1}{40}$

13) $\dfrac{1}{6}, \dfrac{1}{11}, \dfrac{1}{16}, \dfrac{1}{21}$

A) $5\dfrac{1}{5}$ B) $\dfrac{1}{5}$ C) 5   D) Not geometric

Determine whether the given sequence is arithmetic, geometric, or neither. If arithmetic, find the common difference. If geometric, find the common ratio.

14) $\{3n - 2\}$

A) Geometric, $r = 3$ B) Arithmetic, $d = -2$ C) Arithmetic, $d = 3$ D) Neither

15) $\{5n^2\}$

A) Neither   B) Geometric, $r = \dfrac{5}{2}$ C) Geometric, $r = 5$ D) Arithmetic, $d = 5$

16) $\left\{\left(\dfrac{6}{5}\right)^n\right\}$

A) Neither   B) Arithmetic, $d = \dfrac{6}{5}$ C) Geometric, $r = \dfrac{5}{6}$ D) Geometric, $r = \dfrac{6}{5}$

2 Find a Formula for a Geometric Sequence

A geometric sequence is given. Find the common ratio and write out the first four terms.

1) $\left\{6\left(\dfrac{1}{2}\right)^{n-1}\right\}$

A) $a_n = 6 \cdot (2)^{n-1}$
$r = 2;\ 6, 12, 24, 48, \ldots$

B) $a_n = 6\left(\dfrac{1}{4}\right)^{n-1}$
$r = \dfrac{1}{4};\ 6, \dfrac{3}{2}, \dfrac{3}{8}, \dfrac{3}{32}, \ldots$

C) $a_n = \dfrac{1}{2}(6)^{n-1}$
$r = \dfrac{1}{2};\ 6, 12, 24, 48, \ldots$

D) $a_n = 6\left(\dfrac{1}{2}\right)^{n-1}$
$r = \dfrac{1}{2};\ 6, 3, \dfrac{3}{2}, \dfrac{3}{4}, \ldots$

Find the fifth term and the nth term of the geometric sequence whose initial term, a, and common ratio, r, are given.

2) $a = 9;\ r = 2$

A) $a_5 = 288;\ a_n = 9 \cdot (2)^{n-1}$ B) $a_5 = 288;\ a_n = 9 \cdot (2)^n$

C) $a_5 = 144;\ a_n = 9 \cdot (2)^n$ D) $a_5 = 144;\ a_n = 9 \cdot (2)^{n-1}$

3) $a = 4;\ r = \dfrac{1}{2}$

A) $a_5 = \dfrac{1}{8};\ a_n = 4 \cdot \left(\dfrac{1}{2}\right)^{n-1}$ B) $a_5 = \dfrac{1}{8};\ a_n = 4 \cdot \left(\dfrac{1}{2}\right)^n$

C) $a_5 = \dfrac{1}{4};\ a_n = 4 \cdot \left(\dfrac{1}{2}\right)^n$ D) $a_5 = \dfrac{1}{4};\ a_n = 4 \cdot \left(\dfrac{1}{2}\right)^{n-1}$

4) $a = 4;\ r = -4$

A) $a_5 = -256;\ a_n = 4 \cdot (-4)^{n-1}$ B) $a_5 = -256;\ a_n = 4 \cdot (-4)^n$

C) $a_5 = 1024;\ a_n = 4 \cdot (-4)^{n-1}$ D) $a_5 = 1024;\ a_n = 4 \cdot (-4)^n$

5) $a = -5;\ r = -5$

   A) $a_5 = -3125;\ a_n = -5 \cdot (-5)^n$
   B) $a_5 = -3125;\ a_n = -5 \cdot (-5)^{n-1}$
   C) $a_5 = 625;\ a_n = -5 \cdot (-5)^n$
   D) $a_5 = 625;\ a_n = -5 \cdot (-5)^{n-1}$

6) $a = 5;\ r = 4\pi$

   A) $a_5 = 1280\pi^4,\ a_n = 5 \cdot 4^{n-1}\pi^{n-1}$
   B) $a_5 = 1280\pi,\ a_n = 5 \cdot 4^{n-1}\pi$
   C) $a_5 = 5 + 16\pi,\ a_n = 5 + 4\pi(n-1)$
   D) $a_5 = 5120\pi^5,\ a_n = 5 \cdot 4^n \pi^n$

7) $a = \sqrt{7};\ r = \sqrt{7}$

   A) $a_5 = 2401\sqrt{7},\ a_n = 7^{n-1/2}$
   B) $a_5 = 16{,}807,\ a_n = 7^n$
   C) $a_5 = 49\sqrt{7},\ a_n = 7^{n/2}$
   D) $a_5 = 7\sqrt{7},\ a_n = 7^{n/2-1}$

Use the formula for the general term (the nth term) of a geometric sequence to find the indicated term of the sequence with the given first term, a_1, and common ratio, r.

8) Find a_8 when $a_1 = 4,\ r = 3$.

   A) 26,244 B) 25 C) 8752 D) 8748

9) Find a_9 when $a_1 = -4,\ r = 3$.

   A) 20 B) -78,732 C) -26,240 D) -26,244

10) Find a_{11} when $a_1 = 5,\ r = -2$.

    A) -10,240 B) 5120 C) 5124 D) -15

11) Find a_{11} when $a_1 = -3,\ r = -3$.

    A) -33 B) 531,441 C) -177,147 D) -177,143

12) Find a_{11} when $a_1 = 2000,\ r = \dfrac{1}{3}$.

    A) $\dfrac{2000}{59049}$ B) $\dfrac{2000}{177147}$ C) $\dfrac{6010}{3}$ D) $\dfrac{2000}{531441}$

13) Find a_8 when $a_1 = 2000,\ r = -\dfrac{1}{3}$.

    A) $\dfrac{5993}{3}$ B) $-\dfrac{2000}{19683}$ C) $\dfrac{2000}{6561}$ D) $-\dfrac{2000}{2187}$

14) Find a_8 when $a_1 = 40{,}000,\ r = 0.1$.

    A) 0.00004 B) 0.004 C) 0.0004 D) 40,000.7

15) Find a_7 for the sequence 0.6, 0.06, 0.006, . . .

    A) 0.00000001 B) 0.0000006 C) 0.00000006 D) 0.000006

Find the nth term of the geometric sequence.

16) $a = 2;\ r = 3$

    A) $a_n = 3 \cdot 2^{n-1}$
    B) $a_n = 2 \cdot 3^n$
    C) $a_n = 2 \cdot 3^{n-1}$
    D) $a_n = 2 \cdot -3^{n-1}$

17) $a = -2; r = \left(-\dfrac{1}{4}\right)$

A) $a_n = -2\left(\dfrac{1}{4}\right)^{n-1}$ B) $a_n = -2\left(-\dfrac{1}{4}\right)^{n-1}$ C) $a_n = -2\left(-\dfrac{1}{4}\right)^n$ D) $a_n = -\dfrac{1}{4}(-2)^{n-1}$

18) $2, 1, \dfrac{1}{2}, \dfrac{1}{4}, \ldots$

A) $a_n = 2\left(\dfrac{1}{4}\right)^{n-1}$ B) $a_n = 2 \cdot 2^{n-1}$ C) $a_n = 2\left(\dfrac{1}{2}\right)^{n-1}$ D) $a_n = \dfrac{1}{2}(2)^{n-1}$

19) $8, 16, 32, 64, 128$

A) $a_n = 8 \cdot 2^n$ B) $a_n = 8 \cdot 2^{n-1}$ C) $a_n = 8 \cdot 2n$ D) $a_n = a_1 + 2^n$

20) $-2, -6, -18, -54, -162$

A) $a_n = -2 \cdot 3n$ B) $a_n = a_1 + 3^n$ C) $a_n = -2 \cdot 3^n$ D) $a_n = -2 \cdot 3^{n-1}$

21) $2, -6, 18, -54, 162$

A) $a_n = 2 \cdot (-3)n$ B) $a_n = 2 \cdot (-3)^n$ C) $a_n = a_1 - 3^n$ D) $a_n = 2 \cdot (-3)^{n-1}$

22) $4, \dfrac{4}{5}, \dfrac{4}{25}, \dfrac{4}{125}, \dfrac{4}{625}$

A) $a_n = 4 \cdot \left(\dfrac{1}{5}\right)^{n+1}$ B) $a_n = 4 \cdot \left(\dfrac{1}{25}\right)^{n-1}$ C) $a_n = 4 \cdot \left(\dfrac{1}{5}\right)^{n-1}$ D) $a_n = 4 \cdot \left(\dfrac{1}{5}\right)^n$

23) $2, -\dfrac{2}{3}, \dfrac{2}{9}, -\dfrac{2}{27}, \dfrac{2}{81}$

A) $a_n = 2 \cdot \left(-\dfrac{1}{3}\right)^{n+1}$ B) $a_n = 2 \cdot \left(-\dfrac{1}{3}\right)^n$ C) $a_n = 2 \cdot \left(-\dfrac{1}{3}\right)^{n-1}$ D) $a_n = 2 \cdot \left(-\dfrac{1}{9}\right)^{n-1}$

Solve the problem.

24) Find the 10th term of the geometric sequence $\dfrac{1}{3}, 1, 3, \ldots$

A) 177,147 B) 2187 C) 19683 D) 6561

25) For the geometric sequence $2, 1, \dfrac{1}{2}, \dfrac{1}{4}, \ldots$, find a_n.

A) $a_n = 2^{n-2}$ B) $a_n = \left(\dfrac{1}{2}\right)^{1-n}$ C) $a_n = \left(\dfrac{1}{2}\right)^{n-2}$ D) $a_n = 2^{1-n}$

26) For the geometric sequence $64, 16, 4, 1, \ldots$, find a_n.

27) A new piece of equipment cost a company $64,000. Each year, for tax purposes, the company depreciates the value by 25%. What value should the company give the equipment after 7 years?

A) $4 B) $16 C) $11,391 D) $8543

Precalculus Enhanced with Graphing Utilities 841

28) A particular substance decays in such a way that it loses half its weight each day. How much of the substance is left after 9 days if it starts out at 64 grams?

   A) 4 grams   B) $\frac{1}{8}$ gram   C) $\frac{1}{4}$ gram   D) 8 grams

29) A bicycle wheel rotates 300 times in a minute as long as the rider is pedalling. If the rider stops pedalling, the wheel starts to slow down. Each minute it will rotate only 2/3 as many times as in the preceding minute. How many times will the wheel rotate in the 4th minute after the rider's feet leave the pedals? Round your answer to the nearest unit.

   A) 89 times   B) 59 times   C) 0 times   D) 11 times

30) A pendulum bob swings through an arc 80 inches long on its first swing. Each swing thereafter, it swings only 64% as far as on the previous swing. What is the length of the arc after 7 swings? Round your answer to two decimal places, if necessary.

   A) 5.5 inches   B) 307.2 inches   C) 3.52 inches   D) 2.25 inches

31) A football player signs a contract with a starting salary of $830,000 per year and an annual increase of 4.5% beginning in the second year. What will the athlete's salary be, to the nearest dollar, in the sixth year?

   A) $1,034,331   B) $1,032,851   C) $1,035,127   D) $1,036,522

32) Jennifer takes a job with a starting salary of $26,000 for the first year with an annual increase of 6.5% beginning in the second year. What is Jennifer's salary, to the nearest dollar, in the seventh year?

   A) $38,543   B) $37,938   C) $36,624   D) $40,222

3 Find the Sum of a Geometric Sequence

Find the sum.

1) 5, 10, 20, 40, 80

   A) 157   B) 155   C) 165   D) 153

2) 4, −16, 64, −256, 1024

   A) 820   B) 1364   C) −1364   D) −820

3) $\frac{1}{7} + \frac{3}{7} + \frac{3^2}{7} + \frac{3^3}{7} + \cdots + \frac{3^{n-1}}{7}$

   A) $-\frac{2}{7}(1 - 3^n)$ B) $-\frac{1}{14}(1 - 3^n)$ C) $-\frac{1}{7}(1 - 3^n)$ D) $\frac{1}{14}(1 - 3^n)$

4) $\sum_{k=1}^{5} \left(\frac{4}{3}\right)(2)^k$

   A) $\frac{296}{3}$ B) $\frac{308}{3}$ C) $\frac{188}{3}$ D) $\frac{248}{3}$

5) $\sum_{k=1}^{4} \left(\frac{3}{4}\right)^{k+1}$

   A) $\frac{1575}{1024}$ B) $\frac{525}{256}$ C) $\frac{7029}{1024}$ D) $\frac{1579}{1024}$

6) $\sum_{k=1}^{n} 5 \cdot 7^{k-1}$

   A) $-\frac{5}{6}(1 - 7^{n-1})$ B) $-\frac{5}{6}(1 - 7^n)$ C) $-30(1 - 7^n)$ D) $-5(1 - 7^n)$

Use a graphing utility to find the sum of the geometric sequence. Round answer to two decimal places, if necessary.

7) $5 + 15 + 45 + 135 + 405 + \cdots + 5 \cdot 3^{12}$

   A) 3,985,785   B) 3,985,805   C) 3,985,807   D) 3,985,842

8) $-6 - 18 - 54 - 162 - 486 - \cdots - 6 \cdot 3^{11}$

   A) $-1,594,340$ B) $-1,594,318$ C) $-1,594,283$ D) $-1,594,320$

9) $7 - 14 + 28 - 56 + 112 - \cdots + 7 \cdot (-2)^7$

   A) -588 B) -595 C) -597 D) -601

10) $\frac{1}{5} + \frac{2}{5} + \frac{2^2}{5} + \frac{2^3}{5} + \cdots + \frac{2^{13}}{5}$

    A) 3276.6   B) 3276.4   C) 6553.4   D) 6553.2

11) $\sum_{k=1}^{5} 4(4)^k$

    A) 10,280   B) 272   C) 1360   D) 5456

12) $\sum_{k=1}^{5} 3(-4)^k$

    A) -7710 B) -2460 C) 244   D) -156

13) $\sum_{k=1}^{10} \frac{1}{6} \cdot (-2)^{k-1}$

    A) -55.67 B) -57.83 C) -57.17 D) -56.83

Solve the problem.

14) A small business owner made $40,000 the first year he owned his store and made an additional 5% over the previous year in each subsequent year. Find how much he made during his fourth year of business. Find his total earnings during the first four years. (Round to the nearest cent, if necessary.)

    A) $46,305.00; $172,405.00
    B) $135,000.00; $325,000.00
    C) $202,500.00; $527,500.00
    D) $5.00; $42,105.00

15) As Sunee improves her algebra skills, she takes 0.9 times as long to complete each homework assignment as she took to complete the preceeding assignment. If it took her 70 minutes to complete her first assignment, find how long it took her to complete the fifth assignment. Find the total time she took to complete her first five homework assignments. (Round to the nearest minute.)

   A) 41 min; 287 min   B) 41 min; 241 min   C) 46 min; 287 min   D) 46 min; 241 min

16) Initially, a pendulum swings through an arc of 3 feet. On each successive swing, the length of the arc is 0.8 of the previous length. After 10 swings, what total length will the pendulum have swung (to the nearest tenth of a foot)?

17) Joytown has a present population of 40,000 and the population is increasing by 2.5% each year. How long will it take for the population to double? Round your answer to the nearest year.

   A) 40 years   B) 41 years   C) 28 years   D) 29 years

4 Find the Sum of a Geometric Series

Find the sum of the infinite geometric series.

1) $3 + \dfrac{3}{4} + \dfrac{3}{16} + \cdots$

   A) $\dfrac{15}{4}$ B) 3   C) 4   D) $\dfrac{3}{4}$

2) $5 - \dfrac{5}{3} + \dfrac{5}{9} - \cdots$

   A) 5   B) $\dfrac{10}{3}$ C) $\dfrac{15}{4}$ D) $-\dfrac{5}{3}$

3) $150 + 30 + 6 + \cdots$

   A) 186   B) $\dfrac{375}{2}$ C) $-\dfrac{75}{2}$ D) 150

4) $-6 - 3 - \dfrac{3}{2} - \cdots$

   A) $-\dfrac{21}{2}$ B) -6 C) 6   D) -12

5) $\displaystyle\sum_{k=1}^{\infty} 4\left(\dfrac{2}{3}\right)^{k-1}$

   A) 12   B) $\dfrac{8}{3}$ C) 16   D) 4

6) $\displaystyle\sum_{k=1}^{\infty} 10(0.2)^{k-1}$

   A) -9.5 B) 12.5
   C) -12.5 D) The series has no sum.

7) $(0.996) + (0.996)^2 + (0.996)^3 + \ldots$

Express the repeating decimal as a fraction in lowest terms.

8) $0.\overline{6} = \dfrac{6}{10} + \dfrac{6}{100} + \dfrac{6}{1,000} + \dfrac{6}{10,000} \ldots$

A) $\dfrac{33}{50}$ B) $\dfrac{2}{3}$ C) $\dfrac{2}{33}$ D) $\dfrac{333}{500}$

9) $0.\overline{66} = \dfrac{66}{100} + \dfrac{66}{10,000} + \dfrac{66}{1,000,000} + \ldots$

A) $\dfrac{2200}{333}$ B) $\dfrac{3333}{5000}$ C) $\dfrac{2}{3}$ D) $\dfrac{2222}{333}$

Solve the problem.

10) A pendulum bob swings through an arc 80 inches long on its first swing. Each swing thereafter, it swings only 90% as far as on the previous swing. How far will it swing altogether before coming to a complete stop?

A) 89 inches B) 800 inches C) 400 inches D) 178 inches

11) A ball is dropped from a height of 25 feet. Each time it strikes the ground, it bounces up to 0.7 of the previous height. The total distance the ball has travelled before the second bounce is $25 + 2(25 \cdot 0.7)$ feet, and the total distance the ball has travelled before bounce $n + 1$ is

$$25 + \sum_{k=1}^{n} 50\left(0.7^k\right) \text{ feet.}$$

Use facts about infinite geometric series to calculate the total distance the ball has travelled by the time it has stopped bouncing.

12) A ping-pong ball is dropped from a height of 9 ft and always rebounds $\dfrac{1}{3}$ of the distance fallen. Find the total sum of the rebound heights of the ball.

A) 3 ft B) 6 ft C) 13.5 ft D) 4.5 ft

13) Write 0.363636... as a fraction. (Find the sum of the repeating decimal.)

A) $\dfrac{2}{5}$ B) $\dfrac{4}{11}$ C) $\dfrac{8}{11}$ D) $\dfrac{4}{111}$

14) After being struck with a hammer, a gong vibrates 22 vibrations in the first second and in each second thereafter makes $\dfrac{2}{3}$ as many vibrations as in the previous second. Find how many vibrations the gong makes before it stops vibrating.

A) 76 vibrations B) 28 vibrations C) 33 vibrations D) 66 vibrations

11.4 Mathematical Induction

1 Prove Statements Using Mathematical Induction

Use the Principle of Mathematical Induction to show that the statement is true for all natural numbers n.

1) $4 + 9 + 14 + \ldots + (5n - 1) = \dfrac{n}{2}(5n + 3)$

2) $5 + 10 + 15 + \ldots + 5n = \dfrac{5n(n + 1)}{2}$

Precalculus Enhanced with Graphing Utilities 845

3) $1 + 6 + 6^2 + \ldots + 6^{n-1} = \dfrac{6^n - 1}{5}$

4) $n^2 - n + 2$ is divisible by 2

5) $\dfrac{1}{2} + \dfrac{1}{4} + \dfrac{1}{8} + \dfrac{1}{16} + \ldots + \dfrac{1}{2^n} = 1 - \dfrac{1}{2^n}$

6) Use the Principle of Mathematical Induction to show that the statement "5 is a factor of $7^n - 2^n$" is true for all natural numbers. (Hint: $7^{k+1} - 2^{k+1} = 7(7^k - 2^k) + 5 \cdot 2^k$)

7) Show that the formula
$$2 + 4 + 6 + 8 + \cdots + 2n = n^2 + n + 3$$
obeys Condition II of the Principle of Mathematical Induction. That is, show that if the formula is true for some natural number k, it is also true for the next natural number k + 1. Then show that the formula is false for n = 1.

8) Show that the statement "$n^2 - n + 3$ is a prime number" is true for n = 1, but is not true for n = 3.

Use mathematical induction to prove the statement is true for all positive integers n.

9) $6 + 12 + 18 + \ldots + 6n = 3n(n + 1)$

10) $5 + 5 \cdot \dfrac{1}{2} + 5 \cdot \left(\dfrac{1}{2}\right)^2 + \ldots + 5 \cdot \left(\dfrac{1}{2}\right)^{n-1} = \dfrac{5\left[1 - \left(\dfrac{1}{2}\right)^n\right]}{1 - \dfrac{1}{2}}$

11) $1^2 + 4^2 + 7^2 + \ldots + (3n - 2)^2 = \dfrac{n(6n^2 - 3n - 1)}{2}$

12) $1 \cdot 2 + 2 \cdot 3 + 3 \cdot 4 + \ldots + n(n + 1) = \dfrac{n(n + 1)(n + 2)}{3}$

13) $\left(1 - \dfrac{1}{2}\right)\left(1 - \dfrac{1}{3}\right) \cdots \left(1 - \dfrac{1}{n+1}\right) = \dfrac{1}{n+1}$

14) $6 + 2 \cdot 6 + 3 \cdot 6 + \ldots + 6n = \dfrac{6n(n + 1)}{2}$

15) $(2^4)^n = 2^{4n}$

Solve the problem.

16) Let P(n) represent the statement:
$$-3 + 3 + 9 + \ldots + (6n - 9) = 3n^2 - 6n$$
In the proof that P(n) is true for all integers n, n ≥ 1, what term must be added to both sides of P(k) to show P(k +1) follows from P(k)?

A) P(k + 1)   B) 6k + 3   C) 6k - 9   D) 6k - 3

17) Let P(n) represent the statement:
$$-2 + 4 + 10 + \ldots + (6n - 8) = 3n^2 - 5n$$
In the proof that P(n) is true for all integers n, n ≥ 1, what term must be added to both sides of P(k) to show P(k +1) follows from P(k)?

11.5 The Binomial Theorem

1 Evaluate a Binomial Coefficient

Evaluate the expression.

1) $\binom{12}{9}$

   A) 1320   B) 220   C) 1   D) 110

2) $\binom{7}{2}$

   A) 0   B) 7   C) 1   D) 21

3) $\binom{8}{4}$

   A) 70   B) 1680   C) 35   D) 140

4) $\binom{156}{2}$

   A) 12,090   B) $\dfrac{156!}{154!}$ C) 154   D) 24,180

5) $\binom{6}{0}$

   A) 1   B) 720   C) 0   D) 2

6) $\binom{6}{1}$

   A) $\dfrac{6}{5}$ B) 6   C) 1   D) 6!

7) $\binom{10}{10}$

   A) 2   B) 3,628,800   C) 1   D) 0

2 Use the Binomial Theorem

Find the indicated coefficient or term.

1) The coefficient of x in the expansion of $(5x + 3)^3$

   A) 270   B) 135   C) 225   D) 9

2) The coefficient of x in the expansion of $(3x + 4)^5$

   A) 2880   B) 5120   C) 960   D) 3840

3) The coefficient of $\dfrac{1}{x}$ in the expansion of $\left(2x + \dfrac{1}{x}\right)^3$

4) The coefficient of x^8 in the expansion of $(x^2 - 3)^7$
   A) 2835   B) -945   C) -2835   D) 945

5) The 3rd term in the expansion of $(8x + 9)^3$
   A) 1944x   B) 81   C) 3888x   D) $1728x^2$

6) The 5th term in the expansion of $(2x + 3)^5$
   A) $540x^2$ B) 270x   C) 1215   D) 810x

7) The 8th term in the expansion of $(2x - 3y)^9$
   A) $104,976x^2y^8$ B) $104,976x^7y^2$ C) $-314,928x^2y^7$ D) $-157,464x^7y^2$

8) The 8th term in the expansion of $(x + 2y)^9$
   A) $4608x^7y^2$ B) $2304x^2y^8$ C) $4608x^2y^7$ D) $2304x^7y^2$

9) The 4th term in the expansion of $(2x + 3y)^{12}$
   A) $1,520,640x^3y^9$ B) $1,013,760x^3y^9$ C) $1,013,760x^9y^4$ D) $3,041,280x^9y^3$

10) The 8th term in the expansion of $(x - 3y)^{10}$
   A) $87,480x^3y^8$ B) $-262,440x^7y^3$ C) $-262,440x^3y^7$ D) $87,480x^7y^3$

Expand the expression using the Binomial Theorem.

11) $(x - 1)^6$
   A) $x^6 - 6x^5 + 30x^4 - 120x^3 + 30x^2 - 6x + 1$
   B) $x^6 - 6x^5 - 15x^4 - 20x^3 - 15x^2 - 6x - 6$
   C) $x^6 - 6x^5 + 15x^4 - 20x^3 + 15x^2 - 6x + 1$
   D) $x^6 - 6x^5 - 30x^4 - 120x^3 - 360x^2 - 720x + 720$

12) $(x + 7)^5$
   A) $x^5 + 35x^4 + 980x^3 + 6860x^2 + 12,005x + 7$
   B) $x^5 + 35x^4 + 980x^3 + 6860x^2 + 12,005x + 16,807$
   C) $x^5 + 35x^4 + 490x^3 + 3430x^2 + 12,005x + 7$
   D) $x^5 + 35x^4 + 490x^3 + 3430x^2 + 12,005x + 16,807$

13) $(2x + 4)^3$
   A) $4x^6 + 8x^3 + 4096$
   B) $4x^2 + 16x + 16$
   C) $8x^3 + 48x^2 + 96x + 64$
   D) $8x^3 + 48x^2 + 48x + 64$

14) $(4x + 5)^4$
   A) $256x^3 + 1280x^2 + 2400x + 2000$
   B) $(16x^2 + 20x + 25)^4$
   C) $1280x^4 + 6400x^3 + 2400x^2 + 10,000x + 625$
   D) $256x^4 + 1280x^3 + 2400x^2 + 2000x + 625$

15) $(4x + 2)^5$
   A) $1024x^5 + 320x^4 + 1280x^3 + 1280x^2 + 320x + 32$
   B) $1024x^5 + 512x^4 + 256x^3 + 128x^2 + 64x + 32$
   C) $1024x^5 + 2560x^4 + 2560x^3 + 1280x^2 + 320x + 32$
   D) $(16x^2 + 16x + 4)^5$

16) $(2x^2 + 2)^3$
   A) $8x^6 + 24x^4 + 24x^2 + 8$
   B) $8x^3 + 24x^2 + 24x + 8$
   C) $(4x^4 + 8x^2 + 4)^3$
   D) $16x^8 + 8x^6 + 24x^4 + 24x^2 + 8$

17) $(3x - 4y)^3$
   A) $9x^3y - 12x^2y^2 + 16xy^3$
   B) $9x^3y - 24x^2y^2 + 16xy^3$
   C) $27x^3 - 36x^2y + 48xy^2 - 64y^3$
   D) $27x^3 - 108x^2y + 144xy^2 - 64y^3$

18) $(2x + y)^6$
   A) $2x^6 + 12x^5y + 30x^4y^2 + 40x^3y^3 + 12xy^5 + y^6$
   B) $64x^6 + 192x^5y + 480x^4y^2 + 960x^3y^3 + 1440x^2y^4 + 12xy^5 + y^6$
   C) $64x^6 + 192x^5y + 240x^4y^2 + 160x^3y^3 + 60x^2y^4 + 12xy^5 + y^6$
   D) $64x^6 + 192x^5y + 240x^4y^2 + 160x^3y^3 + 240x^2y^4 + 192xy^5 + 64y^6$

19) $(w - s)^6$
   A) $w^6 - 8w^5s + 17w^4s^2 - 22w^3s^3 + 17w^2s^4 - 8ws^5 + s^6$
   B) $w^6 - 6w^5s + 15w^4s^2 - 20w^3s^3 + 15w^2s^4 - 6ws^5 + s^6$
   C) $w^6 - s^6$
   D) $w^6 - 6w^5s - 30w^4s^2 + 120w^3s^3 + 360w^2s^4 - 720ws^5 - 720s^6$

20) $(g - 2h)^3$

21) $\left(x - \dfrac{3}{\sqrt{x}}\right)^4$

Solve the problem.

22) Use the Binomial Theorem to approximate $(1.01)^5 = (1 + 10^{-2})^5$ to 7 decimal places.

23) Use the Binomial Theorem to approximate $(0.98)^6 = (1 - 2(10^{-2}))^6$ to 5 decimal places.

24) A cube with side length s units has a volume equal to s^3. Use the Binomial Theorem to calculate the difference in volume if the side length is increased by h units. That is, expand and simplify the expression $(s + h)^3 - s^3$. How much (to three decimal places) does the volume of a cube of side length 2 inches change if this side is increased by 0.03 of an inch?

Ch. 11 Sequences; Induction; the Binomial Theorem
Answer Key

11.1 Sequences
1 Write the First Several Terms of a Sequence
1) B
2) D
3) D
4) C
5) A
6) B
7) B
8) C
9) D
10) C
11) B
12) D
13) D
14) C
15) A
16) D

2 Write the Terms of a Sequence Defined by a Recursive Formula
1) B
2) A
3) D
4) C
5) 135, 45, 15, 5
6) B
7) A

3 Use Summation Notation
1) A
2) A
3) B
4) B
5) C
6) D
7) B
8) C
9) D
10) A
11) B
12) B
13) C
14) A
15) A
16) D
17) A

4 Find the Sum of a Sequence Algebraically and Using a Graphing Utility
1) B
2) D
3) D
4) C
5) C
6) B
7) B

8) B
9) A
10) A
11) B
12) A
13) B
14) D
15) B
16) A

5 Solve Annuity and Amortization Problems
1) $865
2) $25,924.25
3) 106 deer
4) 299,880 gallons
5) $1824.12
6) B
7) A

6 Demonstrate Additional Understanding and Skills
1) B
2) D
3) C
4) B
5) C
6) C
7) B

11.2 Arithmetic Sequences

1 Determine If a Sequence Is Arithmetic
1) B
2) A
3) B
4) A
5) B
6) A
7) B
8) A
9) C

2 Find a Formula for an Arithmetic Sequence
1) A
2) C
3) D
4) C
5) A
6) C
7) D
8) B
9) B
10) B
11) D
12) C
13) D
14) C
15) D
16) B
17) D
18) C

19) D
20) D

3 Find the Sum of an Arithmetic Sequence
1) C
2) C
3) 341
4) C
5) A
6) B
7) D
8) B
9) A
10) 20 years
11) C

11.3 Geometric Sequences; Geometric Series
1 Determine If a Sequence Is Geometric
1) A
2) B
3) B
4) A
5) B
6) A
7) B
8) C
9) B
10) A
11) C
12) A
13) D
14) C
15) A
16) D

2 Find a Formula for a Geometric Sequence
1) D
2) D
3) D
4) C
5) B
6) A
7) C
8) D
9) D
10) B
11) C
12) A
13) D
14) B
15) B
16) C
17) B
18) C
19) B
20) D
21) D
22) C

23) C
24) D
25) C
26) $a_n = \left(\frac{1}{4}\right)^{n-4}$
27) D
28) B
29) B
30) A
31) A
32) B

3 Find the Sum of a Geometric Sequence

1) B
2) A
3) B
4) D
5) A
6) B
7) B
8) D
9) B
10) A
11) D
12) B
13) D
14) A
15) C
16) approximately 13.4 feet
17) C

4 Find the Sum of a Geometric Series

1) C
2) C
3) B
4) D
5) A
6) B
7) 249
8) B
9) C
10) B
11) $141\frac{2}{3}$ feet
12) D
13) B
14) D

11.4 Mathematical Induction
1 Prove Statements Using Mathematical Induction

1) First we show that the statement is true when n = 1.

For n = 1, we get $4 = \dfrac{(1)}{2}(5(1) + 3) = 4$.

This is a true statement and Condition I is satisfied.

Next, we assume the statement holds for some k. That is,

$4 + 9 + 14 + \ldots + (5k - 1) = \dfrac{k}{2}(5k + 3)$ is true for some positive integer k.

We need to show that the statement holds for k + 1. That is, we need to show that

$4 + 9 + 14 + \ldots + (5(k + 1) - 1) = \dfrac{k + 1}{2}(5(k + 1) + 3)$.

So we assume that $4 + 9 + 14 + \ldots + (5k - 1) = \dfrac{k}{2}(5k + 3)$ is true and add the next term, $5(k + 1) - 1$, to both sides of the equation.

$$4 + 9 + 14 + \ldots + (5k - 1) + 5(k + 1) - 1 = \dfrac{k}{2}(5k + 3) + 5(k + 1) - 1$$

$$= \dfrac{1}{2}\big[k(5(k + 1) - 2) + 10(k + 1) - 2\big]$$

$$= \dfrac{1}{2}\big[5k(k + 1) - 2k + 10(k + 1) - 2\big]$$

$$= \dfrac{1}{2}\big[(k + 1)(5k + 10) - 2(k + 1)\big]$$

$$= \dfrac{k + 1}{2}(5k + 8)$$

$$= \dfrac{k + 1}{2}(5(k + 1) + 3).$$

Conditions II is satisfied. As a result, the statement is true for all natural numbers n.

2) First, we show the statement is true when n = 1.

For n = 1, we get $5 = \dfrac{5(1)((1)+1)}{2} = 5$.

This is a true statement and Condition I is satisfied.

Next, we assume the statement holds for some k. That is,

$5 + 10 + 15 + ... + 5k = \dfrac{5k(k+1)}{2}$ is true for some positive integer k.

We need to show that the statement holds for k + 1. That is, we need to show that

$5 + 10 + 15 + ... + 5(k+1) = \dfrac{5(k+1)(k+2)}{2}$.

So we assume that $5 + 10 + 15 + ... + 5k = \dfrac{5k(k+1)}{2}$ is true and add the next term, $5(k+1)$, to both sides of the equation.

$$\begin{aligned}
5 + 10 + 15 + ... + 5k + 5(k+1) &= \dfrac{5k(k+1)}{2} + 5(k+1) \\
&= 5\left(\dfrac{k(k+1)}{2} + k + 1\right) \\
&= 5\left(\dfrac{k(k+1)}{2} + \dfrac{2(k+1)}{2}\right) \\
&= 5 \cdot \dfrac{k(k+1) + 2(k+1)}{2} \\
&= 5 \cdot \dfrac{(k+1)(k+2)}{2} \\
&= \dfrac{5(k+1)(k+2)}{2}.
\end{aligned}$$

Conditions II is satisfied. As a result, the statement is true for all natural numbers n.

3) First, we show that the statement is true when $n = 1$.

For $n = 1$, we get 1 (or $6[(1) - 1]) = \dfrac{6(1) - 1}{5} = \dfrac{5}{5} = 1$.

This is a true statement and Condition I is satisfied.

Next, we assume the statement holds for some k. That is,

$$1 + 6 + 6^2 + \ldots + 6^{k-1} = \dfrac{6^k - 1}{5}$$ is true for some positive integer k.

We need to show that the statement holds for $k + 1$. That is, we need to show that

$$1 + 6 + 6^2 + \ldots + 6^k = \dfrac{6^{k+1} - 1}{5}.$$

So we assume that $1 + 6 + 6^2 + \ldots + 6^{k-1} = \dfrac{6^k - 1}{5}$ is true and add the next term, 6^k, to both sides of the equation.

$$1 + 6 + 6^2 + \ldots + 6^{k-1} + 6^k = \dfrac{6^k - 1}{5} + 6^k$$

$$= \dfrac{6^k - 1 + 5 \cdot 6^k}{5}$$

$$= \dfrac{6 \cdot 6^k - 1}{5}$$

$$= \dfrac{6^{k+1} - 1}{5}.$$

Conditions II is satisfied. As a result, the statement is true for all natural numbers n.

4) First, we show that the statement is true when $n = 1$.

For $n = 1$, $n^2 - n + 2 = (1)^2 - (1) + 2 = 2$.

This is a true statement and Condition I is satisfied.

Next, we assume the statement holds for some k. That is,

$k^2 - k + 2$ is divisible by 2 is true for some positive integer k.

We need to show that the statement holds for $k + 1$. That is, we need to show that

$(k + 1)^2 - (k + 1) + 2$ is divisible by 2.

So we assume $k^2 - k + 2$ is divisible by 2 and look at the expression for $n = k + 1$.

$(k + 1)^2 - (k + 1) + 2 = k^2 + 2k + 1 - k - 1 + 2$

$= (k^2 - k + 2) + 2k$

Since $k^2 - k + 2$ is divisible by 2, then $k^2 - k + 2 = 2m$ for some integer m. Hence,

$(k + 1)^2 - (k + 1) + 2 = (k^2 - k + 2) + 2k$

$= 2m + 2k$

$= 2(m + k)$.

Conditions II is satisfied. As a result, the statement is true for all natural numbers n.

5) When n = 1, the left side of the statement is $\frac{1}{2n} = \frac{1}{2^1} = \frac{1}{2}$, and the right side of the statement is $1 - \frac{1}{2n} = 1 - \frac{1}{2^1} = 1 - \frac{1}{2} = \frac{1}{2}$, so the statement is true when n = 1.

Assume the statement is true for some natural number k. Then,
$$\frac{1}{2} + \frac{1}{4} + \frac{1}{8} + \frac{1}{16} + \ldots + \frac{1}{2^k} + \frac{1}{2^{k+1}} = \left(1 - \frac{1}{2^k}\right) + \frac{1}{2^{k+1}} = 1 - \frac{1}{2^k}\left(1 - \frac{1}{2}\right) = 1 - \frac{1}{2^{k+1}}.$$

So the statement is true for k + 1. Conditions I and II are satisfied; by the Principle of Mathematical Induction, the statement is true for all natural numbers.

6) When n = 1, $7^n - 2^n = 7^1 - 2^1 = 5$, so the statement is true when n = 1. Assume the statement is true for some natural number k. That is, $7^k - 2^k = 5m$ for some integer m. Then,
$7^{k+1} - 2^{k+1} = 7(7^k - 2^k) + 5 \cdot 2^k = 7(5m) + 5 \cdot 2^k = 5(7m + 2^k)$.

So the statement is true for k + 1. Conditions I and II are satisfied; by the Principle of Mathematical Induction, the statement is true for all natural numbers.

7) Assume the statement is true for some natural number k. Then
$$2 + 4 + 6 + 8 + \cdots + 2k + 2(k+1) = (k^2 + k + 3) + 2(k+1)$$
$$= (k^2 + 3k + 2) + 3$$
$$= (k+2)(k+1) + 3$$
$$= ((k+1) + 1)(k+1) + 3$$
$$= (k+1)^2 + (k+1) + 3$$

So the statement is true for k + 1.

However, when n = 1, the left side of the statement is 2n = 2(1) = 2, and the right side of the statement is $n^2 + n + 3 = 1^2 + 1 + 3 = 5$, so the formula is false for n = 1.

8) When n = 1, $n^2 - n + 3 = 1^2 - 1 + 3 = 3$, which is a prime number, so the statement is true when n = 1. When n = 3, $n^2 - n + 3 = 3^2 - 3 + 3 = 9$, which is not a prime number, so the statement is not true for n = 3.

9) Answers will vary.
10) Answers will vary.
11) Answers will vary.
12) Answers will vary.
13) Answers will vary.
14) Answers will vary.
15) Answers will vary.
16) D
17) 6k - 2

11.5 The Binomial Theorem
1 Evaluate a Binomial Coefficient
1) B
2) D
3) A
4) A
5) A
6) B
7) C

2 Use the Binomial Theorem
1) B
2) D
3) 6
4) B
5) A
6) D
7) C

8) C
9) D
10) C
11) C
12) D
13) C
14) D
15) C
16) A
17) D
18) C
19) B
20) $g^3 - 6g^2h + 12gh^2 - 8h^3$
21) $x^4 - 12x^{5/2} + 54x - \dfrac{108}{\sqrt{x}} + \dfrac{81}{x^2}$

22) 1.0510101
23) 0.88584
24) $3s^2h + 3sh^2 + h^3$; 0.365 cubic inches

Ch. 12 Counting and Probability

12.1 Sets and Counting

1 Find All the Subsets of a Set

Write down all the subsets of the given set.

1) {1, 4, 10, 11}
   A) {1}, {4}, {10}, {11}, {1, 4}, {1, 10}, {1, 11}, {4, 10},
      {4, 11}, {10, 11}, {1, 4, 10}, {1, 4, 11}, {1, 10, 11}, {4, 10, 11}, {1, 4, 10, 11}
   B) {1}, {4}, {10}, {11}, {1, 4}, {1, 10}, {1, 11}, {4, 10},
      {4, 11}, {1, 4, 10}, {1, 4, 11}, {1, 10, 11}, {4, 10, 11},
      {1, 4, 10, 11}, ∅
   C) {1}, {4}, {10}, {11}, {1, 4}, {1, 10}, {1, 11}, {4, 10},
      {4, 11}, {10, 11}, {1, 4, 10}, {1, 4, 11}, {1, 10, 11}, {4, 10, 11}, {1, 4, 10, 11}, ∅
   D) {1}, {4}, {10}, {11}, {1, 4}, {1, 10}, {1, 11}, {4, 10},
      {4, 11}, {10, 11}, {1, 4, 10}, {1, 4, 11}, {1, 10, 11},
      {4, 10, 11}, ∅

2) {2, α, 6, π}
   A) {2}, {α}, {6}, {π}, {2, α}, {2, 6}, {2, π}, {α, 6},
      {α, π}, {6, π}, {2, α, π}, {2, 6, π}, {α, 6, π},
      {2, α, 6, π}, ∅
   B) {2}, {α}, {6}, {π}, {2, α}, {2, 6}, {2, π}, {α, 6},
      {α, π}, {6, π}, {2, α, 6}, {2, α, π}, {2, 6, π}, {α, 6, π},
      {2, α, 6, π}, ∅
   C) {2}, {α}, {6}, {π}, {2, α}, {2, 6}, {2, π}, {α, 6},
      {α, π}, {6, π}, {2, α, 6}, {2, α, π}, {2, 6, π}, {α, 6, π},
      {2, α, 6, π}
   D) {2}, {α}, {6}, {π}, {2, α}, {2, 6}, {2, π}, {α, 6},
      {α, π}, {6, π}, {2, α, 6}, {2, α, π}, {2, 6, π}, {α, 6, π}, ∅

3) {a}
   A) {a}   B) ∅, {a}   C) a   D) {a, b}

4) {p, q, r}
   A) ∅, {p}, {q}, {r}, {p, q}, {p, r}, {q, r}
   B) ∅, {p}, {q}, {r}, {p, q}, {p, r}, {q, r}, {p, q, r}
   C) {p}, {q}, {r}, {p, q}, {p, r}, {q, r}, {p, q, r}
   D) ∅, {p}, {q}, {r}, {p, q}, {p, r}, {q, r}, {p, p}, {q, q}, {r, r}, {p, q, r}

2 Find the Intersection and Union of Sets

Let A = {q, s, u, w, y}, B = {q, s, y, z}, and C = {v, w, x, y, z}. Find the indicated set.

1) A ∪ B
   A) {q, s, u, w, y, z}   B) {q, s, y, z}   C) {q, s, u, y, z}   D) {q, s, y}

2) A ∩ B
   A) {q, s, y}   B) {q, s, u, y, z}   C) {q, s, u, w, y, z}   D) {q, s, y, z}

3) A ∪ (B ∩ C)
   A) {q, s, u, w, y, z}   B) {q, y, z}   C) {q, r, w, y, z}   D) {q, w, y}

4) B ∩ (A ∪ C)
   A) {q, r, w, y, z}   B) {q, w, y}   C) {q, s, u, w, y, z}   D) {q, s, y, z}

5) (A ∩ C) ∪ (A ∩ B)

   A) {q, s, w, y}   B) {y}   C) {q, s, w, y, y}   D) {q, s, u, w, y, z}

Find the indicated set.

6) Let A = {2, 5, 8, 17, 24} and B = { 1, 5, 18, 21, 24}.
   Find A ∩ B.

   A) ∅

   B) {1, 2, 5, 8, 17, 18, 21, 24}

   C) {5, 24}

   D) none of these

7) Let A = {2, 7, 9, 12, 15}, B = {3, 9, 11, 15, 20}, and C = {4, 8, 9, 10, 12, 14}.
   Find (A ∩ B) ∩ C.

   A) {9}

   B) {9, 12, 15}

   C) {2, 3, 4, 7, 8, 9, 10, 11, 12, 14, 15, 20}

   D) {9, 12}

8) Let A = {2, 7, 9, 12, 15}, B = {3, 9, 15, 20}, and C = {4, 9, 12, 15}.
   Find (A ∪ B) ∪ C.

   A) {9, 12, 15}

   B) {9, 15}

   C) {2, 7, 9, 12, 15, 3, 9, 15, 20, 4, 9, 12, 15}

   D) {2, 3, 4, 7, 9, 12, 15, 20}

9) Let A = {0, 3, 5, 8}, B = {1, 4, 6, 9}, and C = {0, 4, 5, 6}.
   Find (A ∪ C) ∩ B.

   A) {0, 1, 4, 5, 6, 9}   B) {0, 1, 3, 4, 5, 6, 8, 9}   C) {4, 6}   D) ∅

10) Let A = {0, 3, 5, 8}, B = {1, 4, 6, 9}, and C = {0, 4, 5, 6, 7}.
    Find (A ∪ C) ∩ (B ∪ C).

    A) {4, 5, 6}   B) {0, 4, 5, 6}   C) {0, 3, 4, 5, 6, 7}   D) {0, 4, 5, 6, 7}

Solve the problem.

11) A group of friends were discussing vegetables they liked. Andy liked only broccoli, beets, spinach, and mushrooms. Brad only liked corn, potatoes, and mushrooms. Carl only liked broccoli, spinach, eggplant, and mushrooms. David only liked beets, spinach, corn, and mushrooms. Which vegetable(s) do all of them like? Which vegetable(s) do both Carl and David like?

    A) {mushrooms}, {corn, spinach, mushrooms}

    B) {mushrooms}, {beets, corn, broccoli, eggplant, spinach, mushrooms}

    C) {mushrooms}, {beets, corn, broccoli, eggplant}

    D) {mushrooms}, {spinach, mushrooms}

3 Find the Complement of a Set

Let U = {q, r, s, t, u, v, w, x, y, z}, A = {q, s, u, w, y}, B = {q, s, y, z}, and C = {v, w, x, y, z}. Find the indicated set.

1) $\overline{(A \cup B)}$

   A) {r, t, v, x}   B) {t, v, x}   C) {r, s, t, u, v, w, x, z}   D) {s, u, w}

2) $\overline{(A \cap B)}$

   A) {s, u, w}   B) {t, v, x}   C) {r, t, u, v, w, x, z}   D) {q, s, t, u, v, w, x, y}

3) $\overline{C} \cup \overline{A}$

   A) {s, t}   B) {w, y}   C) {q, r, s, t, u, v, x, z}   D) {q, s, u, v, w, x, y, z}

4) $\overline{C} \cap \overline{A}$
   A) {q, s, u, v, w, x, y, z}   B) {w, y}   C) {q, r, s, t, u, v, x, z}   D) {r, t}

5) $A \cap \overline{B}$
   A) {u, w}   B) {q, s, t, u, v, w, x, y}   C) {t, v, x}   D) {r, s, t, u, v, w, x, z}

6) $\overline{\overline{B} \cup C}$
   A) {r, t, u, v, w, x, y, z}   B) {r, t, u}   C) {v, w, x}   D) {q, s}

Find the indicated set.

7) Let U = Universal set = {0, 1, 2, 3, 4, 5, 6, 7, 8, 9}, A = {2, 3, 4, 5, 8}, B = {1, 4, 6, 7, 9} and C = {2, 3, 4, 6}.
   Find $\overline{C}$.
   A) {0, 1, 5, 7, 8, 9}   B) {2, 3, 4, 6}   C) {1, 5, 7, 8, 9}   D) {0, 1, 5, 6, 7, 8, 9}

8) Let U = Universal set = {0, 1, 2, 3, 4, 5, 6, 7, 8, 9}, A = {2, 3, 4, 5, 8}, B = {1, 4, 6, 7, 9} and C = {2, 3, 4, 6}.
   Find $\overline{A \cap B}$.
   A) {0}
   B) {0, 1, 2, 3, 5, 6, 7, 8, 9}
   C) {1, 2, 3, 4, 5, 6, 7, 8, 9}
   D) {4}

9) Let U = Universal set = {0, 1, 2, 3, 4, 5, 6, 7, 8, 9}, A = {2, 4, 5, 6}, B = {1, 3, 5, 8} and C = {2, 5, 9}.
   Find $\overline{A \cup B \cup C}$.
   A) {0, 7}
   B) {0, 1, 2, 3, 4, 6, 7, 8, 9}
   C) {0, 7, 9}
   D) {1, 2, 3, 4, 5, 6, 8, 9}

10) Let U = Universal set = {0, 1, 2, 3, 4, 5, 6, 7, 8, 9}, A = {2, 4, 5, 6, 9}, B = {1, 3, 5, 8, 9} and C = {1, 4, 5, 9}.
    Find $\overline{\overline{A} \cap B}$
    A) {1, 2, 3, 4, 5, 6, 8, 9}   B) {1, 3, 8}   C) {0, 2, 4, 5, 6, 7, 9}   D) {0, 1, 2, 3, 4, 6, 7, 8}

11) Let U = Universal set = {0, 1, 2, 3, 4, 5, 6, 7, 8, 9}, A = {2, 4, 5, 6, 9}, B = {1, 3, 5, 8, 9} and C = {1, 4, 5, 9}.
    Find $\overline{A \cap B \cap C}$
    A) {0, 7}
    B) {0, 1, 2, 3, 4, 6, 7, 8}
    C) {1, 2, 3, 4, 5, 6, 7, 8, 9}
    D) {5, 9}

4 Count the Number of Elements in a Set

Solve the problem.

1) If n(A) = 31, n(B) = 28, and n(A ∩ B) = 17, find n(A ∪ B).
   A) 59   B) 42   C) 76   D) 25

2) If n(A) = 37, n(B) = 26, and n(A ∪ B) = 56, find n(A ∩ B).
   A) 63   B) 49   C) 14   D) 7

3) If n(B) = 48, n(A ∩ B) = 9, and n(A ∪ B) = 84, find n(A).
   A) 36   B) 47   C) 45   D) 43

4) If n(A) = 10, n(A ∪ B) = 28, and n(A ∩ B) = 6, find n(B).
   A) 18   B) 24   C) 25   D) 23

5) If $n(A \cup B) = 52$, $n(A \cap B) = 20$, and $n(A) = n(B)$, find $n(A)$.

A) 10   B) 16   C) 26   D) 36

Use the information given in the figure.

6)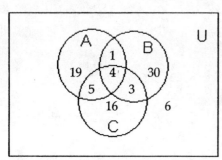

How many are in set A?

A) 19   B) 29   C) 25   D) 35

7)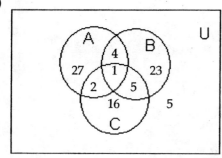

How many are in B or C?

A) 51   B) 6   C) 50   D) 44

8)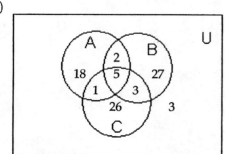

How many are in B and C?

A) 64   B) 56   C) 8   D) 5

9)

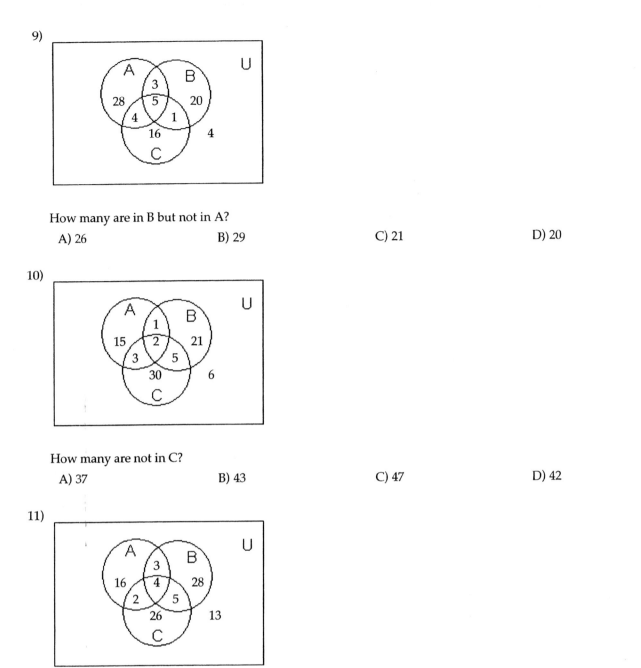

How many are in B but not in A?

A) 26   B) 29   C) 21   D) 20

10)

How many are not in C?

A) 37   B) 43   C) 47   D) 42

11)

How many are in A and B and C?

A) 84   B) 14   C) 4   D) 70

12)

How many are in A or B or C?

A) 10   B) 75   C) 76   D) 4

Solve the problem.

13) In a survey of 50 hospital patients, 25 said they were satisfied with the nursing care, 17 said they were satisfied with the medical treatment, and 6 said they were satisfied with both. How many patients were satisfied with neither? How many were satisfied with only the medical treatment?

A) 14; 17   B) 19; 11   C) 14; 11   D) 20; 17

14) In a survey of 347 computer buyers, 174 put price as a main consideration, 180 put performance as a main consideration, and 57 listed both price and performance. How many computer buyers listed other considerations? How many looked only for performance?

A) 117; 123   B) 107; 180   C) 50; 180   D) 50; 123

15) In survey of 50 households, 25 responded that they have an HDTV television, 35 responded that they had a multimedia personal computer and 15 responded they had both. How many households had neither an HDTV television nor a multimedia personal computer?

A) 25   B) 5   C) 35   D) 15

16) In a student survey, 109 students indicated that they speak Spanish, 33 students indicated that they speak French, 9 students indicated that they speak both Spanish and French, and 129 students indicated that they speak neither. How many students participated in the survey?

A) 253   B) 271   C) 133   D) 262

17) Among a group of 72 investors, 20 owned shares of Stock A, 28 owned shares of Stock B, 44 owned shares of Stock C, 10 owned shares of both Stock A and Stock B, 11 owned shares of Stock A and Stock C, 15 owned shares of Stock B and Stock C, and 6 owned shares of all three. How many investors did not have shares in any of the three? How many owned shares of either Stock A or Stock C but not Stock B?

A) 10; 29   B) 10; 34   C) 16; 29   D) 10; 35

18) In a survey of 188 vacationers in a popular beach resort town, 72 indicated they would consider buying a home there, 91 would consider buying a beach villa, 57 would consider buying a lot, 28 would consider both a home and a villa, 26 would consider both a home and a lot, 28 would consider both a villa and a lot, and 14 would consider all three. How many vacationers would not consider any of the three? How many would consider only a home?

A) 50; 46   B) 36; 17   C) 36; 49   D) 36; 32

19) A survey of 2569 credit card users indicated that 1168 had bought books online, 1149 had bought music online, 511 had bought pet supplies online, 125 had bought both books and music, 229 had bought both books and pet supplies, 163 had bought both music and pet supplies, and 97 had bought all three. How many credit card users did not buy any of the three? How many bought either books or pet supplies but not music?

A) 161; 1224   B) 161; 1127   C) 161; 1259   D) 258; 1127

20) The following data represent the marital status of females 18 years and older in a certain U.S. city in 2002.

Marital Status	Number (in thousands)
Married	308
Widowed	55
Divorced	56
Never married	114

Determine the number of females 18 years old and older who are married or widowed.

A) 419,000   B) 308,000   C) 363,000   D) 364,000

12.2 Permutations and Combinations

1 Solve Counting Problems Using the Multiplication Principle

Solve the problem.

1) A man has 7 shirts and 9 ties. How many different shirt and tie arrangements can he wear?

A) 126   B) 49   C) 63   D) 81

2) A restaurant offers a choice of 4 salads, 8 main courses, and 4 desserts. How many possible 3-course meals are there?

A) 256 possible meals   B) 32 possible meals   C) 16 possible meals   D) 128 possible meals

3) Lisa has 5 skirts, 9 blouses, and 4 jackets. How many 3-piece outfits can she put together assuming any piece goes with any other?

A) 180 possible outfits   B) 360 possible outfits   C) 45 possible outfits   D) 18 possible outfits

4) How many 8-symbol codes can be formed using 3 different symbols? Repeated symbols are allowed.

A) 6720   B) 56   C) 6561   D) 5

5) A certain mathematics test consists of 20 questions. Goldie decides to answer the questions without reading them. In how many ways can Goldie fill in the answer sheet if the possible answers are true and false?

A) 40   B) 1,048,576   C) 400   D) 190

6) A student must choose 1 of 4 mathematics electives, 1 of 6 science electives, and 1 of 9 programming electives. How many possible course selections are there?

A) 19 course selections
B) 216 course selections
C) 432 course selections
D) 24 course selections

7) How many arrangements of answers are possible in a multiple-choice test with 7 questions, each of which has 4 possible answers?

A) 16,384   B) 3   C) 35   D) 210

8) How many 2-letter codes can be formed using the letters A, B, C, D, E, F, G, H, and I. Repeated letters are allowed.

   A) 81 B) 36 C) 72 D) 512

9) How many 3-digit numbers can be formed using the digits 0, 1, 2, 3, 4, 5, 6, 7, 8, and 9 if the first digit cannot be 0? Repeated digits are allowed.

   A) 504 B) 810 C) 900 D) 729

10) How many different license plates can be made using 2 letters followed by 3 digits selected from the digits 0 through 9, if letters and digits may be repeated?

   A) 676,000 B) 6 C) 36 D) 260

2 Solve Counting Problems Using Permutations

Find the value of the permutation.

1) $P(7, 5)$

   A) 42 B) 2520 C) 1260 D) 4

2) $P(7, 0)$

   A) 5040 B) 10,080 C) 4 D) 1

3) $P(5, 1)$

   A) 5 B) 24 C) 1 D) 120

4) $P(5, 5)$

   A) 1 B) 60 C) 120 D) 2

Solve the problem.

5) List all the ordered arrangements of 6 objects a, b, c, d, e, and f choosing 2 at a time without repetition. What is $P(6, 2)$?

   A) ab, ac, ad, ae, af, bc, bd, be, bf, cd, ce, cf, de, df, ef
      $P(6, 2) = 15$

   B) ab, ac, ad, ae, af, ba, bc, bd, be, bf, ca, cb, cd, ce, cf, da, db, dc, de, df, ea, eb, ec, ed, ef
      $P(6, 2) = 25$

   C) aa, ab, ac, ad, ae, af, ba, bb, bc, bd, be, bf, ca, cb, cc, cd, ce, cf, da, db, dc, dd, de, df, ea, eb, ec, ed, ee, ef, fa, fb, fc, fd, fe, ff
      $P(6, 2) = 36$

   D) ab, ac, ad, ae, af, ba, bc, bd, be, bf, ca, cb, cd, ce, cf, da, db, dc, de, df, ea, eb, ec, ed, ef, fa, fb, fc, fd, fe
      $P(6, 2) = 30$

6) List all the ordered arrangements of 4 objects 1, 2, 3, and 4 choosing 3 at a time without repetition. What is P(4, 3)?

   A) 111, 112, 113, 114, 121, 122, 123, 124, 131, 132, 133, 134, 141, 142, 143, 144, 211, 212, 213, 214, 221, 222, 223, 224, 231, 232, 233, 234, 241, 242, 243, 244, 311, 312, 313, 314, 321, 322, 323, 324, 331, 332, 333, 334, 341, 342, 343, 344, 411, 412, 413, 414, 421, 422, 423, 424, 431, 432, 433, 434, 441, 442, 443, 444
   P(4, 3) = 64

   B) 123, 124, 132, 142, 143, 213, 214, 231, 241, 243, 312, 314, 321, 341, 342, 412, 413, 421, 423, 431
   P(4, 3) = 20

   C) 123, 124, 132, 134, 142, 143, 213, 214, 231, 234, 241, 243, 312, 314, 321, 324, 341, 342, 412, 413, 421, 423, 431, 432
   P(4, 3) = 24

   D) 123, 124, 134, 234
   P(4, 3) = 4

7) In how many ways can 7 people be lined up?
   A) 1
   B) 2520
   C) 7
   D) 5040

8) 8 different books are to be arranged on a shelf. How many different arrangements are possible?
   A) 8
   B) 20,160
   C) 5040
   D) 40,320

9) How many different 5-letter codes are there if only the letters A, B, C, D, E, F, G, H, and I can be used and no letter can be used more than once?
   A) 59,049
   B) 15,120
   C) 5
   D) 126

10) How many 2-digit numbers can be formed using the digits 1, 2, 3, 4, 5, 6, 7, 8, 9, and 0? No digit can be used more than once.
    A) 1,814,400
    B) 45
    C) 3,628,800
    D) 90

11) How many different license plates can be made using 3 letters followed by 3 digits selected from the digits 0 through 9, if neither letters nor digits may be repeated?
    A) 11,232,000
    B) 7,862,400
    C) 17,576,000
    D) 312,000

12) How many different license plates can be made using 4 letters followed by 2 digits selected from the digits 0 through 9, if digits may be repeated but letters may not be repeated?
    A) 35,880,000
    B) 41,127,840
    C) 45,697,600
    D) 889.880952

13) In how many ways can 3 people each have different birth months?
    A) 1728
    B) 36
    C) 1320
    D) 220

14) A group of 6 friends goes bowling. How many different possibilities are there for the order in which they play if the youngest person is to bowl first?
    A) 720
    B) 5
    C) 120
    D) 6

3 Solve Counting Problems Using Combinations

Find the value of the combination.

1) C(9, 7)
   A) 72
   B) 4
   C) 36
   D) 90,720

2) C(6, 6)

   A) 0.5 B) 720 C) 1 D) 180

Solve the problem.

3) List all the combinations of 6 objects a, b, c, d, e, and f taken 2 at a time. What is C(6, 2)?

   A) ab, ac, ad, ae, af, ba, bc, bd, be, bf, ca, cb, cd, ce, cf, da, db, dc, de, df, ea, eb, ec, ed, ef, fa, fb, fc, fd, fe
      C(6, 2) = 30

   B) aa, ab, ac, ad, ae, af, ba, bb, bc, bd, be, bf, ca, cb, cc, cd, ce, cf, da, db, dc, dd, de, df, ea, eb, ec, ed, ee, ef, fa, fb, fc, fd, fe, ff
      C(6, 2) = 36

   C) ab, ac, ad, ae, bc, bd, be, cd, ce, cf, de, df
      C(6, 2) = 12

   D) ab, ac, ad, ae, af, bc, bd, be, bf, cd, ce, cf, de, df, ef
      C(6, 2) = 15

4) List all the combinations of 4 objects 1, 2, 3, and 4 taken 3 at a time. What is C(4, 3)?

   A) 123, 124, 132, 134, 142, 143, 213, 214, 231, 234, 241, 243, 312, 314, 321, 324, 341, 342, 412, 413, 421, 423, 431, 432
      C(4, 3) = 24

   B) 111, 112, 113, 114, 121, 122, 123, 124, 131, 132, 133, 134, 141, 142, 143, 144, 211, 212, 213, 214, 221, 222, 223, 224, 231, 232, 233, 234, 241, 242, 243, 244, 311, 312, 313, 314, 321, 322, 323, 324, 331, 332, 333, 334, 341, 342, 343, 344, 411, 412, 413, 414, 421, 422, 423, 424, 431, 432, 433, 434, 441, 442, 443, 444
      C(4, 3) = 64

   C) 123, 124, 134, 234
      C(4, 3) = 4

   D) 123, 124, 134, 234, 321, 432
      C(4, 3) = 6

5) From 8 names on a ballot, a committee of 3 will be elected to attend a political national convention. How many different committees are possible?

   A) 336 B) 56 C) 6720 D) 168

6) A hot dog stand sells hot dogs with cheese, relish, chili, tomato, onion, mustard, or ketchup. How many different hot dogs can be concocted using any 3 of the extras?

   A) 35 B) 105 C) 840 D) 210

7) An exam consists of 9 multiple-choice questions and 6 essay questions. If the student must answer 6 of the multiple-choice questions and 2 of the essay questions, in how many ways can the questions be chosen?

   A) 648 B) 1260 C) 261,273,600 D) 1,814,400

8) Mary finds 9 fish at a pet store that she would like to buy, but she can afford only 5 of them. In how many ways can she make her selection? How many ways can she make her selection if he decides that one of the fish is a must?

   A) 126; 70 B) 3024; 1680 C) 7560; 840 D) 15,120; 1680

9) How many 5-card poker hands consisting of three 7's and two cards that are not 7's are possible in a 52-card deck?

   A) 4512 B) 2652 C) 2256 D) 5304

10) A committee is to be formed consisting of 2 men and 2 women. If the committee members are to be chosen from 8 men and 9 women, how many different committees are possible?

   A) 64   B) 2380   C) 1008   D) 4032

11) How many ways are there to choose a soccer team consisting of 3 forwards, 4 midfield players, and 3 defensive players, if the players are chosen from 5 forwards, 8 midfield players, and 5 defensive players?

   A) 90   B) 43,758   C) 6,048,000   D) 7000

4 Solve Counting Problems Using Permutations Involving n Nondistinct Objects

Solve the problem.

1) How many different 11-letter words (real or imaginary) can be formed from the letters in the word ENGINEERING?

   A) 25,200   B) 277,200   C) 554,400   D) 39,916,800

2) How many different 11-letter words (real or imaginary) can be formed from the letters of the word MISSISSIPPI? Leave your answer in factorial form.

3) How many different vertical arrangements are there of 6 flags if 3 are white, 2 are blue, and 1 is red?

   A) 60   B) 6   C) 20   D) 11

4) An environmental organization has 23 members. Each member will be placed on exactly 1 of 4 teams. Each team will work on a different issue. The first team has 3 members, the second has 4, the third has 7, and the fourth has 9. In how many ways can these teams be formed?

   A) $2.585201674 \times 10^{22}$ B) $3.141770579 \times 10^{18}$ C) $9.816086280 \times 10^{10}$ D) $8.834477652 \times 10^{11}$

12.3 Probability

1 Construct Probability Models

Solve the problem.

1) In a probability model, which of the following numbers could be the probability of an outcome:

   $0, \ 0.2, \ -0.01, \ -\dfrac{1}{3}, \ \dfrac{1}{2}, \ \dfrac{5}{4}, \ 1, \ 1.5$

   A) $0, \ 0.2, \ -0.01, \ 1, \ 1.5$
   B) $0.2, \ \dfrac{1}{2}, \ 1$
   C) $0, \ 0.2, \ \dfrac{1}{2}, \ 1$
   D) $0, \ 0.2, \ -0.01, \ -\dfrac{1}{3}, \ \dfrac{1}{2}, \ 1$

Determine whether the following is a probability model.

2)

Outcome	Probability
Red	0.17
Blue	0.21
Green	0.28
White	0.34

   A) Yes   B) No

3)

Outcome	Probability
Red	0.19
Blue	0.28
Green	0.30
White	0.50

A) Yes   B) No

4)

Outcome	Probability
Red	0.15
Blue	0.16
Green	0.18
White	0.36

A) Yes   B) No

5)

Outcome	Probability
Red	−0.24
Blue	0.29
Green	0.30
White	0.17

A) Yes   B) No

6)

Outcome	Probability
Jim	0
Tom	0
Bill	1
Carl	0

A) Yes   B) No

7)

Outcome	Probability
Golfing	0.05
Skiing	0.15
Swimming	0.05
Biking	0.26
Hiking	0.49

A) Yes   B) No

Construct a probability model for the experiment.

8) Tossing two fair coins once

9) Tossing one fair coin three times

10) Rolling a 6-sided fair die once

11) Rolling a 6-sided fair die once and tossing a fair coin once.

12) Rolling a 6-sided fair die twice

13) Tossing a fair coin twice given that the coin is weighted so that heads is three times as likely as tails to occur.

14) Spinner I has 4 sections of equal area, numbered 1, 2, 3, and 4, and Spinner II has 4 sections of equal area, labeled Red, Yellow, Green, and Blue. Spin Spinner I and then spin Spinner II.
What is the probability of getting a 1 or 3 followed by Red?

15) Spinner I has 4 sections of equal area, numbered 1, 2, 3, and 4. Spinner II has 3 sections of equal area, labeled Red, Yellow, and Green. Spinner III has 2 sections of equal area labeled A and B. Spin Spinner I, then Spinner II, then Spinner III.
What is the probability of getting a 2, followed by Yellow or Green, followed by B?

16) Spinner I has 3 sections of equal area, numbered 1, 2, and 3. Spinner II has 3 sections of equal area, labeled Red, Yellow, and Green. Spin Spinner I twice, then Spinner II.
What is the probability of getting a 2, followed by a 1, followed by Yellow or Red?

Solve the problem.

17) A twelve-sided die is weighted so that only the numbers 1 through 5 will appear and they will occur with the same probability. What probability should be assigned to each face?

18) A die is weighted so that an even-numbered face is three times as likely to occur as an odd-numbered face. What probability should be assigned to each face?

19) A coin is weighted so that heads is 8 times as likely as tails to occur. What probability should be assigned to heads? to tails?

2 Compute Probabilities of Equally Likely Outcomes

Solve the problem.

1) A bag contains 7 red marbles, 9 blue marbles, and 6 green marbles. If one marble is selected at random, determine the probability that it is blue.
   A) $\frac{9}{22}$
   B) $\frac{7}{22}$
   C) $\frac{9}{16}$
   D) $\frac{3}{11}$

2) A 6-sided die is rolled. What is the probability of rolling a number less than 3?
   A) $\frac{1}{3}$
   B) $\frac{1}{6}$
   C) $\frac{1}{2}$
   D) $\frac{2}{3}$

3) Two 6-sided dice are rolled. What is the probability the sum of the two numbers on the dice will be 5?
   A) $\frac{5}{6}$
   B) $\frac{8}{9}$
   C) 4
   D) $\frac{1}{9}$

4) A bag contains 13 balls numbered 1 through 13. What is the probability of selecting a ball that has an even number when one ball is drawn from the bag?
   A) $\frac{2}{13}$
   B) 6
   C) $\frac{13}{6}$
   D) $\frac{6}{13}$

5) What is the probability that the arrow will land on an odd number? Assume that all sectors have equal area.

A) 0   B) $\frac{3}{5}$ C) $\frac{2}{5}$ D) 1

6) Suppose that the sample space is S = {1, 2, 3, 4, 5, 6, 7, 8, 9, 10} and that outcomes are equally likely. Compute the probabaility of the event E = {2, 10}.

A) $\frac{2}{9}$ B) $\frac{1}{10}$ C) 2   D) $\frac{1}{5}$

7) Suppose that the sample space is S = {1, 2, 3, 4, 5, 6, 7, 8, 9, 10} and that outcomes are equally likely. Compute the probabaility of the event E = {1, 2, 4, 5, 7, 9, 10}.

A) $\frac{7}{9}$ B) 7   C) $\frac{7}{10}$ D) $\frac{4}{5}$

8) Suppose that the sample space is S = {1, 2, 3, 4, 5, 6, 7, 8, 9, 10} and that outcomes are equally likely. Compute the probabaility of the event E: "a number divisible by 3".

A) $\frac{1}{3}$ B) $\frac{3}{10}$ C) $\frac{2}{5}$ D) 3

9) Find the probability of getting 2 tails when 3 fair coins are tossed.

A) $\frac{2}{3}$ B) $\frac{3}{8}$ C) $\frac{1}{2}$ D) $\frac{1}{4}$

10) Find the probability of having 4 girls in a 4-child family.

A) $\frac{1}{16}$ B) $\frac{1}{8}$ C) $\frac{1}{32}$ D) $\frac{1}{4}$

3 Use the Addition Rule to Find Probabilities

Solve the problem.

1) Given that P(A) = 0.23, P(B) = 0.27, and P(A ∩ B) = 0.13, find P(A ∪ B).
A) 0.5   B) 0.24   C) 0.37   D) 0.63

2) Given that P(A) = 0.63, P(B) = 0.34, and P(A ∪ B) = 0.74, find P(A ∩ B).
A) 0.51   B) 0.2142   C) 0.97   D) 0.23

3) Given that P(A) = 0.30 and P(B) = 0.26, find P(A ∪ B) if A and B are mutually exclusive.
A) 0   B) 0.482   C) 0.078   D) 0.56

4) Given that P(A) = 0.28 and P(B) = 0.26, find P(A ∩ B) if A and B are mutually exclusive.
A) 0.0728   B) 0.54   C) 0.4672   D) 0

5) Given that P(A) = 1.09, P(A ∪ B) = 1.24, and P(A ∩ B) = 0.41, find P(B).
   A) 0.15   B) 0.56   C) 0.97   D) 0.68

6) The table below shows the results of a consumer survey of annual incomes in 100 households.

Income	Number of households
$0 – 14,999	7
$15,000 – 24,999	25
$25,000 – 34,999	30
$35,000 – 44,999	25
$45,000 or more	13

What is the probability that a household has an annual income of $25,000 or more?
   A) 0.38   B) 0.62   C) 0.68   D) 0.3

7) The table below shows the results of a consumer survey of annual incomes in 100 households.

Income	Number of households
$0 – 14,999	7
$15,000 – 24,999	20
$25,000 – 34,999	25
$35,000 – 44,999	26
$45,000 or more	22

What is the probability that a household has an annual income less than $25,000?
   A) 0.52   B) 0.2   C) 0.27   D) 0.73

8) The table below shows the results of a consumer survey of annual incomes in 100 households.

Income	Number of households
$0 – 14,999	9
$15,000 – 24,999	24
$25,000 – 34,999	28
$35,000 – 44,999	26
$45,000 or more	13

What is the probability that a household has an annual income between $15,000 and $44,999 inclusive?
   A) 0.78   B) 0.5   C) 0.52   D) 0.28

9) In a survey about the number of siblings of college students, the following probability table was constructed:

Number of Siblings	Probability
0	0.27
1	0.33
2	0.21
3	0.10
4 or more	0.09

What is the probability that a student has at least 2 siblings?
   A) 0.81   B) 0.6   C) 0.19   D) 0.4

10) In a survey about the number of siblings of college students, the following probability table was constructed:

Number of Siblings	Probability
0	0.25
1	0.34
2	0.18
3	0.11
4 or more	0.12

What is the probability that a student has at most 2 siblings?

A) 0.41   B) 0.23   C) 0.59   D) 0.77

11) In a survey about the number of siblings of college students, the following probability table was constructed:

Number of Siblings	Probability
0	0.25
1	0.34
2	0.18
3	0.10
4 or more	0.13

What is the probability that a student 3 or more siblings?

A) 0.1   B) 0.13   C) 0.23   D) 0.77

12) In a survey about the number of siblings of college students, the following probability table was constructed:

Number of Siblings	Probability
0	0.26
1	0.30
2	0.21
3	0.10
4 or more	0.13

What is the probability that a student has less than 2 siblings?

A) 0.21   B) 0.23   C) 0.77   D) 0.56

13) In a survey about the number of siblings of college students, the following probability table was constructed:

Number of Siblings	Probability
0	0.26
1	0.32
2	0.21
3	0.12
4 or more	0.09

What is the probability that a student has 1, 2, or 3 siblings?

A) 0.33   B) 0.91   C) 0.65   D) 0.53

14) A bag contains 6 red marbles, 4 blue marbles, and 1 green marble. What is the probability of choosing a marble that is red or green when one marble is drawn from the bag?

A) $\frac{4}{11}$ B) 7 C) $\frac{11}{7}$ D) $\frac{7}{11}$

15) Each of ten tickets is marked with a different number from 1 to 10 and put in a box. If you draw a ticket from the box, what is the probability that you will draw 1, 8, or 7?

A) $\frac{1}{8}$ B) $\frac{3}{10}$ C) $\frac{1}{10}$ D) 1

16) A lottery game has balls numbered 1 through 19. What is the probability of selecting an even numbered ball or a 9?

A) $\frac{9}{10}$ B) $\frac{8}{19}$ C) $\frac{10}{19}$ D) $\frac{9}{19}$

17) A spinner has regions numbered 1 through 18. What is the probability that the spinner will stop on an even number or a multiple of 3?

A) $\frac{2}{3}$ B) 15 C) 1 D) $\frac{1}{3}$

18) The psychology lab at a college is staffed by 6 male doctoral students, 9 female doctoral students, 12 male undergraduates, and 7 female undergraduates. If a person is selected at random from the group, find the probability that the selected person is an undergraduate or a female.

A) $\frac{8}{17}$ B) $\frac{19}{34}$ C) $\frac{21}{34}$ D) $\frac{14}{17}$

19) The faculty at a college consists of 94 full-time teachers and 48 part-time teachers. Of the 94 full-time teachers, 48 are female. Of the 48 part-time teachers, 26 are female. Find the probability that a randomly selected teacher is male or works part-time.

A) $\frac{36}{71}$ B) $\frac{11}{71}$ C) $\frac{47}{71}$ D) $\frac{58}{71}$

4 Use the Complement Rule to Find Probabilities

Solve the problem.

1) A bag contains 5 red marbles, 3 blue marbles, and 1 green marble. What is the probability of choosing a marble that is not blue when one marble is drawn from the bag?

A) $\frac{3}{2}$ B) $\frac{1}{3}$ C) 6 D) $\frac{2}{3}$

2) During July in Jacksonville, Florida, it is not uncommon to have afternoon thunderstorms. On average, 9.8 days have afternoon thunderstorms. What is the probability that a randomly selected day in July will not have a thunderstorm? Round to two decimal places, if necessary.

A) 0.9 B) 0.32 C) 0.68 D) 0.67

3) In the city of Gloomville, the probability of rain on New Year's Day is 40%. What is the probability that next New Year's Day it will not rain in Gloomville?

A) −40% B) 40% C) 16% D) 60%

4) Sam estimates that if he leaves his car parked outside his office all day on a weekday, the chance that he will get a parking ticket is 11%. If Sam leaves his car parked outside his office all day next Tuesday, what is the chance that he will not get a parking ticket?

   A) −11%   B) 11%   C) 89%   D) 1.21%

5) What is the probability that at least 2 people have the same birth month in a group of 5 people?

   A) 0.618   B) 0.637   C) 0.382   D) 0.363

Ch. 12 Counting and Probability
Answer Key

12.1 Sets and Counting
1 Find All the Subsets of a Set
1) C
2) B
3) B
4) B

2 Find the Intersection and Union of Sets
1) A
2) A
3) A
4) D
5) A
6) C
7) A
8) D
9) C
10) D
11) D

3 Find the Complement of a Set
1) A
2) C
3) C
4) D
5) A
6) D
7) A
8) B
9) A
10) C
11) B

4 Count the Number of Elements in a Set
1) B
2) D
3) C
4) B
5) D
6) B
7) A
8) C
9) C
10) B
11) C
12) B
13) C
14) D
15) B
16) D
17) B
18) D
19) C
20) C

12.2 Permutations and Combinations

1 Solve Counting Problems Using the Multiplication Principle
1) C
2) D
3) A
4) C
5) B
6) B
7) A
8) A
9) C
10) A

2 Solve Counting Problems Using Permutations
1) B
2) D
3) A
4) C
5) D
6) C
7) D
8) D
9) B
10) D
11) A
12) A
13) C
14) C

3 Solve Counting Problems Using Combinations
1) C
2) C
3) D
4) C
5) B
6) A
7) B
8) A
9) A
10) C
11) D

4 Solve Counting Problems Using Permutations Involving n Nondistinct Objects
1) B
2) $\dfrac{11!}{4!4!2!}$
3) A
4) C

12.3 Probability

1 Construct Probability Models
1) C
2) A
3) B
4) B
5) B
6) A
7) A

8) S = {HH, HT, TH, TT}; each outcome has the probability of $\frac{1}{4}$

9) S = {HHH, HTH, HHT, HTT, THH, TTH, THT, TTT}; each outcome has the probability of $\frac{1}{8}$

10) S = {1, 2, 3, 4, 5, 6}; each outcome has the probability of $\frac{1}{6}$

11) S = {1H, 2H, 3H, 4H, 5H, 6H, 1T, 2T, 3T, 4T, 5T, 6T}; each outcome has a probability of $\frac{1}{12}$

12) S = {11, 12, 13, 14, 15, 16, 21, 22, 23, 24, 25, 26, 31, 32, 33, 34, 35, 36, 41, 42, 43, 44, 45, 46, 51, 52, 53, 54, 55, 56, 61, 62, 63, 64, 65, 66}; each outcome has a probability of $\frac{1}{36}$

13) S = {HH, HT, TH, TT}

$$\begin{bmatrix} \text{Outcome} & \text{Probability} \\ \text{HH} & 9/16 \\ \text{HT} & 3/16 \\ \text{TH} & 3/16 \\ \text{TT} & 1/16 \end{bmatrix}$$

14) S = {1 Red, 1 Yellow, 1 Green, 1 Blue, 2 Red, 2 Yellow, 2 Green, 2 Blue, 3 Red, 3 Yellow, 3 Green, 3 Blue, 4 Red, 4 Yellow, 4 Green, 4 Blue}; each outcome has a probability of $\frac{1}{16}$

The probability of getting a 1 or 3 followed by Red is $\frac{1}{8}$

15) S = {1 Red A, 1 Red B, 1 Yellow A, 1 Yellow B, 1 Green A, 1 Green B, 2 Red A, 2 Red B, 2 Yellow A, 2 Yellow B, 2 Green A, 2 Green B, 3 Red A, 3 Red B, 3 Yellow A, 3 Yellow B, 3 Green A, 3 Green B, 4 Red A, 4 Red B, 4 Yellow A, 4 Yellow B, 4 Green A, 4 Green B};

Each outcome has a probability of $\frac{1}{24}$.

The probability of getting a 2, followed by Yellow or Green, followed by B is $\frac{1}{12}$

16) S = {11 Red, 11 Yellow, 11 Green, 12 Red, 12 Yellow, 12 Green, 13 Red, 13 Yellow, 13 Green, 21 Red, 21 Yellow, 21 Green, 22 Red, 22 Yellow, 22 Green, 23 Red, 23 Yellow, 23 Green, 31 Red, 31 Yellow, 31 Green, 32 Red, 32 Yellow, 32 Green, 33 Red, 33 Yellow, 33 Green };

Each outcome has a probability of $\frac{1}{27}$.

The probability of getting a 2, followed by a 1, followed by Yellow or Red is $\frac{2}{27}$.

17) The faces numbered 1 through 5 will each have the probability $\frac{1}{5}$; faces numbered higher than 5 will each have probability 0.

18) The faces numbered 1, 3, 5 will each have probability 1/12.
The faces numbered 2, 4, 6 will each have probability 1/4.

19) $P(H) = \frac{8}{9}$; $P(T) = \frac{1}{9}$

2 Compute Probabilities of Equally Likely Outcomes

1) A
2) A
3) D
4) D
5) B
6) D

7) C
8) B
9) B
10) A

3 Use the Addition Rule to Find Probabilities
1) C
2) D
3) D
4) D
5) B
6) C
7) C
8) A
9) D
10) D
11) C
12) D
13) C
14) D
15) B
16) C
17) A
18) D
19) C

4 Use the Complement Rule to Find Probabilities
1) D
2) C
3) D
4) C
5) A

Ch. 13 A Preview of Calculus: The Limit, Derivative, and Integral of a Function

13.1 Finding Limits Using Tables and Graphs

1 Find a Limit Using a Table

Use the TABLE feature of a graphing utility to find the limit.

1) $\lim\limits_{x \to 2} (x^2 + 8x - 2)$

   A) −18   B) 0   C) 18   D) does not exist

2) $\lim\limits_{x \to 0} \dfrac{x+2}{3x+2}$

   A) $\dfrac{2}{3}$ B) 0   C) 1   D) does not exist

3) $\lim\limits_{h \to -3} \left(\dfrac{h^2 - 9}{h^2 + 3h} \right)$

   A) −2   B) 2   C) 0   D) does not exist

4) $\lim\limits_{d \to 3} \left(\dfrac{d^3 - 27}{d - 3} \right)$

   A) 27   B) 9   C) 0   D) does not exist

5) $\lim\limits_{x \to 0} \dfrac{x^2}{\cos x}$

   A) 1   B) 0   C) −1   D) does not exist

6) $\lim\limits_{x \to 0} (e^x - e^{-x})$

   A) 1   B) 0   C) 2   D) does not exist

2 Find a Limit Using a Graph

Use the graph shown to determine if the limit exists. If it does, find its value.

1) $\lim\limits_{x \to 1} f(x)$

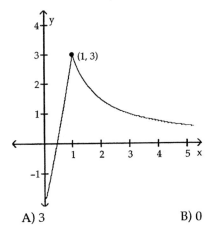

   A) 3   B) 0   C) 1   D) does not exist

2) $\lim_{x \to 4} f(x)$

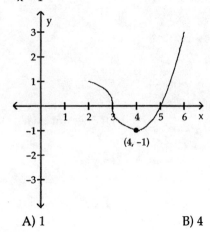

A) 1   B) 4   C) -1   D) does not exist

3) $\lim_{x \to 2} f(x)$

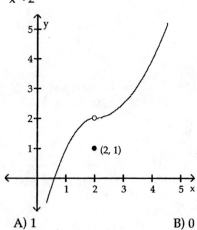

A) 1   B) 0   C) 2   D) does not exist

4) $\lim_{x \to 3} f(x)$

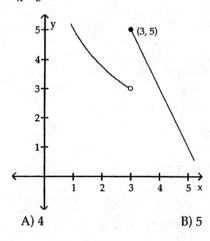

A) 4   B) 5   C) 3   D) does not exist

Use the grid to graph the function. Find the limit, if it exists

5) $\lim_{x \to 5} f(x)$, $f(x) = 4x - 2$

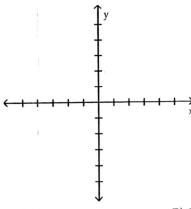

A) 18 B) 20 C) 3 D) does not exist

6) $\lim_{x \to -2} f(x)$, $f(x) = 4x - 1$

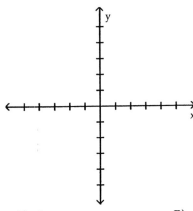

A) –2 B) –8 C) –9 D) does not exist

7) $\lim_{x \to 2} f(x)$, $f(x) = 3 - x^2$

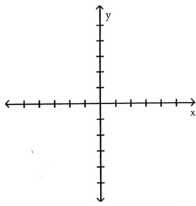

A) 3 B) –1 C) 1 D) does not exist

8) $\lim_{x \to -3} f(x)$, $f(x) = |5x|$

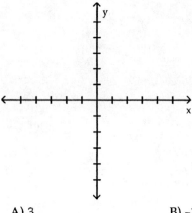

A) 3   B) −15   C) 15   D) does not exist

9) $\lim_{x \to \pi/2} f(x)$, $f(x) = \sin x + 2$

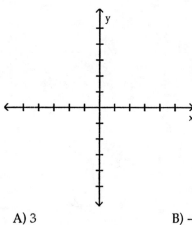

A) 3   B) −1   C) 2   D) does not exist

10) $\lim_{x \to 0} f(x)$, $f(x) = \begin{cases} x + 3 & x < 0 \\ 4x + 3 & x \geq 0 \end{cases}$

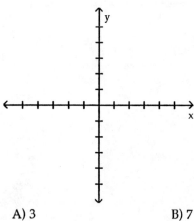

A) 3   B) 7   C) 0   D) does not exist

Use a graphing utility to find the indicated limit rounded to two decimal places.

11) $\lim_{x \to 1} \dfrac{x^3 - x^2 + 3x - 3}{x^4 - x^3 + x - 1}$

A) 2.04   B) 3.00   C) 2.96   D) 2.00

13.2 Algebra Techniques for Finding Limits

1 Find the Limit of a Sum, a Difference, a Product, and a Quotient

Find the limit algebraically.

1) $\lim_{x \to 4} -8$

A) 4   B) -8   C) 8   D) 0

2) $\lim_{x \to 5} x$

A) -5   B) 1   C) 5   D) 0

3) $\lim_{x \to 4} (5x - 4)$

A) 1   B) 4   C) 16   D) does not exist

4) $\lim_{x \to 0} (x - \sqrt{5})(x + \sqrt{5})$

A) 0   B) -5   C) 5   D) does not exist

5) $\lim_{x \to 1} \dfrac{2x - 7}{4x + 5}$

A) $-\dfrac{1}{2}$ B) $-\dfrac{7}{5}$ C) $-\dfrac{5}{9}$ D) does not exist

2 Find the Limit of a Polynomial

Find the limit algebraically.

1) $\lim_{x \to 4} -13x^3$

A) 64   B) -13   C) 4   D) -832

2) $\lim_{x \to 0} (x^2 - 5)$

A) -5   B) 0   C) 5   D) does not exist

3) $\lim_{x \to -9} (3 - 5x^2)$

A) 93   B) 408   C) -402   D) 48

4) $\lim_{x \to -2} (x^2 - 5x - 7)$

A) 7   B) -13   C) 21   D) 1

5) $\lim_{x \to 2} (x^3 + 5x^2 - 7x + 1)$

A) 29   B) 15   C) 0   D) does not exist

6) $\lim_{x \to 0} \dfrac{x^3 + 12x^2 - 5x}{5x}$

A) 5   B) −1   C) 0   D) does not exist

7) $\lim_{x \to -3} \dfrac{x^2 - 2x - 15}{x + 3}$

A) 0   B) 5   C) −8   D) does not exist

8) $\lim_{x \to 5} \dfrac{x^2 - 2x - 15}{x + 3}$

A) 0   B) 5   C) −8   D) does not exist

9) $\lim_{x \to 1} \dfrac{3x^2 + 7x - 2}{3x^2 - 4x + 2}$

A) 0   B) 8   C) $-\dfrac{7}{4}$ D) does not exist

10) $\lim_{x \to 1} \dfrac{x^4 - 1}{x - 1}$

A) 2   B) 4   C) 0   D) does not exist

11) $\lim_{x \to 0} \dfrac{x^3 - 6x + 8}{x - 2}$

A) 4   B) −4   C) 0   D) does not exist

12) $\lim_{x \to 1} \dfrac{x^3 + 5x^2 + 3x - 9}{x - 1}$

A) 0   B) −16   C) 16   D) does not exist

3 Find the Limit of a Power or a Root

Find the limit algebraically.

1) $\lim_{x \to 1} (3x^2 - 19)^2$

A) 144   B) 256   C) −352   D) 100

2) $\lim_{x \to -4} (4x^3 - 2x + 123)^{2/3}$

A) 25   B) −5   C) −125   D) 125

3) $\lim_{x \to 1} (x^2 - 2)^3$

A) 1   B) 3   C) −1   D) −3

4) $\lim\limits_{x \to 1} \sqrt{3x - 2}$

A) $\sqrt{2}$ B) -1 C) 1 D) does not exist

5) $\lim\limits_{x \to 0} (\sqrt{x} - 2)$

A) -2 B) 2 C) 0 D) does not exist

4 Find the Limit of an Average Rate of Change

Find the limit as x approaches c of the average rate of change of the function from c to x.

1) $c = 3;\quad f(x) = 2x + 4$

A) 4 B) -4 C) -2 D) 2

2) $c = 3;\quad f(x) = 3x^2 + 39x$

A) 57 B) 108 C) 30 D) 90

3) $c = -4;\quad f(x) = x^3$

A) 16 B) 0 C) 32 D) 48

13.3 One-sided Limits; Continuous Functions

1 Find the One-sided Limits of a Function

Use the graph of y = g(x) to answer the question.

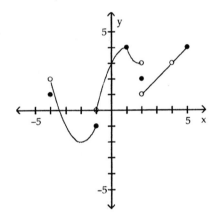

1) What is the domain of g?

A) $\{x \mid -4 < x < -1 \text{ or } -1 < x < 2 \text{ or } 2 < x \leq 5\}$
B) $\{x \mid -4 \leq x < 4 \text{ or } 4 < x \leq 5\}$
C) $\{x \mid -4 \leq x < 5\}$
D) $\{x \mid -4 < x < 5\}$

2) What is the range of g?

A) $\{y \mid -2 \leq y \leq 4\}$ B) $\{y \mid -2 \leq y \leq 5\}$ C) all real numbers D) $\{y \mid 2 \leq y \leq 4\}$

3) Find the y-intercept(s), if any, of g.

A) 0 B) -1 C) 3 D) $-\dfrac{4}{3}$

Precalculus Enhanced with Graphing Utilities 887

4) Find f(-1).
   A) 4   B) -1   C) 0   D) 1

5) Find f(1).
   A) -1   B) 4   C) 3   D) 0

6) Find f(2).
   A) 1   B) 2   C) 3   D) $-\frac{1}{2}$

7) Find f(-4).
   A) -4   B) 0   C) 1   D) 2

8) Find $\lim_{x \to 1^-} g(x)$.
   A) 0   B) 4   C) -4   D) does not exist

9) Does $\lim_{x \to 4} g(x)$ exist? If it does, what is it?
   A) yes; 4   B) yes; 3   C) yes; 0   D) does not exist

Find the one-sided limit.

10) $\lim_{x \to 0^-} (4 - 5x)$
    A) 4   B) 0   C) -1   D) does not exist

11) $\lim_{x \to 2^+} (x^2 - 2x + 1)$
    A) 1   B) 7   C) -1   D) 9

12) $\lim_{x \to 3^-} \frac{x^2 - 9}{x - 3}$
    A) 3   B) -3   C) -6   D) 6

13) $\lim_{x \to 4^+} \frac{16 - x^2}{4 - x}$
    A) 8   B) 0   C) 4   D) does not exist

14) $\lim_{x \to (3\pi/4)^-} \sin x$
    A) $\frac{1}{2}$ B) 0   C) 1   D) $\frac{\sqrt{2}}{2}$

15) $\lim_{x \to 0^+} (2 \cos x)$
    A) -2   B) 2   C) 0   D) does not exist

2 Determine Whether a Function Is Continuous

Determine whether f is continuous at c.

1) $f(x) = \dfrac{5}{x+9}; c = -9$

   A) continuous   B) not continuous

2) $f(x) = \dfrac{5}{x-9}; c = 0$

   A) continuous   B) not continuous

3) $f(x) = \dfrac{2}{x^2 + 3x}; c = 0$

   A) not continuous   B) continuous

4) $f(x) = \dfrac{8}{x^2 - 6x}; c = 6$

   A) continuous   B) not continuous

5) $f(x) = \dfrac{5}{x^2 - 9x}; c = -5$

   A) not continuous   B) continuous

6) $f(x) = \dfrac{x-8}{(x-4)(x-6)}; c = 8$

   A) continuous   B) not continuous

7) $f(x) = \dfrac{x-2}{(x+5)(x-3)}; c = 0$

   A) not continuous   B) continuous

8) $f(x) = \dfrac{x-6}{(x+9)(x+8)}; c = -9$

   A) not continuous   B) continuous

9) $f(x) = \dfrac{x-9}{(x+8)(x+1)}; c = -1$

   A) continuous   B) not continuous

10) $f(x) = \dfrac{x+3}{x-1}; c = -3$

    A) continuous   B) not continuous

11) $f(x) = \dfrac{x-7}{x-4}; c = 0$

    A) continuous   B) not continuous

12) $f(x) = \dfrac{x+2}{x+1}$; $c = -1$

   A) continuous   B) not continuous

13) $f(x) = \dfrac{x^2 - 64}{x - 8}$; $c = -8$

   A) not continuous   B) continuous

14) $f(x) = \dfrac{x^2 - 36}{x - 6}$; $c = 0$

   A) not continuous   B) continuous

15) $f(x) = \dfrac{x^2 - 64}{x - 8}$; $c = 8$

   A) continuous   B) not continuous

16) $f(x) = 8x^4 - 3x^3 + x - 6$; $c = 6$

   A) not continuous   B) continuous

17) $f(x) = 8x^4 - 2x^3 + x - 7$; $c = 0$

   A) not continuous   B) continuous

18) $f(x) = 7x^4 - 2x^3 + x - 3$; $c = 2$

   A) continuous   B) not continuous

19) $f(x) = \begin{cases} x^2 - 5, & x < 0 \\ -4, & x \geq 0 \end{cases}$; $c = -3$

   A) not continuous   B) continuous

20) $f(x) = \begin{cases} 6x - 4, & x < 1 \\ 1, & x = 1 \\ 4x - 5, & x > 1 \end{cases}$; $c = 1$

   A) not continuous   B) continuous

21) $f(x) = \begin{cases} -2x + 9, & x < 1 \\ -4x + 11, & x > 1 \end{cases}$; $c = 1$

   A) not continuous   B) continuous

22) $f(x) = \begin{cases} \dfrac{1}{x+1}, & x > -1 \\ x^2 + 3x, & x \leq -1 \end{cases}$; $c = -1$

   A) not continuous   B) continuous

Find the numbers at which f is continuous. At which numbers is f discontinuous?

23) $f(x) = 5x - 4$

   A) continuous for all real numbers except $x = -\frac{4}{5}$
   B) continuous for all real numbers
   C) continuous for all real numbers except $x = 4$
   D) continuous for all real numbers except $x = \frac{4}{5}$

24) $f(x) = -4x^2 - 2x$

   A) continuous for all real numbers except $x = \frac{1}{4}$
   B) continuous for all real numbers
   C) continuous for all real numbers except $x = -4$
   D) continuous for all real numbers except $x = -\frac{1}{4}$

25) $f(x) = 3 \cos x$

   A) continuous for all real numbers
   B) continuous for all real numbers except $x = 1$
   C) continuous for all real numbers except $x = 3$
   D) continuous for all real numbers except $x = 0$

26) $f(x) = 4 \tan x$

   A) continuous for all real numbers except $x = \frac{k\pi}{4}$ where k is an odd integer
   B) continuous for all real numbers except $x = 0$
   C) continuous for all real numbers except $x = \frac{k\pi}{2}$ where k is an odd integer
   D) continuous for all real numbers

27) $f(x) = \frac{5x - 3}{x^2 - 4}$

   A) continuous for all real numbers except $x = -2$, $x = 2$ and $x = -\frac{3}{5}$
   B) continuous for all real numbers
   C) continuous for all real numbers except $x = 2$
   D) continuous for all real numbers except $x = -2$ and $x = 2$

28) $f(x) = \frac{x + 5}{x^2 - 11x + 24}$

   A) continuous for all real numbers except $x = -5$, $x = 8$, and $x = 3$
   B) continuous for all real numbers except $x = 8$, $x = 3$
   C) continuous for all real numbers except $x = 5$, $x = -8$, and $x = -3$
   D) continuous for all real numbers except $x = -8$, $x = -3$

29) $f(x) = \begin{cases} x - 3 & \text{if } x \leq 3 \\ 2x - 6 & \text{if } x > 3 \end{cases}$

   A) continuous for all real numbers except $x = 3$
   B) continuous for all real numbers except $x = -3$, $x = 3$
   C) continuous for all real numbers
   D) continuous for all real numbers except $x = 0$

30) $f(x) = \begin{cases} \dfrac{x^2 - 16}{x - 4} & \text{if } x \neq 4 \\ 8 & \text{if } x = 4 \end{cases}$

A) continuous for all real numbers except x = 4
B) continuous for all real numbers
C) continuous for all real numbers except x = 4, x = 16
D) continuous for all real numbers except x = 16

31) $f(x) = \begin{cases} 4x & \text{if } x < 3 \\ 11 & \text{if } x = 3 \\ x^2 + 3 & \text{if } x > 3 \end{cases}$

A) continuous for all real numbers except x = 3
B) continuous for all real numbers
C) continuous for all real numbers except x = 0, x = 3
D) continuous for all real numbers except x = 4, x = 3

Determine where the rational function is undefined. Determine whether an asymptote of a hole appears at such numbers.

32) $R(x) = \dfrac{x^3 - x^2 + x - 1}{x^4 - x^3 + 3x - 3}$

13.4 The Tangent Problem; The Derivative

1 Find an Equation of the Tangent Line to the Graph of a Function

Find the slope of the tangent line to the graph at the given point.

1) $f(x) = 4x - 9$ at (4, 7)

   A) $\dfrac{1}{4}$ B) 9   C) -9   D) 4

2) $f(x) = x^2 + 5x$ at (4, 20)

   A) 3   B) 21   C) 9   D) 13

3) $f(x) = 5x^2 + x$ at (-4, 76)

   A) -39   B) -14   C) -41   D) 6

4) $f(x) = -4x^2 + 7x$ at (5, 65)

   A) -13   B) -33   C) 33   D) 3

5) $f(x) = 2x^2 + x - 3$ at (4, 33)

   A) 17   B) 5   C) 19   D) 15

6) $f(x) = x^2 + 11x - 15$ at (1, -3)

   A) 11   B) -9   C) 26   D) 13

Find the equation of the line tangent to the graph of f at the given point.

7) $f(x) = x^2 + 5x$ at (4, 20)

   A) y = 13x - 72   B) y = 2x + 5   C) y = 2x - 5   D) y = 13x - 32

8) $f(x) = 5x^2 + x$ at $(-4, 76)$

   A) $y = 10x + 1$ B) $y = -39x - 232$ C) $y = 10x$ D) $y = -39x - 80$

9) $f(x) = -4x^2 + 7x$ at $(5, 65)$

   A) $y = -4x + 7$ B) $y = 2x - 7$ C) $y = -33x + 230$ D) $y = -8x + 7$

10) $f(x) = 2x^2 + x - 3$ at $x = (4, 33)$

    A) $y = 4x + 3$ B) $y = 4x + 1$ C) $y = 17x - 35$ D) $y = 2x - 3$

11) $f(x) = x^2 + 11x - 15$ at $(1, -3)$

    A) $y = 2x + 11$ B) $y = 13x - 16$ C) $y = 11x$ D) $y = 11x + 15$

2 Find the Derivative of a Function

Find the derivative of the function at the given value of x.

1) $f(x) = x^3 + 1$; $x = 2$

   A) 6   B) 24   C) 3   D) 12

2) $f(x) = 6x + 16$; $x = 7$

   A) 6   B) 42   C) 112   D) 16

3) $f(x) = x^2 + 6x + 9$; $x = 2$

   A) 4   B) 8   C) 10   D) 6

4) $f(x) = x^3 + 4x$; $x = -2$

   A) 12   B) 16   C) -2   D) 4

Find the derivative of the function at the given value of x using a graphing utility. If necessary, round to four decimal places.

5) $f(x) = 4x \cos x$; $x = \dfrac{\pi}{2}$

6) $f(x) = 4x^3 - 6x + 2$; $x = -4$

3 Find Instantaneous Rates of Change

Solve the problem.

1) The function $f(x) = x^3$ describes the volume of a cube, $f(x)$, in cubic inches, whose length, width, and height each measure x inches. Find the instantaneous rate of change of the volume with respect to x when x = 7 inches.

   A) 147 cubic inches per inch
   B) 149.11 cubic inches per inch
   C) 21 cubic inches per inch
   D) 1029 cubic inches per inch

2) The function $V(r) = 3\pi r^2$ describes the volume of a right circular cylinder of height 3 feet and radius r feet. Find the instantaneous rate of change of the volume with respect to the radius when r = 9. Leave answer in terms of π.

   A) 18π cubic feet per feet
   B) 6π cubic feet per feet
   C) 54π cubic feet per feet
   D) 27π cubic feet per feet

3) The volume of a right cylindrical cone of height 6 cm and radius r cm is $V(r) = 2\pi r^2$ cubic centimeters (cm^3). Find the instantaneous rate of change of the volume with respect to the radius r when r = 4 cm.

   A) 16 cm^3/cm B) 4π cm^3/cm C) 16π cm^3/cm D) 32π cm^3/cm

4) The volume of a rectangular box with square base and a height of 5 feet is $V(x) = 5x^2$, where x is the length of a side of the base. Find the instantaneous rate of change of volume with respect to x when x = 3 feet.

   A) 6 ft^3/ft B) 30 ft^3/ft C) 15 ft^3/ft D) 45 ft^3/ft

5) The volume V of a right circular cylinder of height 2 and radius r is $V(r) = 2\pi r^2$. Find the instantaneous rate of change of volume with respect to radius r when r = 7.

   A) 98π ft^3/ft B) 32π ft^3/ft C) 28π ft^3/ft D) 23π ft^3/ft

4 Find the Instantaneous Speed of a Particle

Solve the problem.

1) If an object is thrown straight upward from the ground with an initial speed of 112 feet per second, then its height, h, in feet after t seconds is given by the equation $h(t) = -16t^2 + 112t$. Find the instantaneous speed of the object at t = 6.

   A) 16 ft/sec B) 112 ft/sec C) −464 ft/sec D) −80 ft/sec

2) An explosion causes debris to rise vertically with an initial velocity of 48 feet per second. The function $s(t) = -16t^2 + 48t$ describes the height of the debris above the ground, s(t), in feet, t seconds after the explosion. What is the instantaneous speed of the debris 0.7 second(s) after the explosion?

   A) 22.4 ft/sec B) −22.4 ft/sec C) −25.6 ft/sec D) 25.6 ft/sec

3) An explosion causes debris to rise vertically with an initial velocity of 144 feet per second. The function $s(t) = -16t^2 + 144t$ describes the height of the debris above the ground, s(t), in feet, t seconds after the explosion. What is the instantaneous speed of the debris when it hits the ground?

   A) −144 B) 288 C) −288 D) 144

4) A foul tip of a baseball is hit straight upward from a height of 4 feet with an initial velocity of 112 feet per second. The function $s(t) = -16t^2 + 112t + 4$ describes the ball's height above the ground, s(t), in feet, t seconds after it was hit. What is the instantaneous speed of the ball 0.7 seconds after it was hit?

   A) 89.6 B) −22.4 C) 22.4 D) −89.6

5) A foul tip of a baseball is hit straight upward from a height of 4 feet with an initial velocity of 144 feet per second. The function $s(t) = -16t^2 + 144t + 4$ describes the ball's height above the ground, s(t), in feet, t seconds after it was hit. The ball reaches its maximum height above the ground when the instantaneous speed reaches zero. After how many seconds does the ball reach its maximum height?

   A) $\dfrac{3}{32}$ B) 144 C) $\dfrac{9}{2}$ D) 9

13.5 The Area Problem; The Integral

1 Approximate the Area under the Graph of a Function

Solve the problem.

1) $f(x) = 2x - 2$ is defined on the interval $[1,5]$. In (a) and (b), approximate the area under f as follows:
   (a) Partition $[1,5]$ into four subintervals of equal length and choose u as the left endpoint of each subinterval.
   (b) Partition $[1,5]$ into four subintervals of equal length and choose u as the right endpoint of each subinterval.
   (c) What is the actual area A?

Approximate the area under the curve and above the x-axis using n rectangles. Let the height of each rectangle be given by the value of the function at the right side of the rectangle.

2) $f(x) = 2x^2 + x + 3$ from $x = 0$ to $x = 6$; $n = 6$
   A) 230   B) 221   C) 200   D) 211

3) $f(x) = 2x^3 - 1$ from $x = 1$ to $x = 6$; $n = 5$
   A) 825   B) 850   C) 875   D) 800

4) $f(x) = 2x + 3$ from $x = 0$ to $x = 2$; $n = 4$
   A) 15   B) 17   C) 13   D) 11

5) $f(x) = 5x - 1$ from $x = 1$ to $x = 4$; $n = 2$
   A) 55.75   B) 35.75   C) 45.75   D) 25.75

6) $f(x) = 3x^2 - 2$ from $x = 1$ to $x = 5$; $n = 4$
   A) 140   B) 154   C) 150   D) 144

7) $f(x) = x^2$ from $x = 0$ to $x = 4$; $n = 4$
   A) 33   B) 27   C) 36   D) 30

8) $f(x) = x^2 - 3x + 4$ from $x = 1$ to $x = 5$; $n = 4$
   A) 18   B) 28   C) 48   D) 38

9) $f(x) = 2x^2 + x + 3$ from $x = -2$ to $x = 1$; $n = 3$
   A) 17   B) 21   C) 13   D) 25

10) $f(x) = 2x^2 + x + 3$ from $x = -2$ to $x = 1$; $n = 6$
    A) 13   B) 9   C) 7   D) 11

11) $f(x) = x^2 + 2$ from $x = 1$ to $x = 4$; $n = 6$
    A) 26.875   B) 24.875   C) 30.875   D) 28.875

Solve the problem.

12) Given the function f defined over the interval [a, b], graph the function indicating the area A under f from a to b. Then express the area A as an integral.

$f(x) = x^2 + 9$, $[-1, 4]$

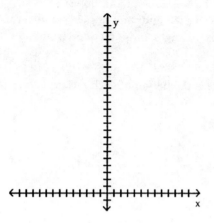

13) Provide a graph that illustrates the area represented by the integral.

$$\int_2^4 x^2\, dx$$

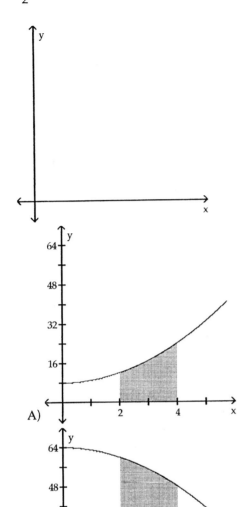

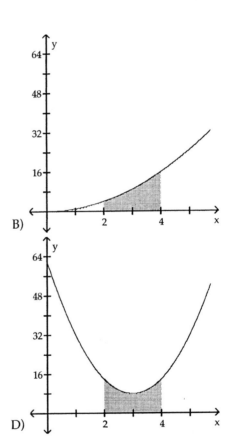

2 Approximate Integrals Using a Graphing Utility

Solve the problem.

1) $f(x) = \sqrt{x^2 + 1}$ is defined on the interval $[0, 4]$. In (a) and (b), approximate the area under f (to three decimal places) as follows:
   (a) Partition $[0, 4]$ into four subintervals of equal length and choose u as the left endpoint of each subinterval.
   (b) Partition $[0, 4]$ into four subintervals of equal length and choose u as the right endpoint of each subinterval.
   (c) Use a graphing utility to approximate the actual area A.

2) $f(x) = \sqrt[3]{x}$ is defined on the interval $[0, 6]$.
   (a) Approximate the area A under the graph of f by partitioning $[0, 6]$ into three subintervals of equal length and choose u as the left endpoint of each subinterval.
   (b) Approximate the area A under the graph of f by partitioning $[0, 6]$ into three subintervals of equal length and choose u as the right endpoint of each subinterval.
   (c) Express the area A as an integral.
   (d) Use a graphing utility to approximate this integral to three decimal places.

3) $f(x) = \tan x$ is defined over the interval $\left[0, \frac{3}{2}\right]$.

   (a) Approximate the area A by partitioning $\left[0, \frac{3}{2}\right]$ into 4 subintervals of equal length and choosing u as the left endpoint of each subinterval.

   (b) Approximate the area A by partitioning $\left[0, \frac{3}{2}\right]$ into 8 subintervals of equal length and choosing u as the right endpoint of each subinterval.
   (c) Express the area A as an integral.
   (d) Use a graphing utility to approximate this integral to three decimal places.

4) (a) What area does the integral $\int_0^{\pi/2} \sin x \cos x \, dx$ represent?

   (b) Use a graphing utility to approximate the area to three decimal places.

Ch. 13 A Preview of Calculus: The Limit, Derivative, and Integral of a Function
Answer Key

13.1 Finding Limits Using Tables and Graphs
1 Find a Limit Using a Table
1) C
2) C
3) B
4) B
5) B
6) B

2 Find a Limit Using a Graph
1) A
2) C
3) C
4) D
5) A
6) C
7) B
8) C
9) A
10) A
11) D

13.2 Algebra Techniques for Finding Limits
1 Find the Limit of a Sum, a Difference, a Product, and a Quotient
1) B
2) C
3) C
4) B
5) C

2 Find the Limit of a Polynomial
1) D
2) A
3) C
4) A
5) B
6) B
7) C
8) A
9) B
10) B
11) B
12) C

3 Find the Limit of a Power or a Root
1) B
2) A
3) C
4) C
5) A

4 Find the Limit of an Average Rate of Change
1) D
2) A
3) D

13.3 One-sided Limits; Continuous Functions
1 Find the One-sided Limits of a Function
1) B
2) A
3) C
4) B
5) B
6) B
7) C
8) B
9) B
10) A
11) A
12) D
13) A
14) D
15) B

2 Determine Whether a Function Is Continuous
1) B
2) A
3) A
4) B
5) B
6) A
7) B
8) A
9) B
10) A
11) A
12) B
13) B
14) B
15) B
16) B
17) B
18) A
19) B
20) A
21) A
22) A
23) B
24) B
25) A
26) C
27) D
28) B
29) C
30) B
31) A
32) $x = -\sqrt[3]{3}$: asymptote
    $x = 1$: hole

13.4 The Tangent Problem; The Derivative
1 Find an Equation of the Tangent Line to the Graph of a Function
1) D

- 2) D
- 3) A
- 4) B
- 5) A
- 6) D
- 7) D
- 8) D
- 9) C
- 10) C
- 11) B

2 Find the Derivative of a Function
- 1) D
- 2) A
- 3) C
- 4) B
- 5) −6.2832
- 6) 186

3 Find Instantaneous Rates of Change
- 1) A
- 2) C
- 3) C
- 4) B
- 5) C

4 Find the Instantaneous Speed of a Particle
- 1) D
- 2) D
- 3) A
- 4) A
- 5) C

13.5 The Area Problem; The Integral

1 Approximate the Area under the Graph of a Function
- 1) (a) 12   (b) 20   (c) 16
- 2) B
- 3) C
- 4) D
- 5) C
- 6) B
- 7) D
- 8) B
- 9) C
- 10) A
- 11) C

12)

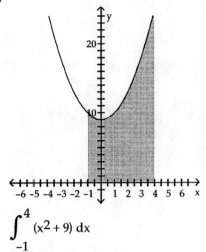

$$\int_{-1}^{4} (x^2 + 9)\, dx$$

13) B

2 Approximate Integrals Using a Graphing Utility

1) (a) 7.813   (b) 10.936   (c) 9.294

2) (a) 5.695   (b) 9.329   (c) $\int_{0}^{6} \sqrt[3]{x}\, dx$ (d) 8.177

3) (a) 1.282   (b) 1.760   (c) $\int_{0}^{3/2} \tan x\, dx$ (d) 2.649

4) (a) The area under the graph of $f(x) = \sin x \cos x$ from 0 to $\dfrac{\pi}{2}$.   (b) 0.500

Ch. 14 (Appendix) Review

14.1 Algebra Review

1 Evaluate Algebraic Expressions

Find the value of the expression using the given values.

1) $x + 15y$ $x = 1, y = 2$
   A) 16   B) 3   C) 31   D) 17

2) $-6xy + 8y - 2$ $x = -4, y = 1$
   A) 34   B) -18   C) 32   D) 30

3) $-12x + y$ $x = 5, y = -3$
   A) 2   B) -15   C) 57   D) -63

4) $\dfrac{9x - 2y}{5}$ $x = 6, y = 4$
   A) $\dfrac{46}{5}$ B) $\dfrac{52}{5}$ C) $\dfrac{24}{5}$ D) $\dfrac{62}{5}$

5) $\dfrac{14x - 11y}{x + 10}$ $x = 7, y = 6$
   A) $\dfrac{7}{17}$ B) $\dfrac{7}{16}$ C) 2   D) $\dfrac{32}{17}$

6) $\dfrac{4xy + 5}{x}$ $x = 5, y = 7$
   A) 5   B) 1   C) 29   D) 33

7) $|x + y|$ $x = -3, y = 8$
   A) -5   B) 5   C) -11   D) 11

8) $|x| + |y|$ $x = -5, y = 3$
   A) 8   B) -8   C) 2   D) -2

9) $|6x - 7y|$ $x = 4, y = 4$
   A) -4   B) 52   C) 4   D) -52

10) $4|x| + 5|y|$ $x = 9, y = -6$
    A) -6   B) 66   C) -66   D) 6

11) $|-8x + 9y|$ $x = 4, y = 2$
    A) 52   B) 50   C) 14   D) 20

12) $|3x| - |-5y|$ $x = -2, y = 3$
    A) -9   B) -1   C) 19   D) 21

2 Determine the Domain of a Variable

Determine which value(s), if any, must be excluded from the domain of the variable in the expression.

1) $\dfrac{3}{x-2}$

   A) $x = -2$ B) $x = 2$ C) $x = 0$ D) none

2) $\dfrac{4}{x+7}$

   A) $x = 7$ B) $x = -7$ C) $x = 0$ D) none

3) $\dfrac{x-9}{3}$

   A) $x = -9$ B) $x = 0$ C) $x = 9$ D) none

4) $\dfrac{x-7}{5-x}$

   A) $x = 5$ B) $x = 5, x = 7$ C) $x = -5$ D) none

5) $\dfrac{7x-9}{x^2 - 36}$

   A) $x = 6, x = -6$ B) $x = 6$ C) $x = \dfrac{9}{7}$ D) $x = 36$

6) $\dfrac{64x^2 - 16}{3x - 18}$

   A) $x = \dfrac{1}{2}, x = -\dfrac{1}{2}$ B) $x = 6$ C) $x = 18$ D) none

7) $\dfrac{x^3 + 7x^4}{x^2 + 9}$

   A) $x = -9$ B) $x = -3$ C) $x = 0, x = -\dfrac{1}{7}$ D) none

8) $\dfrac{x^3 + 3x^2 + 2x}{x^2 - 10x}$

   A) $x = 10$ B) $x = 0, x = 10, x = -2, x = -1$
   C) $x = 0, x = 10$ D) $x = -2, x = -1$

9) $\dfrac{x^2 - 7x - 8}{x + 1}$

   A) $x = -1, x = 8$ B) $x = -1$ C) $x = 1$ D) none

Determine the domain of the variable x in the expression.

10) $\dfrac{x-5}{x+4}$

   A) $\{x \mid x \ne -4\}$ B) $\{x \mid x \ge -4\}$ C) $\{x \mid x \ne 5\}$ D) $\{x \mid x = -4\}$

11) $\dfrac{3x + 4}{x - 1}$

   A) $\{x \mid x \neq 1\}$ B) $\{x \mid x \neq 0, -1\}$ C) $\{x \mid x \neq 0\}$ D) all real numbers

12) $\dfrac{4x}{x(x - 64)}$

   A) $\{x \mid x \neq \pm 64, x \neq 0\}$ B) $\{x \mid x \neq 64, x \neq 0\}$ C) $\{x \mid x \neq \pm 8\}$ D) $\{x \mid x \neq 8\}$

3 Graph Inequalities

Insert <, >, or = to make the statement true.

1) -5 ___ -8

   A) =   B) >   C) <

2) -2.8 ___ -5.0

   A) =   B) <   C) >

3) -87 ___ -49

   A) =   B) <   C) >

4) $|-7|$ ___ $|5|$

   A) =   B) >   C) <

5) $|-9|$ ___ $|9|$

   A) >   B) <   C) =

6) 16 ___ -16

   A) >   B) =   C) <

7) $|-9|$ ___ 0

   A) >   B) =   C) <

8) 70 ___ 700

   A) =   B) >   C) <

9) $\dfrac{15}{3}$ ___ $\dfrac{20}{4}$

   A) =   B) >   C) <

10) $\dfrac{1}{5}$ ___ 0.2

   A) <   B) >   C) =

11) 3.14 ___ π

   A) =   B) <   C) >

Write the statement as an inequality.

12) y is negative

   A) y > 0   B) y ≥ 0   C) y ≤ 0   D) y < 0

13) x is greater than seven

   A) x > 7   B) x ≥ 7   C) x < 7   D) x = 7

14) x is less than or equal to thirteen

   A) x ≤ 13   B) x < 13   C) x ≠ 13   D) x ≥ 13

15) x is greater than seven.

   A) x = 7   B) x > 7   C) x < 7   D) x ≥ 7

16) x is less than or equal to thirteen.

   A) x < 13   B) x ≥ 13   C) x ≠ 13   D) x ≤ 13

Graph the numbers x on the real number line.

17) x ≥ 5

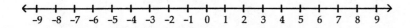

A)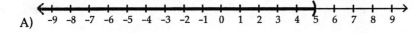

B)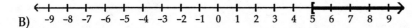

C)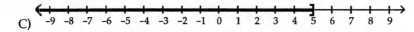

D)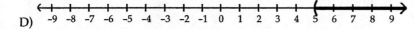

18) x < −1

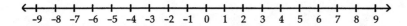

A)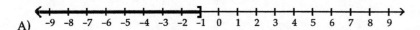

B)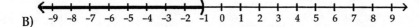

C)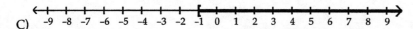

D)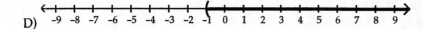

19) $x > -3$

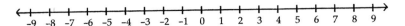

20) $x \leq 2$

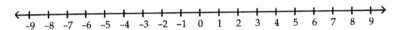

4 Find Distance on the Real Number Line

Use the given real number line to compute the distance.

1) Find $d(A, B)$

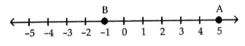

A) 7 B) 5 C) –6 D) 6

5 Use the Laws of Exponents

Simplify the expression.

1) 4^4
   A) –256 B) 256 C) 16 D) –16

2) -4^3
   A) –12 B) 12 C) 64 D) –64

3) $2 \cdot 4^3$
   A) 24 B) 66 C) 512 D) 128

Precalculus Enhanced with Graphing Utilities 907

4) 3^{-2}
   A) 9   B) $\frac{1}{9}$ C) $\frac{1}{6}$ D) -9

5) $(-5)^{-2}$
   A) 25   B) -25   C) $\frac{1}{25}$ D) $-\frac{1}{25}$

6) -4^{-3}
   A) -64   B) $-\frac{1}{64}$ C) $\frac{1}{12}$ D) 64

7) $(-6)^3$
   A) -216   B) 18   C) 216   D) -18

8) -6^1
   A) 0   B) -6   C) 6   D) 1

9) $2^{-1} + 5^{-1}$
   A) 2   B) $-\frac{1}{3}$ C) $\frac{10}{7}$ D) $\frac{7}{10}$

10) $\left(\frac{5}{7}\right)^3$
    A) $\frac{125}{343}$ B) $\frac{343}{125}$ C) $\frac{125}{7}$ D) $\frac{7}{125}$

11) $\left(\frac{1}{6}\right)^2$
    A) $\frac{1}{36}$ B) $\frac{1}{12}$ C) $\frac{1}{8}$ D) $\frac{1}{3}$

12) $\left(\frac{1}{4}\right)^{-3}$
    A) $\frac{1}{12}$ B) $\frac{1}{64}$ C) 64   D) -64

13) $3^{-6} \cdot 3^3$
    A) $\frac{1}{27}$ B) 9   C) $\frac{1}{81}$ D) 27

14) $(5^{-2})^{-1}$
    A) 5   B) $\frac{1}{25}$ C) 25   D) $\frac{1}{5}$

Simplify the expression. Express the answer so that all exponents are positive. Whenever an exponent is 0 or negative, we assume that the base is not 0.

15) x^{-6}

A) $x^{1/6}$ B) $-\dfrac{1}{x^6}$ C) $-x^6$ D) $\dfrac{1}{x^6}$

16) $\dfrac{1}{x^{-6}}$

A) $-x^6$ B) $x^{1/6}$ C) x^6 D) $-x^{1/6}$

17) $\dfrac{x^{-6}}{x^3}$

A) $\dfrac{1}{x^9}$ B) $\dfrac{1}{x^3}$ C) $\dfrac{1}{x^{18}}$ D) x^9

18) $\dfrac{x^7}{x^{-2}}$

A) x^{14} B) $\dfrac{1}{x^9}$ C) $\dfrac{1}{x^{14}}$ D) x^9

19) $-6x^{-3}$

A) $-\dfrac{6}{x^3}$ B) $\dfrac{1}{216x^3}$ C) $\dfrac{-216}{x^3}$ D) $-\dfrac{1}{216x^3}$

20) $\dfrac{(xy^7)(x^4y)}{(x^5y^2)^2}$

A) $\dfrac{x^6}{y^3}$ B) $\dfrac{x^5}{y^4}$ C) $\dfrac{y^3}{x^6}$ D) $\dfrac{y^4}{x^5}$

21) $\left(\dfrac{5x^{-2}}{8y^{-2}}\right)^{-3}$

A) $\dfrac{512y^6}{125x^6}$ B) $\dfrac{512x^6}{125y^6}$ C) $\dfrac{125x^{23}}{512y^{23}}$ D) $\dfrac{125x^6}{512y^6}$

22) $\left(\dfrac{4x^{-2}y^2}{12x^{-4}y^{-1}}\right)^3$

A) $\dfrac{x^6y^9}{27}$ B) $\dfrac{x^6y^9}{3}$ C) $\dfrac{x^8y^6}{9}$ D) $\dfrac{x^2y^3}{3}$

23) $(3x^2)^3(2x)^{-1}$

A) $\dfrac{9x^5}{2}$ B) $\dfrac{27x^5}{2}$ C) $6x^5$ D) $6x^4$

24) $\dfrac{(4xy^{-2})^{-2}}{2xy^3}$

  A) $-\dfrac{8y}{x^3}$
  B) $-\dfrac{4}{x^3y^{-7}}$
  C) $\dfrac{y}{32}$
  D) $\dfrac{y}{32x^3}$

25) $\dfrac{-36x^5y^2}{-9x^4y^4}$

26) $(3xy)^2$

  A) $3x^2y^2$
  B) $9x^2y^2$
  C) $9xy$
  D) $\dfrac{1}{9x^2y^2}$

27) $\left(\dfrac{-4x^6y^{-3}}{5z^7}\right)^{-2}$

  A) $\dfrac{25y^6}{16x^{12}z^{14}}$
  B) $\dfrac{16x^{12}}{25y^6z^{14}}$
  C) $\dfrac{25z^{14}}{16x^{12}y^6}$
  D) $\dfrac{25y^6z^{14}}{16x^{12}}$

28) $(x^{-6}y^3)^{-2}z^6$

  A) $\dfrac{x^{12}}{y^6z^6}$
  B) $\dfrac{y^6z^6}{x^{12}}$
  C) $\dfrac{y^6}{x^{12}z^6}$
  D) $\dfrac{x^{12}z^6}{y^6}$

6 Evaluate Square Roots

Simplify the expression.

1) $\sqrt{4}$

  A) $\dfrac{1}{4}$
  B) 2
  C) 16
  D) not a real number

2) $\sqrt{(-3)^2}$

  A) 81
  B) $\dfrac{1}{9}$
  C) 3
  D) not a real number

7 Demonstrate Additional Understanding and Skills

Express the statement as an equation involving the indicated variables.

1) The area A of a square with sides of length s is the square of s.

  A) $A = \sqrt{s}$
  B) $A = \dfrac{s}{2}$
  C) $A = s^2$
  D) $A = 2s$

2) The area A of a circle with radius r is the product of π and the square of r.

  A) $A = \pi\sqrt{r}$
  B) $A = r\pi^2$
  C) $A = 2\pi r$
  D) $A = \pi r^2$

3) The circumference C of a circle is twice the product of π and its radius r.

  A) $C = 2(\pi + r)$
  B) $C = \pi r^2$
  C) $C = 2\pi r$
  D) $C = \dfrac{\pi}{2r}$

4) The perimeter P of a triangle with sides of length a, b, and c is the sum of the lengths of the sides.

  A) $P = abc$
  B) $P = 3abc$
  C) $P = a + b + c$
  D) $P = a - b - c$

Precalculus Enhanced with Graphing Utilities

5) The volume V of a sphere is $\frac{4}{3}$ times π times the cube of the radius r.

   A) $V = \frac{4}{3}\pi r$
   B) $V = \frac{4}{3}\pi r^2$
   C) $V = \frac{4}{3}\pi r^3$
   D) $V = \frac{4}{3}\pi \sqrt[3]{r}$

6) The surface area S of a sphere is 4 times π times the square of the radius r.

   A) $S = 4\pi\sqrt{r}$
   B) $S = \pi r^2$
   C) $S = 4\pi r^2$
   D) $S = 4\pi r$

7) The volume V of a rectangular box of length l, width w, and height h is the product of the length, width, and height.

   A) $V = l - w - h$
   B) $V = lwh$
   C) $V = (lwh)^3$
   D) $V = l + w + h$

8) The surface area S of a rectangular box of length l, width w, and height h is twice the product of the length and width, plus twice the product of the width and height, plus twice the product of the length and height.

   A) $S = 2lw + 2wh + 2lh$
   B) $S = (lwh)^2$
   C) $S = lwh$
   D) $S = 2(lwh)$

Solve the problem.

9) A breakfast cereal company produces a brand of cereal with a stated net weight of 18 oz. Only boxes with a net weight within 0.02 oz. of this stated amount are acceptable. If x is the net weight of a particular box of cereal brand, a formula describing this situation is $|x - 18| \leq 0.02$. Which of the following cereal boxes is acceptable?

   A) a box with a net weight of 17.03 oz
   B) a box with a net weight of 18.03 oz
   C) a box with a net weight of 18.01 oz
   D) a box with a net weight of 17 oz

10) XYZ Dumbbells, Inc. manufactures free weights varying in size from 2-pound weights to 50-pound weights. Only sets of dumbbells that actually weigh within 5% of the stated weight, are considered acceptable. If y is the actual weight of a 30-pound dumbbell that is unacceptable, then this is represented by the expression $|y - 30| > 0.05(30) = 1.5$. Which of the following 30-pound dumbbells is unacceptable?

   A) one that actually weighs 28.7 pounds
   B) one that actually weighs 31.7 pounds
   C) one that actually weighs 30.1 pounds
   D) one that actually weighs 28.5 pounds

14.2 Geometry Review

1 Use the Pythagorean Theorem and Its Converse

The lengths of the legs of a right triangle are given. Find the hypotenuse.

1) a = 6, b = 8
   A) 10
   B) 8
   C) 9
   D) 7

2) a = 11, b = 20
   A) $\frac{31}{2}$
   B) $\frac{521}{2}$
   C) $\sqrt{521}$
   D) 521

3) a = 5, b = 6
   A) 18
   B) 328
   C) 656
   D) $4\sqrt{41}$

The lengths of the sides of a triangle are given. Determine if the triangle is a right triangle. If it is, identify the hypotenuse.

4) 6, 8, 10
   A) Not right triangle
   B) Right triangle; 10

5) 12, 16, 20
   A) Not a right triangle
   B) Right triangle; 20

6) 12, 13, 5
   A) Not a right triangle
   B) Right triangle; 13

7) 5, 11, 10
   A) Right triangle; 11
   B) Not a right triangle

8) 1, 2, 3
   A) Right triangle; 3
   B) Not a right triangle

9) 4, 11, 12
   A) Not a right triangle
   B) Right triangle; 12

2 Know Geometry Formulas

Solve the problem.

1) Find the area A of a rectangle with length 11 m and width 20 m.
   A) $A = 31 \text{ m}^2$
   B) $A = 44 \text{ m}^2$
   C) $A = 220 \text{ m}^2$
   D) $A = 440 \text{ m}^2$

2) Find the area A of a rectangle with length 6.4 ft and width 10.0 ft.
   A) $A = 64 \text{ ft}^2$
   B) $A = 25.6 \text{ ft}^2$
   C) $A = 16.4 \text{ ft}^2$
   D) $A = 128 \text{ ft}^2$

3) Find the area A of a triangle with height 9 cm and base 7 cm.
   A) $A = \frac{63}{2} \text{ cm}$
   B) $A = 63 \text{ cm}^2$
   C) $A = \frac{63}{2} \text{ cm}^2$
   D) $A = 63 \text{ cm}$

4) Find the area A and circumference C of a circle of radius 11 cm.
   A) $A = 121\pi \text{ cm}^2$; $C = 22\pi \text{ cm}$
   B) $A = 44\pi \text{ cm}^2$; $C = 11\pi \text{ cm}$
   C) $A = 22\pi \text{ cm}^2$; $C = 22\pi \text{ cm}$
   D) $A = 484\pi \text{ cm}^2$; $C = 11\pi \text{ cm}$

5) Find the area A and circumference C of a circle of diameter 20 yd. Use 3.14 for π. Round the result to the nearest tenth.
   A) $A = 1256 \text{ yd}^2$; $C = 31.4 \text{ yd}$
   B) $A = 314 \text{ yd}^2$; $C = 62.8 \text{ yd}$
   C) $A = 62.8 \text{ yd}^2$; $C = 62.8 \text{ yd}$
   D) $A = 125.6 \text{ yd}^2$; $C = 31.4 \text{ yd}$

6) Find the area A and circumference C of a circle of radius 15 ft. Use 3.14 for π. Round the result to the nearest tenth.
   A) $A = 706.5 \text{ ft}^2$; $C = 94.2 \text{ ft}$
   B) $A = 188.4 \text{ ft}^2$; $C = 47.1 \text{ ft}$
   C) $A = 94.2 \text{ ft}^2$; $C = 94.2 \text{ ft}$
   D) $A = 2826 \text{ ft}^2$; $C = 47.1 \text{ ft}$

7) Find the volume V of a rectangular box with length 3 in, width 9 in, and height 8 in.
   A) V = 576 in³   B) V = 216 in³   C) V = 72 in³   D) V = 243 in³

8) Find the volume V and surface area S of a sphere of radius 4 centimeters.
   A) $V = \frac{64}{3}\pi$ cm³; $S = \frac{256}{3}\pi$ cm²   B) V = 256π cm³; S = 4π cm²
   C) V = 64π cm³; S = 16π cm²   D) $V = \frac{256}{3}\pi$ cm³; S = 64π cm²

9) Find the volume V of a right circular cylinder with radius 10 ft and height 17 ft.
   A) V = 425π ft³   B) V = 1700π ft³   C) V = 170π ft³   D) V = 85π ft³

10) Find the volume V of a sphere of radius 4.5 in. Use 3.14 for π. If necessary, round the result to the nearest tenth.
    A) V = 84.8 in³   B) V = 381.5 in³   C) V = 214.6 in³   D) V = 3052.1 in³

11) Find the volume V of a sphere of diameter 3 ft. Use 3.14 for π. If necessary, round the result to the nearest tenth.
    A) V = 7.9 ft³   B) V = 113 ft³   C) V = 9.4 ft³   D) V = 14.1 ft³

12) Find the volume V of a sphere of diameter 2.5 m. Use 3.14 for π. If necessary, round the result to the nearest tenth.
    A) V = 8.2 m³   B) V = 4.6 m³   C) V = 65.4 m³   D) V = 6.5 m³

13) Find the area of the shaded region.

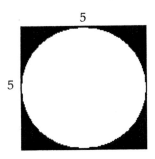

   A) 100 − 25π square units   B) $\frac{25}{4}\pi + 25$ square units
   C) $25 - \frac{25}{4}\pi$ square units   D) $25 - \frac{25}{2}\pi$ square units

14) Find the area of the shaded region in the figure. Round results to the nearest unit.

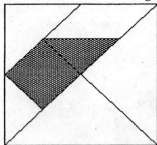

The square portion of the shaded region has sides of length 5.31 m. The hypotenuse of the triangular portion of the shaded region has length 7.5 m. The sides of the outer square have length 15 m.

A) 56 m^2 B) 34 m^2 C) 42 m^2 D) not enough data

15) Find the area of the shaded region in the figure. Round results to the nearest unit.

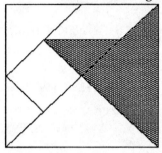

The hypotenuse of the smaller triangular portion of the shaded region has length 4.5 ft. The sides of the outer square have length 9 ft.

A) 20 ft^2 B) 25 ft^2 C) 51 ft^2 D) not enough data

16) Find the area of the shaded region in the figure. Round results to the nearest unit.

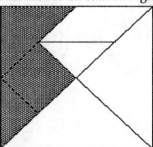

The hypotenuse of the smallest, unshaded triangular region has length 1.06 in. The sides of the outer square have length 3 in. Half the diagonal of the outer square is 2.12 in.

A) 6 in^2 B) 4 in^2 C) 3 in^2 D) not enough data

17) Find the area of the shaded region in the figure. Round results to the nearest unit.

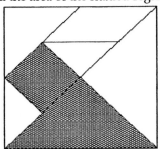

The sides of the square portion of the shaded region have length 2.83 yd. The sides of the outer square have length 8 yd.

A) 20 yd^2 B) 31 yd^2 C) 24 yd^2 D) not enough data

18) Find the area of the shaded region in the figure. Round results to the nearest unit.

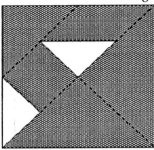

The sides of the outer square have length 29 m.

A) 736 m^2 B) 788 m^2 C) 631 m^2 D) not enough data

19) Find the area of the shaded region in the figure. Round results to the nearest unit.

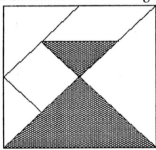

The sides of the outer square have length 8 in. Half the diagonal of the outer square is 5.66 in.

A) 40 in^2 B) 28 in^2 C) 20 in^2 D) not enough data

20) How many inches does a wheel with a diameter of 14 inches travel after 10 revolutions?

21) A rectangular patio has dimensions 10 feet by 15 feet. The patio is surrounded by a border with a uniform width of 3 feet. Find the area of the border.

22) A circular swimming pool, 25 feet in diameter, is enclosed by a circular deck that is 3.5 feet wide. What is the area of the deck?

23) A Norman window consists of a rectangle surmounted by a semicircle. If the width of the window is 5 feet and the total height of the window is 8 feet, how much wood frame is needed to enclose the window?

Precalculus Enhanced with Graphing Utilities 915

24) A person who is 5 feet tall is looking out at the Pacific Ocean while standing on an observation deck of a lighthouse that is 175 feet above sea level. To the nearest tenth of a mile, how far can this person see? Assume the radius of the Earth to be 3960 miles. [Note: 1 mile = 5280 feet]

14.3 Polynomials and Rational Expressions

1 Recognize Special Products

Perform the indicated operations. Express the answer as a polynomial written in standard form.

1) $(x + 13)^2$
   A) $x + 169$
   B) $x^2 + 26x + 169$
   C) $x^2 + 169$
   D) $169x^2 + 26x + 169$

2) $(x - 9)^2$
   A) $x^2 + 81$
   B) $x^2 - 18x + 81$
   C) $81x^2 - 18x + 81$
   D) $x + 81$

3) $(14 + y)^2$
   A) $y + 196$
   B) $196y^2 + 28y + 196$
   C) $y^2 + 28y + 196$
   D) $y^2 + 196$

4) $(b - x)^2$
   A) $b^2 - bx + x^2$
   B) $b^2 - 2bx + x^2$
   C) $b^2 + 2bx + x^2$
   D) $b^2 - 2bx - x^2$

5) $(2x + 11)^2$
   A) $2x^2 + 121$
   B) $2x^2 + 44x + 121$
   C) $4x^2 + 121$
   D) $4x^2 + 44x + 121$

6) $(5x - 7)^2$
   A) $5x^2 - 70x + 49$
   B) $5x^2 + 49$
   C) $25x^2 + 49$
   D) $25x^2 - 70x + 49$

7) $(9x + 7y)^2$
   A) $81x^2 + 126xy + 49y^2$
   B) $9x^2 + 49y^2$
   C) $81x^2 + 49y^2$
   D) $9x^2 + 126xy + 49y^2$

8) $(4x - 9y)^2$
   A) $4x^2 - 72xy + 81y^2$
   B) $4x^2 + 81y^2$
   C) $16x^2 - 72xy + 81y^2$
   D) $16x^2 + 81y^2$

9) $(x + 9)(x - 9)$
   A) $x^2 - 18x - 81$
   B) $x^2 - 18$
   C) $x^2 - 81$
   D) $x^2 + 18x - 81$

10) $(12 + y)(12 - y)$
    A) $24 - y^2$
    B) $144 - 24y - y^2$
    C) $144 + 24y - y^2$
    D) $144 - y^2$

11) $(4z + 11)(4z - 11)$
    A) $16z^2 - 121$
    B) $16z^2 + 88z - 121$
    C) $16z^2 - 88z - 121$
    D) $z^2 - 121$

12) $(1 - 12x)(1 + 12x)$
    A) $1 - 24x - 144x^2$
    B) $1 - 144x^2$
    C) $1 - 12x^2$
    D) $1 + 24x - 144x^2$

13) $(x + 4y)(x - 4y)$
    A) $x^2 - 16y^2$
    B) $x^2 + 8xy - 16y^2$
    C) $x^2 - 8xy - 16y^2$
    D) $x^2 - 8y^2$

14) $(9y + x)(9y - x)$
   A) $81y^2 - x^2$
   B) $81y^2 + 18xy - x^2$
   C) $18y^2 - x^2$
   D) $81y^2 - 18xy - x^2$

15) $(x + \frac{1}{8})(x - \frac{1}{8})$
   A) $x^2 + \frac{1}{4}x - \frac{1}{64}$
   B) $x^2 - \frac{1}{64}$
   C) $x^2 - \frac{1}{4}x - \frac{1}{64}$
   D) $x^2 - 1$

16) $(\frac{1}{3}y - 7)(\frac{1}{3}y + 7)$
   A) $\frac{1}{9}y^2 + \frac{14}{3}y - 49$
   B) $\frac{1}{9}y^2 - \frac{14}{3}y - 49$
   C) $\frac{1}{9}y^2 - 49$
   D) $\frac{1}{9}y^2 - 14$

17) $(6 + \frac{3}{4}z)(6 - \frac{3}{4}z)$
   A) $36 - \frac{3}{2}z^2$
   B) $36 - \frac{9}{16}z^2$
   C) $36 + 9z - \frac{9}{16}z^2$
   D) $36 - 9z - \frac{9}{16}z^2$

18) $(x - 3)^3$
   A) $x^3 - 9x^2 + 27x - 27$
   B) $x^3 - 9x^2 + 15x - 27$
   C) $x^3 - 9x^2 + 9x - 27$
   D) $x^3 - 3x^2 + 15x - 27$

19) $(3x + 4)^3$
   A) $9x^2 + 24x + 16$
   B) $9x^6 + 12x^3 + 4096$
   C) $27x^3 + 108x^2 + 144x + 64$
   D) $27x^3 + 108x^2 + 108x + 64$

20) $(x + 3y)^3$
   A) $x^3 + 27y^3$
   B) $3(x + 3y)$
   C) $x^3 + 3x^2y + 6xy + 9xy^2 + 18y^2 + 27y^3$
   D) $x^3 + 9x^2y + 27xy^2 + 27y^3$

21) $(x^2 + x + 5)(x^2 - x - 5)$
   A) $x^4 - x^2 - 10x + 25$
   B) $x^4 - x^2 - 10x - 25$
   C) $x^4 - 2x^3 - x^2 - 10x - 25$
   D) $x^4 + 2x^3 - x^2 - 10x - 25$

22) $(x^2 - 5x + 1)(x^2 - 5x + 4)$
   A) $x^4 + x^3 + 30x^2 - 25x + 4$
   B) $x^4 + 25x^2 + 4$
   C) $x^4 - 10x^3 + 30x^2 - 25x + 4$
   D) $x^4 - 10x^2 + 4$

23) $(x + 5)^3 - (x - 5)^3$
   A) $2x^3 + 30x^2 + 150x + 250$
   B) $30x^2 + 250$
   C) $2x^3 + 150x$
   D) $x^3 - 30x^2 + 150x - 125$

2 Factor Polynomials

Factor completely. If the polynomial cannot be factored, say it is prime.

1) $x^2 - 9$
   A) $(x + 3)^2$
   B) $(x + 3)(x - 3)$
   C) $(x - 3)^2$
   D) prime

2) $4x^2 - 25$
   A) $(2x + 5)^2$ B) $(2x + 5)(2x - 5)$ C) $(2x - 5)^2$ D) prime

3) $x^3 + 125$
   A) $(x + 5)(x^2 + 25)$
   B) $(x - 5)(x^2 + 5x + 25)$
   C) $(x - 125)(x^2 - 1)$
   D) $(x + 5)(x^2 - 5x + 25)$

4) $729 - x^3$
   A) $(9 + x)(81 - 9x + x^2)$ B) $(9 - x)(81 + 9x + x^2)$ C) $(9 + x)(81 - x^2)$ D) $(9 - x)(81 + x^2)$

5) $x^2 + 20x + 100$
   A) $(x + 10)^2$ B) $(x + 10)(x - 10)$ C) $(x - 10)^2$ D) prime

6) $25x^2 - 20x + 4$
   A) $(5x - 2)(5x + 2)$ B) $(5x + 2)^2$ C) $(5x - 2)^2$ D) $(25x + 1)(x + 4)$

7) $x^2 - x - 42$
   A) $(x + 7)(x - 6)$ B) $(x + 6)(x - 7)$ C) $(x + 1)(x - 42)$ D) prime

8) $x^2 + 11x + 30$
   A) $(x - 5)(x + 1)$ B) $(x - 5)(x + 6)$ C) $(x + 5)(x + 6)$ D) prime

9) $x^2 - x - 63$
   A) $(x - 7)(x + 9)$ B) $(x - 63)(x + 1)$ C) $(x + 7)(x - 9)$ D) prime

10) $20x^2 - 12x - 25x + 15$
    A) $(4x + 5)(5x + 3)$ B) $(4x - 5)(5x - 3)$ C) $(20x + 5)(x + 3)$ D) $(20x - 5)(x - 3)$

11) $12x^6 - 15x^3 + 20x^3 - 25$
    A) $(3x^3 + 5)(4x^3 - 5)$ B) $(3x^6 + 5)(4x - 5)$ C) $(3x^3 - 5)(4x^3 + 5)$ D) $(12x^3 - 5)(x^3 + 5)$

12) $15x^2 + 26x + 8$
    A) $(15x + 4)(x + 2)$ B) $(3x + 4)(5x + 2)$ C) $(3x - 4)(5x - 2)$ D) prime

13) $12z^2 - 7z - 12$
    A) $(10z + 3)(z - 3)$ B) $(2z + 3)(5z - 3)$ C) $(2z - 3)(5z + 3)$ D) prime

14) $21x^2 - 91x - 70$
    A) $7(3x - 2)(x + 5)$ B) $7(3x + 2)(x - 5)$ C) $(21x + 14)(x - 5)$ D) prime

15) $10x^2 - 35x - 20$
    A) $5(2x - 1)(x + 4)$ B) $5(2x + 1)(x - 4)$ C) $(10x - 5)(x + 4)$ D) prime

16) $6y^2 + 27y - 15$
    A) $(6y - 3)(y + 5)$ B) $3(2y + 1)(y - 5)$ C) $3(2y - 1)(y + 5)$ D) prime

17) $48x^9 - 40x^6 + 72x^4$
   A) $8(6x^9 - 5x^6 + 9x^4)$
   B) $x^4(48x^5 - 40x^2 + 72)$
   C) $8x^4(6x^5 - 5x^2 + 9)$
   D) prime

18) $x^2 + 59x + 60$
   A) $(x + 60)(x - 1)$
   B) $(x - 12)(x + 5)$
   C) $(x + 12)(x - 5)$
   D) prime

19) $-18x^2 + 51x - 15$
   A) $(-3x + 3)(2x - 5)$
   B) $(3x - 1)(-6x + 15)$
   C) $-3(3x + 1)(2x + 5)$
   D) $-3(3x - 1)(2x - 5)$

20) $x^2 - 12x + 144$
   A) $(x - 12)^2$
   B) $(x + 12)^2$
   C) $(x + 12)(x - 12)$
   D) prime

21) $x^3 - 25x + 2x^2 - 50$
   A) $(x - 5)^2(x + 2)$
   B) $(x^2 - 25)(x + 2)$
   C) $(x + 5)(x - 5)(x + 2)$
   D) prime

22) $x^4 - 12x^2 - 64$
   A) $(x - 4)^2(x^2 + 4)$
   B) $(x + 4)(x - 4)(x + 2)(x - 2)$
   C) $(x^2 - 16)(x^2 + 4)$
   D) $(x + 4)(x - 4)(x^2 + 4)$

23) $x^4 + 5x^3 + 8x + 40$
   A) $(x + 5)(x + 2)(x^2 - 2x + 4)$
   B) $(x + 5)(x - 2)(x^2 + 2x + 4)$
   C) $(x + 5)(x^3 + 8)$
   D) prime

An expression that occurs in calculus is given. Factor completely.

24) $2(x + 4)(x - 3)^3 + (x + 4)^2 \cdot 4(x - 3)^2$
   A) $(x + 4)(x - 3)^2(3x + 5)$
   B) $2(x + 3)(x - 4)^2(3x + 5)$
   C) $2(x + 4)(x - 3)^2(3x + 5)$
   D) $(x + 4)(x - 3)^2(6x + 10)$

25) $(2x + 1)^2 + 3(2x + 1) - 4$

26) $3(x + 5)^2(2x - 1)^2 + 4(x + 5)^3(2x - 1)$

3 Simplify Rational Expressions

Perform the indicated operations and simplify the result. Leave the answer in factored form.

1) $\dfrac{2x}{4x + 2} \cdot \dfrac{6x + 3}{2}$
   A) $\dfrac{3}{2}$
   B) $\dfrac{3x}{2}$
   C) $\dfrac{3x}{4}$
   D) $\dfrac{x}{2}$

2) $\dfrac{5x - 5}{x} \cdot \dfrac{2x^2}{7x - 7}$
   A) $\dfrac{10x}{7}$
   B) $\dfrac{10x^3 - 10x^2}{7x^2 - 7x}$
   C) $\dfrac{7}{10x}$
   D) $\dfrac{35x^2 + 70x + 35}{2x^3}$

3) $\dfrac{x^3 + 1}{x^3 - x^2 + x} \cdot \dfrac{2x}{-6x - 6}$

A) $-\dfrac{1}{3}$ B) $-\dfrac{x^3 + 1}{3(x + 1)}$ C) $-\dfrac{x^2 + 1}{3}$ D) $\dfrac{x + 1}{3(-x - 1)}$

4) $\dfrac{x^2 + 6x + 8}{x^2 + 13x + 36} \cdot \dfrac{x^2 + 14x + 45}{x^2 + 7x + 10}$

A) $\dfrac{x + 9}{x + 5}$ B) 1   C) $\dfrac{x + 2}{x + 9}$ D) $\dfrac{1}{x + 5}$

5) $\dfrac{x^2 - 5x + 6}{x^2 - 9x + 8} \cdot \dfrac{x^2 - 8x + 7}{x^2 - 7x + 12}$

A) $\dfrac{(x - 2)}{(x - 4)}$ B) $\dfrac{(x - 2)(x - 7)}{(x - 8)(x - 4)}$

C) $\dfrac{(x + 2)(x + 7)}{(x + 8)(x + 4)}$ D) $\dfrac{(x^2 - 5x + 6)(x^2 - 8x + 7)}{(x^2 - 9x + 8)(x^2 - 7x + 12)}$

6) $\dfrac{x^2 + 12x + 35}{x^2 + 16x + 63} \cdot \dfrac{x^2 + 9x}{x^2 - 2x - 35}$

A) $\dfrac{x}{x^2 + 16x + 63}$ B) $\dfrac{1}{x - 7}$ C) $\dfrac{x(x + 9)}{x - 7}$ D) $\dfrac{x}{x - 7}$

7) $\dfrac{9x^4 - 72x}{3x^2 - 12} \cdot \dfrac{x^2 + x - 2}{4x^3 + 8x^2 + 16x}$

A) $\dfrac{3(x - 1)}{4}$ B) $\dfrac{3x(x - 1)(x - 2)^2}{4(x + 2)^2}$ C) $\dfrac{3x(x + 1)}{4}$ D) $\dfrac{3x(x - 1)}{4}$

8) $\dfrac{x^2 + 12x + 35}{x^2 + 14x + 45} \cdot \dfrac{x^2 + 9x}{x^2 + 12x + 35}$

A) $\dfrac{x^2 + 9x}{x + 5}$ B) $\dfrac{x}{x + 5}$ C) $\dfrac{x}{x^2 + 14x + 45}$ D) $\dfrac{1}{x + 5}$

9) $\dfrac{10x^2 - 3x - 4}{5x^2 - x - 4} \cdot \dfrac{15x^2 + 12x}{1 - 4x^2}$

A) $\dfrac{3x(5x + 4)}{(1 - 2x)(x - 1)}$ B) $\dfrac{(5x + 4)(5x - 4)}{(1 - 2x)(x - 1)}$ C) $\dfrac{3x(5x - 4)}{(1 - 2x)(x - 1)}$ D) $\dfrac{3x}{(1 - 2x)(x - 1)}$

An expression that occurs in calculus is given. Reduce the expression to lowest terms.

10) $\dfrac{(x^2 + 4) \cdot 8 - (8x + 9) \cdot 5x}{(x^2 + 4)^3}$

A) $\dfrac{-40x^2 - 45x + 8}{(x^2 + 4)^2}$ B) $\dfrac{-48x^2 - 45x + 32}{(x^2 + 4)^3}$ C) $\dfrac{-32x^2 - 45x + 32}{(x^2 + 4)^3}$ D) $\dfrac{32x^2 + 45x - 32}{(x^2 + 4)^3}$

11) $\dfrac{(2x+5)4x - 2x^2(2)}{(2x+5)^2}$

4 Use the LCM to Add Rational Expressions

Perform the indicated operations and simplify the result. Leave the answer in factored form.

1) $\dfrac{2}{x} + \dfrac{5}{x-8}$

   A) $\dfrac{16x-7}{x(8-x)}$ B) $\dfrac{7x-16}{x(x-8)}$ C) $\dfrac{7x-16}{x(8-x)}$ D) $\dfrac{16x-7}{x(x-8)}$

2) $-\dfrac{9}{32} - \dfrac{8}{8x}$

   A) $\dfrac{-9x-32}{32x}$ B) $\dfrac{9x-32}{32x}$ C) $\dfrac{-9x+32}{32x}$ D) $\dfrac{-17}{32-8x}$

3) $\dfrac{x}{x^2-64} - \dfrac{8}{64-x^2}$

   A) $\dfrac{1}{x+8}$ B) $\dfrac{x-8}{x+8}$ C) $\dfrac{1}{x-8}$ D) $\dfrac{x+8}{x-8}$

4) $\dfrac{x+4}{x^2+15x+56} + \dfrac{4x-1}{x^2+9x+8}$

   A) $\dfrac{5x+3}{2x^2+24x+64}$ B) $5x+3$ C) $\dfrac{5x^2+32x-3}{(x+8)(x+7)(x+1)}$ D) $\dfrac{5x^2+32x-3}{(x-8)(x-7)(x-1)}$

5) $\dfrac{3}{x^2-3x+2} + \dfrac{5}{x^2-1}$

   A) $\dfrac{8x-7}{(x-1)(x+1)(x-2)}$ B) $\dfrac{30x-7}{(x-1)(x+1)(x-2)}$ C) $\dfrac{8x-7}{(x-1)(x-2)}$ D) $\dfrac{7x-8}{(x-1)(x+1)(x-2)}$

6) $\dfrac{x}{x^2-16} - \dfrac{3}{x^2+5x+4}$

   A) $\dfrac{x^2+2x+12}{(x-4)(x+4)(x+1)}$ B) $\dfrac{x^2-2}{(x-4)(x+4)(x+1)}$ C) $\dfrac{x^2-2x+12}{(x-4)(x+4)(x+1)}$ D) $\dfrac{x^2-2x+12}{(x-4)(x+4)}$

7) $\dfrac{7}{x-1} - \dfrac{8}{1-x}$

   A) $\dfrac{15}{1-x}$ B) $\dfrac{-1}{x-1}$ C) $\dfrac{15-15x}{(x-1)(1-x)}$ D) $\dfrac{15}{x-1}$

8) $\dfrac{x+3}{x^2+x-12} - \dfrac{x+4}{x^2-9}$

9) $\dfrac{x-3}{x^2+7x+12} + \dfrac{x-5}{x^2-16}$

   A) $\dfrac{2x^2-x-27}{(x+4)(x-4)(x+3)}$ B) $\dfrac{2x^2-x-3}{(x+4)(x-4)(x+3)}$ C) $\dfrac{2x^2-9x-3}{(x+4)(x-4)(x+3)}$ D) $\dfrac{2x^2-9x-27}{(x+4)(x-4)(x+3)}$

10) $\dfrac{3x}{x^2 - 5x - 36} - \dfrac{x - 1}{x^2 - 16} + \dfrac{1}{x^2 - 13x + 36}$

A) $\dfrac{2x - 1}{(x - 9)(x + 4)}$ B) $\dfrac{2x^2 - x - 5}{(x - 9)(x + 4)(x - 4)}$ C) $\dfrac{2x^2 - 21x + 13}{(x - 9)(x + 4)(x - 4)}$ D) $\dfrac{2x - 1}{(x - 9)(x - 4)}$

14.4 Polynomial Division; Synthetic Division

1 Divide Polynomials Using Long Division

Find the quotient and the remainder.

1) $15x^8 - 25x^5$ divided by $5x$
   A) $3x^7 - 5x^4$; remainder 0
   B) $3x^9 - 5x^6$; remainder 0
   C) $15x^7 - 25x^4$; remainder 0
   D) $3x^8 - 5x^5$; remainder 0

2) $40x^2 + 10x - 12$ divided by $5x$
   A) $8x^2 + 2x - \dfrac{12}{5}$; remainder 0
   B) $40x + 10$; remainder -12
   C) $8x - 10$; remainder 0
   D) $8x + 2$; remainder -12

3) $x^2 + 8x + 15$ divided by $x + 3$
   A) $x + 5$; remainder 0
   B) $x^2 + 5$; remainder 0
   C) $x^3 - 12$; remainder 0
   D) $x - 12$; remainder 0

4) $3x^2 + 21x - 24$ divided by $x + 8$
   A) $3x - 3$; remainder 0
   B) $3x - 3$; remainder 2
   C) $x - 3$; remainder 0
   D) $3x + 3$; remainder 0

5) $x^2 + 4x - 6$ divided by $x + 6$
   A) $x + 2$; remainder 6
   B) $x - 6$; remainder 2
   C) $x - 2$; remainder 0
   D) $x - 2$; remainder 6

6) $x^2 + 14x + 42$ divided by $x + 8$
   A) $x + 6$; remainder 6
   B) $x + 7$; remainder 0
   C) $x + 6$; remainder -6
   D) $x + 6$; remainder 0

7) $8x^3 + 40x^2 - 39x + 54$ divided by $x + 6$
   A) $8x^2 + 8x + 9$; remainder 0
   B) $x^2 + 8x + 8$; remainder 0
   C) $x^2 + 9x + 10$; remainder 0
   D) $8x^2 - 8x + 9$; remainder 0

8) $-8x^3 - 6x^2 + 15x + 14$ divided by $2x + 3$
   A) $-4x^2 + 3x + 3$; remainder 0
   B) $x^2 + 3$; remainder 3
   C) $-4x^2 + 3x + 3$; remainder 5
   D) $-4x^2 + 3x + 3$; remainder 8

9) $5x^3 - 7x^2 + 7x - 8$ divided by $5x - 2$
   A) $x^2 - x + 1$; remainder -6
   B) $x^2 + x - 1$; remainder -6
   C) $x^2 - x + 1$; remainder 6
   D) $x^2 - x + 1$; remainder 10

10) $x^4 + 625$ divided by $x - 5$
    A) $x^3 + 5x^2 + 25x + 125$; remainder 625
    B) $x^3 + 5x^2 + 25x + 125$; remainder 0
    C) $x^3 + 5x^2 + 25x + 125$; remainder 1250
    D) $x^3 - 5x^2 + 25x - 125$; remainder 1250

Precalculus Enhanced with Graphing Utilities 922

© 2006 Pearson Education, Inc., Upper Saddle River, NJ. All rights reserved. This material is protected under all copyright laws as they currently exist.
No portion of this material may be reproduced, in any form or by any means, without permission in writing from the publisher.

11) $8e^4 + 12e^3 - 2e$ divided by $2e^2 + e$

    A) $4e^2 + 4e$; remainder $-6e$ B) $4e^2 + 4e - 2$; remainder 0

    C) $4e^2 + 6e$; remainder $-2e$ D) $4e^2 + 8e + 4$; remainder $2e$

2 Divide Polynomials Using Synthetic Division

Use synthetic division to find the quotient and remainder.

1) $2x^4 - 4x^3 - 12x - 14$ divided by $x - 3$

    A) $2x^2 + 2x + 6$; remainder 4   B) $2x^3 + 2x^2 + 6x + 6$; remainder 4

    C) $2x^2 + 2x + 6$; remainder 6   D) $2x^3 + 2x^2 + 6x$; remainder 6

2) $5x^4 - 13x^2 - 6x + 9$ divided by $x - 3$

    A) $5x^3 + 8x^2 + 18x$; remainder 63   B) $5x^3 - 15x^2 + 32x - 102$; remainder -297

    C) $5x^3 + 15x^2 + 32x + 90$; remainder 279   D) $5x^3 + 8x^2 + 18x + 63$; remainder 0

3) $6x^5 - 5x^4 + x - 4$ divided by $x + \frac{1}{2}$

    A) $6x^4 - 2x^3 - x^2 + \frac{1}{2}x + \frac{5}{4}$; remainder $-\frac{27}{8}$ B) $6x^4 - 2x^3 + x^2 - \frac{1}{2}x + \frac{5}{4}$; remainder $-\frac{37}{8}$

    C) $6x^4 - 8x^3 + 5$; remainder $-\frac{13}{2}$ D) $6x^4 - 8x^3 + 4x^2 - 2x + 2$; remainder -5

Use synthetic division to determine whether x − c is a factor of the given polynomial.

4) $3x^4 - 10x^3 - 59x^2 + 146x + 120$; $x + 5$

    A) Yes   B) No

5) $x^4 + 7x^3 - 19x^2 - 103x + 210$; $x - 3$

    A) Yes   B) No

6) $2x^4 - 4x^3 - 12x - 14$; $x - 3$

    A) Yes   B) No

7) $6x^5 - 5x^4 + x - 4$; $x + \frac{1}{2}$

    A) Yes   B) No

8) $3x^4 - 2x^2 + x - 1$; $x + 1$

    A) Yes   B) No

9) $2x^6 - 18x^4 + x^2 - 9$; $x + 3$

    A) Yes   B) No

14.5 Solving Equations

1 Solve Linear Equations

Solve the equation.

1) $4x = -8$

    A) {4}   B) {−4}   C) {−2}   D) {2}

2) $10x = 1$
   A) $\left\{\dfrac{1}{10}\right\}$ B) $\{-10\}$ C) $\{10\}$ D) $\left\{-\dfrac{1}{10}\right\}$

3) $2x + 2 = 0$
   A) $\{-2\}$ B) $\{2\}$ C) $\{-1\}$ D) $\{1\}$

4) $5x - 11 = 0$
   A) $\left\{\dfrac{5}{11}\right\}$ B) $\left\{-\dfrac{11}{5}\right\}$ C) $\left\{\dfrac{11}{5}\right\}$ D) $\left\{-\dfrac{5}{11}\right\}$

5) $6x + 19 = 0$
   A) $\left\{\dfrac{19}{6}\right\}$ B) $\left\{-\dfrac{19}{6}\right\}$ C) $\left\{\dfrac{6}{19}\right\}$ D) $\left\{-\dfrac{6}{19}\right\}$

6) $\dfrac{3}{7}x = -\dfrac{3}{5}$
   A) $\left\{-\dfrac{21}{5}\right\}$ B) $\left\{\dfrac{7}{5}\right\}$ C) $\left\{-\dfrac{7}{5}\right\}$ D) $\left\{-\dfrac{5}{7}\right\}$

7) $11x + 6 = 5x + 30$
   A) $\{-4\}$ B) $\{5\}$ C) $\{4\}$ D) $\{-5\}$

8) $-8x - 60 = -2x - 18$
   A) $\{7\}$ B) $\{-7\}$ C) $\{-10\}$ D) $\{10\}$

9) $35 - 3x = 7 + 4x$
   A) $\{-5\}$ B) $\{5\}$ C) $\{-4\}$ D) $\{4\}$

10) $9x - (7x - 1) = 2$
    A) $\left\{-\dfrac{1}{16}\right\}$ B) $\left\{\dfrac{1}{2}\right\}$ C) $\left\{-\dfrac{1}{2}\right\}$ D) $\left\{\dfrac{1}{16}\right\}$

11) $5(x + 8) = 6(x - 8)$
    A) $\{88\}$ B) $\{6\}$ C) $\{-8\}$ D) $\{-88\}$

12) $2(2x - 4) = 3(x + 4)$
    A) $\{20\}$ B) $\{6\}$ C) $\{-4\}$ D) $\{-20\}$

13) $\dfrac{1}{5}x - 5 = 1$
    A) $\{-20\}$ B) $\{30\}$ C) $\{-30\}$ D) $\{20\}$

14) $\dfrac{1}{5}x - \dfrac{1}{5} = -6$
    A) $\{31\}$ B) $\{29\}$ C) $\{-31\}$ D) $\{-29\}$

15) $\frac{2}{5}x = 2 + \frac{1}{3}x$

   A) {60}   B) {-60}   C) {-30}   D) {30}

16) $\frac{1}{4}x - 2 = \frac{1}{3}x - 7$

   A) {60}   B) {-60}   C) $\left\{-\frac{5}{12}\right\}$ D) $\left\{\frac{5}{12}\right\}$

17) $\frac{5}{3} - \frac{1}{5}x = \frac{14}{15}$

   A) $\left\{\frac{11}{3}\right\}$ B) $\left\{-\frac{11}{5}\right\}$ C) $\left\{-\frac{11}{3}\right\}$ D) $\left\{\frac{11}{5}\right\}$

18) $-4.4x + 1.1 = -31.9 - 1.4x$

   A) {7.5}   B) {-36}   C) {7.8}   D) {11}

19) $1 - \frac{2}{3x} = \frac{1}{3}$

   A) {1}   B) {-9}   C) $\left\{\frac{1}{3}\right\}$ D) {-1}

20) $\frac{5}{x} + \frac{2}{5} = \frac{7}{x}$

   A) {5}   B) {2}   C) {-5}   D) {-2}

21) $(x+5)(x-1) = (x+1)^2$

   A) $\left\{\frac{6}{5}\right\}$ B) {6}   C) $\left\{\frac{5}{2}\right\}$ D) {3}

22) $x(6x - 1) = (6x + 1)(x - 4)$

   A) $\left\{-\frac{4}{5}\right\}$ B) {2}   C) $\left\{-\frac{4}{23}\right\}$ D) $\left\{-\frac{2}{11}\right\}$

23) $x(x^2 + 5) = 3 + x^3$

   A) $\left\{\frac{5}{3}\right\}$ B) $\left\{\frac{3}{5}\right\}$ C) {3}   D) {5}

2 Solve Rational Equations

Solve the equation.

1) $\frac{x}{x-4} - 3 = \frac{4}{x-4}$

   A) {-4}   B) {4, -4}   C) {4}   D) no solution

2) $\frac{1}{x} + \frac{1}{x-3} = \frac{x-2}{x-3}$

   A) {1}   B) {-1}   C) {3}   D) {-3}

3) $\dfrac{2x}{x^2-9} = \dfrac{2}{x^2-9} - \dfrac{1}{x+3}$

   A) $\{\tfrac{1}{2}\}$ B) $\{-\tfrac{1}{3}\}$ C) $\{-1\}$ D) $\{\tfrac{5}{3}\}$

4) $\dfrac{9}{x+5} - \dfrac{5}{x-5} = \dfrac{2}{x^2-25}$

   A) $\{72\}$ B) $\{-18\}$ C) $\{2\sqrt{17}\}$ D) $\{18\}$

5) $\dfrac{6}{2x-3} = \dfrac{4}{2x+5}$

   A) $\{\tfrac{21}{2}\}$ B) $\{-\tfrac{2}{21}\}$ C) $\{-\tfrac{21}{2}\}$ D) $\{\tfrac{2}{21}\}$

6) $\dfrac{x-2}{x-5} = \dfrac{x+6}{x-7}$

   A) $\{-4\}$ B) $\{-\tfrac{15}{7}\}$ C) $\{\tfrac{8}{5}\}$ D) $\{\tfrac{22}{5}\}$

7) $\dfrac{6x-1}{2x+5} = \dfrac{21x-5}{7x+9}$

   A) $\{-\tfrac{8}{71}\}$ B) $\{\tfrac{1}{3}\}$ C) $\{\tfrac{17}{24}\}$ D) $\{-\tfrac{17}{71}\}$

8) $\dfrac{-2}{x-6} = \dfrac{6}{x-12} - \dfrac{-12}{x^2-18x+72}$

   A) $\{12\}$ B) $\{6\}$ C) $\{6, 12\}$ D) no solution

9) $\dfrac{y+7}{y^2-2y-15} - \dfrac{7}{y^2-10y+25} = \dfrac{y-7}{y^2-2y-15}$

   A) $\{91\}$ B) $\{13\}$ C) $\{-13\}$ D) no solution

10) $\dfrac{x}{2x+2} = \dfrac{2x-3}{x+1} - \dfrac{2x}{4x+4}$

    A) $\{\tfrac{3}{2}\}$ B) $\{-3\}$ C) $\{3\}$ D) no solution

11) $\dfrac{4}{3x} - \dfrac{1}{x+1} = \dfrac{3}{x(2x+2)}$

    A) $\{1\}$ B) $\{\tfrac{1}{6}\}$ C) $\{\tfrac{1}{2}\}$ D) no solution

3 Solve Quadratic Equations by Factoring

Find the real solutions of the equation by factoring.

1) $x^2 + 6x = 0$

   A) $\{0, -6\}$ B) $\{6\}$ C) $\{-6\}$ D) $\{0, 6\}$

2) $8x^2 - 15x = 0$
   A) $\{-\frac{15}{8}, 0\}$
   B) $\{0\}$
   C) $\{\frac{15}{8}, 0\}$
   D) $\{\frac{15}{8}, -\frac{15}{8}\}$

3) $18x^2 + 39x = 0$
   A) $\{-\frac{13}{6}, 0\}$
   B) $\{0\}$
   C) $\{\frac{13}{6}, -\frac{13}{6}\}$
   D) $\{\frac{13}{6}, 0\}$

4) $x^2 - 81 = 0$
   A) $\{9, -9\}$
   B) $\{81\}$
   C) $\{-9\}$
   D) $\{9\}$

5) $x^2 + x - 12 = 0$
   A) $\{-4, 3\}$
   B) $\{-4, -3\}$
   C) $\{-3, 4\}$
   D) $\{3, 4\}$

6) $x^2 - 9x + 20 = 0$
   A) $\{5, -4\}$
   B) $\{5, 4\}$
   C) $\{-5, 4\}$
   D) $\{-5, -4\}$

7) $x^2 - 6x - 7 = 0$
   A) $\{-1, 7\}$
   B) $\{1, -7\}$
   C) $\{-1, -7\}$
   D) $\{1, 7\}$

8) $4x^2 + 15x - 4 = 0$
   A) $\{\frac{1}{4}, -4\}$
   B) $\{-\frac{1}{4}, 4\}$
   C) $\{-\frac{1}{4}, -4\}$
   D) $\{\frac{1}{4}, 4\}$

9) $5x^2 - 15 = 0$
   A) $\{-3, 3\}$
   B) $\{7.5\}$
   C) $\{-\sqrt{3}, \sqrt{3}\}$
   D) $\{4\}$

10) $x(x - 9) + 18 = 0$
    A) $\{-3, -6\}$
    B) $\{-3, 6\}$
    C) $\{3, 6\}$
    D) $\{3, -6\}$

11) $x(x + 3) = 108$
    A) $\{-12, -9\}$
    B) $\{12, 9\}$
    C) $\{-12, 9\}$
    D) $\{12, -9\}$

12) $49x^2 - 28x + 4 = 0$
    A) $\{\frac{7}{2}\}$
    B) $\{-\frac{7}{2}\}$
    C) $\{-\frac{2}{7}\}$
    D) $\{\frac{2}{7}\}$

13) $2x - 7 = \frac{4}{x}$
    A) $\{-\frac{1}{2}, 2\}$
    B) $\{-\frac{1}{2}, 4\}$
    C) $\{-2, 4\}$
    D) $\{\frac{1}{7}, -\frac{1}{2}\}$

14) $36x + \frac{18}{x} = -54$
    A) $\{-1, \frac{1}{2}\}$
    B) $\{36, 2\}$
    C) $\{1, \frac{1}{2}\}$
    D) $\{-1, -\frac{1}{2}\}$

15) $\dfrac{x-3}{x} = \dfrac{8}{x+3}$

A) {9, 1} B) {3, -1} C) {9, -1} D) {3, 1}

4 Solve Quadratic Equations Using the Square Root Method

Solve the equation by the Square Root Method.

1) $x^2 = 81$

A) {9} B) {-9} C) {-6561, 6561} D) {-9, 9}

2) $x^2 - 6 = 0$

A) {$\sqrt{6}, -\sqrt{6}$} B) {$\sqrt{6}$} C) {6} D) {-6, 6}

3) $(x-4)^2 = 36$

A) {40} B) {-6, 6} C) {-10, -2} D) {-2, 10}

4) $(x+4)^2 = 36$

A) {2} B) {-6, 6} C) {-10} D) {-10, 2}

5) $(2x-3)^2 = 81$

A) {-6, 12} B) {-3, 6} C) {-6, 3} D) {-12, 6}

6) $(4x+4)^2 = 64$

A) {-3, 1} B) {0, 1} C) {-17, 17} D) {1, 3}

7) $(x+4)^2 = 15$

A) {$-\sqrt{15}, \sqrt{15}$} B) {11}
C) {$-4-\sqrt{15}, -4+\sqrt{15}$} D) {$4-\sqrt{15}, 4+\sqrt{15}$}

5 Solve Quadratic Equations by Completing the Square

What number should be added to complete the square of the expression?

1) $x^2 + 8x$

A) 16 B) 8 C) 4 D) 32

2) $x^2 - 18x$

A) 162 B) 81 C) -9 D) 41

3) $x^2 + \dfrac{3}{5}x$

A) $\dfrac{9}{100}$ B) $\dfrac{1}{5}$ C) $\dfrac{9}{25}$ D) $\dfrac{9}{50}$

4) $x^2 - \dfrac{2}{5}x$

A) $-\dfrac{1}{5}$ B) $\dfrac{4}{25}$ C) $\dfrac{1}{25}$ D) $-\dfrac{2}{25}$

Solve the equation by completing the square.

5) $x^2 - x = 25$

   A) $\{\dfrac{-1-\sqrt{101}}{2}, \dfrac{-1+\sqrt{101}}{2}\}$
   B) $\{\dfrac{1-\sqrt{101}}{4}, \dfrac{1+\sqrt{101}}{4}\}$
   C) $\{\dfrac{1-\sqrt{101}}{2}, \dfrac{1+\sqrt{101}}{2}\}$
   D) $\{-\dfrac{9}{2}, \dfrac{11}{2}\}$

6) $x^2 + 8x = 5$

   A) $\{4+\sqrt{21}\}$
   B) $\{-1-\sqrt{21}, -1+\sqrt{21}\}$
   C) $\{-4-\sqrt{21}, -4+\sqrt{21}\}$
   D) $\{-4-2\sqrt{21}, -4+2\sqrt{21}\}$

7) $x^2 + 8x - 5 = 0$

   A) $\{-4-2\sqrt{21}, -4+2\sqrt{21}\}$
   B) $\{4+\sqrt{21}\}$
   C) $\{-1-\sqrt{21}, -1+\sqrt{21}\}$
   D) $\{-4-\sqrt{21}, -4+\sqrt{21}\}$

8) $x^2 + 6x - 27 = 0$

   A) $\{\sqrt{6}, -1\}$
   B) $\{-18, -9\}$
   C) $\{3, -9\}$
   D) $\{-3, 9\}$

9) $x^2 + 12x + 22 = 0$

   A) $\{-12+\sqrt{22}\}$
   B) $\{6+\sqrt{14}\}$
   C) $\{6-\sqrt{22}, 6+\sqrt{22}\}$
   D) $\{-6-\sqrt{14}, -6+\sqrt{14}\}$

10) $x^2 - 12x - 7 = 0$

    A) $\{-6-\sqrt{43}, -6+\sqrt{43}\}$
    B) $\{12-\sqrt{151}, 12+\sqrt{151}\}$
    C) $\{6-\sqrt{43}, 6+\sqrt{43}\}$
    D) $\{6-\sqrt{7}, 6+\sqrt{7}\}$

11) $x^2 + 5x - 5 = 0$

    A) $\{\dfrac{-5-3\sqrt{5}}{2}\}$
    B) $\{-5-3\sqrt{5}, -5+3\sqrt{5}\}$
    C) $\{\dfrac{5+3\sqrt{5}}{2}\}$
    D) $\{\dfrac{-5-3\sqrt{5}}{2}, \dfrac{-5+3\sqrt{5}}{2}\}$

12) $x^2 - \dfrac{4}{3}x - \dfrac{5}{9} = 0$

    A) $\{-\dfrac{1}{3}, \dfrac{5}{3}\}$
    B) $\{\dfrac{1}{3}, \dfrac{5}{3}\}$
    C) $\{\dfrac{1}{3}, -\dfrac{5}{3}\}$
    D) $\{-\dfrac{1}{3}, -\dfrac{5}{3}\}$

13) $\dfrac{1}{4}x^2 + \dfrac{1}{16}x - \dfrac{1}{8} = 0$

    A) $\{\dfrac{\sqrt{33}-1}{8}, -\dfrac{\sqrt{33}+1}{8}\}$
    B) $\{\dfrac{1}{8}, -\dfrac{1}{8}\}$
    C) $\{\dfrac{\sqrt{33}-1}{8}, \dfrac{\sqrt{33}+1}{8}\}$
    D) $\{\dfrac{\sqrt{33}}{8}, -\dfrac{\sqrt{33}}{8}\}$

14) $25x^2 + 40x + 12 = 0$

    A) $\{-\dfrac{6}{25}, \dfrac{18}{25}\}$
    B) $\{-\dfrac{2}{25}, -\dfrac{6}{25}\}$
    C) $\{-\dfrac{2}{5}, -\dfrac{6}{5}\}$
    D) $\{\dfrac{2}{5}, \dfrac{6}{5}\}$

15) $3x^2 - 2x - 3 = 0$

    A) $\{\dfrac{3-\sqrt{10}}{9}, \dfrac{3+\sqrt{10}}{9}\}$ B) $\{\dfrac{-1-\sqrt{10}}{3}, \dfrac{-1+\sqrt{10}}{3}\}$

    C) $\{-3, \dfrac{11}{3}\}$ D) $\{\dfrac{1-\sqrt{10}}{3}, \dfrac{1+\sqrt{10}}{3}\}$

6 Solve Quadratic Equations Using the Quadratic Formula

Find the real solutions, if any, of the equation. Use the quadratic formula.

1) $x^2 + 8x - 5 = 0$

    A) $\{-4-\sqrt{21}, -4+\sqrt{21}\}$ B) $\{-1-\sqrt{21}, -1+\sqrt{21}\}$
    C) $\{-4-2\sqrt{21}, -4+2\sqrt{21}\}$ D) $\{4+\sqrt{21}\}$

2) $x^2 - 12x - 5 = 0$

    A) $\{6+\sqrt{5}, 6-\sqrt{5}\}$ B) $\{12+\sqrt{41}, 12-\sqrt{41}\}$
    C) $\{6+\sqrt{41}, 6-\sqrt{41}\}$ D) $\{-6+\sqrt{41}, -6-\sqrt{41}\}$

3) $x^2 + 7x + 4 = 0$

    A) $\{\dfrac{-7-\sqrt{33}}{2}, \dfrac{-7+\sqrt{33}}{2}\}$ B) $\{\dfrac{-7-\sqrt{65}}{2}, \dfrac{-7+\sqrt{65}}{2}\}$
    C) $\{\dfrac{7-\sqrt{33}}{2}, \dfrac{7+\sqrt{33}}{2}\}$ D) $\{\dfrac{-7-\sqrt{33}}{14}, \dfrac{-7+\sqrt{33}}{14}\}$

4) $x^2 + x + 1 = 0$

    A) $\{\dfrac{-1-\sqrt{3}}{2}, \dfrac{1+\sqrt{3}}{2}\}$ B) $\{\dfrac{-1-\sqrt{3}}{2}, \dfrac{-1+\sqrt{3}}{2}\}$ C) $\{\dfrac{1-\sqrt{3}}{2}, \dfrac{1+\sqrt{3}}{2}\}$ D) no real solution

5) $64x^2 + 48x + 9 = 0$

    A) $\{-\dfrac{3}{8}\}$ B) $\{-\dfrac{3}{8}, 24\}$ C) $\{\dfrac{3}{8}\}$ D) no real solution

6) $49x^2 + 98x + 24 = 0$

    A) $\{-\dfrac{2}{7}, -\dfrac{12}{7}\}$ B) $\{-\dfrac{2}{49}, -\dfrac{12}{49}\}$ C) $\{-\dfrac{12}{49}, \dfrac{36}{49}\}$ D) $\{\dfrac{2}{7}, \dfrac{12}{7}\}$

7) $4x^2 - x + 1 = 0$

    A) $\{\dfrac{-1-\sqrt{17}}{8}, \dfrac{1+\sqrt{17}}{8}\}$ B) $\{\dfrac{-1-\sqrt{17}}{8}, \dfrac{-1+\sqrt{17}}{8}\}$
    C) $\{\dfrac{-1+\sqrt{17}}{8}, \dfrac{1+\sqrt{17}}{8}\}$ D) no real solution

8) $2x^2 + x - 4 = 0$

    A) $\{\dfrac{-1-\sqrt{33}}{4}, \dfrac{-1+\sqrt{33}}{4}\}$ B) $\{\dfrac{1-\sqrt{33}}{4}, \dfrac{1+\sqrt{33}}{4}\}$
    C) $\{\dfrac{-1-\sqrt{33}}{2}, \dfrac{-1+\sqrt{33}}{2}\}$ D) no real solution

9) $5x^2 + 12x + 2 = 0$

A) $\{\dfrac{-12-\sqrt{26}}{5}, \dfrac{-12+\sqrt{26}}{5}\}$

B) $\{\dfrac{-6-\sqrt{46}}{5}, \dfrac{-6+\sqrt{46}}{5}\}$

C) $\{\dfrac{-6-\sqrt{26}}{5}, \dfrac{-6+\sqrt{26}}{5}\}$

D) $\{\dfrac{-6-\sqrt{26}}{10}, \dfrac{-6+\sqrt{26}}{10}\}$

10) $3x^2 + 8x = -2$

A) $\{\dfrac{-4-\sqrt{10}}{3}, \dfrac{-4+\sqrt{10}}{3}\}$

B) $\{\dfrac{-8-\sqrt{10}}{3}, \dfrac{-8+\sqrt{10}}{3}\}$

C) $\{\dfrac{-4-\sqrt{22}}{3}, \dfrac{-4+\sqrt{22}}{3}\}$

D) $\{\dfrac{-4-\sqrt{10}}{6}, \dfrac{-4+\sqrt{10}}{6}\}$

11) $2x^2 = -6x - 1$

A) $\{\dfrac{-6-\sqrt{7}}{2}, \dfrac{-6+\sqrt{7}}{2}\}$

B) $\{\dfrac{-3-\sqrt{11}}{2}, \dfrac{-3+\sqrt{11}}{2}\}$

C) $\{\dfrac{-3-\sqrt{7}}{2}, \dfrac{-3+\sqrt{7}}{2}\}$

D) $\{\dfrac{-3-\sqrt{7}}{4}, \dfrac{-3+\sqrt{7}}{4}\}$

12) $4 + \dfrac{12}{x} + \dfrac{6}{x^2} = 0$

A) $\{\dfrac{-3-\sqrt{3}}{2}, \dfrac{-3+\sqrt{3}}{2}\}$

B) $\{\dfrac{3-\sqrt{3}}{2}, \dfrac{3+\sqrt{3}}{2}\}$

C) $\{\dfrac{-3-\sqrt{3}}{8}, \dfrac{-3+\sqrt{3}}{8}\}$

D) no real solution

Solve using the quadratic formula. Round any solutions to two decimal places.

13) $\dfrac{1}{4}x^2 - 2\sqrt{3}x = 3$

A) {-0.21, 14.67}   B) {0.82, -14.67}   C) {0.21, -14.67}   D) {-0.82, 14.67}

Use the discriminant to determine whether the quadratic equation has two unequal real solutions, a repeated real solution, or no real solution without solving the equation.

14) $x^2 + 6x + 5 = 0$

A) repeated real solution   B) two unequal real solutions   C) no real solution

15) $x^2 - 8x + 16 = 0$

A) repeated real solution   B) two unequal real solutions   C) no real solution

16) $x^2 - 2x + 3 = 0$

A) repeated real solution   B) two unequal real solutions   C) no real solution

17) $7x^2 + 2x + 2 = 0$

A) repeated real solution   B) two unequal real solutions   C) no real solution

18) $2x^2 + 8x - 7 = 0$

A) repeated real solution   B) two unequal real solutions   C) no real solution

7 Solve Equations Quadratic in Form

Find the real solutions of the equation.

1) $x^4 - 256 = 0$
   A) $\{-16, 16\}$
   B) $\{-4, 4\}$
   C) $\{-2, 2\}$
   D) no real solution

2) $x^4 - 37x^2 + 36 = 0$
   A) $\{-36, 36\}$
   B) $\{-37, 37\}$
   C) $\{-6, 6\}$
   D) $\{-1, 1, -6, 6\}$

3) $x^4 - 6x^2 - 27 = 0$
   A) $\{-9, 3\}$
   B) $\{-\sqrt{3}, \sqrt{3}\}$
   C) $\{-3, 3\}$
   D) no real solution

4) $3x^4 - 43x^2 - 80 = 0$
   A) $\left\{-\dfrac{\sqrt{15}}{3}, \dfrac{\sqrt{15}}{3}, -4, 4\right\}$
   B) $\{4\}$
   C) $\{-4, 4\}$
   D) no real solution

5) $x^6 + 63x^3 - 64 = 0$
   A) $\{-4, 1\}$
   B) $\{4, -1\}$
   C) $\{64\}$
   D) $\{4\}$

6) $4(x + 1)^2 + 15(x + 1) + 9 = 0$
   A) $\left\{-\dfrac{7}{16}, -3\right\}$
   B) $\left\{-\dfrac{3}{4}, -4\right\}$
   C) $\left\{-\dfrac{7}{4}, -4\right\}$
   D) $\left\{\dfrac{1}{2}, 2\right\}$

7) $(x - 4)^2 + 3(x - 4) - 18 = 0$
   A) $\{-7, 2\}$
   B) $\{-6, 1\}$
   C) $\{-2, 7\}$
   D) $\{-1, 6\}$

8) $(4x - 7)^2 + 2(4x - 7) - 8 = 0$
   A) $\left\{-\dfrac{3}{4}, -\dfrac{9}{4}\right\}$
   B) $\left\{\dfrac{11}{4}, -\dfrac{5}{4}\right\}$
   C) $\left\{-\dfrac{11}{7}, \dfrac{5}{4}\right\}$
   D) $\left\{\dfrac{3}{4}, \dfrac{9}{4}\right\}$

9) $(7x + 1)^2 - 15(7x + 1) + 54 = 0$
   A) $\{9, 6\}$
   B) $\left\{\dfrac{8}{7}, \dfrac{5}{7}\right\}$
   C) $\left\{\dfrac{10}{7}, 1\right\}$
   D) $\left\{-\dfrac{8}{7}, -\dfrac{5}{7}\right\}$

8 Solve Absolute Value Equations

Solve the equation.

1) $|x| = 2$
   A) $\{-2, 2\}$
   B) $\{4\}$
   C) $\{2\}$
   D) $\{-2\}$

2) $|x| = -6$
   A) $\{6, -6\}$
   B) $\{6\}$
   C) $\{-6\}$
   D) no real solution

3) $|x + 6| = 0$
   A) $\{6, -6\}$
   B) $\{-6\}$
   C) $\{6\}$
   D) no real solution

Precalculus Enhanced with Graphing Utilities

4) $|x - 7| = 9$
   A) {2, 16}   B) {-16}   C) {-2, 16}   D) no real solution

5) $|3x + 5| = 4$
   A) $\left\{\frac{1}{3}, 3\right\}$ B) $\left\{-\frac{1}{3}, -3\right\}$ C) $\left\{-\frac{1}{5}, -\frac{9}{5}\right\}$ D) no real solution

6) $|2 - 4x| = 3$
   A) $\left\{-\frac{7}{2}, -\frac{1}{2}\right\}$ B) $\left\{-\frac{1}{4}, \frac{5}{4}\right\}$ C) $\left\{-\frac{5}{4}, \frac{1}{4}\right\}$ D) $\left\{\frac{1}{2}, \frac{7}{2}\right\}$

7) $|-4x| = 9$
   A) $\left\{-\frac{9}{4}, \frac{9}{4}\right\}$ B) {-9, 9}   C) $\left\{-\frac{4}{9}, \frac{4}{9}\right\}$ D) no real solution

8) $|-x| = 5$
   A) {25}   B) {-5}   C) {-5, 5}   D) {5}

9) $6 - |5x| = 4$
   A) $\left\{-\frac{5}{2}, \frac{5}{2}\right\}$ B) $\left\{-\frac{2}{5}, \frac{2}{5}\right\}$ C) {-2, 2}   D) no real solution

10) $\frac{1}{4}|x| = 8$
    A) {-31, 31}   B) {-2, 2}   C) {-32, 32}   D) no real solution

11) $\left|\frac{x}{3} + \frac{1}{5}\right| = 1$
    A) $\left\{\frac{12}{5}\right\}$ B) $\left\{-\frac{12}{5}, \frac{18}{5}\right\}$ C) $\left\{-\frac{18}{5}, \frac{12}{5}\right\}$ D) no real solution

12) $|u - 5| = -\frac{1}{2}$
    A) $\left\{\frac{9}{2}, \frac{11}{2}\right\}$ B) $\left\{\frac{11}{2}\right\}$ C) $\left\{\frac{9}{2}\right\}$ D) no real solution

13) $|x^2 - 81| = 0$
    A) {-9, 9}   B) {9}   C) {-9}   D) {-6561, 6561}

14) $|x^2 - 7x| = 0$
    A) {-7, 0}   B) {0, 7}   C) {-7, 0, 7}   D) no solution

15) $|x^2 + 3x - 2| = 2$
    A) {-4, 4, -1, 1}   B) {-3, -1, 0, 4}   C) {-4, -3, 1}   D) {-4, -3, 0, 1}

9 Solve Equations by Factoring

Find the real solutions of the equation by factoring.

1) $x^3 - x^2 - 42x = 0$
   - A) $\{-6, 0, 7\}$
   - B) $\{-7, 0, 6\}$
   - C) $\{-6, 7\}$
   - D) $\{-7, 6\}$

2) $x^3 + 4x^2 - 12x = 0$
   - A) $\{2, 0, -6\}$
   - B) $\{-2, 6\}$
   - C) $\{-2, 0, 6\}$
   - D) $\{2, -6\}$

3) $x^3 + 5x^2 - x - 5 = 0$
   - A) $\{-5, 5\}$
   - B) $\{1, -5, 5\}$
   - C) $\{25\}$
   - D) $\{-1, 1, -5\}$

4) $x^3 + 5x^2 - 4x - 20 = 0$
   - A) $\{4, -5\}$
   - B) $\{-2, 2, -5\}$
   - C) $\{2, -5\}$
   - D) $\{-2, 2, 5\}$

5) $x^3 + 6x^2 + 25x + 150 = 0$
   - A) $\{6\}$
   - B) $\{-6\}$
   - C) $\{-5, 5, -6\}$
   - D) no real solution

6) $2x^3 + 3x^2 = 50x + 75$
   - A) $\{-5, -\frac{3}{2}, 5\}$
   - B) $\{-\frac{3}{2}, 5\}$
   - C) $\{-\frac{3}{2}, 0\}$
   - D) $\{-5, 5\}$

7) $100x^3 + 3 = 75x^2 + 4x$
   - A) $\{-\frac{1}{25}, \frac{1}{25}, \frac{3}{4}\}$
   - B) $\{-\frac{1}{5}, \frac{1}{5}, \frac{3}{4}\}$
   - C) $\{0, \frac{3}{4}\}$
   - D) $\{-\frac{1}{5}, \frac{1}{5}, \frac{4}{3}\}$

10 Demonstrate Additional Understanding and Skills

Solve the formula for the indicated variable.

1) $PV = nRT$ for R
   - A) $R = \frac{PVT}{n}$
   - B) $R = \frac{PV}{nT}$
   - C) $R = \frac{PV}{T}$
   - D) $R = \frac{nPV}{T}$

2) $S = 2\pi rh + 2\pi r^2$ for h
   - A) $h = 2\pi(S - r)$
   - B) $h = \frac{S}{2\pi r} - 1$
   - C) $h = \frac{S - 2\pi r^2}{2\pi r}$
   - D) $h = S - r$

3) $F = \frac{9}{5}C + 32$ for C
   - A) $C = \frac{F - 32}{9}$
   - B) $C = \frac{9}{5}(F - 32)$
   - C) $C = \frac{5}{9}(F - 32)$
   - D) $C = \frac{5}{F - 32}$

4) $A = P(1 + rt)$ for t
   - A) $t = -\frac{A + P}{rP}$
   - B) $t = \frac{A - P}{rP}$
   - C) $t = \frac{P - A}{rP}$
   - D) $t = \frac{A + P}{rP}$

5) $P - \dfrac{3Q}{7} = \dfrac{P+5}{2} + 1$ for P

   A) $P = \dfrac{21 + 6Q}{7}$
   B) $P = \dfrac{49 - 6Q}{7}$
   C) $P = \dfrac{21 - 6Q}{7}$
   D) $P = \dfrac{49 + 6Q}{7}$

6) $I = PRT$ for T

   A) $T = \dfrac{IP}{R}$
   B) $T = \dfrac{IR}{P}$
   C) $T = \dfrac{I}{PR}$
   D) $T = IPR$

7) $A = \dfrac{1}{2}h(b_1 + b_2)$ for b_1

   A) $b_1 = \dfrac{2A}{h} - b_2$
   B) $b_1 = \dfrac{2A - b_2}{h}$
   C) $b_1 = \dfrac{2A}{h} + b_2$
   D) $b_1 = \dfrac{2A + b_2}{h}$

14.6 Complex Numbers; Quadratic Equations in the Complex Number System

1 Add, Subtract, Multiply, and Divide Complex Numbers

Write the expression in the standard form a + bi.

1) $(4 - 9i) + (7 + 5i)$
   A) $-3 + 14i$
   B) $11 + 4i$
   C) $11 - 4i$
   D) $-11 + 4i$

2) $(4 + 6i) - (-7 + i)$
   A) $-3 + 7i$
   B) $11 + 5i$
   C) $11 - 5i$
   D) $-11 - 5i$

3) $9i(5 - 3i)$
   A) $45i + 27i^2$
   B) $45i - 27$
   C) $27 + 45i$
   D) $45i - 27i^2$

4) $(6 + 9i)(2 - 5i)$
   A) $57 + 12i$
   B) $-45i^2 - 12i + 12$
   C) $-33 + 48i$
   D) $57 - 12i$

5) $(5 + 7i)(5 - 7i)$
   A) $25 - 49i$
   B) 74
   C) $25 - 49i^2$
   D) -24

6) $\dfrac{3}{4i}$
   A) $-\dfrac{3}{4}$
   B) $\dfrac{3}{4}i$
   C) $-\dfrac{3}{4}i$
   D) $\dfrac{3}{4}$

7) $\dfrac{4 - i}{2i}$
   A) $\dfrac{1}{2} + 2i$
   B) $\dfrac{1}{2} - 2i$
   C) $-\dfrac{1}{2} - 2i$
   D) $-\dfrac{1}{2} + 2i$

8) $\dfrac{8}{1 - 8i}$
   A) $-\dfrac{8}{63} - \dfrac{64}{63}i$
   B) $\dfrac{8}{65} + \dfrac{64}{65}i$
   C) $\dfrac{8}{65} - \dfrac{64}{65}i$
   D) $-\dfrac{8}{63} + \dfrac{64}{63}i$

9) $\dfrac{9}{5+i}$

    A) $\dfrac{45}{26} + \dfrac{9}{26}i$ B) $\dfrac{45}{26} - \dfrac{9}{26}i$ C) $\dfrac{15}{8} + \dfrac{3}{8}i$ D) $\dfrac{15}{8} - \dfrac{3}{8}i$

10) $\dfrac{-2 - 16i}{1 - 2i}$

    A) $6 - 4i$ B) $-4 + 6i$ C) $-4 - 6i$ D) $6 + 4i$

11) $\dfrac{9 + 5i}{9 - 4i}$

    A) $\dfrac{61}{97} + \dfrac{81}{97}i$ B) $\dfrac{61}{65} - \dfrac{81}{65}i$ C) $\dfrac{101}{65} - \dfrac{81}{65}i$ D) $\dfrac{101}{97} - \dfrac{9}{97}i$

12) $\left(\dfrac{\sqrt{2}}{2} + \dfrac{\sqrt{2}}{2}i\right)^2$

    A) $-\dfrac{i}{2}$ B) i C) $\dfrac{i}{2}$ D) $-i$

13) $(4 + 7i)^2$

    A) $16 + 56i + 49i^2$ B) $-33 + 56i$ C) $65 + 56i$ D) -33

14) i^{24}

    A) 1 B) -1 C) $-i$ D) i

15) i^{29}

    A) $-i$ B) -1 C) 1 D) i

16) i^{-31}

    A) i B) 1 C) -1 D) $-i$

17) i^{27}

    A) i B) $-i$ C) -1 D) 1

18) $2i^{15} - i^7$

    A) 1 B) i C) -1 D) $-i$

19) $5i^5(1 + i^3)$

    A) $-5 + 5i$ B) $-5 - 5i$ C) $5 - 5i$ D) $5 + 5i$

20) $(1 + i)^3$

    A) $-2 + i$ B) $-2 + 2i$ C) $-2 - 2i$ D) $2 + 2i$

21) $i^{14} + i^{12} + i^{10} + 1$

    A) -1 B) 1 C) i D) 0

Perform the indicated operations and express your answer in the form a + bi.

22) $\sqrt{-16}$

    A) $-4i$ B) $4i$ C) $\{4, -4\}$ D) $-2i$

23) $\sqrt{(5i - 12)(12 + 5i)}$

    A) $\sqrt{119}i$ B) 13 C) $-13i$ D) $13i$

24) $\sqrt{(2 + i)(i - 2)}$

    A) $-\sqrt{5}i$ B) $-\sqrt{5}$ C) $\sqrt{5}$ D) $\sqrt{5}i$

Write the expression in the standard form a + bi.

25) If $z = 7 - 2i$, evaluate $z + \overline{z}$.

    A) 14 B) $-4i$ C) $14 - 4i$ D) $14 + 4i$

26) If $w = 9 + 2i$, evaluate $w - \overline{w}$.

    A) 0 B) $-18 + 4i$ C) 18 D) $4i$

27) If $z = 9 - 4i$, evaluate $z\overline{z}$.

    A) $81 - 16i$ B) 97 C) 65 D) $81 - 16i^2$

28) If $z = 3 + 9i$ and $w = -2 + i$, evaluate $\overline{z - w}$.

    A) $-5 - 8i$ B) $1 + 10i$ C) $5 - 8i$ D) $5 + 8i$

29) If $z = 2 - 5i$ and $w = 7 + 2i$, evaluate $\overline{\overline{z} + \overline{w}}$.

    A) $9 - 3i$ B) $-9 + 3i$ C) $-5 + 7i$ D) $9 + 3i$

2 Solve Quadratic Equations with a Negative Discriminant

Solve the equation in the complex number system.

1) $x^2 + 16 = 0$

    A) $\{-4, 4\}$ B) $\{4i\}$ C) $\{-4i, 4i\}$ D) $\{4\}$

2) $x^2 + 8x + 32 = 0$

    A) $\{-4 - 4i, -4 + 4i\}$ B) $\{-4 - 16i, -4 + 16i\}$ C) $\{-8, 0\}$ D) $\{-4 + 4i\}$

3) $x^2 + x + 3 = 0$

    A) $\{\frac{1}{2} - \frac{\sqrt{11}}{2}i, \frac{1}{2} + \frac{\sqrt{11}}{2}i\}$

    B) $\{\frac{-1 - \sqrt{11}}{2}, \frac{-1 + \sqrt{11}}{2}\}$

    C) $\{-\frac{1}{2} - \frac{\sqrt{11}}{2}i, -\frac{1}{2} + \frac{\sqrt{11}}{2}i\}$

    D) $\{\frac{1 - \sqrt{11}}{2}, \frac{1 + \sqrt{11}}{2}\}$

4) $16x^2 - 7x + 1 = 0$

    A) $\{-\frac{7}{32} - \frac{\sqrt{15}}{32}i, -\frac{7}{32} + \frac{\sqrt{15}}{32}i\}$

    B) $\{\frac{7}{32} - \frac{\sqrt{15}}{32}i, -\frac{7}{32} + \frac{\sqrt{15}}{32}i\}$

    C) $\{\frac{7}{32} - \frac{\sqrt{15}}{32}i, \frac{7}{32} + \frac{\sqrt{15}}{32}i\}$

    D) $\{-\frac{7}{32} - \frac{\sqrt{15}}{32}i, \frac{7}{32} + \frac{\sqrt{15}}{32}i\}$

Precalculus Enhanced with Graphing Utilities 937

5) $x^3 - 729 = 0$

   A) $\{9, -\frac{9}{2} - \frac{9\sqrt{3}}{2}i, -\frac{9}{2} + \frac{9\sqrt{3}}{2}i\}$
   B) $\{9, -\frac{9}{2} - \frac{9\sqrt{3}}{2}, -\frac{9}{2} + \frac{9\sqrt{3}}{2}\}$
   C) $\{9\}$
   D) $\{9, -9i, 9i\}$

6) $x^4 - 625 = 0$

   A) $\{-5, 5, 5i\}$ B) $\{-5, 5, -5i, 5i\}$ C) $\{-5, 5\}$ D) $\{5\}$

7) $x^4 - 6x^2 - 7 = 0$

   A) $\{\sqrt{7}, 7\}$ B) $\{-\sqrt{7}, \sqrt{7}, i, -i\}$ C) $\{-\sqrt{7}i, -i\}$ D) $\{\sqrt{7}i, i\}$

Without solving, determine the character of the solutions of the equation in the complex number system.

8) $x^2 + 3x - 4 = 0$

   A) two unequal real solutions
   B) two complex solutions that are conjugates of each other
   C) a repeated real solution

9) $x^2 + 2x + 1 = 0$

   A) two unequal real solutions
   B) a repeated real solution
   C) two complex solutions that are conjugates of each other

10) $x^2 + 2x + 4 = 0$

    A) two unequal real solutions
    B) two complex solutions that are conjugates of each other
    C) a repeated real solution

11) $x^2 - 6x + 6 = 0$

    A) two complex solutions that are conjugates of each other
    B) a repeated real solution
    C) two unequal real solutions

Solve the problem.

12) $2 - i$ is a solution of a quadratic equation with real coefficients. Find the other solution.

    A) $-2 - i$ B) $-2 + i$ C) $2 + i$ D) $2 - i$

14.7 Problem Solving

1 Translate Verbal Descriptions into Mathematical Expressions

Translate the sentence into a mathematical equation. Be sure to identify the meaning of all symbols.

1) The surface area of a sphere is 4π times the square of the radius.

   A) If S represents the surface area and r the radius, then $S = 4\pi r^2$.
   B) If S represents the surface area and r the radius, then $S = 4\pi r$.
   C) If S represents the surface area and r the radius, then $4\pi S = r^2$.
   D) If S represents the surface area and r the radius, then $S = \pi r^2$.

2) The volume of a right prism is the area of the base times the height of the prism.

   A) If V represents the volume, B the area of the base, and h the height, then $V = B + h$.

   B) If V represents the volume, B the area of the base, and h the height, then $V = \frac{1}{2}Bh$.

   C) If V represents the volume, B the area of the base, and h the height, then $V = \frac{B}{h}$.

   D) If V represents the volume, B the area of the base, and h the height, then $V = Bh$.

3) Speed is measured by distance divided by time.

   A) If S represents speed, d distance, and t time, then $t = \frac{S}{d}$.

   B) If S represents speed, d distance, and t time, then $S = \frac{d}{t}$.

   C) If S represents speed, d distance, and t time, then $d = \frac{S}{t}$.

   D) If S represents speed, d distance, and t time, then $S = \frac{t}{d}$.

4) Momentum is the product of the mass of an object and its velocity.

   A) If M represents momentum, m mass, and v velocity, then $M = \frac{m}{v}$.

   B) If M represents momentum, m mass, and v velocity, then $M = m + v$.

   C) If M represents momentum, m mass, and v velocity, then $M = \frac{1}{2}mv$.

   D) If M represents momentum, m mass, and v velocity, then $M = mv$.

5) The force of gravity between two objects is the gravitational constant times the product of their masses divided by the square of the distance between them.

   A) If F is the force of gravity, G the gravitational constant, m_1 the mass of one object, m_2 the mass of the second, and d the distance between them, then $F = G\frac{m_1 m_2}{d^2}$.

   B) If F is the force of gravity, G the gravitational constant, m_1 the mass of one object, m_2 the mass of the second, and d the distance between them, then $F = G\frac{m_1 + m_2}{d^2}$.

   C) If F is the force of gravity, G the gravitational constant, m_1 the mass of one object, m_2 the mass of the second, and d the distance between them, then $FG = \frac{m_1 m_2}{d^2}$.

   D) If F is the force of gravity, G the gravitational constant, m_1 the mass of one object, m_2 the mass of the second, and d the distance between them, then $F = G\frac{m_1 m_2}{d}$.

6) The total cost of producing refrigerators in one production line is $3600 plus $450 per unit produced.

   A) If C is the total cost and x is the number of units produced, then $C = \dfrac{3600}{450x}$.

   B) If C is the total cost and x is the number of units produced, then $C = 3600x + 450$.

   C) If C is the total cost and x is the number of units produced, then $C = (3600 + 450)x$.

   D) If C is the total cost and x is the number of units produced, then $C = 3600 + 450x$.

7) The profit derived from the sale of x video cameras is $470 per unit less the sum of $2300 costs plus $140 per unit.

   A) If P is profit and x the units sold, then $P = \dfrac{470}{x} - (2300 + \dfrac{140}{x})$ or $P = \dfrac{330}{x} - 2300$.

   B) If P is profit and x the units sold, then $P = 470x - (2300 - 140x)$ or $P = 610x - 2300$.

   C) If P is profit and x the units sold, then $P = 470x + 2300 - 140x$ or $P = 330x + 2300$.

   D) If P is profit and x the units sold, then $P = 470x - (2300 + 140x)$ or $P = 330x - 2300$.

2 Solve Interest Problems

Solve the problem.

1) Don James wants to invest $60,000 to earn $7110 per year. He can invest in B-rated bonds paying 15% per year or in a Certificate of Deposit (CD) paying 8% per year. How much money should be invested in each to realize exactly $7110 in interest per year?

   A) $34,000 in B-rated bonds and $26,000 in a CD   B) $27,000 in B-rated bonds and $33,000 in a CD
   C) $26,000 in B-rated bonds and $34,000 in a CD   D) $33,000 in B-rated bonds and $27,000 in a CD

2) A bank loaned out $58,000, part of it at the rate of 14% per year and the rest at a rate of 6% per year. If the interest received was $5880, how much was loaned at 14%?

   A) $28,000   B) $30,000   C) $27,000   D) $31,000

3) A loan officer at a bank has $94,000 to lend and is required to obtain an average return of 13% per year. If he can lend at the rate of 14% or the rate of 11%, how much can he lend at the 11% rate and still meet his required return?

   A) $846,000.00   B) $6714.29   C) $31,333.33   D) $3760.00

4) A college student earned $9000 during summer vacation working as a waiter in a popular restaurant. The student invested part of the money at 10% and the rest at 9%. If the student received a total of $838 in interest at the end of the year, how much was invested at 10%?

   A) $2800   B) $4500   C) $1000   D) $6200

5) Susan purchased some municipal bonds yielding 7% annually and some certificates of deposit yielding 9% annually. If Susan's investment amounts to $19,000 and the annual income is $1590, how much money is invested in bonds and how much is invested in certificates of deposit?

   A) $13,000 in bonds; $6000 in certificates of deposit   B) $6000 in bonds; $13,000 in certificates of deposit
   C) $13,500 in bonds; $5500 in certificates of deposit   D) $5500 in bonds; $13,500 in certificates of deposit

6) Martin purchased some municipal bonds yielding 8% annually and some certificates of deposit yielding 11% annually. If Martin's investment amounts to $23,000 and the annual income is $2230, how much money is invested in bonds and how much is invested in certificates of deposit?

   A) $10,000 in bonds; $13,000 in certificates of deposit
   B) $8000 in bonds; $15,000 in certificates of deposit
   C) $15,000 in bonds; $8000 in certificates of deposit
   D) $13,000 in bonds; $10,000 in certificates of deposit

7) Kevin invested part of his $10,000 bonus in a certificate of deposit that paid 6% annual simple interest, and the remainder in a mutual fund that paid 11% annual simple interest. If his total interest for that year was $700, how much did Kevin invest in the mutual fund?

   A) $8000   B) $1000   C) $3000   D) $2000

3 Solve Mixture Problems

Solve the problem.

1) The manager of a coffee shop has one type of coffee that sells for $7 per pound and another type that sells for $15 per pound. The manager wishes to mix 30 pounds of the $15 coffee to get a mixture that will sell for $10 per pound. How many pounds of the $7 coffee should be used?

   A) 50 lb   B) 40 lb   C) 80 lb   D) 25 lb

2) The owners of a candy store want to sell, for $6 per pound, a mixture of chocolate-covered raisins, which usually sells for $3 per pound, and chocolate-covered macadamia nuts, which usually sells for $8 per pound. They have a 50-pound barrel of the raisins. How many pounds of the nuts should they mix with the barrel of raisins so that they hit their target value of $6 per pound for the mixture?

   A) 80 lb   B) 75 lb   C) 70 lb   D) 65 lb

3) The manager of a candy shop sells chocolate covered peanuts for $7 per pound and chocolate covered cashews for $15 per pound. The manager wishes to mix 60 pounds of the cashews to get a cashew-peanut mixture that will sell for $10 per pound. How many pounds of peanuts should be used?

   A) 80 lb   B) 100 lb   C) 50 lb   D) 160 lb

4) A chemist needs 50 milliliters of a 29% solution but has only 27% and 32% solutions available. Find how many milliliters of each that should be mixed to get the desired solution.

   A) 20 ml of 27%; 30 ml of 32%
   B) 40 ml of 27%; 10 ml of 32%
   C) 30 ml of 27%; 20 ml of 32%
   D) 10 ml of 27%; 40 ml of 32%

5) How much pure acid should be mixed with 2 gallons of a 50% acid solution in order to get an 80% acid solution?

   A) 1 gal   B) 8 gal   C) 3 gal   D) 5 gal

6) The radiator in a certain make of car needs to contain 70 liters of 40% antifreeze. The radiator now contains 70 liters of 20% antifreeze. How many liters of this solution must be drained and replaced with 100% antifreeze to get the desired strength?

   A) 17.5 L   B) 35 L   C) 23.3 L   D) 28 L

7) How many gallons of a 30% alcohol solution must be mixed with 60 gallons of a 14% solution to obtain a solution that is 20% alcohol?

   A) 27 gal   B) 12 gal   C) 36 gal   D) 7 gal

8) How many liters of 80% hydrochloric acid must be mixed with 40% hydrochloric acid to get 15 liters of 65% hydrochloric acid? Write your answer rounded to three decimals.

   A) 8 L   B) 4.688 L   C) 3.125 L   D) 9.375 L

4 Solve Uniform Motion Problems

Solve the problem.

1) An airplane flies 420 miles with the wind and 340 against the wind in the same length of time. If the speed of the wind is 30, what is the speed of the airplane in still air?

   A) 290 mph   B) 127.5 mph   C) 275 mph   D) 285 mph

2) A boat heads upstream a distance of 30 miles on the Mississippi river, whose current is running at 5 miles per hour. If the trip back takes an hour less, what was the speed of the boat in still water? Give the answer rounded to two decimal places, if necessary.

   A) 18.03 mph   B) 15 mph   C) 16.58 mph   D) 6 mph

3) Two friends decide to meet in Chicago to attend a Cub's baseball game. Rob travels 210 miles in the same time that Carl travels 175 miles. Rob's trip uses more interstate highways and he can average 7 mph more than Carl. What is Rob's average speed?

   A) 35 mph   B) 42 mph   C) 34 mph   D) 50 mph

4) Gary can hike on level ground 3 miles an hour faster than he can on uphill terrain. Yesterday, he hiked 27 miles, spending 2 hours on level ground and 5 hours on uphill terrain. Find his average speed on level ground.

   A) 3 mph   B) 6 mph   C) $3\frac{6}{7}$ mph   D) $6\frac{3}{7}$ mph

5) Two cars start from the same point and travel in the same direction. If one car is traveling 60 miles per hour and the other car is traveling at 46 miles per hour, how far apart will they be after 8.7 hours?

   A) 922.2 mi   B) 400.2 mi   C) 121.8 mi   D) 522 mi

6) Two trains leave a train station at the same time. One travels east at 8 miles per hour. The other train travels west at 7 miles per hour. In how many hours will the two trains be 121.5 miles apart?

   A) 8.1 hr   B) 4.1 hr   C) 8.6 hr   D) 16.2 hr

7) Ken and Kara are 27 miles apart on a calm lake paddling toward each other. Ken paddles at 4 miles per hour, while Kara paddles at 7 miles per hour. How long will it take them to meet?

   A) $2\frac{5}{11}$ hr   B) 16 hr   C) $1\frac{7}{8}$ hr   D) 9 hr

8) A freight train leaves a station traveling at 32 km/h. Two hours later, a passenger train leaves the same station traveling in the same direction at 52 km/h. How long does it takes the passenger train to catch up to the freight train?

   A) 3.2 hr   B) 5.2 hr   C) 4.2 hr   D) 2.2 hr

9) Five friends drove at an average rate of 50 miles per hour to a weekend retreat. On the way home, they took the same route but averaged 75 miles per hour. What was the distance between home and the retreat if the round trip took 10 hours?

   A) 1500 mi   B) 600 mi   C) 6 mi   D) 300 mi

10) During a hurricane evacuation from the east coast of Georgia, a family traveled 240 miles west. For part of the trip, they averaged 60 mph, but as the congestion got bad, they had to slow to 20 mph. If the total time of travel was 8 hours, how many miles did they drive at the reduced speed?

   A) 120 mi B) 125 mi C) 115 mi D) 130 mi

11) Richard works for a company that pays $34 per hour for the first forty hours and $51 per hour for each hour in the week worked above the 40 hours. If he earned $1717 this week, how many overtime hours did he work?

   A) 6 hr B) 47 hr C) 7 hr D) 10.5 hr

12) Jamie sells handcrafted dolls at local art fairs. She sells small dolls for $15 and large dolls for $50. At the end of the Little Town Art Fair, she determined that the total amount she made by selling 15 dolls was $330. Determine the number of small and the number of large dolls that she sold.

   A) 12 small, 3 large B) 11 small, 4 large C) 3 small, 12 large D) 4 small, 11 large

5 Solve Constant Rate Job Problems

Solve the problem.

1) An experienced bank auditor can check a bank's deposits twice as fast as a new auditor. Working together it takes the auditors 10 hours to do the job. How long would it take the experienced auditor working alone?

   A) 20 hr B) 10 hr C) 30 hr D) 15 hr

2) Bob can overhaul a boat's diesel inboard engine in 20 hours. His apprentice takes 60 hours to do the same job. How long would it take them working together assuming no gain or loss in efficiency?

   A) 15 hr B) 12 hr C) 80 hr D) 6 hr

3) Tracy can wallpaper 5 rooms in a new house in 20 hours. Together with her trainee they can wallpaper the 5 rooms in 14 hours. How long would it take the trainee working by herself to do the job?

   A) 100 hr B) 70 hr C) 50 hr D) 20 hr

4) Brandon can paint a fence in 12 hours and Elaine can paint the same fence in 11 hours. How long will they take to paint the fence if they work together? Give your answer rounded to one decimal place, if necessary.

   A) 11.5 hr B) 5.5 hr C) 5.7 hr D) 5.8 hr

5) Sue can sew a precut dress in 3 hours. Helen can sew the same dress in 2 hours. If they work together, how long will it take them to complete sewing that dress? Give your answer rounded to one decimal place, if necessary.

   A) 5 hr B) 1.2 hr C) 1.8 hr D) 2.5 hr

6) Two pumps can fill a water tank in 45 minutes when working together. Alone, the second pump takes 3 times longer than the first to fill the tank. How long does it take the first pump alone to fill the tank?

   A) 67.5 min B) 60 min C) 33.75 min D) 80 min

6 Solve Problems Involving Quadratic Equations

Solve the problem.

1) The area of a vegetable garden is 112 square feet. If the length of the garden is 6 feet longer than its width, what are the dimensions?

   A) 8 ft by 14 ft B) 9 ft by 15 ft C) 7 ft by 15 ft D) 7 ft by 13 ft

2) Find the dimensions of a rectangle whose perimeter is 34 meters and whose area is 66 square meters.

   A) 6 m by 11 m B) 5 m by 10 m C) 7 m by 10 m D) 5 m by 12 m

3) Suppose that an open box is to be made from a square sheet of cardboard by cutting out 5-inch squares from each corner as shown and then folding along the dotted lines. If the box is to have a volume of 180 cubic inches, find the original dimensions of the sheet of cardboard.

   A) 16 in. by 16 in. B) 6 in. by 6 in. C) $6\sqrt{5}$ in. by $6\sqrt{5}$ in. D) $\sqrt{6}$ in. by $\sqrt{30}$ in.

4) A 31-inch-square TV is on sale at the local electronics store. If 31 inches is the measure of the diagonal of the screen, use the Pythagorean theorem to find the length of the side of the screen.

   A) $\sqrt{31}$ in. B) $\dfrac{961}{2}$ in. C) $\dfrac{\sqrt{31}}{2}$ in. D) $\dfrac{31\sqrt{2}}{2}$ in.

5) A ball is thrown vertically upward from the top of a building 144 feet tall with an initial velocity of 128 feet per second. The distance s (in feet) of the ball from the ground after t seconds is $s = 144 + 128t - 16t^2$. After how many seconds does the ball strike the ground?

   A) 11 sec B) 9 sec C) 145 sec D) 8 sec

6) As part of a physics experiment, Ming drops a baseball from the top of a 300-foot building. To the nearest tenth of a second, for how many seconds will the baseball fall? (Hint: Use the formula $h = 16t^2$, which gives the distance h, in feet, that a free-falling object travels in t seconds.)

   A) 18.8 sec B) 75 sec C) 1.1 sec D) 4.3 sec

7) A circular pool measures 10 feet across. One cubic yard of concrete is to be used to create a circular border of uniform width around the pool. If the border is to have a depth of 2 inches, how wide will the border be? Use 3.14 to approximate π. Express your solution rounded to two decimal places. (1 cubic yard = 27 cubic feet)

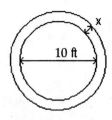

   A) 3.75 ft B) 3.6 ft C) 10.34 ft D) 6.78 ft

8) If a polygon, of n sides has $\dfrac{1}{2}n(n-3)$ diagonals, how many sides will a polygon with 629 diagonals have?

   A) 36 sides B) 39 sides C) 37 sides D) 38 sides

14.8 Interval Notation; Solving Inequalities

1 Use Interval Notation

Express the graph shown using interval notation. Also express it as an inequality involving x.

1)

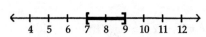

   A) [7, 9)   B) [7, 9]   C) (7, 9)   D) (7, 9]
      $7 \le x < 9$ $7 \le x \le 9$ $7 < x < 9$ $7 < x \le 9$

2)

A) (-2, 3]
  $-2 < x \leq 3$

B) [-2, 3]
  $-2 \leq x \leq 3$

C) (-2, 3)
  $-2 < x < 3$

D) [-2, 3)
  $-2 \leq x < 3$

3)

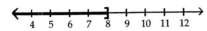

A) (-∞, 8]
  $x \leq 8$

B) (-∞, 8)
  $x < 8$

C) (8, ∞)
  $x > 8$

D) [8, ∞)
  $x \geq 8$

4)

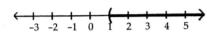

A) (-∞, 1)
  $x < 1$

B) (1, ∞)
  $x > 1$

C) (-∞, 1]
  $x \leq 1$

D) [1, ∞)
  $x \geq 1$

5)

A) [-8, -4]
  $-8 \leq x \leq -4$

B) (-8, -4]
  $-8 < x \leq -4$

C) (-8, -4)
  $-8 < x < -4$

D) [-8, -4)
  $-8 \leq x < -4$

6)

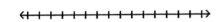

A) [-6, 3]
  $-6 \leq x \leq 3$

B) (-∞, 3)
  $x < 3$

C) [-6, 3)
  $-6 \leq x < 3$

D) (-6, 3]
  $-6 < x \leq 3$

Write the inequality using interval notation, and illustrate the inequality using the real number line.

7) $5 < x < 8$

A) [5, 8)

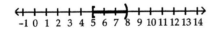

B) (5, 8)

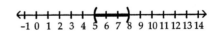

C) (5, 8)

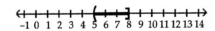

D) [5, 8]

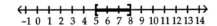

8) $3 \leq x \leq 6$

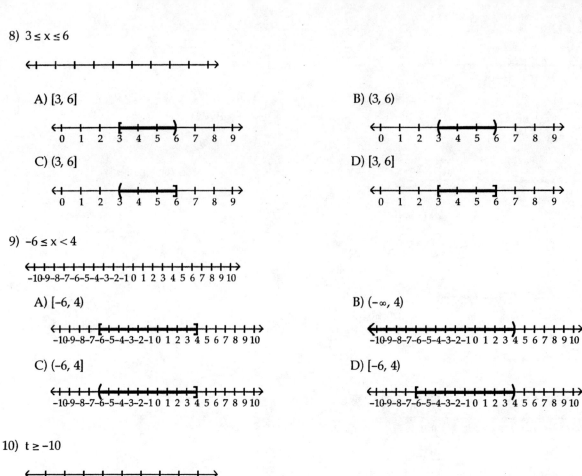

9) $-6 \leq x < 4$

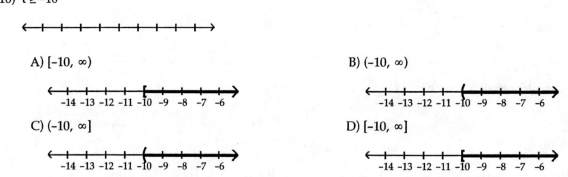

10) $t \geq -10$

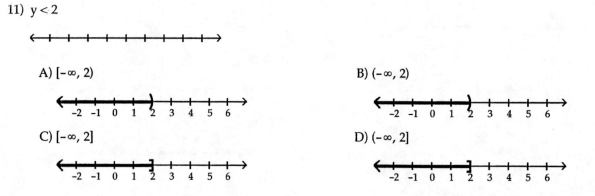

11) $y < 2$

A) $[-\infty, 2)$

B) $(-\infty, 2)$

C) $[-\infty, 2]$

D) $(-\infty, 2]$

Write the interval as an inequality involving x, and illustrate the inequality using the real number line.

12) [−9, 4)

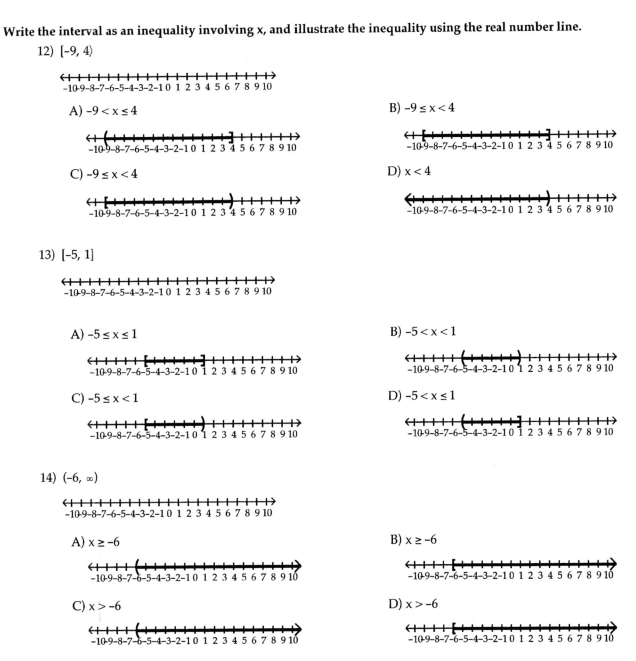

A) −9 < x ≤ 4

B) −9 ≤ x < 4

C) −9 ≤ x < 4

D) x < 4

13) [−5, 1]

A) −5 ≤ x ≤ 1

B) −5 < x < 1

C) −5 ≤ x < 1

D) −5 < x ≤ 1

14) (−6, ∞)

A) x ≥ −6

B) x ≥ −6

C) x > −6

D) x > −6

15) [−6, ∞)

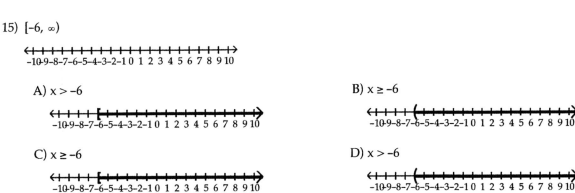

A) x > −6

B) x ≥ −6

C) x ≥ −6

D) x > −6

Precalculus Enhanced with Graphing Utilities 947

16) $(-\infty, 6)$

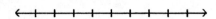

A) $x \leq 6$

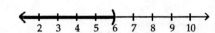

B) $x < 6$

C) $x \leq 6$

D) $x < 6$

2 Use Properties of Inequalities

Write the inequality obtained by performing the indicated operation on the given inequality.

1) Subtract 2 from each side of the inequality $1 - 5x > -1$.

   A) $-1 - 7x < -3$ B) $-1 - 5x < -3$ C) $-1 - 7x > -3$ D) $-1 - 5x > -3$

2) Multiply each side of the inequality $4 - 4x \leq 1$ by -3.

   A) $-12 - 4x \geq -3$ B) $-12 + 12x \leq -3$ C) $-12 + 12x \geq -3$ D) $-12 - 4x \leq -3$

Fill in the blank with the correct inequality symbol.

3) If $x < 14$, then $x - 14$ ___ 0.

   A) $\geq$ B) $>$ C) $\leq$ D) $<$

4) If $x < -15$, then $x + 15$ ___ 0.

   A) $\geq$ B) $<$ C) $>$ D) $\leq$

5) If $x > -6$, then $6x$ ___ -36.

   A) $\leq$ B) $>$ C) $\geq$ D) $<$

6) If $x < 4$, then $-5x$ ___ -20.

   A) $\geq$ B) $>$ C) $\leq$ D) $<$

7) If $x > -8$, then $-2x$ ___ 16.

   A) $\leq$ B) $>$ C) $<$ D) $\geq$

8) If $-2x < 10$, then x ___ -5.

   A) $\geq$ B) $\leq$ C) $>$ D) $<$

3 Solve Linear Inequalities

Solve the inequality. Express your answer using interval notation. Graph the solution set.

1) $x - 3 < 3$

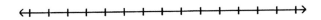

A) $(-\infty, 0]$

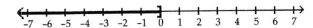

B) $(6, \infty)$

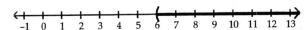

C) $(-\infty, 0)$

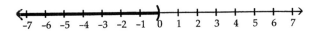

D) $(-\infty, 6)$

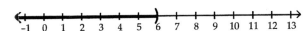

2) $x - 7 < -4$

A) $(-\infty, 3)$

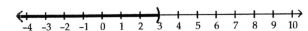

B) $(-\infty, -11)$

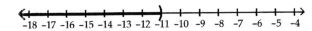

C) $(3, \infty)$

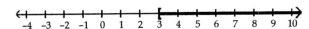

D) $(-\infty, 3]$

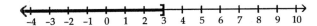

3) $4x + 2 < 38$

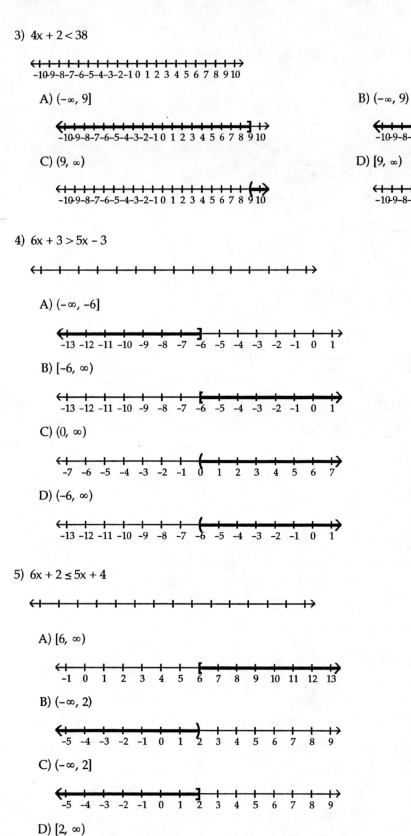

A) $(-\infty, 9]$

B) $(-\infty, 9)$

C) $(9, \infty)$

D) $[9, \infty)$

4) $6x + 3 > 5x - 3$

A) $(-\infty, -6]$

B) $[-6, \infty)$

C) $(0, \infty)$

D) $(-6, \infty)$

5) $6x + 2 \leq 5x + 4$

A) $[6, \infty)$

B) $(-\infty, 2)$

C) $(-\infty, 2]$

D) $[2, \infty)$

6) $-6x - 5 \geq -7x - 11$

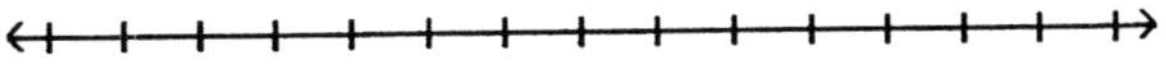

A) $(-16, \infty)$

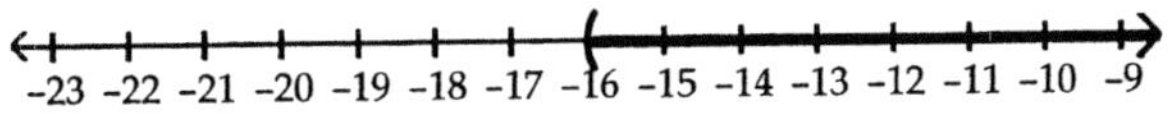

B) $[-6, \infty)$

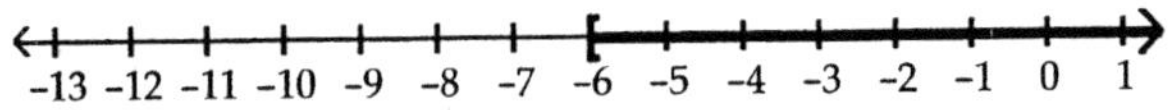

C) $(-\infty, -6)$

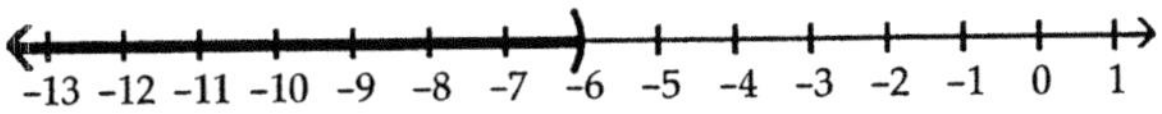

D) $(-\infty, -6]$

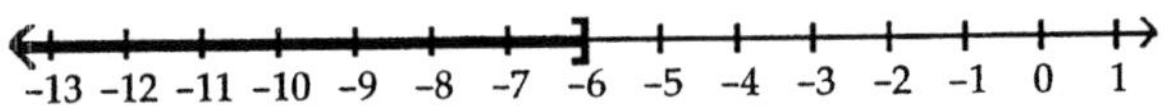

7) $7x - 7 > 6x - 5$

A) $[2, \infty)$

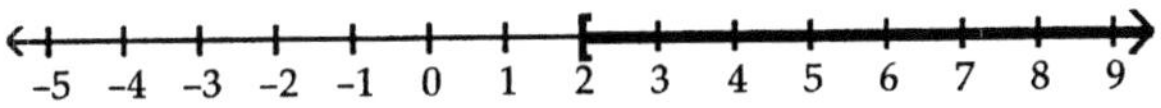

B) $(-\infty, 2]$

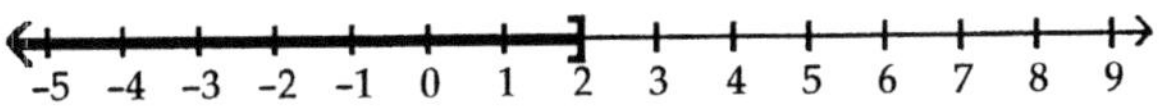

C) $(2, \infty)$

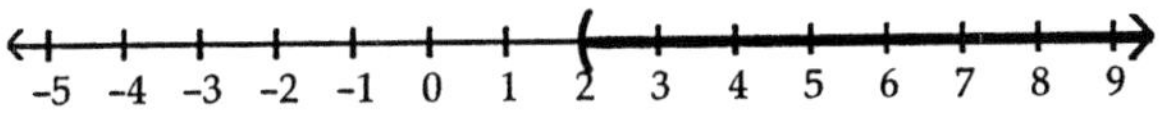

D) $(-12, \infty)$

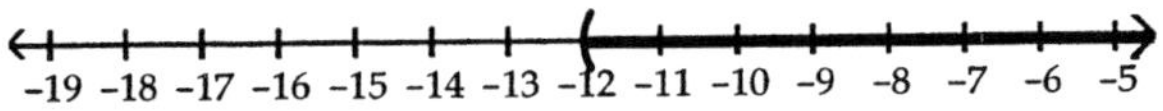

8) $8 - 3(2 - x) \leq 14$

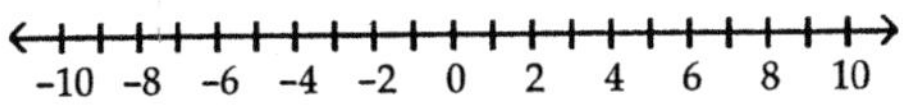

A) $(-\infty, 5]$

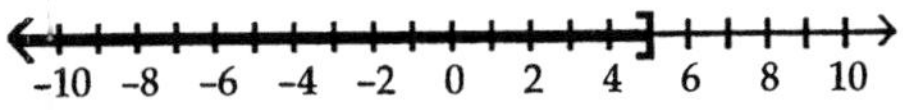

B) $(-\infty, 4]$

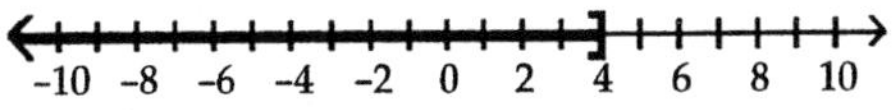

C) $(-\infty, 4)$

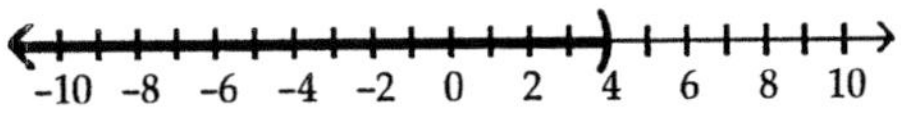

D) $[4, \infty)$

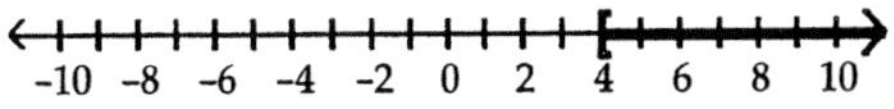

9) $-18x - 3 \leq -3(5x + 4)$

A) $(-\infty, 3)$

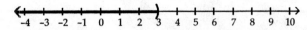

B) $[3, \infty)$

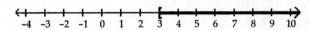

C) $(-\infty, 3]$

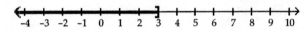

D) $[3, \infty)$

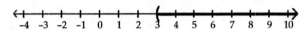

10) $-3(3x - 4) < -12x + 9$

A) $(-\infty, -1]$

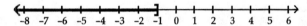

B) $(-\infty, -7]$

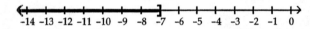

C) $(-\infty, -1)$

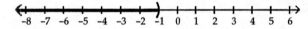

D) $(-1, \infty)$

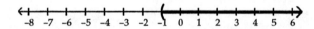

11) $\dfrac{x}{2} \geq 5 + \dfrac{x}{12}$

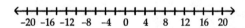

A) $(12, \infty)$

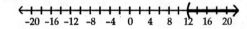

B) $(-\infty, 12]$

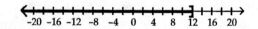

C) $[12, \infty)$

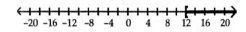

D) $[-12, \infty)$

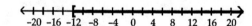

12) $x(16x - 2) \le (4x + 3)^2$

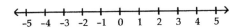

A) $(-\infty, -\frac{9}{26}]$

B) $(-\infty, -\frac{9}{26}]$

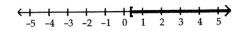

C) $[-\frac{9}{26}, \infty)$

D) $[\frac{9}{26}, \infty)$

4 Solve Combined Inequalities

Solve the inequality. Express your answer using interval notation. Graph the solution set.

1) $15 < 5x \le 25$

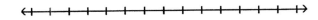

A) $[3, 5)$

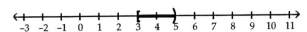

B) $(-5, -3]$

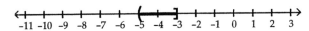

C) $(3, 5]$

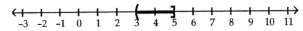

D) $[-5, -3)$

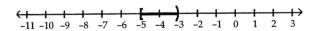

2) $-4 < x + 4 \le 2$

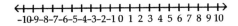

A) $[0, 6)$

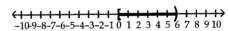

B) $[-8, -2)$

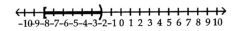

C) $(0, 6]$

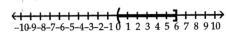

D) $(-8, -2]$

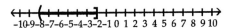

3) $11 \leq 4x - 1 \leq 23$

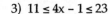

A) (3, 6)

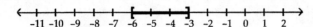

B) [−6, −3]

C) [3, 6]

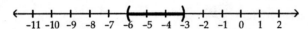

D) (−6, −3)

4) $-27 \leq -4x - 3 < -19$

A) [4, 6)

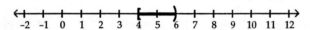

B) [−6, −4)

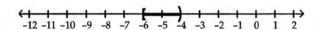

C) (−6, −4]

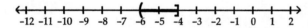

D) (4, 6]

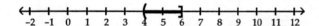

5) $-18 \leq -4x + 2 \leq -14$

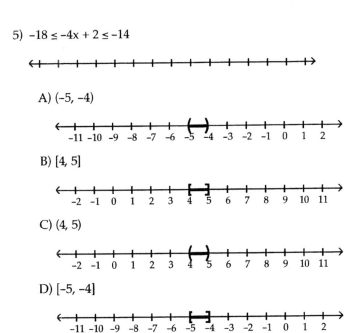

A) $(-5, -4)$

B) $[4, 5]$

C) $(4, 5)$

D) $[-5, -4]$

6) $0 \leq \dfrac{2x+5}{4} < 3$

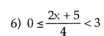

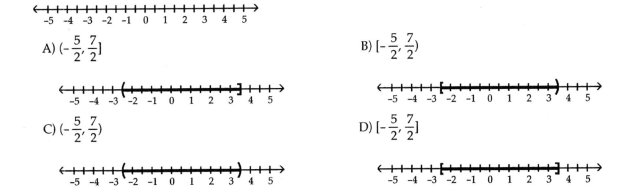

A) $(-\dfrac{5}{2}, \dfrac{7}{2}]$

B) $[-\dfrac{5}{2}, \dfrac{7}{2})$

C) $(-\dfrac{5}{2}, \dfrac{7}{2})$

D) $[-\dfrac{5}{2}, \dfrac{7}{2}]$

7) $-2 \leq \dfrac{2}{3}x - 4 < 0$

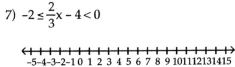

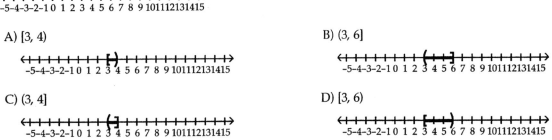

A) $[3, 4)$

B) $(3, 6]$

C) $(3, 4]$

D) $[3, 6)$

8) $-\dfrac{1}{3} \le \dfrac{3x-1}{9} < \dfrac{1}{3}$

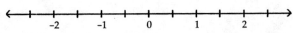

A) $\left(-\dfrac{2}{3}, \dfrac{4}{3}\right]$

B) $\left(-\dfrac{2}{3}, \dfrac{4}{3}\right)$

C) $\left[-\dfrac{2}{3}, \dfrac{4}{3}\right)$

D) $\left[-\dfrac{8}{9}, \dfrac{10}{9}\right)$

5 Solve Absolute Value Inequalities

Solve the inequality. Express your answer using interval notation. Graph the solution set.

1) $|x| < 3$

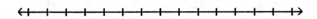

A) $(-\infty, 3]$

B) $[-3, 3]$

C) $(-\infty, -3) \cup (3, \infty)$

D) $(-3, 3)$

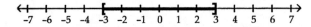

2) |x| > 3

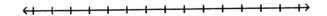

A) [3, ∞)

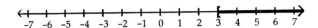

B) (−3, 3)

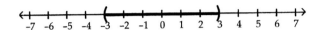

C) (−∞, −3) ∪ (3, ∞)

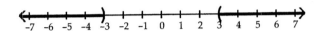

D) [−3, 3]

3) |x| > −2

A) [−2, ∞)

B) (−∞, −2) ∪ (2, ∞)

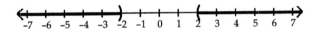

C) (−2, 2)

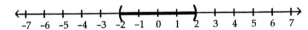

D) (−∞, ∞)

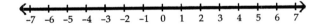

4) $|x| < -2$

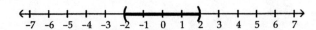

  A) $(-2, 2)$

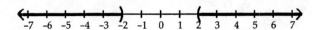

  B) $(-\infty, -2) \cup (2, \infty)$

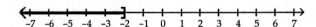

  C) $(-\infty, -2]$

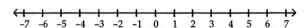

  D) $\emptyset$

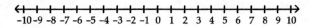

5) $|6x| < 36$

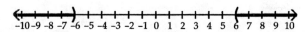

  A) $(-\infty, -6)$ or $(6, \infty)$

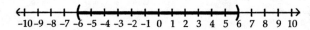

  B) $(-6, 6)$

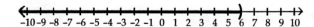

  C) $(-\infty, 6)$

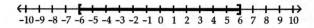

  D) $(-6, 6)$

6) $|5x| > 45$

A) $(-9, 9)$

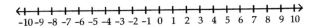

B) $(-\infty, -9]$ or $[9, \infty)$

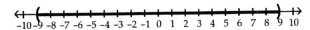

C) $(9, \infty)$

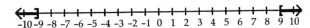

D) $(-\infty, -9)$ or $(9, \infty)$

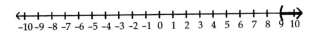

7) $|x - 10| < 17$

A) $(-27, 7)$

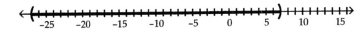

B) $(-\infty, -7)$

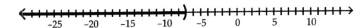

C) $(-\infty, 27)$

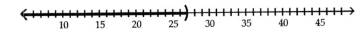

D) $(-7, 27)$

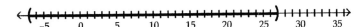

8) $|x + 5| > 2$

A) $(-\infty, -7) \cup (-3, \infty)$

B) $(-7, -3)$

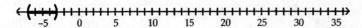

C) $(3, 7)$

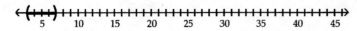

D) $(-3, \infty)$

9) $|2k + 8| \geq 5$

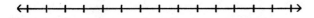

A) $[-\frac{3}{2}, \infty)$

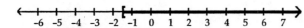

B) $(-\infty, -\frac{13}{2}] \cup [-\frac{3}{2}, \infty)$

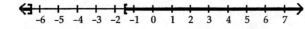

C) $(-\frac{13}{2}, -\frac{3}{2})$

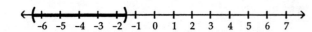

D) $[-\frac{13}{2}, -\frac{3}{2}]$

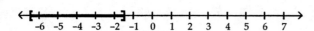

10) $|4k - 5| \le 6$

A) $(-\infty, -\frac{1}{4}] \cup [\frac{11}{4}, \infty)$

B) $(-\infty, \frac{11}{4}]$

C) $[-\frac{1}{4}, \frac{11}{4}]$

D) $(-\frac{1}{4}, \frac{11}{4})$

11) $|x - 4| - 2 \le 7$

A) $[-5, 7]$

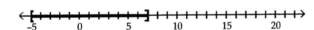

B) $(-5, 13)$

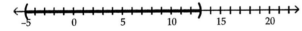

C) $[-5, 13]$

D) ∅

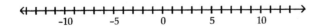

12) $|x - 4| + 8 \geq 12$

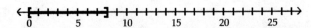

A) [0, 8]

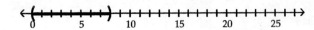

B) (0, 8)

C) [8, ∞)

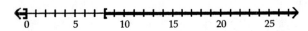

D) (-∞, 0] ∪ [8, ∞)

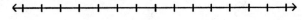

13) $|2k + 6| + 5 > 14$

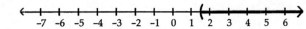

A) $(-\frac{15}{2}, \frac{3}{2})$

B) $(\frac{3}{2}, \infty)$

C) $(-\infty, -\frac{15}{2}] \cup [\frac{3}{2}, \infty)$

D) $(-\infty, -\frac{15}{2}) \cup (\frac{3}{2}, \infty)$

14) $|3k + 4| - 1 < 8$

A) $(-\infty, \frac{5}{3})$

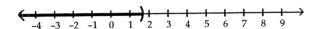

B) $(-\infty, -\frac{13}{3}) \cup (\frac{5}{3}, \infty)$

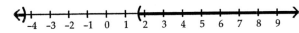

C) $(-\frac{13}{3}, \frac{5}{3})$

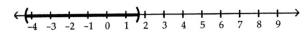

D) $(-\infty, -\frac{13}{3})$

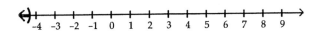

15) $|x - 4| \geq 0$

A) $(-\infty, 4) \cup (4, \infty)$

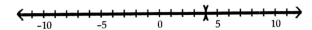

B) $(-\infty, -4) \cup (-4, \infty)$

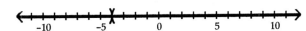

C) 4

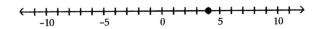

D) $(-\infty, \infty)$

6 Demonstrate Additional Skills and Understanding

Solve the problem.

1) Express the fact that x differs from −7 by more than 3 as an inequality involving absolute value. Solve for x.
   A) $|x + 7| > 3$; $\{x | -10 < x < -4\}$
   B) $|x + 7| > 3$; $\{x | x < -10 \text{ or } x > -4\}$
   C) $|x + 7| < 3$; $\{x | x < -10 \text{ or } x > -4\}$
   D) $|x + 7| < 3$; $\{x | -10 < x < -4\}$

2) A landscaping company sells 40-pound bags of top soil. The actual weight x of a bag, however, may differ from the advertised weight by as much as 0.75 pound. Write an inequality involving absolute value that expresses the relationship between the actual weight x of a bag and 40 pounds. Over what range may the weight of a 40-pound bag of to soil vary?

A) $|x - 40| \geq 0.75$; $\{x | x \leq 39.25 \text{ or } x \geq 40.75\}$

B) $|x - 40| \geq 0.75$; $\{x | 39.25 \leq x \leq 40.75\}$

C) $|x - 40| \leq 0.75$; $\{x | 39.25 \leq x \leq 40.75\}$

D) $|x - 40| < 0.75$; $\{x | 39.25 < x < 40.75\}$

3) In one city, the local cable TV company charges $1.84 for each pay-per-view movie watched. In addition, each monthly bill contains a basic customer charge of $18.50. If last month's bills ranged from a low of $24.02 to a high of $51.62, over what range did customers watch pay-per-view movies?

A) movies watched varied from 3 to 18 inclusive

B) movies watched varied from 4 to 19 inclusive

C) movies watched varied from 2 to 19 inclusive

D) movies watched varied from 2 to 17 inclusive

4) During the first five months of the year, Len earned commissions of $3550, $3650, $2000, $2730, and $3830. If Len must have average monthly earnings of at least $3120 in order to qualify for retirement benefits, what must he earn in the sixth month in order to qualify for benefits?

A) at least $3146   B) at least $3120   C) at least $3152   D) at least $2960

5) A real estate agent agrees to sell an office building according to the following commission schedule: $39,000 plus 25% of the selling price in excess of $800,000. Assuming that the office building will sell at some price between $800,000 and $1,200,000, inclusive, over what range does the agent's commission vary?

A) The commission will vary between $39,000 and $139,000, inclusive.

B) The commission will vary between $239,000 and $339,000, inclusive.

C) The commission will vary between $40,000 and $139,000, inclusive.

D) The commission will vary between $39,000 and $329,000, inclusive.

6) Jim has gotten scores of 85 and 92 on his first two tests. What score must he get on his third test to keep an average of 85 or better?

A) at least 87   B) at least 88.5   C) at least 78   D) at least 76

7) In his algebra class, Rob has scores of 79, 85, 81, and 65 on his first four tests. To get a grade of C, the average of the first five tests must be greater than or equal to 70 and less than 80. Solve an inequality to find the range of scores that Rob can earn on the fifth test to get a C.

A) $40 < x < 90$, where x represents Bob's score on the fifth test

B) $x \geq 40$, where x represents Bob's score on the fifth test

C) $40 \leq x < 90$, where x represents Bob's score on the fifth test

D) $40 \leq x \leq 90$, where x represents Bob's score on the fifth test

8) Marianne is planning a shopping trip to buy birthday gifts for her son. She estimates that the total price of the items she plans to purchase will be between $350 and $400 inclusive. If sales are taxed at a rate of 8.375% in her area, what is the range of the amount of sales tax she should expect to pay on her purchases? If Marianne's budget for the shopping trip is $425, will she necessarily be able to buy all the gifts that she has planned?

A) $\$29.31 \leq x \leq \$33.50$, where x represents the amount of sales tax; No.

B) $\$29.31 < x < \$33.50$, where x represents the amount of sales tax; Yes.

C) $\$29.31 < x < \$33.50$, where x represents the amount of sales tax; No.

D) $\$29.31 \leq x \leq \$33.50$, where x represents the amount of sales tax; Yes.

9) At Bargain Car Rental, the cost of renting an economy car for one day is $19.95 plus 20 cents per mile. At Best Deal Car Rental, the cost of renting a similar car for one day is $24.95 plus 15 cents per mile. Solve the inequality $24.95 + 0.15x < 19.95 + 0.20x$ to find the range of miles driven such that Best Deal is a better deal than Bargain.

A) $x > 100$ mi   B) $x < 100$ mi   C) $x < 10$ mi   D) $x > 10$ mi

14.9 nth Roots; Rational Exponents; Radical Equations

1 Work with nth Roots

Simplify the expression. Assume that all variables are positive when they appear.

1) $\sqrt[3]{1000}$

A) 100   B) ±10   C) 32   D) 10

2) $\sqrt[3]{-1000}$

A) −10   B) 100   C) 32   D) ±10

3) $\sqrt[4]{625}$

A) 5   B) 625   C) −5   D) not a real number

4) $\sqrt[3]{\frac{1}{27}}$

A) $\frac{1}{3}$ B) 3   C) $\frac{1}{\sqrt[3]{3}}$ D) $\frac{1}{9}$

5) $\sqrt[3]{27}$

A) 27   B) 9   C) 33   D) 3

6) $\sqrt[3]{320}$

A) $4\sqrt[3]{5}$ B) 20   C) $4\sqrt[3]{20}$ D) 4

2 Simplify Radicals

Simplify the expression. Assume that all variables are positive when they appear.

1) $\sqrt{6}$

A) $2\sqrt{3}$ B) $\sqrt{6}$ C) $3\sqrt{2}$ D) 2

2) $\sqrt{52}$

A) $4\sqrt{13}$ B) 26   C) $\sqrt{52}$ D) $2\sqrt{13}$

3) $\sqrt{64y^{10}}$

A) $8y^5$ B) $64y^5$ C) $8y^8$ D) $8y^{10}$

4) $\sqrt{567x^2}$

A) $567x$ B) $9x\sqrt{7}$ C) $9\sqrt{7x}$ D) $7x^2\sqrt{9}$

Precalculus Enhanced with Graphing Utilities 965

5) $\sqrt{\dfrac{20x^2y}{49}}$

A) $4x\sqrt{5y}$ B) $\dfrac{2x\sqrt{5y}}{7}$ C) $\dfrac{2\sqrt{5x^2y}}{7}$ D) $x\sqrt{\dfrac{20y}{7}}$

6) $\sqrt{y^{13}}$

A) $y^{12}\sqrt{y}$ B) $y\sqrt{y^{11}}$ C) $\sqrt{y^{13}}$ D) $y^6\sqrt{y}$

7) $\sqrt[3]{\dfrac{4}{y^{21}}}$

A) $\dfrac{\sqrt[3]{4}}{y^7}$ B) $\dfrac{4}{y^7}$ C) $\sqrt[3]{\dfrac{4}{y^{21}}}$ D) $\dfrac{\sqrt[3]{4}}{y^{18}}$

8) $\sqrt[3]{\dfrac{250}{x^{27}}}$

A) $\sqrt[3]{\dfrac{250}{x^{27}}}$ B) $\dfrac{5\sqrt[3]{2}}{x^{24}}$ C) $\dfrac{5\sqrt[3]{2}}{x^9}$ D) $\dfrac{\sqrt[3]{250}}{x^9}$

9) $\sqrt[3]{-216x^{12}}$

A) $-6x^{12}$ B) $-6x^9$ C) $-6x^4$ D) $6x^4$

10) $\sqrt[3]{p^{38}}$

A) $\sqrt[3]{p^{38}}$ B) $p^{12}\sqrt[3]{p^2}$ C) $p\sqrt[3]{p^{35}}$ D) p^{14}

11) $\sqrt{6}\cdot\sqrt{5}$

A) $\sqrt{6+5}$ B) $\sqrt{11}$ C) 30 D) $\sqrt{30}$

12) $\sqrt{3}\cdot\sqrt{4}$

A) $\sqrt{12}$ B) 2 C) $2\sqrt{3}$ D) $\sqrt{6}$

13) $\sqrt{32}\cdot\sqrt{72}$

A) 24 B) $24\sqrt{2}$ C) 48 D) 72

14) $\sqrt[3]{35}\cdot\sqrt[3]{49}$

A) $7\sqrt[3]{5}$ B) $\sqrt[6]{1715}$ C) $\sqrt[3]{1715}$ D) $7\sqrt[3]{35}$

15) $9\sqrt[3]{2}\cdot 7\sqrt[3]{3}$

A) $16\sqrt[3]{6}$ B) $63\sqrt[3]{6}$ C) $63\sqrt[3]{5}$ D) $63\sqrt[3]{2}\cdot\sqrt[3]{3}$

16) $\sqrt{7}(\sqrt{5}-\sqrt{3})$

A) $\sqrt{35}-\sqrt{21}$ B) $7\sqrt{5}+7\sqrt{3}$ C) $8\sqrt{7}$ D) $\sqrt{56}$

17) $(\sqrt{5}+2)(\sqrt{5}-2)$
   A) 3
   B) 1
   C) 9
   D) $5-2\sqrt{2}$

18) $(\sqrt{5}+\sqrt{13})(\sqrt{3}-\sqrt{13})$
   A) $\sqrt{15}+\sqrt{39}-\sqrt{65}+13$
   B) $\sqrt{15}+\sqrt{39}+\sqrt{65}-13$
   C) $\sqrt{15}+\sqrt{39}+\sqrt{65}+13$
   D) $\sqrt{15}+\sqrt{39}-\sqrt{65}-13$

19) $(\sqrt{13}+\sqrt{z})(\sqrt{13}-\sqrt{z})$
   A) $13-z$
   B) $13-2\sqrt{z}$
   C) $13z$
   D) $13-2\sqrt{13z}$

20) $(2\sqrt{2}-5)^2$
   A) $-17-20\sqrt{2}$
   B) $33-20\sqrt{2}$
   C) $3-20\sqrt{2}$
   D) $33+20\sqrt{2}$

21) $(4\sqrt{10})(7\sqrt{100})$
   A) $2800\sqrt{10}$
   B) $280\sqrt{100}$
   C) $280\sqrt{10}$
   D) $28\sqrt{1000}$

22) $-9\sqrt{3}-5\sqrt{75}$
   A) $34\sqrt{3}$
   B) $-34\sqrt{3}$
   C) $-14\sqrt{3}$
   D) $14\sqrt{3}$

23) $-5\sqrt{80}-5\sqrt{20}$
   A) $30\sqrt{5}$
   B) $-30\sqrt{5}$
   C) $10\sqrt{5}$
   D) $-10\sqrt{5}$

24) $\sqrt{5x^2}-4\sqrt{125x^2}-7\sqrt{125x^2}$
   A) $-11x\sqrt{5}$
   B) $-11x\sqrt{76}$
   C) $-54x\sqrt{5}$
   D) $-54x\sqrt{76}$

25) $\sqrt[3]{8y}-\sqrt[3]{128y}$
   A) $2-4\sqrt[3]{2}$
   B) $2\sqrt[3]{y}-4\sqrt[3]{2y}$
   C) $4\sqrt[3]{2y}-2\sqrt[3]{y}$
   D) $6\sqrt[3]{y}$

26) $4\sqrt[3]{8x}+4\sqrt[3]{27x}$
   A) $20x$
   B) $5\sqrt[3]{x}$
   C) $4\sqrt[3]{35x}$
   D) $20\sqrt[3]{x}$

3 Rationalize Denominators

Rationalize the denominator of the expression. Assume that all variables are positive when they appear.

1) $\dfrac{\sqrt{121}}{\sqrt{3}}$
   A) $11\sqrt{3}$
   B) $\dfrac{11\sqrt{3}}{3}$
   C) $\dfrac{121\sqrt{3}}{3}$
   D) 20

2) $\dfrac{8}{\sqrt{5}}$
   A) $\dfrac{8\sqrt{5}}{5}$
   B) $\dfrac{8}{5}$
   C) $\dfrac{64\sqrt{5}}{5}$
   D) $8\sqrt{5}$

3) $\dfrac{3}{9-\sqrt{5}}$

   A) $\dfrac{27+3\sqrt{5}}{-4}$ B) $\dfrac{27+3\sqrt{5}}{76}$ C) $\dfrac{1}{3}-\dfrac{3}{\sqrt{5}}$ D) $\dfrac{27-3\sqrt{5}}{76}$

4) $\dfrac{\sqrt{2}}{\sqrt{3}+7}$

   A) $\dfrac{3\sqrt{6}+3\sqrt{21}}{2}$ B) $\dfrac{\sqrt{6}-7\sqrt{2}}{10}$ C) $\dfrac{\sqrt{6}-7\sqrt{2}}{-46}$ D) $\dfrac{\sqrt{6}+7\sqrt{2}}{-46}$

5) $\dfrac{6-\sqrt{5}}{6+\sqrt{5}}$

   A) 1 B) $\dfrac{41+12\sqrt{5}}{31}$ C) $\dfrac{41-12\sqrt{5}}{31}$ D) $\dfrac{31-12\sqrt{5}}{41}$

6) $\dfrac{2}{8-\sqrt{3}}$

   A) $\dfrac{16-2\sqrt{3}}{61}$ B) $\dfrac{16+2\sqrt{3}}{61}$ C) $\dfrac{2}{8}-\dfrac{2}{\sqrt{3}}$ D) $\dfrac{16+2\sqrt{3}}{-5}$

7) $\dfrac{4}{\sqrt{x+h}-\sqrt{x}}$

   A) $\dfrac{4\sqrt{x+h}+\sqrt{x}}{h}$ B) $\dfrac{4(\sqrt{x+h}+\sqrt{x})}{h}$ C) $\dfrac{4(\sqrt{x+h}-\sqrt{x})}{h}$ D) $\dfrac{4\sqrt{h}}{h}$

4 Solve Radical Equations

Find the real solutions of the equation.

1) $\sqrt[3]{5x+1}=-3$

   A) $\{-\dfrac{32}{5}\}$ B) $\{\dfrac{8}{5}\}$ C) $\{-\dfrac{27}{5}\}$ D) $\{-\dfrac{28}{5}\}$

2) $\sqrt{7x+60}=x$

   A) {-5, 12} B) {-10} C) {12} D) no real solution

3) $\sqrt{20x-20}=x+4$

   A) {6} B) {7} C) {-5} D) {-6}

5 Simplify Expressions with Rational Exponents

Simplify the expression.

1) $64^{1/2}$

   A) 4 B) 8 C) 16 D) 32

2) $256^{1/4}$

   A) 1024 B) 64 C) 4 D) 16

3) $-32^{1/5}$

   A) 32   B) -8   C) 16   D) -2

4) $\left(\dfrac{1}{64}\right)^{1/3}$

   A) $\dfrac{-1}{4}$ B) -4   C) $\dfrac{1}{4}$ D) 4

5) $8^{4/3}$

   A) 128   B) 16   C) 64   D) 32

6) $16^{5/4}$

   A) 32   B) 256   C) 512   D) 128

7) $243^{4/5}$

   A) 2187   B) 19,683   C) 81   D) 6561

8) $\left(\dfrac{1}{27}\right)^{2/3}$

   A) $\dfrac{1}{9}$ B) $\dfrac{1}{13}$ C) $-\dfrac{1}{9}$ D) $\dfrac{1}{4}$

9) $(-27)^{4/3}$

   A) -81   B) 81   C) not a real number   D) 729

10) $16^{-3/2}$

    A) -64   B) $-\dfrac{1}{64}$ C) $\dfrac{1}{64}$ D) 64

11) $8^{-4/3}$

    A) $-\dfrac{1}{16}$ B) not a real number   C) 16   D) $\dfrac{1}{16}$

12) $16^{-5/4}$

    A) not a real number   B) $\dfrac{1}{32}$ C) 32   D) $-\dfrac{1}{32}$

13) $32^{-4/5}$

    A) 16   B) $-\dfrac{1}{16}$ C) not a real number   D) $\dfrac{1}{16}$

14) $-8^{-4/3}$

    A) $\dfrac{1}{16}$ B) 16   C) not a real number   D) $-\dfrac{1}{16}$

15) $-243^{-4/5}$

    A) not a real number   B) 81   C) $-\dfrac{1}{81}$ D) $\dfrac{1}{81}$

16) $125^{-1/3}$

  A) 25 B) 5 C) $\frac{1}{5}$ D) $\frac{1}{25}$

Simplify the expression. Express the answer so that only positive exponents occur. Assume that all variables are positive when they appear.

17) $(64x^{10})^{1/2}$

  A) $8\sqrt[5]{x}$ B) $64x^5$ C) $8x^5$ D) $8x^{10}$

18) $(-343x^9)^{1/3}$

  A) $-343x^3$ B) $-7x^9$ C) $-7\sqrt[3]{x}$ D) $-7x^3$

19) $x^{3/8} \cdot x^{5/8}$

  A) $x^{15/8}$ B) $x^{15/64}$ C) x D) $\frac{1}{x}$

20) $\dfrac{y^{3/4}}{y^{1/4}}$

  A) $\frac{1}{y}$ B) y C) $y^{3/4}$ D) $y^{1/2}$

21) $(x^7)^{2/7}$

  A) x^2 B) $x^{2/49}$ C) $x^{1/2}$ D) $x^{9/7}$

22) $z^{-2/7} \cdot z^{3/7}$

  A) $z^{7/6}$ B) $z^{6/7}$ C) $z^{1/7}$ D) $z^{-1/7}$

23) $(9x^{1/5} \cdot y^{1/5})^2$

  A) $81x^{1/10}y^{1/10}$ B) $81x^{2/5}y^{2/5}$ C) $81x^2y^2$ D) $81x^{1/25}y^{1/25}$

24) $y^{5/9}(y^{3/9} - 5y^{2/9})$

  A) $y^{8/9} - 5y^{7/9}$ B) $y^{15/81} - 5y^{10/81}$ C) $y^{5/3} - 5y^{5/2}$ D) $y^{2/9} - 5y^{3/9}$

25) $(4x^{2/7} - 5x^{3/7})(4x^{2/7} - 5x^{3/7})$

  A) $16x^{4/7} - 40x^{5/7} + 25x^{6/7}$
  B) $8x^{4/7} - 20x^{-10/7} - 10x^{6/7}$
  C) $16x^{4/49} + 5x$
  D) $16x^{4/7} - 20x^{5/7} + 25x^{6/7}$

26) $\dfrac{x^{1/7} \cdot x^{1/2}}{x^{-3/5}}$

  A) $x^{87/70}$ B) $\frac{1}{x^{87/70}}$ C) $\frac{1}{x^{3/70}}$ D) $x^{3/70}$

27) $\dfrac{(3x^{5/2})^5}{x^{2/7}}$

  A) $3x^{179/14}$ B) $243x^{171/14}$ C) $243x^{179/14}$ D) $3x^{171/14}$

28) $(x^2y^4)^{7/3}$
   A) $x^{6/7}y^{12/7}$
   B) $x^{13/3}y^{19/3}$
   C) $x^{28/3}y^{14/3}$
   D) $x^{14/3}y^{28/3}$

29) $(x^{-6}y^3z^{-12})^{-4/3}$
   A) $\dfrac{x^8y^4}{z^{16}}$
   B) $\dfrac{x^8z^{16}}{y^4}$
   C) $\dfrac{y^4z^{16}}{x^8}$
   D) $\dfrac{y^4}{x^8z^{16}}$

30) $(9x^8y^{-6})^{3/2}$
   A) $\dfrac{27x^{12}}{y^9}$
   B) $\dfrac{27}{x^{12}y^9}$
   C) $27x^{12}y^9$
   D) $\dfrac{9x^{12}}{y^9}$

6 Demonstrate Additional Understanding and Skills

An expression that occurs in calculus is given. Write the expression as a single quotient in which only positive exponents and/or radicals appear.

1) $\dfrac{(49-x^2)^{1/2} + 3x^2(49-x^2)^{-1/2}}{49-x^2}$

   A) $\dfrac{49 + 2x^2}{(49-x^2)^{3/2}}$
   B) $\dfrac{3x^2}{\sqrt{49-x^2}}$
   C) $\dfrac{49 - 4x^2}{(49-x^2)^{3/2}}$
   D) $\dfrac{3x^2}{(49-x^2)^{3/2}}$

2) $10x^{3/2}(x^3+x^2) - 12x^{5/2} - 12x^{3/2}$
   A) $10x^{3/2}(x^3-1) + 10x^3 - 2x^{3/2}(x-1)$
   B) $2x^{3/2}(x+1)(5x^2-6)$
   C) $10x^{3/2}(x+1)(5x^2-6)$
   D) $x^{3/2}(x-1)[10(x^2-1) - 2] + 10x^3$

3) $(x^2-1)^{1/2} \cdot \dfrac{3}{2}(2x+5)^{1/2} \cdot 2 + (2x+5)^{3/2} \cdot \dfrac{1}{2}(x^2-1)^{-1/2} \cdot 2x$

4) $\dfrac{(x^2+2)^{1/2} \cdot \dfrac{3}{2}(2x+1)^{1/2} \cdot 2 - (2x+1)^{3/2} \cdot \dfrac{1}{2}(x^2+2)^{-1/2} \cdot 2x}{x^2+2}$

Factor the expression.

5) $x^{5/8} + 3x^{2/8}$
   A) $x^{1/8}(x^{5/8} + 3)$
   B) $x^{1/4}(x^{3/8} + 3)$
   C) $x^{1/8}(x^{3/8} + 3)$
   D) $x^{1/4}(x^{-3/8} + 3)$

6) $6x^{2/7} - 5x^{-4/7}$
   A) $x^{-4/7}(6x^{6/7} - 5x)$
   B) $x^{2/7}(6x^{6/7} - 5x^{6/5})$
   C) $x^{1/3}(6 - 5x^{6/5})$
   D) $x^{-4/7}(6x^{6/7} - 5)$

7) $x^{-2/7} + x^{-4/7}$
   A) $x^{-4/7}(x^{2/7} - 1)$
   B) $x^{-4/7}(x^{2/7} + 1)$
   C) $x^{-4/7}(x^{2/7})$
   D) $x^{-4/7}(x^{2/7} + x)$

8) $x^{1/4} - x^{-5/4}$
   A) $x^{-5/4}(x^{3/2} + 1)$
   B) $x^{-5/4}(x^{3/2} - 1)$
   C) $x^{-5/4}(x^{3/2} - x)$
   D) $x^{-5/4}(x^{3/2})$

Ch. 14 (Appendix) Review
Answer Key

14.1 Algebra Review
1 Evaluate Algebraic Expressions
    1) C
    2) D
    3) D
    4) A
    5) D
    6) C
    7) B
    8) A
    9) C
    10) B
    11) C
    12) A
2 Determine the Domain of a Variable
    1) B
    2) B
    3) D
    4) A
    5) A
    6) B
    7) D
    8) C
    9) B
    10) A
    11) A
    12) B
3 Graph Inequalities
    1) B
    2) C
    3) B
    4) B
    5) C
    6) A
    7) A
    8) C
    9) A
    10) C
    11) B
    12) D
    13) A
    14) A
    15) B
    16) D
    17) B
    18) B
    19) B
    20) A
4 Find Distance on the Real Number Line
    1) D
5 Use the Laws of Exponents
    1) B

2) D
3) D
4) B
5) C
6) B
7) A
8) B
9) D
10) A
11) A
12) C
13) A
14) C
15) D
16) C
17) A
18) D
19) A
20) D
21) B
22) A
23) B
24) D
25) $\dfrac{4x}{y^2}$
26) B
27) D
28) D

6 Evaluate Square Roots
1) B
2) C

7 Demonstrate Additional Understanding and Skills
1) C
2) D
3) C
4) C
5) C
6) C
7) B
8) A
9) C
10) B

14.2 Geometry Review
1 Use the Pythagorean Theorem and Its Converse
1) A
2) C
3) D
4) B
5) B
6) B
7) B
8) B
9) A

2 Know Geometry Formulas

1) C
2) A
3) C
4) A
5) B
6) A
7) B
8) D
9) B
10) B
11) D
12) A
13) C
14) C
15) B
16) C
17) C
18) A
19) C
20) $140\pi \approx 439.8$ inches
21) 186 sq ft
22) $99.75\pi \approx 313.4$ sq ft
23) $16 + 2.5\pi \approx 23.9$ ft
24) 16.4 mi

14.3 Polynomials and Rational Expressions

1 Recognize Special Products

1) B
2) B
3) C
4) B
5) D
6) D
7) A
8) C
9) C
10) D
11) A
12) B
13) A
14) A
15) B
16) C
17) B
18) A
19) C
20) D
21) B
22) C
23) B

2 Factor Polynomials

1) B
2) B
3) D
4) B

5) A
6) C
7) B
8) C
9) D
10) B
11) A
12) B
13) B
14) B
15) B
16) C
17) C
18) D
19) D
20) D
21) C
22) D
23) A
24) C
25) $2x(2x + 5)$
26) $(x + 5)^2(2x - 1)(10x + 17)$

3 Simplify Rational Expressions
1) B
2) A
3) A
4) B
5) B
6) D
7) A
8) B
9) C
10) C
11) $\dfrac{4x(x + 5)}{(2x + 5)^2}$

4 Use the LCM to Add Rational Expressions
1) B
2) A
3) C
4) C
5) A
6) C
7) D
8) $\dfrac{-2x - 7}{(x + 4)(x + 3)(x - 3)}$
9) C
10) B

14.4 Polynomial Division; Synthetic Division
1 Divide Polynomials Using Long Division
1) A
2) D
3) A
4) A
5) D

Precalculus Enhanced with Graphing Utilities 975

6) C
7) D
8) C
9) A
10) C
11) B

2 Divide Polynomials Using Synthetic Division

1) B
2) C
3) D
4) B
5) A
6) B
7) B
8) B
9) A

14.5 Solving Equations
1 Solve Linear Equations

1) C
2) A
3) C
4) C
5) B
6) C
7) C
8) B
9) D
10) B
11) A
12) A
13) B
14) D
15) D
16) A
17) A
18) D
19) A
20) A
21) D
22) D
23) B

2 Solve Rational Equations

1) D
2) A
3) D
4) D
5) C
6) D
7) B
8) D
9) B
10) C
11) C

3 Solve Quadratic Equations by Factoring

1) A

2) C
3) A
4) A
5) A
6) B
7) A
8) A
9) C
10) C
11) C
12) D
13) B
14) D
15) C

4 Solve Quadratic Equations Using the Square Root Method

1) D
2) A
3) D
4) D
5) B
6) A
7) C

5 Solve Quadratic Equations by Completing the Square

1) A
2) B
3) A
4) C
5) C
6) C
7) D
8) C
9) D
10) C
11) D
12) A
13) A
14) C
15) D

6 Solve Quadratic Equations Using the Quadratic Formula

1) A
2) C
3) A
4) D
5) A
6) A
7) D
8) A
9) C
10) A
11) C
12) A
13) D
14) B
15) A
16) C

17) C
18) B

7 Solve Equations Quadratic in Form

1) B
2) D
3) C
4) C
5) A
6) C
7) C
8) D
9) B

8 Solve Absolute Value Equations

1) A
2) D
3) B
4) C
5) B
6) B
7) A
8) C
9) B
10) C
11) C
12) D
13) A
14) B
15) D

9 Solve Equations by Factoring

1) A
2) A
3) D
4) B
5) B
6) A
7) B

10 Demonstrate Additional Understanding and Skills

1) B
2) C
3) C
4) B
5) D
6) C
7) A

14.6 Complex Numbers; Quadratic Equations in the Complex Number System

1 Add, Subtract, Multiply, and Divide Complex Numbers

1) C
2) B
3) C
4) D
5) B
6) C
7) C
8) B
9) B

10) A
11) A
12) B
13) B
14) A
15) D
16) A
17) B
18) D
19) D
20) B
21) D
22) B
23) D
24) D
25) A
26) D
27) B
28) C
29) A

2 Solve Quadratic Equations with a Negative Discriminant

1) C
2) A
3) C
4) C
5) A
6) B
7) B
8) A
9) B
10) B
11) C
12) C

14.7 Problem Solving

1 Translate Verbal Descriptions into Mathematical Expressions

1) A
2) D
3) B
4) D
5) A
6) D
7) D

2 Solve Interest Problems

1) D
2) B
3) C
4) A
5) B
6) A
7) D

3 Solve Mixture Problems

1) A
2) B
3) B
4) C

5) C
6) A
7) C
8) D

4 Solve Uniform Motion Problems
1) D
2) A
3) B
4) B
5) C
6) A
7) A
8) A
9) D
10) A
11) C
12) A

5 Solve Constant Rate Job Problems
1) D
2) A
3) C
4) C
5) B
6) B

6 Solve Problems Involving Quadratic Equations
1) A
2) A
3) A
4) D
5) B
6) D
7) A
8) C

14.8 Interval Notation; Solving Inequalities
1 Use Interval Notation
1) B
2) C
3) A
4) B
5) D
6) C
7) B
8) D
9) D
10) A
11) B
12) C
13) A
14) C
15) C
16) B

2 Use Properties of Inequalities
1) D
2) C
3) D

4) B
5) B
6) B
7) C
8) C

3 Solve Linear Inequalities

1) D
2) A
3) B
4) D
5) C
6) B
7) C
8) B
9) B
10) C
11) C
12) C

4 Solve Combined Inequalities

1) C
2) D
3) C
4) D
5) B
6) B
7) D
8) C

5 Solve Absolute Value Inequalities

1) D
2) C
3) D
4) D
5) B
6) D
7) D
8) A
9) B
10) C
11) C
12) D
13) D
14) C
15) D

6 Demonstrate Additional Skills and Understanding

1) B
2) C
3) A
4) D
5) A
6) C
7) C
8) A
9) A

14.9 nth Roots; Rational Exponents; Radical Equations

1 Work with nth Roots
1) D
2) A
3) A
4) A
5) D
6) A

2 Simplify Radicals
1) B
2) D
3) A
4) B
5) B
6) D
7) A
8) C
9) C
10) B
11) D
12) C
13) C
14) A
15) B
16) A
17) B
18) D
19) A
20) B
21) C
22) B
23) B
24) C
25) B
26) D

3 Rationalize Denominators
1) B
2) A
3) B
4) C
5) C
6) B
7) B

4 Solve Radical Equations
1) D
2) C
3) A

5 Simplify Expressions with Rational Exponents
1) B
2) C
3) D
4) C
5) B
6) A
7) C

8) A
9) B
10) C
11) D
12) B
13) D
14) D
15) C
16) C
17) C
18) D
19) C
20) D
21) A
22) C
23) B
24) A
25) A
26) A
27) B
28) D
29) B
30) A

6 Demonstrate Additional Understanding and Skills

1) A
2) B
3) $\dfrac{(2x+5)^{1/2}(5x^2+5x-3)}{(x^2-1)^{1/2}}$
4) $\dfrac{(2x+1)^{1/2}(x^2-x+6)}{(x^2+2)^{3/2}}$
5) B
6) D
7) B
8) B